BASIC SCIENTIFIC SUBROUTINES, Volume II

BASIC Scientific
Subroutines, Volume II

by F. R. Ruckdeschel

BYTE/McGRAW-HILL
70 MAIN ST
PETERBOROUGH, NH 03458

BASIC SCIENTIFIC SUBROUTINES Vol. II

Copyright © 1981 by BYTE Publications Inc. All rights reserved. Printed in the United States of America. No part of this publication may be reproduced, stored in a retrieval system, or transmitted, in any form or by any means, electronic, mechanical, photocopying, recording, or otherwise, without the prior written permission of the publisher.

The author of the programs provided with this book has carefully reviewed them to ensure their performance in accordance with the specifications described in the book. Neither the author nor BYTE Publications Inc, however, makes any warranties concerning the programs and assumes no responsibility or liability of any kind for errors in the programs, or for the consequences of any such errors. The programs are the sole property of the author and have been registered with the United States Copyright Office.

Library of Congress Cataloging in Publication Data

Ruckdeschel, F. R.
    BASIC scientific subroutines.

    Includes bibliographies and indexes.
    1. Mathematics—Computer programs. 2. Basic (Computer program language) I. Title.
QA76.95.R82   502'.8'5425   80-19582
ISBN 0-07-054201-5 (v. 1)   AACR1
ISBN 0-07-054202-3 (v. 2)

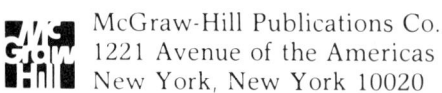

McGraw-Hill Publications Co.
1221 Avenue of the Americas
New York, New York 10020

DEDICATION

Dedicated to four very important people in my life: Anne, Jake, Janet and Nathan.

ACKNOWLEDGMENTS

The author is grateful to BYTE Publications/McGraw-Hill for offering the opportunity to present this information to other computer users. The author also wishes to acknowledge the excellent editing and production skills of Bruce Roberts and Peg Clement at BYTE Books.

Table of Contents

Introduction 1

Chapter 1 Least-Squares Approximation   7

- 1.1   General Introduction to Polynomial Approximations 7
- 1.2   First-Order Least Squares 16
- 1.3   Second-Order Least Squares 24
- 1.4   Nth-Order Least Squares 31
- 1.5   Multidimensional Least Squares 45
- 1.6   Least-Squares Fitting with Orthogonal Polynomials 56
- 1.7   Iterated Regression 65
- 1.8   Parametric Least Squares 75
- 1.9   Extending the Use of Least Squares 82
- 1.10 Summary and Conclusion 87

Chapter 2 Series-Approximation Techniques   89

- 2.1   Introduction 89
- 2.2   Taylor Series and Horner's Rule 93
- 2.3   Asymptotic Series 102
- 2.3.1 The Bessel Function 103
- 2.3.2 The Chi-Square Distribution Functions 108
- 2.3.3 The Gamma Function 118
- 2.3.4 The Error Function 119
- 2.4   Chebyshev Polynomials 123
- 2.5   Economization 130
- 2.6   Reversion, Inversion, and Shifting 142
- 2.6.1 Polynomial Reversion 142
- 2.6.2 Polynomial Inversion 148
- 2.6.3 Shifting the Expansion Point 155
- 2.7   Rational Polynomials 159
- 2.8   Infinite Products 166
- 2.9   Complex Series 169
- 2.10 Summary and Conclusion 174

Chapter 3 Functional Approximations by Iteration and Recursion 175

- 3.1 Introduction 175
- 3.2 Roots by Iteration 177
- 3.3 Tangent Iteration 190
- 3.4 Arctangent by Recursion 194
- 3.5 Arcsine by Recursion 199
- 3.6 Elliptic Integrals by Recursion 202
- 3.7 Natural Logarithm by Recursion 206
- 3.8 Bessel Functions by Recursion 209
- 3.9 Orthogonal Polynomial Coefficients by Recursion 215
- 3.10 Summary and Conclusion 229

Chapter 4 CORDIC Approximation Techniques and Alternatives 231

- 4.1 Introduction 231
- 4.2 The Trigonometric Functions 233
- 4.3 Generating the T_i and P_N Coefficients 243
- 4.4 The Inverse Trigonometric Functions 245
- 4.5 The Exponential Function 254
- 4.6 The Natural-Logarithm Function 261
- 4.7 The Hyperbolic Trigonometric Functions 269
- 4.8 Inverse Hyperbolic Trigonometric Functions 277

Chapter 5 Table Interpolation, Differentiation, and Integration 285

- 5.1 Introduction 285
- 5.2 Lagrange Interpolation 287
- 5.3 Newton Divided-Differences Interpolation and Error Estimates 293
- 5.4 Choosing the Table Values 300
- 5.5 Semi-Spline Interpolation 305
- 5.6 Calculating Derivatives from Tables 311
- 5.7 Table Integration 316
- 5.8 Interpolation and Integration of $2/\sqrt{\pi}\ e^{-x^2}$ 326
- 5.9 Summary and Conclusion 332

Chapter 6 Finding the Real Roots of Functions 333

- 6.1 Introduction 333
- 6.2 Gaining Preliminary Knowledge About the Roots of a Polynomial 335

6.3	Interval Searches	340
6.4	Successive Substitution	348
6.5	Examples of Successive Substitution	352
6.6	Forcing the $x = g(x)$ Form	362
6.7	Formalizing the Generation of $x_{n+1} = g(x_n)$	364
6.8	Newton's Method	366
6.9	The Secant and False-Position Methods	371
6.10	Numerical Comparisons of the Newton, Secant, and False-Position Methods	380
6.11	Aitken Acceleration	383
6.12	Aitken-Steffenson Iteration	393
6.13	Comparison of Algorithms	398
6.14	Finding More Roots: Multiplicity	399
6.15	Finding More Roots: Removal	401
6.16	Conclusion	411

Chapter 7 Finding the Complex Roots of Functions　　413

7.1	Introduction	413
7.2	Review of the Fundamental Properties of Functions in the Complex Domain	415
7.3	Interval Search	418
7.4	Newton's Method in the Complex Domain	443
7.5	Mueller's Method in One Dimension	448
7.6	Two-Dimensional Form of Mueller's Method	457
7.7	Mueller's Method in the Complex Plane	464
7.8	Representing Polynomials in the $\mu(x,y) + i\,\nu(x,y)$ Form and Removing Roots	473
7.9	The Quadratic Formula	488
7.10	Lin's Method	492
7.11	Bairstow's Method	500
7.12	Summary—Comparison of Algorithms	506

Chapter 8 Optimization by Steepest Descent　　509

8.1	Introduction	509
8.2	Steepest Descent with Functional Derivatives	512
8.3	Steepest Descent with Approximate Derivatives	519
8.4	Summary and Conclusions	531

References 533

Appendix IA Software Index by Number 539

Appendix IB Software Index by Function 543

Appendix IIA Full Listings of North Star BASIC Demonstration and
Subroutine Programs 549

Appendix IIB Compacted North Star BASIC Subroutine Listing 645

Appendix III Conversion to Other BASIC Dialects and Microsoft
BASIC Program Listings 687

Index 787

INTRODUCTION

BASIC Scientific Subroutines, Volume II is the second in a continuing series of books dealing with scientific programming in the BASIC language. The object of this series is to provide students, engineers, and scientists with a well-documented library of subroutines that permit the full power of the small computer to be utilized for scientific tasks.

In the past, mathematical calculations, such as those associated with multidimensional nonlinear regression (linear in the coefficients, but nonlinear with respect to the variables), have been considered to require the subroutine packages available in FORTRAN as implemented on a large computer. This has been a self-sustaining tradition. Engineers and scientists have learned FORTRAN in order to capitalize on the wealth of scientific programs in that language, and then have contributed further to the collection. Thus, the FORTRAN library has continued to grow.

The advent of the microcomputer has brought with it the ready availability of significant computing power in an easy-to-use computer language—BASIC. Programmers of FORTRAN have not embraced BASIC for three main reasons. First, the early microcomputer versions of BASIC were very limited with respect to their instruction sets and versatility. This changed dramatically with Microsoft's introduction of Extended BASIC, and with the later releases of North Star, Cromemco, and others. Second, BASIC has historically been an interpreted language. This is an excellent feature in terms of aiding program development, but interpreters execute *very* slowly. The introduction of BASIC compilers has somewhat neutralized the speed disadvantage, however. This leaves the one remaining shortfall of BASIC—its lack of support software for scientific applications. There is an abundance of scientific software written in FORTRAN that is accessible through convenient subroutine calls. Such is not the case for BASIC. The *BASIC Scientific Subroutines* series is designed to help reduce this gap.

The subjects covered in Volume II are coordinated with the material in Volume I (see Ref. 1) both in mechanics and in concept. The line numbers of the subroutines and demonstration programs presented in Volume II sequentially follow those of the previous volume. Some of the subroutines previously developed are used and listed again here. Thus, Volume II is a self-contained sequel to Volume I.

The approximation of functions was the last subject covered in Volume I. This discussion is expanded upon and is divided into several topics with associated subroutines in Volume II.

Chapter 1 of Volume II, of particular interest to those involved in statistics and data processing, deals with the least-squares approximation of functions and the smoothing of data. Linear through Nth-order regression is considered from two algorithmic approaches: simple polynomials in X and Forsythe orthogonal polynomials. Ultimately, the former approach is more general in its applications, and is easily extended to multidimensional nonlinear regression. The orthogonal polynomial method, though more specific, computes more quickly and suffers much less from round-off error. An interesting iterative technique is also provided for optionally reducing the effect of round-off error, although it is at the expense of computer time.

The idea of least-squares regression is subsequently taken beyond the bounds of polynomials into the realm of parametric regression. A very general subroutine for the one-dimensional case is provided. The technique is discussed in enough detail to permit you to apply the methodology to multidimensional problems with little difficulty.

Chapter 2 is directed at approximating functions by series other than least-squares polynomials. Taylor series estimates of functions and examples of *asymptotic* series are included. Chebyshev polynomials are presented and then applied to create near-min-max approximations. General routines that permit Taylor series and least-squares polynomials to be *economized* (*telescoped*) by replacement with truncated Chebyshev series are also supplied. The economization technique is extremely powerful for reasons that become apparent in the discussion.

Chapter 2 continues with subroutines for polynomial reversion and inversion, examples of approximation by rational polynomials, and product sequences. A subroutine for the evaluation of complex-variable polynomial series concludes the chapter.

Functional approximation by iteration and recursion is introduced in Chapter 3. The discussion starts with the classic Newton iteration scheme for finding the Nth-root (where N is an integer) of a variable. The algorithm is subsequently extended to general Nth-root determination. Chapter 3 continues with an examination of the use of recursion formulae. These calculate the series coefficients both for several important functional approximations and for the values of functions themselves.

The concept of recursion is extended in Chapter 4 to the high accuracy approximation of trigonometric, hyperbolic, and exponential functions (and their inverses). The techniques discussed are similar to CORDIC approximation. Because of the dearth of easily readable literature, and because implementation in BASIC is somewhat unique, this chapter is treated in some detail.

The subject of approximation takes a new turn in Chapter 5 where the interpolation, differentiation, and integration of tables are examined. Three interpolation subroutines—Lagrange, Newton and Akima splines—are given. In each case, polynomials are fitted through the table values. The polynomial can then be used not only to derive intermediate values, but also to estimate the derivative. Interpolation is also combined with an enhanced form of trapezoidal numerical integration to form a very general table integration subroutine. Both the intrinsic (*truncation*) and round-off errors of these algorithms are discussed to ensure that you see the practical limitations of such methods.

Chapters 6 and 7 are devoted to the problem of estimating the *roots* of both polynomial and general functions. Chapter 6 contains the introductory sections and algorithms designed for *real* function analysis whereas Chapter 7 gives subroutines applicable to *complex* functions. Most of the techniques discussed are iterative in nature, and are thereby, in principle, capable of great accuracy. Several different methods are considered, however, because practical problems often require specific algorithms to achieve maximum accuracy in minimum computing time.

The algorithms presented in Chapters 6 and 7 include most of the methods taught in introductory numerical methods courses, plus more. This discussion should be particularly interesting to engineering students. Because the basic root-seeking concepts are actually implemented (and tested with examples) in BASIC, the practical values of these methods are put into perspective. For example, the simple classic algorithm taught in all introductory numerical methods courses— the secant technique—is shown to often suffer from convergence problems in real situations. In addition, it can be prone to round-off error instabilities. In this case, *modified* regula-falsi, an alternative subroutine, is provided.

Volume II concludes with Chapter 8, in which the fundamental concept of the gradient is applied to finding the maxima and minima of functions. This discussion should be especially interesting to engineers, business people, and economists who wish to maximize their *objective function*.

Many of the subroutines presented in Volume II are not commonly available in FORTRAN subroutine libraries, nor are they easily obtainable from the literature. Therefore, this text may be useful as an instructional tool, and perhaps may serve as a launching point for even more sophisticated programs.

The style of documentation and coding in *BASIC Scientific Subroutines* represents a compromise between completeness and utility. It is the author's contention that students, engineers, and scientists do not necessarily wish to see an esoteric mathematical treatise followed by an unreadable program. Rather, they want a brief introduction which gives the essential mathematics (or at least the concept), followed by a well-documented program. This documentation should, at the very least, include both a description of the input and output parameters and variables, and at least one example.

BASIC Scientific Subroutines provides a level of documentation that surpasses this minimum. Subroutine-connection diagrams are also given, for example, and subroutine cross-references are supplied in an appendix. In addition, the subroutines themselves internally contain REMarks to aid in understanding the program flow. The programs have been written with an emphasis on clarity rather than on execution speed. For this reason, you will find no concatenated lines or overly subtle programming tricks.

Because of the clarity of the subroutine coding and the decision to use a carefully chosen subset of BASIC commands, very little alteration is needed to translate the subroutines presented in this series to other BASIC dialects. To aid the user, two versions are provided. The programs displayed within the text conform to a language subset of North Star BASIC. Equivalent programs conforming to the Microsoft BASIC syntax are given in an appendix; there are few changes. Several of the peculiarities appearing in the various BASIC dialects have even been circumvented. For example, in some dialects a FOR/NEXT loop is always executed at least once when encountered. But in other cases, the loop arguments are checked first. In situations where this difference in execution can affect the program, a test has been inserted in front of the FOR/NEXT loop. This leads to a small loss in program efficiency in some cases, but the gain in universality more than offsets this loss.

The major dialect problem appears with the format of the PRINT statements in the demonstration programs. Between the two versions supplied—North Star and CP/M Microsoft BASIC—almost all variants are covered. However, even within the Microsoft BASIC category there are variations that cause minor differences in the output format. These will be readily apparent when the demonstration routines are used. The required changes are minor.

North Star and Microsoft BASIC were chosen because of both their popularity and their dissimilarity. Both dialects are widely employed, but they represent near extremes in their syntax. Also, North Star BASIC is much more efficient in memory utilization than Microsoft BASIC, but is much slower in execution (see Ref. 2). Therefore, almost all the differences that you might encounter in using some other BASIC should be apparent in these two.

Although the subroutines here were written with the user in mind, they were not designed to comprehensively check for misinformation or ill-conditioned situations. For this reason, you should have some understanding of the algorithm and the precautions involved in application. There is minimal error trapping in terms of erroneous input. It is assumed that the calling program has the responsibility of assuring that the parameters and variables passed to the subroutine are within the acceptable ranges. However, error checks are often provided within the subroutines so that the most likely failures are not fatal (i.e., the entire program aborts).

The majority of the subroutines require less than two kilobytes of memory above the size of the operating system. In several of the programs involving arrays, the memory demand, which can be quite large, is dictated by the size of the array.

In general, the terminal line width requirements were kept below 64 characters, although this was not possible in some of the tabular displays. All demonstration program input data is assumed to be entered from the keyboard.

All execution time comparisons are based on an IMSAI 8080 running at 2-megahertz with no wait states.

The programs are presented as subroutines *per se*. That is, they are terminated with a RETURN statement. Because it may be desirable to assemble the software into a subroutine library package, the routines are sequentially numbered, starting at statement number 43500. (The range 40000 to 43487 was used in Volume I.) This usually places them high enough in line number to be out of the way of the main calling program.

When using the subroutines, you should be aware of the overlaps between the variable list used by the main program and that employed by the subroutine. In this book, the following conventions have been chosen:

- **FOR/NEXT loops:** Six general indices have been reserved: $I<n>$, $J<n>$, $K<n>$, $L<n>$, $M<n>$, where $<n>$ is optionally a digit in the second location. Included in this convention are their dimensioned extensions. For example, $I4(K)$ might be found in a subroutine.
- **Running variables:** Six general running variables are used: $U<n>$, $V<n>$, $W<n>$, $X<n>$, $Y<n>$, and $Z<n>$, as well as their dimensioned extensions.
- **Parameters:** Five parameters are employed: $A<n>$, $B<n>$, $C<n>$, $D<n>$, and $E<n>$, as well as their dimensioned extensions.

The program execution speed of an interpreter is slow by nature, and the variation among the various BASIC dialects is great. As an example, the following fundamental comparison, as seen in Ref. 2, can be made:

Execution Time in Milliseconds

Function	MITS 8 K (8080)	OSI 8 K (6502)	North Star (8080)	North Star + Floating-Point Board
multiply	4	2	5	2
divide	7	3	16	2

Execution Time in Milliseconds

Function	MITS 8 K (8080)	OSI 8 K (6502)	North Star (8080)	North Star + Floating-Point Board
sine/cosine	23	17	99	11
logarithm	19	14	99	9
exponent	28	22	73	8

The timing comparisons shown for the subroutines given in this book correspond to standard 8-digit North Star BASIC. The 14-digit version runs about 40% slower. By using another dialect of BASIC, you could expect at least a twofold increase in speed. An additional twofold improvement is possible by using a 4-megahertz microprocessor instead of the 2-megahertz one used for the comparisons given in the text. Thus, the values given in each chapter should be considered to be very conservative.

The *BASIC Scientific Subroutines* series is not a complete set of texts on numerical methods, but rather a software tool kit designed to aid users in employing their microcomputers or minicomputers more fully for scientific applications. The degree to which this is accomplished will be a measure of the success of the series.

Chapter 1

Least-Squares Approximation

1.1 General Introduction to Polynomial Approximations

The discussion in this chapter centers on one of the popular concepts in approximation theory: least-squares curve fitting, or regression (the statistical determination of parametric dependencies, e.g., the polynomial coefficients of a least-squares curve fit to a set of data). Although the associated techniques are usually applied to sets of data containing noise, they may also be employed to provide approximations to functions. This is particularly useful when high-accuracy tables that represent the function are available. In the subroutines given in this chapter, there is no operational distinction made between curve-fitting noisy data or fitting function tables. Therefore, these subroutines can be used both for statistical regression and for functional approximations.

The least-squares curve-fitting criterion is simple and is based directly on statistical concepts. We will take the linear correlation between two variables, x and y, as an example. If x is defined as the independent variable, and y the dependent variable, then ideally the linear relationship between the two is

$$y = a + bx \qquad (1.1.1)$$

If we are given a set of data values (x_i, y_i) in which there is some error in the y_i measurement, we have

$$y_i = a + bx_i + E_i \qquad (1.1.2)$$

E_i represents the measurement error. However, the error is usually an unknown quantity. Therefore, a and b cannot be obtained from equation (1.1.2). Instead, the *form* of equation (1.1.1) is assumed, and by some means, the data set is used to obtain approximate values for a and b. We will call these approximations α and β respectively.

If the error, E_i, is normally distributed with variance σ^2 (this is a fairly good assumption and is expected from the Central Limit Theorem—see Ref. 3), then the probability that the estimate, $y'_i = \alpha + \beta x_i$, is within Δy of y_i is

$$P(y_i - \Delta y \leq y_i \leq y_i + \Delta y) = \frac{2\Delta y}{\sigma\sqrt{2\pi}} e^{-(y_i - \alpha - \beta x_i)^2/2\sigma^2} \tag{1.1.3}$$

In addition, if the variance, σ^2, is independent of y_i (i.e., the noise is additive and not signal dependent), and the errors are not *correlated* (i.e., E_i is independent of E_j; $i \neq j$), then the probability that all the predicted values are within some Δy of the measured values is

$$P = \left(\frac{2\Delta y}{\sigma\sqrt{2\pi}}\right)^N \exp \sum_{i=1}^{N}\left[-(y_i - \alpha - \beta x_i)^2\right] \tag{1.1.4}$$

P is called the *likelihood*. The values of α and β that maximize P are called the *maximum-likelihood estimators* (see Ref. 4). It can further be shown that the values of α and β that maximize P are the most precise unbiased estimates of a and b statistically attainable using the given data.

The maximum-likelihood criterion can be restated in the more familiar least-squares form as follows. The "best" values of α and β are those that minimize $S(\alpha,\beta)$:

$$S(\alpha,\beta) = \sum_{i=1}^{N}(y_i - \alpha - \beta x_i)^2 \tag{1.1.5}$$

As we will see later, the above least-squares example can be generalized to an Mth-degree polynomial ($M \leq N - 1$).

The approximation algorithms that result from the least-squares criterion will be considered in subsequent sections. The discussion for the remainder of this section will concentrate on the differences between least-squares curve fitting and several of the other approximation methods reviewed later in this volume, notably Taylor series and min-max polynomials. An understanding of the intrinsic error characteristics of these distinctly different approximation methods is very helpful in choosing the best algorithm for the task.

Perhaps the most frequently used polynomial approximation method is the Taylor series expansion:

$$f(x) = f(x_0) + (x - x_0)\left.\frac{df}{dx}\right|_{x_0} + \frac{(x - x_0)^2}{2!}\left.\frac{d^2f}{dx^2}\right|_{x_0} + \cdots \tag{1.1.6}$$

Its popularity is based on generality; any *analytical* function can be expanded in a Taylor series. The standard mathematical handbooks (e.g., References 5, 6, and 7) contain tables of Taylor series expansions for many functions. It is therefore not surprising to find that functions are often evaluated in computers by means of these expansions. In particular, this seems to be the case with much of the software for small computers in which the basic trigonometric and exponential functions (and their inverses) are implemented in assembly language. The programmers who write this type of software are usually not aware of the significant shortfalls of Taylor series expansions.

As discussed in the last chapter of Volume I of *BASIC Scientific Subroutines*, Taylor series expansions are very convenient, but they are usually among the poorest of the polynomial-approximation techniques.

Figure 1.1.1 shows the *truncation* error associated with using only the first four terms of a Taylor series (Maclaurin series subset, in this case) for approximating $\ln(1 + x)$. The error is very small close to the expansion point, but rapidly grows away from it. The truncation error can usually be reduced by including more terms in the series, but it is difficult to use such a representation for evaluation far from the expansion point. In fact, the series diverges for $x > 1$.

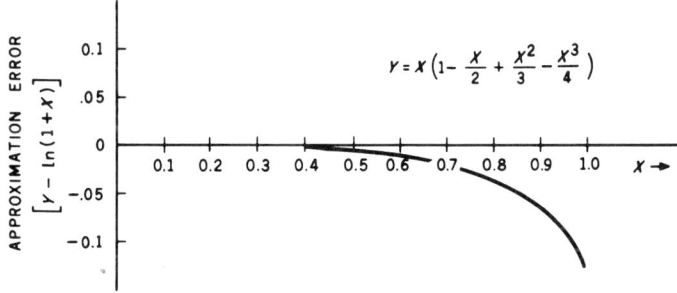

Figure 1.1.1 Truncation error associated with using a fourth-degree Taylor series polynomial to approximate $\ln(1 + x)$ over the range $0 \leq x \leq 1$. The expansion point was $x = 0$. See also the discussion in Hart (Ref. 7).

Table 1.1.1

SUBROUTINE	SUBROUTINE SIZE (bytes) (including support programs)	DEMONSTRATION PROGRAM SIZE (bytes)	EXECUTION SPEED	CONDITIONS/ COMMENTS
LSTSQR1 (one-dimensional linear least squares)	799	1267	1.8 seconds	20 data points

SUBROUTINE	SUBROUTINE SIZE (bytes) (including support programs)	DEMON- STRATION PROGRAM SIZE (bytes)	EXECUTION SPEED	CONDITIONS/ COMMENTS
LSTSQR2 (one-dimensional parabolic least squares)	1164	1333	3.5 seconds	20 data points
LEASTSQR (one-dimensional, Nth-order least squares)	6233	759	48 seconds	20 data points, fourth-degree fit
LEASTSQR (multidimensional, Nth-order least squares)	7367	906	187 seconds	20 data points, parabolic fit in each of two dimensions
LSQRPOLY (multidimensional, Nth-order least squares)	2497	598	20.4 seconds	20 data points, fourth-degree fit
REGITER (iterated, multidimensional, Nth-order least squares)	10080	891	203 seconds/ iteration	20 data points, parabolic fit in each of two dimensions
PARAFIT (iterated, one-dimensional curvilinear regression)	1805	532	30 seconds/ iteration	10 data points, gaussian fit (section 1.8)
CHISQA (χ^2 distribution)	609	197	0.34 seconds/ evaluation	

Table 1.1.1 *A summary of the programs presented in Chapter 1. See also table 1.10.1. The subroutine LEASTSQR can be used in several ways, depending on which coefficient generation subroutine is called. In the first appearance above, POLYCM is called; in the second, MLTNLREG.*

A *much* better approximation can be obtained simply by using a fourth-degree least-squares polynomial (see figure 1.1.2). In this case, the fourth-degree polynomial was generated by accurately calculating values for $[\ln(1+x)]/x$ at 101 points (100 intervals) in the range $0 \leq x \leq 1$. A least-squares cubic polynomial was then fitted to this "data." The result was a much better fourth-degree approximation than that obtained by using the Taylor series, with the maximum error reduced by a factor of 200.

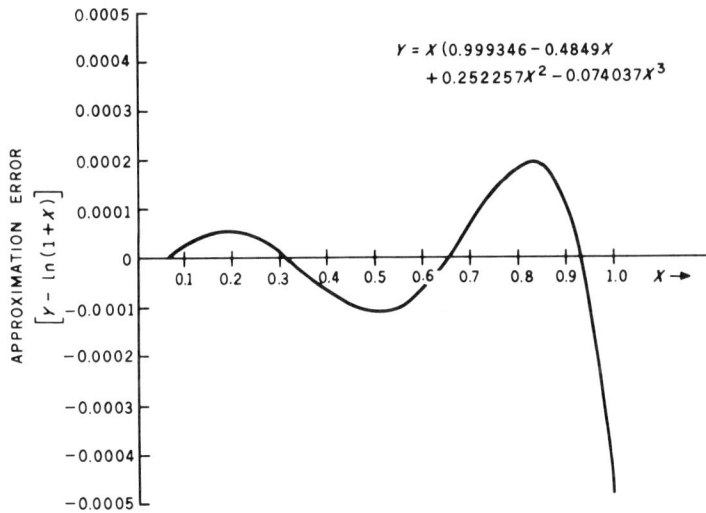

Figure 1.1.2 *Error observed in fitting a fourth-degree least-squares polynomial to the function* $\ln(1+x)$ *over the interval* $0 \leq x \leq 1$. *Note that x was factored out before the calculation such that the function actually approximated was* $[\ln(1+x)]/x$. *Therefore, only four coefficients needed to be calculated to obtain a fourth-degree function.*

The results shown in figure 1.1.2 demonstrate two features that are characteristic of least-squares polynomial fits. First, the error oscillates. Second, the maximum error appears at the right end of the interval. However, the function actually fitted was y/x, and if the error in that approximation is plotted (see figure 1.1.3), then it is apparent that the error in the fit is concentrated at *both* ends of the interval.

The uneven distribution of error in least-squares approximations is characteristic of the method. Becket and Hurt (Ref. 8) point out that there are theoretical reasons for not expecting the error near the endpoints of the interval to go to zero regardless of the degree of the fit. This is reminiscent of the *Gibbs' Phenomenon* in Fourier transform theory; a little *ringing* is always present. Therefore, if least-squares fitting is employed, it is a good practice to approximate the function over a range wider than that eventually needed.

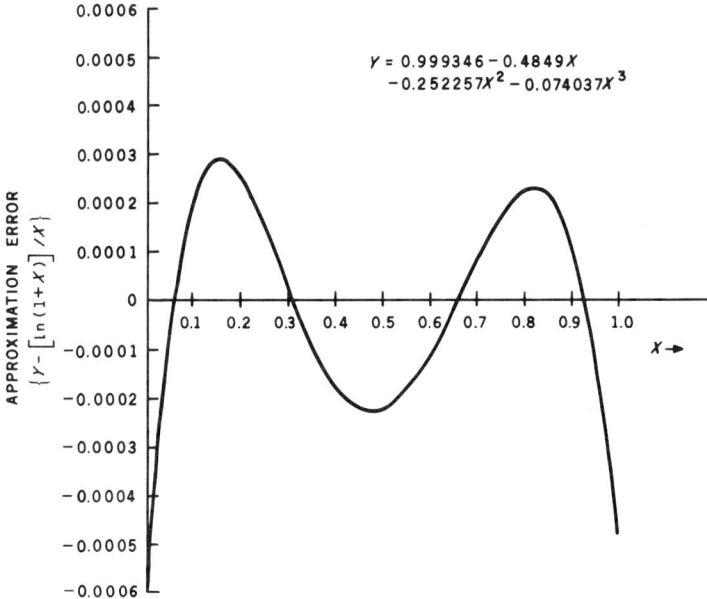

Figure 1.1.3 *Error observed in the least-squares fitting of $[\ln(1+x)]/x$ over the interval $0 \leq x \leq 1$. Note that the error tends to be greater at the ends of the interval than in the interior. Approximating the ratio $[\ln(1+x)]/x$ was chosen in order to give more weight to minimizing the relative error than to minimizing the absolute error.*

It is possible to force the truncation error to be more evenly distributed. This is accomplished by employing *min-max* polynomials. Approximations to these special polynomials can be obtained by using the truncated Chebyshev series discussed in Chapter 2. For the $\ln(1+x)$ example, a four-term truncated Chebyshev series approximation results in the error plot shown in figure 1.1.4. Even though the Chebyshev approximation is only cubic, the maximum error is about the same as that observed for the quartic least-squares fit (figure 1.1.2). If the next term in the Chebyshev series were included, the maximum error would be reduced sevenfold.

Truncated Chebyshev series do not give true min-max polynomials. An exact min-max fit would have equal-ripple error. The ripple pattern appearing in figure 1.1.4 only approximates this behavior. However, not much is to be gained by laboriously determining the precise min-max representation.

In some situations (e.g., Bessel functions), it is desirable to evaluate the function for large values of the argument. One class of approximations is particularly well suited for this application—*asymptotic* series. The type of error behavior characteristic of such approximations is shown in figure 1.1.5. As x increases, the error decreases. In a sense, asymptotic series are similar to Taylor series, but with the expansion "point" being $x = \infty$. Therefore, the larger x is, the better the approximation. This is quite the reverse of the normal Taylor series expansion error behavior. Asymptotic series will be discussed further in Chapter 2.

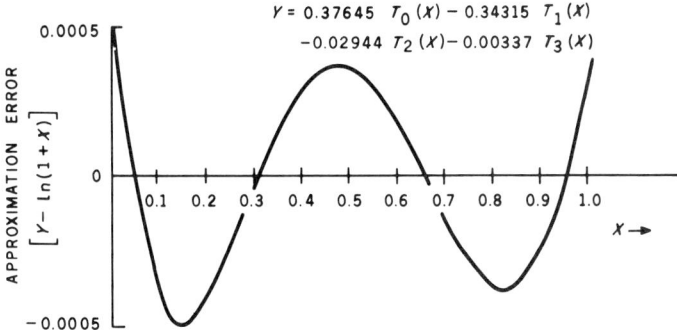

Figure 1.1.4 *Error observed for a truncated Chebyshev series approximation to* $\ln(1+x)$ *over the interval* $0 \le x \le 1$. *This is a near-min-max polynomial having four calculated coefficients. Although the resulting polynomial is only cubic, it has the same maximum error as observed for the fourth-degree least-squares representation. Including one more degree* $[T_4(x)]$ *would reduce the maximum error roughly sevenfold. The fitted equation was obtained from Hastings (Ref. 9).*

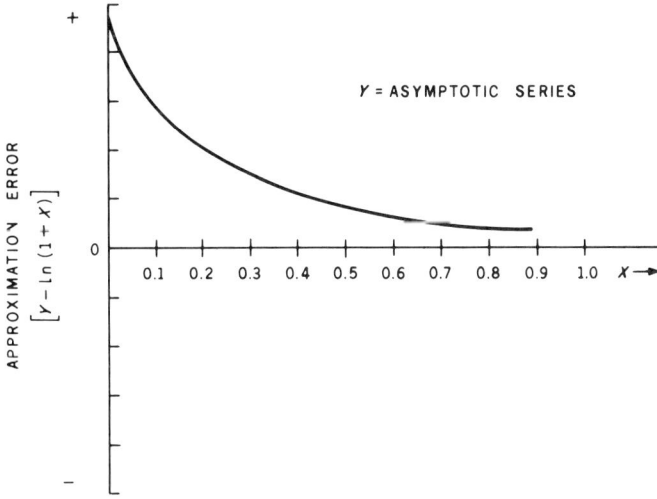

Figure 1.1.5 *Hypothetical error curve for an asymptotic series approximation to* $\ln(1+x)$. *Compare with figure 1.1.1. Whereas the Taylor series approximation gets poorer with increasing x, the asymptotic series gets better.*

Another interesting and important approximation function is the *collocation* polynomial. The error criterion in this case is that the approximating polynomial exactly match the function at N points. This can always be accomplished by a polynomial of degree $N - 1$, and occasionally by one of a lower degree. The type of error observed with such a fit has the characteristic behavior depicted in figure 1.1.6. Because the error curve is continuous and must pass through N zeros, it must alternate in sign.

Collocation polynomials are implicitly used in table-interpolation algorithms. The intermediate error in the fit is related to the $(N + 1)$th derivative of the function being fitted. For further discussion, see Chapter 5.

So far, we have compared various estimation approaches in terms of how the error is distributed over the approximation interval. Now we will briefly discuss the ultimate limits to this error.

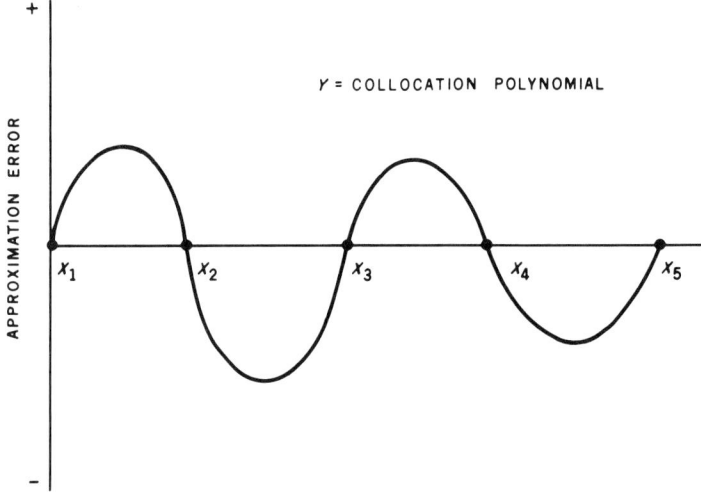

Figure 1.1.6 *Typical error curve associated with using a collocation polynomial. x_1 through x_5 are points where the approximation is chosen to be exact; the polynomial must pass through these points. In between those points there is approximation error, the sign of which oscillates. The number of collocation points is one more than the degree of the polynomial. Collocation polynomials are used extensively in table interpolation algorithms (see Chapter 5).*

Figure 1.1.7 shows a sketch of how the total error is characteristically affected by increasing the number of sequential terms in the fit. The asymptotic series typically diverges eventually *for a given evaluation point* as the number of terms increases. This is not hard to rationalize if we apply the concept that the expansion point is really at $x = \infty$. If the number of terms is kept constant, but x is increased, the error will usually decrease. Therefore, for a given x, there is usually an optimum number of terms to use that will give a minimum error.

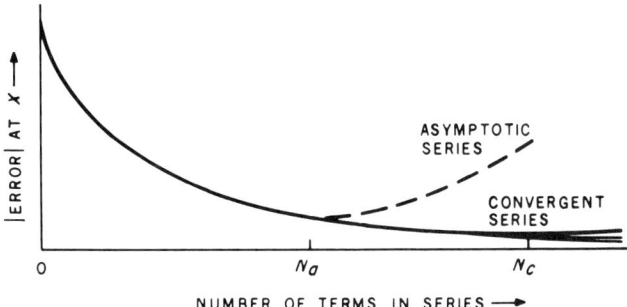

Figure 1.1.7 *Generalized representation of the behavior of the total error observed as the number of terms in the approximation is increased. There is an optimum number, N_a, of terms for an asymptotic series approximation at some x. Using more terms than that increases the error. Convergent approximations (e.g., Taylor, least-square, min-max), in principle, become more accurate as more terms are included. However, in practice, round-off error or other calculation instabilities can cause the error to level out and even to increase as more terms are added. Therefore, an optimum N_c may exist.*

The other series considered (Taylor, least-squares, Chebyshev, collocation) will, in principle, give more accurate results as the number of terms is increased. These series are mathematically convergent except possibly at the endpoints of the interval. However, numerically they may show some behavior similar to asymptotic series due to round-off error. Because of the limitations of the computer, it may be better to use fewer terms to avoid round-off error buildup. This is usually not a problem with quickly convergent series (e.g., sin x over $0 \leq x \leq 1$). Slowly convergent series, however, can demonstrate very bad round-off error problems (e.g., arctan x, especially near $x = \pm 1$). It is therefore wise to test series-approximation subroutines with respect to this error source. The subroutine may be *logically perfect*, but *practically useless*.

Becket and Hurt (Ref. 8) provide a particularly interesting rule of thumb for deciding when to stop adding terms to least-squares polynomials. In essence, they suggest examining the error at the endpoints, where it is expected to be the worst. They propose that if the error is much larger at one endpoint than at the other, round-off is likely to be strongly influencing the results. You may wish to experiment with this idea.

The authors of *Applied Numerical Methods* (Ref. 10) suggest using the *index of correlation*, which is a statistical concept. It is beyond the scope of this book to examine the various statistical criteria by which the fitting error can be analyzed. In this volume, we will consider only the *standard deviation*.

We conclude this section with a brief discussion of the importance of Taylor series in the study of approximations.

As we saw from figure 1.1.1 and from the discussion in Volume I of *BASIC Scientific Subroutines*, Taylor series are usually poor candidates for approximation polynomials. However, they can form the bases for much more efficient polynomial representations. This can be accomplished in two general ways. First, an inefficient Taylor series can be employed to accurately approximate the desired function at several points. Then a much more efficient polynomial can be generated using least-squares fitting. Second, a Taylor series can be *economized* by using the polynomial coefficients (as opposed to the calculated function values) to derive the corresponding Chebyshev series. The resulting Chebyshev series can then be truncated to give a near-min-max polynomial.

Therefore, although Taylor series are not recommended for use in routine evaluation algorithms, they can serve as the bases for other polynomial representations. For this reason, Taylor series expansions are considered in Chapter 2.

1.2 First-Order Least Squares

The simplest and most common fitting function is the straight line. The reasons for its popularity are that real data (e.g., population studies) are often very noisy and do not warrant higher-order fits, and that linear fits are ideal for sensitivity analysis. For example, in many control system models, the key parametric inputs to the analysis are the derivatives (slopes) of the input/output responses at the nominal control point. The derivation of the coefficients for this case is very simple and forms a good basis for understanding the mathematics for higher-order fits.

Recall from the previous section the $S(\alpha,\beta)$ function:

$$S(\alpha,\beta) = \sum_{i=1}^{N} (y_i - \alpha - \beta x_i)^2$$

The parameters α and β are to be chosen such that $S(\alpha,\beta)$ is minimized. At this minimum, it must be the case that $\partial S/\partial \alpha = 0$ and $\partial S/\partial \beta = 0$. Thus, we have the following two equations:

$$\sum_{i=1}^{N} (y_i - \alpha - \beta x_i) = 0$$

$$\sum_{i=1}^{N} x_i(y_i - \alpha - \beta x_i) = 0$$

(See Ref. 4, *Fitting Equations to Data*, for a detailed discussion of this method.) We establish the following definitions:

$$\bar{x} = \frac{1}{N} \sum_{i=1}^{N} x_i$$

$$\bar{y} = \frac{1}{N} \sum_{i=1}^{N} y_i$$

(1.2.1)

The solutions for the coefficients are then

$$\beta = \sum_{i=1}^{N} (x_i - \bar{x})(y_i - \bar{y}) / \sum_{i=1}^{N} (x_i - \bar{x})^2 \qquad (1.2.2)$$

$$\alpha = \bar{y} - \beta \bar{x}$$

The final fitted equation is

$$y \cong \alpha + \beta x \qquad (1.2.3)$$

In addition, the unbiased estimate of the standard deviation is

$$S_D = \left[\frac{\sum_{i=1}^{N} (y_i - \alpha - \beta x_i)^2}{N - 2} \right]^{1/2} \qquad (1.2.4)$$

The above equations can be implemented in BASIC as shown in listing 1.2.1. The associated subroutine-connection diagram appears in figure 1.2.1. The inputs to the subroutine are the number of data points (N) and the data pairs [$X(M)$, $Y(M)$]. The results returned are the coefficients A and B ($Y = A + BX$), and an unbiased estimate of the standard deviation, D.

It is instructive to apply LSTSQR1 to the linear approximation of $\sin x$ in the first octant (see listing 1.2.2). Eleven equally spaced points were used to obtain the following least-squares estimate:

$$\sin x \cong 0.104 + 0.662x \qquad \text{for } 0 \le x \le \pi/2$$

The standard deviation of this fit is $\sigma = 0.08$. This fitted equation and the corresponding truncated Taylor series are compared with $\sin x$ in figure 1.2.2. It is apparent that the least-squares fit is much better on the whole than the Taylor series approximation. As expected from the earlier discussion, the truncated Taylor series polynomial is accurate near the expansion point ($x = 0$), but the error grows rapidly as x increases. The least-squares fit also behaves as expected; the maximum relative errors are at the endpoints of the interval.

The least-squares line-fitting subroutine can be applied to create an even better fit by observing that x is a common factor in the Taylor series expansion. Thus, a more accurate fit is expected by using the function $x(a + bx)$. This is accomplished by replacing $\sin x$ with $\sin x/x$, and approximating the latter function. See listing 1.2.3. The resulting approximation is

$$\sin x \cong x(1.078 - 0.3966x)$$

The standard deviation of this estimate is $\sigma = 0.05$, which is a definite improvement over the previous example. In addition, the *relative error* near $x = 0$ has been greatly reduced.

18 BASIC SCIENTIFIC SUBROUTINES

It is interesting to note that using the first two terms of the Taylor series to approximate sin x, $y = x - x^3/3!$, gives a *higher* standard deviation ($\sigma = 0.31$) than that obtained using just the first term ($\sigma = 0.27$). The addition of the second term causes an overcorrection as measured by the least-squares error metric.

LSTSQR1 is very easy to apply and is not very sensitive to round-off error. It also executes quickly. It requires as input *at least three data points*, two of which must be distinct. The three-point requirement is due to the $M - 2$ divisor in the standard-deviation calculation. The distinct-points requirement is due to a simple geometrical consideration: at least two distinct points are needed to define a line. Because these conditions are seldom violated, the subroutine does not check to see if they are met.

In the next section, we will examine second-order least-squares fitting using the parabola.

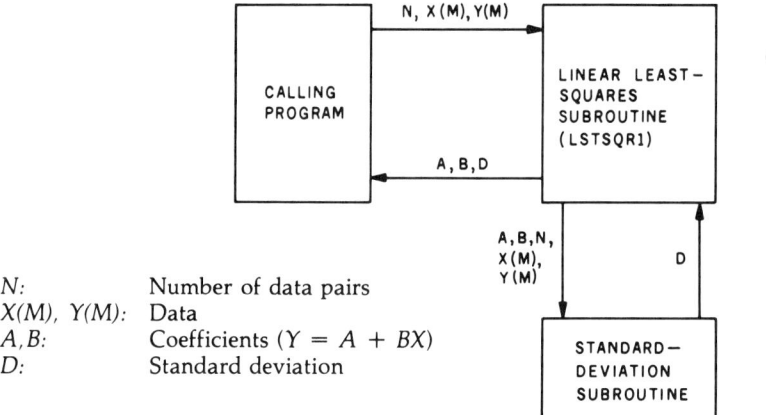

N: Number of data pairs
X(M), Y(M):   Data
A, B: Coefficients ($Y = A + BX$)
D: Standard deviation

Figure 1.2.1 *Subroutine-connection diagram for the least-squares line-fitting subroutine (LSTSQR1) in listing 1.2.1.*

Statements/Functions List

$+, -, *, /$
FOR/NEXT, SQRT

Variables List

A, A1, A2, B, B1, D, D1, M, N, X(M), Y(M)

Variables Passed to Subroutine

N, X(M), Y(M)

Table 1.2.1 *Functions and variables used in the first-order least-squares subroutine (LSTSQR1), given in listing 1.2.1.*

Listing 1.2.1

```
2000 REM LEAST SQUARES DEMONSTRATION PROGRAM
2001 PRINT
2002 PRINT
2003 PRINT "LEAST SQUARES CURVE FIT ROUTINE"
2004 PRINT
2005 PRINT
2006 PRINT "THIS PROGRAM CALCULATES A LINEAR"
2007 PRINT "LEAST SQUARES FIT TO A GIVEN DATA SET. "
2008 PRINT
2009 PRINT "INSTRUCTIONS"
2010 PRINT "----------------"
2011 PRINT
2012 PRINT "THE NUMBER OF DATA COORDINATES PROVIDED "
2013 PRINT "MUST BE GREATER THAN ONE. OTHERWISE, A "
2014 PRINT "DIVIDE BY ZERO ERROR MAY RESULT."
2015 PRINT
2016 PRINT "INPUT THE NUMBER OF DATA POINTS: ",
2017 INPUT N
2018 IF N<2 THEN GOTO 2012
2019 DIM X(N),Y(N)
2020 PRINT
2021 PRINT "THERE ARE TWO INPUT OPTIONS. ONE (1) "
2022 PRINT "INPUTS THE DATA POINTS IN COORDINATE "
2023 PRINT "PAIRS, AND THE OTHER (2) ALLOWS ONE TO "
2024 PRINT "FIRST INPUT THE INDEPENDENT VARIABLE "
2025 PRINT "VALUES, LATER FOLLOWED BY THE DEPENDENT."
2026 PRINT "WHICH MODE DO YOU DESIRE? (1 OR 2): ",
2027 INPUT Z
2028 PRINT
2029 IF Z=2 THEN GOTO 2032
2030 IF Z=1 THEN GOTO 2042
2031 GOTO 2026
2032 FOR M=0 TO N-1
2033 PRINT M+1,
2034 INPUT X(M)
2035 NEXT M
2036 PRINT
2037 FOR M=0 TO N-1
2038 PRINT M+1,
2039 INPUT Y(M)
2040 NEXT M
2041 GOTO 2047
2042 FOR M=0 TO N-1
2043 PRINT M+1,
2044 INPUT X(M),Y(M)
2045 NEXT M
```

```
2046 REM GO TO LINEAR LEAST SQUARES SUBROUTINE
2047 GOSUB 43500
2048 PRINT
2049 PRINT
2050 PRINT "FITTED EQUATION IS: "
2051 PRINT
2052 PRINT "     Y = ",INT(1000000*A)/1000000," ",
2053 IF B>=0 THEN PRINT "+",
2054 PRINT INT(1000000*B)/1000000,"*X"
2055 PRINT
2056 PRINT
2057 PRINT "STANDARD DEVIATION OF FIT: ",
2058 PRINT INT(10000*D)/10000
2059 PRINT
2060 PRINT
2061 END
43491 REM ********************
43492 REM LINEAR LEAST SQUARES SUBROUTINE (LSTSQR1)
43493 REM THE INPUT DATA SET IS (X(M),Y(M)).
43494 REM THE NUMBER OF DATA POINTS IS N.
43495 REM THE NUMBER OF DIFFERENT POINTS MUST BE GREATER THAN ONE.
43496 REM X(M) AND Y(M) MUST BE DIMENSIONED IN THE CALLING PROGRAM.
43497 REM THE SUBROUTINE ALSO CALCULATES THE UNBIASED ESTIMATE
43498 REM OF THE STANDARD DEVIATION, D.
43499 REM THE RETURNED PARAMETERS ARE A,B AND D.
43500 A1=0
43501 A2=0
43502 B0=0
43503 B1=0
43504 FOR M=0 TO N-1
43505 A1=A1+X(M)
43506 A2=A2+X(M)*X(M)
43507 B0=B0+Y(M)
43508 B1=B1+Y(M)*X(M)
43509 NEXT M
43510 A1=A1/N
43511 A2=A2/N
43512 B0=B0/N
43513 B1=B1/N
43514 D=A1*A1-A2
43515 A=A1*B1-A2*B0
43516 A=A/D
43517 B=A1*B0-B1
43518 B=B/D
43519 REM ********************
43520 REM EVALUATION OF STANDARD DEVIATION (UNBIASED ESTIMATE)
43521 D=0
43522 FOR M=0 TO N-1
```

Listing 1.2.1 cont.

```
43523 D1=Y(M)-A-B*X(M)
43524 D=D+D1*D1
43525 NEXT M
43526 D=SQRT(D/(N-2))
43527 RETURN
```

Listing 1.2.1 *Least-squares line-fitting subroutine (LSTSQR1). This program calculates the best-fit line for a given data set, and returns the two coefficients and the unbiased estimate of the standard deviation. Also shown is a demonstration program containing two data input formats. The subroutine-connection diagram appears in figure 1.2.1.*

RUN

Listing 1.2.2

LEAST SQUARES CURVE FIT ROUTINE

THIS PROGRAM CALCULATES A LINEAR
LEAST SQUARES FIT TO A GIVEN DATA SET.

INSTRUCTIONS

THE NUMBER OF DATA COORDINATES PROVIDED
MUST BE GREATER THAN ONE. OTHERWISE, A
DIVIDE BY ZERO ERROR MAY RESULT.

INPUT THE NUMBER OF DATA POINTS: ?11

THERE ARE TWO INPUT OPTIONS. ONE (1)
INPUTS THE DATA POINTS IN COORDINATE
PAIRS, AND THE OTHER (2) ALLOWS ONE TO
FIRST INPUT THE INDEPENDENT VARIABLE
VALUES, LATER FOLLOWED BY THE DEPENDENT.
WHICH MODE DO YOU DESIRE? (1 OR 2): ?1

 1?0,0
 2?.157,.1563558123
 3?.314,.3088655201
 4?.471,.4537776271
 5?.628,.5875275257
 6?.785,.7068251811
 7?.942,.8087360606
 8?1.099,.8907533184
 9?1.256,.9508594605

```
10?1.413,.9875759713
11?1.550,.9997837642

FITTED EQUATION IS:

    Y =  .104584 + .661709*X


STANDARD DEVIATION OF FIT:  .0791


READY
```

Listing 1.2.2 *A sample run of the program shown in listing 1.2.1. An eleven-point (ten-interval) linear approximation to y = sin x over the range of 0 to π/2 was determined. The standard deviation of the fit is 0.08. The standard deviation of the corresponding truncated Taylor series is 0.27. By this metric, the least-squares fit was much better than the truncated Taylor series (see also figure 1.2.2).*

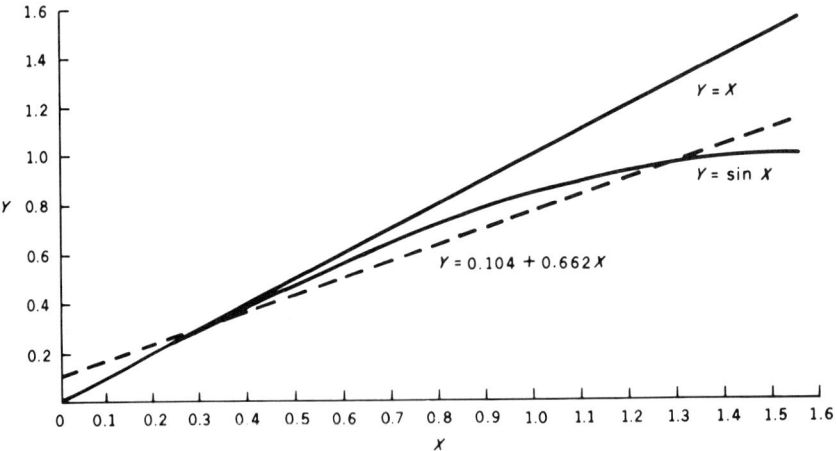

Figure 1.2.2 *Comparison of the linear Taylor series and the eleven-point least-squares approximations to y = sin x over the range $0 \leq x \leq \pi/2$.*

```
RUN

LEAST SQUARES CURVE FIT ROUTINE
```

Listing 1.2.3

Listing 1.2.3 cont.

```
THIS PROGRAM CALCULATES A LINEAR
LEAST SQUARES FIT TO A GIVEN DATA SET.

INSTRUCTIONS
----------------

THE NUMBER OF DATA COORDINATES PROVIDED
MUST BE GREATER THAN ONE. OTHERWISE, A
DIVIDE BY ZERO ERROR MAY RESULT.

INPUT THE NUMBER OF DATA POINTS: ?11

THERE ARE TWO INPUT OPTIONS. ONE (1)
INPUTS THE DATA POINTS IN COORDINATE
PAIRS, AND THE OTHER (2) ALLOWS ONE TO
FIRST INPUT THE INDEPENDENT VARIABLE
VALUES, LATER FOLLOWED BY THE DEPENDENT.
WHICH MODE DO YOU DESIRE? (1 OR 2): ?2

 1?0
 2?.157
 3?.314
 4?.471
 5?.628
 6?.785
 7?.942
 8?1.099
 9?1.256
10?1.413
11?1.57

 1?1
 2?.9918106
 3?.96756368
 4?.92820596
 5?.87526015
 6?.8107458
 7?.73707522
 8?.65693061
 9?.57313037
10?.48849118
11?.40569572


FITTED EQUATION IS:

    Y =   1.07812   -.396575*X
```

STANDARD DEVIATION OF FIT: .0434

READY

Listing 1.2.3 *Repeat of the example given in listing 1.2.2, but for the function $y = \sin x/x$. The resulting approximation is $\sin x = x(1.07812 - 0.396575x)$, and has a standard deviation of $\sigma = 0.05$, which is an improvement over the $\sigma = 0.08$ value observed in the previous run. The corresponding standard deviation for the second-degree truncated Taylor series is $\sigma = 0.31$, which is higher than the $\sigma = 0.27$ value observed in the previous example. Adding another term increased the standard deviation because of overcompensation.*

1.3 Second-Order Least Squares

The concepts of the previous section can easily be extended to parabolic fits. If we assume that the true dependence is $y = a + bx + cx^2$, then the problem is one of determining estimates for the coefficients of that quadratic. Continuing with the notation of the previous section, we will call these estimates α, β, and γ. The function to minimize then is

$$S(\alpha,\beta,\gamma) = \sum_{i=1}^{N} (y_i - \alpha - \beta x_i - \gamma x_i^2)^2 \qquad (1.3.1)$$

The partial derivative approach can again be employed to generate three simultaneous equations. The solution to these equations is a little cumbersome unless some additional definitions are established:

$$\begin{aligned}
S_{xx} &= \frac{1}{N} \sum_{i=1}^{N} (x_i - \overline{x})^2 \\
S_{xy} &= \frac{1}{N} \sum_{i=1}^{N} (x_i - \overline{x})(y_i - \overline{y}) \\
S_{yy} &= \frac{1}{N} \sum_{i=1}^{N} (y_i - \overline{y})^2 \\
S_{xx^2} &= \frac{1}{N} \sum_{i=1}^{N} (x_i - \overline{x})(x_i^2 - \overline{x^2}) \\
S_{x^2x^2} &= \frac{1}{N} \sum_{i=1}^{N} (x_i^2 - \overline{x^2})^2 \\
S_{yx^2} &= \frac{1}{N} \sum_{i=1}^{N} (y_i - \overline{y})(x_i^2 - \overline{x^2})
\end{aligned} \qquad (1.3.2)$$

(For further reference, see any standard statistics text, e.g., Ref. 11.) Using these definitions, it can be shown that the least-squares solutions for the coefficients are

$$\beta = \frac{S_{xy}S_{x^2x^2} - S_{x^2y}S_{xx^2}}{S_{xx}S_{x^2x^2} - (S_{xx^2})^2}$$

$$\gamma = \frac{S_{xx}S_{yx^2} - S_{xx^2}S_{xy}}{S_{xx}S_{x^2x^2} - (S_{xx^2})^2} \qquad (1.3.3)$$

$$\alpha = \overline{y} - \beta \overline{x} - \gamma \overline{x^2}$$

A BASIC subroutine for performing these laborious calculations is shown in listing 1.3.1. The inputs are the number of data points (N) and the data pairs [X(M), Y(M)]. The returned values are the quadratic coefficients (A, B, and C) and the standard deviation (D). Examples of the operation of LSTSQR2 are given in listings 1.3.2 and 1.3.3. These examples correspond to those in section 1.2 in which a linear fit to sin x was obtained. Increasing the order of the fit to a quadratic greatly improves the standard deviation.

The only warning regarding the use of LSTSQR2 is relative to the minimum number of data points. There must be at least four data pairs; at least three of them must be distinct.

The generalization of this least-squares approach is the subject of the next two sections. In those sections, polynomial least-squares fits of arbitrary order are considered by repeating the previous analysis using matrices. The real utility of this generalization is the extension to more than one dimension. For one-dimensional applications, this methodology is *not* recommended because of round-off error. Rather, the orthogonal polynomial algorithm discussed in section 1.6 is suggested.

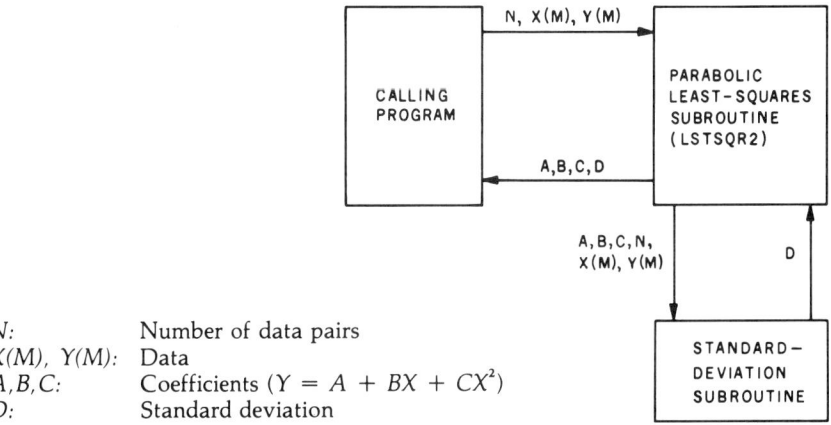

N: Number of data pairs
X(M), Y(M):   Data
A,B,C: Coefficients ($Y = A + BX + CX^2$)
D: Standard deviation

Figure 1.3.1 *Subroutine-connection diagram for LSTSQR2, given in listing 1.3.1.*

Statements/Functions List

+, −, *, /
FOR/NEXT, SQRT

Variables List

A, A0, A1, A2, A3, A4, B, B0, B1, B2, C, D, D1, M, N, X(M), Y(M)

Variables Passed to Subroutine

N, X(M), Y(M)

Table 1.3.1 *Functions and variables used by the second-order least-squares subroutine (LSTSQR2), given in listing 1.3.1.*

Listing 1.3.1

```
2100 REM LEAST SQUARES DEMONSTRATION PROGRAM
2101 PRINT
2102 PRINT
2103 PRINT "LEAST SQUARES CURVE FIT ROUTINE"
2104 PRINT
2105 PRINT
2106 PRINT "THIS PROGRAM CALCULATES A PARABOLIC "
2107 PRINT "LEAST SQUARES FIT TO A GIVEN DATA SET. "
2108 PRINT
2109 PRINT "INSTRUCTIONS"
2110 PRINT "--------------"
2111 PRINT
2112 PRINT "THE NUMBER OF DATA COORDINATES PROVIDED "
2113 PRINT "MUST BE GREATER THAN TWO. OTHERWISE, A "
2114 PRINT "DIVIDE BY ZERO ERROR MAY RESULT."
2115 PRINT
2116 PRINT "INPUT THE NUMBER OF DATA POINTS: ",
2117 INPUT N
2118 IF N<3 THEN GOTO 2112
2119 DIM X(N),Y(N)
2120 PRINT
2121 PRINT "THERE ARE TWO INPUT OPTIONS. ONE (1) "
2122 PRINT "INPUTS THE DATA POINTS IN COORDINATE "
2123 PRINT "PAIRS, AND THE OTHER (2) ALLOWS ONE TO "
2124 PRINT "FIRST INPUT THE INDEPENDENT VARIABLE "
2125 PRINT "VALUES, LATER FOLLOWED BY THE DEPENDENT "
2126 PRINT "WHICH MODE DO YOU DESIRE? (1 OR 2): ",
2127 INPUT Z
2128 PRINT
2129 IF Z=2 THEN GOTO 2132
2130 IF Z=1 THEN GOTO 2142
```

```
2131 GOTO 2126
2132 FOR M=0 TO N-1
2133 PRINT M+1,
2134 INPUT X(M)
2135 NEXT M
2136 PRINT
2137 FOR M=0 TO N-1
2138 PRINT M+1,
2139 INPUT Y(M)
2140 NEXT M
2141 GOTO 2147
2142 FOR M=0 TO N-1
2143 PRINT M+1,
2144 INPUT X(M),Y(M)
2145 NEXT M
2146 REM GO TO PARABOLIC LEAST SQUARES SUBROUTINE
2147 GOSUB 43550
2148 PRINT
2149 PRINT
2150 PRINT "FITTED EQUATION IS: "
2151 PRINT
2152 PRINT "    Y = ",INT(1000000*A)/1000000," ",
2153 IF B>=0 THEN PRINT "+",
2154 PRINT INT(1000000*B)/1000000,"*X ",
2155 IF C>=0 THEN PRINT "+ ",
2156 PRINT INT(1000000*C)/1000000,"*X*X"
2157 PRINT
2158 PRINT
2159 PRINT "STANDARD DEVIATION OF FIT: ",
2160 PRINT INT(10000*D)/10000
2161 PRINT
2162 PRINT
2163 END
43541 REM ********************
43542 REM PARABOLIC LEAST SQUARES SUBROUTINE (LSTSQR2)
43543 REM THE INPUT DATA SET IS (X(M),Y(M)).
43544 REM THE NUMBER OF DATA POINTS IS N.
43545 REM THE NUMBER OF DIFFERENT POINTS MUST BE GREATER THAN THREE.
43546 REM X(M) AND Y(M) MUST BE DIMENSIONED IN THE CALLING PROGRAM.
43547 REM THE SUBROUTINE ALSO CALCULATES THE UNBIASED ESTIMATE
43548 REM OF THE STANDARD DEVIATION, D.
43549 REM THE RETURNED PARAMETERS ARE A,B,C AND D.
43550 A0=1
43551 A1=0
43552 A2=0
43553 A3=0
43554 A4=0
43555 B0=0
43556 B1=0
```

Listing 1.3.1 cont.

```
43557 B2=0
43558 FOR M=0 TO N-1
43559 A1=A1+X(M)
43560 A2=A2+X(M)*X(M)
43561 A3=A3+X(M)*X(M)*X(M)
43562 A4=A4+X(M)*X(M)*X(M)*X(M)
43563 B0=B0+Y(M)
43564 B1=B1+Y(M)*X(M)
43565 B2=B2+Y(M)*X(M)*X(M)
43566 NEXT M
43567 A1=A1/N
43568 A2=A2/N
43569 A3=A3/N
43570 A4=A4/N
43571 B0=B0/N
43572 B1=B1/N
43573 B2=B2/N
43574 D=A0*(A2*A4-A3*A3)-A1*(A1*A4-A3*A2)+A2*(A1*A3-A2*A2)
43575 A=B0*(A2*A4-A3*A3)+B1*(A3*A2-A1*A4)+B2*(A1*A3-A2*A2)
43576 A=A/D
43577 B=B0*(A3*A2-A1*A4)+B1*(A0*A4-A2*A2)+B2*(A2*A1-A0*A3)
43578 B=B/D
43579 C=B0*(A1*A3-A2*A2)+B1*(A1*A2-A0*A3)+B2*(A0*A2-A1*A1)
43580 C=C/D
43581 REM ********************
43582 REM EVALUATION OF STANDARD DEVIATION (UNBIASED ESTIMATE)
43583 D=0
43584 FOR M=0 TO N-1
43585 D1=Y(M)-A-B*X(M)-C*X(M)*X(M)
43586 D=D+D1*D1
43587 NEXT M
43588 D=SQRT(D/(N-3))
43589 RETURN
```

Listing 1.3.1 *Second-order least-squares curve-fitting subroutine (LSTSQR2). This program calculates the coefficients of the parabola that best describes a given data set. Also shown is a program for demonstrating LSTSQR2. See listings 1.3.2 and 1.3.3 for examples.*

```
                   RUN
```
                                                                                            Listing 1.3.2

```
           LEAST SQUARES CURVE FIT ROUTINE


           THIS PROGRAM CALCULATES A PARABOLIC
           LEAST SQUARES FIT TO A GIVEN DATA SET.
```

```
INSTRUCTIONS
-------------------

THE NUMBER OF DATA COORDINATES PROVIDED
MUST BE GREATER THAN TWO. OTHERWISE, A
DIVIDE BY ZERO ERROR MAY RESULT.

INPUT THE NUMBER OF DATA POINTS: ?11

THERE ARE TWO INPUT OPTIONS. ONE (1)
INPUTS THE DATA POINTS IN COORDINATE
PAIRS, AND THE OTHER (2) ALLOWS ONE TO
FIRST INPUT THE INDEPENDENT VARIABLE
VALUES, LATER FOLLOWED BY THE DEPENDENT
WHICH MODE DO YOU DESIRE? (1 OR 2): ?1

1?0,0
2?.157,.1563558123
3?.314,.3088655201
4?.471,.4537776271
5?.628,.5875275257
6?.785,.7068251811
7?.942,.8087360606
8?1.099,.8907533184
9?1.256,.9508594605
10?1.413,.9875759713
11?1.550,.9997837642

FITTED EQUATION IS:

    Y =  -.016427 + 1.180557*X  -.332946*X*X

STANDARD DEVIATION OF FIT:  .0118

READY
```

Listing 1.3.2 *An example of the operation of LSTSQR2 (see listing 1.3.1). Eleven data points representing sin x were employed to find a parabolic approximation to that function. The standard deviation of the resulting fit is $\sigma = 0.01$. Compare with the results shown in listing 1.2.2.*

```
RUN
```

Listing 1.3.3

```
LEAST SQUARES CURVE FIT ROUTINE
```

```
THIS PROGRAM CALCULATES A PARABOLIC
LEAST SQUARES FIT TO A GIVEN DATA SET.

INSTRUCTIONS
-------------------

THE NUMBER OF DATA COORDINATES PROVIDED
MUST BE GREATER THAN TWO. OTHERWISE, A
DIVIDE BY ZERO ERROR MAY RESULT.

INPUT THE NUMBER OF DATA POINTS: ?11

THERE ARE TWO INPUT OPTIONS. ONE (1)
INPUTS THE DATA POINTS IN COORDINATE
PAIRS, AND THE OTHER (2) ALLOWS ONE TO
FIRST INPUT THE INDEPENDENT VARIABLE
VALUES, LATER FOLLOWED BY THE DEPENDENT
WHICH MODE DO YOU DESIRE? (1 OR 2): ?2

 1?0
 2?.157
 3?.314
 4?.471
 5?.628
 6?.785
 7?.942
 8?1.099
 9?1.256
 10?1.413
 11?1.57

 1?1
 2?.9918106
 3?.96756368
 4?.92820596
 5?.87526015
 6?.8107458
 7?.73707522
 8?.65693061
 9?.57313037
 10?.48849118
 11?.40569572


FITTED EQUATION IS:

    Y =   1.013388   -.121702*X   -.175079*X*X
```

```
STANDARD DEVIATION OF FIT:  .011

READY
```

Listing 1.3.3 *The previous example is repeated, but for the function sin x/x. The resulting approximation to sin x is*

$$\sin x \cong x(1.013 - 0.122x - 0.175x^2)$$

The standard deviation of this fit is $\sigma = 0.01$, *which is the same as the error observed in listing 1.3.2. In this exercise, however, the fit is better with respect to relative error.*

1.4 Nth-Order Least Squares

The linear and parabolic least-squares fitting algorithms discussed in the previous sections represent special (and simple) cases of the general least-squares polynomial-approximation procedure. The special case subroutines were presented separately for two reasons. First, those algorithms are easy for the novice to understand. Second, and more importantly, the general case subroutine is slower in execution because much more overhead exists in the calculation. The power of the general approach, however, is that it opens the door to multidimensional curve fitting, as well as to mixed functional dependencies. This will become clear as we go on.

We will develop the mathematics of the general algorithm using matrix algebra. For those who are familiar with (or who simply dislike) this subject, the ensuing discussion can be bypassed.

Consider the following polynomial approximation to $y(x)$:

$$y(x) \cong D_0 x^0 + D_1 x^1 + \cdots + D_N x^N \qquad (1.4.1)$$

If the measurement (or table) pairs are (x_i, y_i), and if there are M of them, then the above approximation can be written in the following matrix form:

$$\mathbf{Y} \cong \mathbf{XD} \qquad (1.4.2)$$

In this notation, **Y** is a column vector of length M; **X** is an M row by $N + 1$ column matrix, and **D** is a column vector of length $N + 1$:

$$\mathbf{Y} = \begin{bmatrix} y_1 \\ y_2 \\ \cdot \\ \cdot \\ \cdot \\ y_M \end{bmatrix} = \text{Value vector}$$

$$X = \begin{bmatrix} x_1^0, & x_1^1, & \ldots, & x_1^N \\ x_2^0 & & & \\ \cdot & & & \\ \cdot & & & \\ \cdot & & & \\ x_M^0, & x_M^1, & \ldots, & x_M^N \end{bmatrix} = \text{Input data array}$$

$$D = \begin{bmatrix} D_0 \\ D_1 \\ \cdot \\ \cdot \\ \cdot \\ D_N \end{bmatrix} = \text{Coefficient vector}$$

Recall that the error function [e.g., $S(\alpha,\beta)$ in section 1.2] had the appearance of the square of the length of a vector in Euclidean space. We can set up the same structure in matrix notation:

$$\text{Error} = (Y - XD)^T(Y - XD) \qquad (1.4.3)$$

The variable in this equation is the coefficient vector, **D**. As we did earlier, we can calculate the differential

$$d(\text{Error}) = (Y - XD)^T X \, dD$$

We want this differential to be zero. Because dD is arbitrary, this equation reduces to

$$(Y - XD)^T X = 0$$

The coefficient vector can then be solved for as

$$D = (X^T X)^{-1} X^T Y \qquad (1.4.4)$$

Equation (1.4.4) has a very simple structure and permits easy solution for the coefficients through very fundamental matrix operations. Unfortunately, most BASIC interpreters and compilers presently available for small computers do not directly support the required matrix calculations. However, a subroutine package exists for such applications (see Ref. 1), and it can be used for calculating equation (1.4.4).

The entire procedure can be implemented as shown in listing 1.4.1. That program consists of several modular parts, as can be seen in figures 1.4.1 and 1.4.2. The calling program is responsible for obtaining the M data pairs in $[X(I), Y(I)]$ form, and the degree of the fit, N. However, a two-dimensional input data array, $Z(I,J)$, is needed. This array corresponds to the matrix X in equation (1.4.4). The POLYCM (POLYnomial Coefficient Matrix) subroutine performs that task and creates the $Z(I,J)$ array using the $X(I)$ data values. This array, along with M, N, and $Y(I)$, is passed to LEASTSQR, a supervisory subroutine that implements equation (1.4.4) by calling several matrix-operation subroutines. Note that $N + 1$, not N, is passed to LEASTSQR. LEASTSQR returns the coefficients to the calling program in the vector $D(I)$. The calling program then obtains the standard deviation from SIGMA. The reason for choosing this modular structure will become apparent in the next section when we consider multidimensional regression.

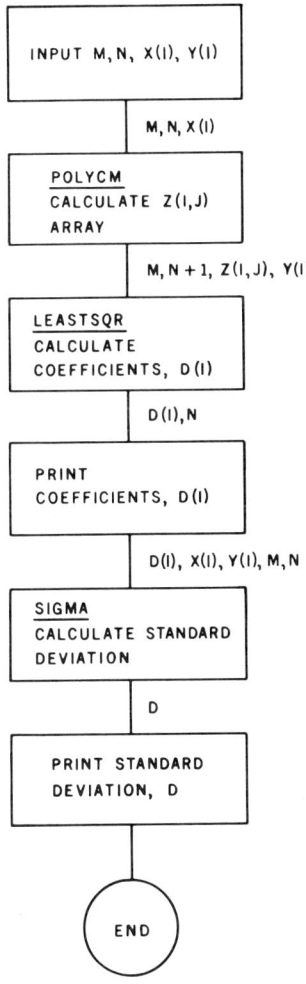

Figure 1.4.1 *Overall flowchart for the program shown in listing 1.4.1. Also displayed are the variables used by the next step in the flowchart.*

The main operational precautions to be observed in using these subroutines center around the validity of the input data. The calling program must ensure that at least $N + 1$ distinct data pairs are passed to POLYCM and LEASTSQR. Also, there are several arrays to be dimensioned.

The practical precaution is simple—watch out for round-off error. Listings 1.4.2 through 1.4.4 show the effect of round-off error on the calculated coefficients when too high a degree of fit is attempted.

In those three examples, the input data set was perfectly linear. By choosing fits higher than linear, larger matrices were processed (in particular, *inverted*) than necessary, leading to increased round-off error. It is therefore suggested that low-order fits be attempted first, and that the standard deviation be examined as higher orders are tried.

Equation (1.4.4) represents a very elegant and powerful solution to the general least-squares problem. However, because matrices are involved, round-off error has a strong influence on the results. In later sections, means by which this source of error can be greatly reduced will be discussed. Before doing so, however, we will examine the extension of the ideas in this section to more than one dimension.

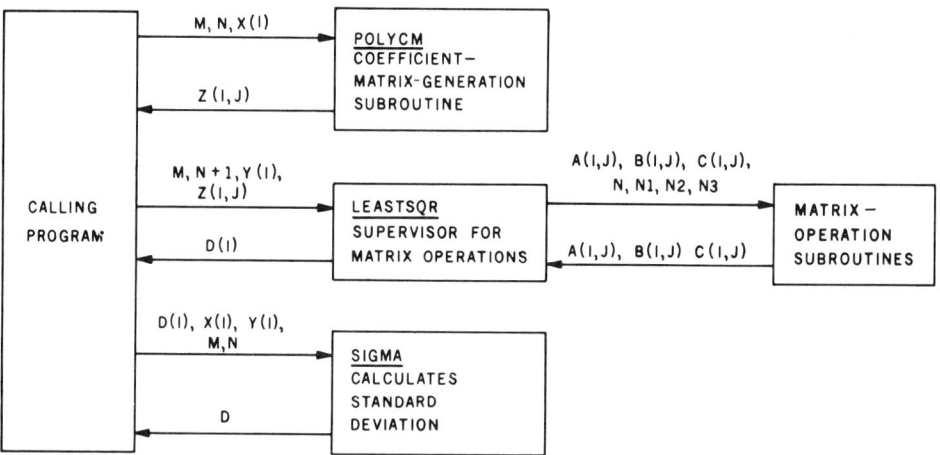

M: Number of data points
X(I), Y(I): Data
N: Degree of fit
Z(I,J):   Input data array
D(I):   Coefficients returned
D: Standard deviation

Figure 1.4.2 *Subroutine-connection diagram associated with applying LEASTSQR (see listing 1.4.1). The calling program is responsible for obtaining the input data array and passing it to LEASTSQR. LEASTSQR itself is largely a supervisory program which calculates equation (1.4.4) using other subroutines.*

Statements/Functions List

+, −, *, /, >
FOR/NEXT, GOSUB, GOTO, IF/THEN, SQRT

Variables List

A(I,J), B, B(I,J), C(I,J), D, D(I), I, I1, I2, I3, J, K, M, M1, M2, M4, N, N1, N2, N3, N4, X(I), Y, Y(I), Z(I,J)

Variables Passed to Subroutine

M, N, X(I) [to POLYCM], Y(I), Z(I,J) [to LEASTSQR]

Table 1.4.1 *Functions and variables used by the general least-squares curve-fitting subroutine (LEASTSQR). Included are all the functions and variables employed by the subroutines that support LEASTSQR (listing 1.4.1). Note that the X(I) values are transformed by POLYCM into the Z(I,J) array, which is subsequently passed to LEASTSQR.*

Listing 1.4.1

```
2200 REM PROGRAM TO DEMONSTRATE ONE DIMENSIONAL
2201 REM OPERATION OF THE MULTI-NONLINEAR REGRESSION
2202 REM SUBROUTINE
2203 PRINT "HOW MANY DATA POINTS ARE THERE: ",
2204 INPUT M
2205 PRINT "WHAT IS THE DEGREE OF THE POLYNOMIAL"
2206 PRINT "TO BE FITTED: ",
2207 INPUT N
2208 DIM X(M),Y(M),Z(M,N+1),D(N+1),A(M,M),B(M,2*M),C(M,M)
2209 PRINT
2210 PRINT "INPUT THE DATA IN (X,Y) PAIRS AS PROMPTED:"
2211 PRINT
2212 FOR I=1 TO M
2213 PRINT I,TAB(5),"X , Y = ",
2214 INPUT X(I),Y(I)
2215 NEXT I
2216 PRINT
2217 PRINT
2218 REM GO TO COEFFICIENTS GENERATION SUBROUTINE
2219 GOSUB 43600
2220 REM GO TO REGRESSION SUBROUTINE
2221 N=N+1
2222 GOSUB 43650
2223 PRINT "THE CALCULATED COEFFICIENTS ARE:"
2224 PRINT
2225 FOR I=1 TO N
2226 PRINT I,TAB(5),INT(1000000*D(I))/1000000
```

```
2227 NEXT I
2228 REM GET STANDARD DEVIATION
2229 N=N-1
2230 GOSUB 43750
2231 PRINT
2232 PRINT
2233 PRINT "STANDARD DEVIATION: ",INT(1000000*D)/1000000
2234 PRINT
2235 PRINT
2236 END
41897 REM ********************
41898 REM MATRIX MULTIPLICATION SUBROUTINE (MATMULT)
41899 REM C=A X B    A IS M1 BY N1    B IS M2 BY N2    C IS M1 BY N2
41900 FOR I=1 TO M1
41901 FOR J=1 TO N2
41902 C(I,J)=0
41903 FOR K=1 TO N1
41904 C(I,J)=C(I,J)+A(I,K)*B(K,J)
41905 NEXT K
41906 NEXT J
41907 NEXT I
41908 RETURN
41947 REM ********************
41948 REM MATRIX TRANSPOSE SUBROUTINE (MATTRANS)
41949 REM B=TRANSPOSE(A)
41950 FOR I=1 TO N
41951 FOR J=1 TO M
41952 B(I,J)=A(J,I)
41953 NEXT J
41954 NEXT I
41955 RETURN
42072 REM ********************
42073 REM MATRIX SAVE (B IN A) SUBROUTINE (MATSAVBA)
42074 REM N1,N2 AND N3 ARE INPUT INDICES
42075 IF N1*N2*N3=0 THEN GOTO 42085
42076 REM CHECK DIMENSION
42077 FOR I1=1 TO N1
42078 FOR I2=1 TO N2
42079 FOR I3=1 TO N3
42080 A(I1,I2,I3)=B(I1,I2,I3)
42081 NEXT I3
42082 NEXT I2
42083 NEXT I1
42084 RETURN
42085 IF N1*N2=0 THEN GOTO 42092
42086 FOR I1=1 TO N1
42087 FOR I2=1 TO N2
42088 A(I1,I2)=B(I1,I2)
42089 NEXT I2
```

Listing 1.4.1 cont.

```
42090 NEXT I1
42091 RETURN
42092 IF N1=0 THEN RETURN
42093 FOR I1=1 TO N1
42094 A(I1)=B(I1)
42095 NEXT I1
42096 RETURN
42097 REM ********************
42098 REM MATRIX SAVE (C IN B) SUBROUTINE (MATSAVCB)
42099 REM N1,N2 AND N3 ARE INPUT INDICES
42100 IF N1*N2*N3=0 THEN GOTO 42110
42101 REM CHECK DIMENSION
42102 FOR I1=1 TO N1
42103 FOR I2=1 TO N2
42104 FOR I3=1 TO N3
42105 B(I1,I2,I3)=C(I1,I2,I3)
42106 NEXT I3
42107 NEXT I2
42108 NEXT I1
42109 RETURN
42110 IF N1*N2=0 THEN GOTO 42117
42111 FOR I1=1 TO N1
42112 FOR I2=1 TO N2
42113 B(I1,I2)=C(I1,I2)
42114 NEXT I2
42115 NEXT I1
42116 RETURN
42117 IF N1=0 THEN RETURN
42118 FOR I1=1 TO N1
42119 B(I1)=C(I1)
42120 NEXT I1
42121 RETURN
42147 REM ********************
42148 REM MATRIX SAVE (A IN C) SUBROUTINE (MATSAVAC)
42149 REM N1,N2 AND N3 ARE INPUT INDICES
42150 IF N1*N2*N3=0 THEN GOTO 42160
42151 REM CHECK DIMENSION
42152 FOR I1=1 TO N1
42153 FOR I2=1 TO N2
42154 FOR I3=1 TO N3
42155 C(I1,I2,I3)=A(I1,I2,I3)
42156 NEXT I3
42157 NEXT I2
42158 NEXT I1
42159 RETURN
42160 IF N1*N2=0 THEN GOTO 42167
42161 FOR I1=1 TO N1
42162 FOR I2=1 TO N2
42163 C(I1,I2)=A(I1,I2)
```

Listing 1.4.1 cont.

```
42164 NEXT I2
42165 NEXT I1
42166 RETURN
42167 IF N1=0 THEN RETURN
42168 FOR I1=1 TO N1
42169 C(I1)=A(I1)
42170 NEXT I1
42171 RETURN
42172 REM ********************
42173 REM MATRIX SAVE (C IN A) SUBROUTINE (MATSAVCA)
42174 REM N1,N2 AND N3 ARE INPUT INDICES
42175 IF N1*N2*N3=0 THEN GOTO 42185
42176 REM CHECK DIMENSION
42177 FOR I1=1 TO N1
42178 FOR I2=1 TO N2
42179 FOR I3=1 TO N3
42180 A(I1,I2,I3)=C(I1,I2,I3)
42181 NEXT I3
42182 NEXT I2
42183 NEXT I1
42184 RETURN
42185 IF N1*N2=0 THEN GOTO 42192
42186 FOR I1=1 TO N1
42187 FOR I2=1 TO N2
42188 A(I1,I2)=C(I1,I2)
42189 NEXT I2
42190 NEXT I1
42191 RETURN
42192 IF N1=0 THEN RETURN
42193 FOR I1=1 TO N1
42194 A(I1)=C(I1)
42195 NEXT I1
42196 RETURN
42394 REM ********************
42395 REM MATRIX INVERSION SUBROUTINE (MATINV)
42396 REM GAUSS-JORDAN ELIMINATION
42397 REM MATRIX A IS INPUT, MATRIX B IS OUTPUT
42398 REM DIM A=N X N     TEMPORARY DIM B=N X 2N
42399 REM FIRST CREATE MATRIX WITH A ON THE LEFT AND I ON THE RIGHT
42400 FOR I=1 TO N
42401 FOR J=1 TO N
42402 B(I,J+N)=0
42403 B(I,J)=A(I,J)
42404 NEXT J
42405 B(I,I+N)=1
42406 NEXT I
42407 REM PERFORM ROW ORIENTED OPERATIONS TO CONVERT THE LEFT HAND
42408 REM SIDE OF B TO THE IDENTITY MATRIX. THE INVERSE OF A WILL
42409 REM THEN BE ON THE RIGHT.
```

Listing 1.4.1 cont.

```
42410 FOR K=1 TO N
42411 IF K=N THEN GOTO 42424
42412 M=K
42413 REM FIND MAXIMUM ELEMENT
42414 FOR I=K+1 TO N
42415 IF ABS(B(I,K))>ABS(B(M,K)) THEN M=I
42416 NEXT I
42417 IF M=K THEN GOTO 42424
42418 FOR J=K TO 2*N
42419 B=B(K,J)
42420 B(K,J)=B(M,J)
42421 B(M,J)=B
42422 NEXT J
42423 REM DIVIDE ROW K
42424 FOR J=K+1 TO 2*N
42425 B(K,J)=B(K,J)/B(K,K)
42426 NEXT J
42427 IF K=1 THEN GOTO 42434
42428 FOR I=1 TO K-1
42429 FOR J=K+1 TO 2*N
42430 B(I,J)=B(I,J)-B(I,K)*B(K,J)
42431 NEXT J
42432 NEXT I
42433 IF K=N THEN GOTO 42441
42434 FOR I=K+1 TO N
42435 FOR J=K+1 TO 2*N
42436 B(I,J)=B(I,J)-B(I,K)*B(K,J)
42437 NEXT J
42438 NEXT I
42439 NEXT K
42440 REM RETRIEVE INVERSE FROM THE RIGHT SIDE OF B
42441 FOR I=1 TO N
42442 FOR J=1 TO N
42443 B(I,J)=B(I,J+N)
42444 NEXT J
42445 NEXT I
42446 RETURN
43590 REM ********************
43591 REM COEFFICIENT MATRIX GENERATION SUBROUTINE FOR THE
43592 REM ONE DIMENSIONAL POLYNOMIAL REGRESSION (POLYCM).
43593 REM THE INPUT DATA SET CONSISTS OF N PAIRS OF
43594 REM (X(I),Y(I)) VALUES.
43595 REM THE REGRESSION ORDER IS N.
43596 REM THE MATRIX RETURNED, Z, IS M ROWS BY N+1 COLUMNS.
43597 REM DIMENSION THIS MATRIX IN THE CALLING PROGRAM.
43598 REM RECALL THAT THE REGRESSION ROUTINE INPUT WILL USE N+1,.
43599 REM YOU MUST SET N TO N+1 BEFORE ENTERING IT.
43600 FOR I=1 TO M
43601 B=1
```

Listing 1.4.1 cont.

```
43602 FOR J=1 TO N+1
43603 Z(I,J)=B
43604 B=B*X(I)
43605 NEXT J
43606 NEXT I
43607 REM Z IS M ROWS BY N+1 COLUMNS
43608 RETURN
43628 REM ********************
43629 REM LEAST SQUARES FITTING SUBROUTINE (LEASTSQR)
43630 REM GENERAL SUBROUTINE FOR MULTIDIMENSIONAL,
43631 REM NONLINEAR REGRESSION.
43632 REM THE EQUATION FITTED HAS THE FORM
43633 REM Y=D(1)X1 + D(2)X2 +  ... + D(N)XN
43634 REM CHANGE IN NOTATION IS DUE TO A DIMENSION CONFLICT.
43635 REM THE COEFFICIENTS ARE RETURNED BY THE PROGRAM IN D(I).
43636 REM THE XI CAN BE SIMPLE POWERS OF X, OR FUNCTIONS.
43637 REM NOTE THAT THE XI ARE ASSUMED TO BE INDEPENDENT.
43638 REM THE MEASURED RESPONSES ARE Y(I)- THERE ARE M OF THEM.
43639 REM Y IS M ROW COLUMN VECTOR, AND Z(I,J) IS A
43640 REM M ROW BY N COLUMN MATRIX.
43641 REM M>=N
43642 REM THE SUBROUTINE INPUTS ARE Y(I), Z(I,J), M AND N.
43643 REM THE WORKING MATRICES WITHIN THE PROGRAM ARE A(I,J),
43644 REM B(I,J) AND C(I,J).
43645 REM THE SUBROUTINE CALLS SEVERAL OTHER MATRIX OPERATION
43646 REM ROUTINES TO PERFORM THE CALCULATION.
43647 REM DIMENSION A,B,C,Y AND Z IN THE CALLING PROGRAM.
43648 REM START PROCEDURE.
43649 REM STORE M AND N
43650 M4=M
43651 N4=N
43652 REM MOVE Z(I,J) TO A(I,J)
43653 FOR I=1 TO M
43654 FOR J=1 TO N
43655 A(I,J)=Z(I,J)
43656 NEXT J
43657 NEXT I
43658 REM A IS M BY N
43659 REM FIND TRANSPOSE OF A AND PUT RESULT IN B
43660 GOSUB 41950
43661 REM B IS N BY M
43662 REM MOVE A TO C, B TO A, AND C TO B
43663 REM WILL HAVE Z(TRANSPOSE) IN A, Z IN B
43664 N1=M
43665 N2=N
43666 N3=0
43667 GOSUB 42150
43668 N1=N
43669 N2=M
```

Listing 1.4.1 cont.

Listing 1.4.1 cont.

```
43670 GOSUB 42075
43671 N1=M
43672 N2=N
43673 GOSUB 42100
43674 REM MULTIPLY A AND B
43675 REM A IS N BY M
43676 REM B IS M BY N
43677 M1=N
43678 N1=M
43679 M2=M
43680 GOSUB 41900
43681 REM RESULT IS IN C, AN N BY N MATRIX.. MOVE TO A AND FIND INVERSE
43682 N1=N
43683 GOSUB 42175
43684 REM A IS N BY N
43685 GOSUB 42400
43686 REM RESTORE M
43687 M=M4
43688 REM INVERSE IS IN B. MOVE TO A, PUT Z(TRANSPOSE) IN B, AND MULTIPLY
43689 REM B IS N BY N
43690 GOSUB 42075
43691 FOR I=1 TO M
43692 FOR J=1 TO N
43693 B(J,I)=Z(I,J)
43694 NEXT J
43695 NEXT I
43696 REM B IS N BY M    A IS N BY N
43697 M2=N
43698 N2=M
43699 GOSUB 41900
43700 REM PRODUCT IS N BY M
43701 REM PRODUCT IS IN C. MOVE TO A, LOAD Y IN B, AND MULTIPLY
43702 N1=N
43703 N2=M
43704 GOSUB 42175
43705 REM B IS A COLUMN VECTOR (M BY 1)
43706 FOR I=1 TO M
43707 B(I,1)=Y(I)
43708 NEXT I
43709 N2=1\N1=M
43710 M2=M
43711 REM A IS N BY M    B IS M BY 1
43712 GOSUB 41900
43713 REM PRODUCT IS N BY 1
43714 REM REGRESSION COEFFICIENTS ARE IN C(I,1)
43715 REM MOVE THEM TO D(I)
43716 FOR I=1 TO N
43717 D(I)=C(I,1)
43718 NEXT I
```

```
43719 RETURN
43742 REM ********************
43743 REM STANDARD DEVIATION SUBROUTINE (SIGMA).
43744 REM THIS SUBROUTINE CALCULATES THE STANDARD
43745 REM DEVIATION FOR A POLYNOMIAL FIT.
43746 REM THE INPUTS ARE THE NUMBER OF DATA POINTS, M,
43747 REM THE DEGREE OF THE FIT, N,
43748 REM THE POLYNOMIAL COEFFICIENTS, D(I),
43749 REM THE ORIGINAL DATA SET, X(I), Y(I).
43750 D=0
43751 FOR I=1 TO M
43752 Y=0
43753 B=1
43754 FOR J=1 TO N+1
43755 Y=Y+D(J)*B
43756 B=B*X(I)
43757 NEXT J
43758 D=D+(Y-Y(I))*(Y-Y(I))
43759 NEXT I
43760 IF M-N-1>0 THEN GOTO 43763
43761 D=0
43762 RETURN
43763 D=D/(M-N-1)
43764 D=SQRT(D)
43765 RETURN
```

Listing 1.4.1 *Subroutine for the matrix solution of general least-squares curve fits (LEASTSQR). Also shown is a program for demonstrating the operation of LEASTSQR, along with several support subroutines. The POLYCM subroutine is used to generate the input matrix, Z(I,J), which is used by LEASTSQR. SIGMA is a general subroutine for calculating the standard deviation of one-dimensional polynomial fits. A series of matrix-operation subroutines supports LEASTSQR and is discussed in Volume I of* BASIC *Scientific Subroutines. The subroutine-connection diagram is shown in figure 1.4.1. Examples are provided in listings 1.4.2 through 1.4.4.*

```
RUN

HOW MANY DATA POINTS ARE THERE: ?11
WHAT IS THE DEGREE OF THE POLYNOMIAL
TO BE FITTED: ?3

INPUT THE DATA IN (X,Y) PAIRS AS PROMPTED:

  1    X , Y = ?1,1
  2    X , Y = ?2,2
  3    X , Y = ?3,3
  4    X , Y = ?4,4
```

Listing 1.4.2

```
5    X , Y = ?5,5
6    X , Y = ?6,6
7    X , Y = ?7,7
8    X , Y = ?8,8
9    X , Y = ?9,9
10   X , Y = ?0,0
11   X , Y = ?1,1
```

THE CALCULATED COEFFICIENTS ARE:

```
1    -.000019
2    1.000039
3    -.000008
4    0
```

STANDARD DEVIATION: .000127

READY

Listing 1.4.2 *An example of the use of the program shown in listing 1.4.1. Eleven values corresponding to the equation $y = x$ were provided to the subroutine. A cubic fit was chosen, and the returned equation was $y = -0.000019 + 1.000039x - 0.000008x^2$, with an associated standard deviation of $\sigma = 0.0001$. Note that there is significant round-off error, but the results are still quite useable.*

```
RUN
```

Listing 1.4.3

```
HOW MANY DATA POINTS ARE THERE: ?11
WHAT IS THE DEGREE OF THE POLYNOMIAL
TO BE FITTED: ?4

INPUT THE DATA IN (X,Y) PAIRS AS PROMPTED:

1    X , Y = ?1,1
2    X , Y = ?2,2
3    X , Y = ?3,3
4    X , Y = ?4,4
5    X , Y = ?5,5
6    X , Y = ?6,6
7    X , Y = ?7,7
8    X , Y = ?8,8
9    X , Y = ?9,9
10   X , Y = ?0,0
11   X , Y = ?1,1
```

THE CALCULATED COEFFICIENTS ARE:

    1   .000151
    2   .999953
    3   -.00007
    4   -.000022
    5   0

STANDARD DEVIATION: .012102

READY

Listing 1.4.3 *The example shown in listing 1.4.2 was repeated, but a fourth-degree fit was sought. Note the increased round-off error.*

RUN

HOW MANY DATA POINTS ARE THERE: ?11
WHAT IS THE DEGREE OF THE POLYNOMIAL
TO BE FITTED: ?1

INPUT THE DATA IN (X,Y) PAIRS AS PROMPTED:

    1   X , Y = ?1,1
    2   X , Y = ?2,2
    3   X , Y = ?3,3
    4   X , Y = ?4,4
    5   X , Y = ?5,5
    6   X , Y = ?6,6
    7   X , Y = ?7,7
    8   X , Y = ?8,8
    9   X , Y = ?9,9
   10   X , Y = ?0,0
   11   X , Y = ?1,1

THE CALCULATED COEFFICIENTS ARE:

    1   -.000001
    2   1

STANDARD DEVIATION: 0

READY

Listing 1.4.4 *If a linear fit is applied to linear data, the round-off error is very small. The conclusion is that round-off error can cause higher-degree fits to be less accurate than expected, in principle; an optimum degree of fit exists. See the discussion in section 1.1.*

1.5 Multidimensional Least Squares

In the previous section we examined the use of matrix equations to find a one-dimensional polynomial fit to a given set of data. We will proceed now to generalize the algorithm for multidimensional use.

The key to the problem is the two-dimensional input data array X [actually, $Z(I,J)$ in the subroutines]. It decides the outcome of equation (1.4.4) along with the measured result vector, Y. The form of this matrix was established by equation (1.4.1):

$$y(x) \cong D_0 x^0 + D_1 x^1 + \cdots + D_N x^N$$

The alteration to treat multiple dimensions can be made at this step.

One approach might be to redefine equation (1.4.1) as follows (for two dimensions, x_1 and x_2):

$$y(x_1, x_2) \cong D_0 x_1^0 + D_1 x_1^1 + \cdots + D_{M_1} x_1^{M_1} + D_{M_1+1} x_2^0 + D_{M_1+2} x_2^1 + \cdots \\ + D_{M_1+M_2+1} x_2^{M_2}$$

The implied assumptions in this approximation are that the degree of the polynomial is M_1 in x_1 and M_2 in x_2, and that there are no cross-dependencies. This is the equivalent of assuming $y(x) \cong P_1(x_1) + P_2(x_2)$, and is very unrealistic. Another approach might be to assume the form $y(x_1, x_2) \cong P_1(x_1) P_2(x_2)$. This permits a measure of the interdependence, but is much too limiting an assumption.

The complete form is much more complicated, but it can be laid out in a logical structure. For example, in two dimensions the structure is

$$\begin{aligned} y(x_1, x_2) \cong\ & (D_1 x_1^0 + D_2 x_1^1 + \cdots + D_{M_1} x_1^{M_1}) x_2^0 \\ & + (D_{M_1+2} x_1^0 + \cdots + D_{2M_1+2} x_1^{M_1}) x_2^1 \\ & + \cdots \\ & + (D_{M_2(M_1+1)+1} x_1^0 + \cdots + D_{(M_2+1)(M_1+1)} x_1^{M_1}) x_2^{M_2} \end{aligned} \quad (1.5.1)$$

The number of coefficients required in this representation is $N = (M_1 + 1)(M_2 + 1)$, where M_1 and M_2 are the degrees of fit in each dimension. If three dimensions were employed, there would be an additional level of brackets required, and the number of coefficients would be $N = (M_1 + 1)(M_2 + 1)(M_3 + 1)$.

Undoubtedly the most difficult problem in using the least-squares curve-fitting algorithm in several dimensions is deciphering the resulting coefficients. Therefore, if you are actually going to use the subroutine that will be presented shortly, the encoding scheme must be clearly understood.

As an example, we will consider the form of the X array [actually, $Z(I,J)$ in the subroutines] for the three-dimensional case in which the degree of fit is linear in the first two dimensions, but quadratic in the third. The number of coefficients to be found is $N = 2 \cdot 2 \cdot 3 = 12$. Each row of the X array corresponds to a particular data point, say $y(I)$. Therefore, we have

Row I of x: $\quad 1, x_1(I), x_2(I), x_1(I)x_2(I), x_3(I), x_1(I)x_3(I), x_2(I)x_3(I), x_1(I)x_2(I)x_3(I),$
$\qquad x_3^2(I), x_1(I)x_3^2(I), x_2(I)x_3^2(I), x_1(I)x_2(I)x_3^2(I)$

Because there are 12 coefficients to be determined, there must be at least 12 distinct data points. Therefore, the minimum size of the X array is 12×12.

As an exercise, try to determine what the X array would look like if the number of dimensions were three, but the fits were quadratic, linear, and linear respectively.

Although generating such an array and processing it according to matrix equation (1.4.4) would be almost impossible by hand, computers are well suited to such tasks.

We can use the computer algorithm given in the previous section for one-dimensional least-squares fitting simply by replacing a few modules. First, the subroutine that generates the X array [actually, $Z(I,J)$ in the program] is replaced with that shown in listing 1.5.1 (MLTNLREG). This subroutine takes the data and generates the $Z(I,J)$ matrix used by LEASTSQR. Also, this replacement subroutine calculates the standard deviation on the second calling, and thereby also replaces SIGMA for multidimensional problems (see figures 1.5.1 and 1.5.2). All the other subroutines used in the previous program (LEASTSQR and all the matrix-operation programs) are kept intact.

The multidimensional least-squares procedure can be demonstrated using the program given in listing 1.5.2. Sample results are shown in listings 1.5.3 and 1.5.4. In each case, the data was generated from the equation $y = x_1 + x_2$. The algorithm clearly identified the linear dependencies, and with reasonably good round-off error. Note, however, that the round-off error was higher for the case in which a parabolic fit was attempted in one of the dimensions. Large arrays lead to increased round-off error.

There are several precautions associated with the use of this collection of programs. First, the degree of fit in each dimension must be linear or higher. This is certainly not a restriction on the utility of the calculation, but it does put the responsibility on the calling program that realistic inputs be supplied. Second, there must be at least $N + 1$ data points, with N of them distinct, where $N = (M_1 + 1)(M_2 + 1) \cdots (M_L + 1)$. Third, the number of dimensions that can be processed is limited to nine. Beyond that point, the coefficient-generation subroutine will abort and set $L = 0$. However, it it very unlikely that this limit would ever be approached since the minimum size of the $Z(I,J)$ array corresponding to nine linear dimensions would be $(2^9 + 1) \times (2^9)$, or 513 rows by 512 columns! Finally, remember to dimension the arrays in the calling program.

In the next section, we will consider an alternative approach to polynomial regression in one dimension. It will have the advantage of low round-off error, high execution speed, and an automatic termination option.

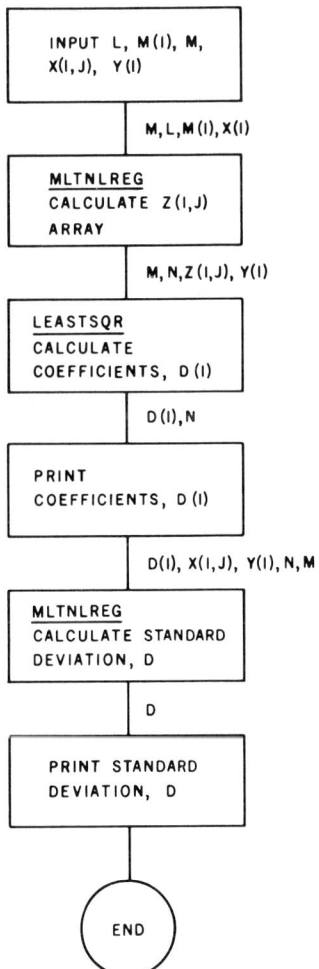

Figure 1.5.1 *Overall flowchart for the multidimensional least-squares algorithm. Also displayed are the variables used by the next step in the flowchart (see also figure 1.5.2).*

48 BASIC SCIENTIFIC SUBROUTINES

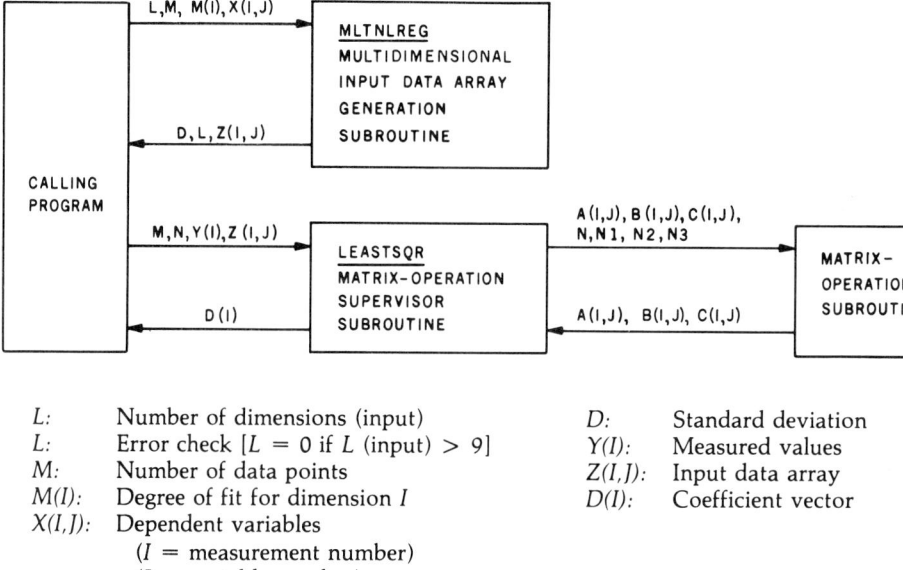

L:	Number of dimensions (input)	D:	Standard deviation
L:	Error check [L = 0 if L (input) > 9]	Y(I):	Measured values
M:	Number of data points	Z(I,J):	Input data array
M(I):	Degree of fit for dimension I	D(I):	Coefficient vector
X(I,J):	Dependent variables		
	(I = measurement number)		
	(J = variable number)		

Figure 1.5.2 *Subroutine-connection diagram for multidimensional curvilinear regression (see also figure 1.5.1).*

Statements/Functions List

+, −, *, /, >
FOR/NEXT, GOSUB, GOTO, IF/THEN

Variables List

A(I,J), B, B(I,J), C, C(I,J), D, D(I), I, I1, I2, I3, J, K, L, M, M(I), M1, M2, M4, N, N1, N2, N3, N4, X(I,J), Y, Y(I), Z(I,J)

Variables Passed to Subroutine

M, M(I), X(I,J) [to MLTNLREG], Y(I), Z(I,J) [to LEASTSQR]

Table 1.5.1 *Functions and variables used by the general least-squares curve-fitting subroutine (LEASTSQR) for multidimensional polynomial regression. Included are all the functions and variables employed by the subroutines that support LEASTSQR. Note that the X(I,J) data values are transformed (by MLTNLREG) into the Z(I,J) array, which is later passed to LEASTSQR.*

```
43783 REM ********************
43784 REM COEFFICIENT MATRIX GENERATION SUBROUTINE FOR MULTIPLE
43785 REM NONLINEAR REGRESSION (MLTNLREG).
43786 REM ALSO CALCULATES THE STANDARD DEVIATION, D, EVEN
43787 REM THOUGH THERE IS SOME REDUNDANT COMPUTING.
43788 REM THE MAXIMUM NUMBER OF DIMENSIONS IS 9.
43789 REM THE INPUT DATA SET CONSISTS OF M DATA SETS OF THE FORM
43790 REM      Y(I), X(I,1), X(I,2),......., X(I,L)
43791 REM THE NUMBER OF DIMENSIONS IS L.
43792 REM THE ORDER OF THE FIT TO EACH DIMENSION IS M(J).
43793 REM THE RESULT IS AN (M1+1)*(M2+1)...*(ML+1)+1
43794 REM COLUMN BY M ROW MATRIX, Z.
43795 REM THIS MATRIX IS ARRANGED AS FOLLOWS ( EXAMPLE- L=2, M(1)=2, M(2)=2)
43796 REM    1  X1  X1*X1  X2  X2*X1  X2*X1*X1  X2*X2  X2*X2*X1  X2*X2*X1*X1
43797 REM THIS MATRIX SHOULD BE DIMENSIONED IN THE CALLING PROGRAM
43798 REM AS SHOULD ALSO THE X(I,J) MATRIX OF DATA VALUES.
43799 REM CALCULATE THE TOTAL NUMBER OF DIMENSIONS
43800 N=1
43801 FOR I=1 TO L
43802 N=N*(M(I)+1)
43803 NEXT I
43804 D=0
43805 FOR I=1 TO M
43806 REM BRANCH ACCORDING TO DIMENSION
43807 REM RETURN IF DIMENSION IS GREATER THAN 9
43808 IF L>0 THEN GOTO 43811
43809 L=0
43810 RETURN
43811 IF L<=9 THEN GOTO 43814
43812 L=0
43813 RETURN
43814 J=0
43815 REM MINIMAL BASIC VERSION, REPLACE FOLLOWING WITH ON/GOTO
43816 IF L=1 THEN GOSUB 43840
43817 IF L=2 THEN GOSUB 43849
43818 IF L=3 THEN GOSUB 43857
43819 IF L=4 THEN GOSUB 43865
43820 IF L=5 THEN GOSUB 43873
43821 IF L=6 THEN GOSUB 43881
43822 IF L=7 THEN GOSUB 43889
43823 IF L=8 THEN GOSUB 43897
43824 IF L=9 THEN GOSUB 43905
43825 REM ARRAY GENERATED FOR ROW I
43826 Y=0
43827 FOR K=1 TO N
43828 Y=Y+D(K)*Z(I,K)
43829 NEXT K
43830 D=D+(Y(I)-Y)*(Y(I)-Y)
43831 NEXT I
```

Listing 1.5.1

```
43832 REM CALCULATE STANDARD DEVIATION
43833 IF M-N>0 THEN GOTO 43836
43834 D=0
43835 RETURN
43836 D=D/(M-N)
43837 D=SQRT(D)
43838 RETURN
43839 RETURN
43840 B=1
43841 C=B
43842 FOR I1=0 TO M(1)
43843 J=J+1
43844 Z(I,J)=B
43845 B=B*X(I,1)
43846 NEXT I1
43847 B=C
43848 RETURN
43849 B=1
43850 C=B
43851 FOR I2=0 TO M(2)
43852 GOSUB 43841
43853 B=B*X(I,2)
43854 NEXT I2
43855 B=C
43856 RETURN
43857 B=1
43858 C=B
43859 FOR I3=0 TO M(3)
43860 GOSUB 43850
43861 B=B*X(I,3)
43862 NEXT I3
43863 B=C
43864 RETURN
43865 B=1
43866 C=B
43867 FOR I4=0 TO M(4)
43868 GOSUB 43858
43869 B=B*X(I,4)
43870 NEXT I4
43871 B=C
43872 RETURN
43873 B=1
43874 C=B
43875 FOR I5=0 TO M(5)
43876 GOSUB 43866
43877 B=B*X(I,5)
43878 NEXT I5
43879 B=C
43880 RETURN
```

Listing 1.5.1 cont.

```
43881 B=1
43882 C=B
43883 FOR I6=0 TO M(6)
43884 GOSUB 43874
43885 B=B*X(I,6)
43886 NEXT I6
43887 B=C
43888 RETURN
43889 B=1
43890 C=B
43891 FOR I7=0 TO M(7)
43892 GOSUB 43882
43893 B=B*X(I,7)
43894 NEXT I7
43895 B=C
43896 RETURN
43897 B=1
43898 C=B
43899 FOR I8=0 TO M(8)
43900 GOSUB 43890
43901 B=B*X(I,8)
43902 NEXT I8
43903 B=C
43904 RETURN
43905 B=1
43906 FOR I9=0 TO M(9)
43907 GOSUB 43898
43908 B=B*X(I,9)
43909 NEXT I9
43910 RETURN
```

Listing 1.5.1 *Input data array generation subroutine for multidimensional nonlinear regression (MLTNLREG). This program generates the Z(I,J) array required by LEASTSQR to calculate the coefficient vector D(I). MLTNLREG is applied as shown in figure 1.5.1. Note that this subroutine also calculates the standard deviation if the D(I) are known. The maximum number of dimensions handled by MLTNLREG is nine.*

```
2250 REM PROGRAM TO DEMONSTRATE MULTI-DIMENSIONAL
2251 REM OPERATION OF THE MULTI-NONLINEAR REGRESSION
2252 REM SUBROUTINE (MLTNLREG)
2253 PRINT "HOW MANY DATA POINTS ARE THERE: ",
2254 INPUT M
2255 PRINT
2256 PRINT "HOW MANY DIMENSIONS ARE THERE: ",
2257 INPUT L
2258 PRINT
2259 FOR I=1 TO L
```

Listing 1.5.2

```
2260 PRINT "WHAT IS THE FIT FOR DIMENSION ",I," ",
2261 INPUT M(I)
2262 NEXT I
2263 N=1
2264 FOR I=1 TO L
2265 N=N*(M(I)+1)
2266 NEXT I
2267 DIM X(M,L),Y(M),Z(M,N),D(N),A(M,M),B(M,2*M),C(M,M)
2268 PRINT
2269 PRINT "INPUT THE DATA AS PROMPTED:"
2270 PRINT
2271 FOR I=1 TO M
2272 PRINT "Y(",I,") = ",
2273 INPUT Y(I)
2274 FOR J=1 TO L
2275 PRINT "X(",I,",",J,") = ",
2276 INPUT X(I,J)
2277 NEXT J
2278 PRINT
2279 NEXT I
2280 REM GOTO COEFFICIENTS GENERATION SUBROUTINE
2281 GOSUB 43800
2282 REM GO TO REGRESSION SUBROUTINE
2283 PRINT
2284 GOSUB 43650
2285 PRINT "THE CALCULATED COEFFICIENTS ARE:"
2286 PRINT
2287 FOR I=1 TO N
2288 PRINT I,TAB(5),INT(1000000*D(I))/1000000
2289 NEXT I
2290 REM GET STANDARD DEVIATION
2291 N=N-1
2292 GOSUB 43800
2293 PRINT
2294 PRINT
2295 PRINT "STANDARD DEVIATION: ",INT(1000000*D)/1000000
2296 PRINT
2297 PRINT
2298 END
```

Listing 1.5.2 *Demonstration program for exercising MLTNLREG and LEASTSQR (listing 1.5.1). This program first calls MLTNLREG in order to generate the Z(I,J) array. LEASTSQR is called next, with the results returned in D(I). MLTNLREG is again called for the standard deviation (see figure 1.5.1).*

```
RUN                                                    Listing 1.5.3

HOW MANY DATA POINTS ARE THERE: ?10

HOW MANY DIMENSIONS ARE THERE: ?2

WHAT IS THE FIT FOR DIMENSION   1 ?2
WHAT IS THE FIT FOR DIMENSION   2 ?1

INPUT THE DATA AS PROMPTED:

Y( 1) = ?7
X( 1, 1) = ?1
X( 1, 2) = ?6

Y( 2) = ?7
X( 2, 1) = ?6
X( 2, 2) = ?1

Y( 3) = ?6
X( 3, 1) = ?3
X( 3, 2) = ?3

Y( 4) = ?8
X( 4, 1) = ?2
X( 4, 2) = ?6

Y( 5) = ?9
X( 5, 1) = ?1
X( 5, 2) = ?8

Y( 6) = ?9
X( 6, 1) = ?7
X( 6, 2) = ?2

Y( 7) = ?6
X( 7, 1) = ?3
X( 7, 2) = ?3

Y( 8) = ?7
X( 8, 1) = ?3
X( 8, 2) = ?4

Y( 9) = ?7
X( 9, 1) = ?4
X( 9, 2) = ?3

Y( 10) = ?2
X( 10, 1) = ?0
```

```
X( 10, 2) = ?2
```

THE CALCULATED COEFFICIENTS ARE:

```
1      .000054
2      .999964
3      .000015
4      .999964
5      .000017
6     -.000002
```

STANDARD DEVIATION: .000406

READY

Listing 1.5.3 *An example of the use of MLTNLREG combined with LEASTSQR. The "data" was generated using the simple function* $y = x_1 + x_2$. *The order of the fit was* $M(1) = 2$ *(parabolic) for* x_1, *and* $M(2) = 1$ *(linear) for* x_2. *The returned coefficients correspond to the following equation:*

$$y = 0.000054 + 0.999964x_1 + 0.000015x_1x_2 + 0.999964x_2 + 0.000017x_1^2 - 0.000002x_1x_2^2$$

The fit was fairly good, although it was influenced by round-off error. The observed standard deviation was $\sigma = 0.0004$.

RUN

Listing 1.5.4

```
HOW MANY DATA POINTS ARE THERE: ?10

HOW MANY DIMENSIONS ARE THERE: ?2

WHAT IS THE FIT FOR DIMENSION  1 ?1
WHAT IS THE FIT FOR DIMENSION  2 ?1

INPUT THE DATA AS PROMPTED:

Y( 1) = ?7
X( 1, 1) = ?1
X( 1, 2) = ?6

Y( 2) = ?7
X( 2, 1) = ?6
X( 2, 2) = ?1

Y( 3) = ?6
X( 3, 1) = ?3
X( 3, 2) = ?3
```

```
Y( 4) = ?8
X( 4, 1) = ?2
X( 4, 2) = ?6

Y( 5) = ?9
X( 5, 1) = ?1
X( 5, 2) = ?8

Y( 6) = ?9
X( 6, 1) = ?7
X( 6, 2) = ?2

Y( 7) = ?6
X( 7, 1) = ?3
X( 7, 2) = ?3

Y( 8) = ?7
X( 8, 1) = ?3
X( 8, 2) = ?4

Y( 9) = ?7
X( 9, 1) = ?4
X( 9, 2) = ?3

Y( 10) = ?2
X( 10, 1) = ?0
X( 10, 2) = ?2


THE CALCULATED COEFFICIENTS ARE:

1      -.000005
2       1.000001
3       1.000001
4      -.000001

STANDARD DEVIATION:  .000002

READY
```

Listing 1.5.4 *Repeat of the example shown in listing 1.5.3, but with a fit linear in both dimensions. The resulting regression equation was*

$$y = -0.000005 + 1.000001x_1 + 1.000001x_2 - 0.000001x_1x_2$$

The fit is very good as evidenced by the standard deviation $\sigma = 0.000002$.

1.6 Least-Squares Fitting With Orthogonal Polynomials

The major problem associated with applying the one-dimensional least-squares polynomial algorithm given in section 1.4 is round-off error. This error is due to the large number of mathematical operations associated with implementing equation (1.4.4):

$$\mathbf{D} = (X^T X)^{-1} X^T \mathbf{Y}$$

By using *Forsythe polynomials* instead of simple powers of x, however, you can still apply the same general matrix equation while drastically reducing the number of calculations. (For a discussion of this technique, see *Applied Numerical Methods*, Ref. 10. It is interesting to note that the very significant round-off error implications of this alternative method are fairly obvious, but appear to have escaped forceful attention in the common literature.)

We will denote the Forsythe polynomial of degree J as $P_J(x)$. The associated approximation equation which is equivalent to equation (1.4.1) is

$$y(x) \cong D_0 + D_1 P_1(x) + D_2 P_2(x) + \cdots + D_N P_N(x) \quad (1.6.1)$$

We will define the matrix P as

$$P = \begin{bmatrix} 1 & P_1(x_1) & P_2(x_1) & \cdots & P_N(x_1) \\ 1 & & & & \\ 1 & & & & \\ 1 & & & & \\ 1 & & & & \\ 1 & P_1(x_M) & P_2(x_M) & \cdots & P_N(x_M) \end{bmatrix}$$

This has the same appearance as the X matrix shown in section 1.4, and we can thereby write the approximation equation as

$$\mathbf{Y} \cong P\mathbf{D} \quad (1.6.2)$$

Again, following the analysis of that section, we will define the error function for the fit to be

$$\text{Error} = (\mathbf{Y} - P\mathbf{D})^T (\mathbf{Y} - P\mathbf{D}) \quad (1.6.3)$$

The corresponding solution obtained for the coefficient vector, $\mathbf{D}$, is

$$\mathbf{D} = (P^T P)^{-1} P^T \mathbf{Y} \quad (1.6.4)$$

The savings in computation (and thus in round-off error) is associated with the form of P^TP. If all the off-diagonal terms of that matrix could somehow be forced to be zero, the total number of calculations would be greatly reduced. $(P^TP)^{-1}$ would then not require a full matrix inversion that is very prone to round-off error. In fact, if that could be accomplished, the coefficients would simply be

$$D_J = \frac{\sum_{i=1}^{M} P_J(x_i)y_i}{\sum_{i=1}^{M} P_J(x_i)^2} \qquad (1.6.5)$$

The ability to force this condition rests in the choice of the $P_J(x)$ polynomials. Forsythe found just a set. The recursion relation for calculating them is

$$P_J(x) = (x - \gamma_J)P_{J-1}(x) - \partial_J P_{J-2}(x) \qquad (1.6.6)$$

where $P_{-1}(x) = 0$

$P_1(x) = 1$

$$\gamma_J = \frac{\sum_{i=1}^{M} x_i [P_{J-1}(x_i)]^2}{\sum_{i=1}^{M} [P_{J-1}(x_i)]^2} \qquad (1.6.7)$$

$$\partial_J = \frac{\sum_{i=1}^{M} x_i P_{J-1}(x_i) P_{J-2}(x_i)}{\sum_{i=1}^{M} [P_{J-2}(x_i)]^2} \qquad (1.6.8)$$

Besides having the nice round-off error property, this formulation also has the advantage that the order of the fit can be increased by one degree without extensive recalculation. The calculation for the additional degree can simply be added on.

A program for performing these calculations is shown in listing 1.6.1. The user inputs the degree of fit (N), the number of data points (M), and the data pairs [$X(I)$, $Y(I)$]. Also, an optional error reduction factor (E) can be included. If $E = 0$, the subroutine will provide a fit to the degree N. If $E > 0$, an increasing sequence of fits will be tried until the standard deviation either is reduced by less than that factor or increases, or until the Nth-degree fit is reached. In the case of termination before the Nth degree, the actual degree of fit is returned in L.

Examples of the use of LSQRPOLY are given in listings 1.6.2 through 1.6.4. It is apparent from these examples that this algorithm is fairly resistant to round-off error.

The precautions involved in using this subroutine are simple. The number of distinct data points must exceed the degree of fit ($M \geq N + 1$). E must be non-negative.

LSQRPOLY is a very effective and fast subroutine, and is highly recommended for polynomial regression.

In the next section, we will examine how the round-off error associated with the multidimensional regression algorithm discussed in section 1.5 can be reduced by a simple application of *iteration*.

M:	Limiting degree of fit
E:	Error reduction factor. Fit terminates when the standard deviation is reduced by E
N:	Number of data points
X(I), Y(I):	Input data pairs
L:	Degree of fit
C(I):	Calculated coefficients
D:	Standard deviation of fit

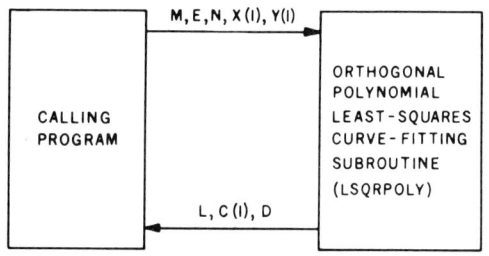

Figure 1.6.1 *Subroutine-connection diagram for the orthogonal polynomial least-squares curve-fitting subroutine (LSQRPOLY).*

Statements/Functions List

$+, -, *, /, <, >$
ABS, FOR/NEXT, GOSUB, GOTO, IF/THEN, SQRT

Variables List

A1, A2, A(I), B1, B2, B(I), C1, C(I), C2(I), D, D1, D(I), E, E(I), F1, F2, F(I), I, L, L2, M, N, N1, V1, V2, V(I), W, X(I), Y(I)

Variables Passed to Subroutine

E, M, N, X(I), Y(I)

Table 1.6.1 *Functions and variables used by LSQRPOLY.*

```
2300 REM PROGRAM TO DEMONSTRATE LSQRPOLY
2301 PRINT
2302 PRINT
2303 PRINT "WHAT IS THE ORDER OF THE FIT: ",
```

Listing 1.6.1

```
2304 INPUT M
2305 PRINT
2306 PRINT "WHAT IS THE ERROR REDUCTION FACTOR: ",
2307 INPUT E
2308 PRINT
2309 PRINT "HOW MANY DATA POINTS ARE THERE: ",
2310 INPUT N
2311 DIM X(N),Y(N),V(N),A(N),B(N),C(N),D(N),C2(N),E(N),F(N)
2312 PRINT
2313 PRINT "INPUT THE DATA POINTS AS PROMPTED: ",
2314 PRINT
2315 FOR I=1 TO N
2316 PRINT I,TAB(5),"X , Y = ",
2317 INPUT X(I),Y(I)
2318 NEXT I
2319 PRINT
2320 PRINT
2321 GOSUB 43950
2322 PRINT "COEFFICIENTS ARE:"
2323 PRINT
2324 FOR I=0 TO L
2325 PRINT I, TAB(5),INT(1000000*C(I))/1000000
2326 NEXT I
2327 PRINT
2328 PRINT
2329 PRINT "STANDARD DEVIATION=",
2330 PRINT INT(100000000*D)/100000000
2331 PRINT
2332 PRINT
2333 END
43932 REM *********************
43933 REM LEAST SQUARES POLYNOMIAL FITTING SUBROUTINE (LSQRPOLY).
43934 REM THIS PROGRAM LEAST SQUARES FITS A POLYNOMIAL TO INPUT DATA.
43935 REM FORSYTHE ORTHOGONAL POLYNOMIALS ARE USED IN THE FITTING.
43936 REM THE NUMBER OF DATA POINTS IS N.
43937 REM THE DATA IS INPUT TO THE SUBROUTINE IN X(I),Y(I) PAIRS.
43938 REM THE COEFFICIENTS ARE RETURNED IN C(I).
43939 REM THE SMOOTHED DATA IS RETURNED IN V(I).
43940 REM THE ORDER OF THE FIT IS SPECIFIED BY M.
43941 REM THE STANDARD DEVIATION OF THE FIT IS RETURNED IN D.
43942 REM THERE ARE TWO OPTIONS AVAILABLE BY USE OF THE PARAMETER E.
43943 REM IF E=0 THE FIT IS TO ORDER M.
43944 REM IF E>0 THE ORDER OF FIT INCREASES TOWARDS M, BUT
43945 REM WILL STOP IF THE RELATIVE STANDARD DEVIATION DOES NOT
43946 REM DECREASE BY MORE THAN E BETWEEN SUCCESSIVE FITS.
43947 REM THE ORDER OF THE FIT THEN OBTAINED IS L.
43948 REM THE ARRAYS X,Y,V,A,B,C,C2,D,E AND F MUST BE DIMENSIONED.
43949 REM A(I) AND B(I) ARE SIMPLY WORK ARRAYS
43950 N1=M+1
```

Listing 1.6.1 cont.

Listing 1.6.1 cont.

```
43951 V1=10000000
43952 REM INITIALIZE THE ARRAYS
43953 FOR I=1 TO N1
43954 A(I)=0
43955 B(I)=0
43956 F(I)=0
43957 NEXT I
43958 FOR I=1 TO N
43959 V(I)=0
43960 D(I)=0
43961 NEXT I
43962 D1=SQRT(N)
43963 W=D1
43964 FOR I=1 TO N
43965 E(I)=1/W
43966 NEXT I
43967 F1=D1
43968 A1=0
43969 FOR I=1 TO N
43970 A1=A1+X(I)*E(I)*E(I)
43971 NEXT I
43972 C1=0
43973 FOR I=1 TO N
43974 C1=C1+Y(I)*E(I)
43975 NEXT I
43976 B(1)=1/F1
43977 F(1)=B(1)*C1
43978 FOR I=1 TO N
43979 V(I)=V(I)+E(I)*C1
43980 NEXT I
43981 M=1
43982 REM SAVE LATEST RESULTS
43983 FOR I=1 TO L
43984 C2(I)=C(I)
43985 NEXT I
43986 L2=L
43987 V2=V
43988 F2=F1
43989 A2=A1
43990 F1=0
43991 FOR I=1 TO N
43992 B1=E(I)
43993 E(I)=(X(I)-A2)*E(I)-F2*D(I)
43994 D(I)=B1
43995 F1=F1+E(I)*E(I)
43996 NEXT I
43997 F1=SQRT(F1)
43998 FOR I=1 TO N
43999 E(I)=E(I)/F1
```

```
44000 NEXT I
44001 A1=0
44002 FOR I=1 TO N
44003 A1=A1+X(I)*E(I)*E(I)
44004 NEXT I
44005 C1=0
44006 FOR I=1 TO N
44007 C1=C1+E(I)*Y(I)
44008 NEXT I
44009 M=M+1
44010 I=0
44011 L=M-I
44012 B2=B(L)
44013 D1=0
44014 IF L>1 THEN D1=B(L-1)
44015 D1=D1-A2*B(L)-F2*A(L)
44016 B(L)=D1/F1
44017 A(L)=B2
44018 I=I+1
44019 IF I<>M THEN GOTO 44011
44020 FOR I=1 TO N
44021 V(I)=V(I)+E(I)*C1
44022 NEXT I
44023 FOR I=1 TO N1
44024 F(I)=F(I)+B(I)*C1
44025 C(I)=F(I)
44026 NEXT I
44027 V=0
44028 FOR I=1 TO N
44029 V=V+(V(I)-Y(I))*(V(I)-Y(I))
44030 NEXT I
44031 REM NOTE THE DIVISION IS BY THE NUMBER OF DEGREES OF FREEDOM
44032 V=SQRT (V/(N-L-1))
44033 L=M
44034 IF E=0 THEN GOTO 44040
44035 REM TEST FOR MIMIMAL IMPROVEMENT
44036 IF ABS(V1-V)/V<E THEN GOTO 44053
44037 REM IF ERROR IS LARGER, QUIT
44038 IF E*V>E*V1 THEN GOTO 44053
44039 V1=V
44040 IF M=N1 THEN GOTO 44043
44041 GOTO 43983
44042 REM SHIFT THE C(I) DOWN SO C(0) IS THE CONSTANT TERM
44043 FOR I=1 TO L
44044 C(I-1)=C(I)
44045 NEXT I
44046 C(L)=0
44047 REM L IS THE ORDER OF THE POLYNOMIAL FITTED
44048 L=L-1
```

Listing 1.6.1 cont.

```
44049 D=V
44050 RETURN
44051 REM SEQUENCE HAS BEEN ABORTED
44052 REM RECOVER LAST VALUES
44053 L=L2
44054 V=V2
44055 FOR I=1 TO L
44056 C(I)=C2(I)
44057 NEXT I
44058 GOTO 44043
```

Listing 1.6.1 *Orthogonal polynomial least-squares curve-fitting subroutine (LSQRPOLY). Also shown is a program for demonstrating LSQRPOLY. Examples appear in listings 1.6.2 through 1.6.4.*

```
RUN
```

Listing 1.6.2

```
WHAT IS THE ORDER OF THE FIT: ?3

WHAT IS THE ERROR REDUCTION FACTOR: ?0

HOW MANY DATA POINTS ARE THERE: ?11

INPUT THE DATA POINTS AS PROMPTED:
  1    X , Y = ?1,1
  2    X , Y = ?2,2
  3    X , Y = ?3,3
  4    X , Y = ?4,4
  5    X , Y = ?5,5
  6    X , Y = ?6,6
  7    X , Y = ?7,7
  8    X , Y = ?8,8
  9    X , Y = ?9,9
 10    X , Y = ?0,0
 11    X , Y = ?1,1

COEFFICIENTS ARE:

 0      0
 1      .999998
 2      0
 3      -.000001
```

```
STANDARD DEVIATION= .00000097

READY
```

Listing 1.6.2 *An example of the use of LSQRPOLY. In this case, eleven data points were generated using the equation $y = x$. The cubic fit found was $y = 0.999998x - 0.000001x^3$, with a standard deviation of $\sigma = 0.000001$. Compare with listing 1.4.2 in which LEASTSQR was directly applied to the same data set. In that case, the cubic fit had a standard deviation of $\sigma = 0.0001$. LSQRPOLY has very good round-off error properties because of the way in which it is structured.*

```
RUN                                                        Listing 1.6.3


WHAT IS THE ORDER OF THE FIT: ?4

WHAT IS THE ERROR REDUCTION FACTOR: ?0

HOW MANY DATA POINTS ARE THERE: ?11

INPUT THE DATA POINTS AS PROMPTED:
 1     X , Y = ?1,1
 2     X , Y = ?2,2
 3     X , Y = ?3,3
 4     X , Y = ?4,4
 5     X , Y = ?5,5
 6     X , Y = ?6,6
 7     X , Y = ?7,7
 8     X , Y = ?8,8
 9     X , Y = ?9,9
10     X , Y = ?0,0
11     X , Y = ?1,1

COEFFICIENTS ARE:

0      0
1      .999999
2      -.000001
3      0
4      -.000001
```

STANDARD DEVIATION= .00000102

READY

Listing 1.6.3 *Repeat of the previous example, but with a fourth-degree fit. The calculated coefficients show very little round-off error, and the standard deviation is exceptionally good ($\sigma = 0.000001$).*

RUN

WHAT IS THE ORDER OF THE FIT? ?4

WHAT IS THE ERROR REDUCTION FACTOR? ?2

HOW MANY DATA POINTS ARE THERE? ?11

INPUT THE DATA POINTS AS PROMPTED:
```
 1    X , Y = ?1,1
 2    X , Y = ?2,2
 3    X , Y = ?3,3
 4    X , Y = ?4,4
 5    X , Y = ?5,5
 6    X , Y = ?6,6
 7    X , Y = ?7,7
 8    X , Y = ?8,8
 9    X , Y = ?9,9
10    X , Y = ?0,0
11    X , Y = ?1,1
```

COEFFICIENTS ARE:

```
0      -.000001
1       1
```

STANDARD DEVIATION= .00000057

READY

Listing 1.6.4 *Repeat of the previous example, but now employing the error-reduction factor. In this case, the degree of fit was terminated at the linear-fit level. As usual with LSQRPOLY, there is very little round-off error.*

1.7 Iterated Regression

The design of numerical algorithms is both a mathematical science and an art. The mathematics forms the background for the methodology, but as in most endeavors, it is the *execution* that decides the quality of the final product.

In section 1.4, we considered an elegant and very general matrix approach to the least-squares fitting of polynomials. Upon execution, however, we found that round-off error could significantly influence the accuracy of the computed results. This difficulty was largely overcome for one-dimensional regression by the Forsythe orthogonal polynomial formulation presented in section 1.6. The trick was to choose a mathematical structure that resulted in matrix operations much less susceptible to round-off error.

The matrix approach was extended to multidimensional applications in section 1.5. Again we encountered round-off error. The objective of this section is to develop a corresponding algorithm that treats this very important case. As you will see, the concept behind this algorithm is fundamental to the use of computers for numerical calculations. This concept is called *iteration*.

We first note that the observed round-off error is mainly associated with the process of subtraction. When two numbers, say A and B, are subtracted, the relative error in their difference is $E/(A - B)$, where E is a measure of the combined numerical (truncation plus round-off) error in A and B. If A and B are comparable in size, the relative error can be quite large. Unfortunately, the matrix operations involved in the LEASTSQR subroutine involve many subtractions of comparable numbers. This is particularly true of the matrix inversion step. Knowing the root of the problem, we can devise a strategy to overcome the difficulty.

The matrix equation given for the coefficient vector was equation 1.4.4:

$$\mathbf{D} = (X^T X)^{-1} X^T Y$$

This has the same form as the classic matrix equation

$$\mathbf{D} = AY$$

Given Y and A, the result $\mathbf{D}$ is *mathematically* determined. However, what we *calculate* is $\mathbf{D}_1$. The residual is

$$\mathbf{r}_1 = \mathbf{D} - \mathbf{D}_1 = \mathbf{D} - AY$$

(See the brief discussion in Ref. 12.) How can we obtain $\mathbf{D}$ when we have $\mathbf{D}_1$? If we knew $\mathbf{r}_1$, even approximately, we could improve upon the estimate $\mathbf{D}_1$ using $\mathbf{D} = \mathbf{D}_1 + \mathbf{r}_1$. Therefore, the problem is reduced to estimating $\mathbf{r}_1$. Recall the original purpose for obtaining $\mathbf{D}$. We want to employ it to approximate (in the least-squares sense) Y:

$$Y = X\mathbf{D}$$

However, applying D_1, we get $Y_1 = XD_1$. Therefore, a reasonable estimate for r_1 is

$$r_1 = A(Y - Y_1)$$

Round-off error will probably also affect the estimate for r_1, so the correction process must be repeated until some error criterion is met:

Step 1	$D_1 = AY$
Step 2	$r_1 = A(Y - XD_1)$
Step 3	$D_2 = D_1 + r_1$
Step 4	$r_2 = A(Y - XD_2)$
Step 5	$D_3 = D_2 + r_2$

and so on. We will use the variance as the criterion for deciding when to stop the iteration sequence. If the variance *increases* upon the next step in the sequence, the round-off error limit has probably been reached. Thus, if $(Y - XD_n)^T(Y - XD_n) < (Y - XD_{n+1})^T(Y - XD_{n+1})$, then D_n is the chosen coefficient vector.

This procedure is restated in figure 1.7.1. It can be implemented in BASIC as shown in listing 1.7.1. See also the subroutine-connection diagram given in figure 1.7.2. The operation of the regression-iteration subroutine (REGITER) is very simple. The calling program supplies the number of data points (M), the number of dimensions (L), the degree of fit for each dimension [M(I)], and the data pairs [X(I,J), Y(I)]. The program then proceeds to iteratively calculate the coefficient vector D(I). The returned results are the coefficients [D(I)], the standard deviation (D), and the number of iterations performed (L1).

An example of the operation of REGITER appears in listing 1.7.2. The coefficients were found with six-digit or better accuracy; the results are very accurate. The corresponding standard deviation of the fit was thousands of times better than the fit found without iteration. Clearly, the method is effective.

REGITER is a fairly reliable program. The input variable precautions are simple and are the same as discussed in section 1.5. There is a possibility (although the author has never encountered such a case) that the iteration may not converge. In that case, the iteration is terminated at the point of divergence. In the worst case, the returned coefficient vector would simply be that calculated on the first pass.

REGITER is a very effective subroutine and should be employed whenever high-accuracy multidimensional least-squares curve fitting is desired. The main disadvantage to using REGITER is that it is slow.

In the next section, we will examine another iterative approach to the least-squares problem. It does not involve matrix operations and it can be used for problems in which the coefficients appear in nonlinear forms.

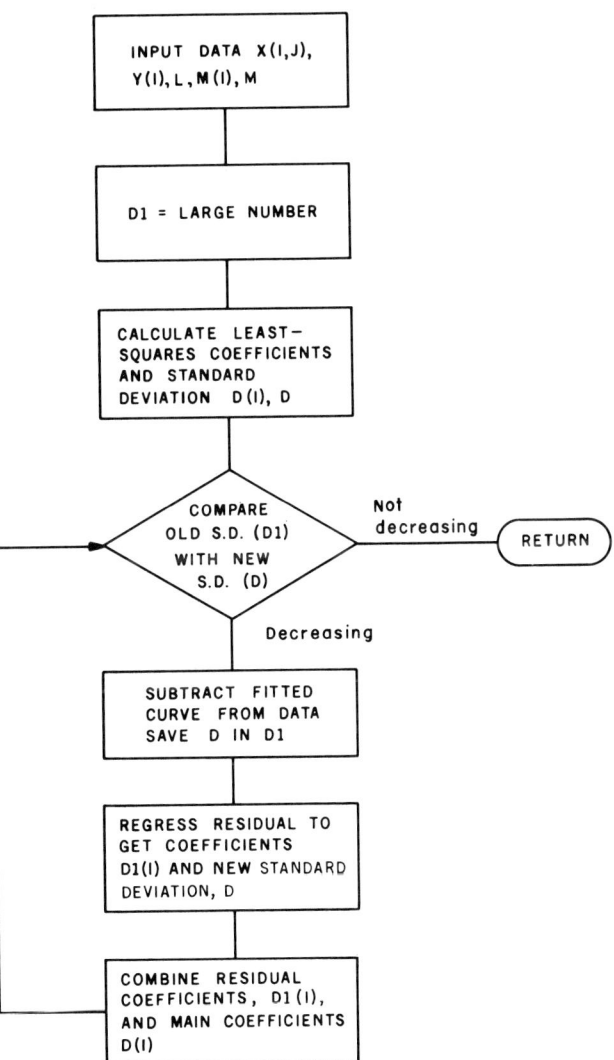

Figure 1.7.1 *Conceptual flowchart for the iterative-regression algorithm.*

68 BASIC SCIENTIFIC SUBROUTINES

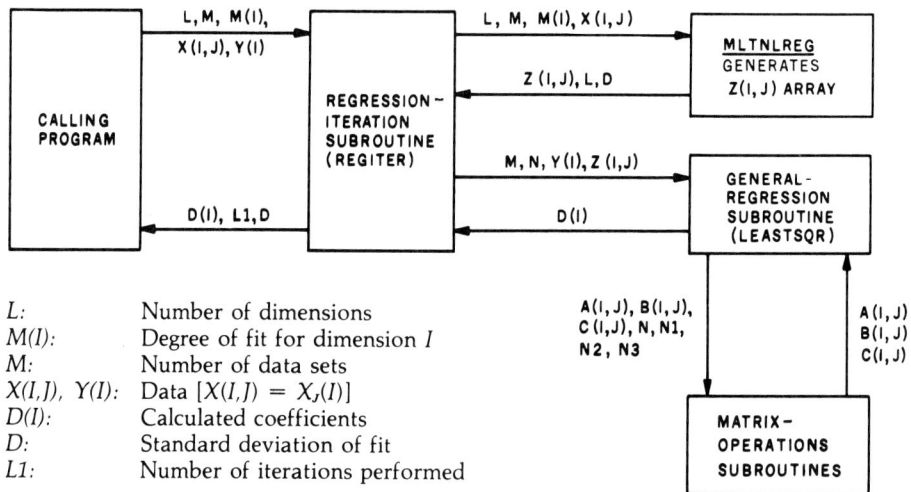

L:	Number of dimensions
M(I):	Degree of fit for dimension I
M:	Number of data sets
X(I,J), Y(I):	Data $[X(I,J) = X_J(I)]$
D(I):	Calculated coefficients
D:	Standard deviation of fit
L1:	Number of iterations performed

Figure 1.7.2 *Subroutine-connection diagram for REGITER. Note that several other subroutines are also called, but these are under the control of REGITER.*

Statements/Functions List

$+, -, *, /, >$
FOR/NEXT, GOSUB, GOTO, IF/THEN

Variables List

A(I,J), B, B(I,J), C, C(I,J), D, D1, D1(I), I, I1, I2, I3, J, K, L, L1, M, M(I), M1, M2, M4, N, N1, N2, N3, N4, X(I,J), Y, Y(I), Y1(I), Z(I,J)

Variables Passed to Subroutine

L, M, M(I), X(I,J) [to MLTNLREG], Y(I), Z(I,J) [to LEASTSQR]

Table 1.7.1 *Functions and variables used by the regression-iteration subroutine (REGITER). Included are the functions and variables employed by all support subroutines.*

Listing 1.7.1

```
2350 REM PROGRAM TO DEMONSTRATE MULTIDIMENSIONAL
2351 REM OPERATION OF THE MULTI-NONLINEAR REGRESSION
2352 REM SUBROUTINE WITH ITERATIVE ERROR REDUCTION
2353 PRINT "HOW MANY DATA POINTS ARE THERE: ",
2354 INPUT M
2355 PRINT
```

```
2356 PRINT "HOW MANY DIMENSIONS ARE THERE: ",
2357 INPUT L
2358 PRINT
2359 FOR I=1 TO L
2360 PRINT "WHAT IS THE FIT FOR DIMENSION ",I," ",
2361 INPUT M(I)
2362 NEXT I
2363 N=1
2364 FOR I=1 TO L
2365 N=N*(M(I)+1)
2366 NEXT I
2367 DIM X(M,L),Y(M),Z(M,N),D(N),A(M,M),B(M,2*M),C(M,M),D1(N),Y1(M)
2368 PRINT
2369 PRINT "INPUT THE DATA AS PROMPTED:"
2370 PRINT
2371 FOR I=1 TO M
2372 PRINT "Y(",I,") = ",
2373 INPUT Y(I)
2374 FOR J=1 TO L
2375 PRINT "X(",I,",",J,") = ",
2376 INPUT X(I,J)
2377 NEXT J
2378 PRINT
2379 NEXT I
2380 REM GO TO ITERATION SUPERVISOR
2381 GOSUB 44100
2382 PRINT
2383 PRINT "THE CALCULATED COEFFICIENTS ARE:"
2384 PRINT
2385 FOR I=1 TO N
2386 PRINT I,TAB(5),INT(1000000*D(I))/1000000
2387 NEXT I
2388 PRINT
2389 PRINT
2390 PRINT "STANDARD DEVIATION: ",INT(1000000*D)/1000000
2391 PRINT
2392 PRINT
2393 PRINT"NUMBER OF ITERATIONS: ",
2394 PRINT L1
2395 PRINT
2396 PRINT
2397 END
44082 REM ********************
44083 REM MULTI-DIMENSIONAL POLYNOMIAL REGRESSION
44084 REM ITERATION SUBROUTINE (REGITER).
44085 REM THIS PROGRAM SUPERVISES THE CALLING OF SEVERAL
44086 REM OTHER SUBROUTINES IN ORDER TO ITERATIVELY
44087 REM FIT LEAST SQUARES POLYNOMIALS IN MORE THAN
44088 REM ONE DIMENSION.
```

Listing 1.7.1 cont.

```
44089 REM THE PROGRAM REPEATEDLY CALCULATES IMPROVED COEFFICIENTS
44090 REM UNTIL THE STANDARD DEVIATION IS NO LONGER REDUCED.
44091 REM THE INPUTS TO THE SUBROUTINE ARE THE NUMBER OF
44092 REM DIMENSIONS, L, THE DEGREE OF FIT FOR EACH
44093 REM DIMENSION, M(I), AND THE INPUT DATA, X(I,J) AND Y(I).
44094 REM THE COEFFICIENTS ARE RETURNED IN D(I), WITH THE
44095 REM STANDARD DEVIATION IN D.
44096 REM ALSO RETURNED IS THE NUMBER OF ITERATIONS TRIED, L1.
44097 REM THE ORIGINAL Y(I) VALUES ARE SAVED IN Y1(I).
44098 REM THE CURRENT COEFFICIENTS ARE STORED IN D1(I).
44099 REM THE PREVIOUS STANDARD DEVIATION IS SAVED IN D1.
44100 L1=0
44101 REM SAVE Y(I)
44102 FOR I=1 TO M
44103 Y1(I)=Y(I)
44104 NEXT I
44105 REM ZERO D1(I)
44106 FOR I=1 TO N
44107 D1(I)=0
44108 NEXT I
44109 REM SET THE INITIAL STANDARD DEVIATION HIGH
44110 D1=10000000
44111 REM GO TO COEFFICIENTS SUBROUTINE
44112 GOSUB 43800
44113 REM GO TO REGRESSION SUBROUTINE
44114 GOSUB 43650
44115 REM GET STANDARD DEVIATION
44116 GOSUB 43800
44117 REM IF STANDARD DEVIATION IS DECREASING, CONTINUE
44118 IF D1>D THEN GOTO 44131
44119 REM TERMINATE ITERATION
44120 FOR I=1 TO N
44121 D(I)=D1(I)
44122 NEXT I
44123 REM RESTORE Y(I)
44124 FOR I=1 TO M
44125 Y(I)=Y1(I)
44126 NEXT I
44127 REM GET THE FINAL STANDARD DEVIATION
44128 GOSUB 43800
44129 RETURN
44130 REM SAVE THE STANDARD DEVIATION
44131 D1=D
44132 L1=L1+1
44133 REM AUGMENT COEFFICIENT MATRIX
44134 FOR I=1 TO N
44135 D(I)=D1(I)+D(I)
44136 D1(I)=D(I)
44137 NEXT I
```

Listing 1.7.1 cont.

```
44138 REM RESTORE Y(I)
44139 FOR I=1 TO M
44140 Y(I)=Y1(I)
44141 NEXT I
44142 REM REDUCE Y(I) ACCORDING TO THE D(I)
44143 GOSUB 44147
44144 REM WE NOW HAVE A SET OF ERROR VALUES
44145 GOTO 44112
44146 REM *********
44147 FOR I=1 TO M
44148 J=0
44149 REM MINIMAL BASIC VERSION. REPLACE FOLLOWING WITH ON/GOTO
44150 IF L=1 THEN GOSUB 44167
44151 IF L=2 THEN GOSUB 44176
44152 IF L=3 THEN GOSUB 44184
44153 IF L=4 THEN GOSUB 44192
44154 IF L=5 THEN GOSUB 44200
44155 IF L=6 THEN GOSUB 44208
44156 IF L=7 THEN GOSUB 44216
44157 IF L=8 THEN GOSUB 44224
44158 IF L=9 THEN GOSUB 44232
44159 REM ARRAY GENERATED FOR ROW I
44160 Y=0
44161 FOR K=1 TO N
44162 Y=Y+D(K)*Z(I,K)
44163 NEXT K
44164 Y(I)=Y(I)-Y
44165 NEXT I
44166 RETURN
44167 B=1
44168 C=B
44169 FOR I1=0 TO M(1)
44170 J=J+1
44171 Z(I,J)=B
44172 B=B*X(I,1)
44173 NEXT I1
44174 B=C
44175 RETURN
44176 B=1
44177 C=B
44178 FOR I2=0 TO M(2)
44179 GOSUB 44168
44180 B=B*X(I,2)
44181 NEXT I2
44182 B=C
44183 RETURN
44184 B=1
44185 C=B
44186 FOR I3=0 TO M(3)
```

```
44187 GOSUB 44177
44188 B=B*X(I,3)
44189 NEXT I3
44190 B=C
44191 RETURN
44192 B=1
44193 C=B
44194 FOR I4=0 TO M(4)
44195 GOSUB 44185
44196 B=B*X(I,4)
44197 NEXT I4
44198 B=C
44199 RETURN
44200 B=1
44201 C=B
44202 FOR I5=0 TO M(5)
44203 GOSUB 44193
44204 B=B*X(I,5)
44205 NEXT I5
44206 B=C
44207 RETURN
44208 B=1
44209 C=B
44210 FOR I6=0 TO M(6)
44211 GOSUB 44201
44212 B=B*X(I,6)
44213 NEXT I6
44214 B=C
44215 RETURN
44216 B=1
44217 C=B
44218 FOR I7=0 TO M(7)
44219 GOSUB 44209
44220 B=B*X(I,7)
44221 NEXT I7
44222 B=C
44223 RETURN
44224 B=1
44225 C=B
44226 FOR I8=0 TO M(8)
44227 GOSUB 44217
44228 B=B*X(I,8)
44229 NEXT I8
44230 B=C
44231 RETURN
44232 B=1
44233 FOR I9=0 TO M(9)
44234 GOSUB 44225
44235 B=B*X(I,9)
```

Listing 1.7.1 cont.

```
44236 NEXT I9
44237 RETURN
```

Listing 1.7.1 *Subroutine for reducing the effect of round-off error through iterative regression on the residual (REGITER). Also shown is a program for demonstrating the operation of REGITER. Note that this routine also calls MLTNLREG, the subroutine for generating the Z(I,J) array for the multidimensional least-squares program, LEASTSQR (which is also called). LEASTSQR in turn uses the matrix-operation subroutines. The entire ensemble results in a ten kilobyte modular program. See also figures 1.7.1 and 1.7.2. A sample run is shown in listing 1.7.2.*

```
RUN

HOW MANY DATA POINTS ARE THERE: ?10

HOW MANY DIMENSIONS ARE THERE: ?2

WHAT IS THE FIT FOR DIMENSION  1 ?2
WHAT IS THE FIT FOR DIMENSION  2 ?1

INPUT THE DATA AS PROMPTED:

Y( 1) = ?7
X( 1, 1) = ?1
X( 1, 2) = ?6

Y( 2) = ?7
X( 2, 1) = ?6
X( 2, 2) = ?1

Y( 3) = ?6
X( 3, 1) = ?3
X( 3, 2) = ?3

Y( 4) = ?8
X( 4, 1) = ?2
X( 4, 2) = ?6

Y( 5) = ?9
X( 5, 1) = ?1
X( 5, 2) = ?8

Y( 6) = ?9
X( 6, 1) = ?7
X( 6, 2) = ?2
```

Listing 1.7.2

```
Y( 7) = ?6
X( 7, 1) = ?3
X( 7, 2) = ?3

Y( 8) = ?7
X( 8, 1) = ?3
X( 8, 2) = ?4

Y( 9) = ?7
X( 9, 1) = ?4
X( 9, 2) = ?3

Y( 10) = ?2
X( 10, 1) = ?0
X( 10, 2) = ?2


THE CALCULATED COEFFICIENTS ARE:
 1      -.000001
 2       .999999
 3      0
 4      1
 5      0
 6      -.000001


STANDARD DEVIATION:  0

NUMBER OF ITERATIONS:  4


READY
```

Listing 1.7.2 *A sample run of the program shown in listing 1.7.1. This is actually a repeat of the example appearing in listing 1.5.3. Observe that the fitted equation is very close to the original:*

$$y = x_1 + x_2$$

versus

$$y \cong -0.000001 + 0.999999 x_1 + x_2 - 0.000001 x_1 x_2^2$$

This was accomplished in four iterations. The standard deviation after the first iteration was $\sigma = 0.0004$; after the second, $\sigma_2 = 0.00000004$; after the third, $\sigma_3 = 0.000000009$; and finally $\sigma_4 = 0.000000008$ (printed out as 0).

1.8 Parametric Least Squares

So far in this chapter, the discussion has been limited to polynomial least-squares fits in which the coefficients appeared in linear form. In fact, this linearity property was key to the iteration procedure discussed in the previous section.

Many problems lend themselves to this form of analysis. For example, the Weibull cumulative-distribution function is often used to analyze the failure modes in manufactured products:

$$P(t) = 1 - e^{-(t/t_0)^w}$$
$$t_0 > 0 \qquad (1.8.1)$$
$$w > 0$$
$$t > 0$$

(See References 1 and 13 for further discussion.) $P(t)$ is the probability that a failure will occur by time t. Often, the procedure is to record the failure times and to fit a Weibull curve graphically in order to determine the characteristic time t_0 and the parameter w (which has implications with respect to the failure mode). This function can be *linearized* for least-squares fitting as follows:

$$1 - P(t) = e^{-(t/t_0)^w}$$
$$\ln[1 - P(t)] = -(t/t_0)^w \qquad (1.8.2)$$
$$\ln\{\ln[1 - P(t)]\} = -w \ln t + w \ln t_0$$

We define the following:

$$X = \ln t$$
$$Y = \ln\{\ln[1 - P(t)]\}$$
$$A = -w$$
$$B = w \ln t_0$$

Equation (1.8.2) then becomes

$$Y = AX + B$$

This last equation has the standard linear form that can easily be treated with the subroutine given in section 1.2 (LSTSQR1). The resulting calculated coefficients, A and B, can be employed to estimate t_0 and w:

$$w = -A$$
$$t_0 = e^{-B/A} \qquad (1.8.3)$$

The linearization method is interesting, but it should not be relied upon for dealing with nonlinear coefficient situations. One problem with the approach is that not all equations can be linearized. Another problem is that the least-squares fit is not with respect to the original equation, but rather with respect to the linearized form. In the case of the Weibull fit, if $P(t)$ is near unity, $1 - P(t)$ is near zero and Y can become large. Thus, it is possible for errors in $P(t)$ in this region to unduly influence the fitted parameters.

An alternative procedure is to iteratively seek the values of the parameters that best satisfy the original equation. We will call this method *parametric least-squares fitting*.

In section 1.1, we found that the values of the parameters that minimized the variance (and thereby the standard deviation) were unbiased estimates. Thus, the minimization of the standard deviation is a good statistical criterion on which to base an iterative procedure.

Figure 1.8.1 shows a heuristic method* for finding the value of a parameter that minimizes the standard deviation. We will call this method the *shuffle*. The basic assumptions are that the sign of the optimum value of the parameter is known, and that there is only one minimum. The basic concept behind the method is that if the estimates are getting better, then you take bigger steps; and when the optimum is passed, you turn around and halve the step size.

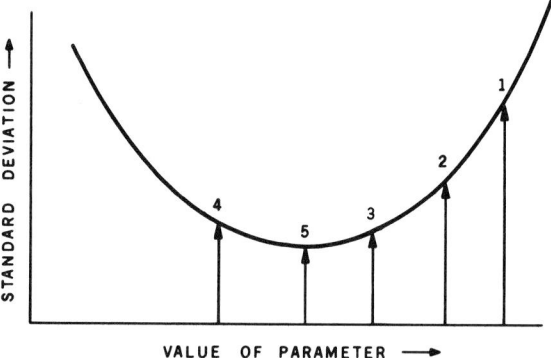

Figure 1.8.1 *An example of the shuffle concept. The second estimate reduced the standard deviation. The next step in that direction is therefore increased—an acceleration. The third estimate has an even lower standard deviation, and so the pace is further increased. However, the fourth estimate indicates that we have overshot the optimum. The direction is changed and the pace reduced.*

---

*The subject of optimization will be considered again in Chapter 8. At that time, the method of *steepest descent* will be presented.

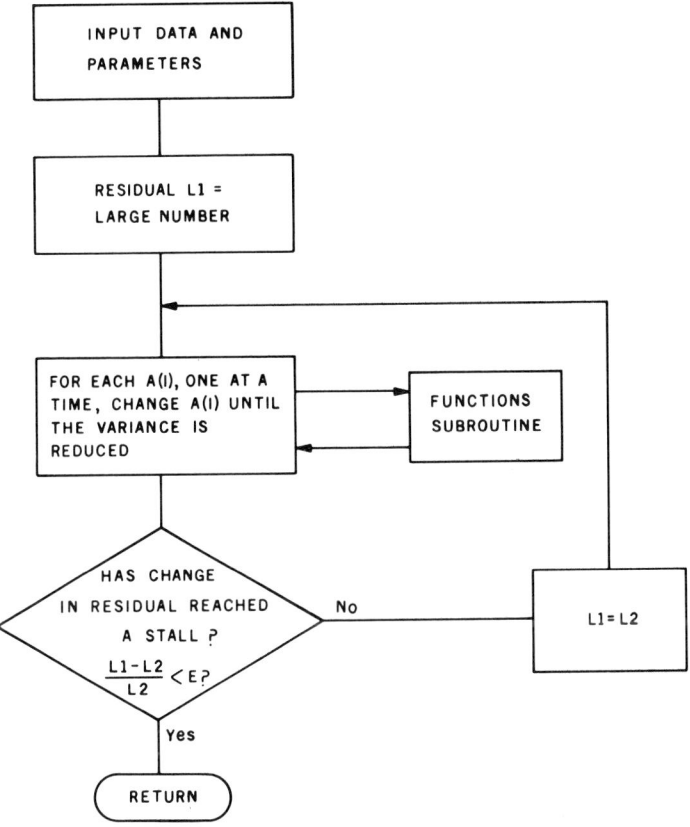

Figure 1.8.2 *General flowchart for the parametric least-squares fitting subroutine (PARAFIT).*

This approach can be generalized to several parameters by taking a step in reducing the standard deviation one parameter at a time, and repeatedly passing through the parameter set. This procedure is outlined in figure 1.8.2.

Assuming that the equation to be fitted can be written in the form $y(x) = f(x;A_1;A_2;A_3;\ldots)$, we can implement this algorithm as shown in listing 1.8.1.

The inputs to PARAFIT are the number of data points (N), the number of coefficients (L), the data pairs [$X(I)$, $Y(I)$], the initial step size ($0 < E1 < 1$), and a convergence factor (E). The returned results are the estimated coefficients [$A(I)$], the standard deviation of the fit using these coefficients (D), and the number of iterations employed (M). It is assumed that the function $Y(X) = f[X;A(I);\ldots]$ is available in the functions subroutine.

$E1$ and E must be specified very carefully. $E1$ is related to the step size, and *must* be in the range $0 < E1 < 1$. If $E1 = 0$, no progress will be made and the algorithm will fail. If $E1 = 1$, the procedure may become unstable. A good compromise choice is $E1 \approx 0.5$, and preferably less then 0.8.

E determines when the iteration is terminated. If the relative change in the standard deviation between two complete successive passes through the parameters is less than E, then the iteration stops.

An example of the use of PARAFIT is shown in listing 1.8.2. In this case, the demonstration program generated a set of ten data pairs using the equation $y(x) = 2e^{-\frac{(x-4.5)^2}{3}}$. This equation has the form of a gaussian, which is a function of considerable importance in statistics, communications, optics, physics, and other sciences. The values of the parameters calculated by PARAFIT are in very good agreement with those used to generate the data set.

PARAFIT is a powerful optimization subroutine. Its chief disadvantage is its slow execution speed. Also, some care is required in the selection of input parameters. PARAFIT does not check on the validity of the inputs. If $L > N$, the subroutine may never terminate. Also, PARAFIT is not always guaranteed to converge. However, it will usually do well if $E \approx 0.1$, $E1 \approx 0.5$, and if the initial guesses for the $A(I)$ are at least correct in sign, and are reasonable overestimates.

Finally, as an exercise, try to apply PARAFIT to the Weibull function discussed earlier in this section.

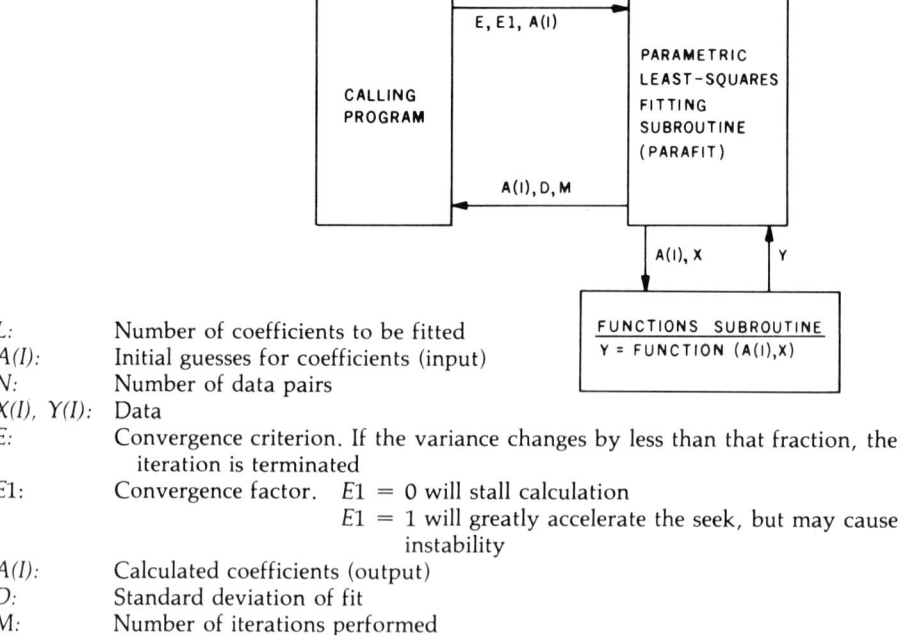

L:	Number of coefficients to be fitted
$A(I)$:	Initial guesses for coefficients (input)
N:	Number of data pairs
$X(I), Y(I)$:	Data
E:	Convergence criterion. If the variance changes by less than that fraction, the iteration is terminated
$E1$:	Convergence factor. $E1 = 0$ will stall calculation
	$E1 = 1$ will greatly accelerate the seek, but may cause instability
$A(I)$:	Calculated coefficients (output)
D:	Standard deviation of fit
M:	Number of iterations performed

Figure 1.8.3 *Subroutine-connection diagram for PARAFIT. Note that the initial estimates, A(I), should be overestimated in magnitude, and at least be of the correct sign.*

Statements/Functions List

$+, -, *, /, <, >$
ABS, GOSUB, GOTO, IF/THEN, SQRT

Variables List

A0, A(I), D, E, E1, E1(I), I, J, L, L1, L2, M, M0, M1, N, X, X(I), Y, Y(I)

Variables Passed to Subroutine

A(I), E, E1, L, N, X(I), Y(I)

Table 1.8.1 *Functions and variables used in the parametric least-squares fitting subroutine (PARAFIT). These are in addition to the functions and variables used in the functions subroutine.*

Listing 1.8.1

```
2400 REM PROGRAM TO DEMONSTRATE THE PARAFIT SUBROUTINE
2401 PRINT
2402 PRINT
2403 N=10
2404 L=3
2405 PRINT "THE INPUT DATA ARE:"
2406 PRINT
2407 FOR I=1 TO N
2408 X(I)=I
2409 Y(I)=2*EXP(-(X(I)-4.5)*(X(I)-4.5)/3)
2410 PRINT "X(",I,") = ",X(I),TAB(15),"Y(",I,") = ",Y(I)
2411 NEXT I
2412 PRINT
2413 PRINT
2414 E=.1
2415 E1=.5
2416 A(1)=10
2417 A(2)=10
2418 A(3)=10
2419 GOSUB 44250
2420 PRINT"THE COEFFICIENTS ARE:"
2421 PRINT A(1)
2422 PRINT A(2)
2423 PRINT A(3)
2424 PRINT
2425 PRINT
2426 PRINT"THE STANDARD DEVIATION OF THE FIT IS",
2427 PRINT INT(10000000*D)/10000000
2428 PRINT
2429 PRINT
2430 PRINT "THE NUMBER OF ITERATIONS WAS",M
```

```
2431 PRINT                                                              Listing 1.8.1 cont.
2432 PRINT
2433 END
44238 REM *********************
44239 REM PARAMETRIC LEAST SQUARES CURVE FIT SUBROUTINE (PARAFIT).
44240 REM THIS PROGRAM LEAST SQUARES FITS A FUNCTION TO A SET OF
44241 REM DATA VALUES BY SUCCESSIVELY REDUCING THE VARIANCE.
44242 REM CONVERGENCE DEPENDS ON THE INITIAL VALUES.- CONVERGENCE IS NOT ASSURED
44243 REM N PAIRS OF DATA VALUES, (X(I),Y(I)), ARE GIVEN.
44244 REM THERE ARE L PARAMETERS, A(J), TO BE OPTIMIZED ACROSS.
44245 REM REQUIRED ARE INITIAL VALUES FOR THE PARAMETER A(L) AND E.
44246 REM ANOTHER IMPORTANT PARAMETER WHICH AFFECTS STABILITY IS E1
44247 REM WHICH IS INITIALLY CONVERTED TO E1(L) FOR THE FIRST INTERVALS.
44248 REM THE PARAMETERS ARE MULTIPLIED BY (1-E1(I)) ON EACH PASS.
44249 REM DIMENSION X(I),Y(I),A(I) AND E1(I) IN THE CALLING PROGRAM
44250 FOR I=1 TO L
44251 E1(I)=E1
44252 NEXT I
44253 M=0
44254 REM SET UP TEST RESIDUAL
44255 L1=1000000
44256 REM MAKE SWEEP THROUGH ALL PARAMETERS
44257 FOR I=1 TO L
44258 A0=A(I)
44259 REM GET VALUE OF RESIDUAL
44260 A(I)=A0
44261 GOSUB 44286
44262 REM STORE RESULT IN M0
44263 M0=L2
44264 REM REPEAT FOR M1
44265 A(I)=A0*(1-E1(I))
44266 GOSUB 44286
44267 M1=L2
44268 REM CHANGE INTERVAL SIZE IF CALLED FOR
44269 REM IF VARIANCE WAS INCREASED, HALVE E1(I)
44270 IF M1>M0 THEN E1(I)=-E1(I)/2
44271 REM IF VARIANCE WAS REDUCED, INCREASE STEP SIZE BY INCREASING E1(
44272 IF M1<M0 THEN E1(I)=1.2*E1(I)
44273 REM IF VARIANCE HAS INCREASED, TRY TO REDUCE IT
44274 IF M1>M0 THEN A(I)=A0
44275 IF M1>M0 THEN GOTO 44261
44276 NEXT I
44277 REM END OF A COMPLETE PASS
44278 REM TEST FOR CONVERGENCE
44279 M=M+1
44280 IF L2=0 THEN RETURN
44281 IF ABS((L1-L2)/L2)<E THEN RETURN
44282 REM IF THIS POINT IS REACHED, ANOTHER PASS IS CALLED FOR
44283 L1=L2
```

```
44284 GOTO 44257
44285 REM RESIDUAL GENERATION SUBROUTINE
44286 L2=0
44287 FOR J=1 TO N
44288 X=X(J)
44289 REM OBTAIN FUNCTION
44290 GOSUB 44300
44291 L2=L2+(Y(J)-Y)*(Y(J)-Y)
44292 NEXT J
44293 D=SQRT(L2/(N-L))
44294 RETURN
44298 REM ********************
44299 REM FUNCTIONS SUBROUTINE
44300 Y=A(1)*EXP(-(X-A(2))*(X-A(2))/A(3))
44301 RETURN
```

Listing 1.8.1 *Parametric least-squares iteration subroutine (PARAFIT). Also shown is a program for demonstrating the operation of PARAFIT for a gaussian form (see listing 1.8.2).*

Listing 1.8.2

```
RUN



THE INPUT DATA ARE:

X( 1) =   1     Y( 1) =   3.3702404E-02
X( 2) =   2     Y( 2) =   .24902896
X( 3) =   3     Y( 3) =   .94473312
X( 4) =   4     Y( 4) =   1.8400889
X( 5) =   5     Y( 5) =   1.8400889
X( 6) =   6     Y( 6) =   .94473312
X( 7) =   7     Y( 7) =   .24902896
X( 8) =   8     Y( 8) =   3.3702404E-02
X( 9) =   9     Y( 9) =   2.3417592E-03
X( 10) =  10    Y( 10) =  8.3539934E-05


THE COEFFICIENTS ARE:
 2.0000001
 4.5
 3.0000002


THE STANDARD DEVIATION OF THE FIT IS 0
```

```
THE NUMBER OF ITERATIONS WAS 19


READY
```

Listing 1.8.2 *Sample run of the program shown in listing 1.8.1. The ten data pairs were generated using the equation*
$$y(x) = 2e^{\frac{-(x-4.5)^2}{3}}$$

The fitted equation is very close to the original. This is a particularly useful example because the gaussian is a very common and important function and must often be fitted to data.

1.9 Extending the Use of Least Squares

Least-squares curve fitting is usually applied to data sets that contain noise in order to provide estimates of the underlying dependencies. The techniques presented in this chapter are well suited for this task. However, they are also applicable to the polynomial fitting of "data" that do not contain noise, e.g., function tables. By using the subroutines given in this chapter, you can generate your own functional approximations. By employing the low round-off error subroutines (LSQRPOLY, iterated LEASTSQR, and PARAFIT), these approximations can be quite accurate.

Tables that exhibit *monotonic* behavior are very good candidates for fitting in this manner. As an example, we will consider the χ^2 cumulative-distribution function, $P(\chi^2)$.

The problem statement is as follows. $P(\chi^2)$ is the probability that the associated random variable will have a value greater than χ^2. This distribution depends on the number of degrees of freedom, M. For $M \geq 100$, we can approximate the interrelation of χ^2 and $P(\chi^2)$ by

$$\chi_M^2(P) \cong M(1 - \frac{2}{9M} + Z_p\sqrt{2/(9M)}\ )^3 \qquad (1.9.1)$$

(See Ref. 14, *Statistics Manual*.) This equation has a form that is simple to encode. However, Z_p is a function of P, and therein lies the complication. In their *Statistics Manual*, Crow, et al, give a table which includes values for Z_p (see table 1.9.1). Two features of the relationship between Z_p and P are apparent from that table. First, only half the table needs to be fitted in that it is asymmetrical. Second, as P becomes very small, Z_p becomes very large (see figure 1.9.1). This example was chosen for discussion because of the particular dependence evidenced in that figure. We will now see where the *art* of curve fitting comes into the analysis.

The curve in figure 1.9.1 does not look like the type that could be easily fitted by a simple polynomial in P (i.e., $Z_p = a_0 + a_1P + \cdots$). The singularity at $P = 0$ creates difficulties. A better choice would be a polynomial in $1/P$ (a truncated Laurent series):

$$Z_p = a_0 + \frac{a_1}{P} + \frac{a_2}{P^2} + \cdots$$

However, $1/P$ has too strong a curvature near $P = 0$. The author stared at this problem for a while and came up with the following polynomial:

$$Z_p = a_0 + a_1(-\ln P) + a_2(-\ln P)^2 + \cdots \qquad (1.9.2)$$

By choosing $y = Z_p$ and $x = -\ln P$, a least-squares fit using LSQRPOLY can be made with the following result:

$$y \cong -0.803 + 1.312x - 0.2118x^2 + 0.016x^3$$

This fitted equation was inserted into equation (1.9.1) and forms the basis of the subroutine shown in listing 1.9.1.

An example of the use of CHISQA is given in listing 1.9.2. The results are accurate to a fraction of one percent. The error is largely due to the inaccuracy associated with the approximation inherent in equation (1.9.1).

The purpose of the above example is to show how it can be advantageous to fit a polynomial in $f(x)$ instead of in x. In this example, the polynomial is in powers of a logarithm. In other situations, an exponential function is more appropriate. For even functions $[Y(x) = Y(-x)]$, and a good choice is often x^2 instead of x:

$$y(x) \cong a_0 + a_1 x^2 + a_2(x^2)^2 + a_3(x^2)^3 + \cdots$$

As you can imagine, considerable trial and error may be required before a good functional form is found.

P	Z_p
0.995	-2.576
0.990	-2.326
0.975	1.960
0.950	-1.645
0.900	-1.282
0.750	-0.6745
0.500	0.0000
0.250	0.6745
0.100	1.282
0.050	1.645
0.025	1.960
0.010	2.326
0.005	2.576

Table 1.9.1 Values for Z_p as a function of P (see Ref. 14).

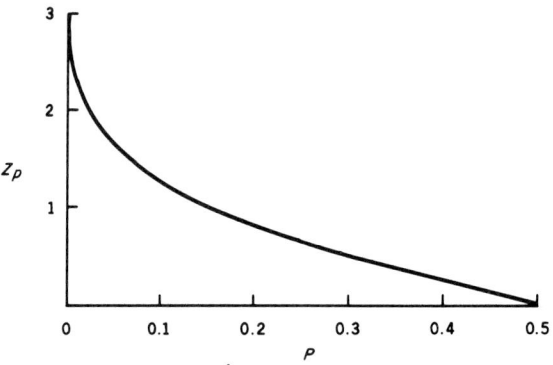

Figure 1.9.1 *Plot of Z_p versus P.*

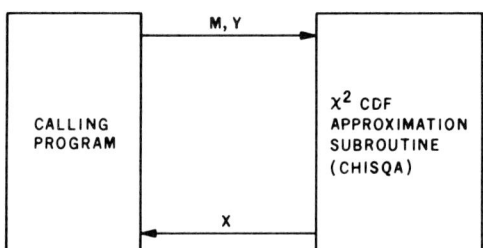

M: Number of degrees of freedom
Y: The cumulative-distribution probability level; probability that the χ^2 value calculated is greater than x
X: Corresponding χ^2_M

Figure 1.9.2 *Subroutine-connection diagram for the χ^2 cumulative-distribution approximation subroutine.*

Statements/Functions List

+, −, *, /, >
EXP, GOTO, IF/THEN, LOG, SQRT

Variables List

M, X, Y, Z

Variables Passed to Subroutine

M, Y

Table 1.9.2 *Functions and variables used by the χ^2 cumulative-distribution function subroutine (CHISQA).*

```
2450 REM PROGRAM TO DEMONSTRATE CHISQA
2451 PRINT
2452 PRINT
2453 PRINT "P(X)",TAB(12),"    X"
2454 PRINT "----",TAB(12),"   ---"
2455 PRINT
2456 M=100
2457 FOR Y=.05 TO 1 STEP .05
2458 GOSUB 44400
2459 PRINT Y,TAB(12),INT(10*X)/10
2460 NEXT Y
2461 PRINT
2462 END
44392 REM ********************
44393 REM CHI-SQUARE CUMMULATIVE DISTRIBUTION APPROXIMATION (CHISQA).
44394 REM GOOD FOR M>100.
44395 REM REFERENCE- STATISTICS MANUAL
44396 REM CROW, MAXFIELD AND DAVIS (DOVER, 1960).
44397 REM THE INPUT VALUE IS Y, THE PROBABILITY.
44398 REM THE OUTPUT VALUE IS THE CORRESPONDING
44399 REM CHI-SQUARE STATISTIC.
44400 X=Y
44401 REM GUARD AGAINST 0 DISCONTINUITY
44402 IF X=0 THEN X=EXP(-100)
44403 IF X>.5 THEN GOTO 44408
44404 X=-LOG(X)
44405 REM REGRESSED TABLE CORRECTION
44406 Z=-.803+1.312*X-.2118*X*X+.016*X*X*X
44407 GOTO 44413
44408 X=1-X
44409 REM GUARD AGAINST 0 DISCONTINUITY
```

Listing 1.9.1

```
44410 IF X=0 THEN GOTO 44415
44411 X=-LOG(X)
44412 Z=.803-1.312**X+.2118**X*X-.016**X*X*X
44413 X=2/(9*M)
44414 X=1-X+Z*SQRT(X)
44415 X=M*X*X*X
44416 RETURN
```

Listing 1.9.1 Subroutine for calculating the X^2 values corresponding to a given $P(X^2)$ (CHISQA). Also shown is a program for demonstrating CHISQA. See listing 1.9.2 for a sample run.

```
RUN

P(X)           X
-----         -----

.05          124.5
.1           118.6
.15          114.5
.2           111.4
.25          108.7
.3           106.5
.35          104.4
.4           102.6
.45          101
.5            99.4
.55           97.6
.6            96
.65           94.3
.7            92.4
.75           90.4
.8            88.1
.85           85.4
.9            82.2
.95           77.7
1              0

READY
```

$P(X^2)$	X^2_{100}	Observed Error
0.05	124.3	+0.2
0.10	118.5	+0.1
0.25	109.1	−0.4
0.50	99.3	+0.1
0.75	90.1	+0.3
0.90	82.4	+0.2
0.95	77.9	+0.2

Listing 1.9.2 A sample run of the program appearing in listing 1.9.1. Compare with the table values shown * (for M = 100 degrees of freedom).

---

*From Ref. 14.

1.10 Summary and Conclusion

The discussion in this chapter started with a rationalization of the use of least squares as a fitting criterion based on statistical concepts. In essence, if the noise in the data is gaussian, additive, and uncorrelated, then the coefficients derived from minimizing the calculated mean-squared error of the fit are unbiased estimates of the true coefficients. The assumption that the noise is gaussian is usually reasonable. The assumption of additivity is often violated by real processes. In electronic systems, the assumption that the noise is uncorrelated implies infinite bandwidth—an impossibility. Finally, the implied assumption that the equation being fitted actually represents the underlying functional dependence is usually tenuous. Despite these possible deficiencies, least-squares curve fitting is popular both because it reflects a clear error criterion and because there are many powerful techniques for performing the calculations.

The distribution of error in the resulting fit was also discussed. It was shown that Taylor series expansions are usually quite inferior to corresponding least-squares polynomials. However, the error in the least-squares fit tends to be largest at the endpoints of the interval. Chebyshev polynomials were demonstrated to be superior in this respect, and are discussed further in Chapter 2.

Several subroutines and collections of subroutines were presented throughout the chapter. These programs are summarized in table 1.10.1. With these subroutines, polynomial approximations of any order for any practical number of dimensions can be made. In addition, very nonlinear equations in one dimension can be fitted (PARAFIT). If particularly high accuracy is required, LSQRPOLY and REGITER are recommended.

In this chapter, two general concepts were introduced and implemented in terms of algorithms. The most pervasive concept was that of matrix algebra to formulate simple solutions to complicated problems. The other concept was iteration. With iteration, we are able to greatly reduce the round-off error in the matrix calculations, as well as to provide a means for dealing with very nonlinear equations. Both concepts would be difficult to implement fully without the aid of a computer. We will discuss many more iterative techniques later in this book.

SUBROUTINE	ADVANTAGES	DISADVANTAGES
LSTSQR1 (one-dimensional linear least squares)	• Short code • Fast • Low round-off error	• Limited use; elementary
LSTSQR2 (one-dimensional parabolic least squares)	• Short code • Fast • Moderate round-off error	
LEASTSQR + POLYCM (one-dimensional, Nth-degree least squares)	• Educational; forms basis for more powerful algorithms	• Susceptible to round-off error for large N • Slow
LSQRPOLY (one-dimensional, Nth-degree least squares)	• Resistant to round-off error • Fast	• Long code
LEASTSQR + MLTNLREG (multidimensional, Nth-degree least squares)	• Very general • Modular; extendable	• Slow • Susceptible to round-off error for large matrices
REGITER + LEASTSQR + MLTNLREG (iterated, multidimensional, Nth-degree least squares)	• Very powerful • Low round-off error • Modular; extendable	• Slow
PARAFIT (iterated, one-dimensional curvilinear regression)	• Very general application to nonlinear functions • Low round-off error	• Slow
CHISQA (χ^2 distribution example)	• Short code • Good example of linearization	

Table 1.10.1 *A summary of the programs given in Chapter 1. See also table 1.1.1.*

Chapter 2

Series Approximation Techniques

2.1 Introduction

The discussion in Chapter 1 was directed at the least-squares technique for fitting polynomials and parametric equations. As indicated in the beginning of that chapter, the least-squares method usually provides much more effective approximations than Taylor series expansions. However, it was also noted that Taylor series are very useful as a means for obtaining the accurate table values required for least-squares fitting.

It would be impossible, as well as inappropriate, both to cover the subject of Taylor series expansions in complete detail and to provide subroutines for a large number of expansions. Instead, an important example is given in section 2.2; you can develop your own subroutines based on that analysis. The particular example is the Bessel function.

One of the chief characteristics of the Taylor series is that the expansion is performed about some finite value of x. When this value is zero, the result is called a Maclaurin series. For numerical calculations, the series summation must be finite, thereby introducing some error. The accuracy of the truncated Taylor series rapidly deteriorates as the evaluation point becomes more remote from the expansion point. An example of this is shown in figure 1.1.1. By contrast, there is another class of series representations in which the accuracy improves as the argument becomes larger; the error *asymptotically* approaches zero. These are called asymptotic series.

The subject of asymptotic series is treated in section 2.3, and several subroutine examples are provided. The extent of the discussion regarding asymptotic series is greater than that for Taylor series for two key reasons. First, the subject of asymptotic series is much less familiar to the average reader and it involves concepts that are somewhat alien to those associated with Taylor expansions. Second, there are many functions that are very effectively approximated by this method, and these functions are not necessarily esoteric. In fact, most of the subroutine examples given in section 2.3 deal with important functions that are frequently found in statistics texts, but only in tabular form. With the approximations given in section 2.3, they can now be directly calculated.

Table 2.1.1

SUBROUTINE	SUBROUTINE SIZE (bytes) (including support programs)	DEMONSTRATION PROGRAM SIZE (bytes)	EXECUTION SPEED	CONDITIONS/ COMMENTS
BESSLSER [Taylor series evaluation of $J_N(x)$]	569	355	0.3 to 1.4 seconds/ evaluation	$E = 10^{-8}$ $X \leq 10$
BESSEL [Taylor series coefficients for $J_N(x)$]	596	334	0.6 seconds	10 terms
BESSEL01 [$J_0(x)$ and $J_1(x)$ for large x]	1163	520	0.5 seconds/ evaluation	$E = 10^{-8}$
LN(X!) (ln $x!$)	442	237	0.3 seconds/ evaluation	
CHI-SQR [$p(\chi^2)$ distribution]	900	417	0.7 seconds/ evaluation	$M = 100$
CHISQ [$P(\chi^2)$ distribution]	1672	435	3.4 seconds/ evaluation	$M = 100$
ASYMERF [erf(x) and erfc(x)]	923	215	0.8 seconds/ evaluation	$X = 3$
CHEBYSER (Chebyshev polynomial coefficients)	558	251	1 second/set	
CHEBECON (Chebyshev economization)	1833	685	45 seconds	16th-degree polynomial economized to the 10th degree

SERIES APPROXIMATION TECHNIQUES 91

SUBROUTINE	SUBROUTINE SIZE (bytes) (including support programs)	DEMONSTRATION PROGRAM SIZE (bytes)	EXECUTION SPEED	CONDITIONS/ COMMENTS
REVERSE (polynomial reversion)	1416	396	1 second	8th degree
RECIPRO (polynomial inversion)	751	478	1.5 seconds	8th degree
HORNER (polynomial shifting)	544	346	0.7 seconds	
INVNORM (inverse of normal distribution)	638	195	0.5 seconds/ evaluation	$0 < X < 0.5$
SINEPROD (sin x by repeated products—for demonstration *only*)	not a subroutine	810	1.5 seconds/ evaluation	10^{-4} accuracy
CMPLXSER (complex series evaluation)	1397	552	3.5 seconds	5th degree

Table 2.1.1 *A summary of the programs presented in Chapter 2.*

Sections 2.4 and 2.5 represent the heart of Chapter 2. In these sections, the subjects of Chebyshev series approximations and *economization* are introduced. The concept of Chebyshev polynomials is very important because it forms the basis of a convenient means for obtaining near-min-max polynomial approximations to many functions.

The essential properties of Chebyshev polynomials are discussed in section 2.4, and a subroutine is provided for calculating the associated coefficients. Using this subroutine and a few decomposition techniques, it is possible to convert a given power series to the corresponding Chebyshev series representation, truncate the latter, and express the results as a near-min-max polynomial. A subroutine for doing that is given in section 2.5. With this economization subroutine, many of the min-max approximations appearing in such classic texts as Hastings (Ref. 9) and Hart

(Ref. 7) can be closely regenerated and improved upon. In addition, by applying the optimization procedure discussed in Chapter 8, the near-min-max polynomial approximations generated using the economization program can be adjusted to give true min-max polynomials. (This is an interesting procedure and is discussed in Hamming, Ref. 12.)

In short, a group of subroutines is provided that can be employed to develop highly efficient min-max polynomial approximations either to sets of data (by economizing the least-squares fitted polynomial) or to functions (by economizing the power series representation).

Section 2.6 deals with some uncommon utility subroutines that can be used to extend the power of polynomial approximations. These programs allow you to manipulate polynomials in three general ways.

First, a polynomial can be reversed. That is, given a polynomial $y = a_0 + a_1x + a_2x^2 + \cdots$, the reversed polynomial $x = b_0 + b_1y + b_2y^2 + \cdots$ can be determined (up to the seventh degree). One example given is $y = \sinh x$ converted to $x = \sinh^{-1} y$. The second useful utility is the inversion subroutine. If the input polynomial is $P(x)$, this routine returns a polynomial approximation to the inverse: $Q(x) \cong 1/P(x)$. The third utility subroutine provided in section 2.6 is one that shifts the expansion point of a quartic polynomial. For example, if the input polynomial is $P(x) = a_0 + a_1x + a_2x^2 + a_3x^3 + a_4x^4$, it can be shifted to $P(x) = b_0 + b_1(x - x_0) + b_2(x - x_0)^2 + b_3(x - x_0)^3 + b_4(x - x_0)^4$. One very important application of this technique is in reducing round-off error. If it is known beforehand that the evaluation points will be in the vicinity of x_0, the round-off error in calculating $P(x)$ can be greatly reduced by using the shifted form of the polynomial. Thus, when very high accuracy is desired, an economized polynomial can be shifted so that accuracy advantages are not lost to round-off error.

In section 2.7, another class of approximations is discussed—the rational polynomial. This form is particularly useful for functions that have both zeros and infinities (and are therefore not analytical), or that have features that crudely resemble such singularities. The particular subroutine example given is for the inverse of the normal distribution function. However, there are many other situations in which simple rational polynomial approximations are very effective.

The text proceeds on to the product-sequence estimation method in section 2.8. Little time is spent on this technique because of convergence and round-off error problems that appear to plague this procedure.

Throughout the discussions in Chapters 1 and 2, it is implicitly assumed that the argument (e.g., x) in the approximating polynomial is real. Section 2.9 remedies this limitation by presenting a subroutine that permits the evaluation of complex series having real coefficients. Engineers will find this subroutine particularly useful.

As you may surmise, Chapter 2 is very broad in scope. The approach in presentation is not to provide a compendium of approximation routines, but rather to supply a set of utility programs and examples.

2.2 Taylor Series and Horner's Rule

The principal properties of the Taylor series approximation method were discussed both in Chapter 1 of this volume, and in Volume I of *BASIC Scientific Subroutines*. Suffice it to say that if a function is analytical in the vicinity of x_0, then a power series in which the coefficients are related to the derivatives of the function being approximated can be constructed:

$$f(x) = a_0 + a_1(x - x_0) + a_2(x - x_0)^2 + \cdots \quad (2.2.1)$$

where $a_0 = f(x_0)$ (2.2.2)

$$a_n = \frac{d^n f(x)}{dx^n}\bigg|_{x=x_0}$$

The structure of the summation shown in equation (2.2.1) can *in principle* be made more efficient using Horner's scheme:

$$\begin{aligned}S_n &= a_n \\ S_k &= (x - x_0)S_{k+1} + a_k \quad \text{for } k = n - 1, n - 2, \ldots, 0\end{aligned} \quad (2.2.3)$$

The approximation is then

$$f(x) \cong S_0$$

(See, for example, Ref. 15.) In effect, the series is summed backwards, starting at $a_n(x - x_0)^n$, and the summation is performed in a manner which reduces the total number of mathematical operations. The final result, S_0, then represents the truncated series, which is an approximation to $f(x)$. This method was discussed in Volume I, and a simple subroutine (SERSUM) was given there for implementing these calculations in an orderly fashion.

In the example in this section, we will not use the Horner rule directly. In effect, it is accomplished by the way in which the problem is approached.

As an example of an interesting Taylor series approximation, we will take the Bessel function of integer order N ($N \geq 0$), $J_N(x)$. Bessel functions, as you may know, are very important in the study of wave propagation (e.g., optics, antenna theory, etc.).

The general expansion for $J_N(x)$ is

$$J_N(x) = \left(\frac{x}{2}\right)^N \sum_{k=0}^{\infty} \frac{(-1)^k (x^2/4)^k}{k! \, \Gamma(N + k + 1)} \quad (2.2.4)$$

(See Ref. 6.) This formula can be simplified. First, $\Gamma(N + k + 1) = (N + k)!$. Second, equation (2.2.4) can be put into the following form:

$$J_N(x) = a \sum_{m=0}^{\infty} b_m$$

$$\text{where} \quad a = \frac{(x/2)^N}{N!} \qquad (2.2.5)$$

$$b_0 = 1$$
$$b_m = \frac{-(x/4)^2}{m(N+m)} b_{m-1} \quad \text{for } m = 1, 2, \ldots$$

(See Ref. 17.) This last equation is called a *recursion* relation. It greatly reduces the number of calculations.

Equation (2.2.5) can be directly implemented in subroutine form as shown in listing 2.2.1 (BESSLSER). The inputs to this subroutine are the order of the Bessel function (N), the argument (X), and a convergence factor (E). The use of this convergence factor is based on a simple property of convergent, alternating-sign series. A point is eventually reached in the series summation where all the succeeding terms monotonically decrease in magnitude. If only the first M terms in this series are summed, then the error in the approximation is bounded by the magnitude of the first term not included. E is compared with the magnitude of this term. If the absolute value of this quantity is larger than E, then the summation is continued; otherwise, it is terminated.

The output of BESSLSER is Y, the approximation to $J_N(X)$, and M, the number of terms included. Listing 2.2.1 also contains a program for exercising BESSLSER, and sample runs are displayed in listing 2.2.2.

In the first two examples, the agreement with published values is excellent. The first example is trivial, but the second shows that the convergence was rapid (only seven terms were summed), and that the results were not influenced by round-off error. In the third example, the convergence rate was much slower (21 terms were included), and the round-off error was significant. This is not unexpected. For large values of x, the terms in the Bessel series initially increase in size. Because the terms alternate in sign, considerable round-off error occurs in calculating the result. We will consider later how the Taylor series can be economized, thereby reducing both the number of terms summed and the round-off error.

There are a few simple precautions involved in using BESSLSER. First, the subroutine does not check on the validity of the inputs. N must be a nonnegative integer, and E must be greater than zero. Also, for $|X| > 5$, there is a strong potential for round-off error. For large values of X, the asymptotic series expansion given in section 2.3 is recommended.

In order to economize the Bessel-function series representation (as will be shown by the example in section 2.5), the coefficients must be evaluated. Also, it is useful to have the results returned in an array so that further processing can be performed without outside intervention. The subroutine shown in listing 2.2.3 (BESSEL) performs this task.

BESSEL is a simple routine and requires only the Bessel-function order (N, a nonnegative integer) and the degree of the truncated series ($M > 0$). No checks are made to test the validity of these inputs. The $M + 1$ coefficients are returned in $A(I)$.

A program for demonstrating BESSEL is also shown in listing 2.2.3. Examples of its use appear in listing 2.2.4. In the first example, the coefficients involved in the evaluation of $J_0(10)$ are shown. The cause of the round-off error observed earlier (in listing 2.2.2) is apparent from that tabulation. For example, the ratio of successive nonzero coefficients near $A(8)$ in this tabulation is roughly -0.01. However, the ratio of nonzero *terms* in this region is roughly -1; the terms are approximately equal in magnitude, but opposite in sign. Round-off error thrives in such an environment. To stay away from this ill-conditioned region of evaluation, x should be less than about 5. Later on (in section 2.5), we will use the coefficients calculated for $J_0(x)$ in the economization subroutine.

The coefficients for the expansion of $J_1(x)$ show the same general behavior as those for $J_0(x)$, and the same comments apply.

In the next section, we will see how the problem in estimating $J_0(x)$ for large x can be circumvented by employing an asymptotic expansion in the region $|x| > 5$.

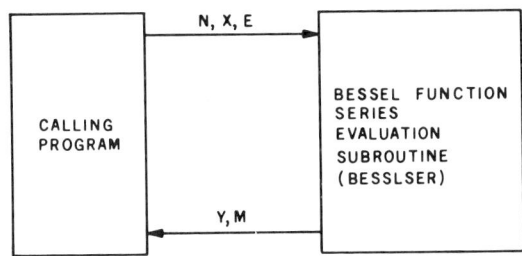

N: Order of the Bessel function (e.g., $J_0, J_1, J_2, \ldots$)
X: Argument [e.g., $J_N(X)$]
E: Convergence criterion; summation terminated at the first term less than E in absolute value
Y: Series sum [e.g., $Y \cong J_0(X)$]
M: Number of terms summed in series

Figure 2.2.1 *Subroutine-connection diagram for the Bessel function series-summation program (BESSLSER).*

Statements/Functions List

$+, -, *, /, <, >$
ABS, GOTO, FOR/NEXT, IF/THEN

Variables List

A, B0, B1, B2, E, I, M, N, X, Y

Variables Passed to Subroutine

E, N, X

Table 2.2.1 *Functions and variables appearing in the Bessel function series-summation subroutine (BESSLSER).*

```
2500 REM PROGRAM TO DEMONSTRATE BESSEL SERIES SUMMATION SUBROUTINE
2501 PRINT
2502 PRINT
2503 PRINT "WHAT IS THE ORDER OF THE BESSEL FUNCTION: ",
2504 INPUT N
2505 PRINT
2506 PRINT "INPUT ARGUMENT",
2507 INPUT X
2508 PRINT
2509 PRINT "INPUT CONVERGENCE CRITERION",
2510 INPUT E
2511 PRINT
2512 PRINT
2513 GOSUB 44425
2514 PRINT "J(",X,") OF ORDER ",N," = ",Y
2515 PRINT
2516 PRINT "NUMBER OF TERMS USED: ",M
2517 PRINT
2518 END
44419 REM ********************
44420 REM BESSEL FUNCTION SERIES SUBROUTINE (BESSLSER)
44421 REM THE ORDER IS N, THE ARGUMENT X.
44422 REM THE RETURNED VALUE IS IN Y.
44423 REM THE NUMBER OF TERMS USED IS RETURNED IN M.
44424 REM E IS THE CONVERGENCE CRITERION
44425 A=1
44426 IF N<=1 THEN GOTO 44431
44427 REM CALCULATE N!
44428 FOR I=1 TO N
44429 A=A*I
44430 NEXT I
```

Listing 2.2.1

```
44431 A=1/A
44432 IF N=0 THEN GOTO 44437
44433 REM CALCULATE MULTIPLYING TERM
44434 FOR I=1 TO N
44435 A=A*X/2
44436 NEXT I
44437 B0=1
44438 B2=1
44439 M=0
44440 REM ASSEMBLE SERIES SUM
44441 M=M+1
44442 B1=-(X*X*B0)/(M*(M+N)*4)
44443 B2=B2+B1
44444 B0=B1
44445 REM TEST FOR CONVERGENCE
44446 IF ABS(B1)>E THEN GOTO 44441
44447 REM FORM FINAL ANSWER
44448 Y=A*B2
44449 RETURN
```

Listing 2.2.1 *Bessel function series-summation subroutine (BESSLSER). Also shown is a program for demonstrating BESSLSER. See listing 2.2.2 for examples.*

RUN

Listing 2.2.2

WHAT IS THE ORDER OF THE BESSEL FUNCTION: ?0

INPUT ARGUMENT?0

INPUT CONVERGENCE CRITERION?.000000001

J(0) OF ORDER   0 =   1

NUMBER OF TERMS USED: 1

READY
RUN

WHAT IS THE ORDER OF THE BESSEL FUNCTION: ?0

INPUT ARGUMENT?2

```
INPUT CONVERGENCE CRITERION?.0000001

J( 2) OF ORDER   0 =   .22389078

NUMBER OF TERMS USED:  7

READY
RUN


WHAT IS THE ORDER OF THE BESSEL FUNCTION:  ?0

INPUT ARGUMENT?10

INPUT CONVERGENCE CRITERION?.000000001

J( 10) OF ORDER  0 =  -.24594148

NUMBER OF TERMS USED:  21

READY
RUN


WHAT IS THE ORDER OF THE BESSEL FUNCTION:  ?4

INPUT ARGUMENT?6

INPUT CONVERGENCE CRITERION?.000000001

J( 6) OF ORDER   4 =   .3576416

NUMBER OF TERMS USED:  12

READY
```

Listing 2.2.2 *Sample runs of the Bessel function series approximation subroutine shown in listing 2.2.1 (BESSLSER). The results are compiled below:*

Example	Calculated value	Error
$J_0(0)$	1.0	0
$J_0(2)$	0.22389078	0
$J_0(10)$	−0.24594148	−0.00000572
$J_4(6)$	0.3576416	< 0.00001

The error was calculated using the tables appearing in Ref. 6. $J_4(6)$ was given to only 5 digits in the tables. The error in the value calculated for $J_0(10)$ is due to round-off.

SERIES APPROXIMATION TECHNIQUES

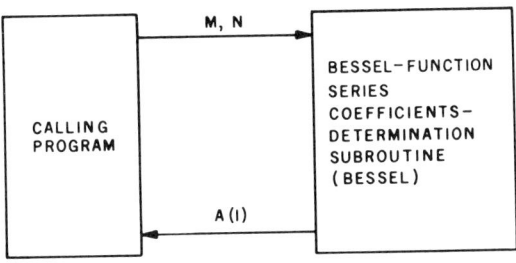

N: Order of Bessel function [e.g., $J_N(x)$]
M: Highest degree coefficient desired
A(I): Returned coefficients; $J_N(x) = a_0 + a_1 x + a_2 x^2 + \cdots$

Figure 2.2.2 Subroutine-connection diagram for the Bessel function series coefficients-determination subroutine (BESSEL).

Statements/Functions List

$+, -, *, /$
FOR/NEXT/STEP

Variables List

A(I), A1, B(I), B1, I, M, N

Variables Passed to Subroutine

M, N

Table 2.2.2 Functions and variables used by the BESSEL subroutine.

```
2520 REM PROGRAM TO DEMONSTRATE THE BESSEL COEFFICIENTS SUBROUTINE
2521 PRINT
2522 PRINT
2523 PRINT "WHAT IS THE BESSEL FUNCTION ORDER: ",
2524 INPUT N
2525 PRINT
2526 PRINT "WHAT DEGREE IS DESIRED: ",
2527 INPUT M
2528 DIM A(M+1),B(M+1)
2529 PRINT
2530 PRINT
2531 GOSUB 44475
```

Listing 2.2.3

```
2532 PRINT"THE COEFFICIENTS ARE:"
2533 PRINT
2534 FOR I=0 TO M
2535 PRINT "A(",I,") = ",A(I)
2536 NEXT I
2537 PRINT
2538 PRINT
2539 END
2540 NEXT I
2541 RETURN
44468 REM ********************
44469 REM BESSEL FUNCTION SERIES COEFFICIENT EVALUATION SUBROUTINE (BESSEL)
44470 REM M+1 IS THE NUMBER OF COEFFICIENTS DESIRED.
44471 REM N IS THE ORDER OF THE BESSEL FUNCTION.
44472 REM THE COEFFICIENTS ARE RETURNED IN A(I).
44473 REM DIMENSION A(I) AND B(I) IN THE CALLING PROGRAM.
44474 REM A1,B1 AND B(I) ARE DUMMY VARIABLES.
44475 A1=1
44476 B1=1
44477 FOR I=1 TO N
44478 B(I-1)=0
44479 B1=B1*I
44480 A1=A1/2
44481 NEXT I
44482 B1=A1/B1
44483 A1=1
44484 FOR I=0 TO M STEP 2
44485 A(I)=A1*B1
44486 A(I+1)=0
44487 A1=-A1/((I+2)*(N+N+I+2))
44488 NEXT I
44489 A1=A1/2
44490 FOR I=0 TO M
44491 B(I+N)=A(I)
44492 NEXT I
44493 FOR I=0 TO N+M
44494 A(I)=B(I)
44495 NEXT I
44496 RETURN
```

Listing 2.2.3 *Subroutine for determining the coefficients of the Taylor series approximation to $J_N(x)$ (BESSEL). Also shown is a program for exercising BESSEL. See listing 2.2.4 for examples.*

RUN

Listing 2.2.4

WHAT IS THE BESSEL FUNCTION ORDER: ?0

Listing 2.2.4 cont.

```
WHAT DEGREE IS DESIRED: ?41

THE COEFFICIENTS ARE:

A(  0) =  1
A(  1) =  0
A(  2) = -.25
A(  3) =  0
A(  4) =  .015625
A(  5) =  0
A(  6) = -4.3402778E-04
A(  7) =  0
A(  8) =  6.7816841E-06
A(  9) =  0
A( 10) = -6.7816841E-08
A( 11) =  0
A( 12) =  4.7095028E-10
A( 13) =  0
A( 14) = -2.4028076E-12
A( 15) =  0
A( 16) =  9.3859672E-15
A( 17) =  0
A( 18) = -2.8969035E-17
A( 19) =  0
A( 20) =  7.2422588E-20
A( 21) =  0
A( 22) = -1.4963345E-22
A( 23) =  0
A( 24) =  2.597803E-25
A( 25) =  0
A( 26) = -3.8429038E-28
A( 27) =  0
A( 28) =  4.901663E-31
A( 29) =  0
A( 30) = -5.4462922E-34
A( 31) =  0
A( 32) =  5.3186447E-37
A( 33) =  0
A( 34) = -4.6009037E-40
A( 35) =  0
A( 36) =  3.55008E-43
A( 37) =  0
A( 38) = -2.4585042E-46
A( 39) =  0
A( 40) =  1.5365651E-49
A( 41) =  0

READY
```

```
RUN


WHAT IS THE BESSEL FUNCTION ORDER: ?1

WHAT DEGREE IS DESIRED: ?15


THE COEFFICIENTS ARE:

A( 0) =   0
A( 1) =   .5
A( 2) =   0
A( 3) =   -.0625
A( 4) =   0
A( 5) =   2.6041667E-03
A( 6) =   0
A( 7) =   -5.425347E-05
A( 8) =   0
A( 9) =   6.781684E-07
A( 10) =  0
A( 11) =  -5.6514035E-09
A( 12) =  0
A( 13) =  3.3639307E-11
A( 14) =  0
A( 15) =  -1.5017548E-13


READY
```

Listing 2.2.4 *Sample runs of the program given in listing 2.2.1. As with the Taylor sin x series, the Bessel function Taylor series rapidly converges for small x. This is evident from the rapidly decreasing coefficient magnitudes. However, from the example shown in listing 2.2.2, terms up to x^{40} (2 times 21 − 1) were required to evaluate $J_0(10)$ for $E = 10^{-9}$. This resulted in considerable round-off error in the Taylor series approximation.*

2.3 Asymptotic Series

The accuracy of Taylor expansions and other *ascending* power series usually falls off rapidly as $x - x_0$ becomes arbitrarily large. As we saw in the previous section for the Bessel function, large values of $x - x_0$ require the summation of many terms before the desired level of truncation error is reached. However, because some series are slowly convergent, the round-off error can build up faster than the truncation error can be reduced.

In this section, we will consider a class of approximations in which the accuracy *improves* as the argument increases in magnitude. However, most of the approximations considered in this section also have the unusual and curious property that the error eventually *increases* as more terms are included. Thus, there is an optimum number of terms that should be used in the summation.

Because asymptotic series offer an important means for circumventing the difficulty associated with large arguments in ascending power series, and because they are so intriguing, several cases will be examined, starting with the Bessel function. You will find the presentation in this section to be oriented for the most part toward examples. For those interested in the theoretical background to this subject, see Erdelyi's book, *Asymptotic Expansions* (Ref. 18).

2.3.1 The Bessel Function

We will start this discussion with approximations for $J_0(x)$ and $J_1(x)$ which involve a group of asymptotic series. Although these approximations are limited to large positive values of x, they are easily extended to negative arguments by noting that $J_0(x) = J_0(-x)$ and $J_1(x) = -J_1(-x)$. The algorithm consists of several relations, as you can see by referring to Refs. 5 and 18:

$$J_0(x) = (2/\pi x)^{1/2} \left[P_0(x) \cos(x - \pi/4) - Q_0(x) \sin(x - \pi/4) \right] \quad (2.3.1)$$

$$J_1(x) = (2/\pi x)^{1/2} \left[P_1(x) \cos(x - 3\pi/4) - Q_1(x) \sin(x - 3\pi/4) \right] \quad (2.3.2)$$

where
$$P_0(x) \sim 1 - \frac{1^2 \cdot 3^2}{2!\,(8x)^2} + \frac{1^2 \cdot 3^2 \cdot 5^2 \cdot 7^2}{4!\,(8x)^4} - \frac{1^2 \cdot 3^2 \cdot 5^2 \cdot 7^2 \cdot 9^2 \cdot 11^2}{6!\,(8x)^6} + \cdots$$

$$Q_0(x) \sim -\frac{1^2}{1!\,(8x)} + \frac{1^2 \cdot 3^2 \cdot 5^2}{3!\,(8x)^3} - \frac{1^2 \cdot 3^2 \cdot 5^2 \cdot 7^2 \cdot 9^2}{5!\,(8x)^5} + \cdots$$

$$P_1(x) \sim 1 + \frac{1^2 \cdot 3 \cdot 5}{2!\,(8x)^2} - \frac{1^2 \cdot 3^2 \cdot 5^2 \cdot 7 \cdot 9}{4!\,(8x)^4} + \frac{1^2 \cdot 3^2 \cdot 5^2 \cdot 7^2 \cdot 9^2 \cdot 11 \cdot 13}{6!\,(8x)^6} - \cdots$$

$$Q_1(x) \sim \frac{1 \cdot 3}{1!\,(8x)} - \frac{1^2 \cdot 3^2 \cdot 5 \cdot 7}{3!\,(8x)^3} + \frac{1^2 \cdot 3^2 \cdot 5^2 \cdot 7^2 \cdot 9 \cdot 11}{5!\,(8x)^5} - \cdots$$

(The "$\sim$" symbol is a conventional approximation sign used in conjuction with asymptotic expansion. See Ref. 15.) The four supporting functions, $P_0(x)$, $Q_0(x)$, $P_1(x)$ and $Q_1(x)$, are represented by asymptotic series. For a given finite x, there is a point in each of these series after which the terms in the summation become larger. If we stop the summation at this point, a crude measure of the error for the truncated asymptotic series is the magnitude of the next term not included in the summation (see Ref. 20).

It is often the case in practical situations that this error estimate is greater than the error limit desired. Thus, the criterion for terminating the summation is not necessarily an error bound chosen by the user (as with the Taylor and Chebyshev approximations), but it may be dictated by the location of the smallest term in the asymptotic series. However, for large x, the convergence is usually rapid and any desired accuracy can be obtained. Thus, there are two error criteria that must be tested for in asymptotic series expansions. If the desired error level is not met, but the optimum point in the series has been reached, the summation should be terminated.

The approximation equations for $J_0(x)$ and $J_1(x)$, along with the convergence tests discussed above, are implemented in the subroutine shown in listing 2.3.1 (BESSEL01). The inputs to the routine are the argument (X) and the desired accuracy ($E3$). The returned results are the approximations to $J_0(X)$ and $J_1(X)$ ($J0$ and $J1$ respectively), an estimate of the error bound (E), and the number of terms summed (N).

Also appearing in listing 2.3.1 is a program for demonstrating BESSEL01, and a sample run is shown in listing 2.3.2. The sample run illustrates the main features of classic asymptotic expansions. First, for small x, an optimum number of summed terms exists for the approximation. As x increases, more of the series is included and the error decreases. Eventually, a point is reached where the accuracy criterion is achieved, and fewer terms are needed to achieve the level as x is further increased.

The results shown in listing 2.3.2 were compared with those given in published tables (see Ref. 6) and the actual errors are displayed in table 2.3.2. On the whole, the actual error decreases with increasing x, but is not in concordance with the error measure computed. This is *not* due to round-off effects, but is due instead to the crude way in which we estimated the error within the program. This represents one of the chief problems in using asymptotic expansions—the error is difficult to predict and, therefore, the correct number of terms to be included in the series can be only roughly estimated.

There are three general precautions related to the use of BESSEL01. First, the argument (X) must be greater than zero. Because X usually represents a *radius* in most physical problems, this is generally not a limitation. Second, the returned error estimate should be used only as a measure of the size of the first term excluded. Any further interpretation is tenuous. Third, the input convergence criterion ($E3$) should be nonnegative. If it is set to zero, the summation will continue until the optimum is found. In most cases, $E3$ should be set much lower than the value actually desired. However, there is no guarantee that the chosen error level will be achieved.

The value of the asymptotic series concept certainly does not rest with its error predictability, but rather with the ability to approximate functions for large values of the argument.

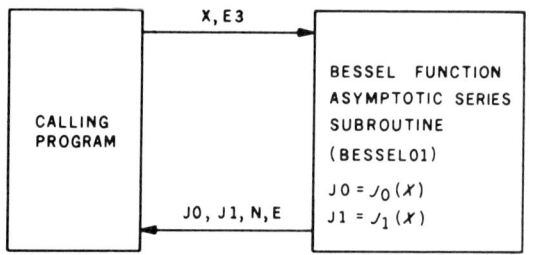

X: Input argument
E3: Desired relative accuracy
J0: $J0 = J_0(X)$
J1: $J1 = J_1(X)$
N: Number of terms in series
E: A measure of the truncation effect on the relative error

Figure 2.3.1 *Subroutine-connection diagram for BESSEL01.*

Statements/Functions List

$+, -, *, /, >$
ABS, COS, GOTO, IF/THEN, SIN, SQRT

Variables List

A, A1, A2, B, C, E, E1, E2, E3, E4, J0, J1, M, M1, M2, N, N1, N2, X, X1

Variables Passed to Subroutine

X, E3

Table 2.3.1 *Functions and variables appearing in BESSEL01.*

Listing 2.3.1

```
2550 REM PROGRAM TO DEMONSTRATE THE BESSEL FUNCTION ASYMPTOTIC SERIES
2551 PRINT
2552 PRINT
2553 PRINT "WHAT IS THE DESIRED ERROR BOUND: ",
2554 INPUT E3
2555 PRINT
2556 PRINT
2557 REM BESSEL FUNCTION SUBROUTINE
2558 PRINT "  X            J0(X)           J1(X)          N        E"
2559 PRINT "-------       -------------   -------------   -----    -----------"
2560 PRINT
2561 FOR X=1 TO 15
2562 GOSUB 44525
2563 PRINT X,TAB(15),INT(100000000*J0)/100000000,TAB(36),INT(100000000*J1)/100000000,
2564 PRINT TAB(55),N,TAB(65),INT(100000000*E)/100000000
2565 NEXT X
2566 PRINT
2567 PRINT
2568 PRINT
2569 END
44515 REM ********************
44516 REM BESSEL FUNCTION ASYMPTOTIC SERIES SUBROUTINE (BESSEL01)
44517 REM THIS PROGRAM CALCULATES THE ZEROTH AND FIRST ORDER BESSEL
44518 REM FUNCTIONS USING AN ASYMPTOTIC SERIES EXPANSION.
44519 REM THE REQUIRED INPUT ARE X AND A CONVERGENCE FACTOR E.
44520 REM RETURNED ARE THE TWO BESSEL FUNCTIONS, J0(X) AND J1(X)
44521 REM REFERENCE-  ALGORITHMS FOR RPN CALCULATORS
44522 REM    BY BALL, J.A., WILEY AND SONS,
44523 REM CALCULATE P AND Q POLYNOMIALS
44524 REM P0(X)=M1   P1(X)=M2   Q0(X)=N1   Q1(X)=N2
44525 A=1
44526 A1=1
```

```
44527 A2=1
44528 B=1
44529 C=1
44530 E1=1000000
44531 M=-1
44532 X1=1/(8*X)
44533 X1=X1*X1
44534 M1=1
44535 M2=1
44536 N1=-1/(8*X)
44537 N2=-3*N1
44538 N=0
44539 M=M+2
44540 A=A*M*M
44541 M=M+2
44542 A=A*M*M
44543 C=C*X1
44544 A1=A1*A2
44545 A2=A2+1
44546 A1=A1*A2
44547 A2=A2+1
44548 E2=A*C/A1
44549 E4=1+(M+2)/M+(M+2)*(M+2)/(A2*8*X)+(M+2)*(M+4)/(A2*8*X)
44550 E4=E4*E2
44551 REM TEST FOR DIVERGENCE
44552 IF ABS(E4)>E1 THEN GOTO 44562
44553 E1=ABS(E2)
44554 M1=M1-E2
44555 M2=M2+E2*(M+2)/M
44556 N1=N1+E2*(M+2)*(M+2)/(A2*8*X)
44557 N2=N2-E2*(M+2)*(M+4)/(A2*8*X)
44558 N=N+1
44559 REM TEST FOR CONVERGENCE CRITERION
44560 IF E1<E3 THEN GOTO 44562
44561 GOTO 44539
44562 A=3.1415926536
44563 E=E2
44564 B=SQRT(2/(A*X))
44565 J0=B*(M1*COS(X-A/4)-N1*SIN(X-A/4))
44566 J1=B*(M2*COS(X-3*A/4)-N2*SIN(X-3*A/4))
44567 RETURN
```

Listing 2.3.1 *Subroutine for the asymptotic series approximation of the zeroeth- and first-order Bessel functions [$J_0(x)$ and $J_1(x)$]. Also shown is a program that calls BESSEL01 to generate a table of Bessel function values. See listing 2.3.2 for a sample run.*

```
RUN

WHAT IS THE DESIRED ERROR BOUND: ?.00000001

    X             J0(X)              J1(X)            N            E
 -------      --------------     --------------     -----      -----------

    1           .73356225          .40223394          1          .1121521
    2           .22148776          .57863407          1          .0070095
    3          -.25995642          .34069879          2          .00078532
    4          -.39682633         -.06588598          2          .00013977
    5          -.17748619         -.32769562          3          .00001554
    6           .15063483         -.27675968          3          .00000361
    7           .3000512          -.00469565          4          .00000038
    8           .17163825          .23464955          5          .00000004
    9          -.09033231          .24532374          5          .00000001
   10          -.24593045          .04347579          6          0
   11          -.17118723         -.17678794          5          0
   12           .04768914         -.2234504           5          0
   13           .20692454         -.0703193           4          0
   14           .17107239          .13337581          4          0
   15          -.0142245           .20510522          4          0

READY
```

Listing 2.3.2 *A sample run of the program given in listing 2.3.1. N is the number of terms included in the series. E is the magnitude of the first divergent term in the asymptotic series expansion of $P_0(x)$.*

X	$E[J_0(x)]$	$E[J_1(x)]$	Error Measure
1	0.03	0.04	0.11
2	0.002	−0.002	0.009
3	0.0001	0.0002	0.0009
4	0.0004	0.0002	0.0001
5	0.0001	−0.0001	0.00001
6	0.00001	0.00008	0.000002
7	0.00003	0.00001	0.0000002
8	0.00001	0.00001	0.00000002
9	0.000001	0.00001	----
10	0.000005	0.000003	----
11	0.000003	0.000003	----
12	0.0000002	0.000003	----
13	0.000002	0.000001	----
14	0.000001	0.0000007	----
15	0.00000003	0.000001	----

Table 2.3.2 *Actual error in the values calculated in listing 2.3.2. Note that the error measure is only effective where the error is large.*

2.3.2 The Chi-Square Distribution Functions

This subsection was inspired by an observation that statisticians appear to be repeatedly thwarted in their work by a lack of table values that cover their particular situation. For those who do not have access to a large computer, this can be very frustrating.

Statistical tables are numerous (e.g., see Refs. 21 and 22). However, the number of possible different situations is infinite, and the tables cover only a few selected cases. This problem appears to be particularly acute for the Chi-Square statistic. In this subsection, therefore, asymptotic series expansions for both the χ^2 probability density function, $p(\chi^2)$, and the χ^2 cumulative distribution function, $P(\chi^2)$, will be presented.

For those not familiar with the subject, the χ^2 statistic is used to test the hypothesis that the measured results either support or disprove a preconceived distribution function. For example, consider the situation where there are N bins into which the results can fall. If E_i is the expected frequency of a result falling into bin i, and O_i is the observed frequency, then the χ^2 statistic is defined to be

$$\chi^2 = \sum_{i=1}^{N} \frac{(O_i - E_i)^2}{E_i} \qquad (2.3.3)$$

(See Ref. 13.) Associated with this statistic is the number of degrees of freedom, $M = N - R$, where R is the number of independent relationships between the O_i and E_i. For example, one relation might be

$$\sum_{i=1}^{N} O_i = \sum_{i=1}^{N} E_i$$

In the extreme case, if there are N independent relationships, and if the expected frequency conjecture is correct, then $\chi^2 = 0$ because the O_i are totally constrained.

The basic relationship we will employ to calculate $p(x)$ is

$$p(x) = \frac{e^{-x/2} \, x^{(M/2 - 1)}}{2^{(M/2)} \, \Gamma(M/2)} \qquad (2.3.4)$$

(See Ref. 24.) In this notation $x = \chi^2$, and M represents the number of degrees of freedom.

The approximation that will be used for $P(x)$ is

$$P(x) = \frac{2x}{M} p(x) \left[1 + \sum_{k=1}^{\infty} \frac{x^k}{(M + 2)(M + 4) \cdots (M + 2k)} \right] \qquad (2.3.5)$$

(See Ref. 14.) To evaluate equation (2.3.5) we need $p(x)$, and to obtain that function we require $\Gamma(M/2)$, the gamma function. Therefore, the starting point involves approximating the gamma function.

One of the problems in evaluating $\Gamma(M/2)$ directly is that there is a strong potential for numeric overflow for large values of M. For example, if $M = 200$, then $\Gamma(M/2) = \Gamma(100) = 99!$, which is beyond the range of many BASIC interpreters. Therefore, it would be better to approximate $\ln[\Gamma(M/2)]$. This can be implemented in equation (2.3.4) by first approximating $\ln[p(x)]$:

$$\ln[p(x)] = -x/2 + (M/2 - 1)\ln x - (M/2)\ln 2 - \ln[\Gamma(M/2)] \qquad (2.3.6)$$

The conversion to $p(x)$ then is simply

$$p(x) = \exp\{\ln[p(x)]\} \qquad (2.3.7)$$

We therefore start with an approximation for $\ln[\Gamma(M/2)]$.

The gamma function is directly related to the generalized factorial

$$\Gamma(x + 1) = x! \qquad (2.3.8)$$

The problem now is one of finding an approximation for $\ln x!$. Since we are doing all of this in order to avoid the numeric overflow problems associated with large arguments in the gamma function, an asymptotic expansion is appropriate:

$$\ln x! \sim (x + \frac{1}{2})\ln x - x + \frac{1}{12x} - \frac{1}{360x^3} + \frac{1}{1260x^5} - \frac{1}{1680x^7}$$
$$+ 0.918938533205 \qquad (2.3.9)$$

(See Ref. 5.) The derivation of this equation will be discussed in the next subsection. For the present, we will note that this expansion is accurate to better than 12 digits for $x > 10$.

Equation (2.3.9) can be implemented in BASIC as shown in listing 2.3.3 [LN(X!)]. This listing also contains a program for exercising LN(X!), and a sample run appears in listing 2.3.4. As you can see from that example, LN(X!) is very accurate even for low values of X.

We can now use the LN(X!) subroutine in conjunction with equation (2.3.6) to provide an approximation to the Chi-Square probability density function, $p(x)$ (see listing 2.3.5). The inputs to CHI-SQR are the number of degrees of freedom (M) and the argument (X). The result is returned in Y. The intrinsic limit to the accuracy of this calculation is the approximation associated with obtaining $\ln(M/2)$. However, that approximation is very good, and therefore the number of degrees of freedom may be as low as $M = 4$ (or even lower).

A sample run of the program appearing in listing 2.3.5 is shown in listing 2.3.6. As expected, the calculated distribution peaks close to the number of degrees of freedom.

There are a few precautions related to the use of CHI-SQR. Because there are no error checks on the inputs, it is the responsibility of the calling program to ensure that M is a nonnegative integer, and that X is positive. Otherwise, CHI-SQR is a very reliable subroutine.

Now that we have a means for accurately evaluating $p(x)$, the cumulative distribution function can be approximated using equation (2.3.5). Since the series summed in this equation contains terms of the same sign, little round-off error is expected in the final result.

The calculations required to evaluate $P(x)$ are performed by the subroutine shown in listing 2.3.7 (CHISQ). This subroutine calls CHI-SQR, which in turn calls LN(X!). The inputs to CHISQ are the number of degrees of freedom (M), the argument (X), and a convergence factor for the series summation (E). The result is returned in Y.

Also appearing in listing 2.3.7 is a program for demonstrating CHISQ. A sample run is shown in listing 2.3.8. The printed results are accurate to all digits shown.

The programs presented in this subsection serve two purposes. First, they are very useful routines for statistical studies. Second, they demonstrate how a set of subroutines can be built one block at a time. This is the overall philosophy behind the *BASIC Scientific Subroutines* series.

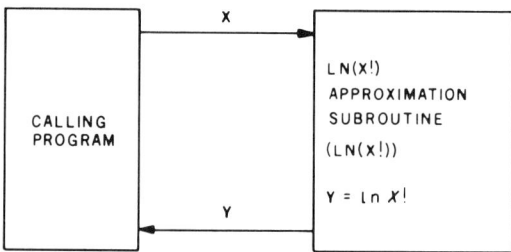

Figure 2.3.2 *Subroutine-connection diagram for the asymptotic series expansion to ln x! [LN(X!)].*

Statements/Functions List

$+, -, *, /$
LOG

Variables List

X, X1, Y

Variable Passed to Subroutine

X

Table 2.3.3 *Functions and variables used by the LN(X!) subroutine.*

```
2575 REM PROGRAM TO DEMOSTRATE LN(X!) SUBROUTINE
2576 PRINT
2577 PRINT
2578 PRINT" X",TAB(8),"LN(X!)",TAB(19),"EXP(LN(X!))"
2579 PRINT "---",TAB(6),"-----------",TAB(19),"------------"
2580 PRINT
2581 FOR X=1 TO 10
2582 GOSUB 44580
2583 PRINT X,TAB(5),Y,TAB(18),EXP(Y)
2584 NEXT X
2585 PRINT
2586 PRINT
2587 END
44572 REM ********************
44573 REM SERIES APPROXIMATION SUBROUTINE FOR LN(X!)  (LN(X!))
44574 REM ACCURACY BETTER THAN 6 PLACES FOR X>=3.
44575 REM ACCURACY BETTER THAN 12 PLACES FOR X>10.
44576 REM ADVANTAGE IS THAT VERY LARGE VALUES OF THE ARGUMENT CAN BE USED
44577 REM WITHOUT FEAR OF OVERFLOW.
44578 REM REFERENCE-  CRC MATH TABLES.
44579 REM X IS THE INPUT, Y IS THE OUTPUT
44580 X1=1/(X*X)
44581 Y=(X+.5)*LOG(X)-X*(1-X1/12+X1*X1/360-X1*X1*X1/1260+X1*X1*X1*X1/1680)
44582 Y=Y+0.918938533205
44583 RETURN
```

Listing 2.3.3 *Subroutine for the asymptotic series approximation to ln x! [LN(X!)]. Also shown is a program that demonstrates the use of LN(X!). See listing 2.3.4 for examples.*

```
RUN

 X      LN(X!)        EXP(LN(X!))
---    -----------    -----------

 1     -.00030754     .99969254
 2      .69314593     1.9999974
 3     1.7917596      6.0000006
 4     3.1780538      23.999999
 5     4.7874915      119.99997
 6     6.5792505      719.99948
 7     8.5251615      5040.0007
 8     10.604604      40320.045
 9     12.801828      362880.18
10     15.104413      3628801.5

READY
```

Listing 2.3.4 *Results of a sample run of the program appearing in listing 2.3.3. The third column indicates the accuracy attainable using the asymptotic expansion. Ideally, exp (ln x!) = x!. Some round-off error is apparent.*

112 BASIC SCIENTIFIC SUBROUTINES

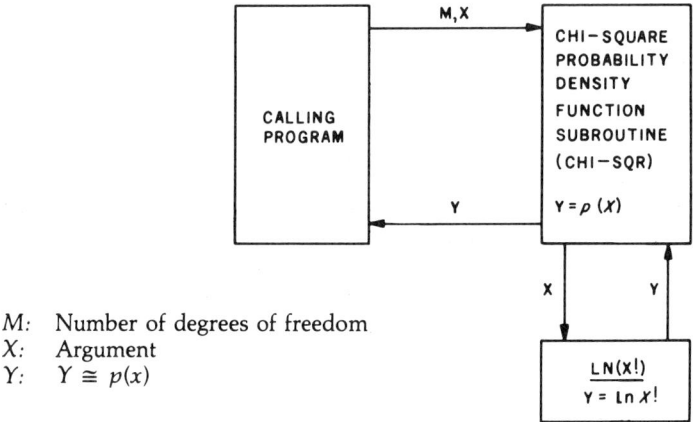

M: Number of degrees of freedom
X: Argument
Y: $Y \cong p(x)$

Figure 2.3.3 *Subroutine-connection diagram for the Chi-Square probability density distribution function subroutine (CHI-SQR). Note that the LN(X!) subroutine is also called.*

Statements/Functions List

$+, -, *, /$
EXP, GOSUB, LOG

Variables List

C, M1, X, X1, Y

Variables Passed to Subroutine

M, X

Table 2.3.4 *Functions and variables employed in both the CHI-SQR and LN(X!) subroutines.*

```
2600 REM PROGRAM TO DEMONSTRATE THE CHI-SQUARE SUBROUTINE
2601 PRINT
2602 PRINT
2603 PRINT "HOW MANY DEGREES OF FREEDOM: ",
2604 INPUT M
2605 PRINT
2606 PRINT "WHAT IS THE RANGE (X1,X2): "
2607 PRINT "X1: ",
2608 INPUT X1
2609 PRINT "X2: ",
2610 INPUT X2
```

Listing 2.3.5

```
2611 PRINT
2612 PRINT "WHAT IS THE TABLE STEP SIZE: ",
2613 INPUT X3
2614 PRINT
2615 PRINT
2616 PRINT "   X",TAB(8),"CHI-SQUARE PDF"
2617 PRINT "   ---",TAB(8),"---------------"
2618 PRINT
2619 FOR X=X1 TO X2 STEP X3
2620 GOSUB 44600
2621 PRINT X,TAB(8),INT(10000*Y)/10000
2622 NEXT X
2623 END
44572 REM ********************
44573 REM SERIES APPROXIMATION SUBROUTINE FOR LN(X!)   (LN(X!))
44574 REM ACCURACY BETTER THAN 6 PLACES FOR X>=3.
44575 REM ACCURACY BETTER THAN 12 PLACES FOR X>10.
44576 REM ADVANTAGE IS THAT VERY LARGE VALUES OF THE ARGUMENT CAN BE USED
44577 REM WITHOUT FEAR OF OVERFLOW.
44578 REM REFERENCE-  CRC MATH TABLES.
44579 REM X IS THE INPUT, Y IS THE OUTPUT
44580 X1=1/(X*X)
44581 Y=(X+.5)*LOG(X)-X*(1-X1/12+X1*X1/360-X1*X1*X1/1260+X1*X1*X1*X1/1680)
44582 Y=Y+0.918938533205
44583 RETURN
44592 REM ********************
44593 REM CHI-SQUARE FUNCTION SUBROUTINE (CHI-SQR)
44594 REM THIS PROGRAM TAKES A GIVEN DEGREE OF FREEDOM, M
44595 REM AND VALUE, X, AND CALCULATES THE CHI-SQUARE
44596 REM DENSITY DISTRIBUTION FUNCTION VALUE, Y.
44597 REM REFERENCE- TEXAS INSTRUMENTS SR-51 OWNERS MANUAL (1974).
44598 REM SUBROUTINE LN(X!) IS ALSO CALLED.
44599 REM SAVE X
44600 M1=X
44601 REM PERFORM CALCULATION
44602 X=M/2-1
44603 REM GOTO LN(X!) SUBROUTINE
44604 GOSUB 44580
44605 X=M1
44606 C=-X/2+(M/2-1)*LOG(X)-(M/2)*LOG(2)-Y
44607 Y=EXP(C)
44608 RETURN
```

Listing 2.3.5 *Chi-Square probability density distribution function subroutine (CHI-SQR). This program calculates $p(X^2)$ using the ln x! asymptotic series approximation subroutine [LN(X!)]. Also shown is a program for demonstrating CHI-SQR. See listing 2.3.6 for a sample run.*

```
RUN


HOW MANY DEGREES OF FREEDOM: ?100

WHAT IS THE RANGE (X1,X2):
X1:?50
X2:?150

WHAT IS THE TABLE STEP SIZE: ?5

   X        CHI-SQUARE PDF
  ---       --------------

  50        0
  55        0
  60        .0001
  65        .0007
  70        .0023
  75        .0057
  80        .011
  85        .0177
  90        .0239
  95        .0277
 100        .0281
 105        .0252
 110        .0202
 115        .0146
 120        .0096
 125        .0058
 130        .0032
 135        .0017
 140        .0008
 145        .0003
 150        .0001


READY
```

Listing 2.3.6 *A sample run of the program given in listing 2.3.5.*

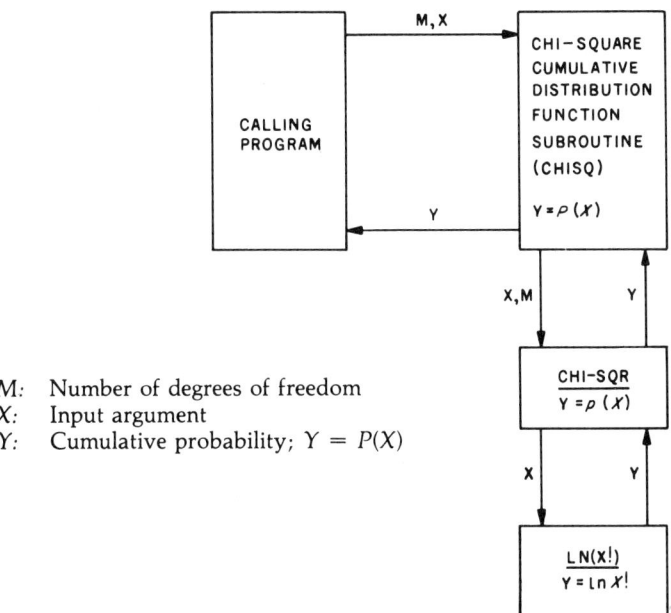

M: Number of degrees of freedom
X: Input argument
Y: Cumulative probability; $Y = P(X)$

Figure 2.3.4 *Subroutine-connection diagram for CHISQ.*

Statements/Functions List

|, −, *, /, <
EXP, GOSUB, GOTO, IF/THEN, LOG

Variables List

C, E, M, M1, M2, X, X1, X2, Y, Y1

Variables Passed to Subroutine

M, X

Table 2.3.5 *Functions and variables employed by CHISQ and its support subroutines, CHI-SQR and LN(X!).*

116 BASIC SCIENTIFIC SUBROUTINES

```
2625 REM PROGRAM TO DEMONSTRATE CHISQ
2626 PRINT
2627 PRINT
2628 PRINT "HOW MANY DEGREES OF FREEDOM: ",
2629 INPUT M
2630 PRINT
2631 PRINT "WHAT IS THE RANGE (X1,X2): "
2632 PRINT "X1: ",
2633 INPUT X1
2634 PRINT "X2: ",
2635 INPUT X2
2636 PRINT
2637 PRINT "STEP SIZE: ",
2638 INPUT X3
2639 PRINT
2640 PRINT "SUMMATION TRUNCATION ERROR BOUND: ",
2641 INPUT E
2642 PRINT
2643 PRINT
2644 PRINT "   X",TAB(8),"CHI-SQUARE CDF"
2645 PRINT " ---",TAB(8),"--------------"
2646 FOR X=X1 TO X2 STEP X3
2647 GOSUB 44625
2648 PRINT X,TAB(9),INT(10000*Y)/10000
2649 NEXT X
2650 END
44572 REM ********************
44573 REM SERIES APPROXIMATION SUBROUTINE FOR LN(X!)   (LN(X!))
44574 REM ACCURACY BETTER THAN 6 PLACES FOR X>=3.
44575 REM ACCURACY BETTER THAN 12 PLACES FOR X>10.
44576 REM ADVANTAGE IS THAT VERY LARGE VALUES OF THE ARGUMENT CAN BE USED
44577 REM WITHOUT FEAR OF OVERFLOW.
44578 REM REFERENCE-  CRC MATH TABLES.
44579 REM X IS THE INPUT, Y IS THE OUTPUT
44580 X1=1/(X*X)
44581 Y=(X+.5)*LOG(X)-X*(1-X1/12+X1*X1/360-X1*X1*X1/1260+X1*X1*X1*X1/1680)
44582 Y=Y+0.918938533205
44583 RETURN
44592 REM ********************
44593 REM CHI-SQUARE FUNCTION SUBROUTINE (CHI-SQR)
44594 REM THIS PROGRAM TAKES A GIVEN DEGREE OF FREEDOM, M
44595 REM AND VALUE, X, AND CALCULATES THE CHI-SQUARE
44596 REM DENSITY DISTRIBUTION FUNCTION VALUE, Y.
44597 REM REFERENCE- TEXAS INSTRUMENTS SR-51 OWNERS MANUAL (1974).
44598 REM SUBROUTINE LN(X!) IS ALSO CALLED.
44599 REM SAVE X
44600 M1=X
44601 REM PERFORM CALCULATION
44602 X=M/2-1
```

Listing 2.3.7

```
44603 REM GOTO LN(X!) SUBROUTINE
44604 GOSUB 44580
44605 X=M1
44606 C=-X/2+(M/2-1)*LOG(X)-(M/2)*LOG(2)-Y
44607 Y=EXP(C)
44608 RETURN
44615 REM ********************
44616 REM CHI-SQUARE CUMMULATIVE DISTRIBUTION (CHISQ)
44617 REM THE PROGRAM IS FAIRLY ACCURATE AND CALLS UPON THE
44618 REM CHI-SQUARE PROBABILITY DENSITY FUNCTION SUBROUTINE (CHI-SQR).
44619 REM THE INPUT PARAMETER IS M, THE NUMBER OF DEGREES OF FREEDOM.
44620 REM ALSO REQUIRED IS THE ORDINATE VALUE. THE PROGRAM RETURNS Y,
44621 REM THE CUMMULATIVE DISTRIBUTION INTEGRAL FROM 0 TO X.
44622 REM REFERENCE- HEWLETT-PACKARD STATISTICS PROGRAMS, 1974.
44623 REM THIS PROGRAM ALSO REQUIRES AN ACCURACY PARAMETER, E, TO
44624 REM DETERMINE THE LEVEL OF SUMMATION.
44625 Y1=1
44626 X2=X
44627 M2=M+2
44628 X2=X2/M2
44629 Y1=Y1+X2
44630 IF X2<E THEN GOTO 44637
44631 M2=M2+2
44632 REM THIS FORM IS USED TO AVOID OVERFLOW
44633 X2=X2*X/M2
44634 REM LOOP TO CONTINUE SUM
44635 GOTO 44629
44636 REM OBTAIN Y, THE PROBABILITY DENSITY FUNCTION
44637 GOSUB 44600
44638 Y=Y1*Y*2*X/M
44639 RETURN
```

Listing 2.3.7 *Chi-Square cumulative distribution function subroutine (CHISQ). Also shown is a program for exercising CHISQ. Note that two other subroutines are called: CHI-SQR and LN(X!) (see figure 2.3.4). An example is given in listing 2.3.8.*

RUN

Listing 2.3.8

```
HOW MANY DEGREES OF FREEDOM: ?100

WHAT IS THE RANGE (X1,X2):
X1: ?50
X2: ?150

STEP SIZE: ?5
```

```
SUMMATION TRUNCATION ERROR BOUND: ?.000001

   X      CHI-SQUARE CDF
  ---     --------------
  50       0
  55       0
  60       .0005
  65       .0026
  70       .0098
  75       .0291
  80       .0703
  85       .142
  90       .2468
  95       .3774
 100       .5188
 105       .6535
 110       .7677
 115       .855
 120       .9155
 125       .954
 130       .9764
 135       .9886
 140       .9948
 145       .9977
 150       .999
READY
```

Listing 2.3.8 *A sample run of the program given in listing 2.3.7.*

2.3.3 The Gamma Function

In the previous subsection, the subroutines for $p(x)$ and $P(x)$ were based on a truncated asymptotic series for $\ln x!$. We will now briefly indicate how this expansion was obtained.

First, we will note that the factorial is directly related to the gamma function:

$$(x-1)! = \Gamma(x) = \int_{-\infty}^{+\infty} e^{-t} t^{x-1} dt \quad \text{for } x > 0 \qquad (2.3.10)$$

(See Ref. 17.) $\Gamma(x)$ can be expressed in the classic Stirling-formula representation as

$$\Gamma(x) = \sqrt{2\pi} \; e^{[(x-1/2) \ln x - x + J(x)]} \qquad (2.3.11)$$

$J(x)$ is the Binet function

$$J(x) = \frac{1}{\pi} \int_0^\infty \frac{x}{t^2 + x^2} \ln\left(\frac{1}{1 - e^{-2\pi t}}\right) dt \qquad (2.3.12)$$

$J(x)$ in turn can be integrated by parts to give the asymptotic expansion

$$J(x) \sim \frac{1}{12x} - \frac{1}{360x^3} + \frac{1}{1260x^5} - \cdots \qquad (2.3.13)$$

Given the above equations, we can then construct the expansion for ln $x!$ shown in equation (2.3.9). As you might imagine, constructing asymptotic expansions is usually not an easy task.

2.3.4 The Error Function

Another classical expression that is amenable to approximation with an asymptotic series is the *error function*, erf(x):

$$\mathrm{erf}(x) = \frac{2}{\sqrt{\pi}} \int_0^x e^{-t^2} dt \qquad (2.3.14)$$

This function [or its complement, erfc(x) = 1 − erf(x)] is often involved in the solutions to heat transfer (see Ref. 26) and ionic diffusion (see Ref. 27) problems. It also appears in statistics (see Ref. 23). In many cases, there is a need to know the value of erf(x) [or erfc(x)] for large values of the argument (e.g., $x > 3$); tables seldom cover this range adequately.

Peirce (see Ref. 25) gives a convenient asymptotic expansion for positive values of x:

$$\mathrm{erf}(x) = 1 - \frac{e^{-x^2}}{x\sqrt{\pi}} \left(1 - \frac{1}{2x^2} + \frac{1 \cdot 3}{(2x^2)^2} - \frac{1 \cdot 3 \cdot 5}{(2x^2)^3} + \cdots \right) \qquad (2.3.15)$$

As with most asymptotic series, the above expression eventually diverges. However, we can again terminate the summation at the smallest term and use the magnitude of the next term as a measure of the error bound. This is done in the subroutine appearing in listing 2.3.9 (ASYMERF). The input to this subroutine is X. The returned values are $Y = \mathrm{erf}(X)$, $Y1 = \mathrm{erfc}(X)$, the error estimate (E), and the number of terms summed (N).

Sample results obtained using ASYMERF are given in listing 2.3.10. It appears that the error-bound estimate returned by the subroutine is conservative, and that erf(x) is accurately approximated for $x > 3$.

For $x \leq 3$, a rapidly convergent Taylor series approximation can be used:

$$\mathrm{erf}(x) = \frac{2}{\sqrt{\pi}} \left(x - \frac{x^3}{3} + \frac{x^5}{5 \cdot 2!} - \frac{x^7}{7 \cdot 3!} + \cdots \right) \qquad (2.3.16)$$

(See Ref. 25.) As a simple exercise, you can now try to develop a subroutine based on equation (2.3.16).

The discussion in this section certainly is not exhaustive. The object is to introduce you to the mechanics of asymptotic series approximations through several useful examples. For those interested in other asymptotic expansions, the author highly recommends the excellent handbook by Abramowitz and Stegun (Ref. 6). In that compendium, for example, you will find expansion pairs such as

$$\text{arcsinh } x = x - \frac{1}{2}\frac{x^3}{3} + \frac{1 \cdot 3}{2 \cdot 4}\frac{x^5}{5} - \frac{1 \cdot 3 \cdot 5}{2 \cdot 4 \cdot 6}\frac{x^7}{7} + \cdots \quad \text{for } |x| \leq 1$$

and

$$\text{arcsinh } x = \log 2x + \frac{1}{2}\frac{1}{2x^2} - \frac{1 \cdot 3}{2 \cdot 4}\frac{1}{4x^4} + \frac{1 \cdot 3 \cdot 5}{2 \cdot 4 \cdot 6}\frac{1}{6x^6} - \cdots$$

$$\text{for } |x| > 1$$

In the next section, we will return to the ascending-power series class of approximations and consider the very important Chebyshev polynomials.

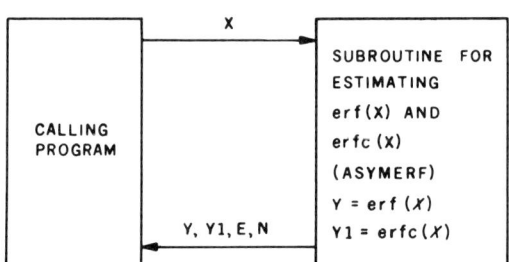

X: Input argument
Y: Returned estimate of erf(X), the error function
Y1: Returned estimate of erfc(X) = 1 − erf(X), the complementary error function
E: Error bound
N: Number of terms summed

Figure 2.3.5 *Subroutine-connection diagram for ASYMERF.*

SERIES APPROXIMATION TECHNIQUES

Statements/Functions List

$+, -, *, /, >$
ABS, GOTO, IF/THEN

Variables List

C1, C2, E, N, X, Y, Y1

Variable Passed to Subroutine

X

Table 2.3.6 *Functions and variables used in the erf(x) asymptotic series approximation subroutine (ASYMERF).*

```
2675 REM PROGRAM TO DEMONSTRATE ASYMERF
2676 PRINT
2677 PRINT
2678 PRINT "INPUT X",
2679 INPUT X
2680 GOSUB 44675
2681 PRINT
2682 PRINT "ERF(X)= ",Y,"  WITH ERROR ESTIMATE= ",INT(100000000*E)/100000000
2683 PRINT "NUMBER OF TERMS EVALUATED WAS",N
2684 PRINT
2685 END
44661 REM *********************
44662 REM ASYMPTOTIC SERIES EXPANSION OF THE INTEGRAL OF
44663 REM   2 EXP(-X*X)/SQRT(PI)  - THE NORMALIZED ERROR FUNCTION (ASYMERF)
44664 REM THIS PROGRAM DETEMINES THE VALUES OF THE ABOVE
44665 REM INTEGRAND USING AN ASYMPTOTIC SERIES WHICH IS
44666 REM EVALUATED TO THE LEVEL OF MAXIMUM ACCURACY.
44667 REM THE INTEGRAL IS FROM 0 TO X.
44668 REM THE INPUT PARAMETER IS X>0. THE RESULTS ARE
44669 REM RETURNED IN Y AND Y1, WITH THE ERROR MEASURE IN E.
44670 REM THE PROGRAM ALSO RETURNS THE NUMBER OF TERMS USED.
44671 REM NOTE- THE ERROR IS ROUGHLY EQUAL TO
44672 REM FIRST TERM NEGLECTED IN THE SERIES SUMMATION.
44673 REM REFERENCE-  A SHORT TABLE OF INTEGRALS BY B.O. PEIRCE
44674 REM    GINN AND COMPANY    1957
44675 N=1
44676 Y=1
44677 C2=1/(2*X*X)
44678 Y=Y-C2
44679 N=N+2
44680 C1=C2
```

Listing 2.3.9

```
44681 C2=-C1*N/(2*X*X)
44682 REM TEST FOR DIVERGENCE
44683 REM THE BREAK POINT IS ROUGHLY N=X*X
44684 IF ABS(C2)>ABS(C1) THEN GOTO 44687
44685 REM CONTINUE SUMMATION
44686 GOTO 44678
44687 N=(N+1)/2
44688 E=EXP(-X*X)/(X*1.772453850905516)
44689 Y1=Y*E
44690 Y=1-Y1
44691 E=E*C2
44692 RETURN
```

Listing 2.3.9 *Asymptotic series approximation subroutine for estimating erf(x) (ASYMERF). Also appearing is a program that demonstrates ASYMERF. See listing 2.3.10 for examples. See figure 2.3.5 for the associated subroutine-connection diagram.*

RUN

Listing 2.3.10

```
INPUT X?1

ERF(X)=  .8962231   WITH ERROR ESTIMATE=  -.15566531
NUMBER OF TERMS EVALUATED WAS 2

READY
RUN




INPUT X?2

ERF(X)=  .9952558   WITH ERROR ESTIMATE=  .000149
NUMBER OF TERMS EVALUATED WAS 5

READY
RUN




INPUT X?3

ERF(X)=  .9999779   WITH ERROR ESTIMATE=  -.00000001
NUMBER OF TERMS EVALUATED WAS 10
```

```
READY
RUN


INPUT X?4

ERF(X)=   1   WITH ERROR ESTIMATE=   0
NUMBER OF TERMS EVALUATED WAS 17

READY
```

Listing 2.3.10 *Examples of the use of the program given in listing 2.3.9. The true values and actual errors are shown below:*

X	erf(x) (True)*	erf(x) (Calculated)	Error	Truncation error-bound estimate
1	0.84270079	0.8962231	−0.05	−0.16
2	0.99532226	0.9952558	0.00007	0.00015
3	?	0.9999779	?	-4×10^{-9}
4	?	1.0000000	?	3×10^{-15}

Entries are missing from this table because the author was not able to find accurate enough values in the literature to compare against. The truncation error estimates appear conservative.

2.4 Chebyshev Polynomials

This is the first of two sections that deal with the subject of Chebyshev polynomials. In this section, we will review some of the special properties of Chebyshev polynomials and why these features may be useful. Also, a subroutine for accurately calculating Chebyshev polynomial coefficients for any order is provided. In the next section, this subroutine is combined with a decomposition program that permits power series to be expressed in terms of Chebyshev polynomial expansions. These expansions in turn facilitate a process called *economization*.

The basic idea behind using Chebyshev polynomials for approximation can easily be summarized by qualitatively examining their basic properties. First, the Chebyshev function of order N [$T_N(x)$] is an Nth-degree polynomial. Second, over the range $-1 \leq x \leq 1$, this polynomial has unity maxima and minima. That is,

$$-1 \leq T_N(x) \leq 1 \qquad (2.4.1)$$

---

*See Handbook of Mathematical Functions (Ref. 6).

Further, any real roots that occur are defined to be in the range $-1 \leq x \leq 1$. The Chebyshev polynomial of order N, therefore, must have N zero crossings in that interval, and $N + 1$ equal (in magnitude) maxima or minima. The shapes of the first few of the unique set of polynomials that satisfy these conditions are given in figure 2.4.1.

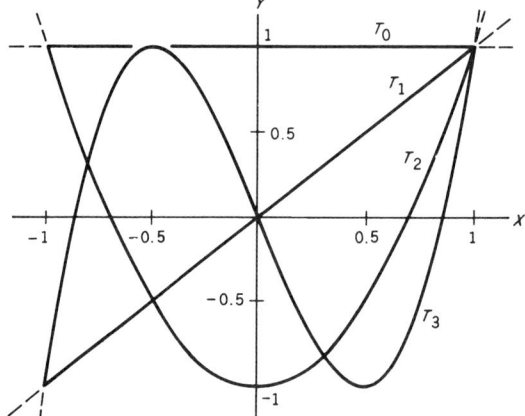

Figure 2.4.1 *Comparative plot of the first four Chebyshev polynomials. Each of the polynomials exists outside the interval $-1 \leq x \leq 1$, but the property that $|T_n(x)| \leq 1$ pertains to only that range. Hastings (Ref. 9) defines the range to be $0 \leq x \leq 1$, and the polynomials he presents are shifted and scaled versions of those given in this section. Although there is no great theoretical significance to the difference, the symmetrical range definition is more convenient for practical use.*

Without getting into mathematical detail, the practical use of this representation can be readily demonstrated. For example, assume that we have an Mth-degree polynomial $P_M(x)$ which exactly (or very accurately) represents some function $f(x)$ over the range $-1 \leq x \leq 1$. (The argument of any function can be shifted and scaled so as to be mapped into this range.) This function can be just as accurately represented by a Chebyshev series:

$$f(x) = \sum_{i=0}^{M} a_i T_i(x) \qquad (2.4.2)$$

If the Chebyshev representation is truncated at the $T_N(x)$ term, we then have the approximation

$$f(x) \cong \sum_{i=0}^{N} a_i T_i(x) \qquad (2.4.3)$$

What can be stated about the error in this estimate? First, because $|T_i(x)| \leq 1$, the error is certainly bounded:

$$|E(x)| \le \sum_{i=N+1}^{M} |a_i| \qquad (2.4.4)$$

Second, if the Chebyshev series is rapidly converging (i.e., $|a_i| \gg |a_{i+1}|$), the error is nearly equal to the first term not included:

$$E(x) \cong a_{N+1} T_{N+1}(x) \qquad (2.4.5)$$

However, $T_{N+1}(x)$ is an oscillatory function. The error therefore oscillates in sign and has nearly equal (in magnitude) maxima and minima. It can be shown (see Ref. 10) that there exists only one polynomial that has equal-ripple error—the *min-max*. The truncated Chebyshev series therefore approximates the min-max polynomial (and hence is called *near-min-max*), and the approximation is better the more rapidly the series converges.

The value in using a min-max polynomial was discussed in Chapter 1, section 1. In essence, the main advantage is that the error is spread out over the interval, and for a given degree of fit, the approximation has the minimum maximum error. This is to be contrasted with the truncated Taylor series which is very accurate near the expansion point, but is inaccurate far from that point.

Chebyshev polynomials can easily be generated using the following defining formula:

$$T_N(x) = \cos N\theta \quad \text{where} \quad N = 0, 1, 2, \ldots \quad \text{and} \quad \theta = \cos^{-1} x \qquad (2.4.6)$$

By employing multiple-angle trigonometric formulae, it can be shown that

$$T_0(x) = 1$$
$$T_1(x) = x$$

Further, because $\cos N\theta = 2 \cos \theta \cos (N - 1)\theta - \cos (N - 2)\theta$, we have the following fundamental *recursion* relation:

$$T_N(x) = 2x\, T_{N-1}(x) - T_{N-2}(x) \qquad (2.4.7)$$

(Recursion, a very powerful technique, is particularly well suited to numerical analysis. This will be discussed more completely in Chapter 3.) It is apparent from equation (2.4.7) that $T_N(x)$ is either an even or an odd function:

$$\begin{array}{ll} T_N(x) = T_N(-x) & N \text{ even} \\ T_N(x) = -T_N(-x) & N \text{ odd} \end{array} \qquad (2.4.8)$$

Therefore, when approximating even functions, we should expect to see only even-order Chebyshev polynomials, and for odd functions, only odd-order $T_N(x)$.

The recursion relation [equation (2.4.7)] can be simply implemented in a computer program for calculating the Chebyshev polynomial coefficients (see listing 2.4.1). The input to CHEBYSER is the order of the polynomial. The outputs are the coefficients $B_N(J) = B(N,J)$. A sample run using the CHEBYSER subroutine and the demonstration program appears in listing 2.4.2. The accuracy of the calculation is very high because the coefficients are all reasonably small integers; there is no round-off error. For very high orders some round-off error can occur. However, high-order Chebyshev approximations are usually not encountered unless there is a significant convergence problem in the original series being economized.

In the next section, the mechanics for achieving the near-min-max approximation indicated by equation (2.4.3) will be considered and an economization subroutine will be developed.

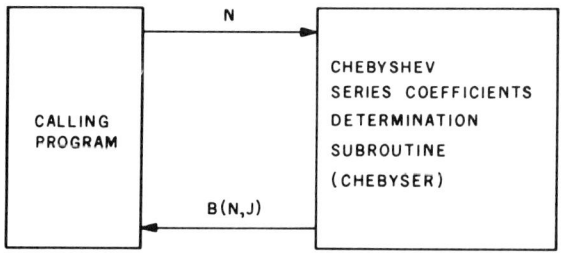

N: Polynomial order
$B(N,J)$: Polynomial coefficients for $T_N(x)$
$T_N(x) = B(N,0) + B(N,1)x + B(N,2)x^2 + \cdots$

Figure 2.4.2 *Subroutine-connection diagram for CHEBYSER.*

Statements/Functions List

$+, -, <$
FOR/NEXT, IF/THEN

Variables List

B(I,J), I, J, N

Variable Passed to Subroutine

N

Table 2.4.1 *Functions and variables employed in the Chebyshev polynomial coefficients-determination subroutine (CHEBYSER).*

```
2700 REM PROGRAM TO DEMONSTRATE CHEBYSER SUBROUTINE
2701 PRINT
2702 PRINT
2703 DIM B(10,10)
2704 FOR N=2 TO 10
2705 GOSUB 44725
2706 PRINT "CHEBYSHEV POLYNOMIAL COEFFICIENTS"
2707 PRINT "FOR DEGREE",N
2708 PRINT
2709 FOR I=0 TO N
2710 PRINT "A(",I,") = ",B(N,I)
2711 NEXT I
2712 PRINT
2713 PRINT
2714 NEXT N
2715 PRINT
2716 END
44717 REM ********************
44718 REM CHEBYCHEV SERIES COEFFICIENT EVALUATION SUBROUTINE (CHEBYSER)
44719 REM THE ORDER OF THE POLYNOMIAL IS N.
44720 REM THE COEFFICIENTS ARE RETURNED IN THE
44721 REM ARRAY B(I,J). I IS THE DEGREE OF THE POLYNOMIAL,
44722 REM J IS THE COEFFICIENT ORDER.
44723 REM DIMENSION B(I,J) IN THE CALLING PROGRAM.
44724 REM ESTABLISH T0 AND T1 COEFFICIENTS
44725 B(0,0)=1
44726 B(1,0)=0
44727 B(1,1)=1
44728 REM RETURN IF ORDER IS LESS THAN TWO
44729 IF N<2 THEN RETURN
44730 FOR I=2 TO N
44731 FOR J=1 TO I
44732 REM BASIC RECURSION RELATION
44733 B(I,J)=B(I-1,J-1)+B(I-1,J-1)-B(I-2,J)
44734 NEXT J
44735 B(I,0)=-B(I-2,0)
44736 NEXT I
44737 RETURN
```

Listing 2.4.1 *Chebyshev polynomial coefficients determination subroutine (CHEBYSER). Also shown is a program for demonstrating the operation of CHEBYSER. A sample run appears in listing 2.4.2.*

RUN   Listing 2.4.2

CHEBYSHEV POLYNOMIAL COEFFICIENTS
FOR DEGREE 2

A(0) =   -1
A(1) =   0
A(2) =   2

CHEBYSHEV POLYNOMIAL COEFFICIENTS
FOR DEGREE 3

A(0) =   0
A(1) =   -3
A(2) =   0
A(3) =   4

CHEBYSHEV POLYNOMIAL COEFFICIENTS
FOR DEGREE 4

A(0) =   1
A(1) =   0
A(2) =   -8
A(3) =   0
A(4) =   8

CHEBYSHEV POLYNOMIAL COEFFICIENTS
FOR DEGREE 5

A(0) =   0
A(1) =   5
A(2) =   0
A(3) = -20
A(4) =   0
A(5) =   16

CHEBYSHEV POLYNOMIAL COEFFICIENTS
FOR DEGREE 6

A(0) =   -1
A(1) =   0
A(2) =   18
A(3) =   0

A(4) = -48
A(5) = 0
A(6) = 32

Listing 2.4.2 cont.

CHEBYSHEV POLYNOMIAL COEFFICIENTS
FOR DEGREE 7

A(0) = 0
A(1) = -7
A(2) = 0
A(3) = 56
A(4) = 0
A(5) = -112
A(6) = 0
A(7) = 64

CHEBYSHEV POLYNOMIAL COEFFICIENTS
FOR DEGREE 8

A(0) = 1
A(1) = 0
A(2) = -32
A(3) = 0
A(4) = 160
A(5) = 0
A(6) = -256
A(7) = 0
A(8) = 128

CHEBYSHEV POLYNOMIAL COEFFICIENTS
FOR DEGREE 9

A(0) = 0
A(1) = 9
A(2) = 0
A(3) = -120
A(4) = 0
A(5) = 432
A(6) = 0
A(7) = -576
A(8) = 0
A(9) = 256

CHEBYSHEV POLYNOMIAL COEFFICIENTS
FOR DEGREE 10

```
A( 0) =   -1
A( 1) =    0
A( 2) =   50
A( 3) =    0
A( 4) = -400
A( 5) =    0
A( 6) = 1120
A( 7) =    0
A( 8) =-1280
A( 9) =    0
A( 10) = 512
```

READY

Listing 2.4.2 *A sample run of the program given in listing 2.4.1. Observe that the Chebyshev polynomials contain either only odd-order terms or only even-order terms. They are either asymmetric $[T_N(x) = -T_N(-x)]$ or symmetric $[T_N(x) = T_N(-x)]$.*

2.5 Economization

The economization (or *telescoping*) of a power series using a truncated Chebyshev representation is very simple. It involves three major steps:

1) The function, $f(x)$, must be approximated by a power series to a level of accuracy in excess of that eventually desired. A truncated Taylor series or least-squares polynomial can be used. The maximum truncation error should be roughly one tenth of that eventually desired in the economized polynomial approximation. Note that x may have to be shifted and scaled so that the argument is within the defined Chebyshev polynomial range.
2) The equivalent Chebyshev representation must be found. This is done by subtracting out Chebyshev polynomial factors starting with the highest degree in the power series. This is called *decomposition*.
3) The Chebyshev series must then be truncated to the desired level of accuracy, and converted back to a polynomial in x.

The decomposition step requires some explanation. Assume we have approximated $f(x)$ with an Mth-degree polynomial, $P_M(x)$:

$$f(x) \cong P_M(x) = a_0 + a_1 x + a_2 x^2 + \cdots + a_{M-1} x^{M-1} + a_M x^M \quad (2.5.1)$$

The first step in the decomposition is to find a value for b_M so that the highest-degree term in $P_M(x)$ can be removed:

$$P_{M-1}(x) = P_M(x) - b_M T_M(x)$$

Next, the highest-degree term in $P_{M-1}(x)$ is removed:

$$P_{M-2}(x) = P_{M-1} - b_{M-1} T_{M-1}(x)$$

This procedure is repeated until we get to

$$0 = P_0(x) - b_0 T_0(x)$$

The original polynomial, $P_M(x)$, can then be replaced:

$$P_M(x) = \sum_{i=0}^{M} b_i T_i(x) \qquad (2.5.2)$$

The Chebyshev series is then truncated:

$$f(x) \cong P_M(x) \cong \sum_{i=0}^{N} b_i T_i(x) \qquad N < M \qquad (2.5.3)$$

If E is the error bound for the original approximation [equation (2.5.1)], then the error bound for the truncated Chebyshev series representation is

$$\text{Error bound} < E + \sum_{i=N+1}^{M} |b_i| \qquad (2.5.4)$$

If the original approximation was very good ($|b_M| \gg E$), and the Chebyshev series coefficients are rapidly decreasing, then

$$\text{Error bound} \approx |b_M| \qquad (2.5.5)$$

If both of those assumptions are reasonable, then the truncated Chebyshev series closely approximates the ideal min-max polynomial. For most practical situations, this is the case.

The above procedure can be readily programmed in BASIC as shown in listing 2.5.1. The economization subroutine given in that listing (CHEBECON) requires as inputs the degree of the original polynomial (M), the degree of the desired economized polynomial ($M1 < M$), and the coefficients of the polynomial to be economized [$C(I)$]. The subroutine also requires a range factor, $X0$. This factor maps the argument range from $-X0 \le X \le X0$ to $-1 \le X \le 1$ for the Chebyshev series decomposition step. The range is restored later.

CHEBECON returns the coefficients of the full Chebyshev series representation [$A(I)$], and the coefficients corresponding to the economized polynomial [$C(I)$]. CHEBYSER is called in the process of performing these calculations.

Listing 2.5.1 also contains a program for demonstrating the operation of

CHEBECON. Using this program, an economized polynomial approximation to sin x in the range $-\pi/2 \leq x \leq \pi/2$ is calculated (see listing 2.5.2). By truncating the Chebyshev series at $T_9(x)$, a ninth-degree polynomial approximation is obtained (five terms) having a truncation error bound of approximately 3×10^{-9}. However, because the input coefficients were to only eight digits, the accuracy is actually limited by the original round-off error in the input coefficients.

We can make two observations regarding this example. First, the coefficients of the economized polynomial bear a close resemblance to those of the original Taylor series. However, they are somewhat altered, and the difference becomes relatively greater for the higher degrees. Second, the decomposition and reconstruction processes involve calculations that are prone to round-off error. Therefore, it is wise to calculate the coefficients using double-precision arithmetic, and then to round off the results. A comparison of the results obtained using both 8- and 14-digit precision calculations is shown in table 2.5.2. In this case, the effect of round-off error was not significant. However, we will soon see an example in which the effect *is* important.

As another exercise for CHEBECON, we will consider the Taylor series expansion for $J_0(x)$ which was discussed in section 2.2. In that particular case, it was found that to ensure low truncation error ($<10^{-9}$) over the range $0 \leq x < 10$, a 40th-degree polynomial was required. However, round-off error drastically reduced the accuracy to five digits, even though 8-digit arithmetic was used. The Taylor expansion can be replaced by an economized polynomial having fewer turns. The result is higher truncation error, but much lower round-off error.

An example of economizing the truncated Taylor series for $J_0(x)$ is shown in listing 2.5.3. By terminating the Chebyshev representation at $T_{24}(x)$, the error introduced by economization is roughly 10^{-8}, which equals the precision of the original coefficients. The result is that the original 21-term series can be replaced by one having only 13 terms. As an exercise, examine the round-off error associated with this shorter polynomial.

The calculations involved in obtaining listing 2.5.3 were performed using 14-digit precision arithmetic. If the computations had been carried out using 8-digit precision, the values shown in table 2.5.3 would have resulted.

The agreement between the two sets of coefficients is on the whole very good, except for the first one. The decomposition proceeded from the high-order terms backwards toward the term that greatly affects $C(0)$, i.e., $A(0)$. $C(0)$ received the brunt of the round-off error buildup.

CHEBECON is a very effective subroutine. With it, many of the approximations appearing in Hart (Ref. 7), Becket and Hurt (Ref. 8), Hastings (Ref. 9), and other references, can be either duplicated or improved upon. It should be noted, however, that the accuracy of the final approximation is dependent on several factors:

- the accuracy of the input polynomial coefficients—the polynomial to be economized. In our examples, these coefficients were accurate to only about eight digits. The economized polynomial coefficients must necessarily be

limited to that accuracy.
- the accuracy of the initial approximation. In our examples, an initial truncation-error limit that exceeded the precision of the coefficients themselves was chosen. In that way, the initial accuracy was clearly limited by round-off error.
- the truncation point in the Chebyshev series. In the examples given in this section, the truncation point was chosen to correspond roughly to the round-off error limit or less. Again, the objective was to have round-off be the main source of error.
- the precision of the calculation used in CHEBECON. In some cases, double precision is required in the determination of the coefficients. These situations are easy to recognize in that they involve slowly convergent series.

So far in the discussions of Chapters 1 and 2, we have concentrated on approximating functions with polynomials. We will now examine how polynomial expressions already calculated can be manipulated in order to extend their utility.

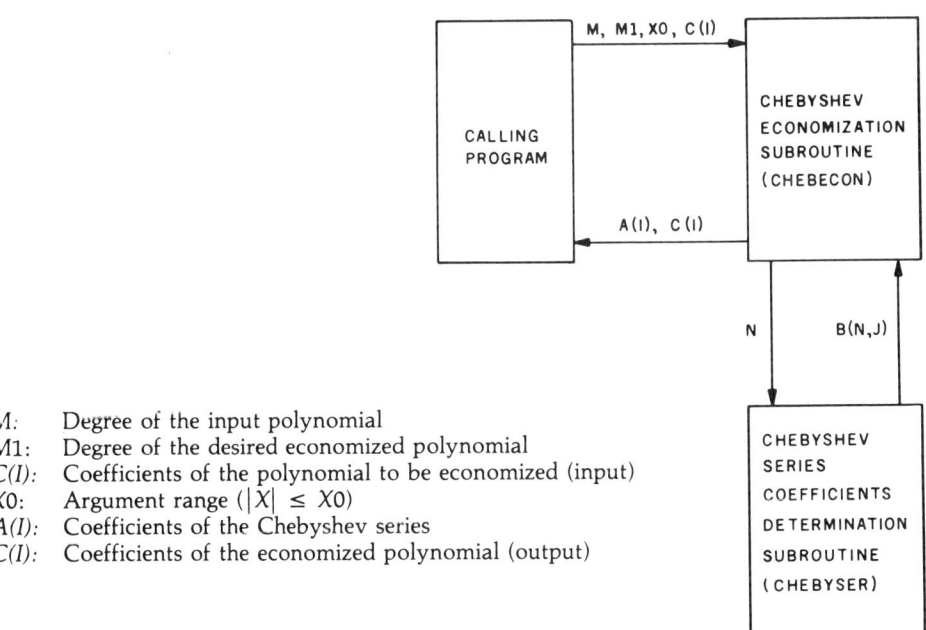

M: Degree of the input polynomial
$M1$: Degree of the desired economized polynomial
$C(I)$: Coefficients of the polynomial to be economized (input)
$X0$: Argument range ($|X| \leq X0$)
$A(I)$: Coefficients of the Chebyshev series
$C(I)$: Coefficients of the economized polynomial (output)

Figure 2.5.1 *Subroutine-connection diagram for the Chebyshev economization subroutine (CHEBECON). Note that the input polynomial coefficients, C(I), are destroyed in the process of calculation.*

Statements/Functions List

+, −, *, /, <
FOR/NEXT, GOSUB, IF/THEN

Variables List

A(N), B, B(N,L), C(L), I, J, L, M, M1, N, X0

Variables Passed to Subroutine

C(I), M, M1, X0

Table 2.5.1 *Functions and variables used in the Chebyshev economization subroutine (CHEBECON). Also included are the functions and variables employed by CHEBYSER, which is called by CHEBECON.*

```
2720 REM PROGRAM TO DEMONSTRATE CHEBYSHEV ECONOMIZATION
2721 PRINT
2722 PRINT "WHAT IS THE DEGREE OF"
2723 PRINT "THE INPUT POLYNOMIAL: ",
2724 INPUT M
2725 PRINT
2726 PRINT "WHAT IS THE DEGREE OF THE"
2727 PRINT "DESIRED ECONOMIZED POLYNOMIAL: ",
2728 INPUT M1
2729 PRINT
2730 PRINT "WHAT IS THE RANGE OF"
2731 PRINT "INPUT POLYNOMIAL: ",
2732 INPUT X0
2733 DIM A(M),B(M,M),C(M)
2734 PRINT
2735 PRINT
2736 PRINT "INPUT THE COEFFICIENTS:"
2737 PRINT
2738 FOR I=0 TO M
2739 PRINT "C(",I,") = ",
2740 INPUT C(I)
2741 NEXT I
2742 PRINT
2743 PRINT
2744 GOSUB 44760
2745 PRINT "THE CHEBYSHEV SERIES COEFFICIENTS ARE:"
2746 PRINT
2747 FOR I=0 TO M
2748 PRINT "A(",I,") = ",A(I)
2749 NEXT I
```

Listing 2.5.1

```
2750 PRINT
2751 PRINT
2752 PRINT "THE ECONOMIZED POLYNOMIAL"
2753 PRINT "COEFFICIENTS ARE:"
2754 PRINT
2755 FOR I=0 TO M1
2756 PRINT "C(",I,") = ",C(I)
2757 NEXT I
2758 PRINT
2759 PRINT
2760 END
44717 REM ********************
44718 REM CHEBYCHEV SERIES COEFFICIENT EVALUATION SUBROUTINE (CHEBYSER)
44719 REM THE ORDER OF THE POLYNOMIAL IS N.
44720 REM THE COEFFICIENTS ARE RETURNED IN THE
44721 REM ARRAY B(I,J). I IS THE DEGREE OF THE POLYNOMIAL,
44722 REM J IS THE COEFFICIENT ORDER.
44723 REM DIMENSION B(I,J) IN THE CALLING PROGRAM.
44724 REM ESTABLISH T0 AND T1 COEFFICIENTS
44725 B(0,0)=1
44726 B(1,0)=0
44727 B(1,1)=1
44728 REM RETURN IF ORDER IS LESS THAN TWO
44729 IF N<2 THEN RETURN
44730 FOR I=2 TO N
44731 FOR J=1 TO I
44732 REM BASIC RECURSION RELATION
44733 B(I,J)=B(I-1,J-1)+B(I-1,J-1)-B(I-2,J)
44734 NEXT J
44735 B(I,0)=-B(I-2,0)
44736 NEXT I
44737 RETURN
44746 REM ********************
44747 REM CHEBYSHEV ECONOMIZATION SUBROUTINE (CHEBECON)
44748 REM ROUTINE TAKES THE INPUT POLYNOMIAL COEFFICIENTS, C(I),
44749 REM AND RETURNS THE CHEBYSCHEV SERIES COEFFICIENTS, A(I).
44750 REM THE DEGREE OF THE SERIES PASSED TO THE ROUTINE IS M.
44751 REM THE DEGREE OF THE SERIES RETURNED IS M1.
44752 REM THE MAXIMUM RANGE OF X IS X0- X0 IS USED FOR SCALING.
44753 REM THE CHEBYSCHEV SERIES COEFFICIENT (B(I,J) SUBROUTINE IS
44754 REM CALLED- I IS THE ORDER OF THE CHEBYSCHEV POLYNOMIAL.
44755 REM NOTE THAT THE INPUT SERIES COEFFICIENTS ARE NULLED DURING THE PROCESS,
44756 REM AND THEN SET EQUAL TO THE ECONOMIZED SERIES COEFFICIENTS.
44757 REM THE CHEBYSHEV SERIES IS VALID ONLY OVER THE RANGE ABS(X/X0)<=1.
44758 REM DIMENSION A(I),B(I,J),C(I) IN THE CALLING PROGRAM.
44759 REM START BY SCALING THE INPUT COEFFICIENTS ACCORDING TO C(I)
44760 B=X0
44761 FOR I=1 TO M
44762 C(I)=C(I)*B
```

```
44763 B=B*X0
44764 NEXT I
44765 REM GET CHEBYSCHEV SERIES COEFFICIENTS.
44766 REM POLYNOMIAL SERIES IS REDUCED FROM THE HIGHEST ORDER DOWN
44767 FOR N=M TO 0 STEP -1
44768 GOSUB 44725
44769 A(N)=C(N)/B(N,N)
44770 FOR L=0 TO N
44771 REM CHEBYSCHEV SERIES OF ORDER L IS SUBTRACTED OUT OF THE POLYNOMIAL
44772 C(L)=C(L)-A(N)*B(N,L)
44773 NEXT L
44774 NEXT N
44775 REM PERFORM TRUNCATION
44776 FOR I=0 TO M1
44777 FOR J=0 TO I
44778 C(J)=C(J)+A(I)*B(I,J)
44779 NEXT J
44780 NEXT I
44781 REM CONVERT BACK TO THE INTERVAL X0
44782 B=1/X0
44783 FOR I=1 TO M1
44784 C(I)=C(I)*B
44785 B=B/X0
44786 NEXT I
44787 RETURN
```

Listing 2.5.1 *Subroutine for economizing a polynomial series (CHEBECON). Also shown is a program for demonstrating the operation of CHEBECON. See listings 2.5.2 and 2.5.3 for examples. CHEBECON uses the Chebyshev series coefficient determination subroutine (CHEBYSER) given in section 2.4. See figure 2.5.1 for the subroutine-connection diagram.*

```
RUN
```

Listing 2.5.2

```
WHAT IS THE DEGREE OF
THE INPUT POLYNOMIAL: ?15

WHAT IS THE DEGREE OF THE
DESIRED ECONOMIZED POLYNOMIAL: ?9

WHAT IS THE RANGE OF
INPUT POLYNOMIAL: ?1.57


INPUT THE COEFFICIENTS:

C( 0) = ?0
```

Listing 2.5.2 cont.

```
C( 1) = ?1
C( 2) = ?0
C( 3) = ?-.166666666
C( 4) = ?0
C( 5) = ?.00833333333
C( 6) = ?0
C( 7) = ?-.00019841270
C( 8) = ?0
C( 9) = ?.000002755732
C( 10) = ?0
C( 11) = ?-.000000025052109
C( 12) = ?0
C( 13) = ?.00000000016059045
C( 14) = ?0
C( 15) = ?-.000000000000076471635


THE CHEBYSHEV SERIES COEFFICIENTS ARE:

A( 0) =    0
A( 1) =    1.1334708
A( 2) =    0
A( 3) =   -.13788415
A( 4) =    0
A( 5) =    4.4798166E-03
A( 6) =    0
A( 7) =   -6.7466708E-05
A( 8) =    0
A( 9) =    5.8648371E-07
A( 10) =   0
A( 11) =  -3.3197271E-09
A( 12) =   0
A( 13) =   1.3197931E-11
A( 14) =   0
A( 15) =  -4.0511212E-14


THE ECONOMIZED POLYNOMIAL
COEFFICIENTS ARE:

C( 0) =    0
C( 1) =    1
C( 2) =    0
C( 3) =   -.16666648
C( 4) =    0
C( 5) =    8.3329005E-03
C( 6) =    0
C( 7) =   -1.9800945E-04
C( 8) =    0
```

```
C( 9) =   2.5905922E-06
```

READY

Listing 2.5.2 *A sample run of the program shown in listing 2.5.1. The input polynomial was a truncated Taylor series representation for sin x: $\sin x \cong x - x^3/3! + x^5/5! - x^7/7! + x^9/9! - x^{11}/11! + x^{13}/13! - x^{15}/15!$. The approximation range was $|x| \le \pi/2$. By terminating the Chebyshev series at $T_9(x)$, a ninth-degree economized approximation polynomial was obtained. The error bound for the economized polynomial is $|A(11)|$; $|error| \le 3 \times 10^{-9}$. If the Taylor series had been truncated at the ninth degree, the truncation error bound would have been $|C(11)| (1.57)^{11}$, or $|error| \le 4 \times 10^{-6}$. Although the Taylor series for sin x is rapidly convergent, Chebyshev economization is still very effective.*

Note that the coefficients of the economized polynomial resemble those of the original Taylor series, but with some modification. Also observe that only 8-digit precision was employed in the calculation, and so the true accuracy is lower than that estimated above.

Coefficient	8-digit Precision Value	14-digit Precision Value
C(1)	1.00000000	0.99999998
C(3)	−0.16666648	−0.16666648
C(5)	$0.83329005 \times 10^{-2}$	$0.83329005 \times 10^{-2}$
C(7)	$-0.19800945 \times 10^{-3}$	$-0.19800945 \times 10^{-3}$
C(9)	$0.25905922 \times 10^{-5}$	$0.25905920 \times 10^{-5}$

Table 2.5.2 *A comparison of the results obtained in listing 2.5.2 using 8-digit precision arithmetic with those obtained using 14 digits (and rounded off to 8). In this case, there is little round-off error in the calculation because there were not many terms in the original series.*

RUN

Listing 2.5.3

```
WHAT IS THE DEGREE OF
THE INPUT POLYNOMIAL: ?40

WHAT IS THE DEGREE OF THE
DESIRED ECONOMIZED POLYNOMIAL: ?24

WHAT IS THE RANGE OF
INPUT POLYNOMIAL: ?10

INPUT THE COEFFICIENTS:
```

Listing 2.5.3 cont.

```
C( 0) = ?1
C( 1) = ?0
C( 2) = ?-.25
C( 3) = ?0
C( 4) = ?.015625
C( 5) = ?0
C( 6) = ?-4.3402778E-4
C( 7) = ?0
C( 8) = ?6.7816841E-6
C( 9) = ?0
C( 10) = ?-6.7816841E-8
C( 11) = ?0
C( 12) = ?4.7095028E-10
C( 13) = ?0
C( 14) = ?-2.4028076E-12
C( 15) = ?0
C( 16) = ?9.3859672E-15
C( 17) = ?0
C( 18) = ?-2.8969035E-17
C( 19) = ?0
C( 20) = ?7.2422588E-20
C( 21) = ?0
C( 22) = ?-1.4963345E-22
C( 23) = ?0
C( 24) = ?2.597803E-25
C( 25) = ?0
C( 26) = ?-3.8429038E-28
C( 27) = ?0
C( 28) = ?4.901663E-31
C( 29) = ?0
C( 30) = ?-5.4462922E-34
C( 31) = ?0
C( 32) = ?5.3186447E-37
C( 33) = ?0
C( 34) = ?-4.6009037E-40
C( 35) = ?0
C( 36) = ?3.55008E-43
C( 37) = ?0
C( 38) = ?-2.4585042E-46
C( 39) = ?0
C( 40) = ?1.5365651E-49


THE CHEBYSHEV SERIES COEFFICIENTS ARE:

A( 0) =   .03153937889659
A( 1) =   0
A( 2) =   -.2146183154965
A( 3) =   0
```

```
A(  4) =   .00433523538825
A(  5) =   0
A(  6) =   -.26620433042028
A(  7) =   0
A(  8) =   .30612528319438
A(  9) =   0
A( 10) =   -.13638882086641
A( 11) =   0
A( 12) =   3.4347534658096E-02
A( 13) =   0
A( 14) =   -5.6980827389343E-03
A( 15) =   0
A( 16) =   6.7750398106601E-04
A( 17) =   0
A( 18) =   -6.0947054446533E-05
A( 19) =   0
A( 20) =   4.3088894795855E-06
A( 21) =   0
A( 22) =   -2.4630010047646E-07
A( 23) =   0
A( 24) =   1.163672445053E-08
A( 25) =   0
A( 26) =   -4.6253681262213E-10
A( 27) =   0
A( 28) =   1.5695501876403E-11
A( 29) =   0
A( 30) =   -4.5998528976889E-13
A( 31) =   0
A( 32) =   1.1811559757991E-14
A( 33) =   0
A( 34) =   -2.6179823856364E-16
A( 35) =   0
A( 36) =   5.7147597435687E-18
A( 37) =   0
A( 38) =   -6.7079898144584E-20
A( 39) =   0
A( 40) =   2.7949956347584E-21
```

Listing 2.5.3 cont.

```
THE ECONOMIZED POLYNOMIAL
COEFFICIENTS ARE:

C(  0) =   .99999999952126
C(  1) =   0
C(  2) =   -.24999999837297
C(  3) =   0
C(  4) =   .01562499908292
C(  5) =   0
C(  6) =   -4.3402757668102E-04
```

```
C( 7) =   0
C( 8) =   6.7816606467589E-06
C( 9) =   0
C( 10) =  -6.7815227896759E-08
C( 11) =  0
C( 12) =  4.7087887510757E-10
C( 13) =  0
C( 14) =  -2.4006814068287E-12
C( 15) =  0
C( 16) =  9.3423756945756E-15
C( 17) =  0
C( 18) =  -2.8349482428351E-17
C( 19) =  0
C( 20) =  6.6374579003594E-20
C( 21) =  0
C( 22) =  -1.1022242672315E-22
C( 23) =  0
C( 24) =  9.761591981951E-26
READY
```

Listing 2.5.3 *In this example, the truncated Taylor series coefficients for $J_0(x)$ are the inputs. Recall from section 2.2 that the truncation-error criterion of $E = 10^{-9}$ required a 40th-degree polynomial to evaluate $J_0(10)$. In the above listing, the economized polynomial degree chosen is 24 (13 terms). The Chebyshev series truncation error associated with this choice is approximately $|A(24)|$, or 10^{-8}. As an exercise, round off the economized series coefficients to eight digits and examine the accuracy (round-off plus truncation) of the resulting approximation.*

Coefficient	8-digit Precision Value	14-digit Precision Value
$C(0)$	0.99999816	1.00000000
$C(2)$	-0.24999999	-0.25000000
$C(4)$	0.15625000	0.15625000
$C(6)$	$-0.43402758 \times 10^{-3}$	$-0.43402758 \times 10^{-3}$
$C(8)$	$0.67816606 \times 10^{-5}$	$0.67816606 \times 10^{-5}$
$C(10)$	$-0.67815228 \times 10^{-7}$	$-0.67815228 \times 10^{-7}$
$C(12)$	$0.47087887 \times 10^{-9}$	$0.47087888 \times 10^{-9}$
$C(14)$	$-0.24006814 \times 10^{-11}$	$-0.24006814 \times 10^{-11}$
$C(16)$	$0.93423756 \times 10^{-14}$	$0.93423757 \times 10^{-14}$
$C(18)$	$-0.28349481 \times 10^{-16}$	$-0.28349482 \times 10^{-16}$
$C(20)$	$0.66374577 \times 10^{-19}$	$0.66374579 \times 10^{-19}$
$C(22)$	$-0.11022242 \times 10^{-21}$	$-0.11022224 \times 10^{-21}$
$C(24)$	$0.97615910 \times 10^{-25}$	$0.97615920 \times 10^{-25}$

Table 2.5.3 *A comparison of the economized polynomial coefficients appearing in listing 2.5.3 (14-digit precision, rounded off to 8 digits) with those calculated using 8-digit precision arithmetic.*

2.6 Reversion, Inversion, and Shifting

It is not unusual for the recipient of results (e.g., from a report) to be in a position where the polynomial approximation provided is not quite what is needed at the moment. For example, a set of experimental data may have been analyzed in the past and a regressed equation $y = P(x)$ obtained. However, what you may really want is $x = Q(y)$. Or perhaps what is required is a polynomial that approximates $1/P(x)$. In this section, numerical techniques that facilitate these forms of polynomial manipulation will be presented.

2.6.1 Polynomial Reversion

We will start with the polynomial reversion problem. Given the polynomial $y = a_0 + a_1x + a_2x^2 + \cdots$, the goal is to determine the coefficients of the reversed polynomial $x = b_0 + b_1y + b_2y^2 + \cdots$. This can be accomplished by direct substitution:

$$\begin{aligned} x &= b_0 + b_1y + b_2y + \cdots \\ &= b_0 + b_1(a_0 + a_1x + a_2x^2 + \cdots) \\ &\quad + b_2(a_0 + a_1x + a_2x^2 + \cdots)^2 \\ &\quad + \cdots \end{aligned} \qquad (2.6.1)$$

By equating like powers of x, equations relating the coefficients can be obtained.

Example: x^0
$$0 = b_0 + b_1a_0 + b_2a_0^2 + b_3a_0^3 + \cdots \qquad (2.6.2)$$

Example: x^1
$$1 = b_1a_1 + 2b_2a_0a_1 + 3b_3a_0^2a_1 + \cdots \qquad (2.6.3)$$

Observe that this particular equation can be satisfied only if $a_1 \neq 0$.

The reversed polynomial is very often infinite in extent. However, if we are willing to accept an approximation, then a reversed polynomial of finite degree N can be chosen. This leads to $N + 1$ coefficient equations in $N + 1$ unknowns (the b_i).

This problem has been solved for $N = 7$. [See Ref. 5. The solution to b_0 is missing in this reference. It can easily be obtained from equation (2.6.2), however, once the other b_i are known.] The resulting expressions are algebraically complicated and are not reproduced here. However, they are implemented in the subroutine (REVERSE) shown in listing 2.6.1. Also appearing in that listing is a program for demonstrating REVERSE (see listing 2.6.2 for examples).

In the first example, the original polynomial was $y = 1 + 5x$. The inputs to REVERSE are N, the degree of the polynomial to be reversed, and $A(I)$, the coefficients. The reversed polynomial coefficients are returned in $B(I)$. The simple reversed equation is $x = -0.2 + 0.2y$.

In the second example, the first seven terms of the Taylor series for $y = \sinh x$ were supplied to the subroutine. It responded with the first seven terms of the Taylor series for $x = \sinh^{-1} y$. This example reveals a very intriguing potential use for REVERSE—approximating the inverse of a function by reversing a polynomial that approximates the function.

The third example demonstrates what happens when the condition $a_1 \neq 0$ is violated. It is the responsibility of the calling program to trap this error.

The remaining two examples demonstrate a key limitation of REVERSE. If the input polynomial is not monotonic, then REVERSE will fail and will return an incorrect set of coefficients. This is not surprising. If there are two values of x that give the same y value, then if y is chosen, what is the correct x value?

Within the restrictions indicated above, REVERSE is a very useful utility program for quickly reversing the dependence of polynomials.

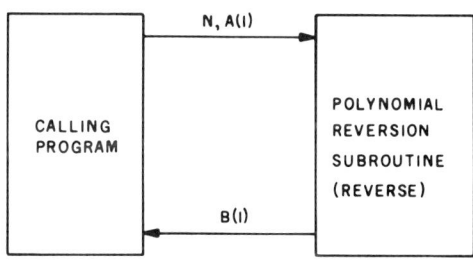

N: Degree of the input polynomial
A(I): Coefficients for the input polynomial [A(1) ≠ 0]
B(I): Coefficients of the reversed polynomial

Figure 2.6.1 *Subroutine-connection diagram for REVERSE.*

Statements/Functions List

+, −, *, /
FOR/NEXT

Variables List

A, A1, A2, A3, A4, A5, A6, A7, A(I), B, B(I), N

Variables Passed to Subroutine

N, A(I)

Table 2.6.1 *Functions and variables appearing in the polynomial reversion subroutine (REVERSE).*

```
2775 REM PROGRAM TO DEMONSTRATE THE SERIES REVERSION SUBROUTINE
2776 PRINT
2777 PRINT
2778 PRINT "WHAT IS THE DEGREE OF"
2779 PRINT "THE INPUT POLYNOMIAL: ",
2780 INPUT N
2781 PRINT
2782 PRINT "INPUT THE COEFFICIENTS AS PROMPTED:"
2783 PRINT
2784 FOR I=0 TO N
2785 PRINT "A(",I,") = ",
2786 INPUT A(I)
2787 NEXT I
2788 PRINT
2789 PRINT
2790 GOSUB 44800
2791 PRINT "THE REVERSED POLYNOMIAL COEFFICIENTS ARE:"
2792 PRINT
2793 FOR I=0 TO 7
2794 PRINT "B(",I,") = ",B(I)
2795 NEXT I
2796 PRINT
2797 PRINT
2798 END
44789 REM ********************
44790 REM SERIES REVERSION SUBROUTINE (REVERSE)
44791 REM THIS PROGRAM TAKES A POLYNOMIAL, Y=A(0) + A(1)X + ..
44792 REM AND RETURNS A POLYNOMIAL X = B(0) + B(1)Y + ...
44793 REM REFERENCE   CRC STANDARD MATHEMATICAL TABLES
44794 REM                 24TH EDITION
44795 REM THE INPUT SERIES COEFFICIENTS ARE A(0),A(1), ETC.
44796 REM A(1) MUST BE NONZERO.
44797 REM THE OUTPUT SERIES COEFFICIENTS ARE B(0),B(1),.....,B(7).
44798 REM THE DEGREE OF REVERSION IS LIMITED TO SEVEN.
44799 REM A1,A2,.... ARE DUMMY VARIABLES.
44800 A1=A(1)
44801 B(1)=1/A1
44802 A=1/A1
44803 B=A*A
44804 A=A*B
44805 B(2)=-A2/A
44806 A3=A(3)
44807 A=A*B
44808 B(3)=A*(2*A2*A2-A1*A3)
44809 A4=A(4)
44810 A=A*B
44811 B(4)=A*(5*A1*A2*A3-A1*A1*A4-5*A2*A2*A2)
44812 A5=A(5)
44813 A=A*B
```

Listing 2.6.1

```
44814 B(5)=6*A1*A1*A2*A4+3*A1*A1*A3*A3+14*A2*A2*A2*A2
44815 B(5)=B(5)-A1*A1*A1*A5-21*A1*A2*A2*A3
44816 B(5)=A*B(5)
44817 A6=A(6)
44818 A=A*B
44819 B(6)=7*A1*A1*A1*A2*A5+7*A1*A1*A1*A3*A4+84*A1*A2*A2*A2*A3
44820 B(6)=B(6)-A1*A1*A1*A1*A6-28*A1*A1*A2*A2*A4
44821 B(6)=B(6)-28*A1*A1*A2*A3*A3-42*A2*A2*A2*A2*A2
44822 B(6)=A*B(6)
44823 A7=A(7)
44824 A=A*B
44825 B(7)=8*A1*A1*A1*A1*A2*A6+8*A1*A1*A1*A1*A3*A5
44826 B(7)=B(7)+4*A1*A1*A1*A1*A4*A4+120*A1*A1*A2*A2*A2*A4
44827 B(7)=B(7)+180*A1*A1*A2*A2*A3*A3+132*A2*A2*A2*A2*A2*A2
44828 B(7)=B(7)-A1*A1*A1*A1*A1*A7-36*A1*A1*A1*A2*A2*A5
44829 B(7)=B(7)-72*A1*A1*A1*A2*A3*A4-12*A1*A1*A1*A3*A3*A3
44830 B(7)=B(7)-330*A1*A2*A2*A2*A2*A3
44831 B(7)=A*B(7)
44832 B(0)=0
44833 A=A(0)
44834 FOR I=1 TO 7
44835 B(0)=B(0)-B(I)*A
44836 A=A*A(0)
44837 NEXT I
44838 RETURN
```

Listing 2.6.1 *Polynomial reversion subroutine (REVERSE). This program takes an input series $y(x) = a_0 + a_1 x + \cdots$, and reverses it to give $x = b_0 + b_1 y + \cdots$. The reversed series is limited to the seventh degree, and the original series must be monotonic.*

Also shown is a program for applying REVERSE. See listing 2.6.2 for examples.

RUN

Listing 2.6.2

```
WHAT IS THE DEGREE OF
THE INPUT POLYNOMIAL: ?1

INPUT THE COEFFICIENTS AS PROMPTED:

A( 0) = ?1
A( 1) = ?5


THE REVERSED POLYNOMIAL COEFFICIENTS ARE:

B( 0) =   -.2
```

```
B( 1) =    .2
B( 2) =    0
B( 3) =    0
B( 4) =    0
B( 5) =    0
B( 6) =    0
B( 7) =    0
```

Listing 2.6.2 cont.

```
READY
RUN



WHAT IS THE DEGREE OF
THE INPUT POLYNOMIAL: ?7

INPUT THE COEFFICIENTS AS PROMPTED:

A( 0) = ?0
A( 1) = ?1
A( 2) = ?0
A( 3) = ?.166666666
A( 4) = ?0
A( 5) = ?.00833333333
A( 6) = ?0
A( 7) = ?.0001984127

THE REVERSED POLYNOMIAL COEFFICIENTS ARE:

B( 0) =    0
B( 1) =    1
B( 2) =    0
B( 3) =   -.16666667
B( 4) =    0
B( 5) =    7.5000004E-02
B( 6) =    0
B( 7) =   -4.4642859E-02

READY
RUN



WHAT IS THE DEGREE OF
THE INPUT POLYNOMIAL: ?3

INPUT THE COEFFICIENTS AS PROMPTED:
```

Listing 2.6.2 cont.

```
A( 0) = ?1
A( 1) = ?0
A( 2) = ?0
A( 3) = ?1


DIVIDE ZERO ERROR IN LINE 44801
READY
RUN



WHAT IS THE DEGREE OF
THE INPUT POLYNOMIAL: ?3

INPUT THE COEFFICIENTS AS PROMPTED:

A( 0) = ?1
A( 1) = ?1
A( 2) = ?1
A( 3) = ?1


THE REVERSED POLYNOMIAL COEFFICIENTS ARE:

B( 0) =    9
B( 1) =    1
B( 2) =    0
B( 3) =   -1
B( 4) =    0
B( 5) =    3
B( 6) =    0
B( 7) =  -12


READY
RUN



WHAT IS THE DEGREE OF
THE INPUT POLYNOMIAL: ?2

INPUT THE COEFFICIENTS AS PROMPTED:

A( 0) = ?0
A( 1) = ?1
A( 2) = ?1
```

THE REVERSED POLYNOMIAL COEFFICIENTS ARE:

```
B( 0) =  0
B( 1) =  1
B( 2) =  0
B( 3) =  0
B( 4) =  0
B( 5) =  0
B( 6) =  0
B( 7) =  0
```

READY

Listing 2.6.2 *Examples of the use of the series reversion subroutine (REVERSE) given in listing 2.6.1. In the first example, the polynomial $y = 1 + 5x$ was reversed to give $x = -0.2 + 0.2y$. In the second example, the truncated Taylor series for sinh x was reversed: $\sinh x \cong y = x + x^3/3! + x^5/5! + x^7/7!$ was reversed to give*

$$x = y - \frac{1}{2}\frac{y^3}{3} + \frac{1\cdot 3}{2\cdot 4}\frac{y^5}{5} - \frac{1\cdot 3\cdot 5}{2\cdot 4\cdot 6}\frac{y^7}{7} \cong \sinh^{-1} y$$

The third example failed because $A(1) = 0$. The fourth example is similar to the second. A reversed polynomial that is accurate for small x was found. In the last example, the input polynomial was $y = x + x^2$. The reversed polynomial found was $x = y$, which clearly is technically incorrect. However, near $x = 0$ the approximation is good.

2.6.2 Polynomial Inversion

The next utility algorithm we will consider is the polynomial approximation to $1/P(x)$. We will denote the approximating polynomial as $Q(x)$. The relationship to satisfy is

$$P(x)\, Q(x) = 1 \qquad (2.6.4)$$

or $\qquad (a_0 + a_1 x + a_2 x^2 + \cdots)(b_0 + b_1 x + b_2 x^2 + \cdots) = 1$

By again equating like powers of x, we obtain the following sequence of equations:

$$\begin{aligned} x^0: &\ a_0 b_0 = 1 \\ x^1: &\ a_0 b_1 + a_1 b_0 = 0 \\ x^2: &\ a_0 b_2 + a_1 b_1 + a_2 b_0 = 0 \\ &\quad \vdots \\ x^N: &\ a_0 b_N + a_1 b_{N-1} + \cdots + a_N b_0 = 0 \end{aligned}$$

By solving the first coefficient equation, enough information is obtained to solve the second, and so on throughout the sequence. There is an infinite number of equations to be solved, but they can be processed only one degree at a time, starting with b_0. Therefore, in principle, any level of approximation is possible. The one restriction on this is that the infinite series for $Q(x)$ will not converge for values of x that correspond to roots of $P(x)$. Also, the truncated approximation to $Q(x)$ behaves in a manner similar to a truncated Taylor series—the error tends to grow with increasing x.

The solution to the coefficient equations can be numerically implemented as shown in listing 2.6.3. The inversion subroutine, RECIPRO, accepts as input the degree (N) of $P(x)$, the corresponding coefficients $[A(I)]$, and the degree (M) of the reciprocal polynomial, $Q(x)$. The $M + 1$ coefficients calculated for $Q(x)$ are returned in the array $B(I)$.

Also appearing in listing 2.6.3 is a program that can be used to exercise RECIPRO. Three examples are given in listing 2.6.4. These particular examples were picked because each displays a different type of approximation error.

In the first example, the full inverse series diverges at $x = \pm 1/2$, even though $P(x)$ has only one root: $x = -1/2$. However, for $|x| < 0.2$, the truncated series accurately approximates $1/P(x)$.

The second example shows an inverse polynomial that diverges at $x = \pm 1$. This is expected in that both points are roots of $P(x)$.

The third example is curious. $P(x)$ has no real roots, but $Q(x)$ clearly diverges at $x = 10$ and roughly $x = -12$. However, the approximation to $1/P(x)$ is very good over the range $-5 < x < 5$.

The error graphs all indicate that $Q(x) \cong 1/P(x)$ to a high degree of accuracy near $x = 0$ (if $a_0 \neq 0$).

RECIPRO is a fairly reliable program. It will always give an approximation to $1/P(x)$ which is good close to $x = 0$ *as long as* $a_0 \neq 0$. A divide-by-zero error will occur if $a_0 = 0$. As we saw from the examples, the approximation to $Q(x)$ will surely diverge at the roots of $P(x)$. However, it may also diverge at other values. Therefore, it is wise to empirically check the range of validity of the calculated polynomial before using it.

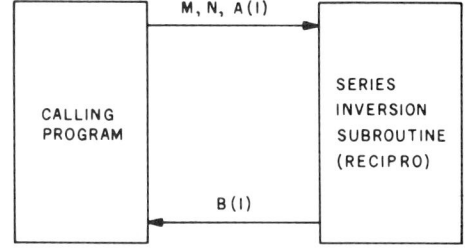

N: Degree of input polynomial, $P(x)$
M: Degree of inverted polynomial, $Q(x)$
$A(I)$: Coefficients of $P(x)$
$B(I)$: Coefficients of the $Q(x)$ truncated series

Figure 2.6.2 *Subroutine-connection diagram for RECIPRO.*

Statements/Functions

$+, -, *, /, <$
GOTO, FOR/NEXT, IF/THEN

Variables List

A(I), B(I), I, J, L, M, N

Variables Passed to Subroutine

A(I), M, N

Table 2.6.2 *Functions and variables used in the series inversion subroutine (RECIPRO).*

```
2800 REM PROGRAM TO DEMONSTRATE THE SERIES INVERSION SUBROUTINE
2801 PRINT
2802 PRINT
2803 PRINT "WHAT IS THE DEGREE OF THE INPUT POLYNOMIAL: ",
2804 INPUT N
2805 PRINT
2806 PRINT "WHAT IS THE DEGREE OF THE INVERTED POLYNOMIAL: ",
2807 INPUT M
2808 DIM A(M),B(M)
2809 PRINT
2810 PRINT "INPUT THE POLYNOMIAL COEFFICIENTS:"
2811 PRINT
2812 FOR I=0 TO N
2813 PRINT "A(",I,") = ",
2814 INPUT A(I)
2815 NEXT I
2816 GOSUB 44850
2817 PRINT
2818 PRINT
2819 PRINT "THE INVERTED POLYNOMIAL COEFFICIENTS ARE:"
2820 PRINT
2821 FOR I=0 TO M
2822 PRINT "B(",I,") = ",B(I)
2823 NEXT I
2824 PRINT
2825 END
44841 REM ********************
44842 REM RECIPROCAL POWER SERIES SUBROUTINE (RECIPRO)
44843 REM REFERENCE- COMPUTATIONAL ANALYSIS BY HENRICI.
44844 REM THE INPUT SERIES COEFFICIENTS ARE A(I).
44845 REM THE OUTPUT SERIES COEFFICIENTS ARE B(I).
44846 REM THE DEGREE OF THE INPUT POLYNOMIAL IS N..
```

Listing 2.6.3

```
44847 REM THE DEGREE OF THE INVERTED POLYNOMIAL IS M.
44848 REM DIMENSION A(I) AND B(I) IN THE CALLING PROGRAM
44849 REM THE PROGRAM WILL TAKE CARE OF THE NORMALIZATION USING L
44850 L=A(0)
44851 FOR I=0 TO N
44852 A(I)=A(I)/L
44853 B(I)=0
44854 NEXT I
44855 REM CLEAR ARRAYS
44856 FOR I=N+1 TO M
44857 A(I)=0
44858 B(I)=0
44859 NEXT I
44860 REM CALCULATE THE B(I) COEFFICIENTS
44861 B(0)=1
44862 FOR I=1 TO M
44863 J=1
44864 B(I)=B(I)-A(J)*B(I-J)
44865 J=J+1
44866 IF J<=I THEN GOTO 44864
44867 NEXT I
44868 REM UN-NORMALIZE THE A(I) AND B(I)
44869 FOR I=0 TO M
44870 A(I)=A(I)*L
44871 B(I)=B(I)/L
44872 NEXT I
44873 RETURN
```

Listing 2.6.3 *Series inversion subroutine (RECIPRO). This program calculates the series coefficients such that $P(x) Q(x) \cong 1$ in the vicinity of $x = 0$. Also shown is a program that demonstrates the operation of RECIPRO. See listing 2.6.4 for examples.*

RUN

Listing 2.6.4

WHAT IS THE DEGREE OF THE INPUT POLYNOMIAL: ?1

WHAT IS THE DEGREE OF THE INVERTED POLYNOMIAL: ?4

INPUT THE POLYNOMIAL COEFFICIENTS:

A(0) = ?1
A(1) = ?2

THE INVERTED POLYNOMIAL COEFFICIENTS ARE:

```
B( 0) =   1
B( 1) =  -2
B( 2) =   4
B( 3) =  -8
B( 4) =  16

READY
RUN


WHAT IS THE DEGREE OF THE INPUT POLYNOMIAL: ?2

WHAT IS THE DEGREE OF THE INVERTED POLYNOMIAL: ?6

INPUT THE POLYNOMIAL COEFFICIENTS:

A( 0) = ?1
A( 1) = ?0
A( 2) = ?1


THE INVERTED POLYNOMIAL COEFFICIENTS ARE:

B( 0) =   1
B( 1) =   0
B( 2) =  -1
B( 3) =   0
B( 4) =   1
B( 5) =   0
B( 6) =  -1

READY
RUN


WHAT IS THE DEGREE OF THE INPUT POLYNOMIAL: ?2

WHAT IS THE DEGREE OF THE INVERTED POLYNOMIAL: ?6

INPUT THE POLYNOMIAL COEFFICIENTS:

A( 0) = ?1
A( 1) = ?.1
A( 2) = ?.01


THE INVERTED POLYNOMIAL COEFFICIENTS ARE:
```

Listing 2.6.4 cont.

```
B( 0) =   1
B( 1) =   -.1
B( 2) =   0
B( 3) =   .001
B( 4) =   -.0001
B( 5) =   0
B( 6) =   .000001
```

READY

Listing 2.6.4 Examples of the operation of RECIPRO. In the first example, the input polynomial was $P(x) = 1 + 2x$. The inverted polynomial calculated was $Q(x) = 1 - 2x + 4x^2 - 8x^3 + 16x^4$. A plot of the product $P(x) Q(x)$ is given in figure 2.6.3. In the second example, the input polynomial was $P(x) = 1 + x^2$. The calculated inverse polynomial was $Q(x) = 1 - x^2 + x^4 - x^6$. A plot of the product $P(x) Q(x)$ is shown in figure 2.6.4. In the last example, $P(x) = 1 + x/10 + x2/100$. The inverse calculated was $Q(x) = 1 - x/10 + x^3/1000 - x^4/10000 + x^6/1000000$ (see figure 2.6.5). Note that $Q(x)$ is actually a truncation of the infinite series which exactly equals $P^{-1}(x)$ if there are no real roots.

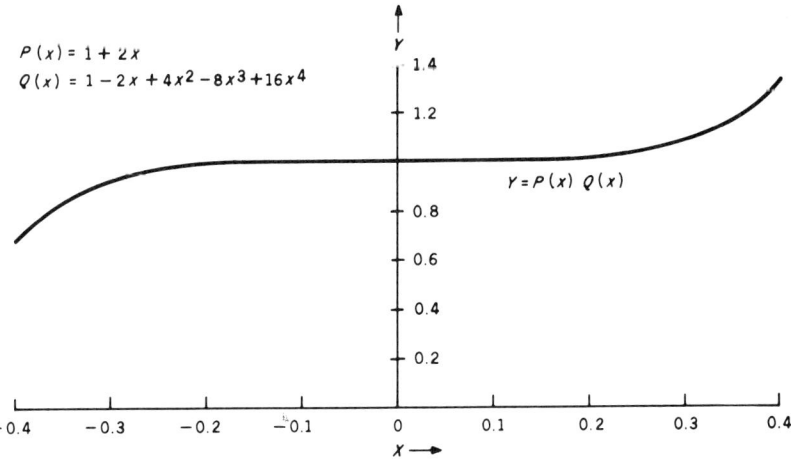

Figure 2.6.3 Plot of the product of $P(x)$ and $Q(x)$ for the first example in listing 2.6.4.

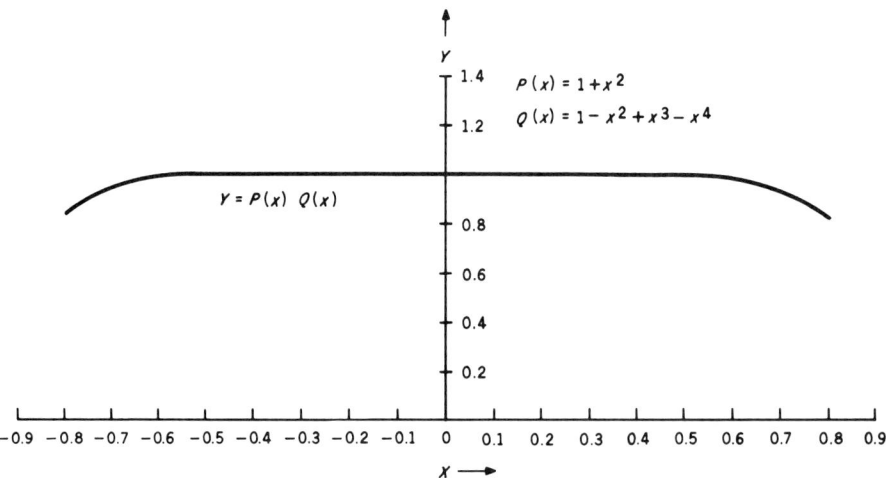

Figure 2.6.4 *Plot of the product of P(x) and Q(x) for the second example in listing 2.6.4.*

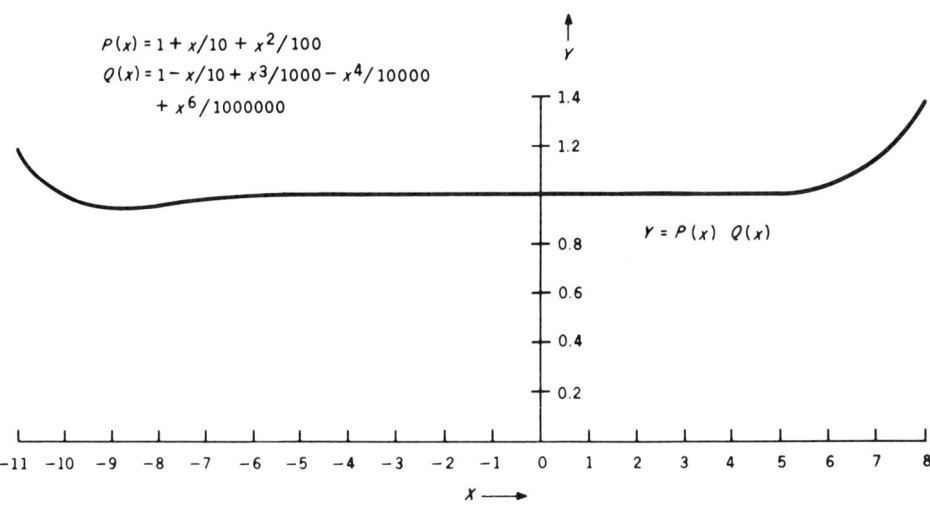

Figure 2.6.5 *Plot of the product of P(x) and Q(x) for the third example in listing 2.6.4.*

2.6.3 Shifting the Expansion Point

The most difficult source of inaccuracy that must be dealt with in numerical analysis is round-off error. In many cases, it is unavoidable. However, it is often possible to restructure the calculation to minimize the influence of round-off error. In this subsection, we will examine how this can be accomplished for some polynomial calculations.

Consider the following quartic polynomial:

$$y(x) = 1 + 4x + 6x^2 + 4x^3 + x^4$$

If this polynomial appeared within a calculation and were to be evaluated near $x = -1$, there is a strong possibility that round-off error would seriously affect the accuracy of the computation. The reason for this is that ideally $y(-1) = 1 - 4 + 6 - 4 + 1 = 0$. However, numerically the result would be E, and the *relative* error would be infinite. If $y(x)$ is a multiplier in an analysis, the results can be useless.

If it were known beforehand that the calculations would involve x values in the vicinity of -1, $y(x)$ could be replaced with a better *conditioned form*:

$$y(x) = (x + 1)^4$$

Now if there is round-off error, its influence is much reduced.

What we have done in the above example is to shift the expansion point so that

$$y(x) = a_0 + a_1 x + a_2 x^2 + \cdots = b_0 + b_1(x - x_0) + b_2(x - x_0)^2 + \cdots$$
$$(2.6.5)$$

The problem is therefore one of determining the coefficients b_i ($i = 0, 1, \ldots$). To solve for the b_i coefficients in terms of the a_i and x_0 values, we again equate like powers:

$$a_0 = b_0 - b_1 x_0 + b_2 x_0^2 - \cdots$$
$$a_1 = b_1 - 2b_2 x_0 + 3b_3 x_0^2 - \cdots$$

and so on. For an Nth-degree polynomial, there are $N + 1$ equations in the $N + 1$ unknowns ($b_0, b_1, \ldots, b_N$), and the problem is, in principle, soluble.

Horner (Ref. 17) developed an algorithm for quartic polynomials that is relatively simple. We will start with the definition

$$b_n^{-1} = a_n \quad \text{for } n = 0, 1, 2, 3, 4$$

Next, for $m = 0, 1, 2, 3, 4$ we will calculate b_0^m:

$$b_0^m = b_0^{m-1}$$
$$b_n^m = x_0 b_{n-1}^m + b_n^{m-1} \quad \text{for } n = 1, 2, 3, 4 - m$$

Finally,

$$b_m = b_m^{4-m} \quad \text{for } m = 0, 1, 2, 3, 4$$

This algorithm can be generalized for polynomials beyond the quartic, but for the purposes of this discussion, we will limit ourselves to the quartic case.

A subroutine for applying Horner's algorithm is given in listing 2.6.5, and is demonstrated in listing 2.6.6. The second example corresponds to the polynomial discussed earlier in this subsection.

HORNER is easy to use and requires no special precautions. As a simple exercise, extend HORNER to polynomials of a higher degree than the quartic.

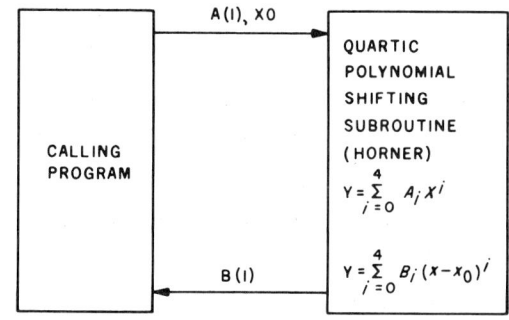

A(I): Coefficients of input polynomial
X0: Expansion point
B(I): Coefficients of shifted polynomial

Figure 2.6.6 *Subroutine-connection diagram for the HORNER shifting subroutine.*

Statements/Functions List

$+, -, *, >$
GOTO, FOR/NEXT, IF/THEN

Variables List

A(I), B(I), C(I,J), I, J

Variables Passed to Subroutine

A(I), X0

Table 2.6.3 *Functions and variables employed by the HORNER subroutine.*

```
2830 REM TEST PROGRAM FOR HORNER'S RULE
2831 DIM C(4,5)
2832 PRINT
2833 PRINT
2834 PRINT "INPUT THE FIVE COEFFICIENTS:"
2835 PRINT
2836 FOR I=0 TO 4
2837 PRINT "A(",I,") = ",
2838 INPUT A(I)
2839 NEXT I
2840 PRINT
2841 PRINT "WHAT IS THE EXPANSION POINT: ",
2842 INPUT X0
2843 PRINT
2844 GOSUB 44900
2845 PRINT
2846 PRINT "THE SHIFTED COEFFICIENTS ARE:"
2847 PRINT
2848 FOR I=0 TO 4
2849 PRINT "B(",I,") = ",B(I)
2850 NEXT I
2851 PRINT
2852 PRINT
2853 END
44892 REM ********************
44893 REM HORNER'S SHIFTING RULE SUBROUTINE (HORNER)
44894 REM THIS SUBROUTINE TAKES A GIVEN QUARTIC
44895 REM POLYNOMIAL AND CONVERTS IT TO A TAYLOR EXPANSION.
44896 REM THE INPUT SERIES COEFFICIENTS ARE A(I).
44897 REM THE EXPANSION POINT IS X0.
44898 REM THE SHIFTED COEFFICIENTS ARE RETURNED IN B(I).
44899 REM C(4,5) MUST BE DIMENSIONED IN THE CALLING PROGRAM.
44900 FOR J=0 TO 4
44901 C(J,0)=A(4-J)
44902 NEXT J
44903 FOR I=0 TO 4
44904 C(0,I+1)=C(0,I)
44905 J=1
44906 IF J>4-I THEN GOTO 44910
44907 C(J,I+1)=X0*C(J-1,I+1)+C(J,I)
44908 J=J+1
44909 GOTO 44906
44910 NEXT I
44911 FOR I=0 TO 4
44912 B(4-I)=C(I,4-I+1)
44913 NEXT I
44914 RETURN
```

Listing 2.6.5 *Subroutine for shifting the expansion point of a quartic polynomial using Horner's algorithm (HORNER). Also shown is a program for demonstrating HORNER. Examples are given in listing 2.6.6.*

```
RUN

INPUT THE FIVE COEFFICIENTS:

A( 0 ) = ?1
A( 1 ) = ?4
A( 2 ) = ?6
A( 3 ) = ?4
A( 4 ) = ?1

WHAT IS THE EXPANSION POINT: ?1

THE SHIFTED COEFFICIENTS ARE:

B( 0 ) =   16
B( 1 ) =   32
B( 2 ) =   24
B( 3 ) =   8
B( 4 ) =   1
READY
RUN

INPUT THE FIVE COEFFICIENTS:

A( 0 ) = ?1
A( 1 ) = ?4
A( 2 ) = ?6
A( 3 ) = ?4
A( 4 ) = ?1

WHAT IS THE EXPANSION POINT: ?-1

THE SHIFTED COEFFICIENTS ARE:

B( 0 ) =   0
B( 1 ) =   0
B( 2 ) =   0
B( 3 ) =   0
B( 4 ) =   1
READY
```

Listing 2.6.6 Examples of the operation of the HORNER subroutine. In the first example, the input quartic was $y = 1 + 4x + 6x^2 + 4x^3 + x^4$. The expansion point was shifted to $X0 = 1$, and the shifted polynomial found was $y = 16 + 32(x - 1) + 24(x - 1)^2 + 8(x - 1)^3 + (x - 1)^4$. As an exercise, show that the two polynomials are equivalent. In the second example, the input polynomial was the same, but the expansion point was $X0 = -1$ instead. The shifted polynomial found was $y = (x + 1)^4$. Observe that $x = -1$ is a multiple root of the polynomial.

2.7 Rational Polynomials

Our attention thus far has been largely directed at polynomial approximations of the form

$$f(x) \cong P(x) = \sum_{i=0}^{N} a_i x^i \qquad (2.7.1)$$

This is the most popular type of approximation partly because it usually works well (after some change of variables perhaps), and partly because very simple and effective techniques exist for determining the a_i. However, there are many cases in which this structure is *not* appropriate. For example, figure 2.7.3 shows a function that has an infinity at some value of x, say x_0. We might approach this problem by replacing the variable x in equation (2.7.1) with $1/(x - x_0)$, thus creating a Laurent series. However, as we saw in Chapter 1, this may also not be appropriate.

There is another polynomial approximation form that is often very useful for dealing with such curve shapes—the *rational polynomial*:

$$R(x) = \frac{P(x)}{Q(x)} \qquad \text{where} \quad Q(x) = \sum_{i=0}^{M} b_i x^i \qquad (2.7.2)$$

We will not go into the theory behind the rational polynomial or the ways in which it can be derived. Instead, a few simple rules of thumb along with some examples will be presented. For a good discussion regarding rational polynomials, see Ref. 15.

We will define the following four parameters:

1) $M(P)$: The lowest degree appearing in $P(x)$.
   For example, for $P(x) = 6x^3 + 7x^4 + 2x^5$, $M(P) = 3$.
2) $N(P)$: The highest degree appearing in $P(x)$.
   For the example above, $N(P) = 5$.
3) $M(Q)$: The lowest degree appearing in $Q(x)$.
4) $N(Q)$: The highest degree appearing in $Q(x)$.

The relative values of these parameters define the asymptotic behavior of $P(x)/Q(x)$ (see figures 2.7.1 through 2.7.4). By examining the behavior of the function to be approximated at $x = 0$ and $x \to \infty$, strong clues as to the form of $P(x)$ and $Q(x)$ can be obtained. For example, if $f(x) =$ constant at $x = 0$, then $M(P) = M(Q)$. If, instead, $f(x) \to \infty$ near $x = 0$, then $M(P) < M(Q)$. If $f(x) = 0$ at $x = 0$, then $M(P) > M(Q)$. There are also three general types of behavior that exist as x becomes infinite: $f(x) \to 0$ [$N(P) < N(Q)$]; $f(x) =$ constant [$N(P) = N(Q)$]; and $f(x) \to \infty$ [$N(P) > N(Q)$]. By examining the shape of $f(x)$, we can usually get an idea of whether or not a rational polynomial is called for, and we can even determine some of the relative properties of $P(x)$ and $Q(x)$.

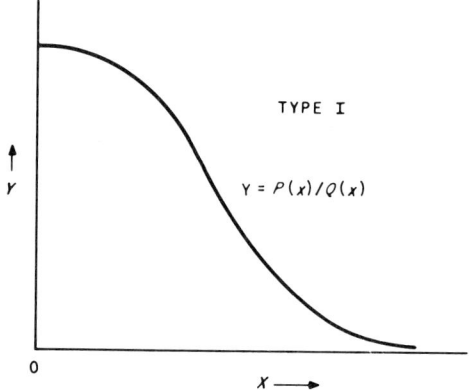

Figure 2.7.1 *A curve form classified as Type I. In this case, a likely rational polynomial approximation would be one in which $M(P) = M(Q)$ and $N(P) < N(Q)$.*

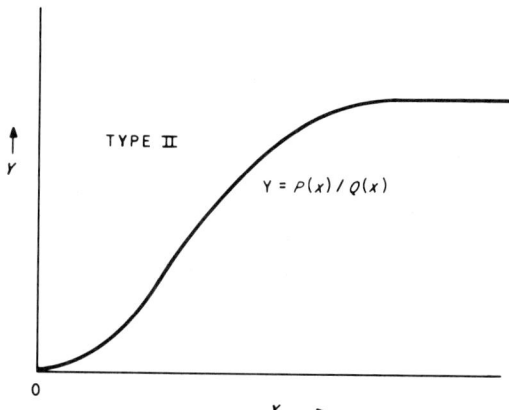

Figure 2.7.2 *A curve form designated as Type II. The appropriate rational polynomial would have $M(P) > M(Q)$ and $N(P) = N(Q)$.*

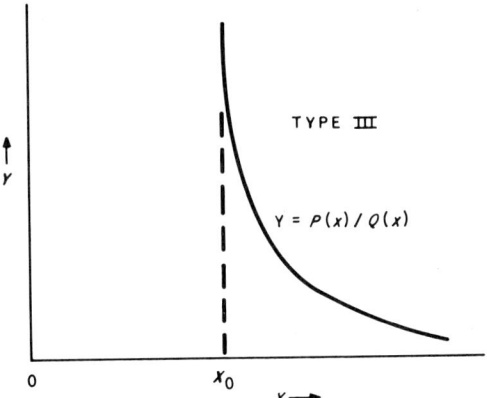

Figure 2.7.3 *This is a Type III curve in which Q(x) has a zero, thus causing a peak in P(x)/Q(x). Note that N(P) < N(Q).*

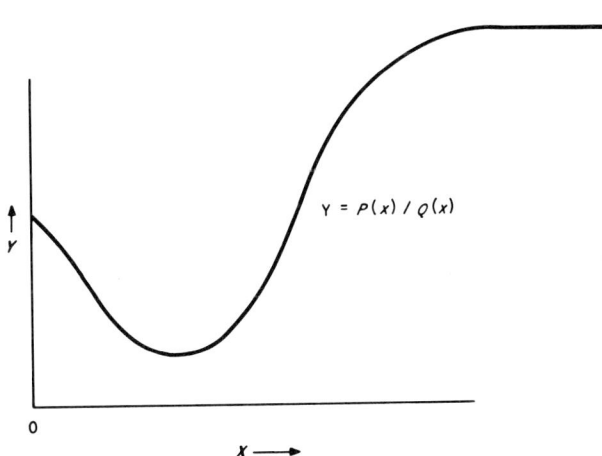

Figure 2.7.4 *A hybrid form in which P(x) has a dip, M(P) = M(Q) and N(P) = N(Q).*

As a numerical example, we will consider an approximation related to the inverse of the complementary error function. We will define $Q(x)$ as

$$Q(x) = \frac{1}{\sqrt{2\pi}} \int_x^\infty e^{-t^2/2} \, dt \qquad (2.7.3)$$

The normal distribution probability density function is

$$p(x) = \frac{1}{\sqrt{2\pi}} e^{-x^2/2} \qquad (2.7.4)$$

Therefore, $Q(x)$ is one minus the cumulative normal distribution function: $Q(x) = 1 - P(x)$. Also, $Q(x)$ is related to the complementary error function: $Q(x) = 1/2 \, \text{erfc}(x)$.

An approximation to the complementary error function was presented in section 2.3. We will now in effect approximate its inverse.

The object is to determine the value of x that corresponds with given Q. Abramowitz and Stegun (Ref. 6) give the following rational polynomial approximation for X_Q ($0 < Q \leq 0.5$):

$$X_Q = t - \frac{c_0 + c_1 t + c_2 t^2}{1 + d_1 t + d_2 t^2 + d_3 t}$$

where $t = (\ln 1/Q^2)^{1/2}$

$c_0 = 2.515517$

$c_1 = 0.802853$

$c_2 = 0.010328$

$d_1 = 1.432788$

$d_2 = 0.189269$

$d_3 = 0.001308$

The error in this approximation is $|E(Q)| < 0.0005$. (Note that the error is referenced to Q, not X_Q. The error in X_Q is relative to how much Q would have to be altered to give a corresponding change in X.) A little ingenuity was applied in choosing the functional form for t. The choice was based on the asymptotic form relating X_Q and Q as $Q \to 0$.

A program for applying this approximation is shown in listing 2.7.1, and a sample run appears in listing 2.7.2. The results are also plotted in figure 2.7.6. This curve is very similar to one encountered in Chapter 1. The approximation associated with that function also involved a logarithmic change of variables.

Other rational polynomial approximations are available from the literature. A few are given below.

Type I (Ref. 15)

$\cos x$ for $-1 \leq x \leq 1$

$$\cos x = \frac{1 + a_2 x^2 + a_4 x^4 + a_6 x^6}{1 + b_2 x^2 + b_4 x^4 + b_6 x^6}$$

$|\text{error}| \leq 2 \times 10^{-11}$

$a_2 = -0.470\ 595\ 788\ 392$
$a_4 = 0.027\ 388\ 289\ 676$
$a_6 = -0.000\ 372\ 342\ 269$
$b_2 = 0.029\ 404\ 211\ 608$
$b_4 = 0.000\ 423\ 728\ 814$
$b_6 = 0.000\ 003\ 235\ 543$

Type II (Ref. 6)

$\text{erf}(x)$ for $0 \leq x < \infty$

$$\text{erf}(x) = 1 - \frac{1}{1 + b_1 x + b_2 x^2 + b_3 x^3 + b_4 x^4 + b_5 x^5 + b_6 x^6}$$

$|\text{error}| \leq 3 \times 10^{-7}$

$b_1 = 0.07052\ 30784$
$b_2 = 0.00927\ 05272$
$b_3 = 0.00027\ 65672$
$b_4 = 0.04228\ 20123$
$b_5 = 0.00015\ 20143$
$b_6 = 0.00004\ 30638$

Type III (Ref. 15)

$\tanh \mu x$ for $-1 \leq x \leq 1$

$\mu = \frac{1}{2} \ln 3$

$$\tanh \mu x = \frac{a_1 x + a_3 x^3}{1 + b_2 x^2 + b_4 x^4}$$

$|\text{error}| \leq 6 \times 10^{-9}$

$a_1 = 0.549306\ 14401$
$a_3 = 0.01574\ 011995$
$b_2 = 0.12923\ 360954$
$b_4 = 0.00085\ 891904$

Although rational polynomials offer an interesting flexibility in functional approximation, they do not appear to be very popular. This is probably due to the difficulty that exists in generating values for the coefficients. This problem will be considered further in Chapter 8.

164 BASIC SCIENTIFIC SUBROUTINES

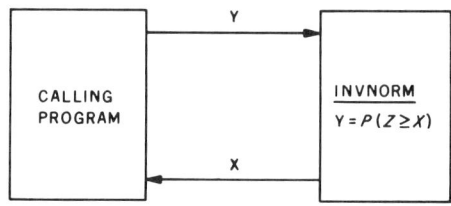

Y: 1 − cumulative normal distribution function [$P(Z \geq X)$]
X: Corresponding X value

Figure 2.7.5 *Subroutine-connection diagram for INVNORM.*

Statements/Functions List

$+, -, *, /$
IF/THEN, LOG, SQRT

Variables List

C0, C1, C2, D1, D2, D3, X, Y, Z

Variable Passed to Subroutine

Y

Table 2.7.1 *Functions and variables employed in INVNORM.*

Listing 2.7.1

```
2860 REM PROGRAM TO DEMONSTRATE INVERSE NORMAL SUBROUTINE
2861 PRINT
2862 PRINT "P(Z>X)",TAB(11),"X"
2863 PRINT "------",TAB(10),"---"
2864 PRINT
2865 FOR Y=.5 TO 0 STEP -.02
2866 GOSUB 44925
2867 PRINT Y,TAB(8),INT(10000*X)/10000
2868 NEXT Y
2869 END
44915 REM ********************
44916 REM INVERSE NORMAL DISTRIBUTION SUBROUTINE (INVNORM)
44917 REM THIS PROGRAM CALCULATES AN APPROXIMATION
44918 REM TO THE INTEGRAL OF THE NORMAL DISTRIBUTION
44919 REM FUNCTION FROM X TO INFINITY (THE TAIL).
```

```
44920 REM A RATIONAL POLYNOMIAL IS USED.
44921 REM THE INPUT IS Y, WITH THE RESULT RETURNED IN X.
44922 REM THE ACCURACY IS BETTER THAN 0.0005 IN THE RANGE 0<Y<=.5
44923 REM REFERENCE- ABRAMOWITZ AND STEGUN
44924 REM DEFINE COEFFICIENTS
44925 C0=2.515517
44926 C1=0.802853
44927 C2=0.010328
44928 D1=1.432788
44929 D2=0.189269
44930 D3=0.001308
44931 IF Y=0 THEN X=10000000000000
44932 IF Y=0 THEN RETURN
44933 Z=SQRT(-LOG(Y*Y))
44934 X=1+D1*Z+D2*Z*Z+D3*Z*Z*Z
44935 X=(C0+C1*Z+C2*Z*Z)/X
44936 X=Z-X
44937 RETURN
```

Listing 2.7.1 *Rational polynomial subroutine for approximating the inverse of the integral of the tail of the normal distribution, i.e., the value of x such that* $y = P(Z \geq x)$ *(INVNORM). Also shown is a program for exercising INVNORM. See listing 2.7.2 for a sample run. Note that y is restricted to the range* $0 < y \leq 0.5$.

RUN

P(Z>X)	X
.5	0
.48	.05
.46	.1001
.44	.1506
.42	.2015
.4	.2529
.38	.305
.36	.358
.34	.412
.32	.4672
.3	.524
.28	.5824
.26	.643
.24	.706
.22	.7719
.2	.8414
.18	.9152

Listing 2.7.2

```
.16       .9944
.14      1.0803
.12      1.175
.1       1.2817
.08      1.4053
.06      1.555
.04      1.751
.02      2.0541
0        1E+13
READY
```

Listing 2.7.2 *A sample run of the program given in listing 2.7.1.*

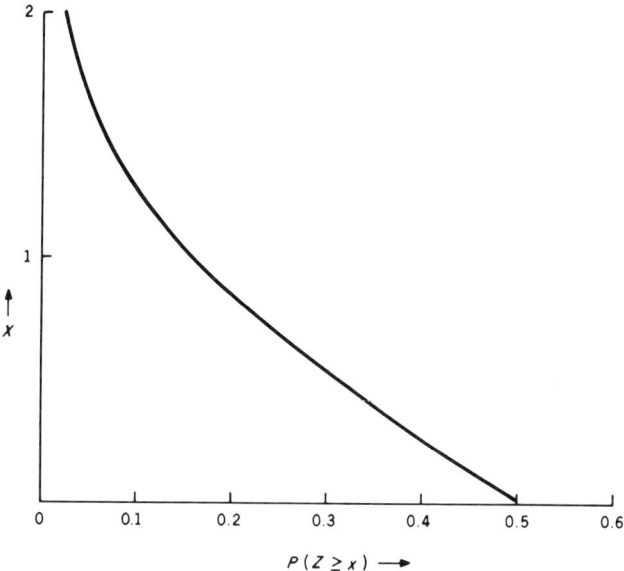

Figure 2.7.6 *A plot of the results shown in listing 2.7.2. The rational polynomial approach is particularly effective in this case. The denominator has a zero at $x = 0$, and the numerator has one at $x = 0.5$.*

2.8 Infinite Products

In this subsection, we will briefly examine another method for approximating some of the standard functions—*infinite products*. The purpose of the discussion is to demonstrate how a seemingly simple approximation scheme can become a numerical disaster. This will help you further understand the conditions that can lead to poor accuracy.

Several infinite product sequences exist for approximating the trigonometric and hyperbolic functions (see Ref. 28). We will take one that approximates sin x as an example:

$$\sin x = x(1 - \frac{x^2}{\pi^2})(1 - \frac{x^2}{2^2\pi^2})(1 - \frac{x^2}{3^2\pi^2}) \cdots \quad (2.8.1)$$

This sequence of products can be programmed very simply and compactly as shown in listing 2.8.1. The inputs to the SINEPROD subroutine are the argument, X, and a convergence criterion, E. The product sequence is terminated when $X^2/N^2\pi^2 < E$. A sample run of the program is shown in listing 2.8.2. The accuracy obtained is not very good even though many terms in the product sequence have been calculated.

SINEPROD is not unusual in its slow convergence. Product sequences of this type are therefore not recommended. Also, because there are so many calculations, round-off error is very likely. Product-sequence approximations should usually be avoided.

Listing 2.8.1

```
2870 REM PROGRAM TO DEMOSTRATE SINEPROD
2871 E=.000001
2872 PRINT
2873 PRINT
2874 PRINT "  X",TAB(10),"SIN(X) CALC.",TAB(25),"SIN(X) TRUE",TAB(42),"K",
2875 PRINT TAB(52),"ERROR"
2876 PRINT " ---",TAB(10),"------------",TAB(25),"-----------",TAB(41),"---"
2877 PRINT TAB(52),"-----"
2878 FOR X=0 TO 2 STEP .05
2879 GOSUB 2888
2880 PRINT X,TAB(10),INT(10000000*Y)/10000000,TAB(25),
2881 PRINT INT(10000000*SIN(X))/10000000,TAB(40),K,
2882 PRINT TAB(49),INT((100000000*(Y-SIN(X))))/100000000
2883 NEXT X
2884 PRINT
2885 PRINT
2886 END
2887 REM ********************
2888 REM SINE PRODUCT SERIES SUBROUTINE (SINEPROD)
2889 REM THIS PROGRAM CALCULATES AN APPROXIMATION TO SIN(X)
2890 REM USING REPEATED PRODUCTS.
2891 REM THE INPUTS TO THE PROGRAM ARE THE ARGUMENT, X
2892 REM AND AN ERROR FACTOR, E.
2893 REM THE APPROXIMATION IS RETURNED IN Y.
2894 Y=X
2895 K=1
2896 L=3.14159265358979323846
```

```
2897 L=X/L
2898 L=L*L
2899 M=L/(K*K)
2900 Y=Y*(1-M)
2901 IF M<E THEN RETURN
2902 K=K+1
2903 GOTO 2899
```

Listing 2.8.1 *Subroutine for determining the sine using repeated products (SINEPROD). This subroutine is not included in the library because of its poor performance. It serves as an example of how a simple algorithm may be extremely poor in its numerical properties. Also shown is a program for demonstrating SINEPROD.*

Listing 2.8.2

X	SIN(X) CALC.	SIN(X) TRUE	K	ERROR
0	0	0	1	0
.05	.0499799	.0499791	16	.00000076
.1	.0998365	.0998334	32	.00000311
.15	.1494451	.1494381	48	.00000702
.2	.1986818	.1986693	64	.00001248
.25	.2474234	.2474039	80	.00001946
.3	.2955481	.2955202	96	.00002792
.35	.3429356	.3428978	112	.00003783
.4	.3894674	.3894183	128	.00004913
.45	.4350272	.4349655	144	.00006176
.5	.4795012	.4794255	160	.00007566
.55	.522778	.5226872	176	.00009077
.6	.56475	.5646424	191	.00010755
.65	.6053112	.6051864	207	.00012486
.7	.6443608	.6442176	223	.00014311
.75	.6818009	.6816387	239	.00016222
.8	.7175381	.717356	255	.00018208
.85	.751483	.7512804	271	.00020259
.9	.7835505	.7833269	287	.00022364
.95	.8136606	.8134155	303	.00024511
1	.8417378	.8414709	319	.00026689
1.05	.867712	.8674232	335	.00028886
1.1	.8915182	.8912073	351	.00031089
1.15	.9130968	.9127639	367	.00033287
1.2	.9323946	.932039	382	.00035558
1.25	.9493617	.9489846	398	.00037708
1.3	.9639563	.9635581	414	.00039813
1.35	.9761419	.9757233	430	.00041861
1.4	.9858881	.9854497	446	.00043839
1.45	.9931703	.9927129	462	.00045734
1.5	.9979703	.9974949	478	.00047535

1.55	1.000276	.9997837	494	.00049227
1.6	1.0000816	.9995736	510	.000508
1.65	.9973874	.996865	526	.00052242
1.7	.9922002	.9916648	542	.0005354
1.75	.9845327	.9839859	558	.00054684
1.8	.9744052	.9738476	573	.0005576
1.85	.9618408	.9612752	589	.00056563
1.9	.9468718	.9463	605	.00057181
1.95	.9295357	.9289597	621	.00057604
2	.9098756	.9092974	637	.00057826

READY

Listing 2.8.2 *A sample run of the program shown in listing 2.8.1. K is the number of products calculated to arrive at "SIN(X) CALC.". Although $E = 10^{-6}$, the resulting error is usually much more than that. If E were decreased to 10^{-9}, the error at x = 0.5 would be 0.000014 (with 5033 products calculated), or one quarter of that appearing above. It is very difficult to get good accuracy from SINEPROD. It is also a very slow subroutine.*

2.9 Complex Series

Once the coefficients of the approximating polynomial have been found, the results can easily be extended into the complex plane

$$P(Z) = a_0 Z + a_1 Z + a_2 Z^2 + \cdots \quad \text{where} \quad Z = x + iy \quad (2.9.1)$$

This can be accomplished by employing a few of the complex algebra subroutines given in the first volume of *BASIC Scientific Subroutines*. The result is the complex series evaluation subroutine (CMPLXSER) shown in listing 2.9.1. This subroutine is supported by three other routines, also given in that listing.

The inputs to CMPLXSER are the degree of the polynomial (M), the corresponding coefficients [A(I); real], and the real and imaginary parts of the argument (X and Y respectively). The result of the summation is returned in two parts, a real (Z1) and an imaginary (Z2) component.

Also appearing in listing 2.9.1 is a program that demonstrates the operation of CMPLXSER (see listing 2.9.2 for examples). As you can see from those examples, CMPLXSER is easy to use and has good round-off error properties.

CMPLXSER can be employed to very effectively extend the application of the small computer to the evaluation of functions having complex arguments. For example, complex functions can often be easily approximated using the techniques discussed in Chapters 1 and 2. To see how this can be accomplished, consider the function H(Z). If H(Z) is analytical, then it can be expanded in a Taylor series:

$$H(Z) = a_0 + a_1 Z + a_2 Z^2 + \cdots \qquad (2.9.2)$$

If the domain over which $H(Z)$ is analytical includes the real axis, we can write

$$H(x) = a_0 + a_1 x + a_2 x^2 + \cdots \qquad (2.9.3)$$

In addition, if $H(x)$ is a real-valued function, then the polynomial-fitting routines discussed so far can be used to find an approximation polynomial $P(x)$. This gives

$$H(x) \cong P(x)$$

It then follows that

$$H(Z) \cong P(Z)$$

Examples of functions that can be treated in this manner are sin Z, cos Z, e^z, ln Z, $J_N(Z)$, and many others. Therefore, although your BASIC interpreter or compiler may already have built-in trigonometric functions, economized (or least-squares fitted) polynomial approximations are still very useful if these functions are to be extended into the complex plane.

Engineers should find this extended ability particularly useful in the analysis of cyclic processes which are often expressed in terms of complex variables.

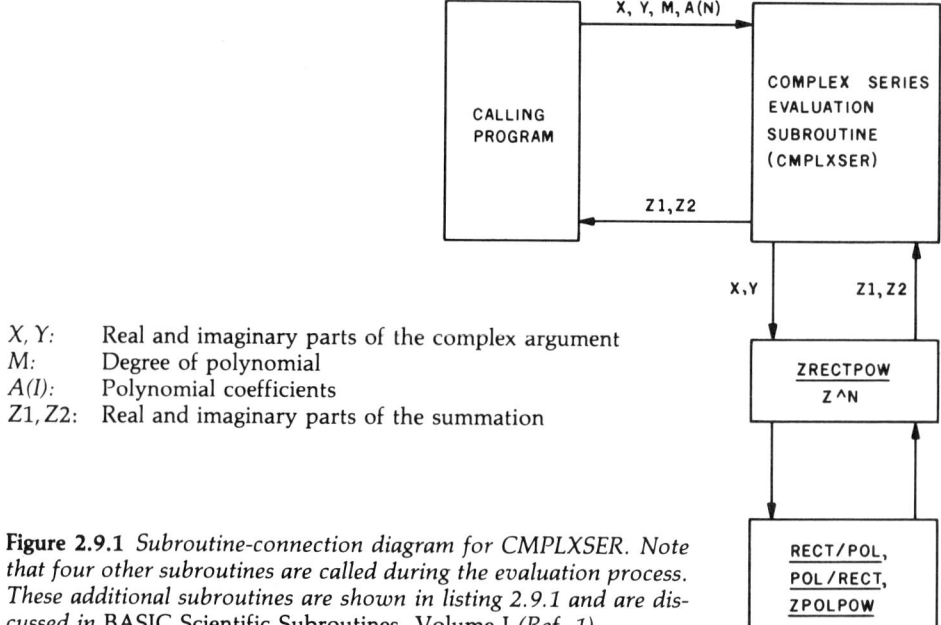

X, Y: Real and imaginary parts of the complex argument
M: Degree of polynomial
$A(I)$: Polynomial coefficients
$Z1, Z2$: Real and imaginary parts of the summation

Figure 2.9.1 *Subroutine-connection diagram for CMPLXSER. Note that four other subroutines are called during the evaluation process. These additional subroutines are shown in listing 2.9.1 and are discussed in* BASIC Scientific Subroutines, Volume I *(Ref. 1).*

Statements/Functions List

+, −, *, /, ^
ATN, COS, GOSUB, FOR/NEXT, IF/THEN, INT, SIN

Variables List

A1, A2, A(N), M, N, U, U1, V, V1, W, X, Y, Z1, Z2

Variables Passed to Subroutine

A(N), M, X, Y

Table 2.9.1 *Functions and variables employed in CMPLXSER and its support subroutines, ZRECTPOW, ZPOLPOW, RECT/POL, and POL/RECT.*

```
3000 REM PROGRAM TO DEMONSTRATE COMPLEX SERIES EVALUATION SUBROUTINE.
3001 REM IT IS ASSUMED THAT THE COEFFICIENTS ARE OBTAINED FROM
3002 REM A SUBROUTINE.
3003 REM GET COEFFICIENTS
3004 GOSUB 3024
3005 REM INPUT COMPLEX NUMBER
3006 PRINT
3007 PRINT
3008 PRINT "INPUT THE COMPLEX NUMBER AS PROMPTED:"
3009 PRINT
3010 PRINT "     REAL PART = ",
3011 INPUT X
3012 PRINT "     COMPLEX PART = ",
3013 INPUT Y
3014 GOSUB 44950
3015 PRINT
3016 PRINT
3017 PRINT "RESULTS ARE:"
3018 PRINT
3019 PRINT "     Z1 = ",Z1
3020 PRINT "     Z2 = ",Z2
3021 PRINT
3022 END
3023 REM COEFFICIENTS SUBROUTINE
3024 M=5
3025 A(0)=1
3026 A(1)=5
3027 A(2)=10
3028 A(3)=10
3029 A(4)=5
3030 A(5)=1
```

Listing 2.9.1

```
3031 RETURN
40398 REM ********************
40399 REM RECTANGULAR TO POLAR CONVERSION SUBROUTINE (RECT/POL)
40400 U=SQRT(X*X+Y*Y)
40401 REM GUARD AGAINST AMBIGUOUS VECTOR
40402 IF Y=0 THEN Y=(.1)^30
40403 REM GUARD AGAINST DIVIDE BY ZERO
40404 IF X=0 THEN X=(.1)^30
40405 REM SOME BASICS REQUIRE A SIMPLE ARGUMENT
40406 W=Y/X
40407 V=ATN(W)
40408 REM CHECK QUADRANT AND ADJUST
40409 IF X<0 THEN V=V+3.1415926535
40410 IF V<0 THEN V=V+6.2831853072
40411 RETURN
40449 REM POLAR TO RECTANGULAR CONVERSION SUBROUTINE (POL/RECT)
40450 X=U*COS(V)
40451 Y=U*SIN(V)
40452 RETURN
41099 REM POLAR POWER SUBROUTINE (ZPOLPOW)
41100 U1=U^N
41101 V1=N*V
41102 V1=V1-6.2831853072*INT(V1/6.2831853072)
41103 RETURN
41198 REM RECTANGULAR COMPLEX NUMBER POWER SUBROUTINE (ZRECTPOW)
41199 REM RECTANGULAR TO POLAR CONVERSION
41200 GOSUB 40400
41201 REM POLAR POWER
41202 GOSUB 41100
41203 REM CHANGE VARIABLE FOR CONVERSION
41204 U=U1
41205 V=V1
41206 REM POLAR TO RECTANGULAR CONVERSION
41207 GOSUB 40450
41208 RETURN
44942 REM ********************
44943 REM COMPLEX SERIES EVALUATION SUBROUTINE (CMPLXSER)
44944 REM THE SERIES COEFFICIENTS ARE A(I), ASSUMED REAL.
44945 REM THE ORDER OF THE POLYNOMIAL IS M.
44946 REM THE SUBROUTINE USES REPEATED CALLS TO THE
44947 REM NTH POWER (Z^N) COMPLEX NUMBER SUBROUTINE.
44948 REM INPUTS TO THE SUBROUTINE ARE X,Y,M, AND THE A(I).
44949 REM OUTPUTS ARE Z1(REAL) AND Z2(IMAGINARY).
44950 Z1=A(0)
44951 Z2=0
44952 REM STORE X AND Y
44953 A1=X
44954 A2=Y
44955 FOR N=1 TO M
```

```
44956 REM RECALL ORIGINAL X AND Y
44957 X=A1
44958 Y=A2
44959 REM GO TO Z^N SUBROUTINE
44960 GOSUB 41200
44961 REM FORM PARTIAL SUM
44962 Z1=Z1+A(N)*X
44963 Z2=Z2+A(N)*Y
44964 NEXT N
44965 REM RESTORE X AND Y
44966 X=A1
44967 Y=A2
44968 RETURN
```

Listing 2.9.1 *Complex series evaluation subroutine (CMPLXSER). Also shown are the support subroutines for the complex algebra operations (RECT/POL, POL/RECT, ZPOLPOW, and ZRECTPOW), and a program for demonstrating CMPLXSER. See listing 2.9.2 for examples.*

RUN

Listing 2.9.2

INPUT THE COMPLEX NUMBER AS PROMPTED:

    REAL PART = ?1
    COMPLEX PART = ?0

RESULTS ARE:

    Z1 = 32
    Z2 = 8E-29

READY
RUN

INPUT THE COMPLEX NUMBER AS PROMPTED:

    REAL PART = ?0
    COMPLEX PART = ?1

RESULTS ARE:

```
            Z1 =   -4.0000009
            Z2 =   -3.9999995

READY
RUN



INPUT THE COMPLEX NUMBER AS PROMPTED:

       REAL PART    = ?1
       COMPLEX PART = ?1


RESULTS ARE:

       Z1 =  -37.999997
       Z2 =   41.000012

READY
```

Listing 2.9.2 *Examples of the use of CMPLXSER for complex polynomial evaluation.*

2.10 Summary and Conclusion

You should leave Chapter 2 with an appreciation for the variety of available polynomial approximation methods, and an awareness of what some of the intrinsic limitations are. Taylor series certainly serve a purpose in obtaining good estimates. However, they should generally not be employed as approximations themselves, but should instead be used as inputs to the economization procedure in order to generate efficient polynomial representations. There are also cases in which Taylor series-based polynomials are not appropriate. For example, when the argument is large, asymptotic series are often much more effective. In addition, some functional forms are better treated using rational polynomials. Therefore, skill and awareness are required in choosing which class of approximations best suits the problem at hand.

Chapter 3

Functional Approximations by Iteration and Recursion

3.1 Introduction

The controversy over the potential limits of artificial intelligence has been brewing for many years. The point that is usually agreed upon is that a well-endowed computer *is* capable of being programmed to apply (though perhaps not ingeniously) the rules of deductive reasoning very effectively to specific classes of problems. In this chapter, we will examine two categories of mathematical approximations that employ deductive reasoning—*iteration* and *recursion*.

Algorithms that use iteration and recursion often appear to be very similar in construction and operation, and the two terms are often used interchangeably. Both iteration and recursion manipulate previously calculated values to obtain new or better estimates. When it comes to the numerical implications, however, there is an important and subtle distinction between the two techniques: round-off error does *not* propagate in an ideally iterative process, but it *does* in a recursive procedure. Two simple examples will serve to make this difference clear.

For the iteration example, we will consider numerically approximating that curious quantity—"zero." The algorithm is simple. Take any number and halve it. Then halve it again, and again, and so on. The eventual result is an arbitrarily accurate approximation to zero. The error in this estimate is limited only by the number of steps taken, and by the resolution of the arithmetic involved in the calculation. Round-off error that occurs at each stage of the iteration does not influence the final result; rather, it *damps* out.

As an example of a recursion algorithm, we will consider calculating an approximation to another interesting constant, π. To do this, an infinite product algorithm will be used:

$$\pi = 4 \prod_{n=1}^{\infty} \left[1 - \frac{1}{(2n+1)^2} \right] \qquad (3.1.1)$$

(See Ref. 28.) The procedure is to start with $4(1 - 1/3^2)$, multiply by $(1 - 1/5^2)$, then by $(1 - 1/7^2)$, and so on. Mathematically, this takes π to any desired level of accuracy, depending upon the number of products taken. Numerically, however, this is a very poor method (see the discussion in section 2.8). Error introduced at any stage directly affects the final result.

As you may already have realized, none of the approximation schemes considered in Chapters 1 and 2 were truly iterative. In fact, they were mostly recursive in nature.

In this chapter, we will examine a few important and clearly iterative algorithms. We will also study several recursive procedures that are very powerful and are only mildly susceptible to round-off error in that they are so rapidly convergent.

The subject of iteration is taken further in Chapters 6 and 7 (finding the zeros of functions), and in Chapter 8 (optimization). The recursion technique is used directly in the discussion in Chapter 5 (the CORDIC method).

Table 3.1.1

SUBROUTINE	SUBROUTINE SIZE (bytes) (including support programs)	DEMON-STRATION PROGRAM SIZE (bytes)	EXECUTION SPEED	CONDITIONS/ COMMENTS
NTHROOT ($y^{1/N}$, N an integer > 1)	574	347	0.18 seconds/ iteration	$(1024)^{1/8}$: 35 iterations
GENROOT (y^N, N arbitrary)	2027	460	11 seconds	32 bit decomposition
TANITER (tangent iteration)	604	273	~ 1 second	
ATANITER (arctangent by Gauss iteration)	805	269	1 second/ iteration	10^{-8} accuracy
ARCSINIT (arcsine)	449	342	0.1 second/ iteration	10^{-7} accuracy
CLIPTIC (complete elliptic integrals by Gauss' method)	927	347	0.1 second/ step	
LOGITER ($\ln x$)	454	316	0.2 seconds/ step	10^{-7} accuracy

SUBROUTINE	SUBROUTINE SIZE (bytes) (including support programs)	DEMONSTRATION PROGRAM SIZE (bytes)	EXECUTION SPEED	CONDITIONS/ COMMENTS
INTBESSL $[J_0(x), J_1(x), J_2(x), J_3(x), J_4(x)]$	620	454	3.7 seconds	$M = 20$ (steps) J_0 through J_4 calculated
LEGNDRE (Legendre polynomial coefficients)	581	271	~1 second/set	
LAGUERR (Laguerre polynomial coefficients)	603	273	~1 second/set	
HERMITE (Hermite polynomial coefficients)	562	271	~1 second/set	

Table 3.1.1 *A summary of the programs appearing in Chapter 3.*

3.2 Roots by Iteration

The classic introductory example for demonstrating iteration is approximating the root of a positive number:

$$x = y^{1/N} \tag{3.2.1}$$

(The reason for breaking with convention and writing the above equation with x on the left side and y on the right will be evident shortly.)

Equation (3.2.1) can be rewritten as

$$y = x^N \tag{3.2.2}$$

We define the function $f(x)$ as

$$f(x) = y - x^N \tag{3.2.3}$$

At the desired value of x, say x_0, $f(x_0) = 0$. However, x_0 is not known *a priori*. Instead, what we may have is a guess, x. This guess and the desired result can be related using a Taylor series expansion about x:

$$f(x_0) = 0 = f(x) + (x_0 - x) \left.\frac{df}{dx}\right|_x + (x_0 - x)^2 \left.\frac{d^2f}{dx^2}\right|_x + \cdots \quad (3.2.4)$$

Because $df/dx = -Nx^{N-1}$, we have the approximation

$$0 \cong y - x^N - (x_0 - x)Nx^{N-1}$$

By rearrangement, this gives

$$x_0 \cong x_0' = \frac{y + (N-1)x^N}{Nx^{N-1}} \quad (3.2.5a)$$

or

$$x_0 \cong x_0' = x + \frac{1}{Nx^{N-1}}(y - x^N) \quad (3.2.5b)$$

(See also Ref. 7.) Given a value for x, equation (3.2.5) can be used to find a better estimate, x_0'. x_0' can then replace x and the process is repeated. Eventually, any desired level of accuracy can, in principle, be achieved. In practice, the accuracy is limited by numerical precision.

For the case $N = 2$, $x = \sqrt{y}$ and the iteration equation is

$$x_{\text{new}} = \frac{1}{2}\left(x_{\text{old}} + \frac{y}{x_{\text{old}}}\right) \quad (3.2.6)$$

This special case equation is called Heron's formula (see Ref. 15). The iteration sequence for finding the square root of 2, which results from applying Heron's formula, is shown in table 3.2.1. As you can see, the convergence is very rapid. Under the right conditions, the convergence of the iteration implied by equation (3.2.5) is quadratic. That is, the error in step $i + 1$ (E_{i+1}) is roughly E_i^2. This root iteration technique is usually very effective.

Iteration Step	Estimate	Error ($\sqrt{2}$ − estimate)
0	1.0	0.4
1	1.5	−0.1
2	1.416666666666667	-2×10^{-3}
3	1.41421568627451	-2×10^{-6}
4	1.41421356237469	-2×10^{-12}
5	1.414213562373095	10^{-15} (all digits shown are correct)

Table 3.2.1 *The sequence of approximations associated with finding the square root of 2 using Heron's Rule. The initial guess was $X = 1$.*

Under the restriction that N be an integer greater than unity, the Nth-root iteration formula can readily be implemented in a BASIC subroutine—NTHROOT—as shown in listing 3.2.1. The inputs to NTHROOT are the argument (Y), the desired root (N), and a convergence criterion (E). The result is returned in X. In addition, the number of iterations is returned in M.

Also appearing in listing 3.2.1 is a program for demonstrating NTHROOT, and sample runs are shown in listing 3.2.2. The corresponding sequences of approximation are displayed in table 3.2.3. The third example, $\sqrt[8]{1024}$, was particularly taxing for NTHROOT. The convergence was slow at first, but it finally homed in on the correct value.

In the examples appearing in listing 3.2.2, it should be noted that the accuracy of the result was always better than the prescribed error criterion. This is because the convergence is so rapid. It is generally expected that the error will be less than E, and is *likely* to be greater than E^2. However, this error estimate must be tempered by the precision of the arithmetic being used.

There are a few important restrictions on the use of NTHROOT. First, Y must be a real number greater than zero. Second, N must be an integer greater than unity. (A direct calculation subroutine for handling complex number arguments and arbitrary N was given in the first volume of BASIC Scientific Subroutines.) Third, E *must* be greater than zero. In many cases, the choosing of E to be much less than the precision of the BASIC being used will give a very accurate estimate. Round-off error can cause some fluctuation in the last digit, however, and an infinite loop may result. Therefore, it is wise to choose E to be at least a few times the precision of the arithmetic being employed. It is always important to keep this thought in mind when setting the error criterion for the iteration and recursion subroutines given in this chapter. This will be discussed further later.

One of the limitations in applying NTHROOT is that it is restricted to positive integer values of N. This could be circumvented by implementing equation (3.2.5) with the exponentiation function (^) available in most dialects of BASIC instead of by employing simple products and division:

$$X_0 \cong X_0^i = X + \frac{1}{N X \char`\^ (N-1)} (Y - X \char`\^ N)$$

However, it is the exponentiation function that is actually being approximated by the algorithm! Therefore, a more fundamental method must be employed to treat arbitrary N.

To develop the algorithm, we redefine the problem:

$$x = y^N \qquad (3.2.7)$$

We need consider only positive N since the result for negative values can be obtained by simple inversion: $y^{-N} = 1/y^N$. Any positive N can be separated into integer and fractional components:

$$N = N_1 + N_2 \qquad (3.2.8)$$
$$N_1 = \text{integer}$$
$$N_2 = \text{fraction} \quad (0 \le N_2 < 1)$$

We then have

$$y^N = y^{N_1} y^{N_2} \qquad (3.2.9)$$

y^{N_1} is simple to calculate; y^{N_2} requires some effort. N_2 can be decomposed into a sum involving powers of 2:

$$N_2 = \sum_{i=1}^{\infty} \frac{a_i}{2^i} \qquad (3.2.10)$$

In this notation, $a_i = 0$ or 1. The above series is actually a *binary* fraction representation of N_2. y^{N_2} can now be expressed as

$$y^{N_2} = y^{\sum_{i=1}^{\infty} a_i/2^i} = y^{a_1/2} y^{a_2/2^2} y^{a_3/2^3} \cdots \qquad (3.2.11)$$

Each factor has the form $y^{a_m/2^m}$, where $a_m/2^m$ is either 0 or $1/2^m$. Thus, $y^{a_m/2^m}$ is either 1 or $y^{1/2^m}$. Since 2^m is clearly an integer, NTHROOT could be called to supply approximations for the required roots. However, using NTHROOT would be wasteful since what is really required is a sequence of square roots (Heron's formula). Therefore, the problem can be broken down into the following steps:

1) Determine N_1 and N_2 ($N = N_1 + N_2$).
2) Decompose N_2 into its binary representation. A limit M must be chosen for the number of *bits*.
3) Apply Heron's formula repeatedly to determine $y^{1/2^m}$.
4) Calculate the factors y_i:

   $y_i = 1$ \qquad if $a_i = 0$

   $y_i = y^{1/2^i}$ \qquad if $a_i = 1$

5) Assemble the results:

$$y^N = y^{N_1} \prod_{i=1}^{M} y_i$$

These steps are encoded in the subroutine shown in listing 3.2.3 (GENROOT). Step 2 in particular is accomplished by a separate support subroutine (RTDECOMP), which accepts N as input and the number of bits in the representation (M). The coefficients (0 or 1) are returned in $A(I)$.

The inputs to GENROOT are the same as to NTHROOT, but with the addition of M. The output, X, is the approximation to the power desired ($X = Y^N$).

Listing 3.2.3 also contains a program for demonstrating both RTDECOMP and GENROOT. Examples are shown in listing 3.2.4 for $M = 32$.

FUNCTIONAL APPROXIMATIONS BY ITERATION AND RECURSION 181

In the first example, the cube root of 27 is computed. The input is $N = 1/3 \cong 0.3333\ldots$. The binary representation of 1/3 is first displayed (32 bits), followed by the result. The error was low (1.3×10^{-6}), but higher than that expected for a pure iteration. The reason is that the decomposition step (RTDECOMP) and reconstruction procedure introduce round-off error which carries through the calculation. In the second example, there was no fractional part to the power, and the result was obtained using simple repeated products. The third example is perhaps the most interesting in that it shows how a simple decimal fraction (0.1) can have a not-so-simple binary representation. The error in the final result was 1.3×10^{-6}, and is entirely due to the round-off error associated with RTDECOMP and the subsequent reconstruction.

GENROOT is a very general subroutine and is capable of fairly good accuracy. However, specific routines exist (see Volume I of *BASIC Scientific Subroutines*) for evaluations such as 10^x and e^x that are more accurate and faster in execution. Therefore, GENROOT should be reserved for values of Y other than e and 10.

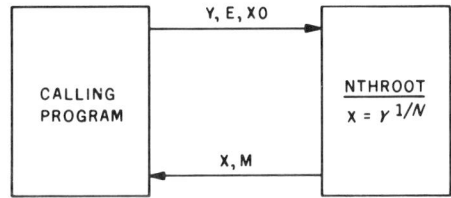

Y: Input value ($Y > 0$)
N: Desired root (integer > 1)
E: Convergence factor. Iteration is terminated when the relative change between successive estimates is less than E in magnitude
X0: Initial guess. $X0 = 1$ will do
X: Estimated root
M: Number of iterations performed

Figure 3.2.1 *Subroutine-connection diagram for the $x = y^{1/N}$ subroutine (NTHROOT).*

Statements/Functions List

$+, -, *, /, <$
ABS, FOR/NEXT, GOTO, IF/THEN, INT

Variables List

E, I, M, N, X0, X1, X2, Y

Variables Passed to Subroutine

E, X0, Y

Table 3.2.2 *Functions and variables employed in the NTHROOT subroutine.*

```
3050 REM PROGRAM TO DEMONSTRATE NTHROOT                Listing 3.2.1
3051 PRINT
3052 PRINT
3053 PRINT "WHAT IS THE NUMBER: ",
3054 INPUT Y
3055 PRINT
3056 PRINT "WHAT ROOT IS DESIRED: ",
3057 INPUT N
3058 PRINT
3059 PRINT "WHAT IS THE CONVERGENCE FACTOR: ",
3060 INPUT E
3061 PRINT
3062 REM THE INITIAL VALUE IS X0.
3063 X0=1
3064 PRINT
3065 GOSUB 45000
3066 PRINT "THE",N,"-TH ROOT OF",Y," IS",X
3067 PRINT
3068 PRINT "THE NUMBER OF ITERATIONS WAS",M
3069 PRINT
3070 END
44991 REM ********************
44992 REM NTH ROOT SUBROUTINE (NTHROOT)
44993 REM USES NEWTON-RAPHSON ITERATION.
44994 REM REFERENCE- HART, COMPUTER APPROXIMATIONS.
44995 REM EXPONENT IS 1/N, INPUT PARAMETER IS N
44996 REM NOTE THAT N MUST BE AN INTEGER.
44997 REM ARGUMENT IS Y, DESIRED ACCURACY IS E.
44998 REM RETURNED VALUE IS X.
44999 REM INITIAL VALUE IS X0.
45000 IF N<=1 THEN RETURN
45001 IF Y<0 THEN RETURN
45002 IF INT(N)<>N THEN RETURN
45003 IF E<=0 THEN RETURN
45004 IF Y>0 THEN GOTO 45007
45005 X=0
45006 RETURN
45007 M=0
45008 REM FIND N-1 POWER OF X0
45009 X2=1
45010 FOR I=1 TO N-1
45011 X2=X2*X0
45012 NEXT I
45013 REM ITERATE
45014 X1=((N-1)*X0+Y/X2)/N
45015 X=X1
45016 M=M+1
45017 IF ABS((X0-X1)/X1)<E THEN RETURN
45018 X0=X1
```

```
45019 GOTO 45009
```

Listing 3.2.1 *Subroutine for determining the Nth root of a positive number. For proper operation, N must be an integer greater than unity. See listing 3.2.2 for examples.*

```
RUN
```
                                        Listing 3.2.2

```
WHAT IS THE NUMBER: ?49

WHAT ROOT IS DESIRED: ?2

WHAT IS THE CONVERGENCE FACTOR: ?.001


THE 2-TH ROOT OF 49 IS 7

THE NUMBER OF ITERATIONS WAS 6

READY
RUN



WHAT IS THE NUMBER: ?27

WHAT ROOT IS DESIRED: ?3

WHAT IS THE CONVERGENCE FACTOR: ?.001


THE 3-TH ROOT OF 27 IS 3.0000005

THE NUMBER OF ITERATIONS WAS 7

READY
RUN



WHAT IS THE NUMBER: ?1024

WHAT ROOT IS DESIRED: ?8

WHAT IS THE CONVERGENCE FACTOR: ?.0001
```

```
THE 8-TH ROOT OF 1024 IS 2.3784143

THE NUMBER OF ITERATIONS WAS 35

READY
```

Listing 3.2.2 *Three examples of the use of NTHROOT. The corresponding approximation sequences that occurred within the subroutine are shown in table 3.2.2. Note that the convergence factor (E) is generally a very conservative measure of the final accuracy. Because the convergence is quadratic, the final error is expected to be roughly between E and E^2.*

Table 3.2.3

Example 1: $(49)^{1/2}$

25
13.48
8.5575075
7.141737
7.0014065
7

Example 2: $(27)^{1/3}$

9.6666667
6.5407583
4.570877
3.4780193
3.0626891
3.0012744
3.0000005

Example 3: $(1024)^{1/8}$

128.875
112.76563
98.669926
86.336185
75.544163
66.101143
57.8385
50.608688
44.282603
38.747278
33.903869
29.665885
25.95765
22.712944
19.873826
17.389598
15.215899
13.313913
11.649676
10.193471
8.9192984
7.8044146
6.8289354
5.9755034
5.229036
4.576604
4.0075723
3.5143355
3.094377
2.754699

Example 3: $(1024)^{1/8}$

2.5167003
2.4022784
2.3792278
2.3784153
2.3784143

Table 3.2.3 *Shown above are the convergent sequences that occurred within the NTHROOT subroutine for the examples given in listing 3.2.2. Note that in the vicinity of the true value, the convergence is very rapid.*

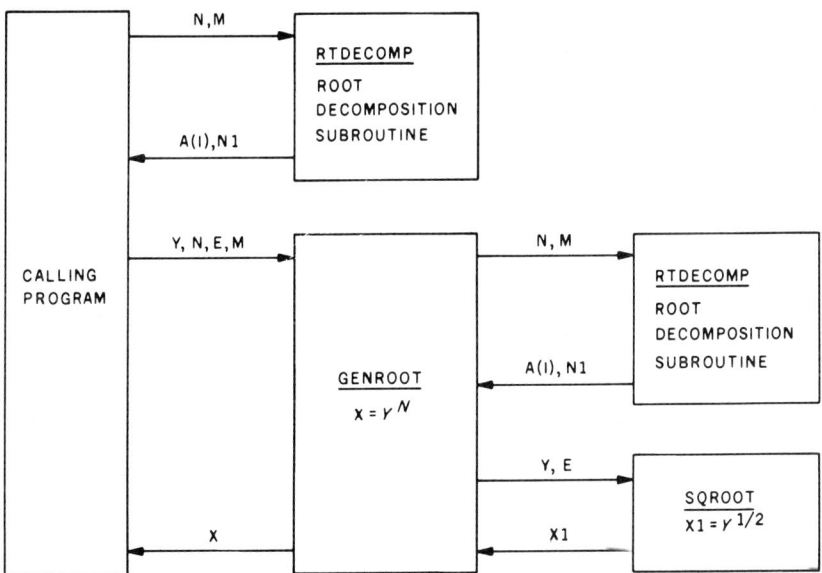

Y: Argument in $X = Y^N$ $(Y > 0)$
N: Exponent
E: Convergence criterion for SQROOT
M: Number of bits to be used in the decomposition which gives $A(I)$
A(I): Binary representation of the fractional part of N
N1: Integer part of N
X: Result $(X = Y^N)$

Figure 3.2.2 *Subroutine-connection diagram for the program shown in listing 3.2.3. The calling program uses RTDECOMP separately to demonstrate the binary decomposition.*

Statements/Functions List

$+, -, *, /, <, >$
ABS, FOR/NEXT, GOSUB, GOTO, IF/THEN

Variables List

A, A(I), E, I, M, N, N1, N2, N3, X, X0, X1, X2, X3, Y

Variables Passed to Subroutine

E, M, N, Y

Table 3.2.4 *Functions and variables employed in GENROOT and its support subroutines, RTDECOMP and SQROOT.*

Listing 3.2.3

```
3075 REM PROGRAM TO DEMONSTRATE GENROOT
3076 PRINT
3077 PRINT
3078 PRINT "WHAT IS THE NUMBER: ",
3079 INPUT Y
3080 PRINT
3081 PRINT "WHAT EXPONENT IS DESIRED: ",
3082 INPUT N
3083 PRINT
3084 PRINT "WHAT IS THE CONVERGENCE FACTOR: ",
3085 INPUT E
3086 PRINT
3087 REM NUMBER OF BITS = M
3088 M=32
3089 DIM A(M)
3090 PRINT "THE BINARY REPRESENTATION OF THE FRACTION IS:"
3091 PRINT
3092 REM GET THE BINARY REPRESENTATION
3093 GOSUB 45060
3094 FOR I=1 TO M
3095 PRINT A(I),
3096 NEXT I
3097 PRINT
3098 PRINT
3099 GOSUB 45030
3100 PRINT "THE",N,"-TH POWER OF",Y," IS",X
3101 PRINT
3102 PRINT
3103 END
45020 REM *******************
45021 REM GENERAL ROOT DETERMINATION SUBROUTINE (GENROOT)
```

```
45022 REM HIGH ACCURACY ITERATION INVOLVING THE SQUARE ROOT.
45023 REM ROUTINE DECOMPOSES EXPONENT INTO A BINARY REPRESENTATION
45024 REM AND THEN APPLIES NEWTON-RAPHSON ITERATION.
45025 REM Y IS THE INPUT, N IS THE EXPONENT, AND X THE RETURNED ROOT.
45026 REM E IS THE DESIRED ACCURACY OF THE ITERATION.
45027 REM M IS THE DESIRED NUMBER OF BITS IN THE REPRESENTATION OF N.
45028 REM SAVE Y FOR RETURNING FROM SUBROUTINE.
45029 REM SAVE N
45030 N3=N
45031 IF Y<0THEN RETURN
45032 IF E<=0 THEN RETURN
45033 IF Y>0 THEN GOTO 45036
45034 X=0
45035 RETURN
45036 X3=Y
45037 REM IF THE EXPONENT IS NEGATIVE, INVERT PROBLEM
45038 IF N>=0 THEN GOTO 45042
45039 N=-N
45040 Y=1/Y
45041 REM BREAK N DOWN INTO POWERS OF 1/2
45042 GOSUB 45060
45043 REM FIND MULTIPLIERS
45044 PRINT
45045 GOSUB 45074
45046 Y=X3
45047 N=N3
45048 RETURN
45053 REM ********************
45054 REM ROOT DECOMPOSITION SUBROUTINE (RTDECOMP)
45055 REM DECOMPOSE ROOT N INTO A BINARY REPRESENTATION.
45056 REM M IS THE NUMBER OF BINARY DIGITS.
45057 REM N IS THE INPUT DECIMAL NUMBER.
45058 REM N1 IS THE INTEGER PART OF N.
45059 REM A(I) IS THE BINARY REPRESENTATION OF THE REMAINING FRACTION.
45060 N1=INT(N)
45061 N2=N-N1
45062 REM N2 IS BETWEEN 0 AND 1
45063 REM DECOMPOSE N2 INTO FRACTIONS
45064 A=.5
45065 FOR I=1 TO M
45066 A(I)=0
45067 IF A<N2 THEN A(I)=1
45068 IF A<N2 THEN N2=N2-A
45069 A=A/2
45070 NEXT I
45071 RETURN
45072 REM ********************
45073 REM FIND MULTIPLYING FACTORS
45074 FOR I=1 TO M
```

Listing 3.2.3 cont.

```
45075 REM FIND SQUARE ROOT OF Y
45076 GOSUB 45109
45077 REM REPLACE Y WITH ITS SQUARE ROOT
45078 Y=X1
45079 IF A(I)=1 THEN GOTO 45083
45080 A(I)=1
45081 GOTO 45084
45082 REM A(I) IS SET EQUAL TO THE LATEST SQUARE ROOT OF Y
45083 A(I)=Y
45084 NEXT I
45085 REM ASSEMBLE RESULTS
45086 REM RETRIEVE Y
45087 Y=X3
45088 REM RETRIEVE N
45089 N=N3
45090 REM TAKE CARE OF N1 MULTIPLICATIONS
45091 X2=1
45092 FOR I=1 TO N1
45093 X2=X2*Y
45094 NEXT I
45095 REM TAKE CARE OF ROOT PORTION
45096 FOR I=1 TO M
45097 X2=X2*A(I)
45098 NEXT I
45099 REM THE FINAL ROOT IS X
45100 X=X2
45101 RETURN
45102 REM ********************
45103 REM SQUARE ROOT SUBROUTINE (SQROOT)
45104 REM USES NEWTON-RAPHSON ITERATION.
45105 REM CALLED HERON'S RULE.
45106 REM REFERENCE- HART, COMPUTER APPROXIMATIONS.
45107 REM ARGUMENT IS Y, RETURNED VALUE IS X1.
45108 REM DESIRED ACCURACY IS E.
45109 X0=1
45110 X1=(X0+Y/X0)/2
45111 IF ABS((X1-X0)/X1)<E THEN RETURN
45112 X0=X1
45113 GOTO 45110
```

Listing 3.2.3 *General subroutine for raising any positive number of the Nth power (GENROOT). N may be any positive or negative value. GENROOT also calls RTDECOMP, a subroutine that returns the binary representation of the fractional part of N, and SQROOT, which is an N = 2 version of NTHROOT. Also shown is a program for demonstrating GENROOT. See listing 3.2.4 for examples.*

FUNCTIONAL APPROXIMATIONS BY ITERATION AND RECURSION 189

WHAT IS THE NUMBER: ?27   Listing 3.2.4

WHAT EXPONENT IS DESIRED: ?.333333333333

WHAT IS THE CONVERGENCE FACTOR: ?.0000001

THE BINARY REPRESENTATION OF THE FRACTION IS:

 0 1 0 1 0 1 0 1 0 1 0 1 0 1 0 1 0 1 0 1 0 1 0 0 0 1 1 0

THE .33333333-TH POWER OF 27 IS 3.0000013

READY
RUN

WHAT IS THE NUMBER: ?100

WHAT EXPONENT IS DESIRED: ?10

WHAT IS THE CONVERGENCE FACTOR: ?.000001

THE BINARY REPRESENTATION OF THE FRACTION IS:

 0 0 0 0 0 0 0 0 0 0 0 0 0 0 0 0 0 0 0 0 0 0 0 0 0 0 0 0

THE 10-TH POWER OF 100 IS 1E+20

READY
RUN

WHAT IS THE NUMBER: ?100

WHAT EXPONENT IS DESIRED: ?.1

WHAT IS THE CONVERGENCE FACTOR: ?.000001

THE BINARY REPRESENTATION OF THE FRACTION IS:

 0 0 0 1 1 0 0 1 1 0 0 1 1 0 0 1 1 0 0 1 1 0 0 1 1 0 0 1

THE .1-TH POWER OF 100 IS 1.5848945

READY

Listing 3.2.4 *Sample runs of the general exponentiation program shown in listing 3.2.3. In the first example, the calculation is $(27)^{1/3}$. The error observed (1.3×10^{-6}) is due to round-off in the decomposition and reconstruction process. The second example demonstrates that N can be a large number. The fractional part of N is zero, and therefore the binary representation is also zero. The calculated value in the third example is high by 1.3×10^{-6}, again due to round-off error.*

3.3 Tangent Iteration

In this subsection, a more general form of equation (3.2.5) will be given. It will then be applied as an example to the determination of the tangent by iteration.

We will start with a truncated Taylor series approximation of equation (3.2.4):

$$0 \cong f(x) + (x_0 - x) \left.\frac{df}{dx}\right|_x \qquad (3.3.1)$$

This can be rearranged to give

$$x_0 \cong x_0' = x - \frac{f(x)}{(df/dx)|_x} \qquad (3.3.2)$$

Equation (3.3.2) is the famous Newton-Raphson iteration formula (see Chapter 6) for finding the zeros of $f(x)$.

The approximation of interest in this section is $x = \tan y$. Proceeding as in section 3.2, we obtain $f(x) = y - \arctan x$. Also, because $df/dx = -1/(1 + x^2)$, equation (3.3.2) results in the following iteration formula:

$$x_{new} = x_{old} + (1 + x_{old}^2)(y - \arctan x_{old}) \qquad (3.3.3)$$

If an ATN (or ATAN) function is available, this iteration formula can be directly implemented [see listing 3.3.1 (TANITER)]. If the ATN (or ATAN) function is accurate, then the low error associated with a full iteration is expected. The examples shown in listing 3.3.2 bear this out.

The main precautions involved in using the TANITER subroutine are: 1) Y must be in the range $-\pi/2 < y < \pi/2$; and 2) E must be greater than zero.

TANITER is not a very useful subroutine unless a routine is also available for evaluating the arctangent, which itself does not depend on the tangent. Such a routine is supplied in the next section. However, TANITER has particular value in that it illustrates a technique for iteratively finding the *inverse* of a given function. This may work quite well under the proper conditions (discussed in Chapter 6). To show

how this can be accomplished, we will simply retrace our steps and change a few definitions:

1) Define $x = g^{-1}(y)$, $y = g(x)$
2) Define $f(x) = y - g(x)$

The iteration formula is then

$$x_{new} = x_{old} + \frac{y - g(x_{old})}{(dg/dx)|_{x_{old}}} \qquad (3.3.4)$$

This technique should be kept in mind in that it offers a quick solution to some problems that are otherwise very difficult to solve. The disadvantage is that convergence may be slow initially. As an exercise, try obtaining the inverses of such functions as $\sin x$ and $\ln x$ by this method.

For the remainder of Chapter 3, the discussion will be directed at approximations via recursion relations. Round-off error may accumulate in recursion calculations, but the calculations are very effective in cases in which the convergence is rapid or the results are explicit (e.g., the Chebyshev polynomial coefficients calculations discussed in Chapter 2).

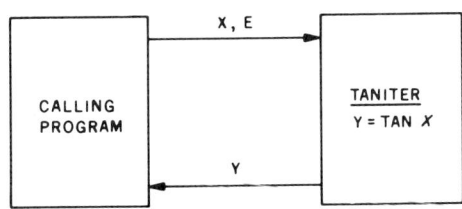

X: Argument ($-\pi/2 < X < \pi/2$)
E: Convergence criterion. If relative change is less than E, then terminate
Y: Result ($Y = \text{TAN } X$)

Figure 3.3.1 *Subroutine-connection diagram for the tangent iteration subroutine (TANITER).*

Statements/Functions List

+, −, *, /, <, >
ABS, ATN, GOTO, IF/THEN

Variables List

E, X, X0, X1, Y

Variables Passed to Subroutine

X, E

Table 3.3.1 *Functions and variables appearing in the tangent iteration subroutine (TANITER).*

```
3125 REM PROGRAM TO DEMONSTRATE TANITER SUBROUTINE
3126 PRINT
3127 PRINT
3128 PRINT "WHAT IS THE ARGUMENT: ",
3129 INPUT X
3130 PRINT
3131 PRINT "WHAT IS THE CONVERGENCE FACTOR: ",
3132 INPUT E
3133 PRINT
3134 PRINT
3135 GOSUB 45125
3136 PRINT "THE TANGENT OF",X," IS",Y
3137 PRINT
3138 PRINT "THE ARCTANGENT OF THE TANGENT IS",ATN(Y)
3139 PRINT
3140 END
45117 REM ********************
45118 REM TANGENT ITERATION SUBROUTINE (TANITER)
45119 REM USES THE INVERSE TANGENT.
45120 REM BASED ON NEWTON-RAPHSON ITERATION.
45121 REM X IS THE ARGUMENT, Y IS THE RESULT.
45122 REM THE DESIRED ACCURACY IS E.
45123 REM NOTE, THE ALLOWABLE RANGE OF THE ARGUMENT IS -PI/2 TO PI/2.
45124 REM INITIAL GUESS IS X0=1
45125 X0=1
45126 REM CHECK FOR DIVIDE BY ZERO
45127 IF X<>0 THEN GOTO 45131
45128 Y=0
45129 RETURN
45130 REM CHECK FOR OUT OF BOUNDS
45131 IF ABS(X)>=3.1415926535/2 THEN RETURN
```

Listing 3.3.1

```
45132 IF E<=0 THEN RETURN
45133 REM CAN CALL ARCTANGENT SUBROUTINE HERE.
45134 X1=X0+(X-ATN(X0))*(1+X0*X0)
45135 REM TEST FOR ACCURACY
45136 IF ABS((X0-X1)/X1)<E THEN GOTO 45139
45137 X0=X1
45138 GOTO 45134
45139 Y=X1
45140 RETURN
```

Listing 3.3.1 *Tangent iteration subroutine (TANITER). TANITER requires the use of the arctangent function. However, the arctangent can also be approximated using an iteration subroutine (see ATANITER in section 3.4). Sample runs of the above program appear in listing 3.3.2.*

```
RUN
```
Listing 3.3.2

```
WHAT IS THE ARGUMENT: ?0

WHAT IS THE CONVERGENCE FACTOR: ?.0000001

THE TANGENT OF 0 IS 0

THE ARCTANGENT OF THE TANGENT IS 0

READY
RUN


WHAT IS THE ARGUMENT: ?1

WHAT IS THE CONVERGENCE FACTOR: ?.000000001

THE TANGENT OF 1 IS 1.5574079

THE ARCTANGENT OF THE TANGENT IS 1

READY
RUN
```

```
WHAT IS THE ARGUMENT: ?1.57

WHAT IS THE CONVERGENCE FACTOR: ?.000000001


THE TANGENT OF 1.57 IS 1255.7746

THE ARCTANGENT OF THE TANGENT IS 1.57

READY
RUN



WHAT IS THE ARGUMENT: ?-1.5705

WHAT IS THE CONVERGENCE FACTOR: ?.000000001


THE TANGENT OF -1.5705 IS -3375.2836

THE ARCTANGENT OF THE TANGENT IS -1.5705

READY
```

Listing 3.3.2 *Sample runs of the program given in listing 3.3.1. The tangent of X is calculated using the ATN (or ATAN) function in BASIC. However, the arctangent can also be calculated using the iteration subroutine ATANITER. The arctangent of the tangent is also evaluated with the ATN (or ATAN) function in BASIC. The arctangent should return the original value, which it does, indicating high accuracy for the calculated tangents.*

3.4 Arctangent by Recursion

There are two general types of recursion. One involves a sequence of calculations that ideally converges on the desired result in the limit. The other involves an explicit calculation based on other results. In this and the following three subsections, we will examine infinite sequence forms of recursion for the arctangent, arcsine, complete elliptic integral and logarithm functions.

We will start with the arctangent recursion sequence. This algorithm was developed by Gauss and is described in Acton (see Ref. 29). It is based on the trigonometric half-angle formulae that result in a recursion relation called the *arithmogeometrical mean*. The function to be approximated is $y = \arctan x$. The evaluation begins with the following definitions:

$$a_0 = \frac{1}{1 + x^2} \tag{3.4.1}$$

$$b_0 = 1$$

The recursion relation that is repeatedly exercised is

$$a_{i+1} = \frac{a_i + b_i}{2} \quad \text{for } i = 1, 2, \ldots \tag{3.4.2}$$

$$b_{i+1} = (a_{i+1}b_i)^{1/2}$$

a_i and b_i eventually converge on one another. If we call this limit L, then

$$\arctan x = \frac{x}{L(1 + x^2)^{1/2}} \tag{3.4.3}$$

Acton assumes that the procedure is carried to completion so that $a_n = b_n$. In the more practical situation, however, the process is repeated until some error criterion is met (e.g., $|(a_n - b_n)/b_n| < E$). In that case, a good estimate for L is

$$L = (a_n b_n)^{1/2} \tag{3.4.4}$$

The only nonelementary function involved in the evaluation procedure is the square root. This can be provided by an efficient iteration using Heron's formula.

The Gauss recursion formula for the arctangent can be programmed in BASIC as shown in listing 3.4.1 (ATANITER). The inputs to this subroutine are the argument (X) and the convergence criterion (E), with the result returned in Y. The meaning of the convergence criterion is not very crisp. Unless there is significant round-off error, the accuracy is generally better than the E.

ATANITER can be exercised using the demonstration program also shown in listing 3.4.1. Sample runs appear in listing 3.4.2. ATANITER works well, but round-off error limits its accuracy to roughly 10 to 30 times the precision of the arithmetic being used.

X: Argument
E: Convergence criterion
Y: Returned result
M: Number of iterations

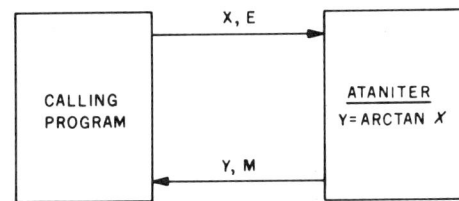

Figure 3.4.1 *Subroutine-connection diagram for the arctangent subroutine (ATANITER).*

Statements/Functions List

$+, -, *, /, <$
ABS, GOSUB, GOTO, IF/THEN

Variables List

A0, A1, B0, B1, E, M, X, X0, X1, X2, Y

Variables Passed to Subroutine

E, X

Table 3.4.1 *Functions and variables used in the arctangent subroutine (ATANITER).*

```
3150 REM PROGRAM TO DEMONSTRATE ATANITER
3151 PRINT
3152 PRINT
3153 PRINT "WHAT IS THE ARGUMENT: ",
3154 INPUT X
3155 PRINT
3156 PRINT "WHAT IS THE CONVERGENCE FACTOR: ",
3157 INPUT E
3158 PRINT
3159 GOSUB 45150
3160 PRINT
3161 PRINT "THE ARCTANGENT OF",X," EQUALS",Y
3162 PRINT
3163 PRINT "THE NUMBER OF ITERATIONS WAS",M
3164 PRINT
3165 PRINT
3166 END
45142 REM *******************
45143 REM INVERSE TANGENT RECURSION SUBROUTINE (ATANITER)
45144 REM USES GAUSS ITERATION
45145 REM REFERENCE- ACTON, NUMERICAL METHODS THAT WORK
45146 REM ARGUMENT IS X, RESULT IS Y
45147 REM DESIRED ACCURACY IS E
45148 REM HERON'S RULE (ITERATION) FOR THE SQUARE ROOT IS ALSO USED
45149 REM A0,A1,B0,B1 ARE DUMMY VARIABLES
45150 IF E<0 THEN RETURN
45151 M=0
45152 Y=1+X*X
45153 REM FIND SQUARE ROOT OF 1/Y
45154 GOSUB 45177
45155 X2=1/X1
45156 A0=X2
```

Listing 3.4.1

```
45157 B0=1
45158 A1=(A0+B0)/2
45159 Y=A1*B0
45160 REM FIND SQUARE ROOT
45161 GOSUB 45177
45162 B1=X1
45163 REM CHECK ACCURACY
45164 M=M+1
45165 IF ABS((A1-B1)/B1)<E THEN GOTO 45170
45166 A0=A1
45167 B0=B1
45168 GOTO 45158
45169 REM COMPUTE FINAL RESULT
45170 Y=A1*B1
45171 REM OBTAIN SQUARE ROOT
45172 GOSUB 45177
45173 Y=X**2/X1
45174 RETURN
45175 REM *******************
45176 REM SQUARE ROOT SUBROUTINE
45177 X0=1
45178 X1=(X0+Y/X0)/2
45179 IF ABS((X0-X1)/X1)<E THEN RETURN
45180 X0=X1
45181 GOTO 45178
```

Listing 3.4.1 *Arctangent recursion subroutine (ATANITER). ATANITER also contains its own square-root (by iteration) subroutine, thereby making it a good candidate for a primary source of arctangent values. See listing 3.4.2 for examples.*

RUN

Listing 3.4.2

WHAT IS THE ARGUMENT: ?1

WHAT IS THE CONVERGENCE FACTOR: ?.01

THE ARCTANGENT OF 1 EQUALS .78600479

THE NUMBER OF ITERATIONS WAS 3

READY
RUN

```
WHAT IS THE ARGUMENT: ?1

WHAT IS THE CONVERGENCE FACTOR: ?.000000001

THE ARCTANGENT OF 1 EQUALS .78539798

THE NUMBER OF ITERATIONS WAS 12


READY
RUN


WHAT IS THE ARGUMENT: ?0

WHAT IS THE CONVERGENCE FACTOR: ?.000000001

THE ARCTANGENT OF 0 EQUALS 0

THE NUMBER OF ITERATIONS WAS 1


READY
RUN


WHAT IS THE ARGUMENT: ?10000000000

WHAT IS THE CONVERGENCE FACTOR: ?.000000001

THE ARCTANGENT OF 1E+10 EQUALS 1.5707958

THE NUMBER OF ITERATIONS WAS 13


READY
```

Listing 3.4.2 *Sample runs of the program given in listing 3.4.1 for the arctangent (ATANITER). The convergence factor (E) is used in both the square-root and main iteration programs. In the first example, $E = 0.01$, and the error in the calculated result is 0.001. In the second example, $E = 10^{-9}$, and the observed error is 2×10^{-7}. Accuracy better than 10^{-8} should not have been expected because only 8-digit precision was used. However, Gauss iteration can suffer from round-off error, as it did in this example.*

3.5 Arcsine by Recursion

The arcsine can also be approximated by a recursion sequence similar to that used in the previous subsection. In this case, there is only one parameter in the sequence, a_i. The sequence starts with the following definitions:

$$a_0 = x(1 - x^2)^{1/2} \qquad (3.5.1)$$
$$a_1 = x$$

(See Ref. 17.) The recursion relation is

$$a_{i+1} = a_i \left(\frac{2a_i}{a_i + a_{i-1}}\right)^{1/2} \quad \text{for } i = 1, 2, \ldots \qquad (3.5.2)$$

Eventually, the sequence of a_i values converges on arcsin x. As with the arctangent algorithm, the square-root function can be replaced with an iteration subroutine.

The arcsine recursion algorithm can be encoded as a BASIC subroutine as shown in listing 3.5.1 (ARCSINIT). Although the square-root function could have been replaced by a call to an iteration subroutine (as was done in ATANITER), that was not done here.

The inputs to ARCSINIT are the argument ($|X| \le 1$) and the convergence criterion (E). The result is $Y \cong \arcsin X$. Also returned is the number of steps, M.

Presented in listing 3.5.1 is a program for demonstrating ARCSINIT, and a sample run appears in listing 3.5.2. As with TANITER, round-off error limits the accuracy to roughly 30 times the precision of the arithmetic.

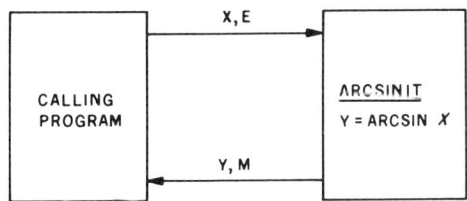

X: Argument ($|X| < 1$)
E: Convergence criterion (absolute error)
Y: Result
M: Number of steps

Figure 3.5.1 *Subroutine-connection diagram for the arcsine recursion subroutine (ARCSINIT). Note that ARCSINIT uses the SQRT function, which can be replaced with an iteration subroutine.*

Statements/Functions List

$+, -, *, /, <, >$
ABS, GOTO, IF/THEN, SQRT

Variables List

E, M, X, Y, U0, U1, U2

Variables Passed to Subroutine

E, X

Table 3.5.1 *Functions and variables appearing in the arcsine recursion subroutine (ARCSINIT).*

```
3175 REM PROGRAM TO DEMONSTRATE ARCSINE RECURSION           Listing 3.5.1
3176 PRINT
3177 PRINT
3178 E=.00000001
3179 PRINT " X",TAB(10),"ARCSIN(X)",TAB(25),"STEPS",TAB(37),"ERROR"
3180 PRINT " ---",TAB(10),"---------",TAB(25),"-----",TAB(35),"---------"
3181 PRINT
3182 FOR X=0 TO 1 STEP .05
3183 GOSUB 45200
3184 PRINT X,TAB(9),INT(10000000*Y)/10000000,TAB(25),
3185 PRINT M,TAB(34),INT(10000000*(SIN(Y)-X)+.5)/10000000
3186 NEXT X
3187 PRINT
3188 PRINT
3189 END
45194 REM ********************
45195 REM ARCSIN(X) RECURSION SUBROUTINE
45196 REM INPUT IS X (-1<X<1)
45197 REM OUTPUT IS Y=ARCSIN(X)
45198 REM CONVERGENCE CRITERIA IS E
45199 REM REFERENCE- COMPUTATIONAL ANALYSIS BY HENRICI
45200 M=0
45201 REM GUARD AGAINST FAILURE
45202 IF E<=0 THEN RETURN
45203 IF X<>0 THEN GOTO 45207
45204 Y=0
45205 RETURN
45206 REM CHECK RANGE
45207 IF ABS(X)>1 THEN RETURN
45208 U0=X*SQRT(1-X*X)
45209 U1=X
```

```
45210 U2=U1*SQRT(2*U1/(U1+U0))
45211 Y=U2
45212 M=M+1
45213 IF ABS(U2-U1)<E THEN RETURN
45214 U0=U1
45215 U1=U2
45216 GOTO 45210
```

Listing 3.5.1 *Subroutine for determining the arcsine by recursion (ARCSINIT). Also appearing is a program for exercising ARCSINIT. See listing 3.5.2 for a sample run.*

RUN

X	ARCSIN(X)	STEPS	ERROR
0	0	0	0
.05	.0500208	7	0
.1	.1001674	8	0
.15	.1505682	9	0
.2	.2013578	9	-.0000001
.25	.2526801	10	-.0000001
.3	.3046925	10	-.0000001
.35	.3575709	10	-.0000001
.4	.4115166	10	-.0000001
.45	.4667652	11	-.0000001
.5	.5235985	11	-.0000002
.55	.582364	11	-.0000001
.6	.6435008	11	-.0000002
.65	.7075842	11	-.0000001
.7	.7753971	11	-.0000003
.75	.8480617	11	-.0000002
.8	.9272949	12	-.0000001
.85	1.0159848	12	-.0000003
.9	1.1197691	12	-.0000002
.95	1.2532356	12	-.0000001
1	1.5707961	13	0

READY

Listing 3.5.2 *Sample run of the program shown in listing 3.5.1. Although 8-digit precision was employed in the calculations, the results are accurate to only a few times 10^{-7}. This is because ARCSINIT is a recursion routine and is affected by round-off error.*

3.6 Elliptic Integrals by Recursion

Elliptic integrals outwardly appear to be very esoteric functions and many practicing scientists and engineers avoid them. But these functions are intimately related to such very common and simple concepts as the period of a pendulum and the circumference of an ellipse.

We will consider the complete elliptic integrals of the first and second kinds. They are defined respectively as

$$K(k) = \int_0^{\pi/2} \frac{1}{\sqrt{1 - k^2 \sin^2 t}} \, dt \qquad (3.6.1)$$

and

$$E(k) = \int_0^{\pi/2} \sqrt{1 + k^2 \sin^2 t} \, dt \qquad (3.6.2)$$

(See Ref. 19.) $K(k)$ can easily be used to accurately calculate the period of an ideal pendulum as follows. If the length of the pendulum is l, the gravitational constant g, and the maximum amplitude of swing α (in radians), then the period is

$$T = 4 \, K(k) \sqrt{l/g} \quad \text{where} \quad k = \sin \alpha/2 \qquad (3.6.3)$$

(See Ref. 30.) In most introductory courses and texts that discuss the pendulum, it is assumed that the amplitude of swing is so small that $\sin \theta$ can be replaced with θ in the defining differential equation. In that case, $k \approx 0$ and $T \cong 2\pi\sqrt{l/g}$. For amplitudes of vibration so small that this approximation is valid, the period is independent of the amplitude and is said to be *isochronous*. However, as the maximum amplitude is increased, the period gets longer (e.g., a church bell stuck upside down). Greenhill (see Ref. 30) gives a practical example that shows that if a pendulum's amplitude is adjusted from a 6° swing to a 10° swing, it will lose 26 seconds a day. As you can imagine, these calculations were particularly important in the days when the pendulum (or a similar clock mechanism) was the primary timepiece.

The complete elliptic integral of the second kind, $E(k)$, also has a simple application. If the major and minor axes of an ellipse are $2a$ and $2b$, respectively, then the circumference of that ellipse is

$$c = 4b \, E(k) \quad \text{where} \quad k^2 = \frac{b^2 - a^2}{b^2} \qquad (3.6.4)$$

(See Ref. 31.) For the special case of a circle, $a = b$ and $c = 2\pi b$.

As you can see, elliptic integrals are involved in some very fundamental calculations. We will now consider how $K(k)$ and $E(k)$ can be approximated.

The steps in the elliptic integral recursion procedure are very similar to those involved in the arctangent calculation (see Ref. 19). First, the starting values are defined:

$$a_0 = 1 + k$$
$$b_0 = 1 - k \tag{3.6.5}$$

Next, the recursion relation is repeatedly exercised until $|a_n - b_n|$ is less than some convergence criterion, E:

$$a_{i+1} = \frac{a_i + b_i}{2}$$
$$b_{i+1} = (a_i b_i)^{1/2} \quad \text{for } i = 1, 2, \ldots \tag{3.6.6}$$

[Note that equation (3.6.6) is nearly identical to equation (3.4.2).] Once the convergence criterion has been met, $K(k)$ is evaluated as

$$K(k) = \frac{\pi}{2a_n} \tag{3.6.7}$$

$E(k)$ requires some further calculation:

$$E(k) = \frac{K(k)}{2}[2 - (a_0^2 - b_0^2) - 2(a_1^2 - b_1^2) - 4(a_2^2 - b_2^2) - \cdots] \tag{3.6.8}$$

Technically, the form of the evaluation for $E(k)$ is prone to round-off error. Convergence is usually rapid, however, and the round-off error tends to be small.

The above recursive procedure can be very compactly programmed in BASIC as shown in listing 3.6.1 (CLIPTIC). The inputs to CLIPTIC are simply the parameter K ($0 \leq K \leq 1$) and the convergence criterion, E. The outputs are $E1 = K(K)$, $E2 = E(K)$, and the number of steps performed, N.

Also appearing in listing 3.6.1 is a program that employs CLIPTIC to prepare a table of elliptic integrals. See listing 3.6.2 for a sample run. As is typical with this kind of recursion calculation, the accuracy is good.

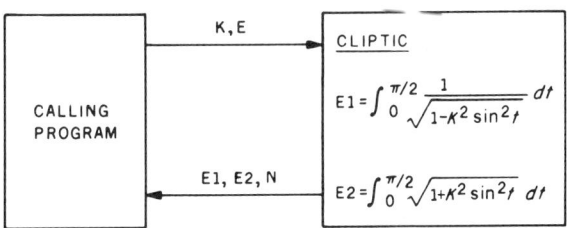

K: Input parameter
E: Convergence criterion (absolute value)
E1: Elliptic integral of the first kind
E2: Elliptic integral of the second kind
N: Number of steps used in the approximation

Figure 3.6.1 *Subroutine-connection diagram for the elliptic integral subroutine (CLIPTIC).*

Statements/Functions List

+, −, *, /, >
ABS, GOTO, IF/THEN, SQRT

Variables List

A(N), B(N), E, E1, E2, K, M, N

Variables Passed to Subroutine

E, K

Table 3.6.1 *Functions and variables appearing in the elliptic integral subroutine (CLIPTIC).*

```
3200 REM PROGRAM TO DEMONSTRATE EVALUATING ELLIPTIC INTEGRALS OF
3201 REM THE FIRST AND SECOND KINDS (COMPLETE)
3202 PRINT
3203 PRINT
3204 DIM A(200),B(200)
3205 E=.0000001
3206 PRINT " K",TAB(14),"K(K)",TAB(29),"E(K)",TAB(41),"STEPS"
3207 PRINT " ---",TAB(10),"------------",TAB(25),"------------",TAB(41),"-----"
3208 PRINT
3209 FOR K=0 TO 1 STEP .05
3210 GOSUB 45250
3211 PRINT K,TAB(10),E1,TAB(25),E2,TAB(42),N
3212 NEXT K
3213 END
45236 REM ********************
45237 REM COMPLETE ELLIPTIC INTEGRAL OF THE FIRST
45238 REM AND SECOND KIND (CLIPTIC)
45239 REM THE INPUT PARAMETER IS K, WHICH SHOULD
45240 REM BE BETWEEN 0 AND 1.
45241 REM TECHNIQUE USES GAUSS' FORMULA FOR THE
45242 REM ARITHMOGEOMETRICAL MEAN.
45243 REM REFERENCE- BALL, ALGORITHMS FOR RPN CALCULATORS.
45244 REM E IS A MEASURE OF THE CONVERGENCE ACCURACY.
45245 REM DEPENDING ON E, A(I) AND B(I) MAY HAVE TO BE DIMENSIONED
45246 REM IN THE CALLING PROGRAM.
45247 REM THE RETURNED VALUES ARE E1, THE ELLIPTIC
45248 REM INTEGRAL OF THE FIRST KIND, AND E2,
45249 REM THE INTEGRAL OF THE SECOND KIND.
45250 A(0)=1+K
45251 B(0)=1-K
45252 N=0
45253 IF K<0 THEN RETURN
```

Listing 3.6.1

```
45254 IF K>1 THEN RETURN
45255 IF E<=0 THEN RETURN
45256 IF K<1 THEN GOTO 45261
45257 E2=1
45258 E1=1000000000
45259 E1=E1*E1*E1*E1
45260 RETURN
45261 N=N+1
45262 REM GENERATE IMPROVED VALUES
45263 A(N)=(A(N-1)+B(N-1))/2
45264 B(N)=SQRT(A(N-1)*B(N-1))
45265 IF ABS(A(N)-B(N))>E THEN GOTO 45261
45266 E1=1.5707963268/A(N)
45267 E2=2
45268 M=1
45269 FOR I=1 TO N
45270 E2=E2-M*(A(I)*A(I)-B(I)*B(I))
45271 M=M*2
45272 NEXT I
45273 E2=E2*E1/2
45274 RETURN
```

Listing 3.6.1 *Subroutine for evaluating complete elliptic integrals of the first and second kinds (CLIPTIC). Also shown is a program for applying CLIPTIC to create a table. See listing 3.6.2 for a sample run.*

RUN

Listing 3.6.2

K	K(K)	E(K)	STEPS
0	1.5707963	1.5707963	1
.05	1.5717794	1.5698141	3
.1	1.5747455	1.5668618	3
.15	1.5797456	1.5619229	3
.2	1.5868679	1.5549688	3
.25	1.5962422	1.5459574	3
.3	1.6080486	1.5348334	3
.35	1.622528	1.5215251	3
.4	1.6399998	1.5059416	3
.45	1.6608862	1.487968	4
.5	1.6857503	1.467462	4
.55	1.7153545	1.4442436	4
.6	1.7507537	1.4180831	4

.65	1.7934541	1.3886865	4
.7	1.845694	1.3556612	4
.75	1.9109897	1.318472	4
.8	1.9953027	1.27635	4
.85	2.1099354	1.2281087	4
.9	2.280549	1.171697	4
.95	2.5900111	1.1027215	5
1	1E+36	1	0

READY

Listing 3.6.2 *Sample run of the elliptic integral program given in listing 3.6.1. K(k) is the elliptic integral of the first kind. E(k) is the elliptic integral of the second kind. Observe that in the notation of Abramowitz and Stegun (Ref. 6), $k^2 = m$ (their tables are given in terms of m). Over the range $0 \leq k < 1$ appearing in the above table, the results are accurate to better than 3×10^{-7}. However, K(1) = infinity.*

3.7 Natural Logarithm by Recursion

The recursion relations for the arcsine and elliptic integral approximations can be modified to give an algorithm for the natural logarithm. Since the procedure development for the logarithm case is particularly easy to follow, it is presented as an example (see also Ref. 17).

We will start with the derivative of x^a with respect to a:

$$\frac{\partial x^a}{\partial a} = x^a \ln x \qquad (3.7.1)$$

(See Ref. 5.) In the limit as a approaches zero, we have

$$\lim_{a \to 0} \frac{\partial x^a}{\partial a} = \ln x \qquad (3.7.2)$$

The numerical approximation for this derivative is

$$\frac{\partial x^a}{\partial a} \cong \frac{x^a - x^{-a}}{2a} \qquad (3.7.3)$$

One approach to estimating $\ln x$ would be to numerically evaluate equation (3.7.3) using a very small value for a. However, the effect of round-off error would be devastating. An alternative is to find a recursion relation which, in effect, evaluates $(x^a - x^{-a})/2a$ as a whole, but halves a on each pass until the limit is reached. Such a relation is given below:

$$s_{i+1} = s_i \left(\frac{2s_i}{s_i + s_{i-1}}\right)^{1/2} \quad \text{for} \quad i = 1, 2, 3, \ldots \qquad (3.7.4)$$

If we take $s_{i-1} = (x^a - x^{-a})/2a$ and $s_i = (x^{a/2} - x^{-a/2})/a$, we get $s_{i+1} = 2(x^{a/4} - x^{-a/4})/a$. If equation (3.7.4) is repeated, we eventually approach the limit desired in equation (3.7.2).

For the logarithm recursion, we will start with

$$s_0 = \frac{1}{4}(x^2 - \frac{1}{x^2})$$
$$s_1 = \frac{1}{2}(x - \frac{1}{x})$$
(3.7.5)

Unless x is near unity, there is little round-off error associated with evaluating these equations. The recursion relation is then exercised until the change in s_i is less than some convergence criterion (E), at which point we take $\ln x = s_{i+1}$.

A subroutine for performing these calculations appears in listing 3.7.1 (LOGITER). The inputs to this subroutine are the argument ($0 < X \le 1$) and the convergence criterion ($E > 0$). The returned results are $Y = \ln X$ and the number of steps used (M). Also provided in listing 3.7.1 is a program for demonstrating LOGITER. See listing 3.7.2 for a sample run. The error observed is due to round-off, and is of the same magnitude as that observed in the earlier recursion subroutines.

If the inputs to LOGITER are not within the allowed ranges, an erroneous result will be obtained. If the returned value of M is zero, then such an error has occurred. Otherwise, LOGITER is a reliable subroutine for approximating $\ln X$.

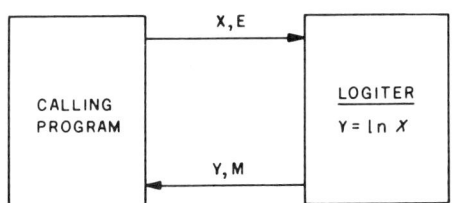

X: Argument ($0 < X \le 1$)
E: Convergence criterion
Y: Result ($Y = \ln X$)

Figure 3.7.1 *Subroutine-connection diagram for the natural logarithm recursion subroutine (LOGITER).*

Statements/Functions List

$+, -, *, /, <$
ABS, GOTO, IF/THEN, SQRT

Variables List

E, M, X, Y, U0, U1, U2

Variables Passed to Subroutine

E, X

Table 3.7.1 *Functions and variables appearing in the LOGITER subroutine.*

```
3225 REM PROGRAM TO DEMONSTRATE LOG BY RECURSION           Listing 3.71
3226 PRINT
3227 PRINT
3228 E=.000001
3229 PRINT "   X",TAB(10),"LOG(X)",TAB(23),"STEPS",TAB(35),"ERROR"
3230 PRINT " ---",TAB(7),"------------",TAB(23),"-----",TAB(32),"------------"
3231 PRINT
3232 FOR X=.05 TO .95 STEP .05
3233 GOSUB 45300
3234 PRINT X,TAB(7),Y,TAB(24),M,TAB(32),INT(100000000*(Y-LOG(X)))/100000000
3235 NEXT X
3236 PRINT
3237 PRINT
3238 END
45294 REM ********************
45295 REM LN(X) RECURSION SUBROUTINE
45296 REM INPUT IS X (0<X<1)
45297 REM OUTPUT IS Y=LN(X)
45298 REM CONVERGENCE CRITERIA IS E
45299 REM REFERENCE-  COMPUTATIONAL ANALYSIS BY HENRICI
45300 M=0
45301 REM GUARD AGAINST FAILURE
45302 IF E<=0 THEN RETURN
45303 IF X<1 THEN GOTO 45307
45304 Y=0
45305 RETURN
45306 REM CHECK RANGE
45307 IF X<=0 THEN RETURN
45308 U0=(X*X-1/(X*X))/4
45309 U1=(X-1/X)/2
45310 U2=U1*SQRT(2*U1/(U1+U0))
45311 Y=U2
```

```
45312 M=M+1
45313 IF ABS(U2-U1)<E THEN RETURN
45314 U0=U1
45315 U1=U2
45316 GOTO 45310
```

Listing 3.7.1 *Subroutine for determining the natural logarithm by recursion (LOGITER). Also shown is a program for demonstrating LOGITER. See listing 3.7.2 for a sample run.*

RUN

X	LOG(X)	STEPS	ERROR
.05	-2.9957324	12	-.0000001
.1	-2.3025852	12	-.0000001
.15	-1.8971203	11	-.0000004
.2	-1.6094381	11	-.0000002
.25	-1.3862944	11	0
.3	-1.203973	10	-.0000003
.35	-1.0498224	10	-.0000003
.4	-.91629079	10	-.00000005
.45	-.79850793	9	-.00000024
.5	-.69314732	9	-.00000015
.55	-.597837	9	0
.6	-.51082568	9	-.00000005
.65	-.43078307	8	-.00000016
.7	-.35667503	8	-.00000008
.75	-.28768227	7	-.00000021
.8	-.22314368	7	-.00000013
.85	-.16251912	6	-.00000019
.9	-.10536069	5	-.00000017
.95	-5.1293402E-02	4	-.00000011

READY

Listing 3.7.2 *Sample run of the program appearing in listing 3.7.1. The results are accurate to better than 3×10^{-7}, and the error is due largely to round-off.*

3.8 Bessel Functions by Recursion

So far in this volume, we have approximated Bessel functions by Taylor, near-min-max, and asymptotic series. In this section, we will examine a very accurate recursive algorithm for determining the values of Bessel functions. This algorithm is particularly interesting in that it applies a recursion relation in an iterative manner that results in very little round-off error. Because this approach is both intriguing and applicable to other similar problems, its development is outlined in detail so that you can extend the technique to other functions (see also Refs. 17 and 32).

The fundamental recursion relation governing Bessel functions is

$$J_{n-1}(x) = \frac{2n}{x} J_n(x) - J_{n+1}(x) \tag{3.8.1}$$

Given $J_0(x)$ and $J_1(x)$, all the other Bessel functions of integer order can be derived.

Equation (3.8.1) is a backward recursion relation that is self-correcting (recessive). That is, if we start with erroneous values for $J_m(x)$ and $J_{m+1}(x)$ (m large), then as we repeatedly use the recursion relation to work toward $J_0(x)$, the calculated results begin to fall in line with the true values (except for some constant multiplier). Therefore, we will use the following recursion sequence:

$$\left. \begin{array}{l} y_{m+1} = 0 \\ y_m = 1 \end{array} \right\} \quad \text{arbitrary}$$

$$y_{n-1} = \frac{2n}{x} y_n - y_{n+1} \quad \text{for} \quad n = m, m-1, \ldots, 1 \tag{3.8.2}$$

At the end of this sequence, we are left with

$$J_0(x) \cong \frac{1}{c} y_0$$

$$J_1(x) \cong \frac{1}{c} y_1$$

and so on. The constant can be evaluated using the following identity:

$$J_0(x) + 2J_2(x) + 2J_4(x) + \cdots = 1 \tag{3.8.3}$$

Therefore, $\quad c \cong y_0 + 2y_2 + 2y_4 + \cdots + 2y_{m \,(or\, m+1)} \tag{3.8.4}$

A subroutine for applying this algorithm is given in listing 3.8.1 (INTBESSL). It follows the flowchart shown in figure 3.8.1. INTBESSL calculates the first five Bessel functions (J_0, J_1, J_2, J_3, J_4), but can easily be extended to higher order. The inputs to INTBESSL are the argument ($X > 0$) and the starting point in the sequence ($M > 4$). The larger the value for M, the better the approximation in terms of the results of the recursion relation [equation (3.8.2)] settling down. However, large values of M can lead to numeric overflow, depending on X. As a rule of thumb, M should be chosen such that the quantity $2M!/X^M$ is comfortably within the numeric range of the BASIC dialect being used.

FUNCTIONAL APPROXIMATIONS BY ITERATION AND RECURSION

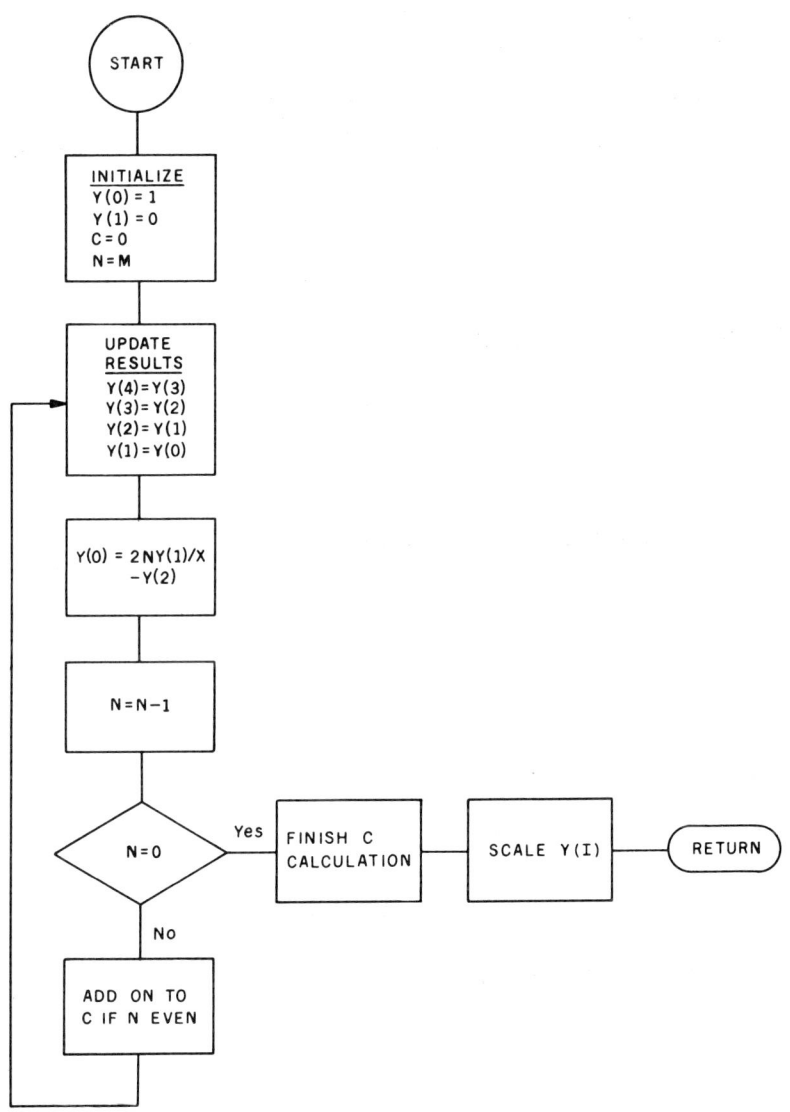

X: Argument $(X > 0)$
M: Number of steps $(M > 4)$
Y(I): Calculated Bessel function value

Figure 3.8.1 *Flowchart for the INTBESSL subroutine. This diagram may also be used as a pattern for other recessive recursion subroutines.*

The outputs of INTBESSL are $Y(0) = J_0(X)$, $Y(1) = J_1(X)$, $Y(2) = J_2(X)$, $Y(3) = J_3(X)$, and $Y(4) = J_4(X)$.

Also shown in listing 3.8.1 is a program for demonstrating INTBESSL. This routine generates a table of Bessel functions as shown in listing 3.8.2. Note that for $M = 20$, the accuracy of the calculations is generally better than 3×10^{-8} over the range $0.1 \leq X \leq 3$. This is near the precision limit of the arithmetic used (eight digits). If M is large enough, INTBESSL is very accurate.

The precautions associated with using INTBESSL are simple. The argument (X) must be positive, and M must be large enough for the error in the arbitrary starting values to damp out.

The two key elements that were required for the Bessel function recursion/iteration algorithm were a recursion relation and a way to normalize the results (find c). Presented below are two classic functions for which these elements are readily available (from Ref. 5). As an exercise, examine approximations to these functions along the lines considered in this section.

Legendre Polynomials: $P_n(x)$ $(-1 \leq x \leq 1)$

Recursion relation: $P_{n-1}(x) = \dfrac{(2n + 1)x}{n} P_n(x) - \dfrac{n + 1}{n} P_{n+1}(x)$

Normalization: $P_n(1) = 1$

$$\dfrac{1}{\sqrt{2 - 2x}} = \sum_{n=0}^{\infty} P_n(x)$$

Chebyshev Polynomials: $T_n(x)$ $(-1 \leq x \leq 1)$

Recursion relation: $T_{n-1}(x) = 2x\, T_n(x) - T_{n+1}(x)$

Normalization: $T_n(1) = 1$

$$\dfrac{1}{2} = \sum_{n=0}^{\infty} T_n(x)$$

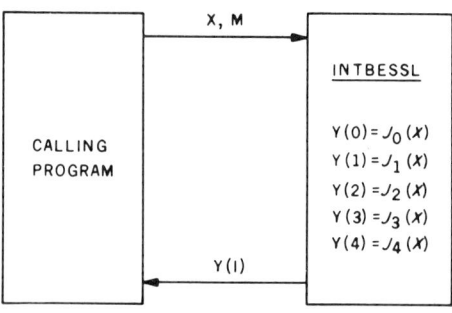

Figure 3.8.2 *Subroutine-connection diagram for INTBESSL.*

Statements/Functions List

$+,-,*,/,<,>$
FOR/NEXT, GOTO, IF/THEN, INT

Variables List

C, I, M, N, X, Y(I)

Variable Passed to Subroutine

X

Table 3.8.1 *Functions and variables appearing in the INTBESSL subroutine.*

Listing 3.8.1

```
3250 REM PROGRAM TO DEMONSTRATE INTBESSL SUBROUTINE
3251 PRINT
3252 PRINT
3253 M=20
3254 PRINT "   X",TAB(10),"J0(X)",TAB(25),"J1(X)",TAB(40),"J2(X)",
3255 PRINT TAB(55),"J3(X)",TAB(70),"J4(X)"
3256 PRINT " ---",TAB(7),"------------",TAB(22),"------------",TAB(37),
3257 PRINT "------------",TAB(52),"------------",TAB(67),"------------"
3258 PRINT
3259 FOR X=.1 TO 3 STEP .1
3260 GOSUB 45350
3261 I=100000000
3262 PRINT X,TAB(7),INT(I*Y(0))/I,TAB(22),INT(I*Y(1))/I,TAB(37),INT(I*Y(2))/I,
3263 PRINT TAB(52),INT(I*Y(3))/I,TAB(67),INT(I*Y(4))/I
3264 NEXT X
3265 PRINT
3266 PRINT
3267 END
45341 REM ********************
45342 REM INTEGER ORDER BESSEL FUNCTION SUBROUTINE (INTBESSL)
45343 REM CALCULATES BESSEL FUNCTIONS OF ORDER 0 THROUGH 4
45344 REM FOR X>0.
45345 REM MILLER'S METHOD USED, SEE HENRICI
45346 REM ARGUMENT IS X
45347 REM NUMBER OF STEPS =M
45348 REM RETURNED RESULTS ARE Y(I)
45349 REM TEST FOR RANGE
45350 IF X<=0 THEN RETURN
45351 IF M<=0 THEN RETURN
45352 Y(0)=1
45353 Y(1)=0
45354 C=0
```

```
45355 N=M
45356 REM UPDATE RESULTS
45357 FOR I=4 TO 1 STEP -1
45358 Y(I)=Y(I-1)
45359 NEXT I
45360 REM APPLY RECURSION RELATION
45361 Y(0)=2*N*Y(1)/X-Y(2)
45362 N=N-1
45363 IF N=0 THEN GOTO 45367
45364 IF INT(N/2)<>N/2 THEN GOTO 45357
45365 C=C+2*Y(0)
45366 GOTO 45357
45367 C=C+Y(0)
45368 REM SCALE THE RESULTS
45369 FOR I=0 TO 4
45370 Y(I)=Y(I)/C
45371 NEXT I
45372 RETURN
```

Listing 3.8.1 *Subroutine for evaluating $J_0(x)$ through $J_4(x)$ using recursion (INTBESSL). Also appearing is a demonstration program that generates a table. See listing 3.8.2 for a sample run.*

Listing 3.8.2

RUN

X	J0(X)	J1(X)	J2(X)	J3(X)	J4(X)
.1	.99750157	.04993752	.00124895	.00002082	.00000026
.2	.99002497	.09950083	.00498335	.00016625	.00000415
.3	.97762625	.14831882	.01116586	.00055934	.00002099
.4	.96039821	.19602658	.01973466	.00132005	.00006613
.5	.93846981	.24226845	.03060402	.00256372	.00016073
.6	.91200486	.28670099	.04366509	.00439965	.00033147
.7	.88120089	.32899573	.05878694	.00692965	.00061009
.8	.84628735	.36884204	.07581776	.01024676	.00103298
.9	.80752379	.40594956	.0945863	.01443402	.00164055
1	.76519768	.44005058	.11490348	.01956335	.00247663
1.1	.71962201	.47090238	.13656415	.02569452	.00358782
1.2	.67113273	.49828906	.15934902	.03287433	.00502266
1.3	.62008598	.52202324	.18302669	.04113582	.00683095
1.4	.56685514	.54194771	.20735589	.05049771	.00906287
1.5	.51208767	.55793651	.23208767	.06096395	.01176813
1.6	.45540214	.56989596	.25696778	.07252345	.01499516
1.7	.39798486	.57776523	.28173894	.08514992	.01879021
1.8	.33998642	.58151696	.30614353	.09880201	.02319651
1.9	.28181855	.58115708	.32992573	.11342341	.02825345
2	.22389076	.57672482	.35283404	.12894325	.03399572

2.1	.16660699	.56829213	.37462363	.14527667	.04045258
2.2	.11036226	.55596304	.39505869	.16232547	.04764714
2.3	.05553976	.5398725	.41391458	.17997893	.05559566
2.4	.00250767	.52018527	.43098005	.1981148	.06430695
2.5	-.0483838	.49709411	.44605907	.21660039	.07378187
2.6	-.09680497	.47081829	.45897289	.23529383	.08401288
2.7	-.14244939	.44160137	.46956151	.2540453	.09498358
2.8	-.18503605	.40970924	.4776855	.27269861	.10666866
2.9	-.22431155	.37542745	.48322703	.29109258	.11903347
3	-.26005193	.33905897	.48609125	.30906271	.13203417

READY

Listing 3.8.2 *A table of Bessel function values generated by the program appearing in listing 3.8.1. The accuracy is better than 3×10^{-8}. In this case, round-off error did not accumulate very quickly.*

3.9 Orthogonal Polynomial Coefficients by Recursion

In section 3.8, we examined the use of recursion relations in iteratively estimating the values of functions. In this section, another use of the recursion relation concept will be investigated for the special class of functions called *orthogonal polynomials*.

Orthogonal polynomials are of importance in several of the sciences. For example, they are encountered in atomic and nuclear physics in the solution of the Schrödinger equation in various coordinate systems. Legendre polynomials in particular occur in that science as well as in electrostatics. In addition, they appear in numerical interpolation and integration schemes (for example, see Ref. 10). The utility of orthogonal polynomials rests in the property that any analytical function can be decomposed into these functions over the range in which they are defined. In many cases, such a decomposition greatly simplifies the solution to a particular problem (e.g., the wave equation).

We will define a set of orthogonal polynomials $P_n(x)$ as follows. First, $P_n(x)$ is of degree n; second, $P_n(x)$ and $P_m(x)$ are orthogonal with respect to a given weighting function $w(x)$ over a range a to b:

$$\int_a^b w(x)\, P_n(x)\, P_m(x)\, dx = 0 \quad \text{for} \quad n \neq m \qquad (3.9.1)$$

It necessarily follows that a recursion relation of the following form must exist:

$$P_{n+1}(x) = (\alpha_n + \beta_n x)\, P_n(x) - \gamma_n\, P_{n-1}(x) \qquad (3.9.2)$$

(See Ref. 6.) If the values of $P_0(x)$ and $P_1(x)$ are known, then so are the values of all higher-order polynomials. Similarly, if the polynomial coefficients of $P_0(x)$ and

$P_1(x)$ are known, then so are those for all $P_n(x)$. The coefficients of $P_{n+1}(x)$ are

$$a_0(n+1) = \alpha_n\, a_0(n) - \gamma_n\, a_0(n-1)$$
$$a_1(n+1) = \alpha_n\, a_1(n) + \beta_n\, a_0(n) - \gamma_n\, a_1(n-1)$$
$$a_2(n+1) = \alpha_n\, a_2(n) + \beta_n\, a_1(n) - \gamma_n\, a_2(n-1) \qquad (3.9.3)$$

$$\cdot$$
$$\cdot$$
$$\cdot$$

$$a_{n+1}(n+1) = \beta_n\, a_n(n)$$

This sequence of calculations can easily be programmed once α_n, β_n, γ_n, $a_0(0)$, $a_0(1)$, and $a_1(1)$ are known. This is done below for three families of orthogonal polynomials: Legendre, Laguerre, and Hermite.

Legendre Polynomials

The Legendre polynomials are defined over the interval $-1 \leq x \leq 1$. They are particularly interesting in that the weighting function is unity [$w(x) = 1$]:

$$\int_{-1}^{1} P_n(x)\, P_m(x)\, dx = 0 \qquad \text{for } n \neq m \qquad (3.9.4)$$

$$= \frac{2}{2n+1} \qquad \text{for } n = m$$

The recursion relation is

$$(n+1)\, P_{n+1}(x) = (2n+1)x\, P_n(x) - n\, P_{n-1}(x) \qquad (3.9.5)$$

$$\text{where } P_0 = 1$$
$$P_1 = x$$

A subroutine for evaluating the Legendre polynomial coefficients is given in listing 3.9.1 (LEGNDRE). The input to this subroutine is simply the order of the polynomial ($N \geq 2$). LEGNDRE returns the $N + 1$ coefficients in $A(I)$. The only precautions involved in using this subroutine are to remember to dimension the A and B arrays in the calling program, and to use only $N \geq 2$.

Also appearing in listing 3.9.1 is a routine for exercising LEGNDRE. This program calls LEGNDRE for $N = 2$ through 10, and prints out the Legendre polynomial coefficients calculated. See listing 3.9.2. Observe that $P_n(x)$ is either an odd or an even function, depending on N.

LEGNDRE also returns another array, $B(I, J)$. This array contains the Jth coefficient of $P_I(x)$ for $I = 0$ to N. However, $B(I, J)$ is not initialized, and for $J > I$ the array values *should* be zero, but may not be. Therefore, if $B(I, J)$ is to be used as an output, it should either be zeroed before entering LEGNDRE or carefully applied afterwards.

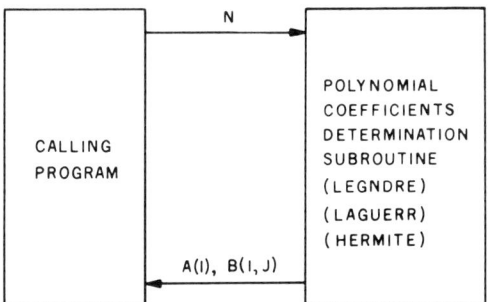

N: The order of the polynomial ($N \geq 2$)
A(I): The returned coefficients $P_n(x)$ [or $L_N(x)$, $H_N(x)$] $= a_0 + a_1 x + a_2 x^2 + \cdots + a_n x^N$
B(I, J): Coefficients for $P_I(x)$

Figure 3.9.1 *Subroutine-connection diagram for the LEGNDRE, LAGUERR and HERMITE polynomial coefficients subroutines.*

Statements/Functions List

$+, -, *, /$
FOR/NEXT, IF/THEN

Variables List

A(I), B(I,J), I, J, N

Variable Passed to Subroutine

N

Table 3.9.1 *Functions and variables employed in the LEGNDRE, LAGUERR, and HERMITE subroutines.*

Listing 3.9.1

```
3275 REM PROGRAM TO DEMONSTRATE THE LEGENDRE COEFFICIENTS SUBROUTINE
3276 PRINT
3277 PRINT
3278 DIM A(10),B(10,10)
3279 FOR N=2 TO 10
```

```
3280 PRINT
3281 PRINT "LEGENDRE POLYNOMIAL COEFFICIENTS"
3282 PRINT "FOR ORDER ",N
3283 PRINT
3284 GOSUB 45400
3285 FOR K=0 TO N
3286 PRINT "A(",K,") = ",A(K)
3287 NEXT K
3288 PRINT
3289 NEXT N
3290 PRINT
3291 PRINT
3292 END
45393 REM ********************
45394 REM LEGENDRE SERIES COEFFICIENT EVALUATION SUBROUTINE (LEGNDRE)
45395 REM BY MEANS OF RECURSION RELATION
45396 REM THE ORDER OF THE POLYNOMIAL IS N
45397 REM THE COEFFICIENTS ARE RETURNED IN A(I)
45398 REM DIMENSION A(I) AND B(I,J) IN THE CALLING PROGRAM
45399 REM ESTABLISH P0 AND P1 COEFFICIENTS
45400 B(0,0)=1
45401 B(1,0)=0
45402 B(1,1)=1
45403 REM RETURN IF ORDER IS LESS THAN TWO
45404 IF N<2 THEN RETURN
45405 FOR I=2 TO N
45406 B(I,0)=-(I-1)*B(I-2,0)/I
45407 FOR J=1 TO I
45408 REM BASIC RECURSION RELATION
45409 B(I,J)=(I+I-1)*B(I-1,J-1)-(I-1)*B(I-2,J)
45410 B(I,J)=B(I,J)/I
45411 NEXT J
45412 NEXT I
45413 FOR I=0 TO N
45414 A(I)=B(N,I)
45415 NEXT I
45416 RETURN
```

Listing 3.9.1 *Legendre polynomial coefficients determination subroutine (LEGNDRE). Also shown is a program that applies LEGNDRE to generate tables of Legendre polynomial coefficients (see listing 3.9.2).*

Listing 3.9.2

```
RUN

LEGENDRE POLYNOMIAL COEFFICIENTS
FOR ORDER   2
```

Listing 3.9.2 cont.

```
A( 0) =   -.5
A( 1) =    0
A( 2) =    1.5


LEGENDRE POLYNOMIAL COEFFICIENTS
FOR ORDER  3

A( 0) =    0
A( 1) =   -1.5
A( 2) =    0
A( 3) =    2.5


LEGENDRE POLYNOMIAL COEFFICIENTS
FOR ORDER  4

A( 0) =    .375
A( 1) =    0
A( 2) =   -3.75
A( 3) =    0
A( 4) =    4.375


LEGENDRE POLYNOMIAL COEFFICIENTS
FOR ORDER  5

A( 0) =    0
A( 1) =    1.875
A( 2) =    0
A( 3) =   -8.75
A( 4) =    0
A( 5) =    7.875


LEGENDRE POLYNOMIAL COEFFICIENTS
FOR ORDER  6

A( 0) =   -.3125
A( 1) =    0
A( 2) =    6.5625
A( 3) =    0
A( 4) =   -19.6875
A( 5) =    0
A( 6) =    14.4375


LEGENDRE POLYNOMIAL COEFFICIENTS
FOR ORDER  7
```

```
A( 0) =   0
A( 1) =   -2.1875
A( 2) =   0
A( 3) =   19.6875
A( 4) =   0
A( 5) =   -43.3125
A( 6) =   0
A( 7) =   26.8125
```

Listing 3.9.2 cont.

```
LEGENDRE POLYNOMIAL COEFFICIENTS
FOR ORDER  8

A( 0) =   .2734375
A( 1) =   0
A( 2) =   -9.84375
A( 3) =   0
A( 4) =   54.140625
A( 5) =   0
A( 6) =   -93.84375
A( 7) =   0
A( 8) =   50.273438


LEGENDRE POLYNOMIAL COEFFICIENTS
FOR ORDER  9

A( 0) =   0
A( 1) =   2.4609376
A( 2) =   0
A( 3) =   -36.09375
A( 4) =   0
A( 5) =   140.76562
A( 6) =   0
A( 7) =   -201.09376
A( 8) =   0
A( 9) =   94.960939


LEGENDRE POLYNOMIAL COEFFICIENTS
FOR ORDER  10

A( 0) =   -.24609375
A( 1) =   0
A( 2) =   13.535156
A( 3) =   0
A( 4) =   -117.30469
A( 5) =   0
A( 6) =   351.91406
```

```
A( 7) =  0
A( 8) =  -427.32423
A( 9) =  0
A( 10) = 180.42578
```

READY

Listing 3.9.2 *A sample run of the program given in listing 3.9.1. The coefficients of $P_N(x)$ for $N = 2$ through 10 are displayed. Note that $P_0(x) = 1$ and $P_1(x) = x$.*

Laguerre Polynomials

Laguerre polynomials are defined over the interval $0 \leq x < \infty$. The weighting function is the exponential [$w(x) = e^{-x}$]:

$$\int_0^\infty e^{-x} L_n(x) L_m(x) \, dx = 0 \quad \text{for } n \neq m \quad (3.9.6)$$
$$= 1 \quad \text{for } n = m$$

The recursion relation is

$$(n + 1) L_{n+1}(x) = (2n + 1 - x) L_n(x) - n L_n(x) \quad (3.9.7)$$

where $L_0(x) = 1$
$L_1(x) = 1 - x$

Simply by changing a few lines in the LEGNDRE subroutine, a program for evaluating Laguerre polynomial coefficients can be generated [see listing 3.9.3 (LAGUERR)]. A sample run appears in listing 3.9.4. Note that the Laguerre polynomial coefficients are all integers, alternate in sign, and have a leading coefficient $a_n = (-1)^n$.

Listing 3.9.3

```
3300 REM PROGRAM TO DEMONSTRATE THE LAGUERRE COEFFICIENTS SUBROUTINE
3301 PRINT
3302 PRINT
3303 DIM A(10),B(10,10)
3304 FOR N=2 TO 10
3305 PRINT
3306 PRINT "LAGUERRE POLYNOMIAL COEFFICIENTS"
3307 PRINT "FOR ORDER ",N
```

```
3308 PRINT
3309 GOSUB 45425
3310 FOR K=0 TO N
3311 PRINT "A(",K,") = ",A(K)
3312 NEXT K
3313 PRINT
3314 NEXT N
3315 PRINT
3316 PRINT
3317 END
45418 REM *******************
45419 REM LAGUERRE POLYNOMIAL COEFFICIENT EVALUATION SUBROUTINE (LAGUERR)
45420 REM BY MEANS OF RECURSION RELATION
45421 REM THE ORDER OF THE POLYNOMIAL IS N
45422 REM THE COEFFICIENTS ARE RETURNED IN A(I)
45423 REM DIMENSION A(I) AND B(I,J) IN THE CALLING PROGRAM
45424 REM ESTABLISH L0 AND L1 COEFFICIENTS
45425 B(0,0)=1
45426 B(1,0)=1
45427 B(1,1)=-1
45428 REM RETURN IF ORDER IS LESS THAN TWO
45429 IF N<2 THEN RETURN
45430 FOR I=2 TO N
45431 B(I,0)=(2*I-1)*B(I-1,0)-(I-1)*(I-1)*B(I-2,0)
45432 FOR J=1 TO I
45433 REM BASIC RECURSION RELATION
45434 B(I,J)=(2*I-1)*B(I-1,J)-B(I-1,J-1)-(I-1)*(I-1)*B(I-2,J)
45435 NEXT J
45436 NEXT I
45437 FOR I=0 TO N
45438 A(I)=B(N,I)
45439 NEXT I
45440 RETURN
```

Listing 3.9.3 *Laguerre polynomial coefficients determination subroutine (LAGUERR). The demonstration program displays the coefficients for $L_2(x)$ through $L_{10}(x)$. See listing 3.9.4 for a sample run.*

Listing 3.9.4

```
RUN



LAGUERRE POLYNOMIAL COEFFICIENTS
FOR ORDER  2
```

Listing 3.9.4 cont.

```
A( 0) =    2
A( 1) =   -4
A( 2) =    1


LAGUERRE POLYNOMIAL COEFFICIENTS
FOR ORDER  3

A( 0) =    6
A( 1) =  -18
A( 2) =    9
A( 3) =   -1


LAGUERRE POLYNOMIAL COEFFICIENTS
FOR ORDER  4

A( 0) =   24
A( 1) =  -96
A( 2) =   72
A( 3) =  -16
A( 4) =    1


LAGUERRE POLYNOMIAL COEFFICIENTS
FOR ORDER  5

A( 0) =  120
A( 1) = -600
A( 2) =  600
A( 3) = -200
A( 4) =   25
A( 5) =   -1


LAGUERRE POLYNOMIAL COEFFICIENTS
FOR ORDER  6

A( 0) =   720
A( 1) = -4320
A( 2) =  5400
A( 3) = -2400
A( 4) =   450
A( 5) =   -36
A( 6) =     1


LAGUERRE POLYNOMIAL COEFFICIENTS
FOR ORDER  7
```

```
A( 0) =    5040
A( 1) =   -35280
A( 2) =    52920
A( 3) =   -29400
A( 4) =    7350
A( 5) =    -882
A( 6) =     49
A( 7) =     -1
```

LAGUERRE POLYNOMIAL COEFFICIENTS
FOR ORDER 8

```
A( 0) =    40320
A( 1) =   -322560
A( 2) =    564480
A( 3) =   -376320
A( 4) =    117600
A( 5) =   -18816
A( 6) =    1568
A( 7) =     -64
A( 8) =      1
```

LAGUERRE POLYNOMIAL COEFFICIENTS
FOR ORDER 9

```
A( 0) =    362880
A( 1) =   -3265920
A( 2) =    6531840
A( 3) =   -5080320
A( 4) =    1905120
A( 5) =   -381024
A( 6) =    42336
A( 7) =    -2592
A( 8) =      81
A( 9) =      -1
```

LAGUERRE POLYNOMIAL COEFFICIENTS
FOR ORDER 10

```
A( 0) =    3628800
A( 1) =   -36288000
A( 2) =    81648000
A( 3) =   -72576000
A( 4) =    31752000
A( 5) =   -7620480
A( 6) =    1058400
```

```
A( 7) =   -86400
A( 8) =    4050
A( 9) =    -100
A( 10) =     1
```

READY

Listing 3.9.4 *The polynomial coefficients for $L_2(x)$ through $L_{10}(x)$ using the LAGUERR subroutine (listing 3.9.3). Observe that $L_0(x) = 1$ and $L_1(x) = 1 - x$.*

Hermite Polynomials

Hermite polynomials are defined over the range $-\infty < x < \infty$. The weighting function is $w(x) = e^{-x^2}$:

$$\int_{-\infty}^{\infty} e^{-x^2} H_n(x) H_m(x)\, dx = 0 \quad \text{for } n \neq m \qquad (3.9.8)$$
$$= f(n) \quad \text{for } n = m$$

The corresponding recursion relation is

$$H_{n+1}(x) = 2x\, H_n(x) - 2n\, H_{n-1}(x) \qquad (3.9.9)$$

$$\text{where } H_0(x) = 1$$
$$H_1(x) = 2x$$

As with the other polynomials, a simple subroutine for evaluating the coefficients can be written [see listing 3.9.5 (HERMITE)]. Also see listing 3.9.6 for examples. Note that Hermite polynomials are either even or odd, depending on N, and the coefficients are integers.

Listing 3.9.5

```
3320 REM PROGRAM TO DEMONSTRATE THE HERMITE COEFFICIENTS SUBROUTINE
3321 PRINT
3322 PRINT
3323 DIM A(10),B(10,10)
3324 FOR N=2 TO 10
3325 PRINT
3326 PRINT "HERMITE POLYNOMIAL COEFFICIENTS"
3327 PRINT "FOR ORDER ",N
3328 PRINT
3329 GOSUB 45450
```

```
3330 FOR K=0 TO N
3331 PRINT "A(",K,") = ",A(K)
3332 NEXT K
3333 PRINT
3334 NEXT N
3335 PRINT
3336 PRINT
3337 END
45443 REM ********************
45444 REM HERMITE POLYNOMIAL COEFFICIENT EVALUATION SUBROUTINE (HERMITE)
45445 REM BY MEANS OF RECURSION RELATION
45446 REM THE ORDER OF THE POLYNOMIAL IS N
45447 REM THE COEFFICIENTS ARE RETURNED IN A(I)
45448 REM DIMENSION A(I) AND B(I,J) IN THE CALLING PROGRAM
45449 REM ESTABLISH H0 AND H1 COEFFICIENTS
45450 B(0,0)=1
45451 B(1,0)=0
45452 B(1,1)=2
45453 REM RETURN IF ORDER IS LESS THAN TWO
45454 IF N<2 THEN RETURN
45455 FOR I=2 TO N
45456 B(I,0)=-2*(I-1)*B(I-2,0)
45457 FOR J=1 TO I
45458 REM BASIC RECURSION RELATION
45459 B(I,J)=2*B(I-1,J-1)-2*(I-1)*B(I-2,J)
45460 NEXT J
45461 NEXT I
45462 FOR I=0 TO N
45463 A(I)=B(N,I)
45464 NEXT I
45465 RETURN
```

Listing 3.9.5 *Subroutine for determining the coefficients of the Hermite polynomial of order N (HERMITE). The demonstration program (also shown above) displays the polynomial coefficents for* $H_2(x)$ *through* $H_{10}(x)$. *See listing 3.9.6 for a sample run.*

```
RUN
```

Listing 3.9.6

```
HERMITE POLYNOMIAL COEFFICIENTS
FOR ORDER   2

A( 0) =  -2
A( 1) =   0
A( 2) =   4
```

HERMITE POLYNOMIAL COEFFICIENTS
FOR ORDER 3

Listing 3.9.6 cont.

A(0) =   0
A(1) =   -12
A(2) =   0
A(3) =   8

HERMITE POLYNOMIAL COEFFICIENTS
FOR ORDER 4

A(0) =   12
A(1) =   0
A(2) =   -48
A(3) =   0
A(4) =   16

HERMITE POLYNOMIAL COEFFICIENTS
FOR ORDER 5

A(0) =   0
A(1) =   120
A(2) =   0
A(3) =   -160
A(4) =   0
A(5) =   32

HERMITE POLYNOMIAL COEFFICIENTS
FOR ORDER 6

A(0) =   -120
A(1) =   0
A(2) =   720
A(3) =   0
A(4) =   -480
A(5) =   0
A(6) =   64

HERMITE POLYNOMIAL COEFFICIENTS
FOR ORDER 7

A(0) =   0
A(1) =   -1680
A(2) =   0
A(3) =   3360

```
A( 4) =   0
A( 5) =   -1344
A( 6) =   0
A( 7) =   128

HERMITE POLYNOMIAL COEFFICIENTS
FOR ORDER  8

A( 0) =   1680
A( 1) =   0
A( 2) =   -13440
A( 3) =   0
A( 4) =   13440
A( 5) =   0
A( 6) =   -3584
A( 7) =   0
A( 8) =   256

HERMITE POLYNOMIAL COEFFICIENTS
FOR ORDER  9

A( 0) =   0
A( 1) =   30240
A( 2) =   0
A( 3) =   -80640
A( 4) =   0
A( 5) =   48384
A( 6) =   0
A( 7) =   -9216
A( 8) =   0
A( 9) =   512

HERMITE POLYNOMIAL COEFFICIENTS
FOR ORDER  10

A( 0) =   -30240
A( 1) =   0
A( 2) =   302400
A( 3) =   0
A( 4) =   -403200
A( 5) =   0
A( 6) =   161280
A( 7) =   0
A( 8) =   -23040
A( 9) =   0
A( 10) =  1024

READY
```

Listing 3.9.6 *The polynomial coefficients of $H_N(x)$ for $N = 2$ through 10 as calculated using the program appearing in listing 3.9.5.*

3.10 Summary and Conclusion

The first two chapters of Volume II pertained exclusively to polynomial approximations. In this chapter, a distinctly different approach was introduced—iteration.

As an initial example of iteration, the Newton-Raphson method for determining the roots of positive numbers was presented. The importance of being able to approximate $y^{1/N}$ by a means other than a polynomial rests in the observation that this function is not analytical at zero for $N > 1$. Therefore, a Taylor series expansion upon which to subsequently build an accurate and efficient approximation does not exist.

The need for a means to evaluate $y^{1/N}$ is related to the appearance of the square-root function in many algorithms. By being able to accurately estimate the square root, an arbitrary general exponentiation subroutine (GENROOT) was developed. Also, the algorithms involving Gauss' formula for the arithmogeometrical mean (ATANITER, ARCSINIT, CLIPTIC, and LOGITER) depend on the square-root function in the recursion sequence. Therefore, if these functions are to be accurately approximated using only elementary operations, it is very useful to have a means for replacing the nonelementary function, $\sqrt{x}$.

The concept of iteration was also shown to be applicable to the task of finding the inverse of functions, again by the Newton-Raphson iteration. The example was the tangent. This technique is useful because it is general and compact. Its main disadvantage is lack of speed.

Recursion was first introduced in Chapter 2 with respect to generating the Chebyshev polynomial coefficients. However, the form of recursion discussed in sections 3.4 through 3.7 was essentially in answer to a round-off error problem associated with numerically estimating derivatives. In principle, each of the functions considered in those four sections could be approximated with a finite difference estimate of a derivative. However, that approach is subject to considerable round-off error in implementation. Thus, a technique was devised for sequentially approximating this derivative by successive refinement using low round-off error operations. Because the accuracy of each step in the procedure affects the next, however, round-off error is still a problem, although to a much reduced degree. For the routines associated with the arithmogeometrical mean, this error appears to be in the $(n-1)$th digit for n-digit precision arithmetic.

The idea of *recessive recursion* was introduced in section 3.8 for the family of functions that are interrelated by three terms. This procedure is interesting in that a recursion relation is combined with a form of iteration in which round-off error *does not* accumulate. For the Bessel functions examined, the actual error observed was only a few times the precision of the arithmetic used. This is a potentially powerful technique and should be considered when attempting to evaluate families of functions that have recursion relations between members, as well as those that have an independent means of establishing the normalization.

Finally, in section 3.9, an interesting class of functions was discussed—orthogonal polynomials. Orthogonal polynomials can be defined through specific recursion relations, and these relations can be used directly to determine the associated polynomial coefficients. We did this first in Chapter 2 for the Chebyshev polynomials. In section 3.9, this was extended to the Legendre, Laguerre, and Hermite polynomials.

Chapter 3 does not represent an exhaustive survey of the iteration and recursion techniques available. However, it does demonstrate by example that with only a little ingenuity, some very powerful algorithms can be implemented in BASIC on a small computer.

Chapter 4

CORDIC Approximation Techniques and Alternatives

4.1 Introduction

So far in Volumes I and II of *BASIC Scientific Subroutines*, we have discussed four basic techniques for finding high-accuracy approximations to functions: polynomial expansions, table interpolation, iteration, and recursion. In this chapter, all four methods will be considered further in the form of the modified coordinate rotational digital computer (CORDIC) algorithm.

The CORDIC technique was devised by J. E. Volder (see Ref. 33). In essence, it is based on vector rotations that can be calculated using the fast register shifting operations available in the hardware or machine language of a binary-based computer. The CORDIC method is very effectively used in calculators such as the Hewlett-Packard series because it not only provides excellent approximations to the trigonometric functions and their inverses, but it also can be used to generate the hyperbolic, inverse hyperbolic, logarithm, and exponent functions.

The following discussion is aimed specifically at computing the CORDIC algorithm in BASIC and requires only the square-root function to build the associated coefficient tables. For a discussion of a routine that uses inverse trigonometric functions to generate the necessary coefficient tables, see "Simple Methods for Calculating Elementary Functions," by J. Rheinstein (in Ref. 34). Also, a more detailed discussion regarding the binary-based computer implementation of the CORDIC algorithm can be found in Schmid (Ref. 35) and Egbert (Ref. 36). As you can see by comparison with the references, the *modified* CORDIC algorithm presented in this text can be implemented without *a priori* knowledge of the inverse tangent table values required by the standard algorithm. This is important when considering very high-accuracy computations, e.g., 14 digits or higher. After all, how are the 14-digit inverse tangent table values to be found initially? The price paid for this modification is a significant reduction in the *potential* computing speed. However, if the true binary CORDIC algorithm were to be implemented in BASIC, the speed advantage, which is largely related to the simple binary register shifts, could not be captured anyway. Therefore, as we will see in a later section, the BASIC implementation of the CORDIC algorithm for the trigonometric and exponential functions is intrinsically wasteful and can easily be improved upon by the addition of a series approximation step.

SUBROUTINE	SUBROUTINE SIZE (bytes) (including support programs)	DEMON-STRATION PROGRAM SIZE (bytes)	EXECUTION SPEED	CONDITIONS/ COMMENTS
TRIGCORD (sine and cosine)	1630	706	~2 seconds	14-digit precision
CORDIC (coefficents program)	----	576	4 minutes (50 line table)	16-digit precision Microsoft program
INVCORD (arcsine, arc-cosine, and arc-tangent)	2161	556	~1.5 seconds	14-digit precision
EXPCORD (e^x)	1122	305	~1.2 seconds	14-digit precision
LNCORDIC (ln x)	1537	352	~1.5 seconds	14-digit precision
SINH COSH TANH (hyperbolic functions)	2403	414	~1.7 seconds ~1.7 seconds ~3.4 seconds	14-digit precision
INVSINH INVCOSH INVTANH (inverse hyperbolic functions)	2898	442	~1 second	8-digit precision

Table 4.1.1 *A summary of the programs appearing in Chapter 4.*

4.2 The Trigonometric Functions

The fundamental mechanics of the general CORDIC algorithm (as modified in this text) can be understood from a discussion of the sine and cosine function approximation case. As this may be your first exposure to the algorithm, the following discussion will be somewhat detailed.

The key concept in the algorithm can be visualized using figure 4.2.1. The vec-

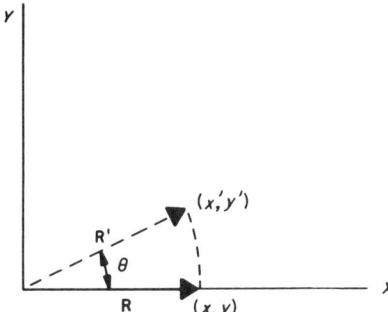

Figure 4.2.1 *The basic rotation used in the CORDIC approximation technique. The vector (x',y') is the result of rotating the vector (x,y) through an angle θ. In the particular example shown above, the original vector lies on the x axis.*

tor $\mathbf{R}'$ is the result of rotating the vector $\mathbf{R}$ through an angle θ. If the end coordinates of the vector $\mathbf{R}$ are (x,y), and those of $\mathbf{R}'$ are (x',y'), then the two vectors are related by the following equations:

$$x' = x \cos \theta - y \sin \theta = \cos \theta \, (x - y \tan \theta)$$
$$y' = y \cos \theta + x \sin \theta = \cos \theta \, (y + x \tan \theta) \quad (4.2.1)$$

To rotate $\mathbf{R}$ to a final angle A, a series of N rotations through the angles θ_i ($i = 1, 2, \ldots, N$) can be performed, and a sequence of equations similar to those shown above would result. The specific requirement on the θ_i values is that they must somehow add up to the total angular rotation desired:

$$A = w_1\theta_1 + w_2\theta_2 + \cdots + w_N\theta_N \quad \text{where} \quad w_i = +1 \text{ or } -1 \quad (4.2.2)$$

(The reason for the choice of $+1$ or -1 for w_i will be explained later.) In actuality, this requirement will not be satisfied exactly because only a finite number of terms is being used. However, the approximation can be very good, as we will now see.

We will restrict our attention to the angular range $-\pi/2 \leq A \leq \pi/2$ radians, or $-90°$ to $+90°$. Most trigonometric series approximations are limited to the range $-45°$ to $+45°$ for convergence reasons. The relation $\sin A = \cos(\pi/2 - A)$ is then employed to extend the results to the $-90°$ to $+90°$ range. This is not necessary for the CORDIC algorithm.

One acceptable choice for the θ_i values is

$$\theta_i = \pi/2^{i+1} \quad \text{for } i = 1, 2, ..., N \quad (4.2.3)$$

Another choice, and the one used in the binary CORDIC algorithm, is to select θ_i so that $\tan \theta_i$ is 2^{-i}. This makes equation (4.2.1) easy to implement with very fast binary shift and add/subtract operations (except for the multiplying cosine term which can be treated separately). However, this very significant speed advantage is lost when computing in BASIC. Instead, the problem of evaluating arctan 2^{-i} arises.

With the choice shown in equation (4.2.3), any angle in the range $-\pi/2 \leq A \leq \pi/2$ can be approximately represented as

$$A \cong \pi \sum_{i=1}^{N} \frac{w_i}{2^{i+1}} \quad (4.2.4)$$

The intrinsic accuracy of this approximation is in excess of $\pi/2^{N+1}$. As we will see, however, a simple addition to the algorithm can increase the accuracy to about $(\pi/2^{N+1})^4$.

The first step in the unabridged CORDIC procedure is to reduce the angle of interest to the range $-\pi/2$ to $\pi/2$, and then to determine the w_i values. The next step involves the repeated application of equation (4.2.1). For example, the $(i + 1)$th rotational step is

$$\begin{aligned} x_{i+1} &= C_i(x_i - T_i y_i) \\ y_{i+1} &= C_i(y_i + T_i x_i) \end{aligned} \quad (4.2.5)$$

where x_{i+1}, y_{i+1} = vector coordinates of the $(i + 1)$th rotated vector

x_i, y_i = vector coordinates of the ith rotated vector

$T_i = \tan w_i \theta_i$

$C_i = \cos w_i \theta_i$

After performing all the required rotations, we obtain

$$\begin{aligned} x_N &\cong |\mathbf{R'}| \cos A \\ y_N &\cong |\mathbf{R'}| \sin A \end{aligned}$$

If we choose $x_0 = 1$, $y_0 = 0$ and $|\mathbf{R'}| = 1$, we have

$$\cos A \cong x_N$$
$$\sin A \cong y_N \qquad (4.2.6)$$

Note that because $w_i \neq 0$ ($w_i = \pm 1$), *all* N rotations must be performed to achieve the final result. An alternative would be to use $w_i = 0$ or $+1$. You should assure yourself that equation (4.2.4) is well satisfied under that choice, although the range of approximation is limited to 0 to $\pi/2$. If the second alternative is used, the rotations corresponding to $w_i = 0$ need not be performed. At first thought, it would appear that this alternative would save some computer time. However, on the average, the first choice ($w_i = \pm 1$) actually reduces the total amount of computation. The reason is that because all rotations are employed, the multiplying cosine factor in equation (4.2.5) can be separated out as

$$P_N = \prod_{i=1}^{N} \cos w_i \theta_i = \prod_{i=1}^{N} \cos \theta_i \qquad (4.2.7)$$

The $w_i = \pm 1$ factor is removable because $\cos(-\theta) = \cos \theta$. For high accuracy values of P_N, see table 4.2.1. As you can surmise, the code is simplified and executes faster by removing the $\cos w_i \theta_i$ multiplier.

The algorithm can now be made more efficient and much more effective by using the following steps:

1) Determine the $w_i = \pm 1$ values given the angle A, and the defining equation (4.2.4).
2) Initialize two dummy variables, u_0 and v_0. We will account for the multiplying cosine constant by removing its effect at this stage:

$$u_0 = P_N$$
$$v_0 = 0$$

3) Ignore the $\cos \theta_i$ multiplier in equation (4.2.5). Recursively calculate equation (4.2.5), but use u_i and v_i instead of x_i and y_i. Note that $\tan w_i \theta_i = w_i \tan \theta_i$ because $w_i = \pm 1$:

$$u_{i+1} = u_i - w_i |T_i| v_i$$
$$v_{i+1} = v_i + w_i |T_i| u_i \qquad (4.2.8)$$

4) We could stop here and extract $\cos A \cong u_N$ and $\sin A \cong v_N$. However, we can account very closely for the inaccuracy of equation (4.2.4) by one last residual rotation through a small angle, z:

$$z = A - \pi \sum_{i=1}^{N} \frac{w_i}{2^{i+1}}$$

Therefore,
$$u_f = (1 - \frac{z^2}{2})[u_N + (z + \frac{z^3}{3})v_N] \quad (4.2.9)$$

$$v_f = (1 - \frac{z^2}{2})[v_N - (z + \frac{z^3}{3})u_N]$$

Note that we have used the idea that to a very high degree of approximation (because z is so small), $\cos z \cong 1 - z^2/2$ and $\tan z \cong z + z^3/3$. A flowchart representing the above procedure is shown in figure 4.2.2.

The addition of the residual rotation is a new twist in the basic CORDIC algorithm and dramatically improves the accuracy of the approximation. The resulting *relative* error is roughly

$$E \sim \frac{z^4}{4!} \quad (4.2.10)$$

For an N-step modified CORDIC sequence, $z \leq \theta_N = \pi/2^{N+1}$, which gives

$$E \lesssim \frac{(\pi/2^{N+1})^4}{24} \quad (4.2.11)$$

Table 4.2.2 gives estimates of the truncation error bound (not including round-off effects) up to $N = 15$. The accuracy is fairly impressive, but not exceptional when compared with the rapidly convergent sine and cosine Taylor series which, for $A \approx \pi/4$, require approximately eight terms. However, the strength of the CORDIC approach will become apparent in a later section in which inverse trigonometric functions are considered.

The BASIC program code for the modified CORDIC subroutine associated with this discussion is shown in listing 4.2.1. The subroutine-connection diagram and the functions/variables list appear in figure 4.2.3 and table 4.2.3, respectively. An example of the use of this subroutine is given in listing 4.2.2.

Listing 4.2.2 shows the sine and cosine values calculated by the CORDIC algorithm compared with the corresponding values generated by the BASIC interpreter (in this case, a 14-digit version of North Star BASIC). The comparison is very good. The differences observed are very likely attributable to round-off error, e.g., the last line in the table corresponds to the angle $\pi/2$. The true value of the cosine of this angle is 0. The subroutine missed by an amount roughly attributable to the number of digits of accuracy in the interpreter.

There are two precautions associated with using the modified CORDIC trigonometric subroutine. The first is the limit on the input angle, A. It must be in the range $-\pi/2 \leq A \leq \pi/2$, or an erroneous result will be returned. The second is that because the w_i weights can take on both positive and negative values, round-off error will occur at a level that may be significantly larger than the value associated with the number of digits accuracy of the BASIC interpreter employed, and thus may exceed that expected from table 4.2.2. In general, the *absolute* error will be less than roughly 10^{-M+1}, where M is the number of digits carried by the interpreter, if N

is sufficiently large. Otherwise, use table 4.2.2.

In the next section, we will discuss how the values in table 4.2.1 were generated and how the range of that table can be extended to even higher approximation levels.

i or N	θ_i (radians)	Tan θ_i	P_N
1	0.7853981633974484	1	0.7071067811865476
2	0.3926990816987242	0.414213562373095	0.6532814824381884
3	0.1963495408493621	0.198912367379658	0.6407288619353766
4	0.09817477042468105	0.09849140335716425	0.6376435773361456
5	0.04908738521234052	0.04912684976946725	0.6368755077217537
6	0.02454369260617026	0.02454862210892544	0.6366836927259824
7	0.01227184630308513	0.01227246237956627	0.6366357516148735
8	6.135923151542565D-03	6.136000157623401D-03	0.6366237671267634
9	3.067961575771283D-03	3.067971201422665D-03	0.6366207710540869
10	1.533980787885641D-03	1.533981991088666D-03	0.6366200220390022
11	7.669903939428207D-04	7.669905443430926D-04	0.6366198347854238
12	3.834951969714103D-04	3.834952157714410D-04	0.6366197879720413
13	1.917475984857052D-04	1.917476008357089D-04	0.6366197762686964
14	9.587379924285259D-05	9.587379953660303D-05	0.6366197733428602
15	4.793689962142629D-05	4.793689965814509D-05	0.6366197726114012
16	2.396844981071315D-05	2.396844981530299D-05	0.6366197724285364
17	1.198422490535657D-05	1.198422490593030D-05	0.6366197723828202
18	5.992112452678286D-06	5.992112452750002D-06	0.6366197723713912
19	2.996056226339143D-06	2.996056226348107D-06	0.6366197723685339
20	1.498028113169572D-06	1.498028113170692D-06	0.6366197723678196
21	7.490140565847858D-07	7.490140565849257D-07	0.6366197723676411
22	3.745070282923929D-07	3.745070282924103D-07	0.6366197723675964
23	1.872535141461964D-07	1.872535141461986D-07	0.6366197723675853
24	9.362675707309822D-08	9.362675707309848D-08	0.6366197723675825
25	4.681337853654911D-08	4.681337853654914D-08	0.6366197723675818
26	2.340668926827456D-08	2.340668926827456D-08	0.6366197723675816

Table 4.2.1 *Fifteen-digit accuracy tables for* tan θ_i *and* P_N *using* $\theta_i = \pi/2^{i+1}$. *Note that* P_N *converges towards a constant as N increases. A program for determining these values is given in the next section.*

238 BASIC SCIENTIFIC SUBROUTINES

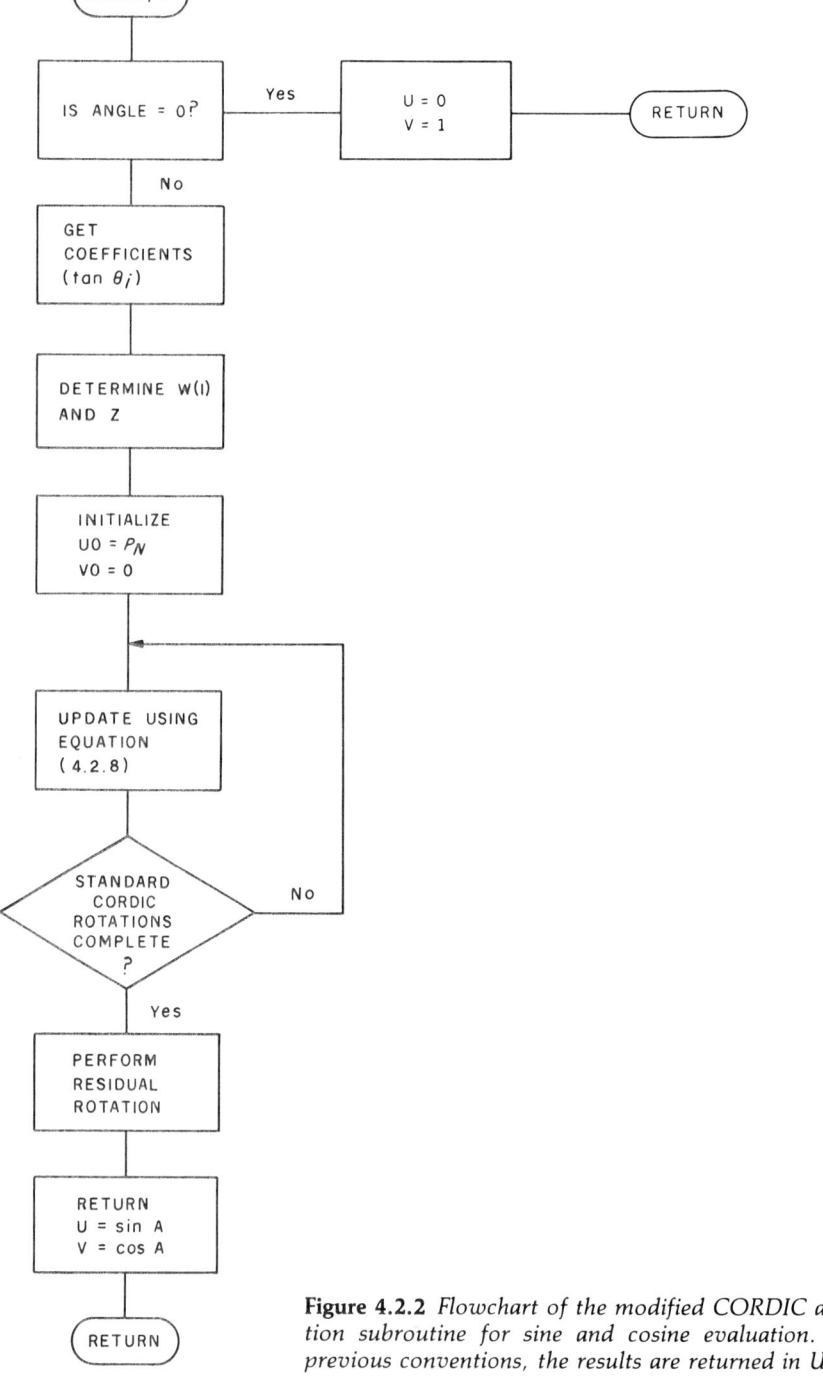

Figure 4.2.2 *Flowchart of the modified CORDIC approximation subroutine for sine and cosine evaluation. To retain previous conventions, the results are returned in U and V.*

N	Error Estimate
1	2×10^{-2}
2	1×10^{-3}
3	6×10^{-4}
4	4×10^{-6}
5	2×10^{-7}
6	2×10^{-8}
7	1×10^{-9}
8	6×10^{-11}
9	4×10^{-12}
10	2×10^{-13}
11	1×10^{-14}
12	1×10^{-15}
13	6×10^{-17}
14	4×10^{-18}
15	2×10^{-19}

Table 4.2.2 *Estimate of the error bound for the modified sine/cosine CORDIC algorithm given in the text as a function of N, the number of angular rotations. This estimate is an upper bound on truncation error, and does not include the effects of round-off inaccuracies.*

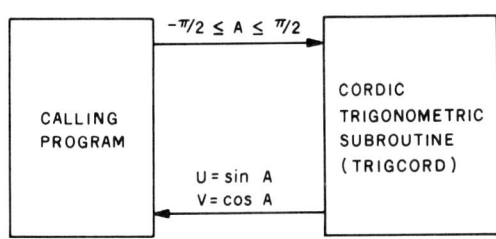

Figure 4.2.3 *Subroutine-connection diagram for the modified trigonometric CORDIC subroutine shown in listing 4.2.1 (see also table 4.2.3).*

240 BASIC SCIENTIFIC SUBROUTINES

Statements/Functions List

$+, -, <, >, *, /$
FOR/NEXT, GOSUB, IF/THEN

Variables List

A, A0, A(I), I, N, P, U, U0, V0, W(I), Z

Variable Passed to Subroutine

A

Table 4.2.3 *Functions and variables used in the CORDIC trigonometric subroutine (TRIGCORD).*

Listing 4.2.1

```
3350 REM PROGRAM TO DEMONSTRATE USE OF THE CORDIC
3351 REM TRIGONOMETRIC APPROXIMATION SUBROUTINE.
3352 PRINT
3353 PRINT
3354 REM PROGRAM PRINTS OUT 21 VALUES OF SINE AND
3355 REM COSINE AND COMPARES THE RESULT WITH VALUES
3356 REM CALCULATED INTERNALLY.
3357 DIM A(12),W(12)
3358 B=3.141592653589793238
3359 PRINT "   ANGLE (RADIANS)",TAB(24),"CALCULATED SINE",TAB(40),
3360 PRINT "      DIFFERENCE",TAB(67),"CALCULATED COSINE",TAB(84),"   DIFFERENCE"
3361 PRINT "   ------------------",TAB(24),"------------------",TAB(40),
3362 PRINT "      ----------",TAB(67),"------------------",TAB(84),"   ----------"
3363 PRINT
3364 B=.025*B
3365 M=1E14
3366 FOR J=0 TO 20
3367 A=J*B
3368 GOSUB 45475
3369 PRINT INT(M*A)/M,TAB(23),INT(M*U)/M,TAB(44),
3370 PRINT INT((U-SIN(A))*M)/M,TAB(66),INT(M*V)/M,TAB(88),INT((V-COS(A))*M)/M
3371 NEXT J
3372 PRINT
3373 PRINT
3374 END
45467 REM ********************
45468 REM TRIGONOMETRIC CORDIC SUBROUTINE (TRIGCORD)
45469 REM THIS SUBROUTINE CALCULATES THE SINE AND COSINE
45470 REM OF AN ANGLE USING THE CORDIC ROTATION METHOD.
45471 REM THE INPUT ANGLE IS A.
45472 REM THE SINE IS RETURNED IN U, AND THE COSINE IN V.
```

```
45473 REM REMEMBER TO DIMENSION W(I) AND A(I) IN THE CALLING PROGRAM
45474 REM IF THE ANGLE IS ZERO, SET FUNCTIONS AND RETURN.
45475 IF A<>0 THEN GOTO 45480
45476 U=0
45477 V=1
45478 RETURN
45479 REM GET THE TANGENT COEFFICIENTS
45480 GOSUB 45518
45481 REM REM DETERMINE THE WEIGHTS, W(I)1060
45482 GOSUB 45506
45483 U0=P
45484 V0=0
45485 REM PERFORM THE ROTATIONS UP TO THE RESIDUAL
45486 FOR I=1 TO N
45487 REM UPDATE U0 AND V0
45488 U1=U0-W(I)*A(I)*V0
45489 V1=V0+W(I)*A(I)*U0
45490 U0=U1
45491 V0=V1
45492 NEXT I
45493 REM PERFORM THE RESIDUAL ROTATION USING Z
45494 REM USE U0 AND V0 AS DUMMY VARIABLES
45495 U0=1-Z*Z/2
45496 V0=Z*(1+Z*Z/3)
45497 REM U AND V ARE THE FINAL RESULTS
45498 V=U0*(U1-V0*V1)
45499 U=U0*(V1+V0*U1)
45500 REM U=SIN(A)
45501 REM V=COS(A)
45502 RETURN
45503 REM TRIG CORDIC WEIGHTS SUBROUTINE
45504 REM THE WEIGHTS ARE W(I)= PLUS OR MINUS 1
45505 REM THE INPUT ANGLE IS A
45506 Z=A
45507 FOR I=1 TO N
45508 W(I)=-1
45509 IF Z>0 THEN W(I)=1
45510 Z=Z-W(I)*A0
45511 A0=A0/2
45512 NEXT I
45513 REM Z IS THE RESIDUAL ANGLE
45514 REM ********************
45515 RETURN
45516 REM TRIG CORDIC COEFFICIENT SUBROUTINE
45517 REM FOR CASE N=12
45518 N=12
45519 REM THE TANGENTS ARE GIVEN IN A(I)
45520 A(1)=1
45521 A(2)=.414213562373095
```

Listing 4.2.1 cont.

```
45522 A(3)=.198912367379658
45523 A(4)=.09849140335716425
45524 A(5)=.04912684976946725
45525 A(6)=.02454862210892544
45526 A(7)=.01227246237956627
45527 A(8)=.006136000157623401
45528 A(9)=.003067971201422665
45529 A(10)=.001533981991088666
45530 A(11)=.0007669905443430926
45531 A(12)=.000383495215771441
45532 REM P REPRESENTS P(N)
45533 P=.6366197879720413
45534 REM A0 IS PI/4
45535 A0=.7853981633974484
45536 RETURN
READY
```

Listing 4.2.1 *The modified CORDIC trigonometric subroutine (TRIGCORD). The case N = 12 is shown. For higher accuracy calculations, N may be increased and more table values [tan θ_i = A(I)] added. Note that if N is increased, P_N must also be changed (see table 4.2.1). In addition, a program that demonstrates the operation of this subroutine is included (see listing 4.2.2).*

RUN

ANGLE (RADIANS)	CALCULATED SINE	DIFFERENCE	CALCULATED COSINE	DIFFERENCE
0	0	0	1	0
.07853981633974	.07845909572785	.00000000000001	.9969173337331	0
.15707963267949	.15643344650024	.00000000000001	.98768834059519	.00000000000009
.23561944901924	.23344536385593	.00000000000002	.97236992039773	.00000000000003
.31415926535898	.30901699437495	0	.95105651629513	.00000000000003
.39269908169873	.3826834323651	.00000000000001	.92387953251127	-.00000000000003
.47123889803847	.45399049973953	-.00000000000002	.89100652418833	-.00000000000007
.54977871437822	.52249856471596	.00000000000001	.85264016435413	.00000000000003
.62831853071796	.58778525229249	.00000000000002	.80901699437499	.00000000000007
.70685834705771	.64944804833015	-.00000000000004	.76040596560001	-.00000000000001
.78539816339745	.70710678118653	-.00000000000002	.70710678118655	.00000000000004
.8639379797372	.76040596560002	-.00000000000001	.64944804833015	-.00000000000002
.94247779607694	.80901699437499	.00000000000004	.58778525229249	-.00000000000001
1.0210176124167	.85264016435411	.00000000000001	.52249856471598	.00000000000005
1.0995574287564	.89100652418829	-.00000000000011	.4539904997396	.00000000000003
1.1780972450962	.92387953251126	-.00000000000004	.38268343236512	.00000000000006
1.2566370614359	.95105651629511	.00000000000001	.30901699437502	.00000000000006
1.3351768777757	.97236992039772	.00000000000002	.23344536385594	.00000000000008
1.4137166941154	.98768834059518	.00000000000008	.1564344650403	.00000000000007
1.4922565104552	.9969173337331	0	.07845909572785	.00000000000006
1.5707963267949	.99999999999999	0	.00000000000006	.00000000000007

READY

Listing 4.2.2 *An example of the operation of the CORDIC trigonometric subroutine. See the demonstration program shown in listing 4.2.1. A 14-digit version of North Star BASIC was employed for the calculations.*

4.3 Generating the T_i and P_N Coefficients

In the previous section, we employed in the CORDIC trigonometric function subroutine a set of 15-digit accuracy T_i and P_N coefficients. In this section, we will consider a numerical means for generating these coefficients which does not require any mathematical operations beyond the simple square-root function. This function itself can be replaced with a sequence of additions and divisions. The discussion also provides an example of the precautions that must often be taken when facing potential round-off error problems.

Recall that the θ_i values were given by equation (4.2.3):

$$\theta_i = \pi/2^{i+1} \quad \text{for} \quad i = 1, 2, \ldots, N$$

To apply the CORDIC algorithm, we require the coefficients

$$T_i = \tan \theta_i \quad \text{and} \quad P_N = \prod_{i=1}^{N} \cos \theta_i = \prod_{i=1}^{N} C_i$$

Fortunately, we have the half-angle formulae

$$\cos \frac{\theta}{2} = \left(\frac{1 + \cos \theta}{2}\right)^{1/2}$$

$$\tan \frac{\theta}{2} = \left(\frac{1 - \cos \theta}{1 + \cos \theta}\right)^{1/2} \quad (4.3.1)$$

Because $\theta_1 = \pi/4$, two values are well known:

$$C_1 = \frac{1}{\sqrt{2}}$$

$$T_1 = 1$$

At this point, we might be tempted to immediately apply the half-angle formulae recursively to calculate C_2, T_2, and so on. However, the term $(1 - \cos \theta)$ in equation (4.3.1) is bound to cause round-off error when θ_i becomes small and $\cos \theta_i$ approaches unity. A means to circumvent this difficulty must be found. One approach is to define $G(\theta) = 1 - \cos \theta$ and to find a way to recursively determine $G(\theta_i)$ directly. The steps given below achieve this:

1) Initialize: $C_1 = 1/\sqrt{2} = 0.7071067811865475$
   $T_1 = 1$
   $G_1 = 1 - C_1$

2) Calculate the next C_i:
   $$C_{i+1} = \left[(1 + C_i)/2\right]^{1/2}$$

3) Calculate the intermediate factor, G_{i+1}, based on G_i and C_{i+1}:
$$G_{i+1} = G_i/[2(1 + C_{i+1})]$$

4) Calculate the next T_i based on G_i and C_i:
$$T_{i+1} = [G_i/(1 + C_i)]^{1/2}$$

5) Return to step 2.

This sequence of operations was used to generate the coefficients shown in table 4.2.1 (see listing 4.3.1). Algebraically-minded readers might try as an exercise to show that the steps listed above are algebraically (though not numerically) equivalent to using the half-angle formulae.

```
3375 REM CORDIC TRIGONOMETRIC COEFFICIENT DETERMINATION PROGRAM
3376 REM WRITTEN IN DOUBLE PRECISION MICROSOFT BASIC
3377 DIM C#(51),T#(51),G#(51)
3378 C#(0)=.7071067811865476#
3379 T#(0)=1#
3380 G#(0)=1#-C#(0)
3381 P#=1#
3382 FOR I=0 TO 50
3383 P#=P#*C#(I)
3384 PRINT I, C#(I),T#(I),P#
3385 Y#=(1#+C#(I))/2#
3386 GOSUB 3395
3387 C#(I+1)=X#
3388 G#(I+1)=G#(I)/(2*(1+C#(I+1)))
3389 Y#=G#(I)/(1#+C#(I))
3390 GOSUB 3395
3391 T#(I+1)=X#
3392 NEXT I
3393 END
3394 REM SQUARE ROOT DETERMINATION SUBROUTINE
3395 X#=1
3396 X1#=X#
3397 X#=(X#+Y#/X#)/2
3398 IF ABS((X#-X1#)/X#)<1E-15 THEN RETURN
3399 GOTO 3396
```

Listing 4.3.1 *Program used to generate the P_N and $\tan \theta_i$ values shown in table 4.2.1. This program was written in a double-precision version of Microsoft BASIC and is not part of the subroutine library. Note that the square root is determined within the program itself.*

4.4 The Inverse Trigonometric Functions

In this section, the previously discussed CORDIC technique will be altered to provide the inverse trigonometric functions of arcsine, arccosine and arctangent. This alteration is accomplished by using the strategy of rotating the vector **R** (see figure 4.4.1) until an adequate approximation to the given sine, cosine, or tangent is achieved. The w_i weights required to achieve this rotation are then used to closely estimate the required angle by employing equation (4.2.4). This angle represents the value of the inverse function.

A flowchart for this algorithm is shown in figure 4.4.2. The subroutine accepts three inputs: U, V, and W. U corresponds to the sine of the desired angle, V to the cosine, and W to the tangent. If U, V, and W are all zero, the subroutine immediately returns a value of zero. Otherwise, the program sorts through the values to determine the desired inverse function. Table 4.4.1 shows the possible (U, V, W) combinations and how they are interpreted.

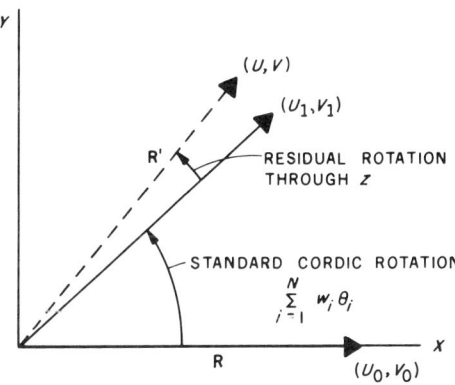

Figure 4.4.1 *Scheme for determining the angle corresponding to a given sine, cosine, or tangent. Starting at 0°, the vector* **R** *is sequentially rotated through the angles* $w_i\theta_i$ ($= \pm\theta_i$), *and finally z, until the sine of the desired angle is achieved, as indicated by* **R'**. *As the required w_i values and z are known, the angle may therefore be determined.*

Once the function is established, U and V are used as the coordinates of the final rotated vector, **R'** (see figure 4.4.1). Next, U and V are checked for the special cases that lead to returned values of $\pm \pi/2$. Following this, the vector **R** is initialized and rotated N steps until a vector having coordinates $(U1, V1)$ is obtained. At this point, equation (4.2.4) gives a reasonably close approximation to the angular rotation required. However, the result is in error by a small angle, z. If we could perform a residual rotation through the angle z, the rotation could be completed. Unfortunately, z is actually an unknown part of what we want to determine. But we can still write the associated equations

$$U = U1 \cos z - V1 \sin z$$
$$V = V1 \cos z + U1 \sin z \qquad (4.4.1)$$

These equations can be solved to give

$$\sin z = \frac{U1 V - V1 U}{U1^2 + V1^2} \qquad (4.4.2)$$

Noting that $U1^2 + V1^2 = 1$, and expanding $\sin z$ in a Maclaurin series, we get

$$z - \frac{z^3}{3!} \cong U1 V - V1 U \qquad (4.4.3)$$

In the spirit of approximation, this last equation can be numerically calculated in two steps:

$$\begin{aligned} 1)\ & z \leftarrow U1 V - V1 U \\ 2)\ & z \leftarrow z + \frac{z^3}{3!} \end{aligned} \qquad (4.4.4)$$

The second step uses the estimate for z obtained in the first step to correct itself. The final angle of rotation is then

$$A = z + \sum_{i=1}^{N} w_i \theta_i$$

From the Maclaurin-series approximation used in equation (4.4.3), we might estimate the error to be roughly $z^5/5!$. However, the second step of equation (4.4.4) introduces an approximation error of magnitude $z^5/12$, which is larger. Following the logic of section 4.2, the approximation error should therefore be

$$E \lesssim \frac{(\pi/2^{n+1})^5}{12} \qquad (4.4.5)$$

Table 4.4.3 shows the approximation error bound calculated using equation (4.4.5). As you can see, the potential truncation error properties of this approximation scheme are exceptional. Unfortunately, there is a strong potential for significant round-off error. This is due to the form of the first step in equation (4.4.4)—taking the difference between two nearly equal numbers. In practice, however, the results are good. The sample runs shown in listing 4.4.2, for example, indicate round-off error to be insignificant.

Other sources of round-off error were directly avoided in the program code. The computational form $(1 + E)^{1/2}$ is encountered in four locations. When E is very small, the square-root function in BASIC can experience difficulties mainly related to round-off error. This problem was circumvented by using series approximations in the danger region.

The utility of the modified CORDIC technique is very apparent in the inverse trigonometric function application. For the inverse sine and cosine cases, any argument in the range from -1 to $+1$ can be used. For the inverse tangent, the acceptable range is from minus infinity to plus infinity (except for the obvious range limitation in the computer). With 12 rotations plus a final extrapolation, 18-digit accuracy is, in principle, possible. If we apply Maclaurin series approximations instead, the series diverges (or at least converges very slowly) for input arguments of unity. Otherwise, they often require a great many terms to be included before any significant truncation accuracy is achieved, and by then round-off error has taken its toll. The least-squares and min-max series are better in this respect, but the CORDIC method is still superior.

There are two precautions involved in using the modified CORDIC subroutine discussed in this section. First, the input value must be encoded according to table 4.4.1. Second, the input value must be in the following range:

$$-1 \leq U \leq 1$$
$$-1 \leq V \leq 1$$
$$-(\text{largest number})^{1/4} < W < (\text{largest number})^{1/4}$$

Because W^4 is used in the program, an overflow can occur if W is too large.

As with the CORDIC subroutine discussed in section 4.2, the truncation accuracy of the program can be increased by using a larger number of rotations. However, round-off error should be carefully evaluated to ensure true accuracy.

Function Assumed	Inputs		
	U	V	W
None: returns 0	0	0	0
arcsin U	$U \neq 0$	anything	anything
arccos V	$U = 0$	$V \neq 0$	anything
arcsin W	$U = 0$	$V = 0$	$W \neq 0$

Table 4.4.1 *Protocol table for the inverse trigonometric CORDIC subroutine.*

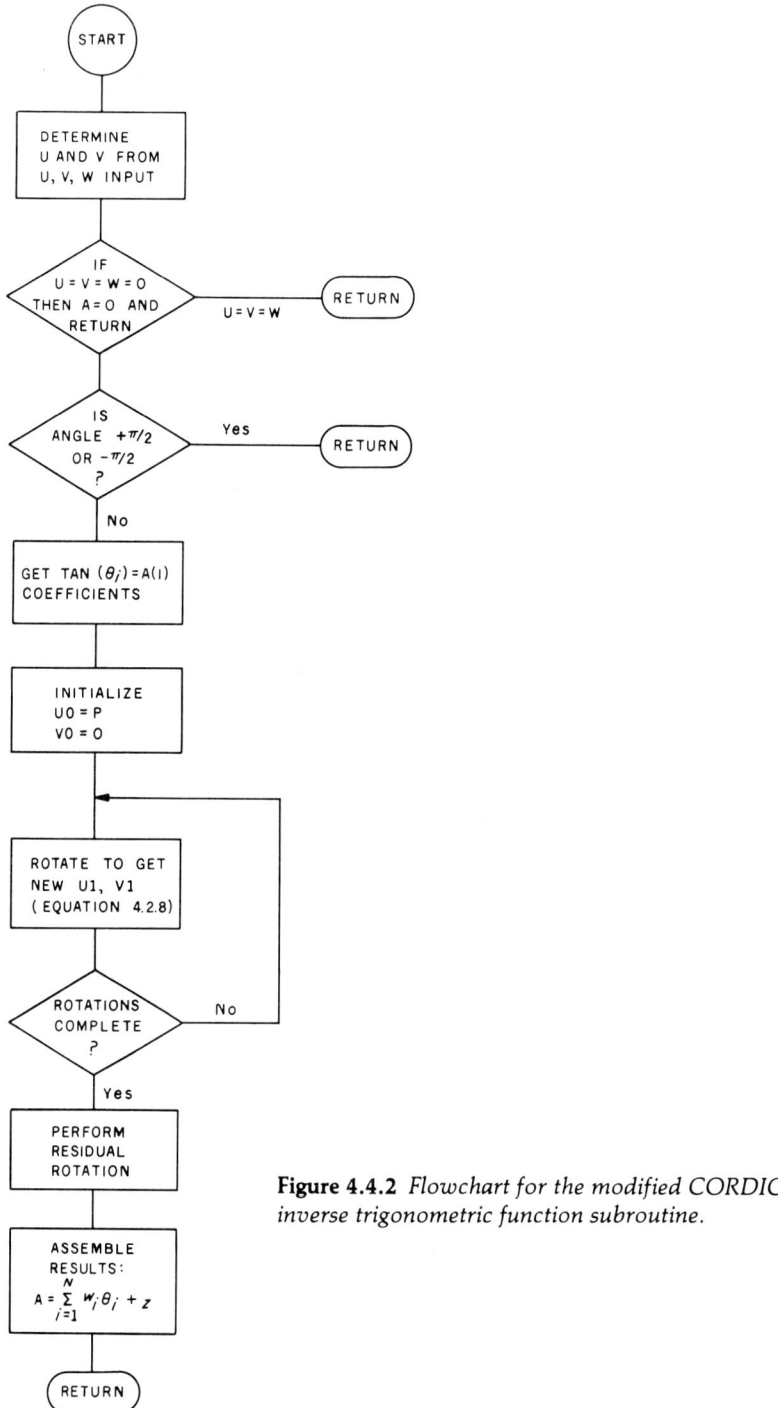

Figure 4.4.2 *Flowchart for the modified CORDIC inverse trigonometric function subroutine.*

CORDIC APPROXIMATION TECHNIQUES AND ALTERNATIVES

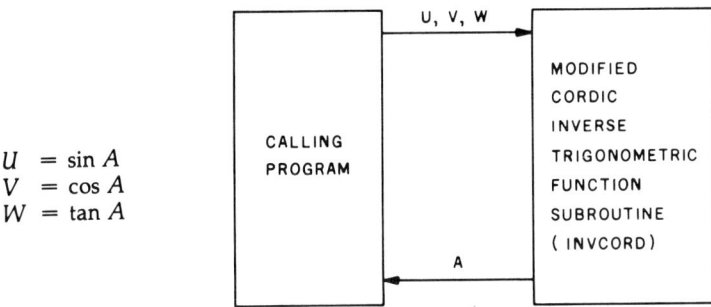

$U = \sin A$
$V = \cos A$
$W = \tan A$

Figure 4.4.3 *Subroutine-connection diagram for the modified CORDIC inverse trigonometric function subroutine. See also table 4.4.1 for the U, V, and W protocol.*

Statements/Functions List

$+,-,<,>,*,/$
FOR/NEXT, GOSUB, IF/THEN

Variables List

A, A0, A(I), I, N, P, U, U0, V, V0, W, W(I), Z

Variables Passed to Subroutine

U, V, W

Table 4.4.2 *Functions and variables used in the modified CORDIC inverse trigonometric function subroutine given in this section.*

```
3425 REM PROGRAM TO DEMONSTRATE THE USE OF THE INVERSE
3426 REM TRIGONOMETRIC CORDIC SUBROUTINE
3427 PRINT
3428 PRINT
3429 PRINT "DO YOU WANT THE INVERSE SINE (S)"
3430 DIM A(12),W(12)
3431 PRINT "                        COSINE (C)"
3432 PRINT "                        OR TANGENT (T): ",
3433 INPUT A$
3434 PRINT
3435 PRINT "WHAT IS THE VALUE: ",
3436 INPUT Z1
3437 U=0
3438 V=0
```

Listing 4.4.1

```
3439 W=0
3440 IF A$="S" THEN U=Z1
3441 IF A$="C" THEN V=Z1
3442 IF A$="T" THEN W=Z1
3443 REM GET RESULT
3444 GOSUB 45550
3445 PRINT
3446 PRINT
3447 IF A$="S" THEN PRINT "ARCSIN(",
3448 IF A$="C" THEN PRINT "ARCCOS(",
3449 IF A$="T" THEN PRINT "ARCTAN(",
3450 PRINT Z1,") = ",A," RADIANS"
3451 PRINT
3452 PRINT
3453 END
45541 REM ********************
45542 REM INVERSE TRIGONOMETRIC CORDIC SUBROUTINE (INVCORD).
45543 REM THIS SUBROUTINE CALCULATES THE ANGLE CORRESPONDING
45544 REM TO A GIVEN SINE, COSINE OR TANGENT USING THE
45545 REM CORDIC ROTATION METHOD.
45546 REM THE INPUT IS U=SIN(A), V=COS(A), OR W=TAN(A).
45547 REM THE RETURNED VALUE IS A.
45548 REM REMEMBER TO DIMENSION W(I) AND A(I) IN THE CALLING PROGRAM
45549 REM TRANSLATE THE U,V,W INPUTS
45550 IF U=0 THEN GOTO 45555
45551 REM INVERSE SINE IS WANTED
45552 IF ABS(U)>=.0001 THEN V=SQRT(1-U*U)
45553 IF ABS(U)<.0001 THEN V=1-U*U/2-U*U*U*U/8
45554 GOTO 45571
45555 IF V=0 THEN GOTO 45560
45556 REM INVERSE COSINE IS WANTED
45557 IF ABS(V)>=.0001 THEN U=SQRT(1-V*V)
45558 IF ABS(V)<.0001 THEN U=1-V*V/2-V*V*V*V/8
45559 GOTO 45571
45560 IF W=0 THEN GOTO 45568
45561 REM INVERSE TANGENT IS WANTED
45562 IF ABS(W)<=10000 THEN U=1/SQRT(1+1/(W*W))
45563 IF ABS(W)>10000 THEN U=1-1/(2*W*W)+3/(8*W*W*W*W)
45564 IF ABS(U)>=.0001 THEN V=SQRT(1-U*U)
45565 IF ABS(U)<.0001 THEN V=1-U*U/2-U*U*U*U/8
45566 U=U*ABS(W)/W
45567 GOTO 45571
45568 A=0
45569 RETURN
45570 REM GET COEFFICIENTS
45571 GOSUB 45616
45572 REM TEST FOR SPECIAL VALUES
45573 IF ABS(U)<1 THEN GOTO 45578
45574 IF ABS(V)>0 THEN GOTO 45578
```

Listing 4.4.1 cont.

Listing 4.4.1 cont.

```
45575 REM SPECIAL CASE FOUND
45576 A=2*A0*ABS(U)/U
45577 RETURN
45578 IF ABS(V)<1 THEN GOTO 45583
45579 IF ABS(U)>0 THEN GOTO 45583
45580 A=0
45581 RETURN
45582 REM SWITCH U WITH V AND INITIALIZE
45583 A=U
45584 U=V
45585 V=A
45586 U0=P
45587 U1=U0
45588 V0=0
45589 V1=V0
45590 REM PERFORM THE ROTATIONS UP TO THE RESIDUAL
45591 FOR I=1 TO N
45592 REM IS ROTATION TO BE PLUS OR MINUS?
45593 W(I)=-1
45594 IF V0<V THEN W(I)=1
45595 REM UPDATE U0 AND V0
45596 U1=U0-W(I)*A(I)*V0
45597 V1=V0+W(I)*A(I)*U0
45598 U0=U1
45599 V0=V1
45600 NEXT I
45601 REM THE SET OF W(I) WEIGHTS HAVE NOW BEEN DETERMINED
45602 REM PERFORM THE RESIDUAL ANGLE APPROXIMATION
45603 Z=V*U1-U*V1
45604 Z=Z+Z*Z*Z/6
45605 REM ASSEMBLE RESULTS
45606 FOR I=1 TO N
45607 Z=Z+W(I)*A0
45608 A0=A0/2
45609 NEXT I
45610 REM RESULT IS IN Z
45611 A=Z
45612 RETURN
45613 REM ********************
45614 REM TRIG CORDIC COEFFICIENT SUBROUTINE
45615 REM FOR CASE N=12
45616 N=12
45617 REM THE TANGENTS ARE GIVEN IN A(I)
45618 A(1)=1
45619 A(2)=.414213562373095
45620 A(3)=.198912367379658
45621 A(4)=.09849140335716425
45622 A(5)=.04912684976946725
45623 A(6)=.02454862210892544
```

```
45624 A(7)=.01227246237956627
45625 A(8)=.006136000157623401
45626 A(9)=.003067971201422665
45627 A(10)=.001533981991088666
45628 A(11)=.0007669905443430926
45629 A(12)=.0003834952157714410
45630 REM P REPRESENTS P(N)
45631 P=.6366197879720413
45632 REM A0 IS PI/4
45633 A0=.7853981633974484
45634 RETURN
```

Listing 4.4.1 *The modified CORDIC inverse trigonometric function subroutine (INVCORD). The inputs are U, V, and W. The output is A = arcsin U, arccos V, or arctan W. The subroutine-connection diagram is shown in figure 4.4.3. See also table 4.4.1 and 4.4.2. Also shown is a demonstration program. See listing 4.4.2 for a sample run.*

Listing 4.4.2

```
RUN


DO YOU WANT THE INVERSE SINE (S)
                        COSINE (C)
             OR TANGENT (T): ?T

WHAT IS THE VALUE: ?1


ARCTAN( 1) =   .7853981633974 RADIANS


READY
RUN


DO YOU WANT THE INVERSE SINE (S)
                        COSINE (C)
             OR TANGENT (T): ?T

WHAT IS THE VALUE: ?-1


ARCTAN( -1) =  -.7853981633974 RADIANS


READY
RUN
```

```
DO YOU WANT THE INVERSE SINE (S)
                      COSINE (C)
               OR TANGENT (T): ?T

WHAT IS THE VALUE: ?.0000001

ARCTAN( .0000001) =  9.999998435E-08 RADIANS


READY
RUN



DO YOU WANT THE INVERSE SINE (S)
                      COSINE (C)
               OR TANGENT (T): ?S

WHAT IS THE VALUE: ?.0000001

ARCSIN( .0000001) =  9.999998435E-08 RADIANS


READY
RUN



DO YOU WANT THE INVERSE SINE (S)
                      COSINE (C)
               OR TANGENT (T): ?C

WHAT IS THE VALUE: ?.0000001

ARCOS( .0000001) =  1.5707962267949 RADIANS


READY
```

Listing 4.4.2 *Sample runs using the program shown in listing 4.4.1. Note that small values represent stress cases for the calculation in terms of possible round-off error. A 14-digit version of North Star BASIC was employed for the calculations.*

N	Error Estimate
1	2×10^{-2}
2	8×10^{-4}
3	2×10^{-5}
4	8×10^{-7}
5	2×10^{-8}
6	7×10^{-10}
7	2×10^{-11}
8	7×10^{-13}
9	2×10^{-14}
10	7×10^{-16}
11	2×10^{-17}
12	7×10^{-19}
13	2×10^{-20}
14	7×10^{-22}
15	2×10^{-23}

Table 4.4.3 *Estimate of the truncation error bound for the modified CORDIC inverse trigonometric function algorithm given in the text as a function of N, the number of rotations. The numbers shown below are upper bounds on the truncation error. The round-off error is likely to be much larger than the truncation error for the large N values.*

4.5 The Exponential Function

The concept of the modified CORDIC method can be extended to the approximation of the exponential function, e^x. In this case, the associated algorithm is simple and very easy to understand.

As with the CORDIC evaluation of the trigonometric functions, applying this method to the exponential function *in BASIC* does not lead to great advantages in speed and accuracy when compared with the Maclaurin series approximation. In this case, though, the difference between the two is not great. As before, the computational advantage appears with the inverse function, which for the exponential is the natural logarithm.

The algorithm developed in this section is *not* the same as the CORDIC method discussed in the previously cited references. The reason for heading in a new direction is that the CORDIC technique for calculating the exponential as presented in the literature loses all its speed and computational advantages when implemented in BASIC. The method treated below attempts to combine the advantages of the rapidly convergent Maclaurin series for e^x and the CORDIC concept to provide a useful and competitive subroutine.

We will restrict our attention to the range $0 \le x < 1$. A simple later addition to the algorithm will take care of arguments outside this range.

For reasons that will become apparent shortly, we can write e^x in a factored form:

$$e^x = e^{c_1}e^{c_2}e^{c_2}e^{c_3} \cdots e^{c_N} \quad (4.5.1)$$

where
$$x = c_1 + c_2 + c_3 + \cdots + c_N \quad (4.5.2)$$

The important property to notice in equation (4.5.1) is that e^x is expressed as a series of products. We can take advantage of this by observing that x can be approximated in the range $0 \le x < 1$ by

$$x \cong \sum_{i=1}^{N} \frac{W_i}{2^i} \quad \text{where} \quad W_i = 0 \text{ or } +1 \quad (4.5.3)$$

[You will find this approach to be very similar to that discussed in Chapter 3 for GENROOT $(X=Y^N)$]. We now define the following coefficient:

$$A_i = e^{1/2^i} \quad (4.5.4)$$

e^x can then be approximated as

$$e^x \cong A_1^{W_1} A_2^{W_2} \cdots A_n^{W_n} = \prod_{i=1}^{N} A_i^{W_i} \quad (4.5.5)$$

Note that if $W_i = 1$, $A_i^{W_i} = A_i$; if $W_i = 0$, $A_i^{W_i} = 1$. The problem is therefore reduced to finding the W_i weights according to equation (4.5.3), an easy numerical task.

A flowchart for the complete algorithm is shown in figure 4.5.1. The first step is to get N (the number of "rotations" to be performed), obtain $E = e$, and transfer the coefficients defined by equation (4.5.4). See table 4.5.1. As you will see shortly, choosing $N = 9$ gives roughly 16-digit truncation accuracy.

The next step is to take the input value, X, and separate it into integer (K) and noninteger ($X - K$) parts. X is thereby reduced to the range $0 \le X < 1$. The W_i coefficients are then determined using equation (4.5.3). However, there is a "residual" exponent of the form

$$Z = X - \sum_{i=1}^{N} \frac{W_i}{2^i} \quad (4.5.6)$$

A much better approximation to e^x than that given in equation (4.5.5) can be constructed using this residual:

$$e^x = e^z + \prod_{i=1}^{N} A_i^{W_i} \quad (4.5.7)$$

Therefore, after e^x is approximated by the N products, an extra multiplication by e^z should be performed. Because Z is small, a series approximation to e^z can be used:

$$e^Z \cong 1 + Z + \frac{Z^2}{2!} + \frac{Z^3}{3!} + \frac{Z^4}{4!} \qquad (4.5.8)$$

The final relative error due to the truncation of the series is then of the order $Z^5/5!$. Because $|Z| < 1/2^{N+1}$, the truncation-error bound is

$$E < \frac{(1/2^{N+1})^5}{5!} \qquad (4.5.9)$$

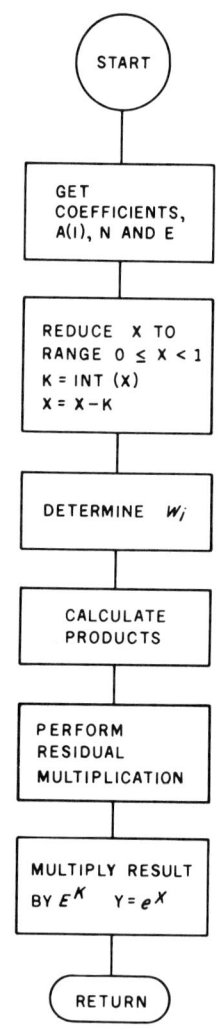

Figure 4.5.1 *Flowchart for the exponential subroutine (EXPCORD).*

i	A_i
1	0.648721270700128
2	1.284025416687742
3	1.133148453066826
4	1.064494458917859
5	1.031743407499103
6	1.015747708586686
7	1.007843097206448
8	1.003913889338348
9	1.001955033591003
10	1.000977039492417
11	1.000488400478694
12	1.000244170429748
13	1.000122077763384
14	1.000061037018933
15	1.000030518043791

Table 4.5.1 A_i coefficients for the following exponential approximation subroutine: $A_i = e^{1/2^i}$.

N	Error Estimate	Approximate Number of Maclaurin series Terms Required
1	3×10^{-4}	7
2	8×10^{-6}	9
3	3×10^{-7}	10
4	8×10^{-9}	12
5	2×10^{-10}	13
6	8×10^{-12}	14
7	2×10^{-13}	15
8	8×10^{-15}	17
9	2×10^{-16}	18
10	7×10^{-18}	19
11	2×10^{-19}	20
12	7×10^{-21}	21

Table 4.5.2 Error associated with the modified "CORDIC" algorithm for determing e^x. Observe that this is an estimate of the upper bound on the truncation error. It does not include round-off effects. Also shown is the approximate number of Maclaurin series terms required to achieve similar accuracy.

Table 4.5.2 shows this error as a function of the number of pseudo-rotations, N. Choosing $N = 9$ gives the 16-digit accuracy mentioned earlier.

The very last step is to multiply the calculated number by e^K (K multiplications by E), and to RETURN.

This algorithm is implemented as shown in listing 4.5.1. Listing 4.5.2 gives a sample run of that program. The final values calculated by the subroutine are limited by round-off error to about 20 times the precision associated with the number of digits carried by the BASIC interpreter.

The exponential subroutine is very easy to use. The only precaution is that you should not try a value of X that will cause an overflow.

Although EXPCORD requires only nine "rotations" plus a 4-term series to give results comparable to an 18-term Maclaurin series, it is not nearly as efficient in computation. It does offer a means to get extremely high-accuracy e^x evaluations, however, and forms the basis for the natural logarithm subroutine discussed in the next section.

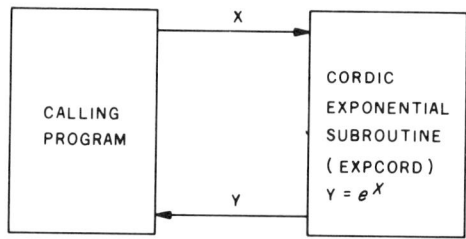

Figure 4.5.2 *Subroutine-connection diagram for the exponential subroutine based on the CORDIC concept.*

Statements/Functions List

$+, -, *, /, <, >$
ABS, FOR/NEXT, GOSUB, GOTO, IF/THEN, INT

Variables List

A, A(I), E, I, K, N, X, Y, W(I), Z

Variable Passed to Subroutine

X

Table 4.5.3 *Functions and variables used by the modified CORDIC subroutine for determining e^x.*

```
3460 REM PROGRAM TO DEMONSTRATE CORDIC EXPONENTIAL SUBROUTINE (EXPCORD)
3461 PRINT
3462 PRINT
3463 PRINT"   X",TAB(20),"    EXP(X)",TAB(50),"RELATIVE DIFFERENCE"
3464 PRINT"   -",TAB(20),"    ------",TAB(50),"-------------------"
3465 PRINT
3466 M=1E14
3467 FOR X=-1 TO 4 STEP .1
3468 GOSUB 45650
3469 PRINTX,TAB(20),Y,TAB(50),INT(M*(Y-EXP(X))/Y)/M
3470 NEXT X
3471 PRINT
3472 PRINT
3473 END
45644 REM ********************
45645 REM MODIFIED CORDIC EXPONENTIAL SUBROUTINE (EXPCORD).
45646 REM THIS PROGRAM TAKES AN INPUT VALUE AND RETURNS Y=EXP(X).
45647 REM X MAY BE ANY POSITIVE OR NEGATIVE VALUE.
45648 REM REMEMBER TO DIMENSION A(I) AND W(I) IN THE CALLING PROGRAM.
45649 REM GET COEFFICIENTS
45650 GOSUB 45684
45651 REM REDUCE THE RANGE OF X
45652 K=INT(X)
45653 X=X-K
45654 REM DETERMINE THE WEIGHTING COEFFICIENTS, W(I)
45655 GOSUB 45673
45656 REM CACLULATE PRODUCTS
45657 Y=1
45658 FOR I=1 TO N
45659 IF W(I)>0 THEN Y=Y*A(I)
45660 NEXT I
45661 REM PERFORM RESIDUAL MULTIPLICATION
45662 Y=Y*(1+Z*(1+Z/2*(1+Z/3*(1+Z/4))))
45663 REM ACCOUNT FOR FACTOR EXP(K)
45664 IF K<0 THEN E=1/E
45665 IF ABS(K)<1 THEN GOTO 45671
45666 FOR I=1 TO ABS(K)
45667 Y=Y*E
45668 NEXT I
45669 REM RESTORE X
45670 X=X+K
45671 RETURN
45672 REM WEIGHT DETERMINATION SUBROUTINE
45673 A=.5
45674 Z=X
45675 FOR I=1 TO N
45676 W(I)=0
45677 IF Z>A THEN W(I)=1
45678 Z=Z-W(I)*A
```

Listing 4.5.1

```
45679 A=A/2
45680 NEXT I
45681 RETURN
45682 REM *********************
45683 REM EXPONENTIAL COEFFICIENTS SUBROUTINE
45684 N=9
45685 E=2.718281828459045
45686 A(1)=1.648721270700128
45687 A(2)=1.284025416687742
45688 A(3)=1.133148453066826
45689 A(4)=1.064494458917859
45690 A(5)=1.031743407499103
45691 A(6)=1.015747708586686
45692 A(7)=1.007843097206448
45693 A(8)=1.003913889338348
45694 A(9)=1.001955033591003
45695 RETURN
```

Listing 4.5.1 *The modified CORDIC subroutine for determining e^x (EXPCORD). See also figure 4.5.2 and table 4.5.3. The coefficients used are sufficient for 15-digit accuracy. Also provided is a demonstration program. See listing 4.5.2 for sample results.*

Listing 4.5.2

RUN

X	EXP(X)	RELATIVE DIFFERENCE
-1	.36787944117145	0
-.9	.40656965974063	.00000000000004
-.8	.44932896411721	-.00000000000003
-.7	.49658530379142	-.00000000000005
-.6	.54881163609405	-.00000000000002
-.5	.60653065971256	-.0000000000001
-.4	.67032004603561	-.00000000000008
-.3	.74081822068163	-.0000000000001
-.2	.81873075307797	-.00000000000002
-.1	.90483741803592	-.00000000000005
0	1	0
.1	1.1051709180757	.00000000000009
.2	1.2214027581601	-.00000000000009
.3	1.349858807576	0
.4	1.4918246976413	0
.5	1.6487212706999	-.00000000000013
.6	1.8221188003904	-.00000000000006

.7	2.0137527074702	-.00000000000015
.8	2.2255409284924	0
.9	2.4596031111568	-.00000000000005
1	2.718281828459	0
1.1	3.0041660239465	.00000000000003
1.2	3.3201169227363	-.0000000000001
1.3	3.6692966676192	-.00000000000006
1.4	4.0551999668447	-.00000000000005
1.5	4.4816890703374	-.00000000000016
1.6	4.9530324243947	-.00000000000009
1.7	5.4739473917264	-.00000000000017
1.8	6.0496474644127	-.00000000000002
1.9	6.6858944422788	-.00000000000006
2	7.3890560989304	-.00000000000003
2.1	8.1661699125677	.00000000000001
2.2	9.0250134994333	-.00000000000008
2.3	9.9741824548144	-.00000000000005
2.4	11.023176380641	-.0000000000001
2.5	12.182493960701	-.00000000000025
2.6	13.463738035	-.00000000000008
2.7	14.87973172487	-.00000000000021
2.8	16.444646771096	-.00000000000007
2.9	18.174145369441	-.00000000000012
3	20.085536923187	-.00000000000005
3.1	22.197951281441	0
3.2	24.532530197107	-.00000000000009
3.3	27.112638920657	0
3.4	29.964100047395	-.00000000000007
3.5	33.115451958685	-.00000000000022
3.6	36.598234443673	-.00000000000017
3.7	40.447304360059	-.0000000000002
3.8	44.701184493297	-.00000000000009
3.9	49.402449105524	-.00000000000019
4	54.598150033142	-.0000000000001

READY

Listing 4.5.2 *Sample run (using 14-digit North Star BASIC) of the program shown in listing 4.5.1. "EXP(X)" is the calculated value. The third column is the relative difference between the value calculated by the subroutine and that calculated internally by BASIC.*

4.6 The Natural Logarithm Function

The subject of this section is approximating the natural logarithm, $\ln x$. We will use the ideas of the previous treatment regarding the exponential function in much the same way as the trigonometric function algorithm was enlisted to develop the inverse trigonometric subroutines.

The discussion is initially restricted to the argument range $1 \le x < e$. The extension to $0 \le x < \infty$ is simple and will be added later. The equation to evaluate is

$$y = \ln x \tag{4.6.1}$$

Exponentiating both sides gives

$$e^y = x \tag{4.6.2}$$

Proceeding as with equation (4.5.7), the above equation becomes

$$x = e^z \prod_{i=1}^{N} A_i^{W_i} \tag{4.6.3}$$

If we can find the Z and W_i values that satisfy this last equation, the problem is nearly solved:

$$y = \ln x = Z + \sum_{i=1}^{N} W_i \ln A_i \tag{4.6.4}$$

Because $A_i = e^{1/2^i}$, we have $\ln A_i = 1/2^i$, and the desired evaluation is

$$y = Z + \sum_{i=1}^{N} \frac{W_i}{2^i} \tag{4.6.5}$$

The problem is clearly one of finding the appropriate W_i and Z values.

The W_i weights are easily determined by successively attempting to divide x (or the result of the last step) by A_i, with the stipulation that if the result is less than unity, the division is skipped and W_i is set equal to zero. Otherwise, the division is performed and $W_i = 1$. In effect, the relevant A_i factors in equation (4.6.3) are being found by a sequence of tests.

The next step is to determine the value of the residual, Z, in equation (4.6.3). This is simply

$$Z = \ln \left(x / \prod_{i=1}^{N} A_i^{W_i} \right) \tag{4.6.6}$$

Because Z is small, and therefore the argument in the logarithm near unity, a short Taylor series expansion can be used for the logarithm function. After the residual is determined, the results can be assembled according to equation (4.6.5).

Arguments outside the range of $1 \le x < e$ are handled by factoring out integral powers of e. For example, if x is the input value, it can be expressed as $x = e^K x'$. The *integer* (positive or negative) K is chosen such that $1 \le x' < e$. We then have $\ln x = K + \ln x'$, and the problem is reduced to evaluating $\ln x'$ as discussed above.

A subroutine for all of this is shown in listing 4.6.1. The associated subroutine-connection diagram appears in figure 4.6.2. In principle, the accuracy of this program is limited only by the series approximation to the residual logarithm term (which gives Z—see listing). The associated error bound is shown in table 4.6.1. However, round-off error has a very significant effect in this algorithm. For example, according to table 4.6.1, nine "rotations" plus the 4-term Taylor series should be sufficient for 14-digit accuracy if it were not for round-off error. Using a 14-digit BASIC, the observed relative error is 10^{-10}, which indicates the presence of considerable round-off error. This error can be tracked down in the code to the determination of Z and $1 - Z$. There are numerical errors associated with calculating Z (which is very near unity in this part of the code), and these are greatly amplified by the $1 - Z$ subtraction. The practical solution to this problem is to use more "rotations" so that the sensitive residual calculations become less important. For this reason, 15 "rotations" are used in the subroutine.

The results of operating this subroutine (i.e., running the demonstration program) are shown in listing 4.6.2. As you can see, the round-off error has been reduced to a reasonable level.

The advantage this algorithm provides over simple series approximations is apparent by inspection of a typical logarithmic Taylor series expansion:

$$\ln x = 2\left[\left(\frac{x-1}{x+1}\right) + \frac{1}{3}\left(\frac{x-1}{x+1}\right)^3 + \cdots + \frac{1}{n}\left(\frac{x-1}{x+1}\right)^n + \cdots\right] \quad (4.6.7)$$

This series converges for all $x > 0$. The truncation error associated with using only the first N of these terms is

$$E \sim \frac{1}{2N+1}\left(\frac{x-1}{x+1}\right)^{2N+1} \quad (4.6.8)$$

For $x = e$, this gives

$$E \sim \frac{1}{(2N+1)}(0.46)^{2N+1}$$

For a truncation error of approximately 10^{-14}, 18 Taylor series terms are required, as opposed to nine "rotations" for the CORDIC method. However, each of these terms contains a subtraction between two comparably sized numbers, which causes round-off error of the same form as encountered (and circumvented) in the CORDIC program.

The three precautions that should be observed when using the modified CORDIC logarithm subroutine are: (1) Do not try to evaluate $\ln(-x)$, for it is not a valid calculation and an incorrect value will be returned; (2) Remember to dimension $A(I)$ and $W(I)$ in the calling program; and (3) Be wary of round-off error!

264 BASIC SCIENTIFIC SUBROUTINES

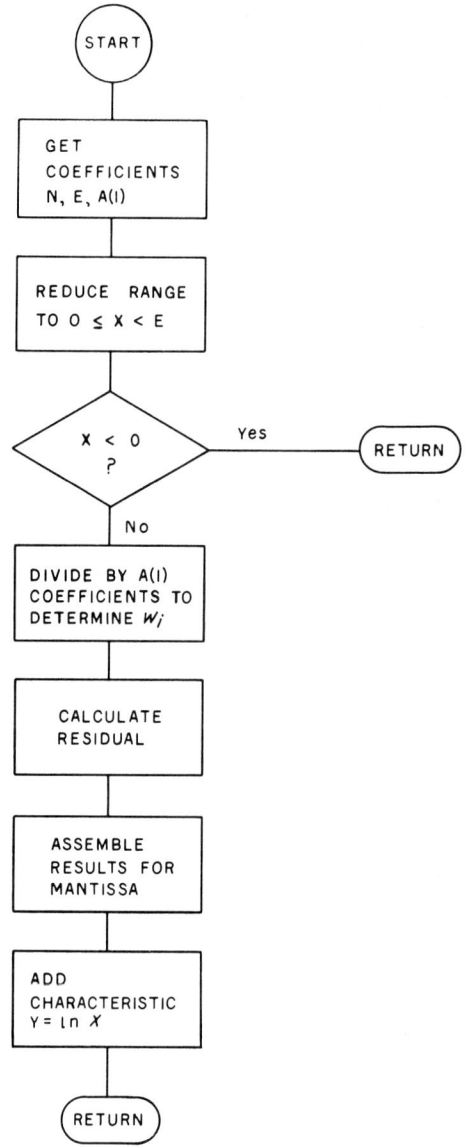

Figure 4.6.1 *Flowchart for natural logarithm subroutine (LNCORDIC).*

CORDIC APPROXIMATION TECHNIQUES AND ALTERNATIVES

N	Error Bound
1	2×10^{-2}
2	4×10^{-4}
3	8×10^{-6}
4	2×10^{-7}
5	5×10^{-9}
6	2×10^{-10}
7	6×10^{-12}
8	2×10^{-13}
9	6×10^{-15}
10	2×10^{-16}
11	6×10^{-18}
12	2×10^{-19}
13	5×10^{-21}
14	2×10^{-22}
15	5×10^{-24}

Table 4.6.1 *Estimated upper bound on the truncation error in the natural logarithm subroutine. Note that this does not include round-off error, which can be very significant.*

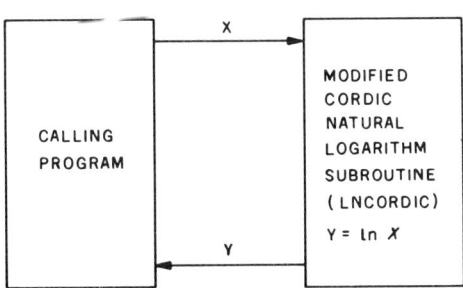

Figure 4.6.2 *Subroutine-connection diagram for the natural logarithm subroutine (LNCORDIC).*

266 BASIC SCIENTIFIC SUBROUTINES

Statements/Functions List

$+, -, *, /, <, >$
FOR/NEXT, GOTO, GOSUB, IF/THEN

Variables List

A, A(I), E, I, K, N, X, X1, Y, W(I), Z

Variable Passed to Subroutine

X

Table 4.6.2 *Functions and variables used by the natural logarithm subroutine.*

Listing 4.6.1

```
3475 REM PROGRAM TO DEMONSTRATE THE MODIFIED CORDIC NATURAL LOGARITHM SUBROUTINE
3476 PRINT
3477 PRINT
3478 PRINT" X", TAB(20),"     LN(X)",TAB(50),"RELATIVE DIFFERENCE"
3479 PRINT" -", TAB(20),"     ------",TAB(50),"-------------------"
3480 PRINT
3481 DIM A(15),W(15)
3482 M=1E14
3483 FOR X=.1 TO 3 STEP .1
3484 GOSUB 45725
3485 PRINTX,TAB(20),INT(M*Y)/M,TAB(49),INT(M*(Y-LOG(X))/(Y+.000001))/M
3486 NEXT X
3487 PRINT
3488 PRINT
3489 END
45719 REM ********************
45720 REM MODIFIED CORDIC NATURAL LOGARITHM SUBROUTINE (LNCORDIC).
45721 REM THIS PROGRAM TAKES AN INPUT VALUE AND RETURNS Y=LN(X).
45722 REM X MAY BE ANY POSITIVE VALUE.
45723 REM REMEMBER TO DIMENSION A(I) AND W(I) IN THE CALLING PROGRAM.
45724 REM GET COEFFICIENTS
45725 GOSUB 45770
45726 REM IF X<=0 THEN AN ERROR EXISTS, RETURN
45727 IF X<=0 THEN RETURN
45728 K=0
45729 REM SAVE X
45730 X1=X
45731 REM REDUCE THE RANGE OF X
45732 IF X<E THEN GOTO 45739
45733 REM DIVIDE OUT A POWER OF E
45734 K=K+1
45735 X=X/E
```

Listing 4.6.1 cont.

```
45736 GOTO 45732
45737 REM TEST IF X>=1. IF SO GO TO NEXT STEP
45738 REM OTHERWISE, BRING X TO >1
45739 IF X>=1 THEN GOTO 45744
45740 K=K-1
45741 X=X*E
45742 GOTO 45739
45743 REM DETERMINE THE WEIGHTING COEFFICIENTS, W(I)
45744 GOSUB 45761
45745 REM CALCULATE RESIDUAL FACTOR BASED ON Z
45746 REM WANT LN(Z), WHERE Z IS NEAR UNITY
45747 Z=Z-1
45748 Z=Z*(1-(Z/2)*(1+(Z/3)*(1-Z/4)))
45749 REM ASSEMBLE RESULTS
45750 A=1/2
45751 FOR I=1 TO N
45752 Z=Z+W(I)*A
45753 A=A/2
45754 NEXT I
45755 REM Z IS NOW THE MANTISSA, K THE CHARACTERISTIC
45756 Y=K+Z
45757 REM RESTORE X
45758 X=X1
45759 RETURN
45760 REM WEIGHT DETERMINATION SUBROUTINE
45761 Z=X
45762 FOR I=1 TO N
45763 W(I)=0
45764 IF Z>A(I) THEN W(I)=1
45765 IF W(I)=1 THEN Z=Z/A(I)
45766 NEXT I
45767 RETURN
45768 REM *********************
45769 REM EXPONENTIAL COEFFICIENTS SUBROUTINE
45770 N=15
45771 E=2.718281828459045
45772 A(1)=1.648721270700128
45773 A(2)=1.284025416687742
45774 A(3)=1.133148453066826
45775 A(4)=1.064494458917859
45776 A(5)=1.031743407499103
45777 A(6)=1.015747708586686
45778 A(7)=1.007843097206448
45779 A(8)=1.003913889338348
45780 A(9)=1.001955033591003
45781 A(10)=1.000977039492417
45782 A(11)=1.000488400478694
45783 A(12)=1.000244170429748
45784 A(13)=1.000122077763384
```

```
45785 A(14)=1.000061037018933
45786 A(15)=1.000030518043791
45787 RETURN
```

Listing 4.6.1 *Natural logarithm subroutine based on the modified CORDIC concept. See also figure 4.6.2 and table 4.6.2. Also shown is a demonstration program. See listing 4.6.2 for a sample run.*

```
RUN

   X                    LN(X)                   RELATIVE DIFFERENCE
   -                    -----                   -------------------

  .1              -2.3025850929941              0
  .2              -1.6094379124339              -.00000000000013
  .3              -1.2039728043258              -.00000000000009
  .4               -.9162907318742               .00000000000004
  .5               -.6931471805597              -.00000000000037
  .6               -.5108256237658              -.0000000000004
  .7               -.3566749439386              -.00000000000037
  .8               -.2231435513142              -.00000000000005
  .9               -.1053605156578              -.00000000000019
 1                  0                            0
 1.1                .09531017980421             -.00000000000113
 1.2                .1823215567939              -.00000000000033
 1.3                .26236426446764              .00000000000053
 1.4                .336647223662122             0
 1.5                .40546510810837              .00000000000051
 1.6                .47000362924573              0
 1.7                .53062825106238              .00000000000041
 1.8                .58778666490231              .0000000000003
 1.9                .64185388617242              .00000000000003
 2                  .69314718056011              .00000000000023
 2.1                .74193734472947              .00000000000012
 2.2                .78845736036442              .0000000000002
 2.3                .83290912293526              .00000000000021
 2.4                .87546873735407              .0000000000002
 2.5                .91629073187433              .00000000000018
 2.6                .9555114450275               .0000000000008
 2.7                .99325177301045              .00000000000016
 2.8               1.02961941718140              .00000000000019
 2.9               1.06471073699250              .00000000000009
 3                 1.0986122886682               .00000000000009

READY
```

Listing 4.6.2 *Sample run using the natural-logarithm subroutine (LNCORDIC) and the demonstration program shown in listing 4.6.1. A 14-digit version of North Star BASIC was employed for the calculations.*

4.7 The Hyperbolic Trigonometric Functions

The three basic hyperbolic trigonometric functions are defined as follows:

$$\sinh x = \frac{e^x - e^{-x}}{2}$$

$$\cosh x = \frac{e^x + e^{-x}}{2} \qquad (4.7.1)$$

$$\tanh x = \frac{e^x - e^{-x}}{e^x + e^{-x}}$$

They are closely related to the ordinary trigonometric functions. If x is a real number, and $i = \sqrt{-1}$, then

$$\begin{aligned} \sin ix &= i \sinh x \\ \cos ix &= \cosh x \\ \tan ix &= i \tanh x \end{aligned} \qquad (4.7.2)$$

The direct connection between the two sets suggests that the CORDIC method used for the trigonometric functions might also be applicable to their hyperbolic relatives. In fact, the analogy is so close that, as we will see, the relevant equations are almost identical!

To probe the relationship, we will first examine the hyperbolic function analogy to the rotation equations [compare with equation (4.2.1)]:

$$\begin{aligned} x' &= x \cosh \theta \pm y \sinh \theta = \cosh \theta \, (x \pm y \tanh \theta) \\ y' &= y \cosh \theta \pm x \sinh \theta = \cosh \theta \, (y + r \tanh \theta) \end{aligned} \qquad (4.7.3)$$

The sign ambiguity is decided by θ. If θ is negative, the minus sign is chosen; if positive, the plus sign is used. Also in accordance with the previous discussion, the total "rotation" (A) can be broken up into steps [compare with equation (4.2.2)]:

$$A = w_1\theta_1 + w_2\theta_2 + \cdots + w_n\theta_n \qquad \text{where} \quad w_i = +1 \text{ or } -1 \qquad (4.7.4)$$

The difference in this case is that A is not limited to $-\pi/2$ to $+\pi/2$, but rather to minus infinity to plus infinity. The net effect in the approximation is that roughly twice as many pseudo-rotations are required to cover the argument range for the hyperbolic functions as were required for their trigonometric counterparts.

As in equation (4.2.3), a simple choice for the θ_i values is possible:

$$\theta_i = 1/2^{N-i+1} \quad \text{for} \quad i = 1, 2, ..., 2N \tag{4.7.5}$$

This definition is slightly different than that employed earlier. The range covered is from 2^{-N} to 2^N. A is thereby approximated as

$$A \cong \sum_{i=1}^{2N} \frac{W_i}{2^{N-i+1}} \tag{4.7.6}$$

[Compare with equation (4.2.4).] Even the required half-angle formula analogies exist [compare with equations (4.3.1)]:

$$\cosh \frac{\theta}{2} = \left(\frac{\cosh \theta + 1}{2}\right)^{1/2}$$
$$\tanh \frac{\theta}{2} = \frac{\sinh \theta}{\cosh \theta + 1} \tag{4.7.7}$$

With all these direct relations, it would seem that the hyperbolic trigonometric functions could be calculated using the modified CORDIC algorithm just as easily as could the ordinary trigonometric functions. But, there are some differences that make the hyperbolic CORDIC method, as implemented in BASIC, highly inefficient.

First, as was discussed earlier, the huge computational advantages of the binary CORDIC algorithms are largely lost when implemented in BASIC. The redeeming quality in the trigonometric function case was that the CORDIC algorithm for the *inverse* trigonometric functions of arcsine, arccosine, and arctangent circumvented the convergence problems of the corresponding Taylor series.

Second, CORDIC approximations to the normal and inverse hyperbolic functions are not required because of the definitions [equations (4.7.1)] and the fact that the inverse hyperbolic trigonometric functions are very simple to calculate:

$$\sinh^{-1} x = \ln\left(x + \sqrt{x^2 + 1}\right)$$
$$\cosh^{-1} x = \ln\left(x + \sqrt{x^2 - 1}\right) \tag{4.7.8}$$
$$\tanh^{-1} x = \frac{1}{2} \ln\left(\frac{1+x}{1-x}\right)$$

The conclusion is that the hyperbolic CORDIC method is not directly suitable to computing in BASIC. However, we can use the definitions [equations (4.7.1)] of these functions, along with the modified CORDIC approximation to e^x to generate the desired functions. That approach will be the subject of the remainder of this section. The corresponding inverse hyperbolic functions will be considered in the following section.

There is a potential problem associated with immediately applying equation (4.7.1)—round-off error. For example, if x is near 0, the hyperbolic sine and tangent relations involve finding the difference between two nearly equal numbers. This is a classic source of round-off error and should therefore be avoided. For values of x that cause this occurrence, an alternate scheme such as a Maclaurin series expansion could be used. Fortunately, for x in the vicinity of zero, a rapidly convergent series exists:

$$\sinh x = x + \frac{x^3}{3!} + \frac{x^5}{5!} + \cdots \qquad (4.7.9)$$

We can determine the maximum number of terms required in the expansion by examining how close to zero x must be before enough round-off error occurs to justify switching to the series approximation. To do this, we will rewrite the sinh x definition:

$$\sinh x = \frac{e^x(1 - e^{-2x})}{2} \qquad (4.7.10)$$

For $e^{-2x} \lesssim 1/2$, the relative round-off error effect attributable to this cause would be small to insignificant. Therefore, a reasonable and conservative estimate of the value of x below which the Maclaurin series should be used is $x \approx 1/2 \ln 2 = 0.35$.

Table 4.7.1 shows the Maclaurin series truncation error that results when using $x = 1/2 \ln 2$. The rapid convergence is apparent.

Number of Terms Kept Beyond the First	Relative Truncation Error
1	2×10^{-2}
2	1×10^{-4}
3	3×10^{-7}
4	6×10^{-10}
5	6×10^{-13}
6	5×10^{-16}
7	3×10^{-19}
8	1×10^{-22}

Table 4.7.1 *Relative truncation error in the sinh x Maclaurin series for $x = \frac{1}{2} \ln 2$.*

The strategy for calculating the hyperbolic sine is therefore the following. For values of x below 0.35, the Maclaurin series approximation is used. Above that, equation (4.7.1) is employed.

For calculating the hyperbolic cosine, the exponential definition is applied directly and without modification since the particular round-off error problem discussed above does not exist.

The approach to handling the potential round-off error problem for the hyperbolic tangent is slightly different. The Maclaurin series for the hyperbolic tangent is not nearly as quickly convergent as that for the hyperbolic sine. For this case, the identity $\tanh x = \sinh x/\cosh x$ is implemented.

The subroutines associated with the above discussion are shown in listing 4.7.1. They are simple to access and work with, and are demonstrated in listing 4.7.2. It is expected that the relative accuracy of the computed results will be limited by round-off error to roughly 10^{-m+1}, where $m = 14$ in this case. Unfortunately, when performing such high-accuracy calculations, it is often difficult to find "true" values to compare against. Such is the case here.

It is not necessary to use the EXPCORD subroutine (see figure 4.7.1) for these calculations. One of the exponential-series evaluation subroutines given in the earlier chapters could also have been employed. The other exponential subroutines connect with the same parameter structure.

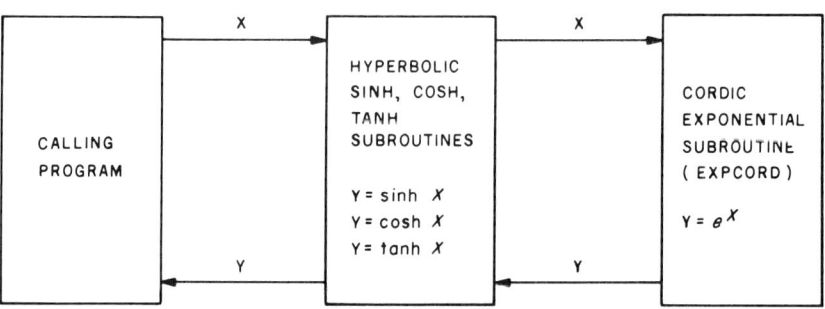

Figure 4.7.1 *Subroutine-connection diagram for the hyperbolic sine, cosine, and tangent subroutines. See table 4.7.2 for the functions and variables used.*

Statements/Functions List

+, −, *, /, <, >
ABS, FOR/NEXT, GOTO, GOSUB, IF/THEN, INT

Variables List

A, A(I), I, V, X, Y, Z

Variable Passed to Subroutine

X

Table 4.7.2 *Functions and variables used by the hyperbolic sine, cosine, and tangent subroutines. Note that these include the functions and variables shown in table 4.5.3 (EXPCORD subroutine).*

Listing 4.7.1

```
3500 REM PROGRAM TO DEMONSTRATE THE HYPERBOLIC SUBROUTINES
3501 PRINT
3502 PRINT
3503 PRINT "   X ",TAB(15)," SINH(X)",TAB(42)," COSH(X)",TAB(71)," TANH(X)"
3504 PRINT"   ---",TAB(15),"-------------",TAB(42),"-------------",TAB(71),"-------------"
3505 PRINT
3506 FOR X=-5 TO 5 STEP .2
3507 PRINT X,
3508 REM GET SINH(X)
3509 GOSUB 45800
3510 PRINT TAB(12),Y,
3511 REM GET COSH(X)
3512 GOSUB 45825
3513 PRINT TAB(40),Y,
3514 REM GET TANH(X)
3515 GOSUB 45840
3516 PRINT TAB(68),Y
3517 NEXT X
3518 PRINT
3519 PRINT
3520 END
45644 REM ********************
45645 REM MODIFIED CORDIC EXPONENTIAL SUBROUTINE (EXPCORD).
45646 REM THIS PROGRAM TAKES AN INPUT VALUE AND RETURNS Y=EXP(X).
45647 REM X MAY BE ANY POSITIVE OR NEGATIVE VALUE.
45648 REM REMEMBER TO DIMENSION A(I) AND W(I) IN THE CALLING PROGRAM.
45649 REM GET COEFFICIENTS
45650 GOSUB 45684
45651 REM REDUCE THE RANGE OF X
45652 K=INT(X)
```

```
45653 X=X-K
45654 REM DETERMINE THE WEIGHTING COEFFICIENTS, W(I)
45655 GOSUB 45673
45656 REM CACLULATE PRODUCTS
45657 Y=1
45658 FOR I=1 TO N
45659 IF W(I)>0 THEN Y=Y*A(I)
45660 NEXT I
45661 REM PERFORM RESIDUAL MULTIPLICATION
45662 Y=Y*(1+Z*(1+Z/2*(1+Z/3*(1+Z/4))))
45663 REM ACCOUNT FOR FACTOR EXP(K)
45664 IF K<0 THEN E=1/E
45665 IF ABS(K)<1 THEN GOTO 45671
45666 FOR I=1 TO ABS(K)
45667 Y=Y*E
45668 NEXT I
45669 REM RESTORE X
45670 X=X+K
45671 RETURN
45672 REM WEIGHT DETERMINATION SUBROUTINE
45673 A=.5
45674 Z=X
45675 FOR I=1 TO N
45676 W(I)=0
45677 IF Z>A THEN W(I)=1
45678 Z=Z-W(I)*A
45679 A=A/2
45680 NEXT I
45681 RETURN
45682 REM ********************
45683 REM EXPONENTIAL COEFFICIENTS SUBROUTINE
45684 N=9
45685 E=2.718281828459045
45686 A(1)=1.648721270700128
45687 A(2)=1.284025416687742
45688 A(3)=1.133148453066826
45689 A(4)=1.064494458917859
45690 A(5)=1.031743407499103
45691 A(6)=1.015747708586686
45692 A(7)=1.007843097206448
45693 A(8)=1.003913889338348
45694 A(9)=1.001955033591003
45695 RETURN
45790 REM ********************
45791 REM HYPERBOLIC SINE SUBROUTINE (SINH).
45792 REM THIS PROGRAM USES THE DEFINITION OF THE
45793 REM HYPERBOLIC SINE AND THE MODIFIED CORDIC
45794 REM EXPONENTIAL SUBROUTINE TO APPROXIMATE
45795 REM ARCSINH(X) OVER THE ENTIRE RANGE OF REAL X.
```

Listing 4.7.1 cont.

```
45796 REM THE INPUT TO THE SUBROUTINE IS X.
45797 REM THE RETURNED VALUE IS Y=ARCSINH(X).
45798 REM START CALCULATION
45799 REM IS X SMALL ENOUGH TO CAUSE ROUND OFF ERROR?
45800 IF ABS(X)<.35 THEN GOTO 45808
45801 REM CALCULATE SINH(X) USING EXPONENTIAL DEFINITION
45802 REM GET EXP(X)
45803 GOSUB 45650
45804 REM CALCULATE SINH(X)
45805 Y=(Y-(1/Y))/2
45806 RETURN
45807 REM SERIES APPROXIMATION
45808 Z=1
45809 Y=1
45810 FOR I=1 TO 8
45811 Z=Z*X*X/((2*I)*(2*I+1))
45812 Y=Y+Z
45813 NEXT I
45814 Y=X*Y
45815 RETURN
45816 REM ********************
45817 REM HYPERBOLIC COSINE SUBROUTINE (COSH).
45818 REM THIS PROGRAM USES THE DEFINITION OF THE
45819 REM HYPERBOLIC COSINE AND THE MODIFIED CORDIC
45820 REM EXPONENTIAL SUBROUTINE TO APPROXIMATE
45821 REM ARCOSH(X) OVER THE ENTIRE RANGE OF REAL X.
45822 REM THE RETURNED VALUE IS Y=ARCOSH(X).
45823 REM START CALCULATION
45824 REM GET EXP(X)
45825 GOSUB 45650
45826 Y=(Y+(1/Y))/2
45827 RETURN
45828 REM ********************
45829 REM HYPERBOLIC TANGENT SUBROUTINE (TANH).
45830 REM THIS PROGRAM USES THE DEFINITION
45831 REM    TAN(X)=SINH(X)/COSH(X)
45832 REM TO CALCULATE THE HYPERBOLIC TANGENT.
45833 REM THE INPUT IS X.
45834 REM THE OUTPUT IS Y=TANH(X).
45835 REM START CALCULATION
45836 REM GET SINH(X)
45840 GOSUB 45800
45841 V=Y
45842 REM GET COSH(X)
45843 GOSUB 45825
45844 Y=V/Y
45845 RETURN
```

Listing 4.7.1 *The hyperbolic sine (SINH), cosine (COSH), and tangent (TANH) subroutines. Note that all these subroutines call the modified CORDIC exponential evaluation subroutine (EXPCORD). The subroutine-connection diagram is shown in figure 4.7.1. Also shown is a demonstration program, the results of which appear in listing 4.7.2.*

X	SINH(X)	COSH(X)	TANH(X)
-5	-74.20321057778	74.20994852478	-.99990920426258
-4.8	-60.751093885835	60.759323632885	-.99986455170074
-4.6	-49.737131903088	49.747183738833	-.99979794161218
-4.4	-40.719295662531	40.731573002434	-.99969857928388
-4.2	-33.33566773205	33.35066330887	-.99955036645955
-4	-27.289917197125	27.308232836014	-.99932929973906
-3.8	-22.339406860721	22.361777632577	-.99899959778585
-3.6	-18.285455360613	18.31277908306	-.99850794233234
-3.4	-14.965363388718	14.998736658678	-.9977749279343
-3.2	-12.245883996565	12.286646200543	-.99668239783968
-3	-10.017874927409	10.067661995777	-.99505475368672
-2.8	-8.1919183542355	8.2527284168605	-.99263152020115
-2.6	-6.694732228393	6.769005806607	-.98902740220115
-2.4	-5.4662292136765	5.5569471669655	-.98367485769375
-2.2	-4.4571051705357	4.567908328898	-.97574313003146
-2	-3.6268604078469	3.7621956910835	-.96402758007582
-1.8	-2.9421742880956	3.1074731763172	-.94680601284626
-1.6	-2.3755679532001	2.5774644711948	-.92166855440645
-1.4	-1.9043015014516	2.1508984653932	-.88535164820227
-1.2	-1.5094613554122	1.8106555673244	-.83365460701216
-1	-1.1752011936438	1.5430806348153	-.76159415595574
-.8	-.88810598218765	1.3374349463049	-.66403677026784
-.6	-.63665358214815	1.1854652182423	-.53704956699794
-.4	-.41075232580285	1.0810723718385	-.37994896225524
-.2	-.2013360025411	1.0200667556191	-.19737532022491
0	0	1	0
.2	.2013360025411	1.0200667556191	.19737532022491
.4	.41075232580285	1.0810723718385	.37994896225524
.6	.63665358214815	1.1854652182423	.53704956699794
.8	.8881059821876	1.3374349463048	.66403677026785
1	1.1752011936438	1.5430806348153	.76159415595574
1.2	1.5094613554121	1.8106555673243	.83365460701215
1.4	1.9043015014516	2.1508984653932	.88535164820227
1.6	2.3755679532	2.5774644711947	.92166855440645
1.8	2.9421742880956	3.1074731763172	.94680601284626
2	3.6268604078469	3.7621956910835	.96402758007582
2.2	4.4571051705355	4.567908328898	.97574313003146
2.4	5.466229213676	5.556947166965	.98367485769375
2.6	6.694732228393	6.769005806607	.98902740220115
2.8	8.1919183542355	8.2527284168605	.99263152020115
3	10.01787492741	10.067661995778	.99505475368672
3.2	12.245883996565	12.286646200543	.99668239783968
3.4	14.965363388718	14.998736658678	.9977749279343
3.6	18.285455360613	18.31277908306	.99850794233234
3.8	22.339406860721	22.361777632577	.99899959778585
4	27.289917197127	27.308232836016	.99932929973906
4.2	33.335667732049	33.350663308869	.99955036645955
4.4	40.719295662529	40.731573002432	.99969857928388
4.6	49.737131903087	49.747183738832	.99979794161218
4.8	60.751093885835	60.759323632885	.99986455170074
5	74.203210577785	74.209948524785	.99990920426258

READY

Listing 4.7.2 *Sample application of the hyperbolic sine, cosine, and tangent subroutines. A 14-digit version of North Star BASIC was employed for the calculations.*

4.8 The Inverse Hyperbolic Trigonometric Functions

This section concludes the chapter on CORDIC approximations. The following discussion is actually not related to the CORDIC method itself, but rather merely completes the analysis of the hyperbolic trigonometric functions.

Recall from the previous section the simple functional relations for the inverses of the hyperbolic functions:

$$\sinh^{-1} x = \ln (x + \sqrt{x^2 + 1})$$
$$\cosh^{-1} x = \ln (x + \sqrt{x^2 - 1})$$
$$\tanh^{-1} x = \frac{1}{2} \ln \left(\frac{1 + x}{1 - x}\right)$$

At first glance, it appears that the evaluation is straightforward. The modified CORDIC natural-logarithm function of section 4.6 can be used in conjunction with a good square-root function. (Even the square-root iteration algorithm discussed earlier can be included if no internal square-root function is available.) As with the earlier discussion, however, there is a potential round-off error problem. For example, in the $\sinh^{-1} x$ case, large negative values of x lead to taking the difference between two large and nearly equal numbers. However, because $\sinh -x = -\sinh x$, the evaluation can be adjusted to always involve *addition*:

$$\sinh^{-1} x = \text{sign } x \ln (|x| + \sqrt{x^2 + 1}) \tag{4.8.1}$$

The $\cosh^{-1} x$ evaluation has potential round-off difficulty near $x = \pm 1$. The function can be rewritten as

$$\cosh^{-1} x = \ln (x + \sqrt{x + 1} \sqrt{x - 1})$$

The round-off error due to the $x \pm 1$ factors is unavoidable.

The $\tanh^{-1} x$ evaluation has the same problem as $\cosh^{-1} x$, but with an extra twist. If x is so near unity that the $x - 1$ factor evaluates to zero, a divide-by-zero error will result. This must be guarded against.

The subroutines associated with the above discussion are shown in listing 4.8.1 and demonstrated in listing 4.8.2. The accuracy of the calculated results is mainly limited by the round-off error that occurs in the natural-logarithm subroutine.

There are two situations to be wary of while using these subroutines. First, the argument transferred to the $\cosh^{-1} x$ subroutine is restricted to the range $1 \leq x < \infty$. If $x < 1$, an incorrect value will be returned. Second, the argument in the $\tanh^{-1} x$ subroutine is tested to avoid a divide-by-zero error. Thus, all x values sufficiently close to unity will result in the same large value returned. Also, the valid argument range for $\tanh^{-1} x$ is $0 \leq x < 1$. If x is outside this range, an incorrect value will be returned.

278 BASIC SCIENTIFIC SUBROUTINES

In the preceding chapters, we considered several methods for approximating functions. These techniques included iteration, least-squares approximation, minmax polynomial approximation, and the modified CORDIC algorithm. In the next chapter, we will see how to interpolate between known values of a function or data set. It will rapidly become apparent that the interpolation schemes treated all have their bases in polynomial approximations over a limited range.

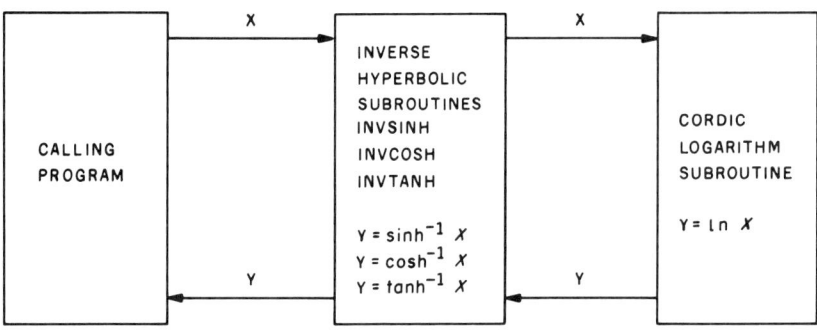

Figure 4.8.1 *Subroutine-connection diagram for the inverse hyperbolic functions discussed in this section.*

Statements/Functions List

$+, -, *, /, <, >$
ABS, FOR/NEXT, GOSUB, GOTO, IF/THEN, SQRT

Variables List

A, A(I), E, I, K, N, W(I), X, X2, Y, Z

Variable Passed to Subroutine

X

Table 4.8.1 *Functions and variables used in the inverse hyperbolic trigonometric function subroutines. These include the functions and variables used by LNCORDIC (see table 4.6.2).*

```
3525 REM PROGRAM TO DEMONSTRATE THE INVERSE HYPERBOLIC FUNCTION SUBROUTINES
3526 PRINT
3527 PRINT
3528 DIM A(15),W(15)
3529 PRINT "   X ",TAB(12),"ARCSINH(X)",TAB(32),"ARCCOSH(X)",TAB(52),"ARCTANH(X)"
3530 PRINT "   ---",TAB(12),"----------",TAB(32),"----------",TAB(52),"----------"
3531 PRINT
3532 FOR X=-3 TO 3 STEP .2
3533 PRINT " ",X,
3534 REM GET ARCSINH(X)
3535 GOSUB 45875
3536 PRINT TAB(11),Y,
3537 REM GET ARCOSH(X)
3538 GOSUB 45900
3539 PRINT TAB(31),Y,
3540 REM GET ARCTANH(X)
3541 GOSUB 45925
3542 PRINT TAB(51),Y
3543 NEXT X
3544 PRINT
3545 PRINT
3546 END
45719 REM ********************
45720 REM MODIFIED CORDIC NATURAL LOGARITHM SUBROUTINE (LNCORDIC).
45721 REM THIS PROGRAM TAKES AN INPUT VALUE AND RETURNS Y=LN(X).
45722 REM X MAY BE ANY POSITIVE VALUE.
45723 REM REMEMBER TO DIMENSION A(I) AND W(I) IN THE CALLING PROGRAM.
45724 REM GET COEFFICIENTS
45725 GOSUB 45770
45726 REM IF X<=0 THEN AN ERROR EXISTS, RETURN
45727 IF X<=0 THEN RETURN
45728 K=0
45729 REM SAVE X
45730 X1=X
45731 REM REDUCE THE RANGE OF X
45732 IF X<E THEN GOTO 45739
45733 REM DIVIDE OUT A POWER OF E
45734 K=K+1
45735 X=X/E
45736 GOTO 45732
45737 REM TEST IF X>=1. IF SO GO TO NEXT STEP
45738 REM OTHERWISE, BRING X TO >1
45739 IF X>=1 THEN GOTO 45744
45740 K=K-1
45741 X=X*E
45742 GOTO 45739
45743 REM DETERMINE THE WEIGHTING COEFFICIENTS, W(I)
45744 GOSUB 45761
45745 REM CALCULATE RESIDUAL FACTOR BASED ON Z
```

Listing 4.8.1

```
45746 REM WANT LN(Z), WHERE Z IS NEAR UNITY
45747 Z=Z-1
45748 Z=Z*(1-(Z/2)*(1+(Z/3)*(1-Z/4)))
45749 REM ASSEMBLE RESULTS
45750 A=1/2
45751 FOR I=1 TO N
45752 Z=Z+W(I)*A
45753 A=A/2
45754 NEXT I
45755 REM Z IS NOW THE MANTISSA, K THE CHARACTERISTIC
45756 Y=K+Z
45757 REM RESTORE X
45758 X=X1
45759 RETURN
45760 REM WEIGHT DETERMINATION SUBROUTINE
45761 Z=X
45762 FOR I=1 TO N
45763 W(I)=0
45764 IF Z>A(I) THEN W(I)=1
45765 IF W(I)=1 THEN Z=Z/A(I)
45766 NEXT I
45767 RETURN
45768 REM ********************
45769 REM EXPONENTIAL COEFFICIENTS SUBROUTINE
45770 N=15
45771 E=2.718281828459045
45772 A(1)=1.648721270700128
45773 A(2)=1.284025416687742
45774 A(3)=1.133148453066826
45775 A(4)=1.064494458917859
45776 A(5)=1.031743407499103
45777 A(6)=1.015747708586686
45778 A(7)=1.007843097206448
45779 A(8)=1.003913889338348
45780 A(9)=1.001955033591003
45781 A(10)=1.000977039492417
45782 A(11)=1.000488400478694
45783 A(12)=1.000244170429748
45784 A(13)=1.000122077763384
45785 A(14)=1.000061037018933
45786 A(15)=1.000030518043791
45787 RETURN
45866 REM ********************
45867 REM ARCSINH(X) SUBROUTINE (INVSINH).
45868 REM THIS ROUTINE CALCULATES THE INVERSE
45869 REM HYPERBOLIC SINE USING THE MODIFIED
45870 REM CORDIC NATURAL LOGARITHM SUBROUTINE.
45871 REM THE INPUT IS X.
45872 REM THE OUTPUT IS Y=ARCSINH(X).
```

Listing 4.8.1 cont.

Listing 4.8.1 cont.

```
45873 REM START CALCULATION
45874 REM TEST FOR ZERO ARGUMENT
45875 IF X<>0 THEN GOTO 45879
45876 Y=0
45877 RETURN
45878 REM SAVE X
45879 X2=X
45880 X=ABS(X)
45881 X=X+SQRT(X*X+1)
45882 REM GET LOGARITHM
45883 GOSUB 45725
45884 REM INSERT SIGN
45885 Y=(X2/ABS(X2))*Y
45886 REM RESTORE X
45887 X=X2
45888 RETURN
45891 REM ********************
45892 REM ARCCOSH(X) SUBROUTINE (INVCOSH).
45893 REM THIS ROUTINE CALCULATES THE INVERSE
45894 REM HYPERBOLIC COSINE USING THE MODIFIED
45895 REM CORDIC NATURAL LOGARITHM SUBROUTINE.
45896 REM THE INPUT IS X.
45897 REM THE OUTPUT IS Y=ARCOSH(X).
45898 REM BEGIN CALCULATION
45899 REM TEST FOR ARGUMENT LESS THAN OR EQUAL TO UNITY
45900 IF X>1 THEN GOTO 45904
45901 Y=0
45902 RETURN
45903 REM SAVE X
45904 X2=X
45905 X=ABS(X)
45906 X=X+SQRT(X-1)*SQRT(X+1)
45907 REM GET LOGARITHM
45908 GOSUB 45725
45909 REM RESTORE X
45910 X=X2
45911 RETURN
45916 REM ARCTANH(X) SUBROUTINE (INVTANH)
45917 REM ********************
45918 REM THIS PROGRAM CALCULATES THE INVERSE
45919 REM HYPERBOLIC TANGENT USING THE MODIFIED
45920 REM CORDIC NATURAL LOGARITHM SUBROUTINE.
45921 REM THE INPUT IS X.
45922 REM THE OUTPUT IS Y=ARCTANH(X).
45923 REM START CALCULATION
45924 REM TEST FOR X>= +/- 1
45925 IF ABS(X)<1 THEN GOTO 45929
45926 Y=(X/ABS(X))*1000000*1000000*1000000
45927 RETURN
```

```
45928 REM TEST FOR ZERO ARGUMENT
45929 IF X<>0 THEN GOTO 45933
45930 Y=0
45931 RETURN
45932 REM SAVE X
45933 X2=X
45934 X=(1+X)/(1-X)
45935 REM GET LOGARITHM
45936 GOSUB 45725
45937 REM RESTORE X
45938 X=X2
45939 RETURN
```

Listing 4.8.1 *Inverse hyperbolic function subroutines. Each of these programs calls the modified CORDIC natural-logarithm subroutine. See also figure 4.8.1 and table 4.8.1. Also shown is a demonstration program, the results of which are displayed in listing 4.8.2.*

Listing 4.8.2

RUN

X	ARCSINH(X)	ARCCOSH(X)	ARCTANH(X)
-3	-1.8184464	0	-1E+18
-2.8	-1.753229	0	-1E+18
-2.6	-1.683743	0	-1E+18
-2.4	-1.6094379	0	-1E+18
-2.2	-1.5296605	0	-1E+18
-2	-1.4436355	0	-1E+18
-1.8	-1.3504408	0	-1E+18
-1.6	-1.2489834	0	-1E+18
-1.4	-1.1379821	0	-1E+18
-1.2	-1.0159732	0	-1E+18
-1	-.88137356	0	-1E+18
-.8	-.73266822	0	-2.1972246
-.6	-.56882485	0	-1.3862944
-.4	-.39003525	0	-.8472978
-.2	-.19869024	0	-.4054651
0	0	0	0
.2	.19869024	0	.40546525
.4	.39003525	0	.84729784
.6	.56882485	0	1.3862945
.8	.73266822	0	2.1972247
1	.88137356	0	1E+18
1.2	1.0159732	.62236237	1E+18

1.4	1.1379821	.8670147	1E+18
1.6	1.2489834	1.046968	1E+18
1.8	1.3504408	1.1929108	1E+18
2	1.4436355	1.3169579	1E+18
2.2	1.5296605	1.4254169	1E+18
2.4	1.6094379	1.5220793	1E+18
2.6	1.683743	1.6094379	1E+18
2.8	1.753229	1.6892356	1E+18
3	1.8184464	1.7627472	1E+18

READY

Listing 4.8.2 *Demonstration of the programs given in listing 4.8.1.*

Chapter 5

Table Interpolation, Differentiation, and Integration

5.1 Introduction

The advent of the scientific pocket calculator dealt such a harsh blow to the manufacturers of slide rules that very few of these classic instruments are to be seen anymore. The consequences of this revolutionary event have been compounded by the appearance of inexpensive small computers. The publishers of function tables have been affected also, but not to such an extreme. The trigonometric functions are available to a higher accuracy on a 10-digit calculator or a small computer than in most mathematical handbooks (see Ref. 6, the *Handbook of Mathematical Functions* for 23-place sine and cosine tables), but they represent only a few of the important scientific functions. Many engineers and scientists use the Bessel, Legendre, elliptic, error, and other functions that have not yet found their way to standard pocket calculators. There is simply not enough demand to justify the production costs. Thus, when such functions are needed, the choice is either to use a published table or to implement a generating routine, such as polynomial expansion, on a computer.

In the previous chapters, several techniques for deriving such polynomial expansions were discussed. However, instead of determining the approximations from specialized and sometimes poorly conditioned expansions (e.g., slowly convergent or subject to round-off error), the desired values can be interpolated (inferred) from a handbook table that need not be very extensive, but must be accurate. The associated interpolation schemes can be implemented as user-defined functions or subroutine calls, the same approach as for the polynomial expansions. In this manner, an extensive function evaluation package that uses only very *general* and *easily applied* interpolation techniques can be developed for a computer system. Also, depending on the particular interpolation algorithm used, relatively fast execution speed is possible.

Dealing with function tables is intrinsically quite a bit different from working with experimental data. Data contain noise, whereas tables are assumed to be perfect. This essential difference is reflected in the types of algorithms that are employed to fit curves to the number set. In both cases, these fitted curves can be later used to determine intermediate values. In the noisy data situation, smoothing of the data is intentionally performed by constructing a curve that passes near the data points, but that does not go through extreme contortions in order to follow the noise. This type of curve fitting is commonly called *regression*, with *least squares* being one of the more common fitting criterion.

In the case of function tables, the value listed in the tabulation is assumed to be accurate to the last digit shown. The fitted curve is then expected to pass by the given table values within the error associated with that last digit. The techniques applied usually assume that the table is perfectly accurate, and the fitted curve is thereby required to pass *through* all the given table values. This latter requirement is the basis for the following interpolation formulae.

The interpolation routines presented in this chapter have two fundamental properties which should be remembered. First, the curves from which the interpolations are made are *piece-wise* approximations. That is, between each two neighboring table values a particular curve fit based on those and surrounding table values is found. The coefficients of the polynomial constructed for the coordinate interval x_i to x_{i+1} may not be the same as those determined for the previous interval x_{i-1} to x_i. Instead, the interpolation routines piece together an overall curve fit that is based on a collection of curvilinear segments. Basically, this is what occurs when a fit using a *french curve* or a *shipbuilder's curve* is made by hand.

The second interpolation property to be remembered is that a table originally generated by a linear, square, cubic, or higher-order function will not only be fitted table value-by-table value, but every interpolated value will also be exactly correct (in principle) *if* the interpolating polynomial is of that order or higher. The implication is that the better behaved (i.e., smoother) a function is, the fewer the number of table values required for a given level of interpolation accuracy. More will be said later about the magnitude of the error that occurs with interpolation.

Once the table values have been manipulated to generate a fitting polynomial, that polynomial in turn can be used to calculate the derivative of the function represented by those table values. Again, the calculated derivative can be very accurate under the conditions of the second interpolation property.

Finally, the integral of the function represented by the table values can be determined using a few simple geometrical concepts coupled with interpolation. This will be considered toward the end of this chapter.

In the following sections, we will explore table-based interpolation, differentiation, and integration.

SUBROUTINE	SUBROUTINE SIZE (bytes) (including support programs)	DEMON-STRATION PROGRAM SIZE (bytes)	EXECUTION SPEED	CONDITIONS/ COMMENTS
LAGRANGE (Nth-order Lagrange interpolation)	1028	846	1 second	Cubic interpolation
NEWTON (cubic divided-differences interpolation)	1107	768	1.8 seconds	
sin x table-spacing program	----	1722	1.1 minutes	Not part of the subroutine library
AKIMA (cubic spline interpolation)	1169	936	2.4 seconds	
D/LAGRNG (Nth-order derivative interpolation)	1174	892	4 seconds	$N = 1$
ITEG (general table integration)	3756	753	10-20 seconds	AKIMA spline subroutine used for interpolating endpoints
e^{-x^2} table-spacing program	----	594	40 seconds	Not a part of the subroutine library

Table 5.1.1 *A summary of the programs presented in Chapter 5.*

5.2 Lagrange Interpolation

Lagrange interpolation is based on the simple premise that for every set of sequential table values, there exists a unique polynomial curve that passes through each and all of those table values (see figure 5.2.1). The Lagrange polynomial for the cubic case is

288 BASIC SCIENTIFIC SUBROUTINES

$$f(x) \cong L_3(x) = \frac{f_1(x - x_2)(x - x_3)(x - x_4)}{(x_1 - x_2)(x_1 - x_3)(x_1 - x_4)}$$
$$+ \frac{f_2(x - x_1)(x - x_3)(x - x_4)}{(x_2 - x_1)(x_2 - x_3)(x_2 - x_4)} \quad (5.2.1)$$
$$+ \frac{f_3(x - x_1)(x - x_2)(x - x_4)}{(x_3 - x_1)(x_3 - x_2)(x_3 - x_4)}$$
$$+ \frac{f_4(x - x_1)(x - x_2)(x - x_3)}{(x_4 - x_1)(x_4 - x_2)(x_4 - x_3)}$$

The other orders of the Lagrange formula follow the same general pattern. In all cases, $N + 1$ table values are required for an Nth-degree polynomial fit.

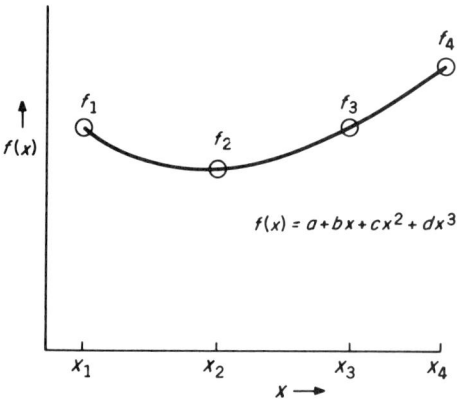

Figure 5.2.1 *Cubic polynomial fitted to four table values. Note that the x coordinates of the table values need not be equally spaced. The cubic polynomial fitted through these four points is unique, i.e., there exists no other cubic polynomial having different coefficients that also passes through the same points.*

Although the above equation may look formidable, we can easily see that when $x = x_1$, much canceling occurs and $f(x) = f_1$ is obtained. The same procedure applies for x_2, x_3, and x_4. Therefore, this polynomial is the required cubic because it passes through all the points. The uniqueness of this particular curve fit follows from a theorem in polynomial theory. Note also that the independent variables (the x_i) need not be equally spaced, a freedom that can be used to advantage.

The Lagrange interpolation formulae for all orders are easy to program because of their simple and repetitive structure (see listing 5.2.1). The associated subroutine-connection diagram appears in figure 5.2.2, and sample runs are shown in listing 5.2.2.

TABLE INTERPOLATION, DIFFERENTIATION, and INTEGRATION

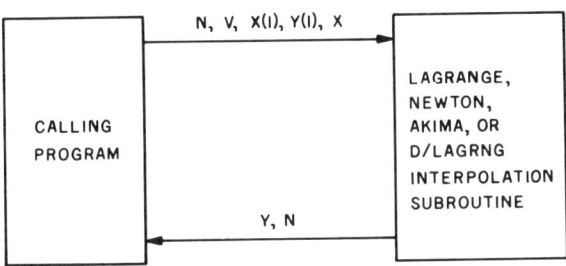

N:	Input – degree of the fit ($N = 1$ for linear, $N = 2$ for parabolic, $N = 3$ for cubic, etc.)
	Output – error check ($N = 0$ indicates insufficient table information)
V:	Total number of table values
X(I), Y(I):	The V table elements, where $Y(I) = f[X(I)]$
X:	Interpolation point
Y:	Calculated result $[Y = f(X)]$

Figure 5.2.2 *Subroutine-connection diagram for the Lagrange interpolation program given in listing 5.2.1. This diagram also applies to the Newton divided-differences subroutine (section 5.3); the Akima spline routine (section 5.5); and the Lagrange derivative interpolation program (section 5.6). Note that N is not required as input to the Newton or Akima cubic interpolation subroutines.*

The inputs to LAGRANGE are the number of table values, V, the table values themselves, X(I) and Y(I), the interpolation point, X, and the degree of the interpolation, N. The subroutine returns the interpolated value Y, and an error check, N. If N is returned as zero, then a condition that called for more table values than were available to the subroutine was encountered. This is demonstrated in the last example shown in listing 5.2.2.

The interpolation is expected to become more accurate as the number of table points employed in a given interval is increased. In the sin x cubic interpolation performed by the demonstration program, the accuracy obtained is fairly good (better than 0.000001) using only 14 table values. This can be seen from table 5.2.2.

The table values used for the sin x interpolation example were not equally spaced in x. For small values of x, sin x ≅ x and there is not much curvature to the function. Therefore, table values were concentrated in the high x region where the function has considerably more curvature. The technique employed for obtaining the table values shown in listing 5.2.1 will be discussed in section 5.4 of this chapter. However, as you can see from the observed errors in table 5.2.2, the choice was roughly optimal in terms of minimizing the maximum error.

Perhaps the most important precaution associated with using LAGRANGE is the restriction on the allowable range of the input interpolation point, X. The interpolation point *must* obey the following rule:

$$X(1) \le X \le X(V - N) \tag{5.2.2}$$

For example, if the number of table values is 14, and a cubic fit is chosen, then $X(1) \le X \le X(11)$. It is also very important that the input table values be in ascending $X(I)$ order, and that no two table positions be equal $[X(I) \ne X(I + 1)]$.

In the next section, we will consider another algorithm called Newton divided-differences interpolation. The mechanics of this algorithm will appear very different from the Lagrange formulation, but the resulting curve fits are, in principle, identical.

Statements/Functions List

$+, -, *, /, <, >$
FOR/NEXT, GOTO, IF/THEN

Variables List

I, J, K, L(J), N, X, X(I), Y(I), V

Variables Passed to Subroutine

N, X, X(I), Y(I), V

Table 5.2.1 *Functions and variables used in the Lagrange interpolation subroutine (LAGRANGE).*

Listing 5.2.1

```
3550 REM DEMONSTRATION PROGRAM FOR LAGRANGE INTERPOLATION OF SIN(X)
3551 PRINT
3552 PRINT
3553 DIM L(9),Y(15),X(15)
3554 V=14
3555 REM INPUT TABLE
3556 FOR I=1 TO V
3557 READ X(I),Y(I)
3558 NEXT I
3559 REM SINE TABLE VALUES FROM
3560 REM HANDBOOK OF MATHEMATICAL FUNCTIONS
3561 REM BY ABRAMOWITZ, M., AND STEGUN, I.A.
3562 REM NBS, JUNE 1964
3563 DATA 0,0,.125,.12467473
3564 DATA .217,.21530095,.299,.29456472
3565 DATA .376,.36720285,.450,.43496553
3566 DATA .520,.49688014,.589,.55552980
3567 DATA .656,.60995199,.721,.66013615
```

```
3568 DATA .7853981634,0.7071067812
3569 DATA .849,.75062005,.911,.79011709
3570 DATA .972,.82601466
3571 REM INPUT INTERPOLATION POINT
3572 PRINT "INPUT X: ",
3573 INPUT X
3574 PRINT "INPUT THE ORDER OF THE INTERPOLATION: ",
3575 INPUT N
3576 REM GOTO INTERPOLATION SUBROUTINE
3577 GOSUB 46000
3578 PRINT "SIN(X)= : ",Y
3579 PRINT "ERROR CHECK: ",N
3580 PRINT
3581 GOTO 3572
3582 END
3583 IF J=K THEN GOTO 3585
3584 L(K)=L(K)*(X-X(J+I))/(X(I+K)-X(J+I))
3585 NEXT J
3586 Y=Y+L(K)*Y(I+K)
3587 NEXT K
3588 RETURN
45988 REM ********************
45989 REM LAGRANGE INTERPOLATION SUBROUTINE (LAGRANGE)
45990 REM N IS THE LEVEL OF THE INTERPOLATION (EG., N=2 IS QUADRATIC).
45991 REM V IS THE TOTAL NUMBER OF TABLE VALUES.
45992 REM (X(I),Y(I)) ARE THE COORDINATE TABLE VALUES, Y(I) BEING THE
45993 REM DEPENDENT VARIABLE. THE X(I) MAY BE ARBITRARILY SPACED.
45994 REM X IS THE INTERPOLATION POINT WHICH IS ASSUMED TO BE IN THE
45995 REM INTERVAL WITH AT LEAST ONE TABLE VALUE TO THE LEFT, AND N TO THE RIGHT.
45996 REM IF THIS IS VIOLATED, N WILL BE SET TO ZERO.
45997 REM IT IS ASSUMED THAT THE TABLE VALUES ARE IN ASCENDING X(I) ORDER.
45998 REM X(I), Y(I) AND L(I) MUST BE DIMENSIONED IN THE CALLING PROGRAM.
45999 REM CHECK TO SEE IF INTERPOLATION POINT IS IN APPROPRIATE RANGE
46000 IF X<X(1) THEN GOTO 46003
46001 IF X<=X(V-N) THEN GOTO 46006
46002 REM AN ERROR HAS BEEN ENCOUNTERED
46003 N=0
46004 RETURN
46005 REM FIND THE RELEVANT TABLE INTERVAL
46006 I=0
46007 I=I+1
46008 IF X>X(I) THEN GOTO 46007
46009 I=I-1
46010 REM BEGIN INTERPOLATION
46011 FOR J=0 TO N
46012 L(J)=1
46013 NEXT J
46014 Y=0
46015 FOR K=0 TO N
```

Listing 5.2.1 cont.

```
46016 FOR J=0 TO N
46017 IF J=K THEN GOTO 46019
46018 L(K)=L(K)*(X-X(J+I))/(X(I+K)-X(J+I))
46019 NEXT J
46020 Y=Y+L(K)*Y(I+K)
46021 NEXT K
46022 RETURN
```

Listing 5.2.1 *Lagrange interpolation subroutine (LAGRANGE). Also shown is a program for demonstrating the use of Lagrange for sin x interpolation. Sample results appear in listing 5.2.2.*

Listing 5.2.2

```
RUN



INPUT X: ?.5
INPUT THE ORDER OF THE INTERPOLATION: ?1
SIN(X)= :  .47919025
ERROR CHECK:  1

INPUT X: ?.7853981634
INPUT THE ORDER OF THE INTERPOLATION: ?1
SIN(X)= :  .70710678
ERROR CHECK:  1

INPUT X: ?.785
INPUT THE ORDER OF THE INTERPOLATION: ?1
SIN(X)= :  .70681637
ERROR CHECK:  1

INPUT X: ?.785
INPUT THE ORDER OF THE INTERPOLATION: ?2
SIN(X)= :  .70682538
ERROR CHECK:  2

INPUT X: ?.785
INPUT THE ORDER OF THE INTERPOLATION: ?3
SIN(X)= :  .70682522
ERROR CHECK:  3

INPUT X: ?.785
INPUT THE ORDER OF THE INTERPOLATION: ?4
SIN(X)= :  .70682522
```

```
ERROR CHECK:   0

INPUT X:   ?
```

Listing 5.2.2 *Sample results for the Lagrange subroutine as applied to sin x interpolation. The first example was arbitrary. For the second example, the input X value corresponded to a table entry, and that number was returned. The next four examples show the effect of increasing the degree of the fit. In the last case, the desired fit required more table values than were available, and the error check (N = 0) indicated failure of the subroutine. The value of Y returned was simply that which existed before the subroutine was called.*

x (in radians)	sin x exact	sin x interpolated	Observed Error	Error Estimate
0.00	0.000000	0.0000000	0.0000000	0.0000000
0.05	0.0499792	0.0499800	−0.0000008	−0.0000003
0.10	0.0998334	0.0998338	−0.0000004	−0.0000002
0.15	0.1494381	0.1494387	−0.0000006	−0.0000004
0.20	0.1986693	0.1986695	−0.0000002	−0.0000002
0.25	0.2474040	0.2474045	−0.0000005	−0.0000004
0.30	0.2955202	0.2955203	−0.0000001	−0.0000000
0.35	0.3428978	0.3428982	−0.0000004	−0.0000003
0.40	0.3894183	0.3894189	−0.0000006	−0.0000004
0.45	0.4349655	0.4349655	0.0000000	0.0000000
0.50	0.4794255	0.4794258	−0.0000003	−0.0000003
0.55	0.5226872	0.5226877	−0.0000005	−0.0000005
0.60	0.5646425	0.5646428	−0.0000003	−0.0000003
0.65	0.6051864	0.6051865	−0.0000001	−0.0000001
0.70	0.6442177	0.6442180	−0.0000003	−0.0000003
0.75	0.6816388	0.6816393	−0.0000005	−0.0000005

Table 5.2.2 *Comparison of Lagrange interpolated values for sin x with corresponding handbook figures. The interpolated values were calculated using the program shown in listing 5.2.1 with N = 3 (cubic interpolation). The error estimates were obtained from the Newton divided-differences interpolation program discussed later in this chapter. Note that the observed error is in rough agreement with the error estimate. Note also that this agreement includes the sign of the error, not just its magnitude.*

5.3 Newton Divided-Differences Interpolation and Error Estimates

Another approach to obtaining the Nth-degree polynomial interpolation curve required to piece-wise fit a given table is to consider the Taylor series expansion about a particular nearby table value:

$$f(x) = f(x_i) + (x - x_i) \frac{df}{dx}\bigg|_{x_i} + \frac{(x - x_i)^2}{2!} \frac{d^2f}{dx^2}\bigg|_{x_i} + \frac{(x - x_i)^3}{3!} \frac{d^3f}{dx^3}\bigg|_{x_i}$$
$$+ \frac{(x - x_i)^4}{4!} \frac{d^4f}{dx^4}\bigg|_{x_i} + \cdots + \frac{(x - x_i)^N}{N!} \frac{d^Nf}{dx^N}\bigg|_{x_i} + \cdots \quad (5.3.1)$$

By truncating the series at the $(x - x_i)^N$ term, the function $f(x)$ is approximated with an Nth-degree polynomial. The resulting error is contained largely in the first term neglected. It can be shown that the error is

$$\frac{(x - x_i)^{N+1}}{(N + 1)!} \frac{d^{N+1}f}{dx^{N+1}}\bigg|_{x'}$$

where x' is some value in the interval between x and x_i.

To generate an algorithm that implements this expansion, table values neighboring the position of interest are employed to find estimates for the $(1/N!) \, d^Nf/dx^N$ terms in a form called *divided differences*. These factors can be derived very methodically as follows:

$$\begin{aligned}
\text{First divided difference} &= F_i^{(1)} = \frac{f_{i+1} - f_i}{x_{i+1} - x_i} \\
\text{Second divided difference} &= F_i^{(2)} = \frac{F_{i+1}^{(1)} - F_i^{(1)}}{x_{i+1} - x_i} \\
\text{Third divided difference} &= F_i^{(3)} = \frac{F_{i+1}^{(2)} - F_i^{(2)}}{x_{i+1} - x_i} \\
\text{Nth divided difference} &= F_i^{(N)} = \frac{F_{i+1}^{(N-1)} - F_i^{(N-1)}}{x_{i+1} - x_i}
\end{aligned} \quad (5.3.2)$$

The polynomial fitted between x_i and x_{i+1} is then

$$f(x) \cong p(x) = f_i + (x - x_i)F_i^{(1)} + (x - x_i)^2 F_i^{(2)} + (x - x_i)^3 F_i^{(3)} + \cdots \quad (5.3.3)$$

Observe that the divided differences automatically include the factorials appearing in equation (5.3.1). Note also that the divided difference corresponding to d^Nf/dx^N is calculated by employing f_i values somewhat remote from the region of interest: $F_i^{(1)}$ uses f_i and f_{i+1}; $F_i^{(2)}$ uses f_i, f_{i+1} and f_{i+2}; etc. In fact, all but one of the points involved is forward of the interpolation position. Therefore, the required derivatives are not being estimated at the exact position desired, but rather a little after it. This same bias is built into the Lagrange formulation. The effect can be reduced by applying a forward/backward, or *central* divided-differences hybrid formula instead. However, this alteration to improve the accuracy is not worth the effort when the increased complexity (and slower execution speed) of the resulting program is considered along with the round-off error, which appears to dominate the accuracy anyway.

There are two sources of error in the Newton divided-differences interpolation scheme. The first and most obvious source is the intrinsic error due to truncating the Taylor series. If Nth-order interpolation is employed, then the truncation error is associated with the first term not included:

$$\text{Error} \approx -\frac{(x - x_i)^{N+1}}{(N + 1)!} \left.\frac{d^{(N+1)}f}{dx^{(N+1)}}\right|_{x_i} \qquad (5.3.4)$$

(The error for the actual numerical calculation has this same form, but it is a little different. It will be discussed shortly.)

Truncation error is, in principle, amenable to estimation. However, in the specific case of Newton interpolation, it is usually not the dominant source of inaccuracy. The Nth divided difference involves N calculations of the differences between nearly equal numbers. This leads to significant round-off error. Unfortunately, this round-off error cannot easily be estimated. We will see its very large effect in a later example.

A subroutine for third-order (cubic) Newton interpolation appears in listing 5.3.1. The associated subroutine-connection diagram and input/output variables are the same as those for the Lagrange routine (see the previous section), with the exception of N. Because the interpolation is cubic, the order input, N, is superfluous.

The NEWTON subroutine can easily be extended to higher orders of interpolation, but this is not wise. Such an extension leads to an increase in the round-off error which, as we will soon see, is bad even at the cubic level.

A priori, the only obvious advantages to the Lagrange interpolation routine are that it is shorter in length, and that it does not require several additional large-dimensioned arrays. Mathematically, Lagrange and Newton interpolation of the same type (forward, backward, etc.) and order are equivalent—only the evaluation forms are different. Therefore, they have the same intrinsic accuracy. In fact, the truncation error for the cubic case in each is

$$\text{Error} \approx \frac{(x - x_i)(x - x_{i+1})(x - x_{i+2})(x - x_{i+3})}{4!} \left.\frac{d^4f}{dx^4}\right|_{x'} \qquad (5.3.5)$$

where x' is an evaluation point somewhere in the interval between x_i and x_{i+3}. Equations (5.3.4) and (5.4.5) appear to be different, with the latter giving larger values for the error estimate. The difference between the two equations can be rationalized as follows. Equation (5.3.4) assumes implicitly that we know the value of all the required derivatives at the position x_i. However, such is not the case in the numerical calculation. Rather, a group of table values is used to estimate the required derivatives, thereby increasing the error. For example, the divided-difference third derivative calculation involves using the points x_i, x_{i+1}, x_{i+2}, and x_{i+3}. The value found is the third derivative at *some point* in that interval (by the *mean value theorem*). This introduces into the algorithm an intrinsic inaccuracy beyond that expected from equation (5.3.4). For a further discussion of this issue, see *Applied Numerical Methods*, Ref. 10.

From the form of the error estimate, it is apparent that a cubic polynomial is fitted exactly because the fourth derivative of a cubic is necessarily zero throughout the interval. This error estimate is also calculated in the Newton program (see listing 5.3.1).

Examples of the operation of the Newton interpolation subroutine and its associated demonstration program are given in listing 5.3.2. The interpolation results for repeated runs are collected in table 5.3.2. The estimated errors are given in both table 5.3.2 and table 5.2.2 (the Lagrange interpolation tabulation). These error estimates were included on that earlier table for comparison with the observed error for the intrinsically equivalent Lagrange interpolation. The comparison for the Lagrange case gives some faith in using this error-estimating technique. However, as you can see from table 5.3.2, the Newton interpolation scheme suffers greatly from round-off error. The Newton method is two orders of magnitude less accurate than the Lagrange method in this particular example. It is not sufficient to simply reduce the truncation error by using more closely spaced table values or higher-order interpolation; *round-off* error must also be guarded against.

It is obvious that Lagrange interpolation is superior to Newton's divided-differences technique in terms of final accuracy.

The precautions associated with using NEWTON are identical to those encountered with LAGRANGE (subsection 5.3).

Statements/Functions List

+, −, *, /, <, >
FOR/NEXT, GOTO, IF/THEN

Variables List

A, B, C, E, I, N, X, Y, Y1, Y1(I), Y2(I), Y3(I), V

Variables Passed to Subroutine

X, X(I), Y(I), V

Table 5.3.1 *Functions and variables used in the Newton divided-differences interpolation subroutine (NEWTON).*

Listing 5.3.1

```
3600 REM PROGRAM TO DEMONSTRATE THE NEWTON INTERPOLATION
3601 PRINT
3602 PRINT
3603 REM SUBROUTINE FOR SIN(X)
3604 V=14
3605 DIM X(V+1),Y(V+1),Y1(V),Y2(V-1),Y3(V-2)
3606 REM INPUT TABLE
```

```
3607 FOR I=1 TO V
3608 READ X(I),Y(I)
3609 NEXT I
3610 REM SIN TABLE VALUES FROM
3611 REM HANDBOOK OF MATHEMATICAL FUNCTIONS
3612 REM BY ABRAMOWITZ, M., AND STEGUN, I.A.
3613 REM NBS, JUNE 1964
3614 DATA 0,0,.125,.12467473,.217,.21530095
3615 DATA .299,.29456472,.376,.36720285
3616 DATA .450,.43496553,.520,.49688014
3617 DATA .589,.55552980,.656,.60995199
3618 DATA .721,.66013615,.7853981634,0.7071067812
3619 DATA .849,.75072005
3620 DATA .911,.79011709,.972,.82601466
3621 REM INPUT INTERPOLATION POINT
3622 PRINT "INPUT X :",
3623 INPUT X
3624 REM GO TO INTERPOLATION SUBROUTINE
3625 GOSUB 46050
3626 PRINT "SIN(",X,") = ",Y
3627 PRINT"ERROR ESTIMATE (W/O ROUNDOFF): ",E
3628 PRINT "ERROR CHECK: ",N
3629 PRINT
3630 PRINT
3631 GOTO 3622
3632 END
46039 REM ********************
46040 REM NEWTON DIVIDED DIFFERENCES INTERPOLATION SUBROUTINE (NEWTON)
46041 REM CALCULATES CUBIC INTERPOLATIONS FOR A GIVEN TABLE.
46042 REM (X(I),Y(I)) ARE THE V COORDINATE PAIRS OF DATA.
46043 REM X(I) IS THE INDEPENDENT VARIABLE, Y(I) THE DEPENDENT.
46044 REM THE INTERPOLATION POINT IS X. IT IS ASSUMED THAT THERE
46045 REM IS AT LEAST ONE DATA POINT TO THE LEFT, AND THREE TO THE RIGHT.
46046 REM IF THIS IS VIOLATED, THEN N IS SET TO 0.
46047 REM E IS THE ERROR ESTIMATE.
46048 REM X,Y,Y1,Y2 AND Y3 ARE ASSUMED DIMENSIONED IN THE CALLING PROGRAM
46049 REM CHECK TO SEE IF X IS IN THE INTERVAL
46050 N=1
46051 IF X>=X(1) THEN GOTO 46054
46052 N=0
46053 RETURN
46054 IF X<=X(V-3) THEN GOTO 46058
46055 N=0
46056 RETURN
46057 REM GENERATE DIVIDED DIFFERENCES
46058 FOR I=1 TO V-1
46059 Y1(I)=(Y(I+1)-Y(I))/(X(I+1)-X(I))
46060 NEXT I
46061 FOR I=1 TO V-2
```

Listing 5.3.1 cont.

```
46062 Y2(I)=(Y1(I+1)-Y1(I))/(X(I+1)-X(I))
46063 NEXT I
46064 FOR I=1 TO V-3
46065 Y3(I)=(Y2(I+1)-Y2(I))/(X(I+1)-X(I))
46066 NEXT I
46067 REM FIND RELEVANT TABLE INTERVAL
46068 I=0
46069 I=I+1
46070 IF X>X(I) THEN GOTO 46069
46071 I=I-1
46072 REM BEGIN INTERPOLATION
46073 A=X-X(I)
46074 B=A*(X-X(I+1))
46075 C=B*(X-X(I+2))
46076 Y=Y(I)+A*Y1(I)+B*Y2(I)+C*Y3(I)
46077 REM CALCULATE NEXT TERM IN THE EXPANSION FOR AN ERROR ESTIMATE
46078 E=C*(X-X(I+3))*Y/24
46079 RETURN
```

Listing 5.3.1 *Subroutine for cubic interpolation using the Newton divided-differences algorithm (NEWTON). Also shown is a program for demonstrating the application of NEWTON to sin x table interpolation. See figure 5.2.2 for the subroutine-connection diagram, and listing 5.3.2 for sample runs.*

Listing 5.3.2

```
RUN

INPUT X :?.5
SIN( .5) =   .47961461
ERROR ESTIMATE (W/O ROUNDOFF):    -2.7745705E-07
ERROR CHECK:   1


INPUT X :?.7853981634
SIN( .78539816) =   .70710678
ERROR ESTIMATE (W/O ROUNDOFF):   0
ERROR CHECK:   1


INPUT X :?.785
SIN( .785) =   .70683078
ERROR ESTIMATE (W/O ROUNDOFF):    -6.0519083E-09
ERROR CHECK:   1


INPUT X :?.9
SIN( .9) =   .70683078
```

```
ERROR ESTIMATE (W/O ROUNDOFF):  -6.0519083E-09
ERROR CHECK:  0


INPUT X :?0
SIN( 0) = 0
ERROR ESTIMATE (W/O ROUNDOFF):  0
ERROR CHECK:  1


INPUT X :?
```

Listing 5.3.2 *Examples of using NEWTON for sin x interpolation. In the second and fifth examples, the interpolation position corresponded to a table entry, and that value was returned. In the fourth example, the interpolation point was beyond the allowable table range, as indicated by the N = 0 error check. In this case, the returned result is not correct.*

x	sin x (Handbook)	Newton Interpolation	Observed Error	Error Estimate
0.00	0.0000000	0.0000000	0.00000	0.0000000
0.05	0.0499792	0.0497300	−0.00025	−0.0000003
0.10	0.0998334	0.0997483	−0.00008	−0.0000002
0.15	0.1494381	0.1494106	−0.00003	−0.0000004
0.20	0.1986693	0.1986957	0.00003	−0.0000002
0.25	0.2474040	0.2474723	0.00007	−0.0000004
0.30	0.2955202	0.2955260	0.00001	−0.0000000
0.35	0.3428978	0.3430434	0.00005	−0.0000003
0.40	0.3894183	0.3895480	0.00013	−0.0000004
0.45	0.4349655	0.4349655	0.00000	0.0000000
0.50	0.4794255	0.4796146	0.00019	−0.0000003
0.55	0.5226872	0.5229237	0.00024	−0.0000005
0.60	0.5646425	0.5647715	0.00013	−0.0000003
0.65	0.6051864	0.6052761	0.00009	−0.0000001
0.70	0.6442177	0.6445039	0.00029	−0.0000003
0.75	0.6816388	0.6818061	0.00017	−0.0000005

Table 5.3.2 *Comparison of sin x handbook values with the corresponding Newton interpolations. The Lagrange and Newton interpolation algorithms are mathematically equivalent. The Newton interpolation method, however, is much more susceptible to round-off error in implementation, and thus has lower net accuracy. Recall that the corresponding Lagrange interpolation results shown in table 5.2.2 were accurate to better than 10^{-6}, and were in rough agreement with the estimated error. In this case, the Lagrange subroutine is two orders of magnitude more accurate than the Newton subroutine.*

5.4 Choosing the Table Values

When implementing an interpolation routine, a table size and spacing must be chosen. The closer the spacing between the table values (and therefore the larger the table), the better the intrinsic accuracy of the interpolation. For *evenly* spaced tables in which $x_{i+1} - x_i = h$, the average truncation error for a cubic interpolation (forward or backward divided differences) is

$$\text{Average truncation error} \approx \frac{(h/2)(3h/2)(5h/2)(7h/2)}{24} \left.\frac{d^4 f}{dx^4}\right|_x \approx \frac{h^4}{4} \left.\frac{d^4 f}{dx^4}\right|_x \tag{5.4.1}$$

In the case of sin x interpolation, the error estimate is approximately $h^4/(4\sqrt{2})$ at $x = \pi/4$. For 11 equally spaced table entries over the range $0 \le x \le \pi/4$, this gives an error estimate of about 7×10^{-6}. However, as you can see from table 5.2.2, the inaccuracy actually observed with the Lagrange program was generally much less than this ($<10^{-6}$), even though there were only 11 table values in the same range. The reason for this is that the sin x table spacings were chosen to be *unequal* so that the size of the ith interval resulted in the same approximate error as that for any other interval. For example, near $x_i = 0$, where the fourth derivative is small, the spacing was chosen to be large. In general, as x_i increases, the chosen interval size decreases. In this manner, the average interpolation error associated with each interval is made roughly the same. This is necessarily less than the error estimated in the beginning of this paragraph. (You may recognize this manipulation to be another application of the *min-max* principle.)

The desired unequal table spacings can be found in approximation using equation (5.4.1). If the magnitude of the average error resulting from the chosen uneven table spacings is E, then the table spacing itself is

$$H_x \cong \left[\frac{4E}{d^4 f(x)/dx^4 \big|_x} \right]^{1/4} \tag{5.4.2}$$

For sin x interpolation, the corresponding equation would be

$$H_x \cong \left[\frac{4E}{\sin x} \right]^{1/4} \tag{5.4.3}$$

The above approximation can be applied in two different ways. First, the average error can be chosen and the table positions calculated sequentially by choosing a starting position, calculating the distance to the next location (the first spacing), determining the next spacing, and so on. The calculation would be considered complete when the desired range is covered. The resulting number of table values is thereby an implied function of the chosen average error, E. The *error function* example given in subsection 5.8 of this chapter uses this procedure.

The second (and more complicated) way of applying equation (5.4.3) is to choose the number of table values that are to be within a given range, and to let E

become whatever is consistent with that choice.

The first approach—choosing the average truncation error *a priori*—is simple and easy to program. As an exercise, determine the table positions (and thereby the number of values) for sin x in the range $0 \le x \le \pi/4$ for $E = 10^{-6}$, $E = 10^{-7}$, and $E = 10^{-8}$. Observe that sin $x = 0$ at $x = 0$, and that it might be better to work backwards from $x = \pi/4$.

It is possible to start with the number of table values. This is a more difficult approach in that E is not known beforehand. However, the powerful method of *successive substitution* can be employed. (This general technique will be discussed in Chapter 6 with respect to finding the roots of equations.) The procedure is first to guess the size of the first interval, and then to apply equation (5.4.1) to find the corresponding E. Using this value, a sequence of table spacings and positions up to the desired number chosen is found. If the location of the last table value is beyond the desired range, then the size of the first interval is scaled down accordingly. If the location of the last table value is short of the range limit, then the initial interval size is scaled up. The calculation is then repeated and the last table position calculated is again examined. The size of the first interval is again scaled, and the process is repeated. The procedure is terminated when the correction factor (scaling) becomes sufficiently close to unity.

A program that performs this sequence of calculations for sin x in the first octant $(0 \le x \le \pi/4)$ is shown in listing 5.4.1. This program is only an example and is not part of the basic subroutine library. It can be used as a pattern to develop similar programs for finding the table locations to be used in interpolating other functions. The associated flowchart is given in figure 5.4.1. Although the flowchart is self-explanatory, three particular features should be highlighted.

The first feature deals with the evaluation of equation (5.4.3) and where the evaluation is made. To improve accuracy, d^4f/dx^4 should be evaluated near the *middle* of each interval. Therefore, the first evaluation point is at $x = H/2$. If $x = 0$ were chosen, a divide-by-zero error would occur, implying an infinite first interval. One of the precautions in following this procedure, therefore, is to watch out for the zeros of d^4f/dx^4. Also, note that problems can arise if d^4f/dx^4 changes *sign*. Thus, although not necessary for this example, absolute values should be used.

The second feature to be aware of is the number of table positions calculated. When dealing with a small number of table intervals, the calculated spacings can be in considerable error. A good rule of thumb is to start with a number of intervals at least twice (and preferably four times) greater than desired. Then pick every second (or fourth) point at the end of the calculation. This was done in the current example, and the results are shown in listing 5.4.2.

The third feature to note is the normalization step. Eleven positions (ten intervals) in the range $0 \le x \le \pi/4$ are desired. However, if a cubic interpolation is to be employed, three additional values beyond $x = \pi/4$ are needed. The program shown in listing 5.4.1 calculates 21 locations (20 intervals) in the range $0 \le x \le \pi/4$, and seven more beyond that. When every other point is chosen, the desired 14 positions are obtained.

302 BASIC SCIENTIFIC SUBROUTINES

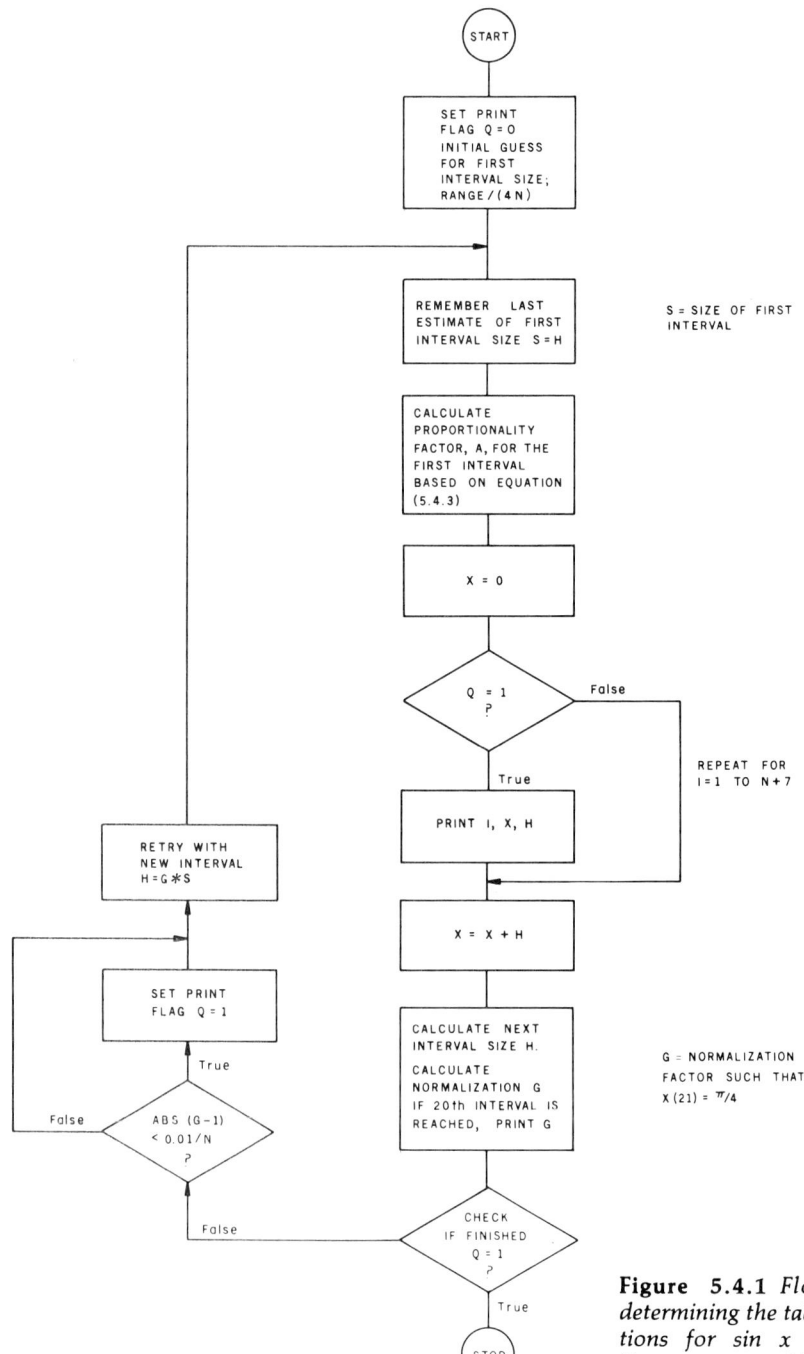

Figure 5.4.1 *Flowchart for determining the table value locations for sin x interpolation. Twenty-seven table locations are determined.*

It is easy to work with equal table spacings. However, although uneven table spacings require an initial investment of effort, the resulting interpolation error is much better distributed. Also, a clear error bound which is considerably smaller than the error bound associated with equal spacing (an order of magnitude effect for the sin x interpolation example) is established.

The subject of unequal table spacing will be considered again for the *error function* (section 5.8). We will now proceed on to examine a third interpolation scheme of particular importance to the engineering sciences—splines.

Listing 5.4.1

```
3650 REM TABLE SPACING PROGRAM FOR SIN(X) IN THE FIRST OCTANT
3651 PRINT
3652 PRINT
3653 REM PROGRAM GENERATES TABLE LOCATIONS WHICH APPROXIMATELY
3654 REM MINIMIZE THE MAXIMUM ERROR FOR A GIVEN NUMBER OF TABLE POINTS.
3655 REM A CUBIC FIT IS ASSUMED.
3656 REM G=CORRECTION TO GUESS.
3657 REM L IS USED TO ESTABLISH THE RANGE OF INTERPOLATION, IN THIS CASE PI/4.
3658 REM FOR ANOTHER FUNCTION, CHANGE NEXT TWO LINES.
3659 L1=3.141595/4
3660 L=SIN(L1)
3661 REM Q=PRINT FLAG
3662 Q=0
3663 REM ESTABLISH THE INITIAL GUESS FOR QUARTER INTERVAL SPACING
3664 REM NOTE THAT THE FINAL NUMBER OF INTERPOLATION INTERVALS IN THE
3665 REM INTERPOLATION RANGE IS N
3666 N=20
3667 H=L1/(4*N)
3668 PRINT "CONVERGENCE FACTOR:"
3669 REM S STORES THE LENGTH OF THE FIRST INTERVAL
3670 S=H
3671 REM PRINT HEADING
3672 IF Q=0 THEN GOTO 3681
3673 PRINT
3674 PRINT
3675 PRINT "INTERVAL",TAB(12),"POSITION",TAB(25),"INTERVAL SIZE"
3676 PRINT "--------",TAB(12),"------------",TAB(25),"---------------"
3677 REM A IS A MEASURE OF THE ERROR WHICH IS TO BE CONSTANT
3678 REM IT IS USED TO DETERMINE THE SPACING FOR ALL SUBSEQUENT INTERVALS
3679 REM THE FUNCTION IN THE NEXT LINE IS THE FOURTH DERIVATIVE (ABSOLUTE VALUE)
3680 REM OF THE FUNCTION TO BE EVALUATED
3681 A=H*(SIN(H/2))^(1/4)
3682 REM START INTERVAL SECTIONING
3683 X=0
3684 FOR I=1 TO N+7
3685 REM IF CONVERGENCE HAS BEEN ACHIEVED, PRINT RESULTS
3686 IF Q=0 THEN GOTO 3689
```

```
3687 PRINT I,TAB(10),INT(100000000*X)/100000000,TAB(26),INT(100000000*H)/100000000
3688 REM MOVE ON TO THE NEXT INTERVAL
3689 X=X+H
3690 REM CALCULATE THE NEW SPACING FOR THAT INTERVAL
3691 H=A*(1/ABS(SIN(X)))^(1/4)
3692 REM AT THIS POINT A CORRECTION IS APPLIED TO MAKE THE SUM
3693 REM OF THE INTERVALS (N OF THEM) EQUAL L1, THE RANGE)
3694 REM NOTE THAT THE FUNCTION ITSELF IS USED ON THE NEXT LINE,
3695 REM NOT ITS DERIVATIVE
3696 IF I=N THEN G=L/ABS(SIN(X))
3697 NEXT I
3698 IF Q=1 THEN GOTO 3705
3699 PRINT G
3700 REM CHECK FOR CONVERGENCE
3701 IF ABS(G-1)<.01/N THEN Q=1
3702 REM ADJUST SPACING OF THE FIRST INTERVAL AND TRY AGAIN
3703 H=G*S
3704 GOTO 3670
3705 PRINT
3706 PRINT
3707 END
```

Listing 5.4.1 *A program for finding 27 table-entry locations (21 in the range 0 to $\pi/4$) for sin x cubic interpolation. The resulting positional accuracy is approximately 0.001. The associated flowchart is shown in figure 5.4.1. Every other point, starting with the first, is used for the table positions appearing in the sin x interpolation programs given in sections 5.2 and 5.3. If the full 27-element table were used, the truncation error would be reduced by $\frac{1}{2}^4$ (to about 10^{-8}).*

Listing 5.4.2

```
RUN


CONVERGENCE FACTOR:
 6.243001
 1.0811704
 1.0141187
 1.0027805
 1.0005601
 1.0001132


INTERVAL        POSITION         INTERVAL SIZE
--------        --------         -------------
   1             0                 .06743335
   2             .06743335         .05671253
```

3	.12414589	.04870924
4	.17285513	.04486793
5	.21772307	.04238365
6	.26010673	.04057463
7	.30068136	.03916785
8	.33984922	.03802699
9	.37787621	.03707441
10	.41495063	.03626187
11	.4512125	.03555737
12	.48676987	.03493868
13	.52170856	.03438976
14	.55609832	.03389864
15	.58999697	.03345621
16	.62345318	.03305533
17	.65650852	.03269036
18	.68919889	.03235675
19	.72155565	.03205076
20	.75360642	.03176933
21	.78537575	.03150987
22	.81688562	.03127022
23	.84815584	.03104853
24	.87920438	.03084324
25	.91004763	.03065299
26	.94070062	.03047659
27	.97117722	.03031304

READY

Listing 5.4.2 *Sample run of the program shown in listing 5.4.1. Twenty-seven table locations are calculated. Every other position is chosen to obtain the 14 values desired.*

5.5 Semi-Spline Interpolation

The *spline* curve-fitting technique has been in use for quite some time and has found considerable application in the design of ship hulls and airplane structures. The method as originally practiced consisted of plotting the values, laying that plot on a board, and sticking pins in the calculated points (the table values, or *knots*). A thin strip of wood was then bent so as to conform to the pins and to span the space between them. Because the natural curve shape for a simply supported, end-loaded light beam is a cubic, you would expect a spline to fit a cubic function perfectly. Therefore, this form of spline fitting can be considered a third-order interpolation algorithm in some ways similar to the Lagrange and Newton methods.

The spline-fitting algorithm also has the subtle property that it minimizes the following integral:

$$D = \int \left[\frac{d^2p}{dx^2} - \frac{d^2f}{dx^2} \right]^2 dx \quad \text{where} \quad \begin{aligned} f(x) &= \text{true function} \\ p(x) &= \text{fitted curve} \end{aligned} \quad (5.5.1)$$

If the true function, $f(x)$, is a cubic, then the spline fit obviously minimizes D because $f(x) = p(x)$, but then so do the other cubic interpolation formulae that we have already discussed. For functions beyond the cubic, the spline fit is a little different in its behavior. It tends to minimize *net curvature* (elastic beams do not like kinks) while still passing through all the table values.

As might be expected intuitively, the bending of the spline is influenced by distant points, although the *coupling* is weaker the further away those points are. This distant coupling is necessary to minimize D. In summary, the spline technique is a cubic interpolation scheme which gives the same results (in principle) as the corresponding Lagrange and Newton cubic formulas for functions of the third degree and lower. However, for higher-order functions, the spline fit *does not* simply reflect the Taylor series expansion, but minimizes net curvature instead. A later example will clearly demonstrate and reinforce this difference.

Akima (see Applied Numerical Methods course notes, University of Michigan, 1978) developed a cubic interpolation technique that approximates the spline properties. It is calculated in a manner vaguely similar to the Newton divided-differences method. It is less coupled to distant points than a true spline fit, but the effect is not severe. Its chief advantage is that it is *much* faster in execution than exact spline interpolation algorithms, which involve time- and memory-consuming matrix operations.

The development of the background mathematics for Akima's algorithm is beyond the scope of this discussion. The resulting subroutine is given in listing 5.5.1 and the associated subroutine-connection diagram appears in figure 5.2.2. The in-

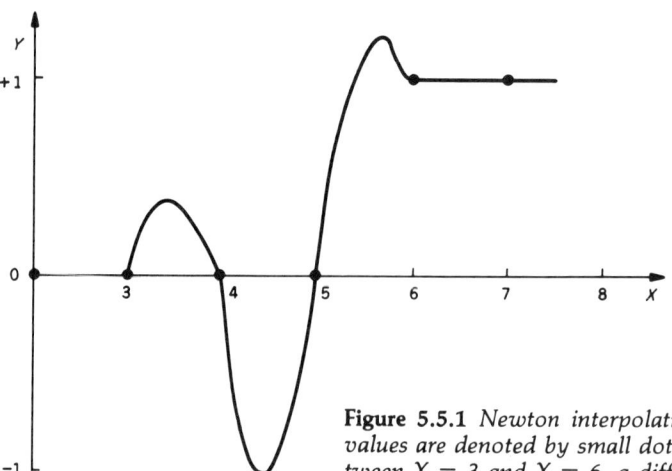

Figure 5.5.1 *Newton interpolation for a step function. The table values are denoted by small dots. In each of the three intervals between X = 3 and X = 6, a different cubic curve is used for interpolation. The overshoots are not symmetrical because the interpolation formula is not; forward differences are employed.*

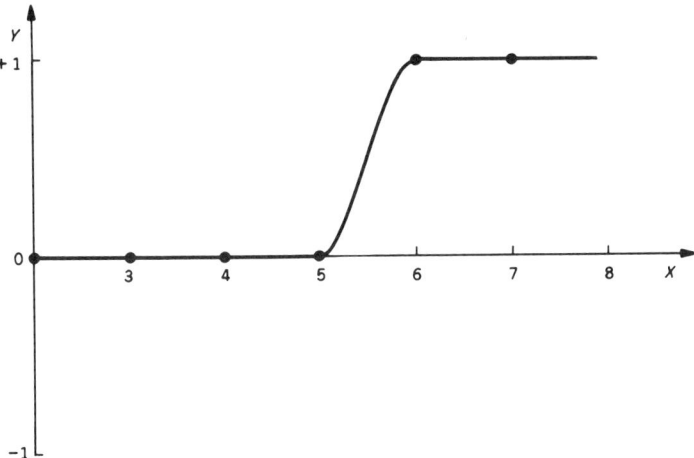

Figure 5.5.2 *Akima spline interpolation for a step function. The table values are denoted by small dots. Observe that the transition between the two levels is smooth. Contrast this with the ringing apparent in figure 5.5.1. This is an extreme example of how the Taylor series based and spline interpolation algorithms perform quite differently when confronted with tables representing functions more complex than a cubic.*

put/output characteristics of the Akima subroutine are very similar to those of the earlier two interpolation programs. Again, the inputs are the number of table values, V, the values themselves, $X(I)$ and $Y(I)$, and the interpolation point, X. The interpolated result is returned in Y. An error flag, N, is also returned. If N is zero, the interpolation point was outside the allowable table range.

Listing 5.5.2 shows an example of the operation of the Akima interpolation subroutine for the test function, $\sin x$. The calculated results are approximately one order of magnitude more accurate than those observed for the corresponding Newton divided-differences example. The results are also approximately one order of magnitude less accurate than those seen with the Lagrange interpolation example. It appears that the Akima subroutine suffers from round-off error, but the effect is moderate.

The key advantage to using AKIMA is its net curvature-minimization property. This property is particularly apparent when dealing with discontinuous functions. For example, if the table represents a step function, then the results obtained using Newton and Akima interpolation are quite different (see figures 5.5.1 and 5.5.2). The Akima spline interpolation in this example is *much* better behaved than the corresponding Newton calculation. It is this particular feature of spline interpolation that has made the technique so popular.

The precautions regarding the use of AKIMA are few. First, the interpolation requires three table values beyond the interpolation point; the range of the table must be wider than that of the interpolation. If this condition is violated, $N = 0$ will be returned. The second precaution is to be aware of round-off error. If round-off error is suspected to be a problem, then you should consider using Lagrange interpolation.

In the next section, we will consider the calculation of derivatives from tables using the Lagrange interpolation formula.

Statements/Functions List

$+, -, *, /, <, >$
ABS, FOR/NEXT, GOTO, IF/THEN

Variables List

A, B, I, M(I), N, V, X, X(I), Y, Y(I), Z(I)

Variables Passed to Subroutine

X, X(I), Y(I), V

Table 5.5.1 *Functions and variables used by the Akima spline interpolation subroutine.*

Listing 5.5.1

```
LIST
3725 REM PROGRAM TO DEMONSTRATE THE AKIMA SPLINE FITTING SUBROUTINE
3726 PRINT
3727 PRINT
3728 REM V=NUMBER OF TABLE VALUES
3729 V=14
3730 DIM X(V+1),Y(V+1),M(V+4),Z(V+1)
3731 REM INPUT TABLE
3732 FOR I=1 TO V
3733 READ X(I),Y(I)
3734 NEXT I
3735 REM SIN TABLE VALUES FROM
3736 REM HANDBOOK OF MATHEMATICAL FUNCTIONS
3737 REM BY ABRAMOWITZ, M., AND STEGUN, I.A.
3738 REM NBS, JUNE 1964
3739 DATA 0,0,.125,.12467473,.217,.21530095
3740 DATA .299,.29456472,.376,.36720285
3741 DATA .450,.43496553,.520,.49688014
3742 DATA .589,.55552980,.656,.60995199
3743 DATA .721,.66013615,.7853981634,0.7071067812
3744 DATA .849,.75072005
3745 DATA .911,.79011709,.972,.82601466
```

```
3746 PRINT
3747 PRINT " X ",TAB(10),"SIN(X) HANDBOOK",TAB(30),"AKIMA INTERPOLATION",
3748 PRINT TAB(54),"OBSERVED ERROR"
3749 PRINT "---",TAB(10),"-----------------",TAB(30),"--------------------",
3750 PRINT TAB(54),"----------------"
3751 PRINT
3752 FOR X=0 TO .75 STEP .05
3753 GOSUB 46100
3754 PRINT X,TAB(10),INT(10000000*SIN(X))/10000000,TAB(32),INT(10000000*Y)/10000000,
3755 PRINT TAB(56),INT(1000000*(SIN(X)-Y))/1000000
3756 NEXT X
3757 PRINT
3758 PRINT
3759 END
46090 REM ********************
46091 REM AKIMA SPLINE FITTING SUBROUTINE (AKIMA)
46092 REM THE INPUT TABLE IS (X(I),Y(I)), WHERE Y(I)
46093 REM IS THE DEPENDENT VARIABLE.
46094 REM THE INTERPOLATION POINT IS X, WHICH IS ASSUMED
46095 REM TO BE IN THE RANGE OF THE TABLE WITH AT LEAST
46096 REM ONE TABLE POINT TO THE LEFT, AND THREE TO THE RIGHT.
46097 REM Y IS RETURNED AS THE INTERPOLATED VALUE.
46098 REM N IS RETURNED AS AN ERROR CHECK (N=0 IMPLIES ERROR).
46099 REM DIMENSION M,X,Y AND Z IN THE CALLING PROGRAM
46100 N=1
46101 REM CHECK TO SEE IF X IS IN THE TABLE RANGE
46102 IF X>=X(1) THEN GOTO 46105
46103 N=0
46104 RETURN
46105 IF X<=X(V-3) THEN GOTO 46108
46106 N=0
46107 RETURN
46108 X(0)=2*X(1)-X(2)
46109 REM CALCULATE AKIMA COEFFICIENTS
46110 FOR I=1 TO V-1
46111 REM SHIFT I TO I+2
46112 M(I+2)=(Y(I+1)-Y(I))/(X(I+1)-X(I))
46113 NEXT I
46114 M(V+2)=2*M(V+1)-M(V)
46115 M(V+3)=2*M(V+2)-M(V+1)
46116 M(2)=2*M(3)-M(4)
46117 M(1)=2*M(2)-M(3)
46118 FOR I=1 TO V
46119 A=ABS(M(I+3)-M(I+2))
46120 B=ABS(M(I+1)-M(I))
46121 IF A+B<>0 THEN GOTO 46124
46122 Z(I)=(M(I+2)+M(I+1))/2
46123 GOTO 46125
46124 Z(I)=(A*M(I+1)+B*M(I+2))/(A+B)
```

Listing 5.5.1 cont.

```
46125 NEXT I
46126 REM FIND RELEVANT TABLE INTERVAL
46127 I=0
46128 I=I+1
46129 IF X>=X(I) THEN GOTO 46128
46130 I=I-1
46131 REM BEGIN INTERPOLATION
46132 B=X(I+1)-X(I)
46133 A=X-X(I)
46134 Y=Y(I)+Z(I)*A+(3*M(I+2)-2*Z(I)-Z(I+1))*A*A/B
46135 Y=Y+(Z(I)+Z(I+1)-2*M(I+2))*A*A*A/(B*B)
46136 RETURN
```

Listing 5.5.1 *The Akima spline interpolation subroutine (AKIMA). Also shown is a program for demonstrating AKIMA. See listing 5.5.2 for sample results. The subroutine-connection diagram appears in figure 5.2.2.*

RUN

X	SIN(X) HANDBOOK	AKIMA INTERPOLATION	OBSERVED ERROR
0	0	0	0
.05	.0499791	.0500401	-.000061
.1	.0998334	.0998435	-.000011
.15	.1494381	.1494309	.000007
.2	.1986693	.1986458	.000023
.25	.2474039	.2474157	-.000012
.3	.2955202	.2955218	-.000002
.35	.3428978	.3428915	.000006
.4	.3894183	.3894264	-.000009
.45	.4349655	.4349655	0
.5	.4794255	.4794204	.000005
.55	.5226872	.5226893	-.000003
.6	.5646424	.5646492	-.000007
.65	.6051864	.6051821	.000004
.7	.6442176	.6442177	-.000001
.75	.6816387	.6816266	.000012

READY

Listing 5.5.2 *Sample results for the program given in listing 5.5.1. The Akima subroutine is not as susceptible to round-off error as the Newton divided-differences interpolation, but some error is apparent.*

5.6 Calculating Derivatives from Tables

In this section, we will consider briefly how the derivative of a function represented by a table can be estimated from that same table. The technique discussed is based on the Lagrange formulation given in section 5.2, although the concept is more akin to the Newton divided-differences analysis presented in section 5.3. The Lagrange approach is preferable to the Newton technique because of its superior round-off error properties.

The Lagrange formula given in section 5.2 [equation (5.2.1)] is a polynomial in x. For a given order of fit, N, we can write the estimate of the derivative of $f(x)$ as

$$\frac{df}{dx} \cong \frac{dL_N(x)}{dx} \qquad (5.6.1)$$

The derivative indicated on the right-hand side can be determined algebraically and programmed relatively simply. As an exercise, perform the required algebraic manipulations.

Before examining the subroutine that encodes these results, it is important to understand the geometrical significance of the derivative estimation process. Consider figures 5.6.1 and 5.6.2. In the first figure, the derivative is estimated simply by calculating the slope of the line between the two surrounding table values. This corresponds to $N = 1$ in the Lagrange formula [equation (5.2.1)] because two table values are used. The next higher-order interpolation for the derivative uses three table values ($N = 2$). In essence, two slopes (M_1 and M_2) are employed to estimate the slope at x (see figure 5.6.2). The algebraic form of this estimate is evident from the Newton divided-differences structure discussed in section 5.3. In the Lagrange formulation, this *mathematically equivalent* end result is achieved by fitting a parabola to the three table values, and then by finding the slope at x.

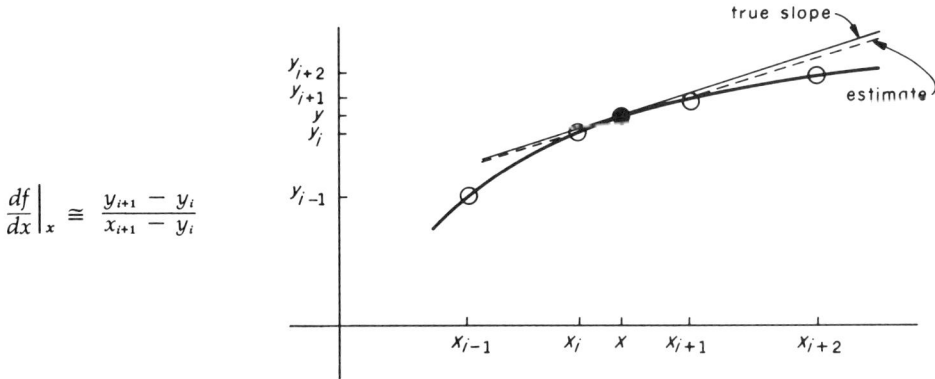

Figure 5.6.1 *Linear estimate ($N = 1$) of the derivative at x. Two table values are used—one to the left of x and one to the right.*

312 BASIC SCIENTIFIC SUBROUTINES

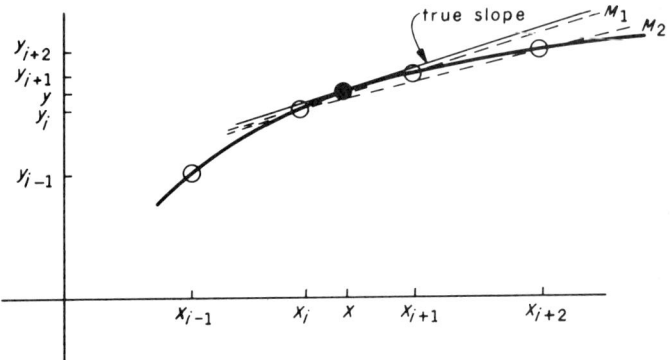

Figure 5.6.2 *Parabolic estimate (N = 2) of the derivative at x. Three table values are used—one to the left of x, and two to the right.*

With this introduction, we can generalize as follows. The Lagrange derivative algorithm for order N employs $N + 1$ table values (one to the left of x, and N to the right) to derive an Nth-degree polynomial fit. The slope of this polynomial, $L_N(x)$, at x is the returned estimate of the derivative at x. The subroutine given in listing 5.6.1 performs this calculation.

At this point you may be wondering specifically why the Lagrange formula is being used to find the derivative rather than the Newton divided-differences equations, which seem more to the point. There are two reasons. First, as we saw earlier, the Lagrange subroutine is much less susceptible to round-off error than the corresponding divided-differences program. Derivatives obtained from tables are *very* prone to round-off error, and an algorithm having round-off error problems with simple interpolation is likely to have even more difficulty when altered to interpolate derivatives. Second, the Lagrange formulation is very amenable to programming for general Nth-order interpolation. The result is a general, fairly accurate, but relatively slow subroutine: D/LAGRNG.

An example of the use of D/LAGRNG is shown in listing 5.6.2 for the test function sin x. The errors in the calculated derivatives are fairly small and can be compared with the generally larger errors observed with NEWTON for the simple interpolation of sin x.

The inputs to D/LAGRNG are identical to those for LAGRANGE, and the precautions are the same with regards to the inputs, with one important exception. LAGRANGE required the interpolation point, X, to be *greater than or equal to* the first table position, $X(1)$. D/LAGRNG always requires the interpolation point to be *greater than* $X(1)$; there *must* be a table value to the left of the point of interest. For this reason, an additional value was added to the front of the sin x table appearing in listing 5.6.1.

D/LAGRNG has an important weakness intrinsic to the mathematics of Taylor series based interpolation. A clue to this weakness appears in figure 5.5.1 (in the preceding section). Although the input table represented a step function, the cubic interpolation appeared quite different in the region of the step. Because this interpolation is used as the basis for the derivative evaluation, very strange results occur for the calculated derivative near the transition region. As an exercise, apply D/LAGRNG to this step-function table and compare your results with figure 5.5.1.

This weakness is the basis for the following warning: unless you are sure that the Nth-order LAGRANGE interpolation is accurate, do not use D/LAGRNG with order $N > 1$. Instead, use $N = 1$. The linear estimate of the derivative is often more realistic than that obtained using higher-order interpolation.

In the next section, we will consider the antithesis of table differentiation—integration.

Statements/Functions List

$+, -, *, /, <, >$
FOR/NEXT, GOTO, IF/THEN

Variables List

I, J, K, L, L(J), K, M(J,K), N, X, X(I), Y, Y(I), V

Variables Passed to Subroutine

N, X, X(I), Y(I), V

Table 5.6.1 *Functions and variables used in the Lagrange derivative interpolation subroutine (D/LAGRNG).*

```
3775 REM DEMONSTRATION PROGRAM FOR LAGRANGE DERIVATIVE INTERPOLATION OF SIN(X)
3776 PRINT
3777 PRINT
3778 V=15
3779 N=3
3780 DIM L(9),M(9,9),X(V),Y(V)
3781 REM INPUT TABLE
3782 FOR I=1 TO V
3783 READ X(I),Y(I)
3784 NEXT I
3785 REM SINE TABLE VALUES FROM
3786 REM HANDBOOK OF MATHEMATICAL FUNCTIONS
3787 REM BY ABRAMOWITZ, M., AND STEGUN, I.A.
3788 REM NBS, JUNE 1964
3789 DATA -.125,-.12467473
```

Listing 5.6.1

```
3790 DATA 0,0,.125,.12467473                                    Listing 5.6.1 cont.
3791 DATA .217,.21530095,.299,.29456472
3792 DATA .376,.36720285,.450,.43496553
3793 DATA .520,.49688014,.589,.55552980
3794 DATA .656,.60995199,.721,.66013615
3795 DATA .7853981634,0.7071067812
3796 DATA .849,.75062005,.911,.79011709
3797 DATA .972,.82601466
3798 PRINT
3799 PRINT " X ",TAB(10),"COS(X) HANDBOOK",TAB(30),"LAGRANGE DERIVATIVE
3800 PRINT TAB(54),"OBSERVED ERROR"
3801 PRINT "---",TAB(10),"----------------",TAB(30),"------------------
3802 PRINT TAB(54),"----------------"
3803 PRINT
3804 FOR X=0 TO .75 STEP .05
3805 GOSUB 46150
3806 PRINT X,TAB(10),COS(X),TAB(32),Y,TAB(56),INT(100000000*(COS(X)-Y))/10000000
3807 NEXT X
3808 END
46138 REM ********************
46139 REM LAGRANGE DERIVATIVE INTERPOLATION SUBROUTINE (D/LAGRNG)
46140 REM N IS THE LEVEL OF THE INTERPOLATION (EG., N=2 IS QUADRATIC).
46141 REM V IS THE TOTAL NUMBER OF TABLE VALUES.
46142 REM (X(I),Y(I)) ARE THE COORDINATE TABLE VALUES, Y(I) BEING THE
46143 REM DEPENDENT VARIABLE. THE X(I) MAY BE ARBITRARILY SPACED.
46144 REM X IS THE INTERPOLATION POINT WHICH IS ASSUMED TO BE IN THE
46145 REM INTERVAL WITH AT LEAST ONE TABLE VALUE TO THE LEFT, AND N TO THE RIGHT.
46146 REM Y IS RETURNED AS THE DESIRED DERIVATIVE.
46147 REM N IS RETURNED AS THE ERROR CHECK (N=0 IMPLIES ERROR).
46148 REM DIMENSION L(I),M(I,J),X(I) AND Y(I) IN THE CALLING PROGRAM
46149 REM CHECK TO SEE IF X IS IN INTERVAL
46150 IF X>X(1) THEN GOTO 46153
46151 N=0
46152 RETURN
46153 IF X<=X(V-N) THEN GOTO 46157
46154 N=0
46155 RETURN
46156 REM FIND THE RELEVANT TABLE INTERVAL
46157 I=0
46158 I=I+1
46159 IF X>X(I) THEN GOTO 46158
46160 I=I-1
46161 REM BEGIN INTERPOLATION
46162 FOR J=0 TO N
46163 L(J)=0
46164 FOR K=0 TO N
46165 M(J,K)=1
46166 NEXT K
46167 NEXT J
```

```
46168 Y=0
46169 FOR K=0 TO N
46170 FOR J=0 TO N
46171 IF J=K THEN GOTO 46179
46172 FOR L=0 TO N
46173 IF L=K THEN GOTO 46178
46174 IF L<>J THEN GOTO 46177
46175 M(L,K)=M(L,K)/(X(I+K)-X(I+J))
46176 GOTO 46178
46177 M(L,K)=M(L,K)*(X-X(J+I))/(X(I+K)-X(I+J))
46178 NEXT L
46179 NEXT J
46180 FOR L=0 TO N
46181 IF L=K THEN GOTO 46183
46182 L(K)=L(K)+M(L,K)
46183 NEXT L
46184 Y=Y+L(K)*Y(I+K)
46185 NEXT K
46186 RETURN
```

Listing 5.6.1 *Table differentiation subroutine (D/LAGRNG). Also shown is a program for demonstrating D/LAGRNG for the case d sin x/dx = cos x (see listing 5.6.2). The subroutine-connection diagram appears in figure 5.2.2.*

Listing 5.6.2

RUN

X	COS(X) HANDBOOK	LAGRANGE DERIVATIVE	OBSERVED ERROR
0	1	.9999938	.0000062
.05	.9987502	.9987485	.0000017
.1	.9950042	.9949889	.0000153
.15	.9887711	.98877895	-.00000785
.2	.9800666	.9800528	.0000138
.25	.9689124	.9689116	.0000008
.3	.9553365	.9553746	-.0000381
.35	.9393727	.9393607	.000012
.4	.921061	.92106544	-.00000444
.45	.9004471	.9004338	.0000133
.5	.8775826	.8775677	.0000149
.55	.8525245	.85252037	.00000413
.6	.82533562	.8253555	-.00001988
.65	.79608381	.7960669	.00001691

.7	.7648422	.7648268	.0000154
.75	.73168888	.73168413	.00000475

READY

Listing 5.6.2 *Sample run of the program shown in listing 5.6.1. D/LAGRNG is used to find interpolated values for d sin x/dx given a sin x table. The true value is cos x. The interpolation results and true values are in fairly good agreement.*

5.7 Table Integration

The task of accurately estimating the integral of a function represented by a set of table values presents a challenge not easily addressed by the standard numerical integration algorithms. First of all, the integration limits (endpoints) may not correspond to table entries. We can interpolate to circumvent that problem. Second, the highly effective integration algorithms, which require specific evaluation points to achieve high-order polynomial fits (e.g., Gauss-Legendre quadrature), may not be usable because the available table values most likely do not match the required grid points. Even the powerful Newton-Cotes formulae, which call for equal-interval table spacings, may not be applicable because the available table may not comply with that requirement. (See the several texts on this subject, including references 10, 16, 20 and 38.) For example, the 14-value sin x table employed throughout this chapter has a table spacing chosen to minimize the cubic interpolation error. The resulting table positions are neither equally spaced nor matched to any common quadrature points. As we will see shortly, however, the integration of short tables is limited for the most part by the interpolation needed to determine the integration endpoints. Therefore, the table spacing should be dictated by the interpolation algorithm, not by the integration method.

The determination of the integration endpoints is an important step in the procedure. Of the three interpolation methods already discussed (Lagrange, Newton, and Akima), the author prefers the Akima spline-fitting technique. The reasoning is as follows. As we saw, the Newton subroutine suffers severely from round-off error problems, and has only speed (relative to AKIMA) in its favor. The interpolation time factor is less important in the integration routine because the interpolation is only part of the exercise. Of the remaining two choices, Lagrange and Akima interpolation, the former is roughly one order of magnitude better than the spline program in terms of round-off error, but the spline fit is *safer*. For reasons that will become apparent later, if the Lagrange scheme is chosen, then the cubic fit would be used. The table to be interpolated is likely to be much more complicated than a cubic. Otherwise, a short and simple Taylor series would be used instead of the table. As we saw in the section on spline interpolation, a Lagrange cubic interpolation of a complicated function, particularly one with discontinuities, can contain undesirable *ringing*. The spline fit is much better in this respect, and is the preferred interpolation method for use with the integration program to be presented.

The second part of the procedure is to numerically integrate over a set of table values, including the two interpolated endpoints. The algorithm employed must be independent of the spacing of the table values, and must have reasonable accuracy. A candidate for this algorithm is a hybrid form of trapezoidal integration.

Trapezoidal integration is accomplished by finding the average value of the function in an incremental segment using the two boundary values, then multiplying this average by the interval length to obtain an incremental area, and finally summing all such areas. This gives an approximation to area under the functional curve, or the desired integral. See figure 5.7.1. In this example, the approximate integral is

$$\int_{x_1}^{x_4} f(x)dx \cong \frac{f_1 + f_2}{2}(x_2 - x_1) + \frac{f_2 + f_3}{2}(x_3 - x_2) + \frac{f_3 + f_4}{2}(x_4 - x_3) \tag{5.7.1}$$

If the interval sizes were all equal, then the error in this integration form would be proportional to H^2, where H is the width of each interval. Therefore, if $I1$ is the approximation obtained using M equal intervals, and $I2$ is the value obtained using $2M$ intervals, a better approximation might be

$$\int f(x)dx \approx \frac{4}{3} I2 - \frac{1}{3} I1 \tag{5.7.2}$$

The simple proof of this correction formula is left to you. (It can be found in Reference 10 under "Richardson's extrapolation.")

This formula can significantly improve the accuracy of the integration, and under the right conditions, can give results equivalent to procedures that will integrate a cubic exactly. Because the spline interpolation algorithm is also cubic in accuracy, the two schemes are well matched.

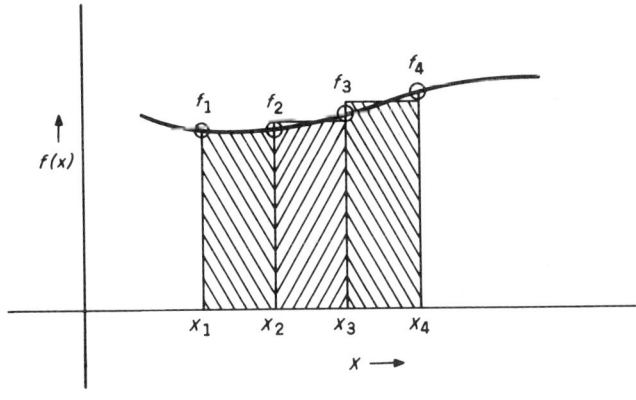

Figure 5.7.1 *Graphic representation of trapezoidal integration.*

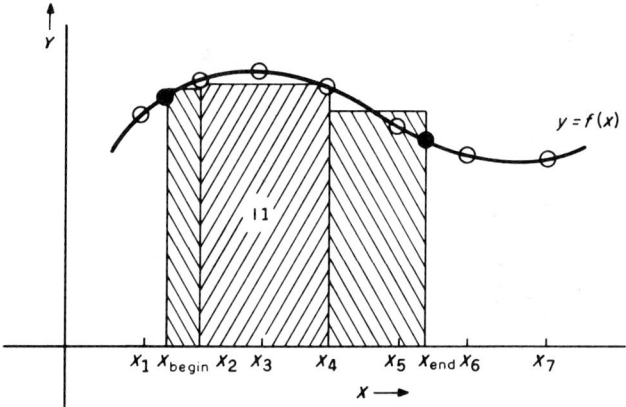

Figure 5.7.2 *First trapezoidal integration. The table values used are indicated by small circles.*

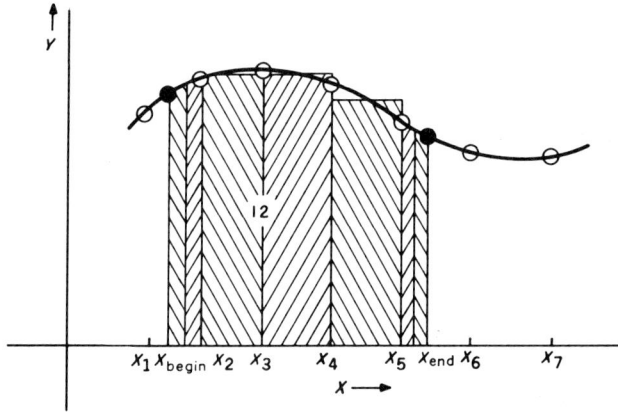

Figure 5.7.3 *Second trapezoidal integration. The table values used are indicated by small circles.*

Although this form of Richardson's extrapolation is theoretically valid only for equally spaced tables, it can be used in approximation for unequal interval tables as shown in figures 5.7.2 and 5.7.3. In the first figure, four points (three intervals) are used, the two endpoints and every other table value between these endpoints. In the second figure, every table value between the integration limits is used, along with two new divisions, giving roughly twice as many intervals (seven). Seven is not twice three, but this leads to little error in the correction equation; the effect is small.

The reason is that for integrals spanning only short lengths of the table, the interpolation error dominates. For spans wide enough to have the integration error become important, the number of intervals is large enough for $M/(2M + 1) \cong 1/2$. Later examples will confirm this assertion.

A subroutine that applies the above concepts is shown in listing 5.7.1 (ITEG). The flow of the program is apparent from both the comments contained within the listing, and the modular structure.

The inputs to ITEG are the table values, $X(I)$ and $Y(I)$; the number of values, V; and the integration limits, $X1$ and $X2$. The integral is returned in Z. Also returned is an error check, $Z1$, and the intermediate results, $I1$ and $I2$.

Observe that at the very end of the integration program, you are given the opportunity to branch to some other interpolation subroutine by changing the GOSUB target line. Keep in mind, however, that the replacement subroutine may require additional input. For example, if LAGRANGE is to be used, set $N = 3$ (for cubic interpolation) before branching to that subroutine.

Examples of the operation of ITEG appear in listing 5.7.2. The fifth example is particularly interesting. The calculated result of 5.018×10^{-9} is in error by roughly 2×10^{-11} in that the true value is $1 - \cos 0.0001 \cong 5.000 \times 10^{-9}$. The relative error is roughly 0.004, and is very likely a reflection of round-off error in the calculation. However, if $1 - \cos 0.0001$ is evaluated directly (in 8-digit North Star BASIC), the result is zero. The interpolation/integration subroutine gives a much more accurate result than a direct computer evaluation! (Round-off error is probably also occurring in the cosine subroutine contained within BASIC; the effect is worse.)

The results of repeated runs of ITEG for the $\int \sin x \, dx$ example are displayed in table 5.7.2. In general, the accuracy of these calculations is better than 10^{-5}, which is not bad considering how few table values are employed.

The two intermediate trapezoidal integration results, $I1$ and $I2$, are also available for inspection upon return from ITEG. These can be used to derive a crude measure of the accuracy of ITEG in the case of low round-off error. Assume $I2$ to be exact. The error in $I1$ is then $E = I2 - I1$. Richardson's extrapolation formula becomes

$$\int f(x) \approx \frac{4}{3} I2 - \frac{1}{3} (I2 - E) = I2 + \frac{E}{3}$$

Therefore, the error estimate is

$$\text{Integration error bound} \sim \frac{I2 - I1}{3} \qquad (5.7.3)$$

In practice, this error-bound estimate is usually somewhat larger than what is actually observed, excluding round-off error effects.

We conclude this section with a brief discussion of the precautions that should be observed when calling ITEG. First, it is very easy to forget the allowable range of

320 BASIC SCIENTIFIC SUBROUTINES

integration for a given table. Therefore, the error flag, Z1, should always be examined after calling ITEG. Second, do not expect great relative accuracy for very short integration spans; round-off and interpolation error can be significant in this range. Third, when integrating functions that contain discontinuities, use the spline interpolation algorithm (AKIMA), not the Lagrange or Newton methods.

In the next section, we will apply the techniques presented in this chapter to the interpolation and integration of the one-sided normal probability distribution function. This function is of particular significance in statistics, and tabulations still appear in textbooks. The integral of this probability distribution, the *error function*, is difficult to calculate using simple numerical methods, thereby explaining the continued popularity of tables for this function.

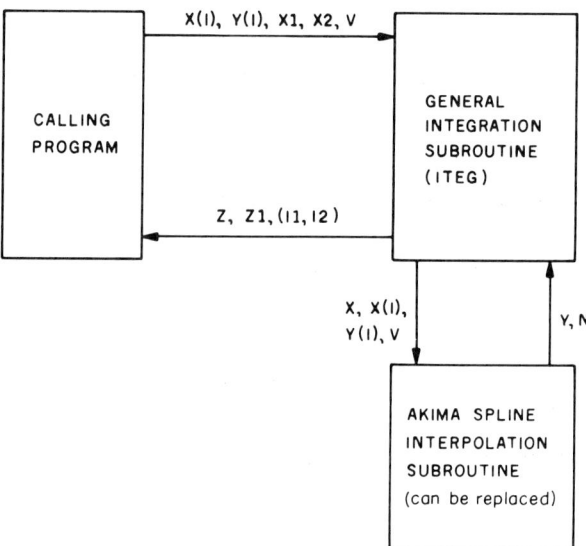

X(I), Y(I):   Table values
V: Number of table values
X1, X2:   Integration range
Z: Calculated integral
Z1:   Error check. Z1 = 0 implies an integration point is outside of the allowable table range
I1, I2:   Results of the two trapezoidal integrations

Figure 5.7.4 *Subroutine-connection diagram for the general integration program (ITEG). This routine also calls an interpolation subroutine, preferably AKIMA. However, this call can easily be changed.*

Statements/Functions List

+, −, *, /, <, >
GOSUB, GOTO, IF/THEN

Variables List

D, I, I1, I2, J1, N1, X, X1, X2, X3, X(I), Y, Y(I), V, Z, Z1

Variables Passed to Subroutine

X(I), X1, X2, Y(I), V

Table 5.7.1 *Functions and variables used by the general integration subroutine (ITEG). This list does not include the functions and variables used by the interpolation subroutine (e.g., AKIMA) which is called by ITEG.*

Listing 5.7.1

```
3825 REM PROGRAM TO DEMONSTRATE THE GENERAL INTEGRATION SUBROUTINE
3826 REM EXAMPLE IS THE INTERGRAL OF SIN(X) FROM X1 TO X2
3827 REM V=NUMBER OF TABLE VALUES
3828 V=14
3829 DIM X(V),Y(V),Z(V),M(V+3)
3830 REM INPUT TABLE
3831 FOR I=1 TO V
3832 READ X(I),Y(I)
3833 NEXT I
3834 REM SIN TABLE VALUES FROM
3835 REM HANDBOOK OF MATHEMATICAL FUNCTIONS
3836 REM BY ABRAMOWITZ, M., AND STEGUN, I.A.
3837 REM NBS, JUNE 1964
3838 DATA 0,0,.125,.12467473,.217,.21530095
3839 DATA .299,.29456472,.376,.36720285
3840 DATA .450,.43496553,.520,.49688014
3841 DATA .589,.55552980,.656,.60995199
3842 DATA .721,.66013615,.7853981634,0.7071067812
3043 DATA .849,.75072005
3844 DATA .911,.79011709,.972,.82601466
3845 PRINT "INPUT START POINT, END POINT: ",
3846 INPUT X1,X2
3847 GOSUB 46200
3848 PRINT "INTEGRAL FROM ",X1," TO ",X2," EQUALS ",Z
3849 PRINT "ERROR CHECK: ",Z1
3850 PRINT
3851 PRINT
3852 GOTO 3845
46090 REM ********************
46091 REM AKIMA SPLINE FITTING SUBROUTINE (AKIMA)
```

```
46092 REM THE INPUT TABLE IS (X(I),Y(I)), WHERE Y(I)
46093 REM IS THE DEPENDENT VARIABLE.
46094 REM THE INTERPOLATION POINT IS X, WHICH IS ASSUMED
46095 REM TO BE IN THE RANGE OF THE TABLE WITH AT LEAST
46096 REM ONE TABLE POINT TO THE LEFT, AND THREE TO THE RIGHT.
46097 REM Y IS RETURNED AS THE INTERPOLATED VALUE.
46098 REM N IS RETURNED AS AN ERROR CHECK (N=0 IMPLIES ERROR).
46099 REM DIMENSION M,X,Y AND Z IN THE CALLING PROGRAM
46100 N=1
46101 REM CHECK TO SEE IF X IS IN THE TABLE RANGE
46102 IF X>=X(1) THEN GOTO 46105
46103 N=0
46104 RETURN
46105 IF X<=X(V-3) THEN GOTO 46108
46106 N=0
46107 RETURN
46108 X(0)=2*X(1)-X(2)
46109 REM CALCULATE AKIMA COEFFICIENTS
46110 FOR I=1 TO V-1
46111 REM SHIFT I TO I+2
46112 M(I+2)=(Y(I+1)-Y(I))/(X(I+1)-X(I))
46113 NEXT I
46114 M(V+2)=2*M(V+1)-M(V)
46115 M(V+3)=2*M(V+2)-M(V+1)
46116 M(2)=2*M(3)-M(4)
46117 M(1)=2*M(2)-M(3)
46118 FOR I=1 TO V
46119 A=ABS(M(I+3)-M(I+2))
46120 B=ABS(M(I+1)-M(I))
46121 IF A+B<>0 THEN GOTO 46124
46122 Z(I)=(M(I+2)+M(I+1))/2
46123 GOTO 46125
46124 Z(I)=(A*M(I+1)+B*M(I+2))/(A+B)
46125 NEXT I
46126 REM FIND RELEVANT TABLE INTERVAL
46127 I=0
46128 I=I+1
46129 IF X>=X(I) THEN GOTO 46128
46130 I=I-1
46131 REM BEGIN INTERPOLATION
46132 B=X(I+1)-X(I)
46133 A=X-X(I)
46134 Y=Y(I)+Z(I)*A+(3*M(I+2)-2*Z(I)-Z(I+1))*A*A/B
46135 Y=Y+(Z(I)+Z(I+1)-2*M(I+2))*A*A*A/(B*B)
46136 RETURN
46189 REM ********************
46190 REM GENERAL INTEGRATION SUBROUTINE (ITEG)
46191 REM INTERPOLATION BY AKIMA (OR OTHER).
46192 REM INTEGRATION BY ENHANCED TRAPAZOIDAL RULE.
```

Listing 5.7.1 cont.

```
46193 REM REM WITH RICHARDSON EXTRAPOLATION TO GIVE CUBIC ACCURACY.
46194 REM CAN BE USED UNDER VERY GENERAL CONDITIONS.
46195 REM THE INTEGRATION RANGE IS (X1,X2).
46196 REM IT IS ASSUMED THAT X1<X2, AND THAT THERE IS AT LEAST ONE TABLE
46197 REM VALUE TO THE LEFT OF X1, AND THREE TO THE RIGHT OF X2.
46198 REM THE RESULT IS RETURNED IN Z
46199 REM AN ERROR CHECK IS RETURNED IN Z1- Z1=0 IMPLIES ERROR.
46200 Z=0
46201 Z1=0
46202 REM CHECK TO SEE IF END POINTS ARE IN ALLOWABLE RANGE
46203 IF X1<X(1) THEN RETURN
46204 IF X2>X(V-3) THEN RETURN
46205 REM IF X1>X2 THEN SWITCH AND SET FLAG
46206 IF X1<X2 THEN GOTO 46211
46207 X3=X1
46208 X1=X2
46209 X2=X3
46210 Z1=1
46211 IF X2=X1 THEN RETURN
46212 REM START TRAPAZOIDAL INTEGRATIONS
46213 REM FIRST INTEGRATION TO GET I1
46214 GOSUB 46232
46215 REM SECOND ROUND TO GET I2
46216 GOSUB 46267
46217 REM RICHARDSON EXTRAPOLATION
46218 Z=4*I2/3-I1/3
46219 REM CHECK TO SEE IF THE END POINTS HAVE BEEN REVERSED
46220 IF Z1=0 THEN GOTO 46225
46221 Z=-Z
46222 X2=X1
46223 X1=X3
46224 REM RESET ERROR FLAG
46225 Z1=1
46226 RETURN
46227 REM Z IS THE INTEGRAL DESIRED
46228 RETURN
46229 REM ********************
46230 REM ROUTINE FOR THE FIRST TRAPAZOIDAL INTEGRATION, I1
46231 REM N1 KEEPS TRACK OF THE NUMBER OF INTERVALS
46232 I1=0
46233 N1=0
46234 X=X1
46235 REM FIND THE BEGINNING OF THE INTERVAL
46236 REM GO TO BRANCH WHICH CALLS THE INTERPOLATION SUBROUTINE
46237 REM FIND THE INTERVAL, I, AND THE LEFT END POINT, Y
46238 GOSUB 46299
46239 REM IS THERE AT LEAST ONE TABLE INTERVAL?
46240 IF X2>X(I+1) THEN GOTO 46250
46241 REM IF NOT, INTEGRAL IS SIMPLE
```

Listing 5.7.1 cont.

```
46242 N1=N1+1                                        Listing 5.7.1 cont.
46243 D=Y
46244 X=X2
46245 REM FIND END POINT Y VALUE
46246 GOSUB 46299
46247 I1=(Y+D)*(X2-X1)/2
46248 RETURN
46249 REM AT LEAST ONE TABLE INTERVAL MUST BE SUMMED OVER
46250 J1=I
46251 I1=I1+(Y+Y(I+1))*(X(I+1)-X)/2
46252 REM ANY MORE INTERVALS? IF NOT, FINISH INTEGRAL WITH END POIN
46253 IF X2<X(J1+3) THEN GOTO 46259
46254 REM OTHERWISE, KEEP SUMMING
46255 N1=N1+1
46256 I1=I1+(Y(J1+1)+Y(J1+3))*(X(J1+3)-X(J1+1))/2
46257 J1=J1+2
46258 GOTO 46253
46259 X=X2
46260 REM FIND LAST Y VALUE
46261 GOSUB 46299
46262 I1=I1+(Y+Y(J1+1))*(X2-X(J1+1))/2
46263 N1=N1+1
46264 RETURN
46265 REM ********************
46266 INTEGRATION FOR I2
46267 I2=0
46268 X=X1
46269 GOSUB 46299
46270 D=Y
46271 IF X2>X(I+1) THEN GOTO 46280
46272 X=X1+(X2-X1)/2
46273 GOSUB 46299
46274 I2=I2+(D+Y)*(X2-X1)/4
46275 D=Y
46276 X=X2
46277 GOSUB 46299
46278 I2=I2+(D+Y)*(X2-X1)/4
46279 RETURN
46280 X=X1+(X(I+1)-X1)/2
46281 J1=I
46282 GOSUB 46299
46283 I2=I2+(Y+D)*(X-X1)/2
46284 I2=I2+(Y+Y(J1+1))*(X(J1+1)-X)/2
46285 IF X2<X(J1+2) THEN GOTO 46289
46286 I2=I2+(Y(J1+1)+Y(J1+2))*(X(J1+2)-X(J1+1))/2
46287 J1=J1+1
46288 GOTO 46285
46289 X=X2-(X2-X(J1+1))/2
46290 GOSUB 46299
```

```
46291 D=Y
46292 I2=I2+(Y(J1+1)+D)*(X2-X)/2
46293 X=X2
46294 GOSUB 46299
46295 I2=I2+(D+Y)*(X2-X(J1+1))/4
46296 RETURN
46297 REM BRANCH TO AN INTERPOLATION SUBROUTINE FROM HERE
46298 REM GO TO AKIMA SPLINE INTERPOLATION SUBROUTINE
46299 GOSUB 46100
46300 REM JUST RETURNED FROM INTERPOLATION SUBROUTINE
46301 REM RETURN TO PROGRAM
46302 RETURN
```

Listing 5.7.1 *General table integration subroutine (ITEG). Also shown is a program for demonstrating ITEG. Sample results appear in listing 5.7.2.*

Listing 5.7.2

```
RUN

INPUT START POINT, END POINT: ?0,.5
INTEGRAL FROM   0 TO   .5 EQUALS   .1224225
ERROR CHECK:  1


INPUT START POINT, END POINT: ?.5,0
INTEGRAL FROM  .5 TO   0 EQUALS  -.1224225
ERROR CHECK:  1


INPUT START POINT, END POINT: ?0,.8
INTEGRAL FROM   0 TO  .8 EQUALS  -.1224225
ERROR CHECK:  0


INPUT START POINT, END POINT: ?-.1,.5
INTEGRAL FROM  -.1 TO  .5 EQUALS  -.1224225
ERROR CHECK:  0


INPUT START POINT, END POINT: ?0,.0001
INTEGRAL FROM   0 TO  .0001 EQUALS   5.0177955E-09
ERROR CHECK:  1


INPUT START POINT, END POINT: ?.0001,.0002
INTEGRAL FROM  .0001 TO  .0002 EQUALS   1.5053307E-08
ERROR CHECK:  1
```

```
INPUT START POINT, END POINT: ?
```

Listing 5.7.2 *Sample runs of the general integration subroutine (ITEG) shown in listing 5.7.1. In the second example, the integration is from 0.5 to 0. Both of these values are within the allowable table range, although in reverse order. The calculated integral is simply the reverse of the first example. The third and fourth examples show what happens when one of the integration endpoints is outside the allowable range; the error check value is zero.*

x	$\int_0^x \sin x \, dx$	Approximation	Observed Error
0.0	0.00000000	0.00000000	0.00000000
0.01	0.00005000	0.00005016	0.00000016
0.025	0.00031248	0.00031331	0.00000083
0.05	0.00124974	0.00125211	0.00000237
0.10	0.00499583	0.00500007	0.00000424
0.15	0.01122892	0.01123318	0.00000426
0.20	0.01993342	0.01993663	0.00000321
0.25	0.03108758	0.03108732	−0.00000026
0.30	0.04466351	0.04466823	0.00000472
0.35	0.06062729	0.06063234	0.00000505
0.40	0.07893901	0.07893666	−0.00000235
0.45	0.09955290	0.09955772	0.00000482
0.50	0.12241744	0.12242250	0.00000506
0.55	0.14747548	0.14747559	0.00000011
0.60	0.17466439	0.17466928	0.00000489
0.65	0.20391620	0.20392111	0.00000491
0.70	0.23515781	0.23516473	0.00000692
0.75	0.26831113	0.26831591	0.00000478

Table 5.7.2 *Integral of sin x as calculated by the general integration program (listing 5.7.1). Observed errors are also shown. Note that $\int_0^x f(x) = 1 - \cos x$. The worst error observed was 7×10^{-6}. Recall that the error for the AKIMA spline interpolation of sin x was of the same order, thereby partially identifying that step as a major contributor to loss of accuracy.*

5.8 Interpolation and Integration of $(2/\sqrt{\pi}) \, e^{-x^2}$

One of the most elementary and important functions in statistics is the one-sided *normal probability distribution function* (pdf):

$$p(x) = \frac{2}{\sqrt{\pi}} e^{-x^2} \quad \text{for} \quad 0 \le x \qquad (5.8.1)$$

(See any introductory text on statistics. Also, see in Volume I of *BASIC Scientific Subroutines* the discussion that describes how normally distributed random numbers can be generated.) This function has the following meaning. The probability of the random variable, x, taking on a value in the incremental range x to $x + dx$ is $p(x)\, dx$. The *cumulative distribution function* (cdf) is defined as

$$P(0 \leq x < X) = \int_0^X p(x)\, dx \quad \text{for } 0 \leq X < \infty \tag{5.8.2}$$

The cumulative distribution function is the probability that x will take on a value in the range $0 \leq x < X$. In the case of the one-sided normal cdf, a special name—the *error function*—is applied:

$$\text{erf}(X) = \frac{2}{\sqrt{\pi}} \int_0^X e^{-x^2}\, dx \tag{5.8.3}$$

Tables of both the error function and its derivative (the normal pdf) can be found in almost every statistics textbook. However, the error function is not common enough to merit implementation in every calculator. Also, the error function is generally not part of any math package directly supplied with BASIC interpreters or compilers. It is therefore an interesting and useful example for table interpolation and integration.

The table interpolation/integration subroutines require table position and value pairs as input. The first step in preparation, therefore, is to determine the table positions. An equal-interval table can easily be generated by directly consulting a handbook. However, as we saw in section 5.4, the maximum interpolation error can be minimized by choosing an appropriate set of *unequal* intervals.

Recall from section 5.4.2 the interval size formula for cubic interpolation:

$$H_x \cong \left[\frac{4E}{d^4 f(x)/dx^4 \big|_x} \right]^{1/4}$$

It can be shown that the fourth derivative of $p(x)$ is

$$\frac{d^4 P(x)}{dx^4} = \frac{8}{\sqrt{\pi}} e^{-x^2} (3 - 6x^2 + 4x^4) \tag{5.8.4}$$

Therefore, the approximate spacing formula is

$$H \cong \text{constant} \left(\frac{e^{x^2}}{3 - 6x^2 + 4x^4} \right)^{1/4} \tag{5.8.5}$$

The multiplying constant can be chosen so that the size of the first interval is of some predetermined length.

Equation (5.8.5) can be applied recursively to find a sequence of table positions for $(2/\sqrt{\pi})\, e^{-x^2}$. A program for doing this is given in listing 5.8.1. In this program, the multiplying constant is chosen so that the initial interval size is 0.025, although only every fourth position is printed out. This is in keeping with the rule of thumb given in section 5.4, which called for at least a twofold finer grid than desired to maintain accuracy. The results of the computation are displayed in listing 5.8.2.

Observe that the calculation performed by the program in listing 5.8.1 is significantly different from that performed by the corresponding program in section 5.4. In that earlier program, the *number* of table positions in a predetermined range was chosen *a priori*, and the computer program distributed the intervals accordingly by iteration. In the present example, the first table interval is chosen beforehand, and the subsequent positions are calculated recursively; each depends on the already calculated values.

Table 5.8.1 shows the 24 table values chosen for the interpolation based on the results appearing in listing 5.8.2. By simply inserting these values as DATA into the demonstration program shown in listing 5.7.1, and by setting $V = 24$, $p(x)$ can be integrated to give erf(x). Also, the same can be done with any of the three interpolation programs to obtain intermediate values for $p(x)$.

As an exercise, make the simple changes in the programs to use the other interpolation subroutines and then examine your results. Table 5.8.2 was constructed by using the Akima spline-fitting subroutine for both the interpolation and the integration. Better results would be expected for Lagrange cubic interpolation, and poorer results for Newton cubic interpolation.

This particular example subtly demonstrates a difficulty associated with determining the table spacings for arbitrary functions. The first indication of a potential problem appears in the interval sizes displayed in listing 5.8.2. The interval size starts at 0.1, increases to 0.17, *decreases* to 0.13, and thereafter increases (without bound). These variations are simply a reflection of $[d^4f(x)/dx^4]^{-1/4}$. That presents a problem, however. What if the fourth derivative changes sign (i.e., becomes negative)? Unless the absolute value is used, an error will result. Therefore, unless you are sure that the sign of the fourth derivative is positive, use absolute values.

A related problem is the case in which $d^4f(x)/dx^4$ approaches zero (or worse, contains a root!). In that situation, the intervals may become very large. The example in this section represents a particularly sensitive condition in which the calculation becomes unstable after $x = 10$. The reason for this is that the table values themselves become much smaller than the prescribed error bound, and there is no appropriate table interval size. For example, if the function is less than E beyond some point, then there is no set of interpolation interval choices in that range that can lead to an intrinsic error greater than that. There is no solution for the interval size. The only way to circumvent this difficulty is to use a finer grid, and thereby reduce the implied error.

As a final note, remember that *three* table values beyond the interpolation point (or integration endpoint) are required. This means that, given the entries shown in table 5.8.1, the argument range is $0 \leq x$ (or $X2$) ≤ 2.32.

```
3875 REM PROGRAM TO CALCULATE THE ERROR FUNCTION TABLE POSITIONS
3876 PRINT
3877 PRINT
3878 REM STARTING POSITION IS X=0
3879 X=0
3880 D=0
3881 I=0
3882 REM INITIAL INTERVAL SIZE IS H0=.025
3883 H0=.025
3884 REM TERMINATION POSITION IS E=10
3885 E=20
3886 PRINT
3887 PRINT
3888 PRINT " POSITION NUMBER",TAB(21),"POSITION",TAB(35),"INTERVAL SIZE"
3889 PRINT " ----------------",TAB(21),"--------",TAB(35),"-------------"
3890 PRINT
3891 REM START ITERATION
3892 IF INT(I/4)<>I/4 THEN GOTO 3895
3893 PRINT TAB(6),I/4+1,TAB(22),INT(1000*X)/1000,TAB(36),INT(1000*D)/1000
3894 D=0
3895 H=EXP(-X*X/4)*SQRT(SQRT((3-6*X*X+4*X*X*X*X)/3))
3896 H=H0/H
3897 I=I+1
3898 X=X+H
3899 D=D+H
3900 IF X<E THEN GOTO 3892
3901 PRINT
3902 PRINT
3903 END
```

Listing 5.8.1 *Program for finding the interpolation table spacings and positions for the function* $f(x) = (2/\sqrt{\pi})\, e^{-x^2}$. *The scaling factor has been chosen so that the first spacing is* $H = 0.025$. *Every fourth position found is printed out until* $x > 10$.

Listing 5.8.2

RUN

POSITION NUMBER	POSITION	INTERVAL SIZE
1	0	0
2	.1	.1
3	.201	.101
4	.306	.104

5	.416	.109
6	.534	.118
7	.665	.131
8	.816	.151
9	.986	.17
10	1.15	.163
11	1.297	.147
12	1.434	.136
13	1.565	.131
14	1.695	.129
15	1.825	.13
16	1.959	.134
17	2.1	.14
18	2.25	.15
19	2.415	.164
20	2.602	.186
21	2.824	.221
22	3.108	.283
23	3.53	.421
24	4.556	1.026

READY

Listing 5.8.2 *Table positions for interpolation (and integration) of the function $f(x) = (2/\sqrt{\pi})\, e^{-x^2}$ as printed out by the program shown in listing 5.8.1. The "interval size" column gives the increment from the previous table position.*

Table 5.8.1

Position Number	Position x	Table Value $(2/\sqrt{\pi})\, e^{-x^2}$
1	0	1.1283791671
2	0.1	1.1171516068
3	0.2	1.0841347871
4	0.31	1.0249893657
5	0.42	0.9459006256
6	0.53	0.8520434444
7	0.67	0.7202781930
8	0.82	0.5760171973
9	0.99	0.4234508779
10	1.15	0.3006772759
11	1.30	0.2082079868
12	1.43	0.1460045107
13	1.57	0.0959317995
14	1.70	0.0627110405

15	1.83	0.0396320255
16	1.96	0.0242141583
17	2.10	0.013715650
18	2.25	0.0071423190
19	2.42	0.003386700
20	2.60	0.0013080500
21	2.82	0.00039698274
22	3.11	7.1107499×10^{-5}
23	3.53	4.3727530×10^{-6}
24	4.56	1.0517423×10^{-9}

Table 5.8.1 Table values for the derivative of the error function, $(2/\sqrt{\pi})\,e^{-x^2}$, for use in the interpolation and integration subroutines (see Handbook of Mathematical Functions, Ref. 6).

X	$(2/\sqrt{\pi})\,e^{-x^2}$ (exact)	$(2/\sqrt{\pi})\,e^{-x^2}$ interpolated	Error	erf(x) exact	erf(x) integrated	Error
0.0	1.12838	1.12838	0.00000	0.00000	0.00000	0.00000
0.1	1.11715	1.11715	0.00000	0.11246	0.11246	0.00000
0.2	1.08413	1.08413	0.00000	0.22270	0.22252	−0.00018
0.3	1.03126	1.03135	0.00009	0.32862	0.32878	0.00016
0.4	0.96154	0.96170	0.00016	0.42839	0.42839	0.00000
0.5	0.87878	0.87911	0.00033	0.52049	0.52051	0.00002
0.6	0.78724	0.78657	−0.00067	0.60385	0.60382	−0.00003
0.7	0.69127	0.69155	0.00028	0.67780	0.67772	−0.00008
0.8	0.59499	0.59514	0.00015	0.74210	0.74210	0.00000
0.9	0.50197	0.50224	0.00027	0.79691	0.79692	0.00001
1.0	0.41511	0.41503	−0.00008	0.84270	0.84296	0.00026
1.1	0.33648	0.33663	0.00015	0.88021	0.88013	−0.00008
1.2	0.26734	0.26726	−0.00008	0.91031	0.91032	0.00001
1.3	0.20821	0.20821	0.00000	0.93401	0.93430	0.00029
1.4	0.15894	0.15880	−0.00014	0.95229	0.95227	−0.00002
1.5	0.11893	0.11898	0.00005	0.96611	0.96613	0.00002
2.0	0.02067	0.02074	0.00007	0.99532	0.99535	0.00003

Table 5.8.2 Calculated values for $(2/\sqrt{\pi})\,e^{-x^2}$ and its integral, erf(x). The integration range was from $X1 = 0$ to $X2 = X$. Observe that the average magnitude of the interpolation error is 0.0002, and that of the integration is 0.0001; they are comparable. The implication is that the dominant source of error is the interpolation step. Note that the Akima spline-interpolation subroutine was used.

5.9 Summary and Conclusion

In the previous sections, we discussed three particular interpolation schemes: Lagrange, Newton, and Akima. The first two were based on the Taylor series expansion, and although they are mathematically equivalent, they evaluate quite differently. The Newton divided-differences subroutine was demonstrated to be subject to considerable round-off error. The Lagrange algorithm was seen to be much less sensitive to round-off error effects, and faster in execution.

The third interpolation method, the Akima spline fit, was introduced as an alternative to the other two. It responds to discontinuities in a much smoother manner than the others and it is not as subject to *ringing*. The Akima algorithm is mathematically equivalent to the Lagrange and Newton methods for linear, quadratic, and cubic functions. Its round-off error behavior is between that of the Lagrange and Newton programs.

For general applications, the Akima spline fit is recommended over the others, although there are many cases (such as the sin x interpolation example) in which the Lagrange method is preferable.

In connection with the interpolation algorithm, it was also shown that equal-interval tables are not optimum in terms of minimizing the maximum error for a given number of table values. The two examples of sin x and $(2/\sqrt{\pi})\,e^{-x^2}$, in which the optimum table spacings were approximately calculated, were considered.

The Lagrange interpolation formula was also used to develop a table differentiation routine (D/LAGRNG). We found that the resulting subroutine worked well, but that it was limited by round-off error. Round-off error generally plagues table differentiation, and calculations beyond the first derivative are not recommended.

Finally, the interpolation concept was combined with an enhanced form of trapezoidal integration to provide a *general* table integration subroutine. Two examples were discussed: $\int \sin x$ and $\int (2/\sqrt{\pi})\,e^{-x^2}$

The above techniques and their associated subroutines form a core library of programs for dealing with the interpolation, differentiation, and integration of tables. The utility of the techniques presented rests in their application to complicated functions that cannot be easily expressed in series expansions over extended regions, but that still must be evaluated accurately.

In the next chapter, we will examine algorithms for finding the roots of functions. In a sense, this is similar to *inverse interpolation* in that polynomial fits can be applied to estimate the position of $f(x) = 0$ according to the values of $f(x)$ at other locations. Inverse interpolation can be performed with tables simply by switching the $f(x_i)$ and x_i columns. However, the range of x_i values and the order must be such that $f(x_i)$ is *monotonically increasing* over the span of the table. Otherwise, the procedure is the same.

Chapter 6

Finding the Real Roots of Functions

6.1 Introduction

Ever since the development of the concept of mathematical equations, there has been both an academic and a practical interest in establishing methods for finding the zeros, maxima, and minima of functions. When the appropriate mathematical models exist, it is possible to determine the maximum range of a ballistic projectile, the maximum force on a bouncing ball, the minimum time between two events, the minimum velocity in a gas stream, or even the optimum inventory for a given supply-and-demand business environment. For the beginning student in numerical methods, the search for zeroes, maxima, and minima becomes an end in itself. In many cases, the key to finding maxima and minima is the ability to find the zeros of the derivative of the function of interest. In the following discussion, a group of simple methods for determining positions of the zeros of a function is presented. As you will see, the techniques do not involve complicated mathematical structures, but rather they apply some elementary thoughts regarding geometry, *iteration*, and *successive substitution*. We will start with a background discussion of polynomials, and we will then review some basic interval search techniques.

Table 6.1.1

SUBROUTINE	SUBROUTINE SIZE (bytes) (including support programs)	DEMON-STRATION PROGRAM SIZE (bytes)	EXECUTION SPEED	CONDITIONS/ COMMENTS
ROOTTEST (information regarding polynomial roots)	-----	1609	-----	Not a subroutine

SUBROUTINE	SUBROUTINE SIZE (bytes) (including support programs)	DEMONSTRATION PROGRAM SIZE (bytes)	EXECUTION SPEED	CONDITIONS/ COMMENTS
BISECT (bisection method)	921	422	0.17 seconds/ iteration	
Z-NEWTON (Newton-Raphson iteration)	703	321	0.25 seconds/ iteration	
SECANT (secant method)	932	375	0.22 seconds/ iteration	
REGULA (modified false-position method)	1063	399	0.25 seconds/ iteration	
AITKEN (Aitken acceleration method)	1176	395	0.25 seconds/ iteration	
A/SITER (Aitken-Steffenson iteration)	1193	389	0.52 seconds/ iteration	
RSYNDIV (polynomial synthetic division)	454	667	3 seconds	20th-degree polynomial divided by a 10th-degree polynomial
NEXTROOT (Newton-Raphson iteration with root removal)	737	504	0.4 seconds/ iteration	5 roots predetermined

Table 6.1.1 *A summary of the programs presented in Chapter 6. Note that the execution speed is dependent on the function being evaluated. For this tabulation, the functions displayed with the subroutine listings in the text were used.*

6.2 Gaining Preliminary Knowledge About the Roots of a Polynomial

There are two classes of functions that will be considered: polynomials and transcendentals. This section will deal with polynomial theorems.

In this book, only polynomials that have real coefficients are treated. These are the most common type encountered. The general form of a polynomial, $p(x)$, is shown in equation (6.2.1). The locations of the zeros of $p(x)$ are called *roots*. The coefficients of $p(x)$ contain all the information possible regarding these roots. Therefore, it is not surprising that general rules for estimating these roots have been formulated based on some simple examinations of the coefficients.

For example, although a polynomial may have real coefficients, the roots may be complex numbers. If a root $r_1 = x + iy$ is found, then it necessarily follows that the complex conjugate of r_1, $r_2 = r_1^* = x - iy$, is also a root. (See Ref. 37.) Complex roots can come only in complex conjugate pairs.

If a polynomial is of degree n, then it has at most n distinct roots. Assume that the polynomial is written in the form

$$p(x) = a_0 + a_1 x + \cdots + a_{n-1} x^{n-1} + a_n x^n \qquad (6.2.1)$$

We can apply the Newton (or Birge-Vieta) relations (see Ref. 37), three of which are:

$a_0 = (-1)^n \times$ product of all roots
$a_{n-2} =$ sum of all products of roots taken two at a time
$a_{n-1} = -$ sum of all roots

Note that because the product of complex roots is positive, a_0 can be used to test for the possible existence of a negative root. Also, it can be shown that

$$r_1^2 + r_2^2 + \cdots + r_n^2 = (a_{n-1})^2 - 2(a_{n-2})$$

It can then be argued that the largest root is bounded in absolute value by

$$|r_{max}| \leq \sqrt{(a_{n-1})^2 - 2a_{n-2}} \qquad (6.2.2)$$

Other helpful rules, which are due to Descartes, relate the number of sign changes between the coefficients, m, to the number of positive and negative roots:

- The number of positive real roots of $p(x)$ is equal to, or is an even integer less than, m.
- The number of negative real roots of $p(x)$ is equal to, or is an even integer less than, m', where m' is the number of coefficient sign changes in $p(-x)$.

The general rules given above can be used to develop a simple computer program that can teach you what to look for in roots. Such a program is shown in listing 6.2.1, and examples of its application are shown in listing 6.2.2. This program supplies only clues as to where the roots can be located. However, these clues are very important in that nearly all the zero-search programs to be presented require at least one or two initial guesses for the start of iteration.

Listing 6.2.1

```
3925 REM ROOT TESTING PROGRAM (ROOTTEST)
3926 PRINT
3927 PRINT
3928 PRINT "THIS PROGRAM WILL HELP YOU"
3929 PRINT "DETERMINE WHERE TO LOOK FOR"
3930 PRINT "ROOTS OF A POLYNOMIAL."
3931 PRINT
3932 PRINT "WHAT IS THE DEGREE OF THE POLYNOMIAL: ",
3933 INPUT N
3934 DIM A(N)
3935 PRINT
3936 PRINT "INPUT THE POLYNOMIAL COEFFICIENTS"
3937 PRINT "AS PROMPTED:"
3938 PRINT
3939 FOR I=0 TO N
3940 PRINT "A(",I,") = ",
3941 INPUT A(I)
3942 NEXT I
3943 REM NORMALIZE SO A(N)=1
3944 FOR I=0 TO N
3945 A(I)=A(I)/A(N)
3946 NEXT I
3947 PRINT
3948 PRINT
3949 PRINT "THERE ARE",N," ROOTS."
3950 PRINT
3951 PRINT
3952 REM FIND THE MAXIMUM VALUE OF ROOT
3953 A=A(N-1)*A(N-1)-2*A(N-2)
3954 IF A>0 THEN GOTO 3958
3955 PRINT "THERE ARE AT LEAST TWO COMPLEX"
3956 PRINT "ROOTS. THE ANALYSIS ENDS."
3957 GOTO 3998
3958 A=SQRT(A)
3959 PRINT
3960 PRINT "THE MAGNITUDE OF THE LARGEST ROOT"
3961 PRINT "IS NOT GREATER THAN",A
3962 PRINT
3963 PRINT
3964 A=-1
3965 IF INT(N/2)=N/2 THEN A=-A
3966 A=A(0)/A
```

```
3967 REM B WILL FLAG A NEGATIVE ROOT
3968 B=0
3969 IF A>0 THEN GOTO 3972
3970 PRINT "THERE IS AT LEAST ONE NEGATIVE REAL ROOT."
3971 B=1
3972 PRINT
3973 REM TEST FOR DESCARTES RULE NUMBER 1
3974 C=0
3975 FOR I=1 TO N
3976 IF A(I-1)*A(I)<0 THEN C=C+1
3977 NEXT I
3978 IF C=1 THEN PRINT "THERE IS AT MOST ONE POSITIVE ",
3979 IF C>1 THEN PRINT "THERE ARE AT MOST",C," POSITIVE ",
3980 IF C=1 THEN PRINT "REAL ROOT."
3981 IF C>1 THEN PRINT "REAL ROOTS."
3982 PRINT
3983 REM TEST FOR DESCARTES RULE NUMBER 2
3984 D=0
3985 FOR I=1 TO N
3986 IF A(I-1)*A(I)>0 THEN D=D+1
3987 NEXT I
3988 IF D=1 THEN PRINT "THERE IS AT MOST ONE NEGATIVE ",
3989 IF D>1 THEN PRINT "THERE ARE AT MOST",D," NEGATIVE ",
3990 IF D=1 THEN PRINT "REAL ROOT."
3991 IF D>1 THEN PRINT "REAL ROOTS."
3992 PRINT
3993 IF INT(C/2)=C/2 THEN GOTO 3995
3994 PRINT "THERE IS AT LEAST ONE POSITIVE REAL ROOT."
3995 IF INT(D/2)=D/2 THEN GOTO 3998
3996 IF B=1 THEN GOTO 3998
3997 PRINT "THERE IS AT LEAST ONE NEGATIVE REAL ROOT."
3998 PRINT
3999 PRINT
4000 END
```

Listing 6.2.1 *Program for aiding in finding the roots of a polynomial having real coefficients.*

Listing 6.2.2

RUN

THIS PROGRAM WILL HELP YOU
DETERMINE WHERE TO LOOK FOR
ROOTS OF A POLYNOMIAL.

WHAT IS THE DEGREE OF THE POLYNOMIAL: ?2

```
INPUT THE POLYNOMIAL COEFFICIENTS
AS PROMPTED:

A( 0) = ?-1
A( 1) = ?0
A( 2) = ?1


THERE ARE 2 ROOTS.


THE MAGNITUDE OF THE LARGEST ROOT
IS NOT GREATER THAN 1.4142136

THERE IS AT LEAST ONE NEGATIVE REAL ROOT.



READY
RUN



THIS PROGRAM WILL HELP YOU
DETERMINE WHERE TO LOOK FOR
ROOTS OF A POLYNOMIAL.

WHAT IS THE DEGREE OF THE POLYNOMIAL: ?2

INPUT THE POLYNOMIAL COEFFICIENTS
AS PROMPTED:

A( 0) = ?1
A( 1) = ?0
A( 2) = ?1


THERE ARE 2 ROOTS.


THERE ARE AT LEAST TWO COMPLEX
ROOTS. THE ANALYSIS ENDS.


READY
```

Listing 6.2.2 cont.

RUN   Listing 6.2.2 cont.

THIS PROGRAM WILL HELP YOU
DETERMINE WHERE TO LOOK FOR
ROOTS OF A POLYNOMIAL.

WHAT IS THE DEGREE OF THE POLYNOMIAL: ?3

INPUT THE POLYNOMIAL COEFFICIENTS
AS PROMPTED:

A(0) = ?-2
A(1) = ?-1
A(2) = ?2
A(3) = ?1

THERE ARE 3 ROOTS.

THE MAGNITUDE OF THE LARGEST ROOT
IS NOT GREATER THAN 2.4494897

THERE IS AT MOST ONE POSITIVE REAL ROOT.

THERE ARE AT MOST 2 NEGATIVE REAL ROOTS.

THERE IS AT LEAST ONE POSITIVE REAL ROOT.

READY
RUN

THIS PROGRAM WILL HELP YOU
DETERMINE WHERE TO LOOK FOR
ROOTS OF A POLYNOMIAL.

WHAT IS THE DEGREE OF THE POLYNOMIAL: ?5

INPUT THE POLYNOMIAL COEFFICIENTS
AS PROMPTED:

```
A( 0) = ?1
A( 1) = ?5
A( 2) = ?10
A( 3) = ?10
A( 4) = ?5
A( 5) = ?1
```

THERE ARE 5 ROOTS.

THE MAGNITUDE OF THE LARGEST ROOT
IS NOT GREATER THAN 2.236068

THERE IS AT LEAST ONE NEGATIVE REAL ROOT.

THERE ARE AT MOST 5 NEGATIVE REAL ROOTS.

Listing 6.2.2 *Sample runs of the root testing program shown in listing 6.2.1. In the first example, $p(x) = (x - 1)(x + 1) = x^2 - 1$ was examined. It has roots $+1$ and -1. In the second example, the function was $p(x) = (x + i)(x - i) = x^2 + 1$. It has no real roots. The polynomial tested in the third example was $p(x) = (x + 1)(x + 2)(x - 1) = x^3 + 2x - x - 2$, which has roots $+1$, -1, and -2. The conclusion from the program is that there is exactly one positive root and two negative roots, and that these roots lie in the range -2.45 to $+2.45$. That is quite a clue! In the last example, the polynomial $p(x) = (x + 1)^5 = x^5 + 5x^4 + 10x^3 + 10x^2 + 5x + 1$ was probed. It has a fivefold multiple root of -1. In this case, the clue is helpful, but not highly specific.*

6.3 Interval Searches

Perhaps the oldest and the most elementary method for numerically locating the position of a zero is to apply simple trial and error calculations. This approach can be organized to ensure locating a zero to any degree of accuracy (under the right conditions) by *interval halving*, or *bisection*. Although this technique is simple, it requires specific *a priori* knowledge. It is also both slow and subject to failure. However, this technique is usually less likely to fail than the more sophisticated algorithms to be discussed later. The real utility of the bisection method is that it can be applied to locate the approximate position of a zero, and then a faster converging algorithm, such as modified regula-falsi, can be used to finish the iteration.

The basic idea behind the bisection method is shown in figure 6.3.1. The object is to find the position at which $y(x)$ crosses the x axis [$y(x) = 0$]. We *assume* that the

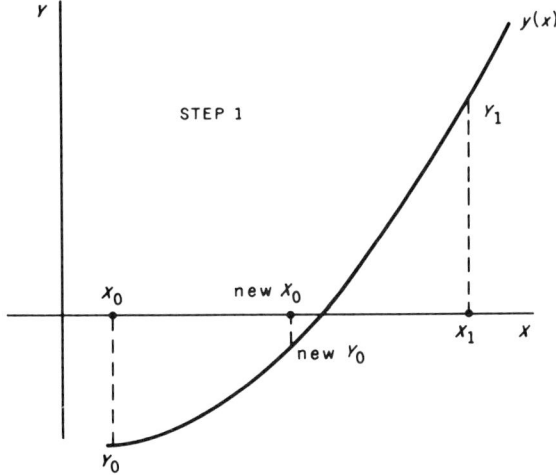
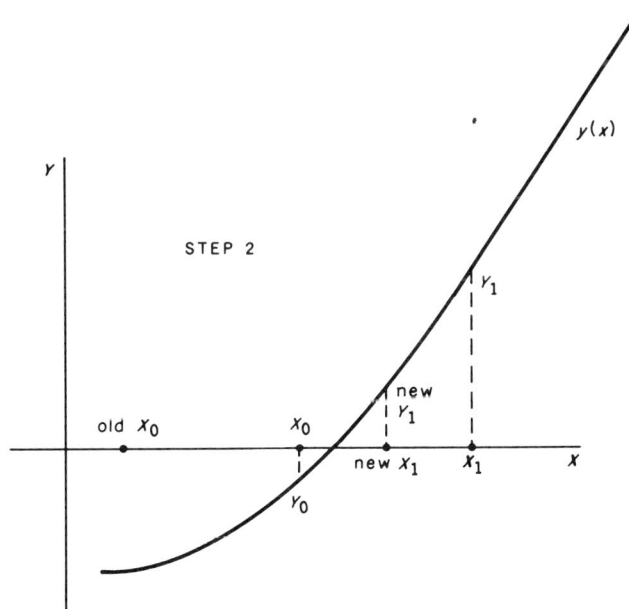

Figure 6.3.1 *Two steps in the bisection method.*

initial guesses (x_0 and x_1) in step 1 of figure 6.3.1 bracket the zero. The next estimate of the position of the zero is $(x_0 + x_1)/2$, the middle of the interval between x_0 and x_1. This new estimate is used to replace either x_0 or x_1 so that the zero is contained within the new interval. In step 1 of figure 6.3.1, x_0 is replaced, whereas in step 2, x_1 is replaced. The choice can be made mathematically as follows:

- If (old y_0) × (new y) < 0, then replace x_1
- If (old y_0) × (new y) > 0, then replace x_0

This type of conditional decision is easily made by a computer and can be used to develop a simple algorithm (BISECT) as shown in listing 6.3.1. The connection diagram for this subroutine is given in figure 6.3.2. The inputs are the two initial bracketing guesses, X1 and X2, and the convergence (error) criterion, E. Returned are the estimate of the root, X, and Y = y(X). Also returned is the number of bisections used, M.

Listing 6.3.2 shows three sets of results obtained for the cubic polynomial example $y(x) = (x - 2)(x + 1)(x + 10)$. The display of the intermediate results for the first example (see listing 6.3.3) was generated by inserting a print statement into the subroutine itself. The two initial guesses in the first example were $x_0 = 0$ and $x_1 = 10$. These bracket the root $x = 2$, and exclude the other two roots, $x = -1$ and $x = -10$. Twenty-four steps were required for convergence to the $x = 2$ root given the chosen error criterion of $E = 10^{-6}$. Note, however, that the final result was accurate to 10^{-7}, which is much better than E. It should be kept in mind that the chosen error criterion represents a maximum, and that the final result is likely to be better than implied by that error bound.

We can easily estimate the number of required iteration steps to attain an accuracy of E. The initial uncertainty in the location of the zero is $x_1 - x_0$. Each step in the iteration halves the width of the interval in which the zero must be located. In n steps, the uncertainty becomes $|x_1 - x_0|2^{-n}$. We therefore require the number of steps, n, to be such that

$$E \gtrsim |x_1 - x_0|2^{-n}$$

Taking logarithms, this gives

$$n \gtrsim 3.3\ (\log_{10} |x_1 - x_0| - \log_{10} E) \tag{6.3.1}$$

For the case $x_1 - x_0 = 10$ and $E = 10^{-6}$, we get $n \gtrsim 23$, which was observed ($n = 24$) in the corresponding example.

The predictability of the required number of iteration steps is a very nice feature of the bisection method. However, the convergence is *slow*. Therefore, the method is often used only to approximately locate a zero, at which point a faster algorithm is applied.

The bisection method is powerful, but it does have limitations. Listing 6.3.2

shows an example in which no roots were contained in the initial range. The method failed by converging on one of the extremes of the interval. The procedure can also fail with double roots. Generally, this will happen if, during the halving process, a bisection point *between* the two roots is not encountered. You will need some luck when dealing with ranges containing more than one root. For example, if the initial guesses were $x_0 = 0$ and $x_1 = -15$, convergence to one of the roots ($x = -1$ and $x = -10$) would have occurred, even though two roots existed in the interval. As an exercise, try to determine which root is found in this case. What would happen if the initial guesses were 0 and -30, or 0 and $+30$?

The above observation regarding intervals containing more than one root immediately implies that real-valued double roots, e.g., $y(x) = (x - a)^2$, cannot be directly found by the bisection method because an intermediate bisection point will never fall *between* the two roots! In fact, real roots of even order [$(x - a)^{2m}$] are invisible to the bisection method.

Another situation that can arise with the bisection method is the occurrence of simple *poles* of the form $y = 1/(x - a)^n$. An example of an odd-order pole is shown in figure 6.3.3. If the initial guesses bracket the pole, the interval halving will converge about $x = a$. If the chosen error criterion, E, is small enough, it is possible for $y(x)$ to exceed the allowable numeric range available in the computer [$y(x)$ becomes $1/E^n$]. If the pole is of even order, it is invisible to the algorithm.

In the next section, we will discuss a different approach to root searching—one that is based on an iteration scheme that is somewhat more complex than interval halving, but that is usually faster (when it works).

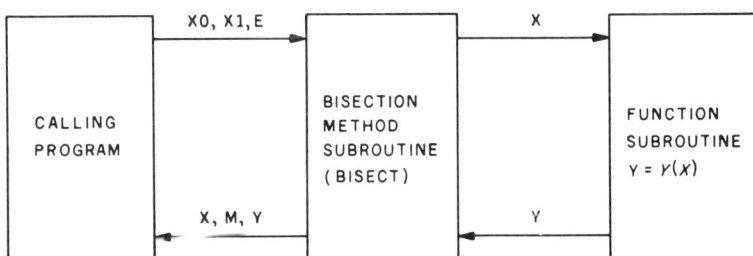

X0, X1:   Initial bracketing guesses
E: Convergence criterion
X: Calculated root
M: Number of iterations

Figure 6.3.2 *Subroutine-connection diagram for the bisection method interval search routine. The functions and variables used are shown in table 6.3.1. The function Y(X) is evaluated in a separate subroutine.*

Statements/Functions List

+, −, *, /, <, >
ABS, GOSUB, GOTO, IF/THEN

Variables List

E, M, X, X0, X1, Y

Variables Passed to Subroutine

E, X0, X1

Table 6.3.1 *Functions and variables appearing in the bisection method subroutine.*

Listing 6.3.1

```
4025 REM PROGRAM TO DEMONSTRATE THE BISECTION SUBROUTINE
4026 PRINT
4027 PRINT
4028 PRINT "WHAT IS THE INITIAL RANGE (X0,X1):"
4029 PRINT
4030 PRINT "X0 = ",
4031 INPUT X0
4032 PRINT "X1 = ",
4033 INPUT X1
4034 PRINT
4035 PRINT "WHAT IS THE CONVERGENCE CRITERION: ",
4036 INPUT E
4037 REM GO TO BISECTION METHOD SUBROUTINE
4038 GOSUB 46350
4039 PRINT
4040 PRINT
4041 PRINT "THE CALCULATED ZERO IS X =",X
4042 PRINT
4043 PRINT "THE ASSOCIATED Y VALUE IS Y = ",Y
4044 PRINT
4045 PRINT "THE NUMBER OF STEPS WAS",M
4046 PRINT
4047 PRINT
4048 END
44298 REM ********************
44299 REM FUNCTION SUBROUTINE
44300 Y=(X-2)*(X+1)*(X+10)
44301 RETURN
46333 REM ********************
46334 REM BISECTION METHOD SUBROUTINE (BISECT)
46335 REM THIS PROGRAM ITERATIVELY SEEKS THE ZERO
46336 REM OF A FUNCTION USING THE METHOD OF INTERVAL
```

```
46337 REM HALVING UNTIL THE INTERVAL IS LESS THAN
46338 REM E IN WIDTH.
46339 REM IT IS ASSUMED THAT THE FUNCTION Y=Y(X)
46340 REM IS AVAILABLE FROM THE FUNCTION SUBROUTINE
46341 REM LOCATED AT 44300.
46342 REM THIS SUBROUTINE REQUIRES AS INPUT THE INITIAL
46343 REM RANGE VALUES (X0 AND X1), AS WELL AS THE
46344 REM CONVERGENCE CRITERION, E.
46345 REM THE ZERO MUST BE WITHIN THE RANGE SPECIFIED
46346 REM OR AN ERRONEOUS VALUE WILL BE RETURNED IN X.
46347 REM THIS SUBROUTINE RETURNS THE ESTIMATE OF THE ROOT
46348 REM IN X, AND THE CORRESPONDING Y VALUE.
46349 REM ALSO RETURNED IS THE NUMBER OF STEPS (M).
46350 M=0
46351 X=X0
46352 GOSUB 44300
46353 Y0=Y
46354 X=X1
46355 GOSUB 44300
46356 X=(X0+X1)/2
46357 GOSUB 44300
46358 M=M+1
46359 IF Y*Y0=0 THEN RETURN
46360 IF Y*Y0<0 THEN X1=X
46361 IF Y*Y0>0 THEN X0=X
46362 IF ABS(X1-X0)>E THEN GOTO 46351
46363 RETURN
```

Listing 6.3.1 *Bisection method subroutine (BISECT).*

Listing 6.3.2

```
RUN

WHAT IS THE INITIAL RANGE (X0,X1):

X0 = ?0
X1 = ?10

WHAT IS THE CONVERGENCE CRITERION: ?.000001


THE CALCULATED ZERO IS X = 2.0000001

THE ASSOCIATED Y VALUE IS Y =   3.6000001E-06

THE NUMBER OF STEPS WAS 24
```

```
READY
RUN


WHAT IS THE INITIAL RANGE (X0,X1):

X0 = ?-5
X1 = ?5

WHAT IS THE CONVERGENCE CRITERION: ?.000001

THE CALCULATED ZERO IS X = -.9999998

THE ASSOCIATED Y VALUE IS Y =  -5.3999996E-06

THE NUMBER OF STEPS WAS 24

READY
RUN


WHAT IS THE INITIAL RANGE (X0,X1):

X0 = ?5
X1 = ?15

WHAT IS THE CONVERGENCE CRITERION: ?.000001

THE CALCULATED ZERO IS X = 14.999999

THE ASSOCIATED Y VALUE IS Y =  5199.999

THE NUMBER OF STEPS WAS 23

READY
```

Listing 6.3.2 Sample runs of the bisection method subroutine for the function $y(x) = (x - 2)(x - 1)(x + 10)$. In the first two examples, a root was contained within the initial search interval, and proper convergence occurred. In the third example, the initial interval did not include a root, and the iteration closed in on one of the endpoints of that interval. The error is apparent in the last value calculated for y.

```
RUN


WHAT IS THE INITIAL RANGE (X0,X1):

X0 = ?0
X1 = ?10

WHAT IS THE CONVERGENCE CRITERION: ?.000001
 1          5                  1760
 2          2.5                270
 3          1.25               21.875
 4          1.875              21.875
 5          2.1875             21.875
 6          2.03125            7.2839355
 7          1.953125           1.139679
 8          1.9921875          1.139679
 9          2.0117188          1.139679
10          2.0019532          .42393837
11          1.9970704          7.0372431E-02
12          1.9995118          7.0372431E-02
13          2.0007325          7.0372431E-02
14          2.0001222          .02637805
15          1.999817           4.3994239E-03
16          1.9999696          4.3994239E-03
17          2.0000459          4.3994239E-03
18          2.0000078          1.6524317E-03
19          1.9999887          2.8080092E-04
20          1.9999983          2.8080092E-04
21          2.0000031          2.8080092E-04
22          2.0000007          1.1160014E-04
23          1.9999995          2.5200008E-05
24          2.0000001          2.5200008E-05


THE CALCULATED ZERO IS X = 2.0000001

THE ASSOCIATED Y VALUE IS Y =   3.6000001E-06

THE NUMBER OF STEPS WAS 24

READY
```

Listing 6.3.3 *Repeat of the first example appearing in listing 6.3.2, but with a PRINT statement inserted within the BISECT subroutine so that intermediate values may be viewed.*

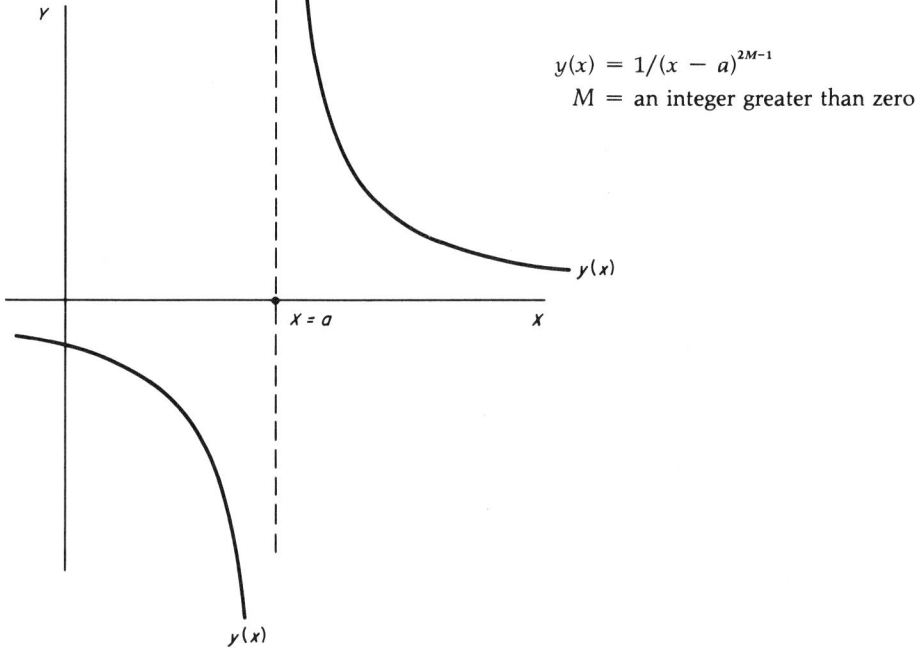

Figure 6.3.3 *A function that causes problems for the bisection method.*

6.4 Successive Substitution

The method to be discussed in this section is based on the ability to write the function of interest, $y(x)$, in a special form:

$$y(x) = x - g(x) \qquad (6.4.1)$$

The object then is to determine the one or more values of x so that $y(x) = 0$. In the case of equation (6.4.1), when $y(x) = 0$ we have

$$x - g(x) = 0$$

or
$$x = g(x) \qquad (6.4.2)$$

The values of x that satisfy this equation are called *fixed points*. As you can see, polynomial equations [e.g., $y(x) = a_0 + a_1 x + a_2 x^2 + a_3 x^3 + \cdots$] can always be written in a form that satisfies equation (6.4.2).

Figure 6.4.1 shows a hypothetical plot of the functions x and $g(x)$. The desired values of x are where x equals $g(x)$, or where the curves intersect. As shown in the figure, this can occur in more than one place: x_A and x_B.

Figure 6.4.2 shows the region surrounding intersection at x_A in more detail.

If we approximate the position of the sought-after intersection with the guess x_1, and use x_1 in equation (6.4.2) to perhaps get a better estimate, x_2, the result will be

$$x_2 = g(x_1)$$

If x_2 is then used as the basis of approximation to get another estimate, we will have

$$x_3 = g(x_2)$$

In general, the approximation sequence is

$$x_{n+1} = g(x_n) \tag{6.4.3}$$

The sequence is graphically shown in figure 6.4.2. Geometrically, we start with a value x_1, go up vertically to the $y = g(x)$ curve, then horizontally to the $y = x$ line, and then reflect down to the new estimate, x_2. The procedure is repeated using x_2. If all goes well, a good approximation to x_A, the intersection and desired fixed point, will eventually be obtained.

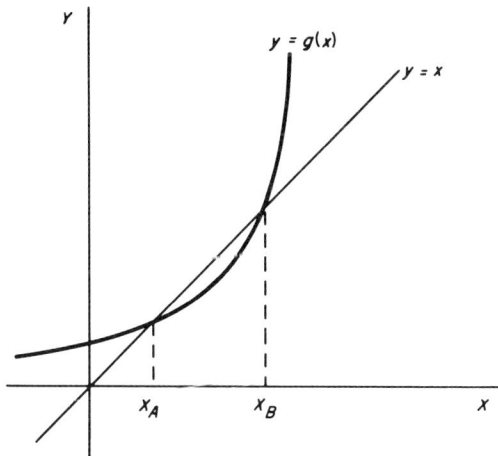

Figure 6.4.1 *Hypothetical plot of the functions $y = x$ and $y = g(x)$.*

If you understand geometrically how the approximation sequence proceeds, then you will see what the conditions are for convergence of this scheme. Figure 6.4.3 shows a case in which the slope of $g(x)$ (or, in mathematical terms, dg/dx) is greater than unity. The successive approximations head away from the true value. Figure 6.4.4 shows a situation in which the slope is less than negative unity ($dg/dx < -1$). Again, there is a failure in the convergence of this method.

In general, convergence is to be expected if, in the region over which the successive approximations are being tried, the derivative (slope) of $g(x)$ obeys the following constraint:

$$\left|\frac{dg}{dx}\right| < 1 \qquad (6.4.4)$$

As we will see in the next section, the closer dg/dx is to zero, the faster the convergence will be.

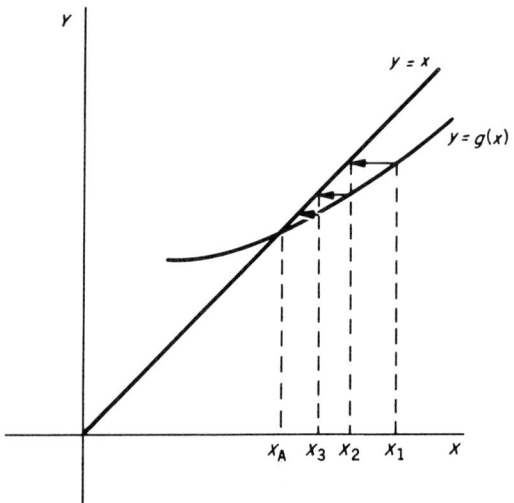

Figure 6.4.2 *Graphic representation of the successive-substitution iteration scheme.*

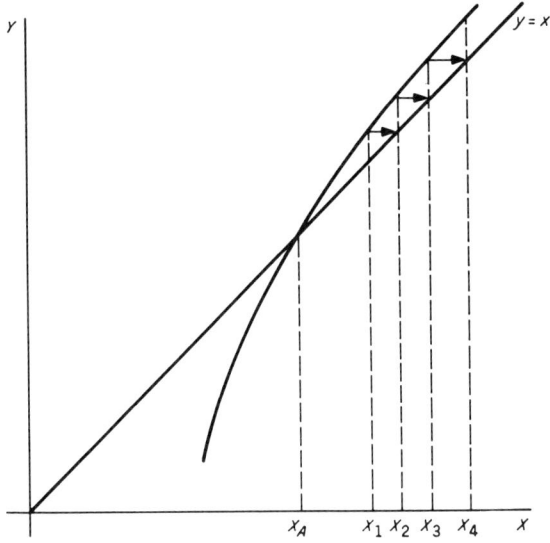

Figure 6.4.3 *Case in which the slope of $g(x)$ is greater than unity, and the sequence diverges.*

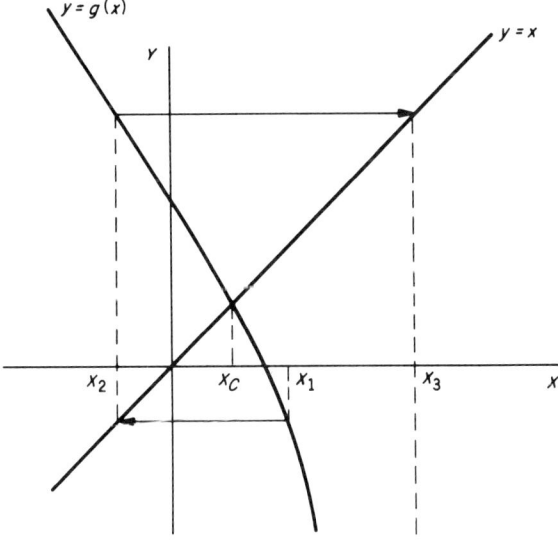

Figure 6.4.4 *Case in which the slope of $g(x)$ is less than negative unity, and the sequence diverges.*

6.5 Examples of Successive Substitution

The purpose of this section is to help you develop some familiarity with the concept of successive substitution, in that it forms the basis for more advanced techniques.

Programming successive-substitution problems can be very easy. In this section, a few simple examples that demonstrate the basic procedure will be presented. These examples also point out weaknesses in the technique that can be partially overcome as shown in the following section.

For the first example, consider finding the roots of

$$y(x) = x - 2 \sin x$$

In this case, $g(x) = 2 \sin x$ and we can write the successive-approximation equation as

$$x_{n+1} = 2 \sin x_n$$

There are three roots to be found: X_A, X_B, and X_C. From figure 6.5.1, it is apparent that $X_B = 0$, and by symmetry $X_A = -X_B$. As a first estimate, we will try $X_1 = \pi/2$. Listing 6.5.1 shows the program used for the iteration. This program calculates the first 40 steps in the successive-substitution sequence. It is apparent that the convergence is rapid in terms of time (i.e., the number of statements evaluated), but it is slow with respect to the number of iterations required (32) before the round-off error level is reached.

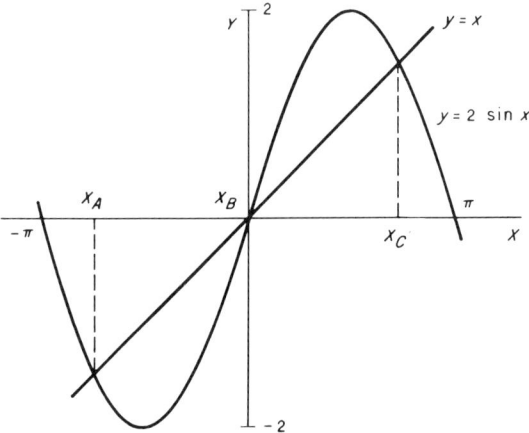

Figure 6.5.1 *Finding the roots of $y = x - 2 \sin x$.*

The number of steps required to obtain a given number of digits in accuracy can easily be approximated. With a little trigonometry and examination of figure 6.4.2, it can be shown that if the error in the initial guess is $\bar{E}_0$, then, if $g(x)$ has little curvature in the vicinity of the fixed point, this error is reduced roughly by the factor $|dg(x)/dx|$ by the first iteration. In n iterations the error is approximately $E_n \approx E_0 |dg(x)/dx|^n$. The number of iterations required to achieve an accuracy of 10^{-m} can then be shown to be

$$n \approx \frac{m + \log_{10} E_0}{-\log_{10} |dg(x)/dx|} \qquad (6.5.1)$$

Note that if $|dg(x)/dx| \geq 1$, m can become infinite or even negative, indicating a failure in convergence.

In the example given, the initial guess was 1.57 and the final result was 1.89. Therefore, $E_0 \cong 0.32$. The number of digits accuracy attempted was $m = 7$. The derivative at the final value is

$$\left.\frac{dg}{dx}\right|_{fixed\ point} = \left.\frac{d(2 \sin x)}{dx}\right|_{x=1.895} = 2 \cos x \Big|_{x=1.895} = -0.64$$

Using equation (6.5.1), we then obtain $n = 37$, which is consistent with the results shown in listing 6.5.1.

For the second example, we will consider the polynomial

$$y(x) = x^3 + 2x^2 - x - 2$$

This polynomial has the known roots of $+1$, -1, and -2, but we will forget this for the moment. We will try to find the values by iteration. The successive approximation form to be used is

$$x_{n+1} = x_n^3 + 2x_n^2 - 2 = g(x_n)$$

The coding of this iteration formula is very simple. However, what do we use as an initial guess? Graphing $y(x)$ is an excellent way to get an idea of where the zeros are. We can also use the program given in section 6.2. However, even if we do get a rough approximation for the locations of the zeros of $y(x)$ for use as starting points in the iteration, convergence is not assured unless $|dg/dx| < 1$ near the root. Therefore, for the purpose of enabling the iteration scheme, a plot of dg/dx is more useful. The desired derivative is

$$\frac{dg}{dx} = 3x^2 + 4x$$

A plot of this function is shown in figure 6.5.2.

Because it is only worth using the successive-substitution iteration scheme in the regions where it stands a chance of converging, i.e., $|dg/dx| < 1$ (and preferably near zero), two initial guesses are suggested: $x_0 = 0$ and $x_0 \cong -1.5$. The iteration results for these two guesses are shown in listing 6.5.2. The first guess, $x_1 = 0$, fortuitously led to one exact root in one iteration! As shown in listing 6.5.3, using $x_1 = \pm 0.001$ as an initial guess does not work at all. Also, as seen in listing 6.5.2, the iteration using $x_1 = -1.5$ as a beginning point does not converge, and it has a pattern similar to that appearing in listing 6.5.3. The source of the convergence problem is apparent from figure 6.5.2. The root $x = -2$ is in a region of divergence. If an estimate at any stage in the iteration is in the divergence region, the error is likely to grow instead of shrink. Using $x_1 = 0$ fortunately led to a "better" guess of $x_2 = -2$, which had no error. However, using a starting position slightly different from 0 (± 0.001) resulted in a second guess of -1.9999998, which had a very small error. Because this value was in a region of divergence, however, the small error grew until the iteration reentered a region of convergence. The iteration then attempted to close in on the root $x = -1$. Unfortunately, that root is just on the border of a region of weak divergence (weak because $|dg/dx| \gtrsim 1$). The iteration then repeatedly crossed the convergence/divergence boundary near $x = -1$, bracketing, but never converging on, the root $x = -1$.

The general result is that, except for the initial guess $x_1 = 0$ (and also the starting points $x_1 = 1, -1,$ and -2), the iteration scheme used does *not* converge. In fact, starting points such as $x_1 = 2$ will cause numeric overflows.

It is apparent that successive-substitution iteration, although simple in implementation, has no universal guarantee of convergence. This is a general problem with root-seeking algorithms.

Since the successive-substitution method is so simple and so easy to code, there is a reluctance to abandon it if it does not work on the first try. For example, although in the last case none of the roots were in the narrow region of possible convergence, it is possible to force the existence of a large convergence interval with the hope that a root will be contained in it. As an example, an alternate form for the polynomial equation just examined is

$$y(x) = x^2 \left(x + 2 - \frac{1}{x} - \frac{2}{x^2} \right)$$

The iteration formula then becomes

$$x_{n+1} = -2 + \frac{1}{x_n} + \frac{2}{x_n^2} = g(x_n)$$

This iteration equation will not progressively lead to large values of x_{n+1} because a large x_n will give x_{n+1} near -2. It is self-limiting. If this equation is used for the iteration, we get the results shown in listing 6.5.4. The root $x = -2$ is now in the region of convergence, which is rapid. By examining dg/dx, you may be satisfied that $x = -2$ is the only root in a convergence region.

As a third example, consider finding the maxima and minima of

$$y(x) = x \cos x$$

We expect an infinity of local maxima and minima near, but not exactly at, the extremes of cos x, which occur at 0, $\pm \pi$, $\pm 2\pi$, and so on. The problem can be converted into one of looking for zeros by examining the derivative dy/dx:

$$\frac{dy}{dx} = \cos x - x \sin x$$

This quickly leads to the iteration form

$$x_{n+1} = \frac{1}{\tan x_n} = g(x_n)$$

We can determine the regions of convergence using dg/dx:

$$dg/dx = \frac{-1}{\sin^2 x}$$

If the extremes of $y(x)$ are near-multiples of π, then sin $x \approx 0$ and there is a convergence problem. The one exception might be the maxima near $x = 0$ (but not at $x = 0$ because $x \cos x$ is zero there). The results of two sample iteration runs are shown in listing 6.5.5. Two extremes were found. The negative value ($x = -1.16234$) corresponds to a local minimum.

The examples presented in this section have served to demonstrate that direct successive-substitution iteration can be very efficient in implementation *when it works*. The method is so easy to use that perhaps it is worth trying on the chance that it will succeed. If direct successive substitution fails, then there are other approaches that can be employed.

Listing 6.5.1

```
1 X1=3.14159/2
2 FOR I=1 TO 40
3 X2=2*SIN(X1)
4 PRINT I,TAB(8),X1
5 X1=X2
6 NEXT I
7 PRINT
8 END

RUN

1       1.570795
2       2
3       1.818595
```

4	1.9389094
5	1.866016
6	1.9134766
7	1.8837148
8	1.9028784
9	1.8907312
10	1.8985118
11	1.8935604
12	1.8967246
13	1.8947078
14	1.8959956
15	1.8951742
16	1.8956984
17	1.895364
18	1.8955774
19	1.8954412
20	1.8955282
21	1.8954726
22	1.8955082
23	1.8954854
24	1.8955
25	1.8954906
26	1.8954966
27	1.8954928
28	1.8954952
29	1.8954936
30	1.8954946
31	1.895494
32	1.8954944
33	1.8954942
34	1.8954944
35	1.8954942
36	1.8954944
37	1.8954942
38	1.8954944
39	1.8954942
40	1.8954944

READY

Listing 6.5.1 *Program for locating the right-hand zero of $y = x - 2 \sin x$. Execution results are also shown.*

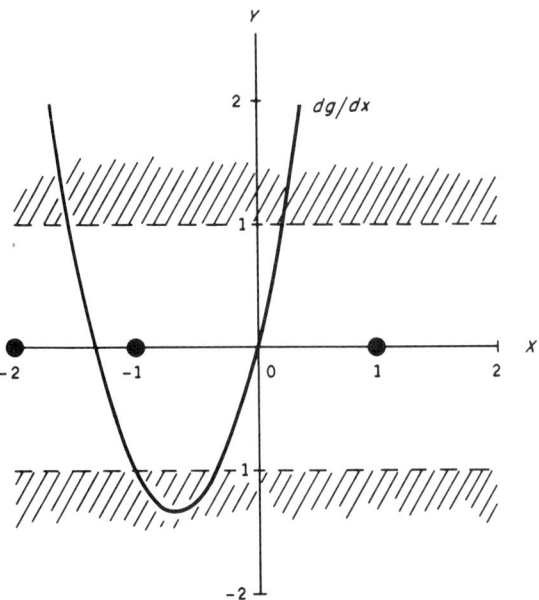

Figure 6.5.2 *Plot of $dg/dx = 3x^2 + 4x$. The shaded regions show where divergence of the iteration scheme is expected. The actual positions of roots are shown as solid circles.*

Listing 6.5.2 part A

```
1 X1=0
2 FOR I=1 TO 10
3 X2=X1*X1*X1+2*X1*X1-2
4 PRINT I,TAB(8),X1
5 X1=X2
6 NEXT I
7 PRINT
8 END
```

Listing 6.5.2 part B

```
1 X1=-1.5
2 FOR I=1 TO 20
3 X2=X1*X1*X1+2*X1*X1-2
4 PRINT I,TAB(8),X1
5 X1=X2
6 NEXT I
7 PRINT
8 END
```

Listing 6.5.2 part A, cont.

RUN

1	0
2	-2
3	-2
4	-2
5	-2
6	-2
7	-2
8	-2
9	-2
10	-2

READY

Listing 6.5.2 part B, cont.

RUN

1	-1.5
2	-.875
3	-1.1386719
4	-.8832246
5	-1.1288195
6	-.8899127
7	-1.1208724
8	-.8955036
9	-1.1142749
10	-.9002762
11	-1.1086769
12	-.9044174
13	-1.1038454
14	-.9080584
15	-1.0996177
16	-.9112945
17	-1.0958762
18	-.9141973
19	-1.0925331
20	-.9168217

READY

Listing 6.5.2 *Successive-substitution iteration results for $x_{n+1} = x_n^3 + 2x_n^2 - 2$ using the initial guesses $x_1 = 0$ and $x_1 = -1.5$.*

Listing 6.5.3 part A

```
1 X1=.001
2 FOR I=1 TO 20
3 X2=X1*X1*X1+2*X1*X1-2
4 PRINT I,TAB(8),X1
5 X1=X2
6 NEXT I
7 PRINT
8 END
```

Listing 6.5.3 part B

```
1 X1=-.001
2 FOR I=1 TO 20
3 X2=X1*X1*X1+2*X1*X1-2
4 PRINT I,TAB(8),X1
5 X1=X2
6 NEXT I
7 PRINT
8 END
```

Listing 6.5.3 part A, cont.

```
RUN

1       .001
2      -1.999998
3      -1.999992
4      -1.999968
5      -1.999872
6      -1.9994881
7      -1.9979535
8      -1.9918307
9      -1.9675891
10     -1.8745243
11     -1.5590984
12      -.9282618
13     -1.0765154
14      -.9297872
15     -1.0747965
16      -.9312165
17     -1.0731893
18      -.9325595
19     -1.071682
20      -.9338246

READY
```

Listing 6.5.3 part B, cont.

```
RUN

1      -.001
2      -1.999998
3      -1.999992
4      -1.999968
5      -1.999872
6      -1.9994881
7      -1.9979535
8      -1.9918307
9      -1.9675891
10     -1.8745243
11     -1.5590984
12      -.9282618
13     -1.0765154
14      -.9297872
15     -1.0747965
16      -.9312165
17     -1.0731893
18      -.9325595
19     -1.071682
20      -.9338246

READY
```

Listing 6.5.3 *Successive-substitution iteration results for $x_{n+1} = x_n^3 + 2x_n^2 - 1$ using $x_1 = \pm 0.001$ as the initial guess.*

Listing 6.5.4 part A

```
1 X1=10
2 FOR I=1 TO 20
3 X2=-2+1/X1+2/(X1*X1)
4 PRINT I,TAB(8),X1
5 X1=X2
6 NEXT I
7 PRINT
8 END
```

Listing 6.5.4 part B

```
1 X1=-10
2 FOR I=1 TO 20
3 X2=-2+1/X1+2/(X1*X1)
4 PRINT I,TAB(8),X1
5 X1=X2
6 NEXT I
7 PRINT
8 END
```

Listing 6.5.4 part A, cont.

```
RUN

 1      10
 2      -1.88
 3      -1.966048
 4      -1.9912163
 5      -1.9977846
 6      -1.999445
 7      -1.9998612
 8      -1.9999653
 9      -1.9999913
10      -1.9999978
11      -1.9999995
12      -1.9999998
13      -2
14      -2
15      -2
16      -2
17      -2
18      -2
19      -2
20      -2

READY
```

Listing 6.5.4 part B, cont.

```
RUN

 1      -10
 2      -2.08
 3      -2.0184911
 4      -2.0045385
 5      -2.0011294
 6      -2.000282
 7      -2.0000705
 8      -2.0000176
 9      -2.0000044
10      -2.0000011
11      -2.0000002
12      -2.0000001
13      -2
14      -2
15      -2
16      -2
17      -2
18      -2
19      -2
20      -2

READY
```

Listing 6.5.4 *Successive-substitution iteration for* $x_{n+1} = -2 + 1/x_n + 2/x_n^2$ *using* $+10$ *and* -10 *as the initial guesses.*

Listing 6.5.5 part A

```
1 X1=10
2 FOR I=1 TO 35
3 X2=1/ATN(X1)
4 PRINT I,TAB(8),X1
5 X1=X2
6 NEXT I
7 PRINT
8 END
```

Listing 6.5.5 part B

```
1 X1=-1
2 FOR I=1 TO 35
3 X2=1/ATN(X1)
4 PRINT I,TAB(8),X1
5 X1=X2
6 NEXT I
7 PRINT
8 END
```

Listing 6.5.5 part A, cont.

RUN

1	10
2	.67975069
3	1.6750246
4	.96844693
5	1.2997646
6	1.0928804
7	1.2051848
8	1.138722
9	1.176236
10	1.1544627
11	1.166902
12	1.15973
13	1.1638436
14	1.161477
15	1.1628361
16	1.1620548
17	1.1625037
18	1.1622457
19	1.162394
20	1.1623087
21	1.1623578
22	1.1623296
23	1.1623458
24	1.1623364
25	1.1623418
26	1.1623387
27	1.1623405
28	1.1623395
29	1.1623401
30	1.1623397
31	1.16234
32	1.1623398
33	1.16234
34	1.1623398
35	1.16234

READY

Listing 6.5.5 part B, cont.

RUN

1	−1
2	−1.2732396
3	−1.1049448
4	−1.1972991
5	−1.142923
6	−1.1737159
7	−1.1558753
8	−1.1660786
9	−1.1601993
10	−1.1635726
11	−1.1616323
12	−1.1627468
13	−1.1621061
14	−1.1624743
15	−1.1622625
16	−1.1623843
17	−1.1623143
18	−1.1623545
19	−1.1623314
20	−1.1623447
21	−1.1623371
22	−1.1623414
23	−1.162339
24	−1.1623404
25	−1.1623395
26	−1.1623401
27	−1.1623397
28	−1.16234
29	−1.1623398
30	−1.16234
31	−1.1623398
32	−1.16234
33	−1.1623398
34	−1.16234
35	−1.1623398

READY

Listing 6.5.5 *Iterative solutions to finding the minima and maxima of $y(x) = x \cos x$. The two initial estimates used were $x_1 = 10$ and $x_2 = -1$.*

6.6 Forcing the $x = g(x)$ Form

The successive-substitution method at first thought appears limited to functional forms that contain a separable x (or x^2, x^3, etc.) factor. However, this is not necessarily the case. For example, consider finding the locations of the zeros of the function, $\sin x$. If x satisfies $\sin x = 0$, then it also satisfies $x = x + \sin x$. The suggested iteration form then is

$$x_{n+1} = x_n + \sin x_n = g(x_n)$$

Computer examples are shown in listing 6.6.1. The convergence is very rapid and an infinite number of solutions can be expected. However, as it turns out, not all of the roots are accessible. We can see this by examining dg/dx:

$$dg/dx = 1 + \cos x$$

The zeros of $\sin x$ correspond to values of ± 1 for $\cos x$. In the vicinity of half of the zeros of $\sin x$, $dg/dx \approx 0$, thus implying fast convergence. However, near the other roots of $\sin x$, $dg/dx \approx 2$, and the iteration sequence will avoid those roots. The other set of roots can be determined by a small change in the iteration equation:

$$x_{n+1} = x_n - \sin x_n$$

This form is justified by noting that we obtained the earlier equation somewhat arbitrarily *by adding* x to each side of the equation $\sin x = 0$. Now we simply subtract. You may wish to check that the remaining roots of $\sin x$ can be found using this second iteration equation.

We can also apply the form-forcing method to the earlier examples (see section 6.5). For the example $y(x) = x - 2 \sin x$, one possible iteration equation is

$$x_{n+1} = 2x_n - 2 \sin x_n$$

The results of using this equation are shown in listing 6.6.2. The root $x = 0$ was quickly found. However, the other two roots are in regions of divergence. We will see later how a small change in the concept of forcing the form $x = g(x)$ can be used to greatly reduce convergence problems.

The second example given in section 6.5 was $y(x) = x^3 + 2x^2 - x - 2$. A possible iteration form (by subtracting x) is

$$x_{n+1} = x_n^3 - 2x_n^2 + 2x_n + 2$$

Using this equation with the initial guesses $x = 0$ and $x = -1.5$, the root $x = 2$ is quickly obtained. As an exercise, perform this exercise, and then compare your results with listing 6.5.2.

The conclusion that can be drawn from this section is that the form $x_{n+1} = g(x_n)$ can always be forced and, therefore, roots obtained with some success. However, the general procedures for choosing the specific form appear very nebulous. As we will see in the next section, several classical root-seeking algorithms can be directly derived by applying some very simple criteria for how $x_{n+1} = g(x_n)$ is obtained.

Listing 6.6.1 part A

```
1 X1=1
2 FOR I=1 TO 7
3 X2=X1+SIN(X1)
4 PRINT I,TAB(8),X1
5 X1=X2
6 NEXT I
7 PRINT
8 END
```

Listing 6.6.1 part B

```
1 X1=20
2 FOR I=1 TO 7
3 X2=X1+SIN(X1)
4 PRINT I,TAB(8),X1
5 X1=X2
6 NEXT I
7 PRINT
8 END
```

Listing 6.6.1 part C

```
1 X1=-4
2 FOR I=1 TO 7
3 X2=X1+SIN(X1)
4 PRINT I,TAB(8),X1
5 X1=X2
6 NEXT I
7 PRINT
8 END
```

```
RUN
1    1
2    1.841471
3    2.8050617
4    3.1352763
5    3.1415926
6    3.1415927
7    3.1415927

READY
```

```
RUN
1    20
2    20.912945
3    21.794055
4    21.989875
5    21.991149
6    21.991149
7    21.991149

READY
```

```
RUN
1    -4
2    -3.2431975
3    -3.1417674
4    -3.1415927
5    -3.1415927
6    -3.1415927
7    -3.1415927

READY
```

Listing 6.6.1 Results using $x_{n+1} = x_n + \sin x_n$ with the initial guesses $x_1 = 1$, 20 and -4.

```
1 X1=3.14159/2
2 FOR I=1 TO 10
3 X2=2*X1-2*SIN(X1)
4 PRINT I,TAB(8),X1
5 X1=X2
6 NEXT I
7 PRINT
8 END
```

```
RUN
 1      1.570795
 2      1.14159
 3       .4645874
 4       .03306684
 5      1.2052E-05
 6      0
 7      0
 8      0
 9      0
10      0

READY
```

Listing 6.6.2 *Successive-substitution iteration of* $x_{n+1} = 2x_n - 2 \sin x_n$ *with an initial guess of* $X1 = \pi/2$.

6.7 Formalizing the Generation of $x_{n+1} = g(x_n)$

Recall that the original goal was to determine the values of x so that $y(x) = 0$. In the last section, we arbitrarily used the iteration formula

$$x_{n+1} = x_n \pm y(x_n)$$

We could just as arbitrarily have used the following more general iteration relation:

$$x_{n+1} = x_n + cy(x_n) = g(x_n)$$

From the previous discussion (section 6.4), this iteration formula will converge if, in the immediate vicinity of the root, $|dg/dx| < 1$. This convergence criterion can be used to derive a choice for c. We find that

$$dg/dx = 1 + c\, dy/dx$$

We also know that the convergence is most rapid if $dg/dx \approx 0$. Thus, a seemingly good choice for c is

$$c = \frac{-1}{dy/dx} \qquad (6.7.1)$$

The iteration formula then becomes

$$x_{n+1} = x_n - \frac{y(x_n)}{dy(x_n)/dx} \qquad (6.7.2)$$

The above equation is called *Newton's method* for finding the locations of the zeros of a function. The method is illustrated in figure 6.7.1.

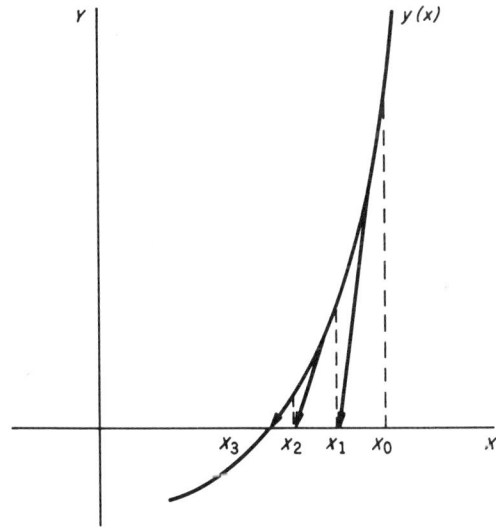

Figure 6.7.1 *Graphic example of Newton's method. In essence, the tangent line to $y(x)$ at x_n is calculated and used to project to the $y = 0$ axis. The intersection with this axis is taken as the new estimate for the root. The procedure is repeated until the desired level of convergence is achieved.*

When Newton iteration is applied to the examples given previously, the convergence is extremely good if the initial guess is reasonably close to a root. Newton's method will be considered in more detail in section 6.8.

The implementation of Newton's method requires that both the function, $y(x)$, and its derivative, dy/dx, be known and numerically evaluated in the program. This is often an inconvenience, especially if the derivative is difficult to obtain. The problem can be circumvented by numerically approximating the derivative as

$$\left.\frac{dy}{dx}\right|_{x_n} \approx \frac{y(x_n) - y(x_{n-1})}{x_n - x_{n-1}}$$

The resulting numerical iteration formula is then

$$x_{n+1} = x_n - y(x_n) \frac{x_n - x_{n-1}}{y(x_n) - y(x_{n-1})}$$

or
$$x_{n+1} = \frac{x_{n-1} y(x_n) - x_n y(x_{n-1})}{y(x_n) - y(x_{n-1})} \tag{6.7.3}$$

The latter equation embodies a concept commonly called the *secant method*. As you can see, the secant method is just an approximation to Newton's method, which itself is just a specific case of the general iteration form $x_{n+1} = x_n + cy(x_n)$. The mathematical structure of equation (6.7.3) is also very similar to the *method of false position*, or *regula-falsi*.

In the following discussion, we will examine the methods introduced in this section more carefully and develop useful subroutines for finding the locations of the zeros of functions. After presentation of the subroutines, the techniques will be compared in their ability to find the roots associated with the previous examples.

6.8 Newton's Method

If the function and its derivative are known analytically, then a simple routine can be written for Newton's method. As Hamming notes (see Ref. 16), when Newton's method closes in on a zero, the number of digits accuracy roughly doubles with each iteration step. This is termed *quadratic convergence*. However, there are conditions under which the direct application of Newton's method does not converge. This may seem confusing in that we used the idea of maximizing the convergence rate to obtain Newton's method (see section 6.7). We can examine the problem by recalling that, for Newton's method,

$$g(x) = x - \frac{y}{dy/dx}$$

The derivative of $g(x)$ is

$$\frac{dg(x)}{d(x)} = 1 - \frac{dy/dx}{dy/dx} + \frac{y(d^2y/dx^2)}{(dy/dx)^2}$$

$$= \frac{y(d^2y/dx^2)}{(dy/dx)^2} \qquad (6.8.1)$$

If $y(x)$ is nearly a straight line near the location of the zero, we have $dy/dx \approx 0$, and convergence is guaranteed. However, if there is considerable curvature to $y(x)$ $[d^2y/dx^2 \geq (dy/dx)^2/y]$, the method may converge poorly, or simply diverge. Cases in which convergence problems occur are illustrated in figures 6.8.1 and 6.8.2. See also listing 6.8.2.

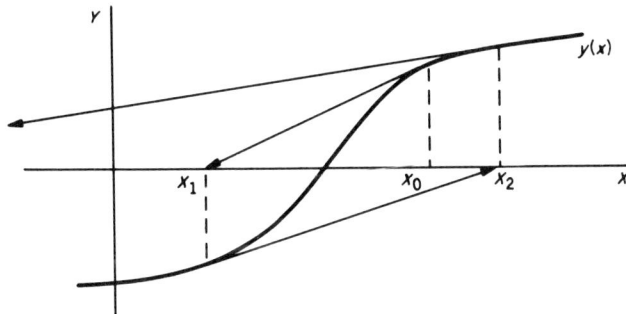

Figure 6.8.1 *Divergence of Newton's method when the root is near a strong inflection point.*

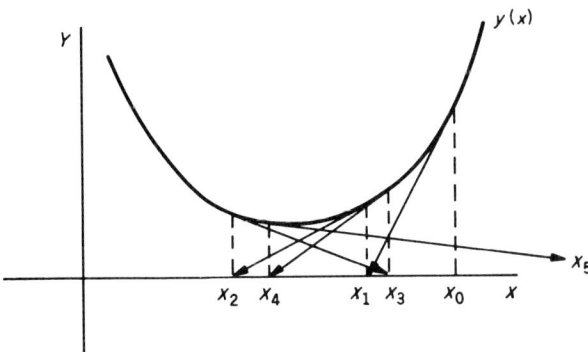

Figure 6.8.2 *Erratic behavior of Newton's method near a local minimum with $y(x) > 0$.*

368 BASIC SCIENTIFIC SUBROUTINES

Although Newton's method is very good when it works, there is a problem in making it converge quickly. The convergence rate problem can usually be overcome if the starting guess is sufficiently near the root so that $dg/dx \approx 0$ in equation (6.8.1). Thus, Newton's method is most applicable as the final routine in finding a root, where the first routine approximately locates the position of that root beforehand.

A subroutine for applying Newton's method is given in listing 6.8.1 and it is connected to the calling program as shown in figure 6.8.3. The variables passed to the subroutine are the initial guess, X0, the maximum number of iterations, M, and the desired accuracy, E. Returned is the approximate root, X, the number of iterations, N, and the last value of Y calculated. The subroutine itself calls another routine which contains the evaluation of $y(x)$, Y, and its derivative, Y1. An example of a function subroutine is also shown in listing 6.8.1 with a sample run appearing in listing 6.8.2. More will be discussed later.

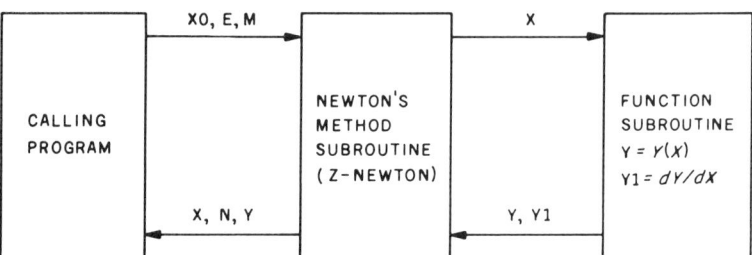

X0: Initial guess
E: Convergence criterion
M: Limit on number of iterations
X: Returned estimate of root
N: Number of iterations used
Y: Last value of Y calculated

Figure 6.8.3 *Subroutine-connection diagram for Newton's method. See table 6.8.1 for the functions and variables used.*

Statements/Functions List

−, /, >
ABS, GOSUB, GOTO, IF/THEN

Variables List

E, M, N, X, X0, Y, Y1

Variables Passed to Subroutine

E, M, M0

Table 6.8.1 *Functions and variables used in the Newton's method subroutine. These are in addition to those employed in the function subroutine.*

Listing 6.8.1

```
4050 REM PROGRAM TO DEMONSTRATE NEWTON'S METHOD
4051 PRINT
4052 PRINT
4053 PRINT "INPUT THE INITIAL GUESS: ",
4054 INPUT X0
4055 PRINT
4056 PRINT "INPUT THE CONVERGENCE FACTOR: ",
4057 INPUT E
4058 PRINT
4059 PRINT "MAXIMUM NUMBER OF ITERATIONS: ",
4060 INPUT M
4061 PRINT
4062 GOSUB 46375
4063 PRINT "THE CALCULATED ROOT IS X =",X
4064 PRINT
4065 PRINT "NUMBER OF ITERATIONS: ",N
4066 PRINT
4067 PRINT
4068 END
44298 REM ********************
44299 REM FUNCTION SUBROUTINE
44300 Y=1+5*X+10*X*X+10*X*X*X+5*X*X*X*X+X*X*X*X*X
44301 Y1=5+20*X+30*X*X+20*X*X*X+5*X*X*X*X
44302 RETURN
46364 REM ********************
46365 REM NEWTON'S METHOD SUBROUTINE (Z-NEWTON)
46366 REM THIS PROGRAM CALCULATES THE ZEROS OF A
46367 REM FUNCTION BY NEWTON'S METHOD.
46368 REM THE ROUTINE REQUIRES AN INITIAL GUESS, X0,
46369 REM AND A CONVERGENCE FACTOR, E.
46370 REM ALSO REQUIRED IS A LIMIT ON THE NUMBER
```

```
46371 REM OF ITERATIONS, M. THE NUMBER USED IS
46372 REM RETURNED IN N.
46373 REM IT IS ASSUMED THAT THE FUNCTION AND ITS
46374 REM DERIVATIVE ARE IN THE SUBROUTINE AT 44300.
46375 N=0
46376 REM GET Y AND Y1
46377 X=X0
46378 GOSUB 44300
46379 REM UPDATE ESTIMATE
46380 X0=X0-Y/Y1
46381 N=N+1
46382 IF N>=M THEN RETURN
46383 IF ABS(Y/Y1)>E THEN GOTO 46377
46384 X=X0
46385 RETURN
```

Listing 6.8.1 *Subroutine implementation of Newton's method for finding the roots of a function (Z-NEWTON).*

Listing 6.8.2

```
RUN


INPUT THE INITIAL GUESS: ?3

INPUT THE CONVERGENCE FACTOR: ?.000001

MAXIMUM NUMBER OF ITERATIONS: ?10

THE CALCULATED ROOT IS X = -.46312904

NUMBER OF ITERATIONS:   10


READY
RUN


INPUT THE INITIAL GUESS: ?3

INPUT THE CONVERGENCE FACTOR: ?.000001

MAXIMUM NUMBER OF ITERATIONS: ?100

THE CALCULATED ROOT IS X = -1.1239209
```

```
NUMBER OF ITERATIONS:   100


READY
RUN



INPUT THE INITIAL GUESS:  ?3

INPUT THE CONVERGENCE FACTOR:  ?.000001

MAXIMUM NUMBER OF ITERATIONS:  ?1000

DIVIDE ZERO ERROR IN LINE 46005
READY
```

Listing 6.8.2 *Examples of poor convergence using Newton's method (Z-NEWTON) for finding the roots of the function $y = (x + 1)^5$. This function has a fivefold multiple root at $x = -1$, thereby leading to considerable curvature in that region. According to equation (6.8.1), convergence problems are expected. Another way of viewing the problem is that Newton's method uses a division by dy/dx, and that slope is zero at $x = -1$.*

6.9 The Secant and False-Position Methods

As shown in section 6.7, the secant algorithm is, in essence, a numerical approximation to Newton's method. In principle, therefore, it works as well and converges as quickly as Newton's method. However, the secant method suffers from a numerical flaw that is obvious from its structure:

$$x_{n+1} = \frac{x_{n-1}y(x_n) - x_n y(x_{n-1})}{y(x_n) - y(x_{n-1})}$$

The Newton and secant algorithms tend to approach from one side of the zero (see figure 6.7.1). Thus, $y(x_n)$ and $y(x_{n-1})$ have the same sign. This can cause a round-off error problem if $y(x_n)$ and $y(x_{n-1})$ are nearly equal, which eventually happens. An alternative is to use the same form for the iteration equation, but to change the procedure slightly so that $y(x_n)$ and $y(x_{n-1})$ have opposite signs. The effect on the iteration is shown in figure 6.9.1. This means that x_{n+1} and x_n bracket the zero of interest. This is a better computational structure, but requires that *two* initial guesses, X0 and X1, be supplied as input to the associated subroutine. These two estimates *must* bracket the zero. Two initial guesses are also required with the secant method, but the bracketing restriction does not apply. The secant method subroutine is shown in listing 6.9.1. The false-position (regula-falsi) method subroutine is shown in listing

6.9.3. Note that the false-position subroutine contains a modification, suggested by Hamming, that speeds up convergence. The geometrical significance of this modification is shown in figure 6.9.2. The connection diagram for both the secant and the false-position subroutines is shown in figure 6.9.3. The functions and variables used are shown in table 6.9.1. Numerical comparisons of the Newton, secant, and false-position subroutines are given in the next section. Also, examples of seeking the roots of $y = (x + 1)^5$ are shown in listings 6.9.2 and 6.9.4. The Newton, secant, and false-position algorithms all have difficulty with multiple roots, but the false-position method usually is more effective in such difficult situations.

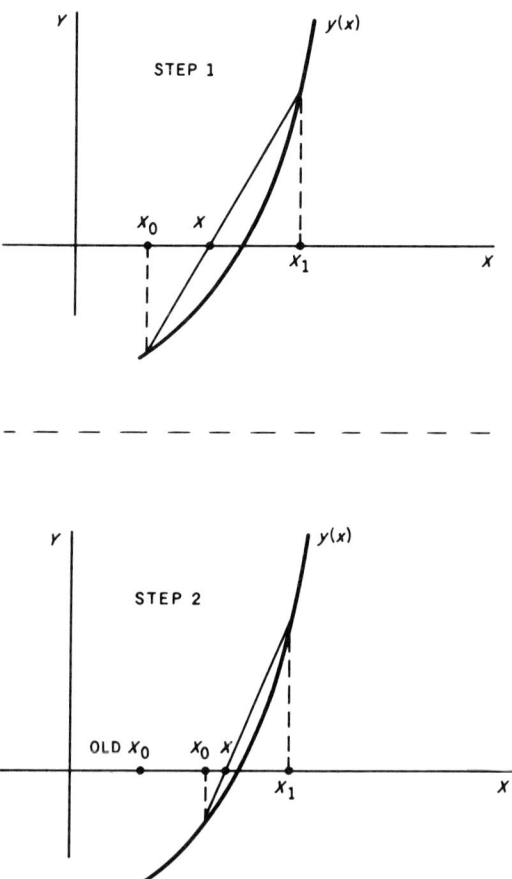

Figure 6.9.1 *An example of two steps in the false-position (regula falsi) iteration sequence. The two estimates, X_0 and X_1, bracket the root at all times.*

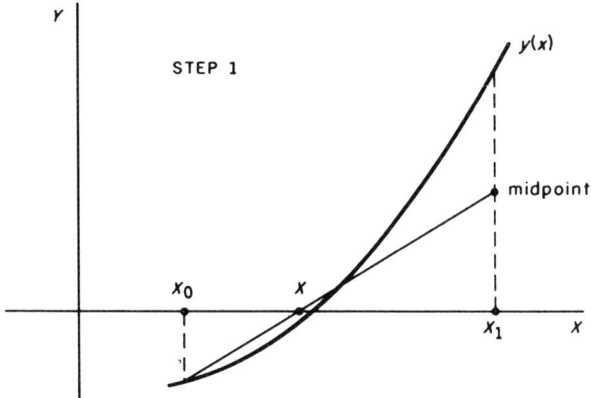

Figure 6.9.2 *Geometrical interpretation of Hamming's modification of the false-position method. Compare the position of the new estimate, X, with the corresponding result shown in figure 6.9.1.*

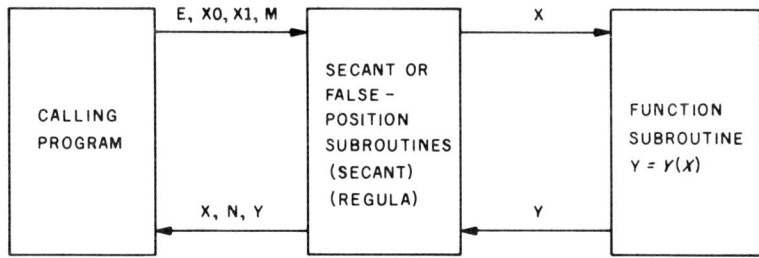

$X0, X1$: Initial guesses
E: Convergence criterion
M: Maximum number of iterations
X: Estimated root
N: Number of iterations
Y: Last Y value calculated

Figure 6.9.3 *Subroutine-connection diagram for the secant and false-position algorithms. See table 6.9.1 for the functions and variables used.*

Statements/Functions List

−, *, /
ABS

Variables List

E, X, X0, X1, Y0, Y1

Variables Passed to Subroutine

E, X0, X1

Table 6.9.1 *Functions and variables used in the secant and false-position subroutines.*

Listing 6.9.1

```
4075 REM PROGRAM TO DEMONSTRATE THE SECANT SUBROUTINE
4076 PRINT
4077 PRINT
4078 PRINT "INPUT THE TWO INITIAL GUESSES:"
4079 PRINT
4080 PRINT "X0 = ",
4081 INPUT X0
4082 PRINT "X1 = ",
4083 INPUT X1
4084 PRINT
4085 PRINT "INPUT THE CONVERGENCE FACTOR: ",
4086 INPUT E
4087 PRINT
4088 PRINT "MAXIMUM NUMBER OF ITERATIONS: ",
4089 INPUT M
4090 PRINT
4091 GOSUB 46425
4092 PRINT "THE CALCULATED ROOT IS X =",X
4093 PRINT
4094 PRINT "NUMBER OF ITERATIONS: ",N
4095 PRINT
4096 PRINT
4097 END
44298 REM *******************
44299 REM FUNCTION SUBROUTINE
44300 Y=1+5*X+10*X*X+10*X*X*X+5*X*X*X*X+X*X*X*X*X
44301 RETURN
46413 REM *******************
46414 REM SECANT METHOD SUBROUTINE (SECANT)
46415 REM THIS SUBROUTINE CALCULATES THE ZEROES OF A
46416 REM FUNCTION USING THE SECANT METHOD.
46417 REM TWO INITIAL GUESSES ARE REQUIRED, X0 AND X1.
```

```
46418 REM THE CONVERGENCE CRITERION IS E.
46419 REM THE MAXIMUM NUMBER OF ITERATIONS IS M.
46420 REM THE NUMBER OF ITERATIONS PERFORMED IS
46421 REM RETURNED IN N.
46422 REM THE RESULT IS RETURNED IN X.
46423 REM IT IS ASSUMED THAT THE FUNCTION, Y(X),
46424 REM IS IN THE SUBROUTINE AT 44300.
46425 N=0
46426 REM START ITERATION
46427 X=X0
46428 GOSUB 44300
46429 Y0=Y
46430 X=X1
46431 REM GET NEXT POINT
46432 GOSUB 44300
46433 Y1=Y
46434 REM CALCULATE NEW ESTIMATE
46435 REM IF Y1=Y0 THEN THERE WILL BE AN OVERFLOW
46436 REM GUARD AGAINST THIS ARTIFICIALLY
46437 IF Y1=Y0 THEN Y1=Y1+.001
46438 X=(X0*Y1-X1*Y0)/(Y1-Y0)
46439 N=N+1
46440 REM TEST FOR CONVERGENCE
46441 IF N>=M THEN RETURN
46442 IF ABS(X1-X0)<E THEN RETURN
46443 REM UPDATE POSITIONS
46444 X0=X1
46445 X1=X
46446 GOTO 46427
```

Listing 6.9.1 *Secant method subroutine (SECANT). See also table 6.9.1 and figure 6.9.3. Sample runs appear in listing 6.9.2.*

RUN

Listing 6.9.2

INPUT THE TWO INITIAL GUESSES:

X0 = ?2
X1 = ?4

INPUT THE CONVERGENCE FACTOR: ?.000001

MAXIMUM NUMBER OF ITERATIONS: ?10

THE CALCULATED ROOT IS X = -.24303569

```
NUMBER OF ITERATIONS:   10


READY
RUN



INPUT THE TWO INITIAL GUESSES:

X0 = ?2
X1 = ?4

INPUT THE CONVERGENCE FACTOR: ?.000001

MAXIMUM NUMBER OF ITERATIONS: ?100

THE CALCULATED ROOT IS X = -.96475163

NUMBER OF ITERATIONS:   47


READY
RUN



INPUT THE TWO INITIAL GUESSES:

X0 = ?2
X1 = ?4

INPUT THE CONVERGENCE FACTOR: ?.00000001

MAXIMUM NUMBER OF ITERATIONS: ?1000

THE CALCULATED ROOT IS X = -.96475135

NUMBER OF ITERATIONS:   1000


READY
```

Listing 6.9.2 Sample runs of the program (SECANT) given in listing 6.9.1. Compare with listing 6.8.2. Although SECANT only numerically approximates the derivative, it performs better than Z-NEWTON in this case. However, the function $y = (x + 1)^5$ is very difficult for methods similar to Z-NEWTON and SECANT, and the poor convergence is again apparent.

```
4100 PROGRAM TO DEMONSTRATE THE MODIFIED REGULA FALSI SUBROUTINE
4101 PRINT
4102 PRINT
4103 PRINT "INPUT THE TWO BRACKETTING GUESSES:"
4104 PRINT
4105 PRINT "X0 = ",
4106 INPUT X0
4107 PRINT "X1 = ",
4108 INPUT X1
4109 PRINT
4110 PRINT "INPUT THE CONVERGENCE FACTOR: ",
4111 INPUT E
4112 PRINT
4113 PRINT "MAXIMUM NUMBER OF ITERATIONS: ",
4114 INPUT M
4115 PRINT
4116 GOSUB 46475
4117 PRINT
4118 PRINT "THE CALCULATED ROOT IS X =",X
4119 PRINT
4120 PRINT "NUMBER OF ITERATIONS: ",N
4121 PRINT
4122 PRINT
4123 END
44298 REM ********************
44299 REM FUNCTION SUBROUTINE
44300 Y=1+5*X+10*X*X+10*X*X*X+5*X*X*X*X+X*X*X*X*X
44301 RETURN
46463 REM ********************
46464 REM MODIFIED FALSE POSITION SUBROUTINE (REGULA)
46465 REM SUBROUTINE USES HAMMING'S MODIFICATION TO
46466 REM SPEED CONVERGENCE.
46467 REM IT IS ASSUMED THAT THE FUNCTION Y(X) IS
46468 REM IN THE SUBROUTINE AT 44300.
46469 REM THE TWO INTITIAL GUESSES ARE X0 AND X1.
46470 REM THESE TWO GUESSES MUST BRACKET THE ZERO.
46471 REM THE CONVERGENCE CRITERION IS E.
46472 REM THE MAXIMUM NUMBER OF GUESSES IS M.
46473 REM THE RESULT IS RETURNED IN X.
46474 REM THE NUMBER OF ITERATIONS IS RETURNED IN N.
46475 N=0
46476 REM MAKE X0<X1
46477 IF X0<X1 THEN GOTO 46481
46478 X=X0
46479 X0=X1
46480 X1=X
46481 X=X0
46482 REM GET Y0 AND Y1
46483 GOSUB 44300
```

Listing 6.9.3

```
46484 Y0=Y
46485 X=X1
46486 REM INITIAL GUESSES FOR A AND B ARE REQUIRED
46487 GOSUB 44300
46488 Y1=Y
46489 REM CALCULATE A NEW ESTIMATE, X
46490 X=(X0*Y1-X1*Y0)/(Y1-Y0)
46491 REM TEST FOR CONVERGENCE
46492 N=N+1
46493 IF N>=M THEN RETURN
46494 IF ABS(X1-X)<E THEN RETURN
46495 REM GET A NEW Y(X) VALUE
46496 GOSUB 44300
46497 REM APPLY HAMMING'S MODIFICATION
46498 IF Y1*Y=0 THEN RETURN
46499 IF Y0*Y>0 THEN GOTO 46504
46500 X1=X
46501 Y1=Y
46502 Y0=Y0/2
46503 GOTO 46490
46504 X0=X
46505 Y0=Y
46506 Y1=Y1/2
46507 GOTO 46490
```

Listing 6.9.3 *Modified false-position subroutine (REGULA). See also table 6.9.1 and figure 6.9.3. Sample runs appear in listing 6.9.4.*

Listing 6.9.4

```
RUN


INPUT THE TWO BRACKETTING GUESSES:

X0 = ?-4
X1 = ?4

INPUT THE CONVERGENCE FACTOR: ?.000001

MAXIMUM NUMBER OF ITERATIONS: ?10


THE CALCULATED ROOT IS X = -1.543147

NUMBER OF ITERATIONS:   10
```

```
READY
RUN


INPUT THE TWO BRACKETTING GUESSES:

X0 = ?-4
X1 = ?4

INPUT THE CONVERGENCE FACTOR: ?.000001

MAXIMUM NUMBER OF ITERATIONS: ?100


THE CALCULATED ROOT IS X = -.99433649

NUMBER OF ITERATIONS:   44

READY
RUN


INPUT THE TWO BRACKETTING GUESSES:

X0 = ?-4
X1 = ?4

INPUT THE CONVERGENCE FACTOR: ?.00000001

MAXIMUM NUMBER OF ITERATIONS: ?1000


THE CALCULATED ROOT IS X = -.99433649

NUMBER OF ITERATIONS:   44


READY
```

Listing 6.9.4 *The difficult examples attempted with Z-NEWTON and SECANT were tried again with the modified false-position subroutine (REGULA). Although the convergence is not rapid, it is better than both of the other two methods. Round-off error kept it from proceeding beyond 44 iterations. REGULA is, on the whole, superior to the other two subroutines when dealing with such poorly convergent iterations.*

6.10 Numerical Comparisons of the Newton, Secant, and False-Position Methods

The objective of this section is to numerically compare the performances of the Newton, secant, and modified false-position subroutines when applied to the following three examples:

1) $y(x) = x - 2 \sin x$

    $dy/dx = 1 - 2 \cos x$

    initial guesses $\begin{cases} x_0 = -1 \\ x_1 = +1 \end{cases}$ and $\begin{cases} x_0 = -0.5 \\ x_1 = +0.5 \end{cases}$

2) $y(x) = x^3 + 2x^2 - x - 2$

    $dy/dx = 3x^2 + 4x - 1$

    initial guesses $\begin{cases} x_0 = 0 \\ x_1 = 2 \end{cases}$ and $\begin{cases} x_0 = -3 \\ x_1 = -1 \end{cases}$

3) $y(x) = \sin x$

    $dy/dx = \cos x$

    initial guesses $\begin{cases} x_0 = 2 \\ x_1 = 4 \end{cases}$

The comparison is in terms of the number of iterations required to achieve a given level of accuracy. The number of iterations is recorded by inserting a counter in the subroutines, and then by printing out its value when control is returned to the calling program. The results of running the above limited set of cases are shown in tables 6.10.1 through 6.10.3. There are several observations to be made regarding these results.

Although the secant algorithm is merely a numerical approximation of Newton's method in terms of estimating the required derivative by finite differences, the two algorithms respond quite differently. For example, in table 6.10.1A (initial guesses $X0 = -1$, $X1 = 1$) 37 iterations were required before the Newton subroutine locked onto the target root. The secant routine quickly converged on it in six iterations. However, when the initial estimates were $X0 = -0.5$ and $X1 = +0.5$, the Newton routine converged in only three iterations, whereas the secant routine still required six. Estimating the derivative, dy/dx, by finite differences greatly influences the behavior of the secant algorithm compared with the corresponding Newton routine. As a further example, see table 6.10.2. The two algorithms converged on different roots!

Another general observation is that the accuracy of the Newton and secant approximations is usually *much* better than E, the accuracy criterion. The reason for this is that Newton/secant convergence is quadratic; the number of digits accuracy tends to double on each iteration. If the function has little curvature in the vicinity of the root, the convergence rate can be phenomenal. This is the case in our examples. Recall that a measure of influence of curvature is given by equation (6.8.1):

$$dg/dx = \frac{y(d^2y/dx^2)}{(dy/dx)^2}$$

The closer dg/dx is to 0, the better. For $y = x - 2 \sin x$, in the vicinity of the root $x = 0$, we have

$$dg/dx \approx \frac{-x^2}{2}$$

At some stage in the iteration, the root will be located to an accuracy on the order of, but greater than, E. For $E = 0.1$, we can estimate that the effect of curvature on the *next* step will be $E^2/2$. For our case, the curvature effect is 0.005. The next estimate launched from x will then hit extremely close to the true value of the root. If the curvature effect were zero, the next estimate would be perfect. From table 6.10.1 and from the other examples, it is apparent that this behavior has occurred. Usually, at some stage in a convergent Newton or secant-iteration sequence, a point is reached after which the convergence rate is extremely rapid.

RESULTS

	Initial Estimate(s)	Convergence Criterion E	Newton Estimate	Newton # of Iterations	Secant Estimate	Secant # of Iterations	False-Position Estimate	False-Position # of Iterations
A)	$X0 = -1$ $X1 = 1$	0.1 0.01	0.00010128 0	37 38	0 -----	6 -----	0 -----	3 -----
B)	$X0 = -0.5$ $X1 = 0.5$	0.1	3×10^{-10}	3	0	6	0	3

Table 6.10.1 *A comparison of the Newton, secant, and false-position subroutines for the case* $y(x) = x - 2 \sin x$.

RESULTS

Initial Estimate(s)	Convergence Criterion E	Newton Estimate	Newton # of Iterations	Secant Estimate	Secant # of Iterations	False-Position Estimate	False-Position # of Iterations
A) $\begin{cases} X0 = 0 \\ X1 = 2 \end{cases}$	0.1	−2	2	0.99971755	16	1.0261519	6
	0.01			1.0000016	18	1.0019662	10
	0.001			1	20	0.99974383	13
	0.0001					1.0000318	16
	0.00001					1.0000020	20
	0.000001					0.99999975	23
	0.0000001					1	27
B) $\begin{cases} X0 = -3 \\ X1 = -1 \end{cases}$	0.1	−2.0003749	4	−1	4	−1	2
	0.01	−2.0000002	5				
	0.001	−2.0000002	5				
	0.0001	−2	6				

Table 6.10.2 *A comparison of the Newton, secant, and false-position subroutines for the case* $y(x) = x^3 + 2x^2 - x - 2$.

RESULTS

Initial Estimate(s)	Convergence Criterion E	Newton Estimate	Newton # of Iterations	Secant Estimate	Secant # of Iterations	False-Position Estimate	False-Position # of Iterations
$X0 = 2$	0.1	3.1415927	5	3.1415903	6	3.1242912	4
$X1 = 4$	0.01			3.1415927	8	3.1441193	7
	0.001					3.1412833	10
	0.0001					3.1415733	14
	0.00001					3.1415900	19
	0.000001					3.1415924	20
	0.0000001					3.1415927	25

Table 6.10.3 *A comparison of the Newton, secant, and false-position subroutines for the case* $y(x) = \sin x$.

The false-position method examples also exhibit diverse and interesting features. Sometimes the convergence is very fast, but other times it is slow. The accuracy achieved is occasionally *much* better than the error criterion, but at other times only comparable to it. In a wide range of comparisons, it is usually observed that the false-position routine requires more iterations than the secant method. However, the modified false-position algorithm is less susceptible to some of the slow convergence problems that can plague the other two methods. In a sense, the modified false-position algorithm combines the fast quadratic convergence rate of the Newton/secant method with the slow-but-sure qualities of the bisection algorithm.

In the next section, we will consider a method by which even the influence of curvature in the function is reduced. This is achieved by using previous estimates to reduce the effect of curvature on the subsequent iteration by means of parabolic curve fitting.

6.11 Aitken Acceleration

In this section, we return to an examination of the fundamental iteration form:

$$x = g(x)$$

Recall that in applying this equation numerically, the iteration sequence is

$$x_{n+1} = g(x_n)$$

When convergence occurs, the result is called a *fixed point*. However, as demonstrated in the earlier discussion and examples, convergence is assured only if $|dg/dx| < 1$ in the range of the iteration sequence. The closer dg/dx is to zero, the faster the convergence will be.

The convergence properties of the above iteration formula can be improved upon by using the Aitken Δ^2 method (see Ref. 17). This technique is, in essence, based on accounting for the curvature of the function as the root is approached. As shown in the previous section for the Newton and secant algorithms, the curvature of the function can greatly affect convergence if the iteration is not close to a root. Aitken (see Ref. 10) has provided a means for *accelerating* the convergence by developing an algorithm that employs three earlier calculated values to parabolically predict an improved position. The acceleration formula is

$$x'_n = x_n - \frac{(x_n - x_{n-1})^2}{x_n - 2x_{n-1} + x_{n-2}} \quad (6.11.1)$$

It is important to note that this formula requires *three* previous $x_{n+1} = g(x_n)$ iteration steps to be calculated before the acceleration can be performed. Once the accelerated value is determined, it is used as the first guess in another sequence of three iterations.

To apply this methodology to the problem of finding the locations of the zeros of $y(x)$, we will again examine the following form:

$$x = g(x) = x + cy(x) \tag{6.11.2}$$

We found in section 6.8 that the Newton, secant, and false-position formulae could be derived if $c = -1/(dy/dx)$. However, this form requires knowledge of dy/dx, which we will not resort to. Instead, a constant value will be used for c. In principle, we can always find values of c that are sufficiently small (and of sign opposite to dy/dx) to assure convergence if the iteration is near a root. However, the resulting convergence may be slow. Therefore, c can be considered an empirically chosen convergence factor. This will be discussed more in association with later examples.

A subroutine that applies the concept of Aitken acceleration to finding the roots of a function, $y(x)$, is shown in listing 6.11.1. The associated subroutine-connection diagram is shown in figure 6.11.1. This subroutine requires as input the convergence factor, c, and an error goal that causes a return when the iterative change in X_n is less than E. The subroutine also requires an initial guess, $X0$, and an iteration limit, M. Two values are returned: X and N. X is the estimate of the position of the zero, and N is the number of iterations used to obtain that estimate. The last value calculated for Y is also available.

Aitken acceleration can be very effective. For example, consider the following functions which were used in previous examples:

$y(x) = x - 2 \sin x$ (initial guess $X0 = 1.57 \cong \pi/2$)
$y(x) = x^3 + 2x^2 - x - 2$ (initial guess $X0 = -1.5$)
$y(x) = \sin x$ (initial guess $X0 = 20$)

Sample results are shown in table 6.11.2. Included in the last column of each tabulation are the results obtained using $c = 1/(dy/dx)$, Newton's method coupled with Aitken acceleration.

There are several observations to be made. First, using a constant for c instead of the Newton form leads to mixed results in terms of the number of iterations required for a given E. In the first table (table 6.11.2A), the convergence is somewhat slower for high-precision results when the Newton form for c is used rather than employing a constant. In the second table, the reverse is true. The two are comparable in the third table.

The second observation is that the accuracy of the results should be viewed with caution. The accuracy is *not* E; it can be much better or much worse. However, E can be used to *roughly* estimate the error.

It has been suggested by Hamming that instead of specifying an error limit E, the number of iterations should be used as the iteration termination criterion. However, by doing this, you would have *no* idea of the resulting accuracy. But, as a counterargument, fatal loops (possibly due to round-off error) could be avoided by setting the number of iterations (M). In this book, we use both methods by employing E as the termination criterion and M to avoid infinite loops.

The third observation is that there appears to be considerable freedom in the choice of values for the convergence factor, c. It is usually the case that the smaller c is, the slower the convergence. However, large values of c can result in leaps in the iteration, which in turn lead to different roots. This is evident in all of the tables. Also, the sign of c may or may not affect which root is converged on, but can affect the convergence rate.

The power of the Aitken acceleration method rests in its stable convergence properties. Initial guesses for the locations of the zeros need not be as accurate as those required for the Newton and secant methods. Also, the number of iterations required is reasonably constant. As with any root-seeking method, however, there are limitations.

The numerical implementation of Aitken's method suffers from a potential round-off error problem associated with the denominator of the acceleration formula: $x_n - 2x_{n-1} + x_{n-2}$. When the iteration nears completion, this factor approaches zero, although the individual values, x_n, x_{n-1} and x_{n-2}, are nonzero (i.e., they are all nearly equal). This can lead to a round-off error problem at the very least, and perhaps a divide-by-zero overflow error. The subroutine as written guards against the overflow problem, but the potential round-off error can result in a reduction in the efficiency of the routine.

Another complication is the choice of the convergence factor, c. As discussed earlier, if c is chosen so that convergence occurs, however slowly, the accuracy test (the difference between successive values being less than E) may not work well. That is, successive values may be close to one another but not close to the root, and termination may occur far from the root. Also, choosing too large a value for c can lead to convergence difficulties. Unfortunately, you have no *a priori* knowledge of what the appropriate range for c is! Fortunately, the algorithm is forgiving.

Another more general difficulty exists in the evaluation of $y(x)$ itself. There are some ways to evaluate $y(x)$ that are better than others. For example, consider the function $y(x) = (x + 1)^5$. Two distinctly different computational ways of writing this equation are

$$y_1 = (X + 1)(X + 1)(X + 1)(X + 1)(X + 1)$$

and
$$y_2 = X^5 + 5X^4 + 10X^3 + 10X^2 + 5X + 1$$

The first form is more compact and faster in execution. It is also less prone to round-off error in evaluation. This can be demonstrated as follows.

Consider the evaluation of $y(x)$ near its root, say $x = -1 + E$. Under ideal conditions (i.e., no round-off), the two computations become

$$y_1 = E^5$$

and
$$y_2 = (1 - E)^5 + 5(1 - E)^4 - 10(1 - E)^3 + 10(1 - E)^2 - 5(1 - E) + 1$$

Mathematically, both expressions give $y = 0$ when $E = 0$. However, when evaluating these forms numerically, y_1 will evaluate to zero when $E = 0$, but y_2 may not because of round-off error. Because we know that there must be values of E that numerically lead to $y_2 = 0$, it is apparent that round-off error may (and very likely does) create a "root" that differs from the "true" or mathematical root by some small amount. In effect, by choosing a particular way to evaluate $y(x)$, false roots may unknowingly be introduced. Listing 6.11.3 shows how extreme the effect can be.

Given the above discussion, we can order what may go wrong with the Aitken acceleration algorithm. First, the iteration may diverge or jump around if c is too large, or if it is of the wrong sign. Second, the iteration may converge, but slowly, and the convergence test $|X_n - X_{n-1}| < E$ may prematurely terminate the iteration. Third, the convergence may be fast, but the result may be a false root caused by round-off error. This last problem is not limited to just the Aitken acceleration subroutine.

Despite all these potential problems, Aitken acceleration is a very useful technique. In the next section, we will consider a variation of this method.

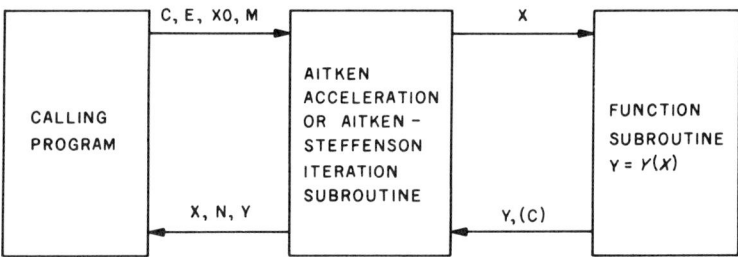

C: Convergence factor ($C \sim -1$)
E: Convergence criterion
X0: Initial guess
M: Limit on the number of iterations
N: Number of steps used
X: Returned result

Figure 6.11.1 *Subroutine-connection diagram for the Aitken acceleration and Aitken-Steffenson iteration subroutines. The functions and variables used are shown in table 6.11.1.*

Statements/Functions List

$+, -, *, /, >, <$
GOSUB, GOTO, IF/THEN

Variables List

C, E, K, M, M1, N, X, X0, X1, X2, Y

Variables Passed to Subroutine

C, E, M, X0

Table 6.11.1 *Functions and variables used by the Aitken acceleration and Aitken-Steffenson iteration subroutines. The M1 in the variables list applies to the Aitken-Steffenson subroutine only.*

Listing 6.11.1

```
4125 REM PROGRAM TO DEMONSTRATE THE AITKEN ACCELERATION SUBROUTINE
4126 PRINT
4127 PRINT
4128 PRINT "INPUT THE INITIAL GUESS: ",
4129 INPUT X0
4130 PRINT
4131 PRINT "INPUT THE CONVERGENCE CRITERION: ",
4132 INPUT E
4133 PRINT
4134 PRINT "INPUT THE CONVERGENCE FACTOR: ",
4135 INPUT C
4136 PRINT
4137 PRINT "MAXIMUM NUMBER OF ITERATIONS: ",
4138 INPUT M
4139 PRINT
4140 GOSUB 46525
4141 PRINT "THE CALCULATED ROOT IS X =",X
4142 PRINT
4143 PRINT "NUMBER OF ITERATIONS: ",N
4144 PRINT
4145 PRINT
4146 END
44298 REM *******************
44299 REM FUNCTION SUBROUTINE
44300 Y=1+5*X+10*X*X+10*X*X*X+5*X*X*X*X+X*X*X*X*X
44301 RETURN
46509 REM *******************
46510 REM AITKEN ACCELERATION SUBROUTINE (AITKEN)
46511 REM THIS ROUTINE CALCULATES THE ZEROS OF A FUNCTION
46512 REM BY ITERATION, AND EMPLOYS AITKEN ACCELERATION TO
```

```
46513 REM SPEED UP CONVERGENCE.
46514 REM REM THE SUBROUTINE REQUIRES AN INITIAL GUESS, X0,
46515 REM AND TWO CONVERGENCE FACTORS, C AND E.
46516 REM E RELATES TO THE ACCURACY OF THE ESTIMATE, AND C
46517 REM IS USED TO AID THE CONVERGENCE.
46518 REM ALSO REQUIRED IS AN ITERATION LIMIT, M.
46519 REM C=-1 IS A NORMAL VALUE. IF DIVERGENCE OCCURS,
46520 REM SMALLER AND/OR POSITIVE VALUES SHOULD BE TRIED.
46521 REM THE RESULT IS RETURNED IN X.
46522 REM THE NUMBER OF ITERATIONS IS RETURNED IN N.
46523 REM IT IS ASSUMED THAT THE FUNCTION Y(X) IS IN
46524 REM THE SUBROUTINE AT 44300.
46525 N=0
46526 X=X0
46527 REM GET Y
46528 GOSUB 44300
46529 Y=X+C*Y
46530 REM ARE THERE ENOUGH POINTS FOR ACCELERATION?
46531 IF N>0 THEN GOTO 46536
46532 X1=Y
46533 X=X1
46534 N=N+1
46535 GOTO 46528
46536 X2=Y
46537 N=N+1
46538 REM GUARD AGAINST A ZERO DENOMINATOR
46539 IF X2-2*X1+X0=0 THEN X0=X0+.001
46540 REM PERFORM ACCELERATION
46541 K=(X2-X1)*(X2-X1)/(X2-2*X1+X0)
46542 X2=X2-K
46543 REM TEST FOR CONVERGENCE
46544 IF N>=M THEN RETURN
46545 IF ABS(K)<E THEN RETURN
46546 X0=X1
46547 X1=X2
46548 X=X1
46549 GOTO 46528
```

Listing 6.11.1 *Aitken acceleration subroutine (AITKEN). The subroutine-connection diagram is shown in figure 6.11.1. See listing 6.11.2 for examples.*

Listing 6.11.2

```
         RUN



         INPUT THE INITIAL GUESS: ?3
```

```
INPUT THE CONVERGENCE CRITERION: ?.000001

INPUT THE CONVERGENCE FACTOR: ?-1

MAXIMUM NUMBER OF ITERATIONS: ?10

THE CALCULATED ROOT IS X = -.9990215

NUMBER OF ITERATIONS:   4

READY
RUN

INPUT THE INITIAL GUESS: ?3

INPUT THE CONVERGENCE CRITERION: ?.00000001

INPUT THE CONVERGENCE FACTOR: ?-1

MAXIMUM NUMBER OF ITERATIONS: ?40

THE CALCULATED ROOT IS X = -.9990215

NUMBER OF ITERATIONS:   4

READY

RUN

INPUT THE INITIAL GUESS: ?3

INPUT THE CONVERGENCE CRITERION: ?0

INPUT THE CONVERGENCE FACTOR: ?-1

MAXIMUM NUMBER OF ITERATIONS: ?50

THE CALCULATED ROOT IS X = -.99902164

NUMBER OF ITERATIONS:   50

READY
```

Listing 6.11.2 *Examples of the behavior of AITKEN for the difficult problem of finding the roots of* $y = (x + 1)^5$. *Although AITKEN also experiences difficulties in convergence for this problem, it is clearly superior in performance relative to Z-NEWTON, SECANT, and REGULA.*

A) $y(x) = x - 2\sin x$ $X0 = 1.57$
roots: 0, 1.8954943

Convergence Factor, c

E	−0.1	+0.1	−1	+1	−1/dy/dx
0.1	1.9829056(3)	2.0200624(3)	1.9999994(2)	−.083621(5)	2.0006853(2)
0.0001	1.8783217(5)	1.9056554(7)	1.8954142(7)	−.001450537(6)	1.8956790(7)
0.0000001	1.8954973(15)	1.8954979(19)	1.8954940(11)	0.00000045(8)	1.8954937(12)
True Value	1.8954943	1.8954943	1.8954943	0	1.8954943

B) $y(x) = x^3 + 2x^2 - x - 2$ $X0 = 1.5$
roots: −2, −1, 1

Convergence Factor, c

E	−0.1	+0.1	−1	+1	−1/dy/dx
0.1	0.167656(4)	0.0533662(3)	−2.0119023(4)	−0.875(2)	1 (2)
0.0001	−2.0038728(15)	−1.43(17)	−2.0001375(8)	−0.99989661(8)	1 (2)
0.0000001	−1.9999989(25)	−2.0000013(97)	−2 (16)	−0.99999989(14)	1 (2)
True Value	−2	−2	−2	−1	1

C) $y(x) = \sin x$ $X0 = 20$
roots: $n\pi$

Convergence Factor, c

E	−0.1	+0.1	−1	+1	−1/dy/dx
0.1	17.967623(3)	17.504558(9)	19.087055(2)	46.180888(3)	18.790797(3)
0.0001	21.952955(9)	12.505906(5)	18.849192(7)	47.123642(7)	18.849757(5)
0.0000001	21.99113(19)	12.566314(9)	18.849556(12)	47.12389(10)	18.849556(8)
True Value	21.99115	12.566371	18.849556	47.12389	18.849556

Table 6.11.2 Examples of Aitken acceleration. Five convergence factors were examined: $c = \pm 0.1, \pm 1$ and $-1/dy/dx$. The last factor corresponds to the term used in Newton's method. Three convergence criteria were used: $E = 0.1, 0.0001,$ and 0.0000001. The numbers in parentheses next to the calculated results represent the number of iterations that occurred.

Listing 6.11.3 part A

```
1 FOR X=-1.00001 TO -.99999 STEP .0000001
2 Y=(X+1)*(X+1)*(X+1)*(X+1)*(X+1)
3 IF Y=0 THEN PRINT X
4 NEXT X
5 PRINT
6 END
```

RUN

 -1

READY

Listing 6.11.3 part B

```
1 FOR X=-1.1 TO -.9 STEP .001
2 Y=X*X*X*X*X+5*X*X*X*X+10*X*X*X+10*X*X+5*X+1
3 IF Y=0 THEN PRINT X
4 NEXT X
5 PRINT
6 END
```

RUN

 -1.063
 -1.058
 -1.056
 -1.052
 -1.05
 -1.049
 -1.048
 -1.047
 -1.046
 -1.045
 -1.044
 -1.043
 -1.042

Listing 6.11.3 part B, cont.

```
-1.041
-1.04
-1.039
-1.038
-1.037
-1.036
-1.035
-1.034
-1.033
-1.032
-1.031
-1.03
-1.029
-1.028
-1.027
-1.026
-1.025
-1.024
-1.023
-1.021
-1.02
-1.019
-1.018
-1.017
-1.016
-1.015
-1.013
-1.012
-1.011
-1.01
-1.009
-1.008
-1.006
-1.005
-1.004
-1.003
-1.002
-1.001
-1
-.999
-.993
-.99
-.987
-.984
-.981
-.98
-.979
-.976
-.975
```

-.974
-.973
-.972
-.969
-.966
-.965
-.963
-.962

READY

Listing 6.11.3 part B, cont.

Listing 6.11.3 *An example of how the form of the evaluation may greatly influence the accuracy of the results through round-off error. The values printed are the numerical roots of y(x) according to the results of direct substitution. An 8-digit version of North Star BASIC was used to execute these programs. In the previous examples, the poor evaluation form for y(x) was intentionally used to demonstrate the susceptibility of the various algorithms to round-off error in y(x). The conclusion is that AITKEN survives this form of error well.*

6.12 Aitken-Steffenson Iteration

Aitken-Steffenson iteration is a variation on the Aitken acceleration scheme. It has very good convergence properties (quadratic) if the initial guess is in the vicinity of the root. This is to be contrasted with some of the algorithms discussed earlier which can diverge even if the initial guess is very close to the root.

As with Aitken acceleration, this method is designed to find the fixed points of $x = g(x)$. If we again define $g(x) = x + cy(x)$, then the Aitken-Steffenson iteration formula is

$$x'_n = x_n - \frac{[g(x_n) - x_n]^2}{g[g(x_n)] - 2g(x_n) + x_n} \qquad (6.12.1)$$

x'_n is employed to directly calculate x_{n+1} using $x_{n+1} = g(x'_n)$. This procedure differs from that of the Aitken acceleration algorithm which requires three prior values in order to perform the acceleration calculation.

A subroutine for applying the Aitken-Steffenson iteration scheme is shown in listing 6.12.1. The associated subroutine-connection diagram was given previously in figure 6.11.1. The results of using this subroutine to find the roots associated with the three following equations are shown in table 6.12.1.

A) $y(x) = x - 2 \sin x$ (initial guess $X0 = 1.57 \cong \pi/2$)
B) $y(x) = x^3 + 2x^2 - x - 2$ (initial guess $X_0 = -1.5$)
C) $y(x) = \sin x$ (initial guess $X0 = 20$)

If we use these examples as a guide, and compare them to the results of section

6.11, we will see that Aitken-Steffenson iteration appears to behave much the same as Aitken acceleration. But it is somewhat slower in convergence—more iterations are required. In 12 of the 60 examples, the iteration did not terminate at all, probably because of round-off error problems in the calculation.

As with Aitken acceleration, convergence can be slow if the chosen c is too small. Also, jumps can occur if c is too large and/or of the wrong sign [see example (A) for the case $E = 0.0000001$, $c = 1$].

Convergence can also be poor if $|dg/dx| \cong 1$ near the root. Because the derivative of $g(x)$ is

$$\frac{dg}{dx} = 1 + \frac{c}{dy/dx}$$

poor convergence can occur if $dy/dx \cong 0$. Geometrically, this result is not surprising.

Even with these limitations, Aitken-Steffenson iteration and its relative, Aitken acceleration, are perhaps among the most powerful of the simple root-finding techniques available.

Listing 6.12.1

```
4150 REM PROGRAM TO DEMONSTRATE AITKEN STEFFENSON ITERATIO
4151 PRINT
4152 PRINT
4153 PRINT "INPUT THE INITIAL GUESS: ",
4154 INPUT X0
4155 PRINT
4156 PRINT "INPUT THE CONVERGENCE CRITERION: ",
4157 INPUT E
4158 PRINT
4159 PRINT "INPUT THE CONVERGENCE FACTOR: ",
4160 INPUT C
4161 PRINT
4162 PRINT "MAXIMUM NUMBER OF ITERATIONS: ",
4163 INPUT M
4164 PRINT
4165 GOSUB 46575
4166 PRINT "THE CALCULATED ROOT IS X =",X
4167 PRINT
4168 PRINT "NUMBER OF ITERATIONS: ",N
4169 PRINT
4170 PRINT
4171 END
44298 REM *******************
44299 REM FUNCTION SUBROUTINE
44300 Y=X-2*SIN(X)
44301 C=1-2*COS(X)
44302 C=-1/C
```

```
44303 RETURN
46559 REM ********************
46560 REM AITKEN-STEFFENSON ITERATION SUBROUTINE (A/SITER)
46561 REM THIS ROUTINE CALCULATES THE ZEROS OF A FUNCTION
46562 REM BY ITERATION, AND EMPLOYS AITKEN ACCELERATION TO
46563 REM SPEED UP CONVERGENCE.
46564 REM REM THE SUBROUTINE REQUIRES AN INITIAL GUESS, X0,
46565 REM AND TWO CONVERGENCE FACTORS, C AND E.
46566 REM E RELATES TO THE ACCURACY OF THE ESTIMATE, AND C
46567 REM IS USED TO AID THE CONVERGENCE.
46568 REM ALSO REQUIRED IS A LIMIT TO THE NUMBER OF ITERATIONS, M.
46569 REM C=-1 IS A NORMAL VALUE. IF DIVERGENCE OCCURS,
46570 REM SMALLER AND/OR POSITIVE VALUES SHOULD BE TRIED.
46571 REM THE RESULT IS RETURNED IN X.
46572 REM THE NUMBER OF ITERATIONS IS RETURNED IN N.
46573 REM IT IS ASSUMED THAT THE FUNCTION Y(X) IS IN
46574 REM THE SUBROUTINE AT 44300.
46575 N=0
46576 M1=0
46577 X=X0
46578 REM GET Y
46579 GOSUB 44300
46580 Y=X+C*Y
46581 REM ARE THERE ENOUGH POINTS FOR ACCELERATION?
46582 IF M1>0 THEN GOTO 46588
46583 N=N+1
46584 M1=M1+1
46585 X=X1
46586 X1=Y
46587 GOTO 46579
46588 X2=Y
46589 REM PERFORM ACCELERATION
46590 REM GUARD AGAINST ZERO DENOMINATOR
46591 K=(X2-2*X1+X0)
46592 IF K=0 THEN K=.001
46593 K=(X1-X0)*(X1-X0)/K
46594 X0=X0-K
46595 REM TEST FOR CONVERGENCE
46596 IF N>=M THEN RETURN
46597 IF ABS(X-X0)<E THEN RETURN
46598 REM REPEAT PROCESS
46599 GOTO 46576
```

Listing 6.12.1 *Aitken-Steffenson iteration subroutine (A/SITER). The associated subroutine-connection diagram is shown in figure 6.11.1. The functions and variables used are shown in table 6.11.1. See listing 6.12.2 for a sample run.*

```
RUN


INPUT THE INITIAL GUESS: ?3

INPUT THE CONVERGENCE CRITERION: ?.00000001

INPUT THE CONVERGENCE FACTOR: ?-1

MAXIMUM NUMBER OF ITERATIONS: ?50

THE CALCULATED ROOT IS X = 1.8954942

NUMBER OF ITERATIONS:  50



READY
RUN


INPUT THE INITIAL GUESS: ?3

INPUT THE CONVERGENCE CRITERION: ?.000001

INPUT THE CONVERGENCE FACTOR: ?-1

MAXIMUM NUMBER OF ITERATIONS: ?50

THE CALCULATED ROOT IS X = 1.8954945

NUMBER OF ITERATIONS:  5



READY
```

Listing 6.12.2 *A sample run of the program shown in listing 6.12.1. In this case, the test function is $y(x) = x - 2 \sin x$, and $c = -1/(dy/dx)$. Note that c is calculated within the subroutine and overrides the value supplied as initial input. This subroutine fails for the previous test function $y(x) = (x + 1)^5$. AITKEN is superior to A/SITER in some cases, particularly those involving round-off error in the calculation of $y(x)$.*

A) $y(x) = x - 2 \sin x$ $\qquad$ X0 = 1.57

Convergence Factor, c

E	−0.1	+0.1	−1	1	$-1/dy/dx$
0.1	1.9105667(4)	1.8999486(4)	1.9066462(4)	1.8623822(5)	2.0006853(2)
0.0001	1.8955203(8)	1.8954486(6)	1.8955338(8)	1.8954390(13)	1.8955769(4)
0.000001	1.8954931(10)	1.8954929(10)	1.8954944(12)	1.8954938(19)	1.8954942(6)
0.0000001	-----	-----	-----	0 (39)	-----
True Value	1.8954942	1.8954942	1.8954942	0	1.8954942

B) $y(x) = x^3 + 2x^2 - x - 2$ $\qquad$ X0 = −1.5

Convergence Factor, c

E	−0.1	+0.1	−1	+1	$-1/dy/dx$
0.1	−1.5646928(3)	-----	−2.0584778(5)	−0.9918552(4)	−1.0536692(3)
0.0001	−1.999971(15)	-----	−2.0000454(13)	−0.9999888(8)	−0.99999869(5)
0.000001	−1.9999996(19)	-----	−2.0000004(18)	−0.9999996(10)	−1 (6)
0.0000001	−1.9999999(21)	-----	−2 (20)	−1 (12)	−1 (6)
True Value	−2		−2	−1	−1

C) $y(x) = \sin x$ $\qquad$ X0 = 20

Convergence Factor, c

E	−0.1	+0.1	−1	1	$-1/dy/dx$
0.1	31.440408(12)	15.659416(26)	18.849381(4)	28.283849(6)	15.813427(4)
0.0001	31.415944(22)	50.265578(104)	18.849556(5)	28.274334(6)	15.707963(6)
0.000001	-----	-----	18.849556(5)	28.274334(6)	15.707963(6)
0.0000001	-----	-----	18.849556(5)	28.274334(6)	15.707963(6)
True Value	31.415927	50.265482	18.849556	28.274334	15.707963

Table 6.12.1 *Examples of Aitken-Steffenson iteration. The numbers in parentheses next to the calculated values are the associated number of iterations. Missing values indicate a fatal loop problem. Compare with table 6.11.2.*

6.13 Comparison of Algorithms

Several algorithms for determining the roots of functions have been presented in this chapter. The purpose of this section is to review the characteristics of these algorithms in a comparative manner. Refer to table 6.13.1 for these comparisons.

It is apparent from the table that there is no perfect algorithm. However, three good candidates do exist: modified false-position, Aitken acceleration, and Aitken-Steffenson iteration. The latter two methods are potentially faster than false-position, but they can suffer from round-off error. Also, they require a convergence factor, c. The false-position scheme clearly avoids the round-off error problem, but in doing so, it requires two initial estimates that bracket the root (and only that root). One possible approach to combining the best properties of each algorithm is to use Aitken acceleration or Aitken-Steffenson iteration to approximately locate the root, and then to apply modified false-position to complete the search. As an exercise, try to implement such combinations.

Table 6.13.1

Algorithm	Advantages	Disadvantages
Interval search (Bisection)	• Most likely to converge • Requires only evaluation of function • Well-defined accuracy	• Can be tricked by close or even-multiple roots, and poles of odd order • Convergence slow • Requires two initial estimates on either side of root
Successive substitution	• Very simple • Short code • Requires only evaluation of function	• Requires algebraic manipulation to set up iteration • May not converge, even when close to root
Newton's method	• Quadratic convergence • Simple geometrical interpretation	• Convergence range limited • Convergence affected by curvature of function • Requires evaluation of function and its derivative
Secant method	• Quadratic convergence • Requires only evaluation of function • Simple geometrical interpretation	• Convergence range limited • Convergence affected by curvature of function • Round-off error problems in evaluation
Modified false-position (regula-falsi)	• Moderate convergence • Overcomes round-off error in the secant method	• Requires two initial estimates on either side of the root • Modification reduces convergence rate in many cases

Algorithm	Advantages	Disadvantages
	• Requires only evaluation of function • Modification counters curvature problem	
Aitken acceleration	• Reasonably stable convergence properties • Accounts for curvature of function • Requires only evaluation of function	• Requires convergence factor, c • Premature termination if convergence is slow • Convergence range limited • Requires three successive substitution steps before acceleration calculation • Round-off problem in evaluation
Aitken-Steffenson iteration	• Theoretically very wide convergence range • Accounts for curvature of function • Requires only evaluation of function	• Requires convergence factor, c • Premature termination if convergence rate is slow • Fatal loops possible because of round-off error

Table 6.13.1 *A comparison of algorithms for finding the zeros of functions. See also table 6.1.1.*

6.14 Finding More Roots: Multiplicity

Once a root has been found using one of the methods discussed earlier in this chapter, we are faced with the subsequent problem of continuing the analysis to find other possible roots. This section deals with a simple means to extract subsequent roots by adjusting the form of the functions $y(x)$ and dy/dx so that the algorithms already presented can be used without modification.

To proceed, we must first determine the multiplicity of the roots already found. To do this, we will write $y(x)$ in the following separable form:

$$y(x) = (x - A)^m f(x) \qquad (6.14.1)$$

This form is based on $y(x)$ being *analytical* in the region surrounding the root, and thus capable of being expanded in a Taylor series. (See Ref. 38.) If this separation is not possible, the method to be presented may fail.

We can determine m by examining the function

$$h_n(x) = \frac{y(x)}{(x - A)^n} = \frac{(x - A)^m f(x)}{(x - A)^n} \qquad (6.14.2)$$

By definition, $f(A) \neq 0$. If $n < m$, we have $h_n(A) = 0$. However, the form $y(x)/(x - A)^n$ cannot be directly evaluated at $x = A$ because the $(x - A)^n$ factor will cause a divide-by-zero error. However, $h_n(x)$ can be evaluated very near A. This can be accomplished numerically as follows.

Consider a point in the close vicinity of A, say $x = A + E$. We then have

$$h_n(A + E) \approx \frac{E^m f(A)}{E^n} = E^{m-n} f(A)$$

If $n < m$, and if E is small, then $h_n(A + E) \approx 0$. When $n = m$, $h_n \approx f(A)$. Thus, by examining the sequence $h_n(A + E)(n = 1, 2, 3, ...)$, the multiplicity of a root can easily be determined by choosing the value of n that leads to a resulting $h_n(A + E)$ of roughly unity. Theoretically, this is not a sufficient grasp on the problem in that the implied assumption is that $f(A) \sim$ unity (within several orders of magnitude). In practical applications, however, this approach is more than adequate. As an example, consider the function

$$y(x) = (x - 1)^4 \tan x + (x - 1)^3 \cos x$$

To probe for the multiplicity of the root at $x = 1$, we will evaluate the sequence $h_n(1.000001)$. The results are shown in table 6.14.1. The triple multiplicity is apparent from the computed values.

n	$h_n(1.000001)$
0	5×10^{-19}
1	5×10^{-13}
2	5×10^{-7}
3	0.5
4	5×10^5

Table 6.14.1 *An example of probing for the multiplicity of a root. The value of h_n closest to unity indicates the multiplicity.*

The heuristic technique described above assumes that the function $y(x)$ can be factored into the form $(x - a)^m f(x)$ in the region near the root. This is a reasonable expectation if the function can be expanded in a Taylor series about the position of the root. For example, $y(x) = \sin^3 x$ is certainly not a simple polynomial containing only a few terms. However, the Taylor series expansion about $x = 0$ (a root) has x^3 as its lowest order term. By our multiplicity test, we would conclude that the multiplicity of this root is three, and that the function can be written as $y(x) = x^3 f(x)$. This is a valid interpretation as long as it is understood that this functional form is *local* in the Taylor series expansion sense.

To summarize, determining the multiplicity of a root is a simple numerical task.

However, the interpretation of the results should be limited to the concept of a local Taylor series unless it is known beforehand that the function is a finite polynomial.

6.15 Finding More Roots: Removal

Consider the situation in which a root, A, has been found, and its multiplicity, m, has been determined. We now want in effect to remove this root from the function and to continue the calculation to find the locations of other zeros. A fairly simple approach would be to replace $y(x)$ with $y(x)/(x - A)^m$. As long as the roots are sufficiently well separated, this is, in principle, a valid procedure.

One possible method for removing roots from polynomials is *synthetic division*. Synthetic division is simply the organized division of one polynomial by another:

$$C(x) = \frac{A(x)}{B(x)}$$

If $A(x)$ is of degree $N1$, and $B(x)$ is of degree $N2$, then the resulting polynomial, $C(x)$, is of degree $N1 - N2$. If $B(x)$ is a factor of $C(x)$, then the division will be even, i.e., $C(x)$ will be a simple polynomial of degree $N2 - N1$ with no fractional remainders.

Synthetic division is easily implemented in a computer algorithm. The first few steps are shown below. In this notation, the polynomial coefficients are A_i, B_i, and C_i, where A_i is the coefficient of x^i, etc.

1) $\alpha = C_{N2-N1} = \dfrac{A_{N2}}{B_{N1}}$

   This gives the leading polynomial coefficient in $C(x)$.
2) $A(x) \rightarrow A(x) - x^{N2-N1} B(x)$
   $N2 \rightarrow N2 - 1$
   $A(x)$ is in effect reduced by one degree.

By repeating the first step with the new $A(x)$ obtained in the second step, the next coefficient in $C(x)$ is obtained.

A program for performing this calculation is shown in listing 6.15.1. Examples of its use are given in listing 6.15.2. Note that the program expects $N2$ to be greater than $N1$, and that it makes no assumption as to whether $B(x)$ is an even factor of $A(x)$ or not.

Caution should be exercised in using synthetic division to remove an estimated root from a polynomial in order to determine other roots. The reason for this is that removing the approximate root by dividing by the factor $(x - A)$ introduces some error (round-off, at least) into the coefficients of the reduced polynomial. The next root determined is thus further in error, as is demonstrated in listing 6.15.2. Dividing by the factor associated with this second root may cause more error in the determination of the next root. In short, the error can quickly grow with each approximate root removed. It is often better to use the following forms if error propagation problems are encountered with synthetic division:

$$y(x) \to \frac{y(x)}{(x - A)^{m_1}}$$

$$y(x) \to \frac{y(x)}{(x - A)^{m_1}(x - B)^{m_2}} \qquad (6.15.1)$$

and so on.

In this approach, directly dividing out the factors (and not using synthetic division) simply tends to keep the iteration away from the area associated with the roots already determined. This algorithm is simple to implement—you merely insert the form $y(x)/(x - A)^m$ into the subroutine where $y(x)$ is normally evaluated.

The implementation for nonmultiple roots can be made more automatic for Newton's method as follows. Assume several simple (not multiple) roots, A_i, have already been found. We are therefore interested in the form

$$F(x) = \frac{y(x)}{[(x - A_1)(x - A_2) \cdots]}$$

The function that Newton's method requires is $F/(dF/dx)$, which can be shown to be

$$F/(dF/dx) = \frac{1}{(dy/dx)/y - 1/(x - A_1) - 1/(x - A_2)} \qquad (6.15.2)$$

A subroutine for applying this concept is shown in listing 6.15.3. You input an initial value, X0, for the root to be found; the number of roots, L, already determined; an error criterion, E; an iteration limit, M; and the previous roots, A(I). The subroutine iterates until the difference between successive estimates of the new root is less than E, and then it returns X as the result.

Examples of the repeated use of this program are shown in listing 6.15.4 and table 6.15.3 for the function $y(x) = \sin(x/2)$. In each step, note how the roots found in the previous steps are removed and a new root determined. Note also that $y(x)$ is not a polynomial, but we are still able to remove roots as if it were.

This program has a problem when it comes to multiple roots. If an attempt is made to automatically remove a multiple root, a divide-by-zero error may occur in the denominator of equation (6.15.2). In that case, the form indicated by equation (6.15.1) should be used.

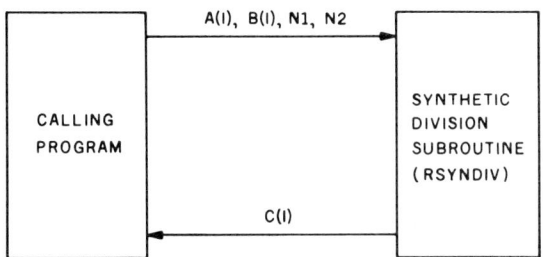

A(I): Coefficients of numerator polynomial
B(I): Coefficients of divisor polynomial
N1: Degree of $A(x)$
N2: Degree of $B(x)$
C(I): Coefficients of resulting polynomial: $C(x) = A(x)/B(x)$

Figure 6.15.1 *Synthetic division subroutine-connection diagram.*

Statements/Functions List

$-, *, /$
FOR/NEXT, GOTO, IF/THEN

Variables List

A(I), B(I), C(I), I, J, N1, N2

Variables Passed to Subroutine

A(I), B(I), N1, N2

Table 6.15.1 *Functions and variables used by the synthetic division subroutine (see listing 6.15.1).*

```
4175 REM PROGRAM TO DEMONSTRATE SYNTHETIC DIVISION (RSYNDIV)
4176 PRINT
4177 PRINT "WHAT IS THE DEGREE OF THE POLYNOMIAL"
4178 PRINT "TO BE DIVIDED INTO: ",
4179 INPUT N1
4180 DIM C(N1)
4181 PRINT
4182 PRINT "INPUT THE POLYNOMIAL COEFFICIENTS AS PROMPTED:"
4183 PRINT
4184 FOR I=0 TO N1
4185 PRINT "C(",I,") = ",
4186 INPUT C(I)
4187 NEXT I
4188 PRINT
4189 PRINT "WHAT IS THE ORDER OF THE POLYNOMIAL "
4190 PRINT "BE DIVIDED BY: ",
4191 INPUT N2
4192 DIM A(N1-N2),B(N2)
4193 PRINT
4194 PRINT "INPUT THE POLYNOMIAL COEFFICIENTS AS PROMPTED:"
4195 PRINT
4196 FOR I=0 TO N2
4197 PRINT "B(",I,") = ",
4198 INPUT B(I)
4199 NEXT I
4200 GOSUB 46625
4201 PRINT
4202 PRINT
4203 PRINT "THE COEFFICIENTS OF THE RESULTING POLYNOMIAL ARE:"
4204 PRINT
4205 FOR I=0 TO N1-N2
4206 PRINT "A(",I,") = ",A(I)
4207 NEXT I
4208 PRINT
4209 PRINT
4210 END
46617 REM ********************
46618 REM SYNTHETIC DIVISION SUBROUTINE (RSYNDIV)
46619 REM ASSUMES REAL POLYNOMIAL COEFFICIENTS.
46620 REM FORM CALCULATED IS A(X)=C(X)/B(X).
46621 REM THE INPUT POLYNOMIAL COEFFICIENTS ARE
46622 REM C(I) AND B(I), THE RESULT IS A(I).
46623 REM C(X) IS OF ORDER N1, B(X) IS OF ORDER N2.
46624 REM RESULT IS OF ORDER N1-N2 (AT MOST).
46625 FOR I=N1 TO N2 STEP -1
46626 A(I-N2)=C(I)/B(N2)
46627 IF I=N2 THEN GOTO 46631
46628 FOR J=0 TO N2
46629 C(I-J)=C(I-J)-A(I-N2)*B(N2-J)
```

Listing 6.15.1

```
46630 NEXT J
46631 NEXT I
46632 RETURN
```

Listing 6.15.1 *Synthetic division subroutine (RSYNDIV). See table 6.15.1 for the functions and variables used, and figure 6.15.1 for the subroutine-connection diagram. See also listing 6.15.2 for examples.*

Listing 6.15.2

```
RUN


WHAT IS THE DEGREE OF THE POLYNOMIAL
TO BE DIVIDED INTO: ?5

INPUT THE POLYNOMIAL COEFFICIENTS AS PROMPTED:

C( 0) = ?1
C( 1) = ?5
C( 2) = ?10
C( 3) = ?10
C( 4) = ?5
C( 5) = ?1

WHAT IS THE ORDER OF THE POLYNOMIAL
BE DIVIDED BY: ?1

INPUT THE POLYNOMIAL COEFFICIENTS AS PROMPTED:

B( 0) = ?1
B( 1) = ?1


THE COEFFICIENTS OF THE RESULTING POLYNOMIAL ARE:

A( 0) =   1
A( 1) =   4
A( 2) =   6
A( 3) =   4
A( 4) =   1


READY
RUN
```

```
WHAT IS THE DEGREE OF THE POLYNOMIAL
TO BE DIVIDED INTO: ?5

INPUT THE POLYNOMIAL COEFFICIENTS AS PROMPTED:

C( 0) = ?1
C( 1) = ?5
C( 2) = ?10
C( 3) = ?10
C( 4) = ?5
C( 5) = ?1

WHAT IS THE ORDER OF THE POLYNOMIAL
BE DIVIDED BY: ?1

INPUT THE POLYNOMIAL COEFFICIENTS AS PROMPTED:

B( 0) = ?.999999
B( 1) = ?1


THE COEFFICIENTS OF THE RESULTING POLYNOMIAL ARE:

A( 0) =   1.000001
A( 1) =   4.000003
A( 2) =   6.000003
A( 3) =   4.000001
A( 4) =   1


READY
```

Listing 6.15.2 *Two examples of synthetic division. In the first example, the exact root, $x = -1$, is used to deflate the original quintic polynomial. The resulting coefficients are correct. In the second example, an approximation to the root is used. The coefficients of the resulting quartic polynomial are therefore slightly in error. This polynomial has the root $x = -1.000003$, which is in even greater error than the first approximate root $x = 0.999999$. Because of round-off error, the roots of the resulting polynomial, $C(x)$, may significantly differ from those of $A(x)$. This error rapidly compounds with repeated division.*

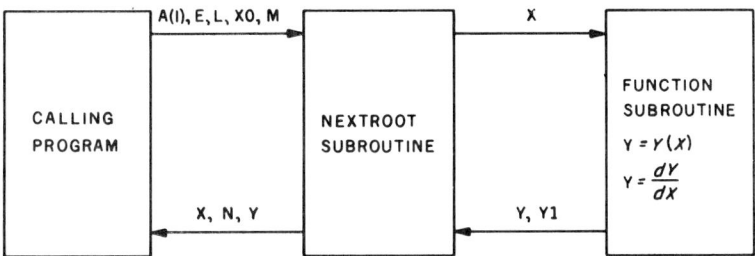

L: Number of roots already found
A(I): Roots already found
X0: Initial guess
E: Convergence criterion
M: Maximum number of iterations
X: Returned estimate of new root
N: Number of iterations performed
Y: Last value of Y calculated

Figure 6.15.2 *Connection diagram for NEXTROOT, a subroutine that seeks the roots of a function, disregarding the previously found values, A(1), A(2), . . . , A(L). The convergence criterion is E, and the initial guess is X0. The maximum number of iterations is M; the number actually performed, N.*

Statements/Functions List

$-, /, <$
ABS

Variables List

A(I), B, E, I, L, M, N, X, X0, X1, Y, Y1

Variables Passed to Subroutine

A(I), E, L, M, X0

Table 6.15.2 *Functions and variables used by the NEXTROOT subroutine (see also listing 6.15.3 and figure 6.15.2).*

```
4225 REM PROGRAM TO DEMONSTRATE NEXTROOT SUBROUTINE
4226 PRINT
4227 PRINT
4228 PRINT "HOW MANY ROOTS HAVE BEEN DETERMINED: ",
4229 INPUT L
4230 PRINT
4231 PRINT
4232 PRINT "INPUT THE ROOTS AS PROMPTED:"
4233 PRINT
4234 FOR I=1 TO L
4235 PRINT "A(",I,") = ",
4236 INPUT A(I)
4237 NEXT I
4238 PRINT
4239 PRINT "WHAT IS THE INITIAL GUESS: ",
4240 INPUT X0
4241 PRINT
4242 PRINT "WHAT IS THE CONVERGENCE CRITERION: ",
4243 INPUT E
4244 PRINT
4245 PRINT "MAXIMUM NUMBER OF ITERATIONS: ",
4246 INPUT M
4247 PRINT
4248 GOSUB 46650
4249 PRINT
4250 PRINT "THE CALCULATED ROOT IS X =",X
4251 PRINT
4252 PRINT "NUMBER OF ITERATIONS: ",N
4253 PRINT
4254 PRINT
4255 END
44298 REM ********************
44299 REM FUNCTION SUBROUTINE
44300 Y=SIN(X/2)
44301 Y1=.5*COS(X/2)
44302 RETURN
46641 REM ********************
46642 REM SUBROUTINE FOR DETERMINING ADDITIONAL ROOTS OF
46643 REM A FUNCTION GIVEN A SET OF ALREADY ESTABLISHED ROOTS (NEXTROOT)
46644 REM USE IS RESTRICTED TO REAL ROOTS.
46645 REM METHOD APPLIED IS NEWTON-RAPHSON ITERATION.
46646 REM THE L ESTABLISHED ROOTS ARE A(I).
46647 REM THE FUNCTION Y AND ITS DERIVATIVE ARE PLACED IN SUBROUTINE 44300.
46648 REM THE INITIAL GUESS IS X0.
46649 REM THE ACCURACY CRITERIA IS E.
46650 N=0
46651 REM GIVEN X0, FIND F/F'.
46652 X=X0
46653 GOSUB 44300
```

Listing 6.15.3

```
46654 B=Y1/Y
46655 FOR I=1 TO L
46656 B=B-1/(X0-A(I))
46657 NEXT I
46658 REM NEWTON-RAPHSON ITERATION
46659 X1=X0-1/B
46660 N=N+1
46661 REM TEST FOR CONVERGENCE
46662 IF N>=M THEN GOTO 46666
46663 IF ABS(X1-X0)<E THEN GOTO 46666
46664 X0=X1
46665 GOTO 46652
46666 X=X1
46667 RETURN
```

Listing 6.15.3 *Newton iteration subroutine for finding the roots of y(x) given L previously determined roots (NEXTROOT). See figure 6.15.2 for the subroutine-connection diagram.*

Listing 6.15.4

```
RUN

HOW MANY ROOTS HAVE BEEN DETERMINED: ?1

INPUT THE ROOTS AS PROMPTED:

A( 1 ) = ?0

WHAT IS THE INITIAL GUESS: ?.1

WHAT IS THE CONVERGENCE CRITERION: ?.001

MAXIMUM NUMBER OF ITERATIONS: ?20

THE CALCULATED ROOT IS X = 119.38052

NUMBER OF ITERATIONS:   4

READY
RUN


HOW MANY ROOTS HAVE BEEN DETERMINED: ?2
```

```
INPUT THE ROOTS AS PROMPTED:

A( 1 ) = ?0
A( 2 ) = ?119.38052

WHAT IS THE INITIAL GUESS: ?119

WHAT IS THE CONVERGENCE CRITERION: ?.001

MAXIMUM NUMBER OF ITERATIONS: ?20


THE CALCULATED ROOT IS X = 75.398223

NUMBER OF ITERATIONS:   4


READY
RUN



HOW MANY ROOTS HAVE BEEN DETERMINED: ?3


INPUT THE ROOTS AS PROMPTED:

A( 1 ) = ?0
A( 2 ) = ?119.38052
A( 3 ) = ?75.398223

WHAT IS THE INITIAL GUESS: ?75

WHAT IS THE CONVERGENCE CRITERION: ?.001

MAXIMUM NUMBER OF ITERATIONS: ?20


THE CALCULATED ROOT IS X = 50.265481

NUMBER OF ITERATIONS:   4


READY
```

Listing 6.15.4 *Sample runs of NEXTROOT. These results are included in table 6.15.3. Note how NEXTROOT very effectively diverts the search away from the previously determined roots, but in a seemingly unpredictable manner.*

Step Number	Initial Guess X0	New Root Found
1	0.1	119.38052
2	119	75.398223
3	75	50.265481
4	50	31.415927
5	31	18.849556
6	18	12.566371
7	12	6.2831852
8	6	−6.2831853

Table 6.15.3 *An example of using NEXTROOT to remove the roots of sin (x/2) and to find others. The convergence criterion used was E = 0.001. The first root removed was $A_1 = 0$. The accuracy of the new roots found is in most cases all the digits shown.*

6.16 Conclusion

This chapter has dealt with several methods for determining the real roots of functions. None of the algorithms presented is ideal in that none could be used under all circumstances. However, this subroutine collection should provide you with an effective set of tools for seeking real roots.

There are other methods too, such as Graeffe's technique for polynomials having real roots, Wegstein's rearrangement iteration scheme, and Mueller's parabolic fitting algorithm. See Reference 10 for a discussion of the first two methods. Mueller's method, which is not limited to real roots, is described in the next chapter.

It should be kept in mind that the problem of finding the real roots of functions is contained in the task of finding complex-valued roots. Many of the techniques presented in the next chapter can therefore be used for determining real roots, but they may be more cumbersome in implementation and slower in execution.

Chapter 7

Finding the Complex Roots of Functions

7.1 Introduction

The concept of complex numbers is a mathematical abstraction that has particular importance to the physical and engineering sciences. It is very useful in analyzing the behavior of cyclic processes such as radio communication, mechanical vibration, and general frequency-dependent phenomenon. As a result, there has been a great deal of study devoted to treating complex-valued functions and to determining specific properties such as *poles* and *zeros*.

In this chapter, we will consider two types of complex-valued functions: polynomials having a finite number of terms, and *analytic* functions. Note that the qualifier "analytic" means that the function can be expanded in a complex Taylor series. This definition naturally includes finite length polynomials, which makes our classification mathematically vague. However, there is an important practical difference involved in this classification that relates to the distinctly different functional forms involved and their influence on the algorithms to be employed. Thus, for the purposes of the following discussion, the function $P(z) = a_0 + a_1 z + a_2 z^2 + a_3 z^3$ will be treated in special ways that may not be compatible with the function $f(z) = \sin z$. However, the techniques that apply to $f(z)$ can generally be applied to $P(z)$.

Finding the complex roots of polynomials is an important goal in engineering analysis. For example, in Volume I of *Basic Scientific Subroutines*, the simultaneous differential equations associated with the compound pendulum were shown to result in a matrix equation. This equation eventually led to the determination of the eigenvalues that describe the modes of oscillation of the pendulum. Finding these eigenvalues required the evaluation of a particular determinant, and naturally resulted in finding the roots of a polynomial. If there were any damping effects involved in the physical formulation of the problem, they would be evident in the complex nature of the roots.

The determination of the roots of a polynomial is also central to the Laplace transform analysis technique commonly employed in electrical engineering and elsewhere. The goal is usually to evaluate the pole (infinity points) and zero structure of the mathematical simulation of a process such as a feedback control system. As you will see later, the procedures for finding poles are very similar to those for locating zeros.

To begin the discussion, we will first briefly review some of the basic and very important computational properties of complex functions.

Table 7.1.1

SUBROUTINE	SUBROUTINE SIZE (bytes) (including support programs)	DEMON-STRATION PROGRAM SIZE (bytes)	EXECUTION SPEED	CONDITIONS/ COMMENTS
ROOTNUM (number of roots within a circle)	1648	450	4 seconds	4 evaluation points/quadrant
ZCIRCLE ("bisection" in the complex plane)	3559	660	6 seconds/ iteration	4 evaluation points/quadrant
CZNEWTON (Newton-Raphson iteration)	1048	482	0.2 - 0.4 seconds/ iteration	
MUELLER (one-dimensional parabolic iteration)	1828	374	0.5 seconds/ iteration	
MUELLER2 (two-dimensional parabolic iteration)	2373	509	1.5 seconds/ iteration	
ZMUELLER (complex plane parabolic iteration)	2577	562	1.3 seconds/ iteration	
ALLROOT (repeated root seeking for functions and polynomials)	5741	754	2 seconds/ iteration	Removes roots already found and continues on to seek others

SUBROUTINE	SUBROUTINE SIZE (bytes) (including support programs)	DEMON- STRATION PROGRAM SIZE (bytes)	EXECUTION SPEED	CONDITIONS/ COMMENTS
QUADRAT (high accuracy quadratic roots)	1029	403	0.3 seconds	
LIN (polynomial roots by Lin's method)	1448	562	0.4 seconds/ iteration	Two root pairs of a 6th-degree polynomial
BAIRSTOW (polynomial roots by Bairstow's method)	1507	566	0.7 seconds/ iteration	Two root pairs of a 6th-degree polynomial

Table 7.1.1 *A summary of the programs given in Chapter 7. See also table 7.12.1.*

7.2 Review of the Fundamental Properties of Functions in the Complex Domain

To establish a theoretical basis for the algorithms to be presented in this chapter, we will first review a few of the fundamental features of complex functions. The complex variable, z, is defined as

$$z = x + iy$$

where x and y are real numbers and $i = \sqrt{-1}$. The complex domain elementary operations of addition, subtraction, multiplication, division, exponentiation, and roots were discussed in detail in Volume I of this series; they will not be discussed further in this section.

Some of the subroutines associated with these mathematical procedures are indirectly used in this chapter. For example, one of the techniques employed to find the complex roots of a polynomial—Newton iteration—requires treatment of the polynomial $P(z)$ and its derivative, dP/dz. Subroutines for evaluating these functions are provided in Chapter 1 of this volume, and those subroutines in turn call others which were described in Volume I. All of the required subroutines are provided in this text.

The analyses to be presented are limited to one general class of complex functions: those that are *analytic*. In simple terms, a function, $f(z)$, is said to be *analytic* in some region of the complex plane if the function and *all* of its derivatives exist in that region. It is immediately apparent that the polynomial $P(z)$ is analytic over the entire complex plane since $d^n P/dz^n$ exists for all n ($n = 1, 2, 3, \ldots$). In fact, $d^n P/dz^n$ is zero for values of n that exceed the degree of the polynomial.

It also follows that if $f(z)$ is analytic in some region, then $f(z)$ can be represented by a polynomial in that same region. The reason for this is that if all $d^n f(z)/dz^n$ exist, a Taylor series polynomial representation can surely be constructed:

$$f(z) = z_0 + (z - z_0) f^1(z) + \frac{(z - z_0)^2}{2!} f^2(z) + \cdots$$

By restricting our attention to analytic functions, we can apply many of the techniques discussed in the previous chapter, such as Newton iteration. This restriction is generally not very encumbering, and as we will see later, can even be ignored to some extent *after* we have developed the algorithms. We will impose another restriction on the analysis which, as it turns out, also does not significantly hamper the utility of the algorithms to be presented.

The polynomial root-seeking algorithms will be restricted to polynomials having real coefficients. In that case, if $z_1 = x_R + iy_R$ is a root of $P(z)$, then so is its complex conjugate $z_2 = z_1' = x_R - iy_R$. In other words, the complex roots of polynomials having real-valued coefficients come in conjugate pairs. As we will see with the Lin and Bairstow polynomial algorithms, this conjugate pair property can be capitalized on.

As in thermodynamics, a resourceful mathematician armed with a few rules can derive a wide spectrum of useful relations. In this case, we can note that if $f(z)$ is analytic in some region, then it can be represented by a Taylor series in that region. By sorting out the terms of the series appropriately, separation of the real and imaginary ($\sqrt{-1}$) parts is possible:

$$f(z) = P(z) = \mu(x,y) + i\nu(x,y) \qquad (7.2.1)$$

where $\mu(x,y)$ and $\nu(x,y)$ are real-valued functions.

The importance of this last result cannot be overemphasized. It is the key to finding the complex roots of general analytic functions using only real-number computations which, after all, are the only ones possible in a computer. We can now trade the problem of finding the complex value of z so that $f(z)$ is zero for that of finding the real values of x and y that *simultaneously* make $\mu(x,y)$ and $\nu(x,y)$ zero:

$$f(z) = 0 \longleftrightarrow \begin{cases} \mu(x,y) = 0 \\ \nu(x,y) = 0 \end{cases}$$

Besides the advantage of both $\mu(x,y)$ and $\nu(x,y)$ being real-valued functions, there are mathematical interrelations that are very useful when interpreted geometrically. These interrelations can be established by taking the derivative of $f(z)$. By recalling the fundamental definition of the derivative ("in the limit of...," etc.*), this derivative can be calculated *two* ways (i.e., from two directions):

$$\frac{df}{dz} = \frac{\partial \mu}{\partial x} + i\frac{\partial \nu}{\partial x} \quad \text{or} \quad \frac{df}{dz} = \frac{\partial \nu}{\partial y} - i\frac{\partial \mu}{\partial y}$$

By equating real and imaginary parts, we get the Cauchy-Riemann differential equations:

$$\frac{\partial \mu}{\partial x} = \frac{\partial \nu}{\partial y} \quad \text{and} \quad \frac{\partial \mu}{\partial y} = \frac{\partial \nu}{\partial x} \qquad (7.2.2)$$

The Cauchy-Riemann equations offer the convenience of reducing the derivative analysis to a consideration of only one function, say $\mu(x,y)$. However, there is also the following powerful geometric interpretation of the Cauchy-Riemann equations.

Consider the two curves $\mu(x,y) = c_1$ and $\nu(x,y) = c_2$ shown in figure 7.2.1. At

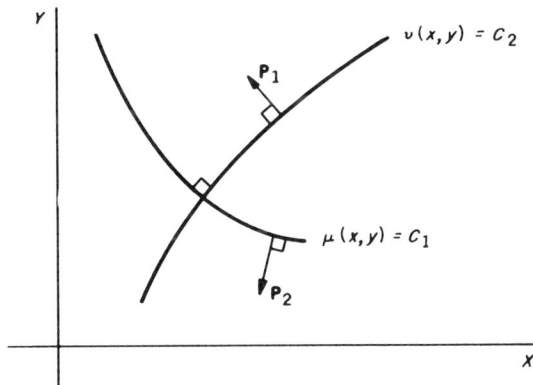

Figure 7.2.1 *If $\mu(x, y)$ and $\nu(x, y)$ are the real and imaginary parts of the analytical function $f(z)$, then the curves $\mu(x, y) = c_1$ and $\nu(x, y) = c_2$ are perpendicular at their intersection.*

every point on each curve a perpendicular to the curve can be constructed: P_1 and P_2. At the intersection of the two curves, it can be shown (by taking the vector dot product of the gradients—see Ref. 38) that the cosine of the angle between the two

---

*See any introductory text on complex variables. A very good discussion is given in Kreysig (Ref. 38).

perpendiculars to these curves at that point is

$$\cos \theta = \frac{\mu_x \nu_x + \mu_y \nu_y}{\text{normalizing factor}}$$

However, using the Cauchy-Riemann relations, it follows that the numerator is zero and thus $\cos \theta = 0$. If the cosine is zero, then the two perpendiculars themselves are orthogonal, and in turn, the $\mu(x,y) = c_1$ and $\mu(x,y) = c_2$ curves are perpendicular at their intersection. This orthogonality property is used as the basis for the root-seeking algorithm given in the next section.

We will conclude this section with a discussion of the classic *mapping* operation $z \to 1/z$. One important use of this transformation is to move the roots of a polynomial to a more convenient region. Consider the polynomial $Q(z)$ which is derived from the Nth-degree polynomial $P(z)$ as follows:

$$Q(z) = z^N P(1/z) \qquad (7.2.3)$$

It can be shown that $Q(z)$ is also a polynomial of degree N, and that its coefficients are the same as those of $P(z)$ but with the order reversed. It can also be shown that the roots of $P(z)$ that are outside the unit circle $|z| = 1$ correspond one-to-one with the roots of $Q(z)$ that are within the circle $|z| = 1$. Thus, a hard-to-find, far-flung root can be moved in a very well-defined region. This holds an advantage when deciding where to start the search for the root. A further advantage is that in the actual computation, the chances of fatal numeric overflows are reduced and replaced by nonfatal numeric underflows. However, a possible disadvantage is that after the transformation, two very large and widely separated roots may both end up close to the origin, and thereby close to one another. This can cause failures in some algorithms. As an exercise, experiment with this mapping procedure using the algorithms presented later in this chapter.

The following sections are divided into two groups. The first group deals with functions of the form $f(z) = \mu(x,y) + i\nu(x,y)$. It quickly tracks the logical sequence of development given in the previous chapter. The next group of sections specifically considers polynomials, both in terms of their treatment as general functions of the form $\mu + i\nu$, and as special finite series. The reason for this partitioning will become apparent as you proceed.

7.3 Interval Search

The first technique considered in Chapter 6 for locating real roots was the interval search (bisection method). Its key advantages were that it was conceptually simple to understand and very easy to code. The main disadvantages found were that it was intrinsically very slow in terms of how many iterations were required to reach a chosen level of accuracy, and that the root had to be initially bracketed with two guesses.

FINDING THE COMPLEX ROOTS OF FUNCTIONS

Interval searching for complex roots is even slower and more complicated. Assuming the existence of an algorithm that would determine whether or not a root was within a given *area*, the basic bisection algorithm could be applied in *two* dimensions. N^2 iterations would be expected for the complex plane calculation. For example, if $N = 20$ ($\sim 10^{-8}$ relative accuracy) were required for the one-dimensional calculation, then $N = 400$ would correspondingly be necessary for the complex plane iteration. Also, determining whether or not a complex root exists within a given region is more complicated and time-consuming than establishing a crossing of the axis in the one-dimensional case. The conclusion is that the classical bisection method for finding complex roots is too inefficient to merit further consideration, especially when *much* faster algorithms exist.

As with all root-seeking algorithms, a starting point must be chosen for the search. Because it is wasteful to probe a region that does not contain a root, it would be advantageous to have a routine that quickly tests for the existence of a root within some prescribed boundary. Such an algorithm can be developed with the help of figure 7.3.1.

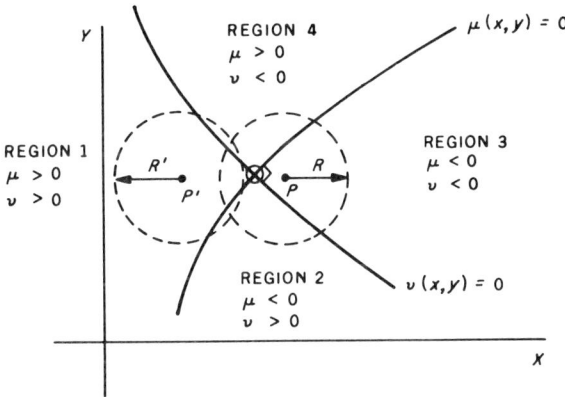

Figure 7.3.1 *The region surrounding a simple root. The desired root is located where the two curves intersect. The area surrounding the root may be broken up into four regions that are defined by the four possible sign combinations of μ and ν. The circle shown with origin P is not centered on the root, but it does pass through each of the four regions, implying that a root is contained within. The circle with origin P' does not enclose the root, nor does it pass through all four regions.*

Recall that the complex roots of $f(z)$ occur where both $\mu(x,y)$ and $\nu(x,y)$ are zero. Thus, a root must be located at the intersection of the curves $\mu(x,y) = 0$ and $\nu(x,y) = 0$. These curves intersect at right angles and thereby divide the nearby space into four quadrants. The $\mu(x,y) = 0$ curve separates the $\mu(x,y) > 0$ and

$\mu(x,y) < 0$ regions. Similarly, the $\nu(x,y) = 0$ curve separates the $\nu(x,y) > 0$ and $\nu(x,y) < 0$ regions. We can number the four quadrants according to the signs of μ and ν as shown in the figure.

We now draw a circle about the point P that has a large enough radius to contain the root. This circle passes through each of the four regions. Note that if we were to travel counterclockwise around the circle, we would pass through each of the four regions in sequence. For example, if the trip started in region 1, the sequence would be (1, 2, 3, 4). If, instead, we were to construct a circle of radius R' (centered on P'), which did not contain the root, the corresponding sequence might be (1, 4, 1, 2), which is quite different.

The algorithm suggested by these observations is the following. We draw a circle about some point in the complex plane and travel around it once counterclockwise, noting the sequence of regions passed through. If a complete cycle appears in this sequence, then an intersection of the $\mu(x,y) = 0$ and $\nu(x,y) = 0$ curves *must* be contained within. The conclusion is inescapable. If, however, no complete cycle is observed, then it is *likely* that no root is contained within, with some exceptions (see figure 7.3.2).

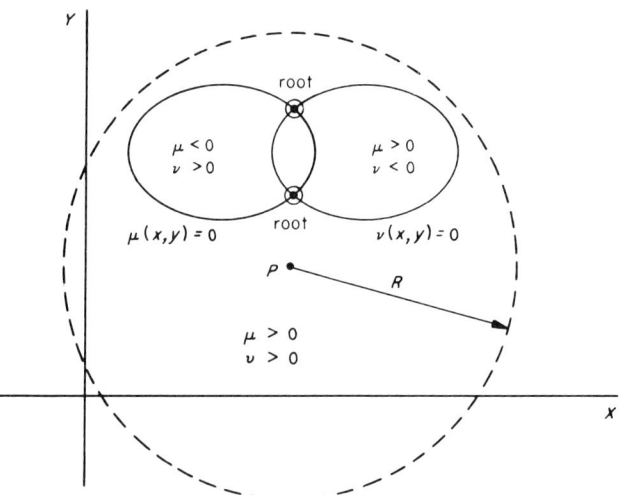

Figure 7.3.2 *A case in which two roots are contained within the test region, but the algorithm fails to note their existence.*

Unfortunately, complications are apparent before we even start to compose a program to implement this algorithm. For example, what if there are multiple roots (more than one at the same point) or several roots (more than one, but not coincident) contained within the search region?

Hamming (see Ref. 16) provides a short but good discussion of this complication. In brief, the $\mu(x,y) = 0$ and $\nu(x,y) = 0$ intersection pattern is very well defined for multiple roots (examples are shown in figure 7.3.3). We can visualize the patterns as follows. To each member of the multiple root there belongs a pair of $\mu = 0$ and $\nu = 0$ curves. When there is more than one root, there are that many pairs of curves. Near the intersection, symmetry is maintained, thereby giving the star patterns apparent in figure 7.3.3.

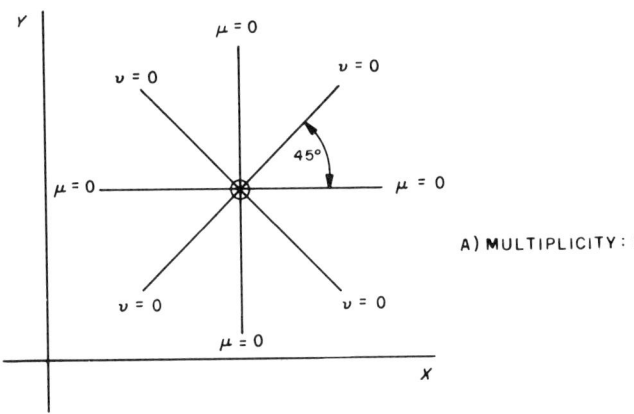

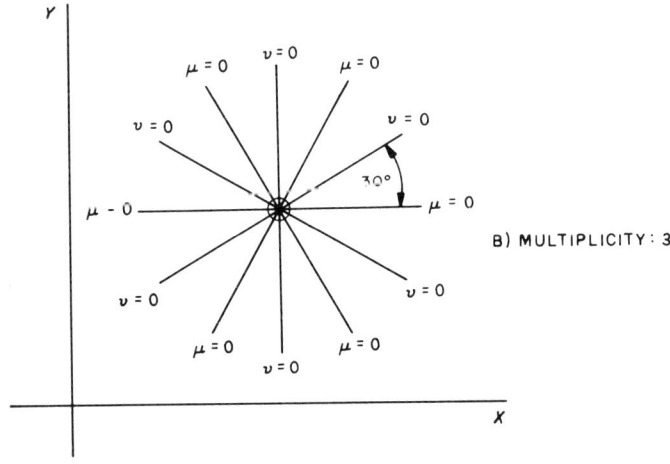

Figure 7.3.3 *Examples of the local intersection patterns associated with multiple roots.*

The pattern corresponding to two very close roots is similar to that for two multiple roots. In the example shown in figure 7.3.4, the pattern for two close roots follows that of the double root pattern shown in figure 7.3.3A when the observation point is some distance away from the roots.

Figure 7.3.4 can also be employed to examine the influence of multiple or several roots on the (μ,ν) sign sequence. Starting at point A on the circle and proceeding counterclockwise, the sequence of regions is

$$1, 2, 3, 4, 1, 2, 3, 4$$

In one trip around the circumference of the circle, two *complete* sequences occur. The modification of the algorithm therefore suggested is that complete cycles should be counted as a measure of the number of roots.

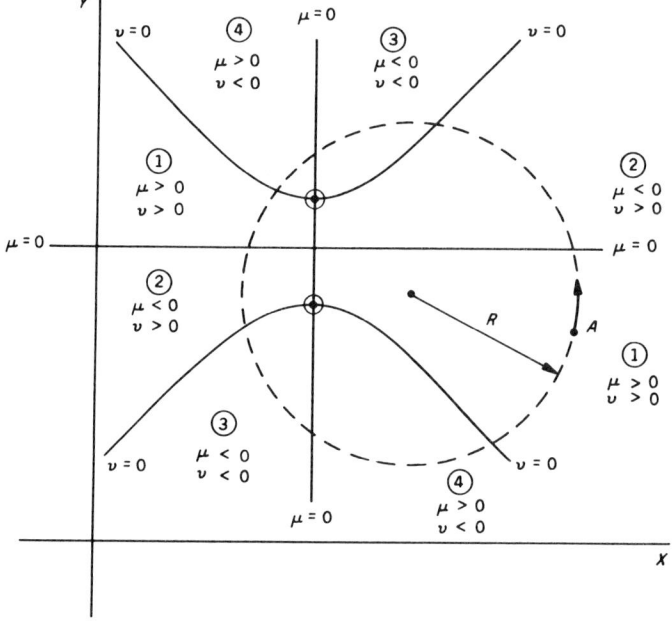

Figure 7.3.4 *Example of two close roots and how they may outwardly appear as a multiple root (compare with figure 7.3.3A).*

The concepts involved in this algorithm can be implemented in BASIC as shown in listing 7.3.1. The associated subroutine-connection diagram appears in figure 7.3.5, and the functions and variables used are listed in table 7.3.1.

The inputs to ROOTNUM are the center of the search circle, $(X0, Y0)$; the radius of the circle, W; and the number of evaluation points per quadrant of the circle, M. The returned values are N, the number of complete cycles (and thus roots) found, and A, an error measure. If $A \neq 0$, then a noninteger number of cycles was encountered and the returned number of roots, N, may be in error. The cause of the error is most likely too few evaluation points, M.

To demonstrate ROOTNUM, the function $f(z) = z^2 + 1 = (x^2 - y^2 + 1) + 2xyi$ is examined (see listing 7.3.2 and figure 7.3.6). This function has two roots—one at $z = +i$ ($x = 0$, $y = 1$), and one at $z = -i$ ($x = 0$, $y = -1$). The first three tests gave easily understood results. For the fourth test, the size of the search circle and its location were chosen so that both roots were on the circumference. The subroutine concluded that there were no roots contained within the circle, which is a reasonable result. When the circle's radius was increased slightly to encompass *both* roots (run #5), the subroutine returned with the answer that only one root was contained within the region, which is clearly incorrect. There was no hint of failure in the error check, A. The source of the discrepancy can be found in the results of run #6. Here the same circle was used, but the number of evaluation points was doubled. The correct value was returned.

The above example points out one of the precautions that should be taken when using ROOTNUM: choose a sufficiently fine sampling about the circumference of the circle to ensure that all the crossings are accounted for. In most cases, four points per circle quadrant are enough. Sometimes eight are needed. Very few functions require 16 or more unless a root is very near the edge of the circle. It should be remembered that situations such as that depicted in figure 7.3.2 can occur, and a root can be missed entirely! It should also be noted that ROOTNUM works only for complex functions.

The fundamental concepts embodied in figure 7.3.1 can be applied to develop a more efficient "bisection" algorithm than the one suggested earlier. The geometry of this new algorithm is shown in figure 7.3.7. (I say "new" because I know of no prior reference to the specific technique discussed in this text, although the analogy with the bisection method in one dimension is fairly obvious.)

It is assumed that a guess of where the root is has already been made and that ROOTNUM has been used to ensure that a root is contained within the search circle. The four intersections of the $\mu(x,y) = 0$ and $\nu(x,y) = 0$ curves with the search circle are then determined. The location of the root within the circular region is then estimated by linearly "interpolating" between the four points. This is done simply by drawing lines between associated pairs of intersection points. This is directly analogous to linear interpolation in two dimensions. The technique also embodies the same concepts as those employed in the regula-falsi method in one dimension.

The new estimate is subsequently used as the center for a new search circle having a smaller radius, and the process is repeated.

The procedure described above is conceptually straightforward. However, the details of implementation are very important to the success of the algorithm. Because no other literature reference appears to be available, some of the important details will be explained more fully below.

The first potential problem that must be guarded against is the possibility that the search circle contains multiple (or several) roots. Figure 7.3.8 shows a case in which a double root is encountered. Four intersection points are found, and a terrible new estimate far outside the search circle is made. A check could be inserted in the subroutine to ignore exterior projections and to go on to find new (μ, ν) intersections, but such a check is actually counterproductive. For example, it may happen that the root is actually outside the search circle, and the exterior estimate is valid (see figure 7.3.9). That situation might occur if the radii of the sequential-search circles are reduced too quickly, or if you simply try potluck in choosing the initial circle.

The alternative procedure is illustrated in figure 7.3.10. In essence, the object is to locate two $\mu(x,y) = 0$ intersections (with the circle) that are as far apart as possible. The same procedure applies for the two $\nu(x,y) = 0$ crossings, but with the added restriction that they be as close as possible to 90° away from the $\mu(x,y) = 0$ crossings.

So far, we have assumed that the intersections (points A, B, C, and D in figure 7.3.10) have been precisely located. However, such is not the case. Only a finite number of evaluation points (M per circle quadrant) is used in searching for the intersections. Therefore, what is actually known are the locations of the two surrounding evaluation points, (x_{i-1}, y_{i-1}) and (x_i, y_i). We can interpolate to find a good estimate of the true intersection. As an example, for the situation depicted in figure 7.3.11, the interpolation equations are

$$x = \frac{x_i \mu_{i-1} + x_{i-1} \mu_i}{\mu_{i-1} + \mu_i}$$
$$y = \frac{y_i \mu_{i-1} + y_{i-1} \mu_i}{\mu_{i-1} + \mu_i} \qquad (7.3.1)$$

Note that these equations have the same form as those used in the secant method in Chapter 6. [Actually, they *look* more like the regula-falsi formulae, but μ_i and μ_{i-1} are necessarily opposite in sign since they bracket $\mu(x,y) = 0$.]

Once the intersection points have been more accurately located, they can be called on to estimate the position of the root using the following equations:

$$x_R = -\frac{b_1 - b_2}{M_1 - M_2} \qquad (7.3.2a)$$

$$y_R = \frac{M_1 b_2 + b_1 M_2}{M_1 + M_2} \qquad (7.3.2b)$$

where
$$M_1 = \frac{y_B - y_A}{x_B - x_A}$$

$$M_2 = \frac{y_D - y_C}{x_D - x_C}$$

$$b_1 = y_A - M_1 x_A$$

$$b_2 = y_c - M_2 x_c$$

Once the next estimate for the root is found, a new circle (having a smaller radius) can be constructed, and the procedure is repeated.

A flowchart describing this iterative process is displayed in figure 7.3.12. It is implemented as shown in listing 7.3.3. The subroutine appearing in that listing contains several safeguards, including checks for divide-by-zero errors, and a special error trap for the situation in which the four required transition points cannot be found. This may happen when there is no root to be found, or when the $\mu(x,y) = 0$ and $\nu(x,y) = 0$ curves do not intersect the search circle. It may also occur when the $\mu(x,y) = 0$ and $\nu(x,y) = 0$ curves *do* cross the circle, but the evaluation-point grid (number of points/quadrant) is not fine enough. This is the same situation as the one encountered in ROOTNUM (see the previous section).

Figures 7.3.14, 7.3.15, and 7.3.16 give three examples of the application of ZCIRCLE. The intermediate results (the values occurring after each iteration) were obtained by inserting PRINT statements within the subroutine.

In the first example (figure 7.3.14), a root was contained within the original search circle, but the radius of the circle was reduced too rapidly and the subroutine failed (gracefully). In the second example, a larger initial circle was used, and the circle radius was reduced less rapidly. The result (after six iterations) was an order of magnitude more accurate than implied by the final size of the search circle. This is usually (but not always) the case after several successful iterations. In the third example (figure 7.3.16), the initial search region contained two roots, but the algorithm unambiguously closed in on only one of them. This situation contained symmetry which could have caused some difficulty in convergence. However, the determination of the $\mu(x,y) = 0$ and $\nu(x,y) = 0$ circle crossings employed a limited number of evaluation points, and this broke the symmetry slightly, thereby favoring one root.

The precautions regarding the use of ZCIRCLE relate to the input parameters. The first search circle should be large enough to contain at least one root. However, if it is too large, the algorithm may fail. (Try an initial radius of 1000 in the third example.) Second, the radius should not be reduced too quickly. A conservative reduction factor of 0.5 will usually work if the initial radius is large enough. Much smaller factors are often possible. Third, the number of evaluation points per quadrant should be at least four, and occasionally eight.

ZCIRCLE is an interesting subroutine. It is reasonably reliable, given the above precautions. As an adventurous exercise, you may wish to further probe the capabilities of this subroutine using the functions given in table 7.3.3.

It should be noted in passing that this subroutine can be used only with analytic functions in the complex domain. It should *not* be used for real-space problems [e.g., $\mu = (x,y)$, $\nu = 0$].

In the next section, we will examine the complex domain analogy of Newton's method. Newton's method is superior to ZCIRCLE in terms of the convergence rate when the initial guess is close to the true root. ZCIRCLE, however, is more stable when the initial guess is not close to the root, or when there is considerable curvature to the functions $\mu(x,y)$ and $\nu(x,y)$.

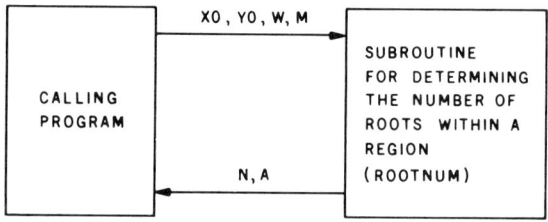

X0, Y0: Center of search circle
W: Radius of search circle
M: Number of evaluation points per quadrant of the circle
N: Number of roots found
A: Residual (error if $A \neq 0$)

Figure 7.3.5 *Subroutine-connection diagram for ROOTNUM, a program that determines the number of complex roots contained within a given circular region in the complex plane.*

Statements/Functions List

$+, -, *, /, <, >$
ABS, COS, FOR/NEXT, GOTO, IF/THEN, INT, SIN

Variables List

A, M, I, N(I), U, V, W, X, X0, Y, Y0

Variables Passed to Subroutine

M, X0, Y0, W

Table 7.3.1 *Functions and variables used in the ROOTNUM subroutine.*

```
4275 REM PROGRAM TO DEMONSTRATE THE COMPLEX ROOT COUNTING SUBROUTINE
4276 PRINT
4277 PRINT
4278 PRINT "WHERE IS THE CENTER OF THE SEARCH CIRCLE (X0,Y0):"
4279 PRINT
4280 PRINT "     X0 = ",
4281 INPUT X0
4282 PRINT "     Y0 = ",
4283 INPUT Y0
4284 PRINT
4285 PRINT "WHAT IS THE RADIUS OF THIS CIRCLE: ",
4286 INPUT W
4287 PRINT "HOW MANY EVALUATION POINTS PER QUADRANT: ",
4288 INPUT M
4289 DIM N(4*M)
4290 GOSUB 46700
4291 PRINT
4292 PRINT
4293 PRINT "NUMBER OF COMPLETE CYCLES FOUND:",
4294 PRINT N
4295 PRINT "RESIDUAL: ",
4296 PRINT A
4297 PRINT
4298 PRINT
4299 END
44298 REM ********************
44299 REM FUNCTIONS SUBROUTINE
44300 U=X*X-Y*Y+1
44301 V=2*X*Y
44302 RETURN
46679 REM ********************
46680 REM COMPLEX ROOT COUNTING SUBROUTINE (ROOTNUM)
46681 REM THIS ROUTINE CALCULATES THE NUMBER OF COMPLEX
46682 REM ROOTS WITHIN A CIRCLE OF RADIUS W CENTERED
46683 REM ON (X0,Y0) BY COUNTING (U,V) TRANSITIONS
46684 REM AROUND THE CIRCUMFERENCE.
46685 REM THE INPUT PARAMETERS ARE
46686 REM     W - THE RADIUS OF THE CIRCLE
46687 REM     (X0,Y0) - THE CENTER OF THE CIRCLE
46688 REM     M - THE NUMBER OF EVALUATION POINTS PER QUADRANT
46689 REM THE ROUTINE RETURNS THE NUMBER OF ROOTS FOUND, N
46690 REM AND THE NUMBER A, WHERE A<>0 INDICATES A FAILURE
46691 REM IN THE ALGORITHM.
46692 REM IT IS ASSUMED THAT THE FUNCTION IS COMPLEX IN THE
46693 REM DOMAIN BEING SEARCHED (U AND V BOTH HAVE TRANSITIONS).
46694 REM IT IS ALSO ASSUMED THAT THE SECTOR SPACING IS CLOSE
46695 REM ENOUGH TO CATCH ALL TRANSITIONS.
46696 REM NOTE THAT N(I) MUST BE DIMENSIONED TO 4M IN THE
46697 REM CALLING PROGRAM.
```

Listing 7.3.1

```
46698 REM OBSERVE THAT U(X,Y) AND V(X,Y) ARE EXPECTED TO BE
46699 REM FOUND IN THE FUNCTIONS SUBROUTINE.
46700 A=3.14159/(2*M)
46701 REM START CALCULATION BY ESTABLISHING THE N(I) ARRAY
46702 FOR I=1 TO 4*M
46703 X=W*COS(A*(I-1))+X0
46704 Y=W*SIN(A*(I-1))+Y0
46705 GOSUB 44300
46706 IF U>=0 THEN IF V>=0 THEN N(I)=1
46707 IF U<0 THEN IF V>=0 THEN N(I)=2
46708 IF U<0 THEN IF V<0 THEN N(I)=3
46709 IF U>=0 THEN IF V<0 THEN N(I)=4
46710 NEXT I
46711 REM COUNT COMPLETE CYCLES COUNTERCLOCKWISE
46712 N=N(1)
46713 A=0
46714 FOR I=2 TO 4*M
46715 IF N=N(I) THEN GOTO 46721
46716 IF N<>4 THEN IF N=N(I)+1 THEN A=A-1
46717 IF N=1 THEN IF N(I)=4 THEN A=A-1
46718 IF N=4 THEN IF N(I)=1 THEN A=A+1
46719 IF N+1=N(I) THEN A=A+1
46720 N=N(I)
46721 NEXT I
46722 REM COMPLETE CIRCLE
46723 IF N<>4 THEN IF N=N(1)+1 THEN A=A-1
46724 IF N=4 THEN IF N(1)=1 THEN A=A+1
46725 IF N=1 THEN IF N(1)=4 THEN A=A-1
46726 IF N+1=N(1) THEN A=A+1
46727 A=ABS(A)
46728 N=INT(A/4)
46729 A=A-4*INT(A/4)
46730 RETURN
```

Listing 7.3.1 *Subroutine for calculating the number of roots contained within a given circular search area (ROOTNUM). Also shown is a program to demonstrate ROOTNUM. See table 7.3.1, figure 7.3.5, and listing 7.3.2.*

Listing 7.3.2

```
RUN

WHERE IS THE CENTER OF THE SEARCH CIRCLE (X0,Y0):
    X0 = ?0
    Y0 = ?1
```

```
WHAT IS THE RADIUS OF THIS CIRCLE: ?.5          Listing 7.3.2 cont.
HOW MANY EVALUATION POINTS PER QUADRANT: ?4


NUMBER OF COMPLETE CYCLES FOUND: 1
RESIDUAL:  0


READY
RUN



WHERE IS THE CENTER OF THE SEARCH CIRCLE (X0,Y0):

     X0 = ?0
     Y0 = ?0

WHAT IS THE RADIUS OF THIS CIRCLE: ?4
HOW MANY EVALUATION POINTS PER QUADRANT: ?4


NUMBER OF COMPLETE CYCLES FOUND: 2
RESIDUAL:  0


READY
RUN



WHERE IS THE CENTER OF THE SEARCH CIRCLE (X0,Y0):

     X0 = ?0
     Y0 = ?0

WHAT IS THE RADIUS OF THIS CIRCLE: ?.5
HOW MANY EVALUATION POINTS PER QUADRANT: ?4


NUMBER OF COMPLETE CYCLES FOUND: 0
RESIDUAL:  0


READY
RUN



WHERE IS THE CENTER OF THE SEARCH CIRCLE (X0,Y0):
```

```
            X0 = ?0
            Y0 = ?0

    WHAT IS THE RADIUS OF THIS CIRCLE: ?1
    HOW MANY EVALUATION POINTS PER QUADRANT: ?4


    NUMBER OF COMPLETE CYCLES FOUND: 0
    RESIDUAL:  0


    READY
    RUN


    WHERE IS THE CENTER OF THE SEARCH CIRCLE (X0,Y0):

            X0 = ?0
            Y0 = ?0

    WHAT IS THE RADIUS OF THIS CIRCLE: ?1.1
    HOW MANY EVALUATION POINTS PER QUADRANT: ?4


    NUMBER OF COMPLETE CYCLES FOUND: 1
    RESIDUAL:  0


    READY
    RUN


    WHERE IS THE CENTER OF THE SEARCH CIRCLE (X0,Y0):

            X0 = ?0
            Y0 = ?0

    WHAT IS THE RADIUS OF THIS CIRCLE: ?1.1
    HOW MANY EVALUATION POINTS PER QUADRANT: ?8


    NUMBER OF COMPLETE CYCLES FOUND: 2
    RESIDUAL:  0

    READY
```

Listing 7.3.2 *Examples of using subroutine ROOTNUM for counting the roots of the function* $f(z) = z^2 + 1 = (x^2 - y^2 + 1) + 2xyi.$

Run	Circle Center	Radius	No. Points/Quadrant	No. Roots Found
1	(0, 1)	0.5	4	1
2	(0, 0)	4.0	4	2
3	(0, 0)	0.5	4	0
4	(0, 0)	1.0	4	0
5	(0, 0)	1.1	4	1
6	(0, 0)	1.1	8	2

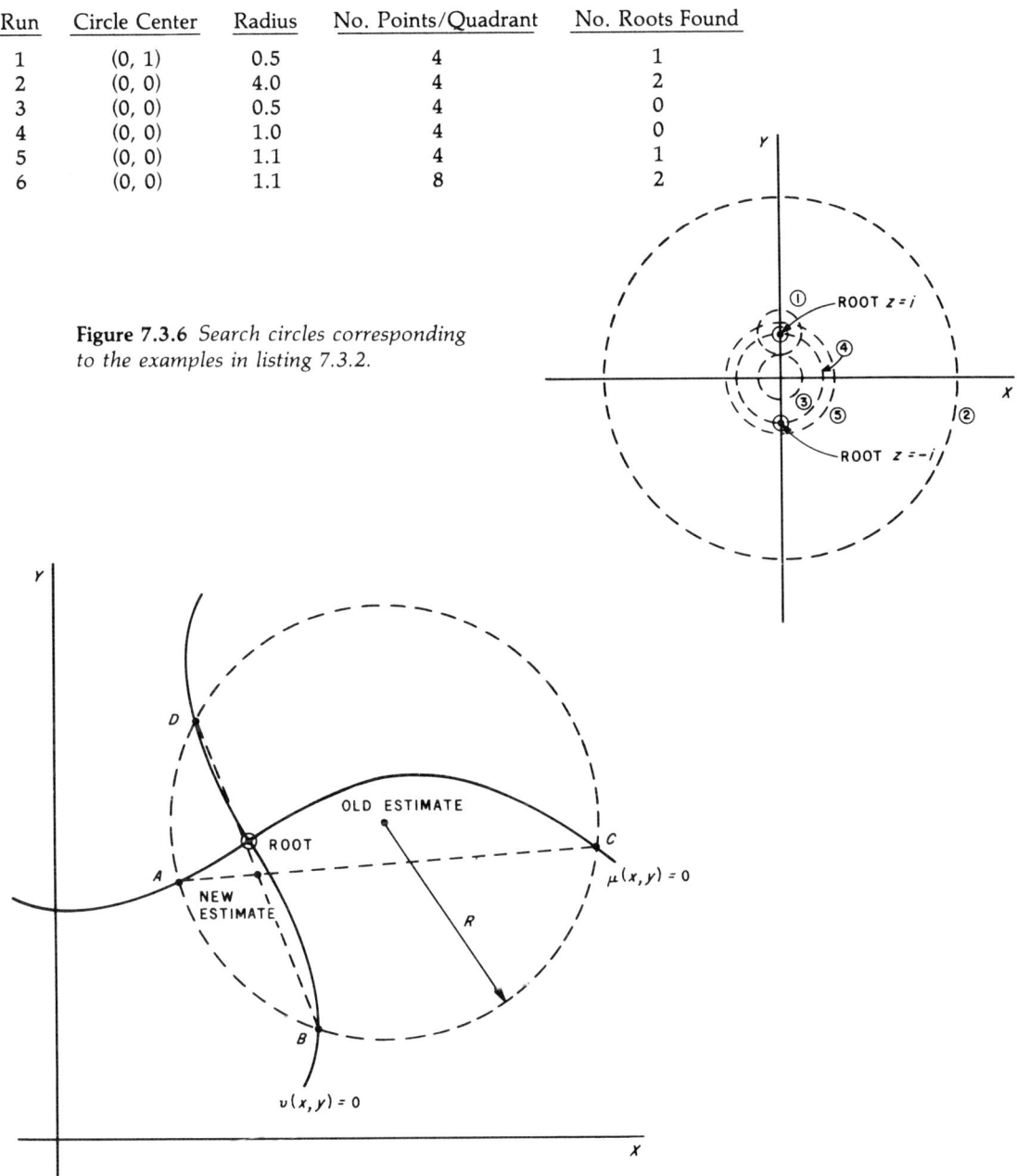

Figure 7.3.6 Search circles corresponding to the examples in listing 7.3.2.

Figure 7.3.7 The complex plane analogy of the secant method. The root is assumed to be contained within the original search circle. The crossings of the $\mu(x, y) = 0$ and $v(x, y) = 0$ curves with the search circle are found and used to estimate where the root is.

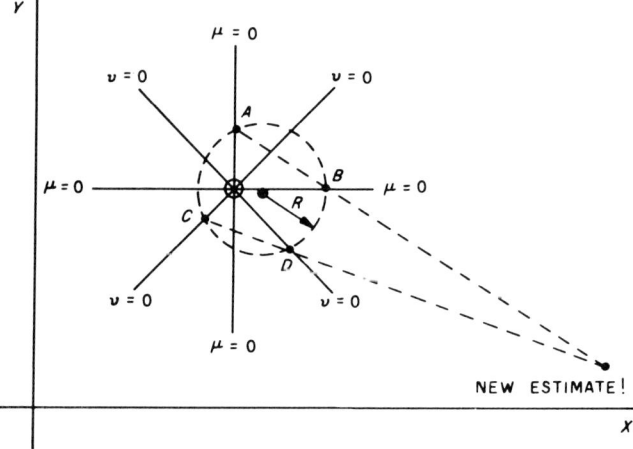

Figure 7.3.8 *A disastrous situation in which the four intersections have been found, but the new estimate is far outside the search circle!*

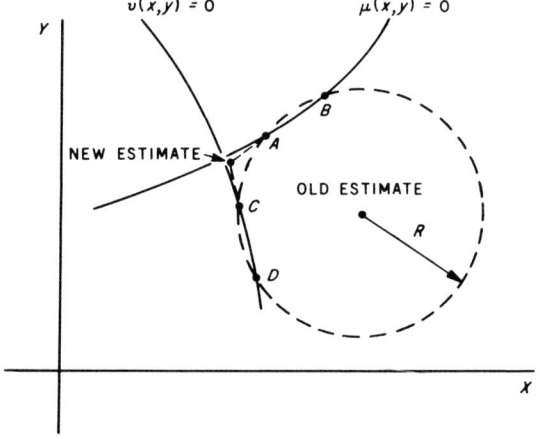

Figure 7.3.9 *Case in which the root is not within the search circle, but the algorithm locates a much better estimate anyway.*

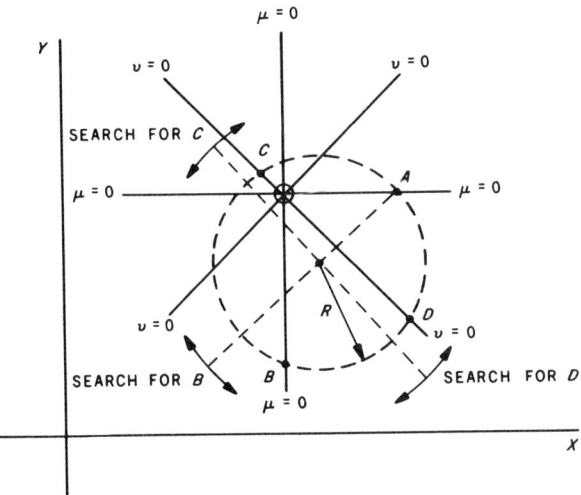

Figure 7.3.10 Procedure for finding the four intersections of the (μ, ν) curves with the search circle when confronted with multiple roots. The first step is to locate one $\mu(x, y) = 0$ crossing with the circle, point A. The second step is to go to the opposite side of the circle and again search for the nearest crossing of the $\mu(x, y) = 0$ curve with the circle, point B. The third step is to move 90° counterclockwise around the circle, starting at point A, and find the nearest $\nu(x, y) = 0$ crossing, point C. Lastly, the side of the circle opposite point C is searched for a $\nu(x, y) = 0$ crossing, point D.

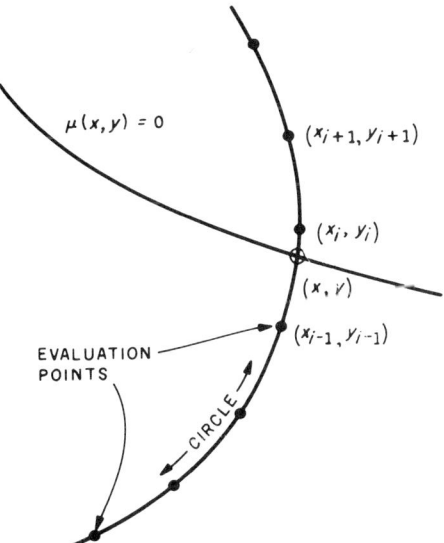

Figure 7.3.11 Expanded view of the region surrounding an intersection of the $\mu(x, y) = 0$ curve with the search circle. The true point is (x, y). The evaluation points are (x_i, y_i), and so on.

434 BASIC SCIENTIFIC SUBROUTINES

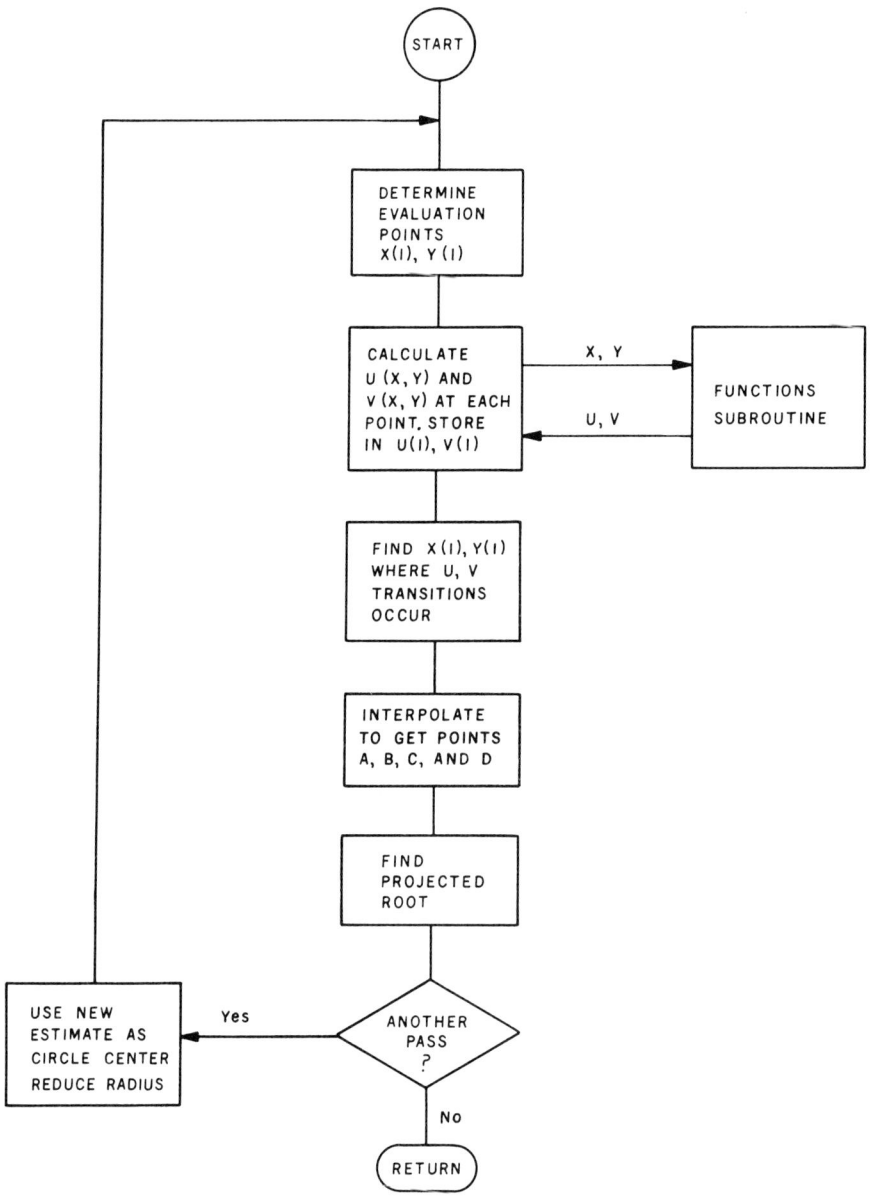

Figure 7.3.12 *General flowchart for hybrid root-searching subroutine called ZCIRCLE (see also listing 7.3.3 and figure 7.3.13).*

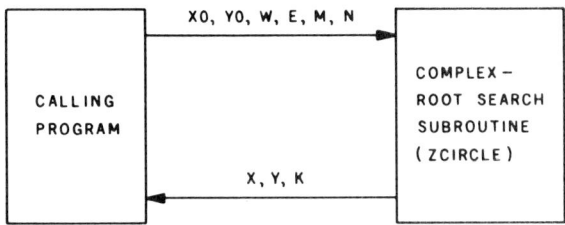

X0, Y0: Initial guess as to the location of the root. Center of first circle
W: Starting radius of circle
E: Factor by which the radius is reduced on each pass
M: Number of evaluation points per quadrant of the circle. M is changed within the subroutine
N: Maximum number of iterations to be performed
X, Y: Returned estimate
K: Actual number of iterations performed. If $K \leq N$, procedure was terminated prematurely

Figure 7.3.13 *Subroutine-connection diagram for ZCIRCLE, a complex-root searching subroutine.*

Statements/Functions List

$+, -, *, /, <, >$
COS, FOR/NEXT, GOSUB, GOTO, IF/THEN, SIN

Variables List

A, B1, B2, E, I, J, K, M, M1, M2, M3, M4, N, X, X0, X1, X2, X3, X4, X(I), Y, Y0, Y1, Y2, Y3, Y4, Y(I), U, U(I), V, V(I), W

Variables Passed to Subroutine

E, M, N, X0, Y0, W

Table 7.3.2 *Functions and variables used in ZCIRCLE, a complex-root searching subroutine.*

```
4300 REM PROGRAM TO DEMONSTRATE THE ZERO SEARCHING ALGORITHM
4301 PRINT
4302 PRINT
4303 PRINT "WHAT IS THE INITIAL GUESS FOR X AND Y: "
4304 PRINT "    X0 = ",
4305 INPUT X0
4306 PRINT "    Y0 = ",
4307 INPUT Y0
4308 PRINT
4309 PRINT "WHAT IS THE RADIUS OF THE FIRST SEARCH CIRCLE: ",
4310 INPUT W
4311 PRINT "BY WHAT FRACTION IS THIS CIRCLE TO BE REDUCED ON EACH ITERATION: ",
4312 INPUT E
4313 PRINT "HOW MANY POINTS ARE TO BE SEARCHED PER QUADRANT: ",
4314 INPUT M
4315 PRINT "HOW MANY ITERATIONS ARE TO BE EMPLOYED: ",
4316 INPUT N
4317 DIM X(4*M),Y(4*M),U(4*M),V(4*M)
4318 GOSUB 46760
4319 PRINT
4320 PRINT
4321 PRINT "THE APPROXIMATE SOLUTION IS Z = ",X,
4322 IF Y>=0 THEN PRINT " +",
4323 IF Y<0 THEN PRINT " ",
4324 PRINT Y," I"
4325 PRINT
4326 PRINT "NUMBER OF ITERATIONS: ",K
4327 PRINT
4328 END
44298 REM ********************
44299 REM FUNCTION SUBROUTINE
44300 U=X*X-Y*Y+1
44301 V=2*X*Y
44302 RETURN
46737 REM ********************
46738 REM COMPLEX ROOT SEARCH SUBROUTINE (ZCIRCLE)
46739 REM THIS PROGRAM SEARCHES FOR THE COMPLEX ROOTS
46740 REM OF AN ANALYTICAL FUNCTION BY ENCIRCLING THE
46741 REM ZERO AND ESTIMATING WHERE IT IS. THE CIRCLE
46742 REM IS SUBSEQUENTLY TIGHTENED BY A FACTOR E, AND
46743 REM A NEW ESTIMATE MADE.
46744 REM THE INPUTS TO THE SUBROUTINE ARE
46745 REM   (X0,Y0) - THE INITIAL GUESSES
46746 REM   W - THE INITIAL RADIUS OF THE SEARCH CIRCLE
46747 REM   E - THE FACTOR BY WHICH THE CIRCLE IS REDUCED
46748 REM   N - THE NUMBER OF ITERATIONS
46749 REM   M - THE NUMBER OF EVALUATION POINTS PER QUADRANT
46750 REM THE RESULTS IS RETURNED IN Z=X+IY (X,Y).
46751 REM ALSO, THE NUMBER OF ITERATIONS PERFORMED, OR
```

Listing 7.3.3

```
46752 REM IN PROGRESS, IS RETURNED IN K.
46753 REM X(I),Y(I),U(I) AND V(I) MUST BE DIMENSIONED
46754 REM IN THE CALLING PROGRAM TO 4M.
46755 REM IT IS ASSUMED THAT THE FUNCTION IS DECOMPOSED
46756 REM INTO ITS REAL AND IMAGINARY PARTS, U(X,Y) AND
46757 REM V(X,Y), AND THAT THESE ARE ACCESSIBLE BY A CALL
46758 REM TO THE FUNCTION SUBROUTINE WHICH RETURNS U AND V.
46759 REM START CALCULATION BY FINDING THE EVALUATION POINTS
46760 M=M*4
46761 K=1
46762 A=6.283185/M
46763 FOR I=1 TO M
46764 X(I)=W*COS(A*(I-1))+X0
46765 Y(I)=W*SIN(A*(I-1))+Y0
46766 NEXT I
46767 REM DETERMINE THE CORRESPONDING U(I) AND V(I)
46768 FOR I=1 TO M
46769 X=X(I)
46770 Y=Y(I)
46771 GOSUB 44300
46772 U(I)=U
46773 V(I)=V
46774 NEXT I
46775 REM FIND THE POSITION AT WHICH U CHANGES SIGN IN THE
46776 REM COUNTERCLOCKWISE DIRECTION
46777 I=1
46778 U=U(I)
46779 GOSUB 46885
46780 IF U*U(I)<0 THEN GOTO 46785
46781 REM GUARD AGAINST INFINITE LOOP
46782 IF I=1 THEN GOTO 46881
46783 GOTO 46779
46784 REM TRANSITION FOUND
46785 M1=I
46786 REM SEARCH FOR THE OTHER TRANSITION, STARTING
46787 REM ON THE OTHER SIDE OF THE CIRCLE
46788 I=M1+M/2
46789 IF I>M THEN I=I-M
46790 J=I
46791 U=U(I)
46792 REM FLIP DIRECTIONS ALTERNATELY
46793 GOSUB 46885
46794 IF U*U(I)<0 THEN GOTO 46801
46795 IF U*U(J)<0 THEN GOTO 46804
46796 REM TEST FOR INFINITE LOOP
46797 IF I=M1+M/2 THEN GOTO 46881
46798 IF J=M1+M/2 THEN GOTO 46881
46799 GOTO 46793
46800 REM TRANSITION FOUND
```

Listing 7.3.3 cont.

```
46801 M3=I                                          Listing 7.3.3 cont.
46802 GOTO 46808
46803 REM TRANSITION FOUND
46804 IF J=M THEN J=0
46805 M3=J+1
46806 REM M1 AND M3 HAVE BEEN DETERMINED. NOW FOR M2 AND M4.
46807 REM NOW FOR THE V TRANSITIONS
46808 I=M1+M/4
46809 IF I>M THEN I=I-M
46810 J=I
46811 V=V(I)
46812 GOSUB 46885
46813 IF V*V(I)<0 THEN GOTO 46820
46814 IF V*V(J)<0 THEN GOTO 46822
46815 REM AGAIN, GUARD AGAINST THE INFINITE LOOP
46816 IF I=M1+M/4 THEN GOTO 46881
46817 IF J=M1+M/4 THEN GOTO 46881
46818 GOTO 46812
46819 REM M2 HAS BEEN FOUND
46820 M2=I
46821 GOTO 46825
46822 IF J=M THEN J=0
46823 M2=J+1
46824 REM M2 HAS BEEN FOUND. NOW FOR M4
46825 I=M2+M/2
46826 IF I>M THEN I=I-M
46827 J=I
46828 V=V(I)
46829 GOSUB 46885
46830 IF U*V(I)<0 THEN GOTO 46836
46831 IF V*V(J)<0 THEN GOTO 46838
46832 REM GUARD AGAINST THE INFINITE LOOP AGAIN
46833 IF I=M2+M/2 THEN GOTO 46881
46834 IF J=M2+M/2 THEN GOTO 46881
46835 GOTO 46829
46836 M4=I
46837 GOTO 46842
46838 IF J=M THEN J=0
46839 M4=J+1
46840 REM ALL THE INTERSECTIONS HAVE BEEN DETERMINED
46841 REM INTERPOLATE TO FIND THE FOUR (X,Y) COORDINATES
46842 I=M1
46843 GOSUB 46891
46844 X1=X
46845 Y1=Y
46846 I=M2
46847 GOSUB 46891
46848 X2=X
46849 Y2=Y
```

```
46850 I=M3
46851 GOSUB 46891
46852 X3=X
46853 Y3=Y
46854 I=M4
46855 GOSUB 46891
46856 X4=X
46857 Y4=Y
46858 REM CALCULATE THE INTERSECTION OF THE LINES
46859 REM GUARD AGAINST A DIVIDE BY ZERO
46860 IF X1<>X3 THEN GOTO 46864
46861 X=X1
46862 Y=(Y1+Y3)/2
46863 GOTO 46868
46864 M1=(Y3-Y1)/(X3-X1)
46865 IF X2<>X4 THEN GOTO 46868
46866 M2=100000000
46867 GOTO 46869
46868 M2=(Y2-Y4)/(X2-X4)
46869 B1=Y1-M1*X1
46870 B2=Y2-M2*X2
46871 X=-(B1-B2)/(M1-M2)
46872 Y=(M1*B2+M2*B1)/(M1+M2)
46873 REM IS ANOTHER ITERATION IN ORDER?
46874 IF K=N THEN RETURN
46875 X0=X
46876 Y0=Y
46877 K=K+1
46878 W=W*E
46879 GOTO 46763
46880 REM INFINITE LOOP ENCOUNTERED, RETURN
46881 X=0
46882 Y=0
46883 RETURN
46884 REM AUXILLIARY SUBROUTINE
46885 I=I+1
46886 J=J-1
46887 IF I>M THEN I=I-M
46888 IF J<1 THEN J=J+M
46889 RETURN
46890 REM AUXILLIARY SUBROUTINE FOR INTERPOLATION
46891 J=I-1
46892 IF J<1 THEN J=J+M
46893 REM REGULA FALSI INTERPOLATION FOR THE ZERO
46894 X=(X(I)*U(J)+X(J)*U(I))/(U(I)+U(J))
46895 Y=(Y(I)*V(J)+Y(J)*V(I))/(V(I)+V(J))
46896 RETURN
```

Listing 7.3.3 *Subroutine (ZCIRCLE) for iteratively searching for the roots of the function $f(z) = \mu(x,y) + i\,\nu(x,y)$. Also shown is a program for demonstrating ZCIRCLE. See also figures 7.3.12 and 7.3.13, as well as table 7.3.2.*

```
RUN                                                          Listing 7.3.4


    WHAT IS THE INITIAL GUESS FOR X AND Y:
        X0 = ?.2
        Y0 = ?1.2

    WHAT IS THE RADIUS OF THE FIRST SEARCH CIRCLE: ?.5
    BY WHAT FRACTION IS THIS CIRCLE TO BE REDUCED ON EACH ITERATION: ?.3
    HOW MANY POINTS ARE TO BE SEARCHED PER QUADRANT: ?4
    HOW MANY ITERATIONS ARE TO BE EMPLOYED: ?6


    THE APPROXIMATE SOLUTION IS Z =   0 + 0 I

    NUMBER OF ITERATIONS:   2

    READY
    RUN


    WHAT IS THE INITIAL GUESS FOR X AND Y:
        X0 = ?.2
        Y0 = ?1.2

    WHAT IS THE RADIUS OF THE FIRST SEARCH CIRCLE: ?1
    BY WHAT FRACTION IS THIS CIRCLE TO BE REDUCED ON EACH ITERATION: ?.5
    HOW MANY POINTS ARE TO BE SEARCHED PER QUADRANT: ?4
    HOW MANY ITERATIONS ARE TO BE EMPLOYED: ?6


    THE APPROXIMATE SOLUTION IS Z =   -2.9131803E-03 + .99911998 I

    NUMBER OF ITERATIONS:   6

    READY
    RUN


    WHAT IS THE INITIAL GUESS FOR X AND Y:
        X0 = ?0
        Y0 = ?0

    WHAT IS THE RADIUS OF THE FIRST SEARCH CIRCLE: ?5
    BY WHAT FRACTION IS THIS CIRCLE TO BE REDUCED ON EACH ITERATION: ?.5
    HOW MANY POINTS ARE TO BE SEARCHED PER QUADRANT: ?4
    HOW MANY ITERATIONS ARE TO BE EMPLOYED: ?10
```

THE APPROXIMATE SOLUTION IS Z = 5.4907034E-04 + .99991877 I

NUMBER OF ITERATIONS: 10

READY

Listing 7.3.4 *Sample runs of the program shown in listing 7.3.3. See also figures 7.3.14 through 7.3.16.*

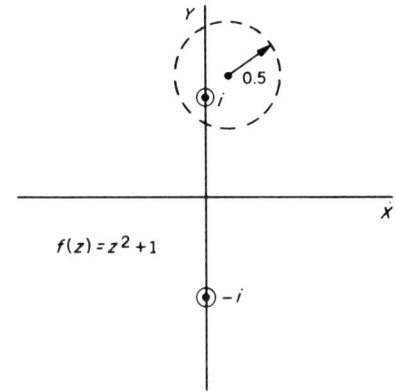

$f(z) = z^2 + 1$

center of first circle:	X0 = 0.2
	Y0 = 1.2
radius:	W = 0.5
reduction factor:	E = 0.3
points/quadrant:	M = 4
max. number of iterations:	N = 6
returned result:	X = 0
	Y = 0
number of iterations:	2
reason:	The first circle contained the root. The new estimate after the first pass was: X = −0.086
	Y = 1.16
	Convergence was occurring. However, the radius of the next search circle was 0.3 × 0.5 = 0.15.
	The actual root lay outside this second search circle and a potential infinite loop was encountered, causing a return to the calling program.

Figure 7.3.14 *An example in which the radius of the search circle was reduced too quickly. The clue that there was a failure is that a six-iteration limit was allowed, but only two passes occurred.*

442 BASIC SCIENTIFIC SUBROUTINES

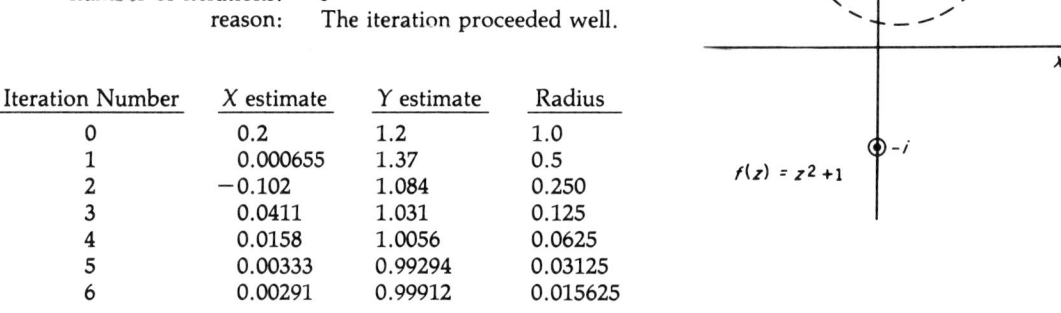

center of first circle:	X0 = 0.2
	Y0 = 1.2
radius:	W = 1.0
reduction factor:	E = 0.5
points/quadrant:	M = 4
max. number of iterations:	N = 6
returned result:	X = −0.0029
	Y = 0.99912
number of iterations:	6
reason:	The iteration proceeded well.

Iteration Number	X estimate	Y estimate	Radius
0	0.2	1.2	1.0
1	0.000655	1.37	0.5
2	−0.102	1.084	0.250
3	0.0411	1.031	0.125
4	0.0158	1.0056	0.0625
5	0.00333	0.99294	0.03125
6	0.00291	0.99912	0.015625

Figure 7.3.15 *Repeat of the previous example, but starting with a larger search circle and reducing the size of the circle by a factor of only 0.5 on each pass.*

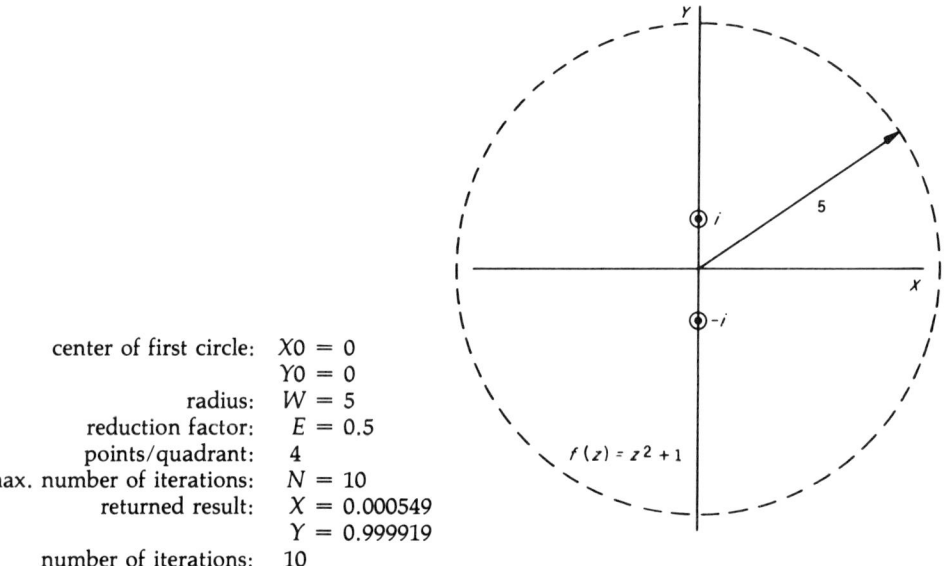

center of first circle:	X0 = 0
	Y0 = 0
radius:	W = 5
reduction factor:	E = 0.5
points/quadrant:	4
max. number of iterations:	N = 10
returned result:	X = 0.000549
	Y = 0.999919
number of iterations:	10

Figure 7.3.16 *A case in which the initial search circle contained two roots. The center of the circle moved and the circle itself closed down onto one of them.*

Function f(Z)	μ, ν Representation
$e^z - 1$	$\mu = e^x \cos y - 1$
	$\nu = e^x \sin y$
$z^2 + 2z + 4$	$\mu = x^2 + 2x - y^2 + 4$
	$\nu = 2y(1 + x)$
$z^2 - 2(1 - i)z + (1 - i)^2$	$\mu = (x - 1)^2 - (y + 1)^2$
	$\nu = 2(x - 1)(y + 1)$

Table 7.3.3 *Functions that you may wish to examine using the techniques presented in this chapter.*

7.4 Newton's Method in the Complex Domain

When the general location of a root is known *reasonably* well, it is possible to apply the rapidly convergent Newton technique discussed in Chapter 6 to very accurately determine its value.

Recall the form of the Newton iteration equation in real space:

$$x_{i+1} = x_i - \frac{f(x_i)}{f'(x_i)}$$

The analogous form in the complex plane is

$$z_{i+1} = z_i - \frac{f(z_i)}{f'(z_i)} \tag{7.4.1}$$

If we represent $f(z)$ with the form $\mu(x,y) + i\nu(x,y)$, then the ratio in the above equation becomes

$$\frac{f}{f'} = \frac{\mu + i\nu}{\mu_x + \mu_y} = \frac{\mu\mu_x + \nu\mu_y}{\mu_x^2 + \mu_y^2} + i\frac{\nu\mu_x - \mu\mu_y}{\mu_x^2 + \mu_y^2} \tag{7.4.2}$$

where $\mu_x = \dfrac{\partial \mu}{\partial x}$

$\mu_y = \dfrac{\partial \mu}{\partial y}$

Equation (7.4.1) can then be broken down into x and y components as follows:

$$x_{i+1} = x_i - \frac{\mu\mu_x + \nu\mu_y}{\mu_x^2 + \mu_y^2}$$

$$y_{i+1} = y_i - \frac{\nu\mu_x - \mu\mu_y}{\mu_x^2 + \mu_y^2} \tag{7.4.3}$$

If $\mu(x,y)$, $\nu(x,y)$, $\partial\mu/\partial x$, and $\partial\mu/\partial y$ are available, then Newton's method can be simply applied using equation (7.4.3).

A program that performs this operation is displayed in listing 7.4.1. The associated subroutine-connection diagram appears in figure 7.4.1, and the functions and variables employed are shown in table 7.4.1.

Two examples of the use of this program are shown in table 7.4.2. In the first example, the initial guess was close to one of the roots and the convergence was very rapid (quadratic after the second iteration). In the second example, the initial guess was not close to the root, and the error roughly halved on each iteration up to the fourth, after which the algorithm quickly closed in on the root. Undoubtedly, there are other functions for which the initial convergence rate is even slower, justifying the use of an algorithm such as ZCIRCLE.

Newton's method in the complex plane suffers from the same instabilities or weaknesses as the corresponding algorithm in real space. High curvature can very dramatically slow convergence, and zero derivatives can be disastrous, especially when round-off error is factored in. In any case, it is best to start the iteration near the root, and to avoid regions where $df/dz \approx 0$.

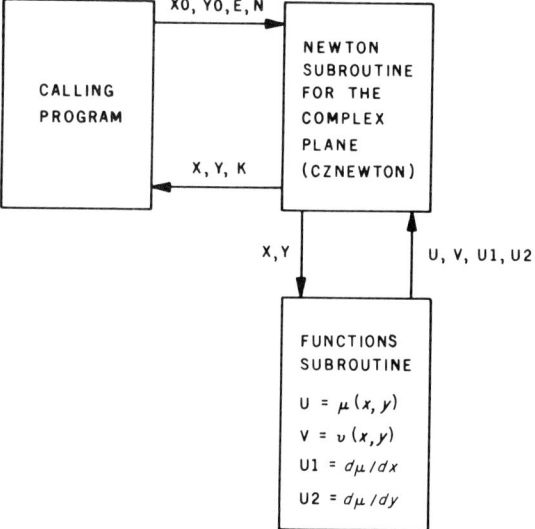

X0, Y0:   Initial guess
E: Convergence criterion. Return on $(x_{i+1} - x_i)^2 + (y_{i+1} - y_i)^2 < E^2$
N: Maximum number of iterations
X, Y: Calculated root
K: Actual number of iterations performed

Figure 7.4.1 *Subroutine-connection diagram for Newton's method in the complex plane (CZNEWTON).*

Statements/Functions List

$+, -, *, /, <, >$
GOSUB, GOTO, IF/THEN

Variables List

A, K, N, X, X0, Y, Y0, U, U1, U2, V

Variables Passed to Subroutine

E, N, X0, Y0

Table 7.4.1 *Functions and variables used by the complex plane Newton iteration subroutine (CZNEWTON).*

Listing 7.4.1

```
4350 REM PROGRAM TO DEMONSTRATE THE NEWTON ROOT
4351 REM PROGRAM IN THE COMPLEX DOMAIN
4352 PRINT
4353 PRINT
4354 PRINT "WHAT IS THE INITIAL GUESS:"
4355 PRINT "     X0 = ",
4356 INPUT X0
4357 PRINT "     Y0 = ",
4358 INPUT Y0
4359 PRINT "WHAT IS THE CONVERGENCE CRITERION: ",
4360 INPUT E
4361 PRINT "WHAT IS THE LIMIT ON THE NUMBER OF ITERATIONS: ",
4362 INPUT N
4363 GOSUB 46925
4364 PRINT
4365 PRINT
4366 PRINT "THE ROOT ESTIMATE IS:"
4367 PRINT "     X0 = ",X
4368 PRINT "     Y0 = ",Y
4369 PRINT
4370 PRINT "THE NUMBER OF ITERATIONS PERFORMED WAS: ",
4371 PRINT K
4372 PRINT
4373 PRINT
4374 END
44298 REM *******************
44299 REM FUNCTIONS SUBROUTINE
44300 U=X*X-Y*Y+1
44301 V=2*X*Y
44302 U1=2*X
44303 U2=-2*Y
```

```
44304 V1=2*Y
44305 V2=2*X
44306 RETURN
46912 REM ********************
46913 REM COMPLEX ROOT SEEKING USING NEWTON'S METHOD (CZNEWTON)
46914 REM THIS ROUTINE USES THE COMPLEX DOMAIN FORM OF
46915 REM NEWTON'S METHOD FOR ITERATIVELY SEARCHING FOR ROOTS.
46916 REM IT IS ASSUMED THAT THE FUNCTION AND ITS FIRST PARTIAL
46917 REM DERIVATIVES ARE AVAILABLE FROM THE FUNCTIONS SUBROUTINE
46918 REM IN THE FORM  F(Z) = U(X,Y) + I V(X,Y).
46919 REM THE REQUIRED DERIVATIVES ARE DU/DX AND DU/DY.
46920 REM THE INPUTS TO THE SUBROUTINE ARE THE INITIAL GUESS, X0, Y0,
46921 REM THE CONVERGENCE CRITERIA, E, AND THE MAXIMUM NUMBER OF
46922 REM ITERATIONS TO BE PERFORMED, N.
46923 REM THE RESULTING APPROXIMATION TO THE ROOT IS RETURNED IN
46924 REM (X,Y), AND THE NUMBER OF ITERATIONS IN K.
46925 K=0
46926 K=K+1
46927 REM GET U, V AND THE DERIVATIVES
46928 X=X0
46929 Y=Y0
46930 GOSUB 44300
46931 A=U1*U1+U2*U2
46932 X=X0+(V*U2-U*U1)/A
46933 Y=Y0-(V*U1+U*U2)/A
46934 REM CHECK FOR CONVERGENCE IN EUCLIDEAN SPACE
46935 IF (X0-X)*(X0-X)+(Y0-Y)*(Y0-Y)<=E*E THEN RETURN
46936 IF K>=N THEN RETURN
46937 X0=X
46938 Y0=Y
46939 GOTO 46926
```

Listing 7.4.1 *Subroutine for applying Newton's method in the complex plane (CZNEWTON). Also shown is a program for demonstrating CZNEWTON.*

Listing 7.4.2

```
          RUN


          WHAT IS THE INITIAL GUESS:
               X0 = ?.2
               Y0 = ?1.2
          WHAT IS THE CONVERGENCE CRITERION: ?.00000001
          WHAT IS THE LIMIT ON THE NUMBER OF ITERATIONS: ?10
```

THE ROOT ESTIMATE IS:
    X0 = 0
    Y0 = 1

THE NUMBER OF ITERATIONS PERFORMED WAS: 5

READY
RUN

WHAT IS THE INITIAL GUESS:
    X0 = ?10
    Y0 = ?20
WHAT IS THE CONVERGENCE CRITERION: ?.00000001
WHAT IS THE LIMIT ON THE NUMBER OF ITERATIONS: ?10

THE ROOT ESTIMATE IS:
    X0 = 0
    Y0 = 1

THE NUMBER OF ITERATIONS PERFORMED WAS: 9

READY

Listing 7.4.2 *Sample runs of the program shown in listing 7.4.1. See also table 7.4.2.*

Table 7.4.2

A) Initial guess: $X0 = 0.2$
$Y0 = 1.2$

Iteration Number	X	Y
1	0.03243243	1.0054054
2	1.90574×10^{-4}	0.9994976
3	9.582×10^{-8}	1.0000001
4	9×10^{-15}	1
5	0	1

B) Initial guess: $X0 = 10$
$Y0 = 20$

Iteration Number	X	Y
1	4.99	10.02
2	2.4750878	5.049984
3	1.1984162	2.6048251
4	0.5263234	1.4608316
5	0.15401333	1.0333615
6	0.00645907	0.990024
7	6.52729×10^{-5}	1.0000288
8	1.88×10^{-9}	1
9	0	1

Table 7.4.2 *Two examples of the convergence properties of CZNEWTON for the function $f(z) = z^2 + 1$. In the first case, the initial guess is close to the root in the upper half plane. In the second case, it is not.*

7.5 Mueller's Method in One Dimension

In Chapter 6 we found that Newton's method could be improved upon by employing Aitken acceleration, and subsequently Aitken-Steffenson iteration. These improvements were based on the observation of a weak point in Newton's method — the assumption that the function was *linear* near the root. However, many nontrivial functions have curvature which causes error, or at least slows convergence, in Newton's algorithm. The solution is to at least partially account for this curvature. That was the basis of the Aitken acceleration concept.

In this section, we will consider an algorithm called Mueller's method. This algorithm assumes the function to be *parabolic* in the neighborhood of the root. Because the complex-plane implementation of this technique is algebraically complicated, we will first examine the procedure for one-dimensional functions $[f(x)]$, then extend it *iteratively* to two dimensions, and finally proceed on to the full complex plane formulation. As you will see, Mueller's method is very powerful in terms of both convergence rate and stability.

The geometrical interpretation of Mueller's method is shown in figure 7.5.1. The idea is to find three evaluation points near the root, fit a parabola through those points, and then determine the roots of the corresponding second degree equation. The root closest to X_3 is then taken as a new evaluation point:

$$\begin{aligned} X_1 \text{ is dropped} \\ X_2 &\to X_1 \\ X_3 &\to X_2 \\ X_{new} &\to X_3 \end{aligned}$$

The procedure is repeated until the change in X_3 is less than some chosen criterion

$$|X_{new} - X_3| < E$$

Note that conceptually this is a simple upgrade of the secant method in which two points are used to generate a linear equation for the intersection with the X axis. This new value is then used to replace one of the previous two points.

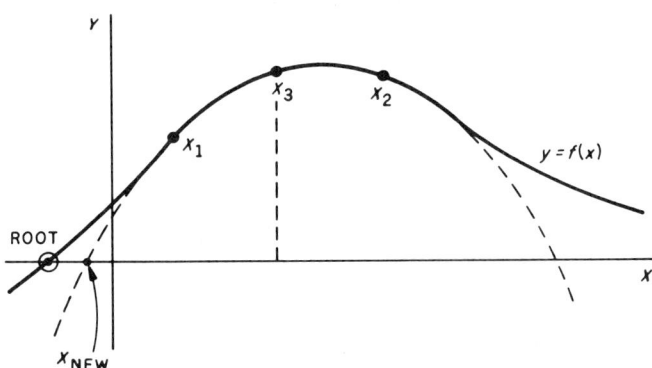

Figure 7.5.1 *The basic geometry of Mueller's method. The three points, X_1, X_2, and X_3, are used to generate a parabola. This parabola is expected to intersect the X axis at two points. The intersection point closest to X_3 is chosen as the new estimate, and X_1 is replaced.*

The algebra associated with this method is somewhat complicated and is only briefly described below. (For a more complete discussion, refer to the text by Becket and Hurt, Ref. 8.)

We will start by defining two dimensionless parameters, L and D:

$$L = \frac{X_3 - X_2}{X_2 - X_1}$$

$$D = \frac{X_3 - X_1}{X_2 - X_1}$$

These are used to calculate G and C:

$$G = L^2 f(X_1) - D^2 f(X_2) + (L + D) f(X_3)$$
$$C = L[L f(X_1) - D f(X_2) + f(X_3)]$$

The above definitions are then used to calculate the parameter λ:

$$\lambda = \frac{-2D\, f(X_3)}{G \pm \sqrt{G^2 - 4DC\, f(X_3)}}$$

The sign in the denominator is chosen so that the magnitude of λ is minimized. The new estimate for the location of the root is then

$$X_{new} = X_3 + \lambda(X_3 - X_2)$$

X_1 is subsequently dropped from the group and the process is repeated until some convergence criterion is met. This procedure can be implemented as shown in listing 7.5.1. The associated subroutine-connection diagram is displayed in figure 7.5.2, and the functions/variables list appears in table 7.5.1.

Two sets of sample runs are shown in listings 7.5.2 and 7.5.3. In the first set, the function examined was $y = x(x - 1)(x - 2)(x - 3)(x - 4)$. The first two runs show very good convergence. The third run terminated in six iterations according to the error criteria, $E = 0.001$, but it was in error by much more than that. Thus the warning: do not assume that the error in the final result is less than the error criterion! Convergence may be slow in some situations. Such was also the case for the fourth run in which the initial guess was far from any root. After ten iterations, the error was very large. However, extending the iteration limit to 20 (fifth run) and then 30 (sixth run) indicates that despite the initial slow convergence, the algorithm eventually, and very accurately, found a root.

Why was the convergence initially so slow? The answer to this question rests in the difference between what the algorithm assumes the form of the function to be, and what the form actually is. The algorithm assumes the function to be locally parabolic. But for an initial guess of $X0 = -30$, which is far from the cluster of five roots, the form of the function is approximately $y = x^5$. The algorithm has difficulty coping with this, but at least it does not fail.

Listing 7.5.3 gives an additional example of this convergence problem. In this case, the function is $y = (x + 1)^5$. Whether the evaluation point is near to or far from the multiple root, the functional form always has the high curvature associated with a quintic function. The sample runs displayed in listing 7.5.3 all show how the algorithm is struggling to find the root, and how it is converging slowly.

These were two particularly difficult examples. Most functions do not create as much trouble for Mueller's method as these do, especially when the initial guess is reasonably near a root. You can compare the results for the $y = (x + 1)^5$ case with a similar example given in Chapter 6 using Newton's method. The comparison clearly shows the superior properties of the Mueller algorithm.

The implementation of the one-dimensional Mueller algorithm as shown in listing 7.5.1 contains several checks to avoid errors in execution. For the most part, these error checks involve avoiding divide-by-zero failures. However, one particular check is associated with the potentially fatal situation indicated in figure 7.5.3. In this case, the three evaluation points lead to a parabolic fit that does not cross the X axis; the roots are imaginary. This is not helpful to the algorithm and the error check substitutes an artificial set of intersections to at least keep the iteration going.

In the next section, we will investigate the extension of Mueller's method to two dimensions by using successive substitution.

FINDING THE COMPLEX ROOTS OF FUNCTIONS 451

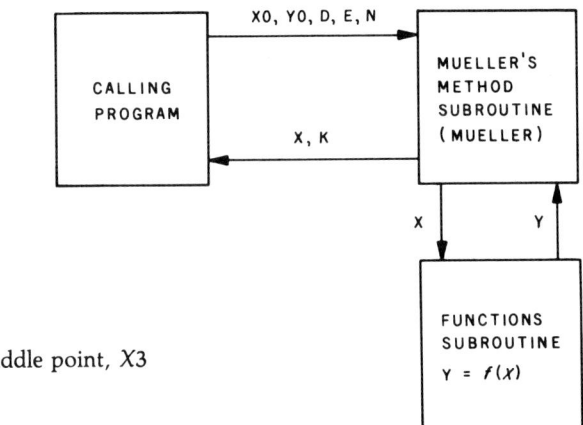

X0: Initial guess for the middle point, X3
D: X1 = X3 − D
    X2 = X3 + D
E: Convergence criterion. Return on $|X - X_3| < E$
N: Maximum number of iterations
X: Returned estimate of root
K: Number of iterations actually performed

Figure 7.5.2 *Subroutine-connection diagram for the one-dimensional Mueller's method algorithm (MUELLER).*

Statements/Functions List

+, −, *, /, <, >
ABS, GOSUB, GOTO, IF/THEN, SQRT

Variables List

A1, B, C1, D, D1, E, E1, E2, E3, K, L, N, X, X1, X2, X3, Y, Y1, Y2, Y3

Variables Passed to Subroutine

D, E, N, X0, Y0

Table 7.5.1 *Functions and variables used in the one-dimensional Mueller's method subroutine (MUELLER).*

Listing 7.5.1

```
4375 REM PROGRAM TO DEMONSTRATE MUELLER'S METHOD
4376 PRINT
4377 PRINT
4378 PRINT "WHAT IS THE INITIAL GUESS: ",
4379 INPUT X0
4380 PRINT "WHAT IS THE BOUND ON THIS GUESS: ",
4381 INPUT D
4382 PRINT "WHAT IS THE ERROR CRITERION: ",
4383 INPUT E
4384 PRINT "HOW MANY INTERATIONS (MAXIMUM): ",
4385 INPUT N
4386 PRINT
4387 GOSUB 46960
4388 PRINT
4389 PRINT "THE ESTIMATED ROOT IS: ",X
4390 PRINT "THE NUMBER OF ITERATIONS PERFORMED WAS: ",K
4391 PRINT
4392 END
44298 REM ********************
44299 REM FUNCTION SUBROUTINE
44300 Y=(X+1)*(X+1)*(X+1)*(X+1)*(X+1)
44301 RETURN
46944 REM ********************
46945 REM PARABOLIC ROOT SEEKING SUBROUTINE (MUELLER)
46946 REM THIS PROGRAM ITERATIVELY SEEKS THE ROOT OF A
46947 REM FUNCTION BY FITTING A PARABOLA TO THREE POINTS
46948 REM AND CALCULATING THE NEAREST ROOT AS DESCRIBED IN
46949 REM BECKET AND HURT, NUMERICAL CALCULATIONS AND ALGORITHMS.
46950 REM THE SUBROUTINE INPUTS ARE
46951 REM     X0 - THE INITIAL GUESS
46952 REM     D - A BOUND ON THE ERROR IN THIS GUESS
46953 REM     E - THE CONVERGENCE CRITERIA
46954 REM     N - THE MAXIMUM NUMBER OF ITERATIONS
46955 REM THE PROGRAM RETURNS THE VALUE OF THE ROOT FOUND, X,
46956 REM AND THE NUMBER OF ITERATIONS PERFORMED, K.
46957 REM IT IS ASSUMED THAT THE FUNCTION Y(X) IS AVAILABLE
46958 REM IN THE FUNCTIONS SUBROUTINE.
46959 REM SET UP THE THREE EVALUATION POINTS
46960 K=1
46961 X3=X0
46962 X1=X3-D
46963 X2=X3+D
46964 REM CALCULATE MUELLER PARAMETERS
46965 REM GUARD AGAINST DIVIDE BY ZERO
46966 IF X2-X1=0 THEN X2=X2*(1.0000001)
46967 IF X2-X1=0 THEN X2=X2+.0000001
46968 L1=(X3-X2)/(X2-X1)
46969 D1=(X3-X1)/(X2-X1)
46970 IF K>1 THEN GOTO 46978
```

```
46971 REM GET VALUES OF FUNCTION
46972 X=X1
46973 GOSUB 44300
46974 E1=Y
46975 X=X2
46976 GOSUB 44300
46977 E2=Y
46978 X=X3
46979 GOSUB 44300
46980 E3=Y
46981 A1=L1*L1*E1-D1*D1*E2+(L1+D1)*E3
46982 C1=L1*(L1*E1-D1*E2+E3)
46983 B=A1*A1-4*D1*C1*E3
46984 REM TEST FOR COMPLEX ROOT, MEANING THE PARABOLA IS INVERTED
46985 IF B<0 THEN B=0
46986 REM CHOOSE CLOSEST ROOT
46987 IF A1<0 THEN A1=A1-SQRT(B)
46988 IF A1>0 THEN A1=A1+SQRT(B)
46989 REM GUARD AGAINST A DIVIDE BY ZERO
46990 IF ABS(A1)+ABS(B)=0 THEN A1=4*D1*E3
46991 REM CALCULATE RELATIVE DISTANCE OF NEXT GUESS
46992 REM GUARD AGAINST DIVIDE BY ZERO
46993 IF A1=0 THEN A1=.0000001
46994 L=-2*D1*E3/A1
46995 REM CALCULATE NEXT ESTIMATE
46996 X=X3+L*(X3-X2)
46997 REM TEST FOR CONVERGENCE
46998 IF ABS(X-X3)<E THEN RETURN
46999 REM TEST FOR NUMBER OF ITERATIONS
47000 IF K>=N THEN RETURN
47001 REM OTHERWISE, MAKE ANOTHER PASS
47002 K=K+1
47003 REM SAVE SOME CALCULATIONS:
47004 X1=X2
47005 X2=X3
47006 X3=X
47007 E1=E2
47008 E2=E3
47009 GOTO 46967
```

Listing 7.5.1 *One-dimensional Mueller's method subroutine (MUELLER). Also shown is a program for demonstrating the operation of MUELLER, sample results of which appear in listings 7.5.2 and 7.5.3.*

```
RUN                                          Listing 7.5.2


WHAT IS THE INITIAL GUESS: ?0
WHAT IS THE BOUND ON THIS GUESS: ?10
WHAT IS THE ERROR CRITERION: ?.001
HOW MANY INTERATIONS (MAXIMUM): ?10


THE ESTIMATED ROOT IS:  0
THE NUMBER OF ITERATIONS PERFORMED WAS:  1

READY
RUN


WHAT IS THE INITIAL GUESS: ?20
WHAT IS THE BOUND ON THIS GUESS: ?40
WHAT IS THE ERROR CRITERION: ?.001
HOW MANY INTERATIONS (MAXIMUM): ?10


THE ESTIMATED ROOT IS:  4
THE NUMBER OF ITERATIONS PERFORMED WAS:  10

READY
RUN


WHAT IS THE INITIAL GUESS: ?20
WHAT IS THE BOUND ON THIS GUESS: ?40
WHAT IS THE ERROR CRITERION: ?.001
HOW MANY INTERATIONS (MAXIMUM): ?6


THE ESTIMATED ROOT IS:  4.6643002
THE NUMBER OF ITERATIONS PERFORMED WAS:  6

READY
RUN


WHAT IS THE INITIAL GUESS: ?-30
WHAT IS THE BOUND ON THIS GUESS: ?1
WHAT IS THE ERROR CRITERION: ?.001
```

```
HOW MANY INTERATIONS (MAXIMUM): ?10


THE ESTIMATED ROOT IS:  -6.662084
THE NUMBER OF ITERATIONS PERFORMED WAS:   10

READY
RUN



WHAT IS THE INITIAL GUESS: ?-30
WHAT IS THE BOUND ON THIS GUESS: ?1
WHAT IS THE ERROR CRITERION: ?.001
HOW MANY INTERATIONS (MAXIMUM): ?20


THE ESTIMATED ROOT IS:   4.0086337
THE NUMBER OF ITERATIONS PERFORMED WAS:   20

READY
RUN



WHAT IS THE INITIAL GUESS: ?-30
WHAT IS THE BOUND ON THIS GUESS: ?1
WHAT IS THE ERROR CRITERION: ?0
HOW MANY INTERATIONS (MAXIMUM): ?30


THE ESTIMATED ROOT IS:   4
THE NUMBER OF ITERATIONS PERFORMED WAS:   30

READY
```

Listing 7.5.2 *Sample runs of the demonstration program shown in listing 7.5.1 for the case* $y = x(x-1)(x-2)(x-3)(x-4)$.

Listing 7.5.3

```
RUN



WHAT IS THE INITIAL GUESS: ?0
WHAT IS THE BOUND ON THIS GUESS: ?2
WHAT IS THE ERROR CRITERION: ?.001
```

```
HOW MANY INTERATIONS (MAXIMUM): ?10                     Listing 7.5.3 cont.


THE ESTIMATED ROOT IS:  -.97847292
THE NUMBER OF ITERATIONS PERFORMED WAS:   6

READY
RUN



WHAT IS THE INITIAL GUESS: ?0
WHAT IS THE BOUND ON THIS GUESS: ?2
WHAT IS THE ERROR CRITERION: ?0
HOW MANY INTERATIONS (MAXIMUM): ?10


THE ESTIMATED ROOT IS:  -1.0301292
THE NUMBER OF ITERATIONS PERFORMED WAS:  10

READY
RUN



WHAT IS THE INITIAL GUESS: ?0
WHAT IS THE BOUND ON THIS GUESS: ?2
WHAT IS THE ERROR CRITERION: ?0
HOW MANY INTERATIONS (MAXIMUM): ?20


THE ESTIMATED ROOT IS:  -.99839055
THE NUMBER OF ITERATIONS PERFORMED WAS:  20

READY
RUN



WHAT IS THE INITIAL GUESS: ?0
WHAT IS THE BOUND ON THIS GUESS: ?2
WHAT IS THE ERROR CRITERION: ?0
HOW MANY INTERATIONS (MAXIMUM): ?40


THE ESTIMATED ROOT IS:  -.99995649
THE NUMBER OF ITERATIONS PERFORMED WAS:  40

READY
```

```
RUN


WHAT IS THE INITIAL GUESS: ?0
WHAT IS THE BOUND ON THIS GUESS: ?2
WHAT IS THE ERROR CRITERION: ?0
HOW MANY INTERATIONS (MAXIMUM): ?80


THE ESTIMATED ROOT IS:   -.99995649
THE NUMBER OF ITERATIONS PERFORMED WAS:   80

READY
```

Listing 7.5.3 *Sample runs of the demonstration program shown in listing 7.5.1 for the case $y = (x + 1)^5$.*

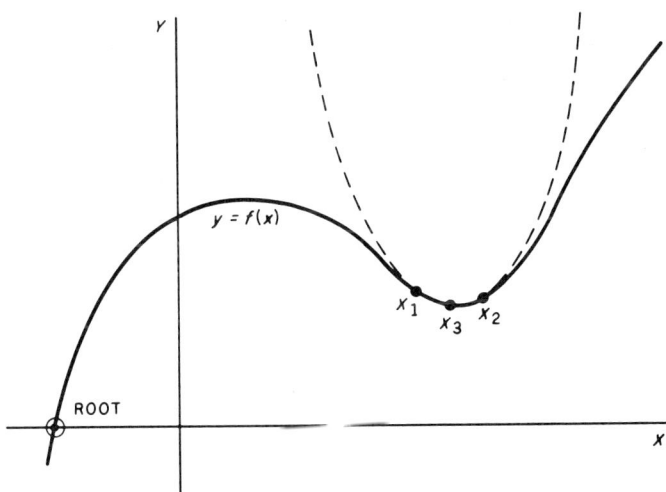

Figure 7.5.3 *A potentially fatal situation for Mueller's method. The subroutine has a check for this situation and gives the algorithm another chance. However, convergence may not occur.*

7.6 Two-Dimensional Form of Mueller's Method

In this section, Mueller's method for one dimension is combined with the concept of successive substitution to produce an algorithm that can be called on to find a root of the function $\mu(x,y)$. We will construct this algorithm by considering each dimension separately.

The starting point is the initial guess and *two* bounds: B_1 for the X direction, and B_2 for the Y direction. The first of these bounds is used to generate three evaluation points:

$$X_3 = X_0$$
$$X_1 = X_0 - B_1$$
$$X_2 = X_0 + B_1$$
$$Y_1, Y_2, Y_3 = Y_0$$

Mueller's method for one dimension is then applied to find a better estimate for the X coordinate of the root, X_{new}. B_1 is replaced with $|X_{new} - X_3|$.

For the iteration in the Y direction, the three evaluation points are

$$Y_3 = Y_0$$
$$Y_1 = Y_0 - B_2$$
$$Y_2 = Y_0 + B_2$$
$$X_1, X_2, X_3 = X_{new}$$

The one-dimensional form of Mueller's method is again used to find a new estimate for the Y coordinate of the root, Y_{new}. As with the X iteration, B_2 is replaced with $|Y_{new} - Y_3|$. If the change in X and Y satisfies the prechosen error criterion,

$$|B_1| + |B_2| < E$$

then the iteration is ended. Otherwise, $X_0 \rightarrow X_{new}$, $Y_0 \rightarrow Y_{new}$, and the procedure is repeated.

A program for performing these calculations is shown in listing 7.6.1 (MUELLER2). It is based primarily on the routine given in the previous section. It requires as inputs an initial guess for the location of the root, $(X0, Y0)$; two measures of the bound on the range of the starting two-dimensional parabola ($B1$ for the X direction and $B2$ for the Y direction); a convergence criterion, E; and a bound on the maximum number of iterations to be performed, N. The outputs of the subroutine are the estimate of the location of the root (X, Y), and the actual number of iterations performed, K.

To demonstrate the basic operation of MUELLER2, the following function was examined:

$$\mu(x,y) = (x + 1)^5(y - 1)^5$$

Sample results appear in listing 7.6.2.

As expected from the example in the previous section, convergence occurs, but it is not rapid. The reason for this is that the curvature of the function is high (quintic) because of the multiple roots.

Another function one might inadvertently attempt to solve is

$$\mu(x,y) = x^2 - y^2 + 1$$

This is the real part of the complex function $f(z) = z^2 + 1$. MUELLER2 completely fails in this exercise by diverging toward infinity, and eventually reaching a numeric overflow. The reason for the failure is that the function has no unique root! Such errors in the key inputs are not guarded against by this subroutine.

The precautions regarding the use of MUELLER2 center on the two-dimensional shape of the function. If the calculation is started in the wrong place, the iteration can diverge. Also, the algorithm is particularly vulnerable to the divergence associated with roots that occur at or near *saddle* points or local minima, although recovery is possible because of the error checks within the subroutine.

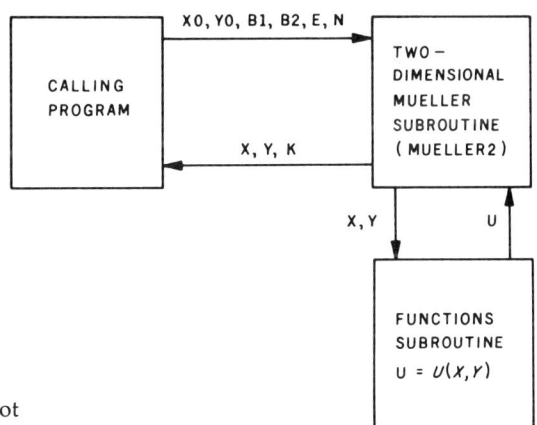

X0, Y0:	Inital guess for location of root
B1:	X1 = X0 − B1
	X2 = X0 + B1
B2:	Y1 = Y0 − B2
	Y2 = Y0 + B2
E:	Convergence criterion. Return on ABS (ΔX) + ABS(ΔY) < E
N:	Maximum number of iterations
X, Y:	Returned estimate of root
K:	Actual number of iterations performed

Figure 7.6.1 *Subroutine-connection diagram for the two-dimensional form of Mueller's method (MUELLER2).*

Statements/Functions List

+, −, *, /, <, >
ABS, GOSUB, GOTO, IF/THEN

Variables List

A1, B, B1, B2, C1, D1, E, E1, E2, E3, K, L, L1, N, X0, X1, X2, X3, Y0, Y1, Y2, Y3, U

Variables Passed to Subroutine

B1, B2, E, N, X0, Y0

Table 7.6.1 *Functions and variables used in the two-dimensional, successive substitution Mueller's method subroutine (MUELLER2).*

Listing 7.6.1

```
4400 REM PROGRAM TO DEMONSTRATE A TWO DIMENSIONAL VERSION OF MUELLER'S METHOD
4401 PRINT
4402 PRINT
4403 PRINT "WHAT ARE THE INITIAL GUESSES AND THEIR BOUNDS:"
4404 PRINT "     X0 = ",
4405 INPUT X0
4406 PRINT "     BOUND ON X0 = ",
4407 INPUT B1
4408 PRINT "     Y0 = ",
4409 INPUT Y0
4410 PRINT "     BOUND ON Y0 = ",
4411 INPUT B2
4412 PRINT "WHAT IS THE ERROR CRITERION: ",
4413 INPUT E
4414 PRINT "HOW MANY INTERATIONS (MAXIMUM): ",
4415 INPUT N
4416 PRINT
4417 GOSUB 47050
4418 PRINT
4419 PRINT "THE ESTIMATED ROOT IS (X,Y) = (",X,",",Y,")"
4420 PRINT "THE NUMBER OF ITERATIONS PERFORMED WAS: ",K
4421 PRINT
4422 END
44298 REM ********************
44299 REM FUNCTIONS SUBROUTINE
44300 W=(X+1)*(X+1)*(X+1)*(X+1)*(X+1)*(Y-1)*(Y-1)*(Y-1)*(Y-1)*(Y-1)
44301 RETURN
47033 REM ********************
47034 REM PARABOLIC ROOT SEEKING SUBROUTINE (MUELLER2)
47035 REM FOR A TWO DIMENSIONAL FUNCTION, W(X,Y).
```

```
47036 REM THIS PROGRAM ITERATIVELY SEEKS THE ROOT OF A
47037 REM FUNCTION BY FITTING A PARABOLA TO THREE POINTS
47038 REM AND CALCULATING THE NEAREST ROOT AS DESCRIBED IN
47039 REM BECKET AND HURT, NUMERICAL CALCULATIONS AND ALGORITHMS.
47040 REM THE SUBROUTINE INPUTS ARE
47041 REM     X0,Y0 - THE INITIAL GUESS
47042 REM     B1,B2 - A BOUND ON THE ERROR IN THIS GUESS
47043 REM     E - THE CONVERGENCE CRITERIA
47044 REM     N - THE MAXIMUM NUMBER OF ITERATIONS
47045 REM THE PROGRAM RETURNS THE VALUE OF THE ROOT FOUND, (X,Y),
47046 REM AND THE NUMBER OF ITERATIONS PERFORMED, K.
47047 REM IT IS ASSUMED THAT THE FUNCTION U(X,Y) IS AVAILABLE
47048 REM IN THE FUNCTIONS SUBROUTINE.
47049 REM SET UP THE THREE EVALUATION POINTS
47050 K=1
47051 X3=X0
47052 X1=X3-B1
47053 X2=X3+B1
47054 REM CALCULATE MUELLER PARAMETERS
47055 REM GUARD AGAINST DIVIDE BY ZERO
47056 IF X2-X1=0 THEN X2=X2*(1.0000001)
47057 IF X2-X1=0 THEN X2=X2+.0000001
47058 L1=(X3-X2)/(X2-X1)
47059 D1=(X3-X1)/(X2-X1)
47060 REM GET VALUES OF FUNCTION
47061 Y=Y0
47062 X=X1
47063 GOSUB 44300
47064 E1=W
47065 X=X2
47066 GOSUB 44300
47067 E2=W
47068 X=X3
47069 GOSUB 44300
47070 E3=W
47071 GOSUB  47111
47072 REM CALCULATE NEW X ESTIMATE
47073 B1=L*(X3-X2)
47074 X=X3+B1
47075 REM TEST FOR CONVERGENCE
47076 IF ABS(B1)+ABS(B2)<E THEN RETURN
47077 X0=X
47078 REM REPEAT FOR THE Y DIRECTION
47079 Y3=Y0
47080 Y1=Y3-B2
47081 Y2=Y3+B2
47082 REM CALCULATE MUELLER PARAMETERS
47083 REM GUARD AGAINST A DIVIDE BY ZERO
47084 IF Y2-Y1=0 THEN Y2=Y2*(1.0000001)
```

Listing 7.6.1 cont.

```
47085 IF Y2-Y1=0 THEN Y2=Y2+.0000001
47086 L1=(Y3-Y2)/(Y2-Y1)
47087 D1=(Y3-Y1)/(Y2-Y1)
47088 REM GET VALUES OF FUNCTION
47089 Y=Y1
47090 GOSUB 44300
47091 E1=W
47092 Y=Y2
47093 GOSUB 44300
47094 E2=W
47095 Y=Y3
47096 GOSUB 44300
47097 E3=W
47098 GOSUB 47111
47099 REM CALCULATE NEW Y ESTIMATE
47100 B2=L*(Y3-Y2)
47101 Y=Y3+B2
47102 REM TEST FOR CONVERGENCE
47103 IF ABS(B1)+ABS(B2)<E THEN RETURN
47104 REM TEST FOR NUMBER OF ITERATIONS
47105 IF K>=N THEN RETURN
47106 Y0=Y
47107 K=K+1
47108 REM START ANOTHER PASS
47109 GOTO 47051
47110 REM UTILITY SUBROUTINE
47111 A1=L1*L1*E1-D1*D1*E2+(L1+D1)*E3
47112 C1=L1*(L1*E1-D1*E2+E3)
47113 B=A1*A1-4*D1*C1*E3
47114 REM TEST FOR COMPLEX ROOT, MEANING THE PARABOLA IS INVERTED
47115 IF B<0 THEN B=0
47116 REM CHOOSE CLOSEST ROOT
47117 IF A1<0 THEN A1=A1-SQRT(B)
47118 IF A1>0 THEN A1=A1+SQRT(B)
47119 REM GUARD AGAINST A DIVIDE BY ZERO
47120 IF ABS(A1)+ABS(B)=0 THEN A1=4*D1*E3
47121 REM CALCULATE RELATIVE DISTANCE OF NEXT GUESS
47122 REM GUARD AGAINST DIVIDE BY ZERO
47123 IF A1=0 THEN A1=.0000001
47124 L=-2*D1*E3/A1
47125 RETURN
```

Listing 7.6.1 *Successive substitution form of Mueller's one-dimensional method as applied to two dimensions (MUELLER2). Also shown is a program to demonstrate this subroutine. Sample results are given in listing 7.6.2.*

RUN   Listing 7.6.2

```
WHAT ARE THE INITIAL GUESSES AND THEIR BOUNDS:
     X0 = ?0
     BOUND ON X0 = ?2
     Y0 = ?0
     BOUND ON Y0 = ?2
WHAT IS THE ERROR CRITERION: ?.001
HOW MANY INTERATIONS (MAXIMUM): ?10


THE ESTIMATED ROOT IS (X,Y) = ( -.95127621, .95127625)
THE NUMBER OF ITERATIONS PERFORMED WAS:  10

READY
RUN


WHAT ARE THE INITIAL GUESSES AND THEIR BOUNDS:
     X0 = ?0
     BOUND ON X0 = ?2
     Y0 = ?0
     BOUND ON Y0 = ?2
WHAT IS THE ERROR CRITERION: ?0
HOW MANY INTERATIONS (MAXIMUM): ?20


THE ESTIMATED ROOT IS (X,Y) = ( -.99787724, .99787719)
THE NUMBER OF ITERATIONS PERFORMED WAS:  20

READY
RUN


WHAT ARE THE INITIAL GUESSES AND THEIR BOUNDS:
     X0 = ?0
     BOUND ON X0 = ?2
     Y0 = ?0
     BOUND ON Y0 = ?2
WHAT IS THE ERROR CRITERION: ?0
HOW MANY INTERATIONS (MAXIMUM): ?40


THE ESTIMATED ROOT IS (X,Y) = ( -.99999738, .9999976)
THE NUMBER OF ITERATIONS PERFORMED WAS:  40
```

```
READY
RUN


WHAT ARE THE INITIAL GUESSES AND THEIR BOUNDS:
    X0 = ?0
    BOUND ON X0 = ?2
    Y0 = ?0
    BOUND ON Y0 = ?2
WHAT IS THE ERROR CRITERION: ?0
HOW MANY INTERATIONS (MAXIMUM): ?80


THE ESTIMATED ROOT IS (X,Y) = ( -.99999977, .9999996)
THE NUMBER OF ITERATIONS PERFORMED WAS:  80

READY
```

Listing 7.6.2 *Sample runs of MUELLER2.*

7.7 Mueller's Method in the Complex Plane

Mueller's one-dimensional parabolic method, as discussed in section 7.5, can be applied directly to complex plane calculations by replacing the real variables and functions contained in the subroutine by their complex plane counterparts. The difficulty in doing this resides in the large number of complex plane algebraic manipulations which must be performed.

This problem can be approached in two ways. The first way would be to perform all the complex plane calculations by calling the subroutines given in Volume I of this series. However, the resulting program would not be efficient in the total length of code, and it would also execute slowly. The second approach is to algebraically rewrite the basic procedure so that the complex plane calculations are broken down into a series of real-number operations. The result is some complicated algebra, but a reasonably short and fast subroutine. The latter approach is preferable.

Listing 7.7.1 shows a real-number, algebraic implementation of Mueller's method in the complex plane. The associated subroutine-connection diagram appears in figure 7.7.1. The subroutine requires as input an initial guess, $(X0, Y0)$; two initial bounds ($B1$ for $X0$; $B2$ for $Y0$); a convergence criterion, E; and a limit on the number of iterations, N. The subroutine returns the new estimate, (X, Y), and the number of iterations actually performed, K.

We can use this program to examine the test function $f(z) = z^2 + 1$ (see listing 7.7.2). The convergence rate is very impressive. It appears that only two iterations are required no matter what the initial guess is! The reason for this high convergence rate is simple: ZMUELLER, in effect, fits a complex-plane parabola to the test function. However, $f(z)$ is only of the second degree, and in principle, the fitted parabola is coincident with $f(z)$ throughout the complex plane. Therefore, the roots are located on the first pass (again, in principle). Thus, we would expect only one iteration to be necessary, if it were not for round-off error. The second iteration is, in essence, a cleanup operation.

$f(z) = z^2 + 1$ was an ideal function for treatment with ZMUELLER. Functions that do not have more than two roots close together are usually very amenable to solution by this algorithm. The high curvature associated with functions like $f(z) = (z + 1)^5$ can slow down convergence, as we saw in the previous sections, but the root is eventually found. ZMUELLER is a fairly reliable subroutine.

The utility of ZMUELLER is not strictly limited to analytic functions of z. This routine can also be employed to search for the roots of other functions of x and y simply by using $\mu(x,y)$ to represent the function (in the "functions" subroutine), and setting $\nu(x,y) = 0$. A difficult example of this (difficult because of the high curvature in the functional form) is demonstrated in listing 7.7.3. Again the fifth-degree function, $(x + 1)^5$, is probed and a root found. This particular example was chosen in part to show that the convergence properties of ZMUELLER for one- and two-dimensional problems are similar to those of the previous two Mueller method subroutines. Also, this example is actually one-dimensional: $f(x,y) = f(x)$. However, the routine followed a path through the (x,y) plane. The calculated values of y have no significance in this case.

In conclusion, ZMUELLER is a very versatile algorithm that can be used for finding the roots of one- and two-dimensional functions in the complex plane. The subroutine can also be extended to the solution of *simultaneous* two-dimensional functions. Simply represent one function with $\mu(x,y)$, and the other with $\nu(x,y)$. [Note that it is not required that $\mu(x,y) + i\,\nu(x,y)$ be an analytic function of z; the Cauchy-Riemann equations need not be satisfied.] This is left as an exercise for you.

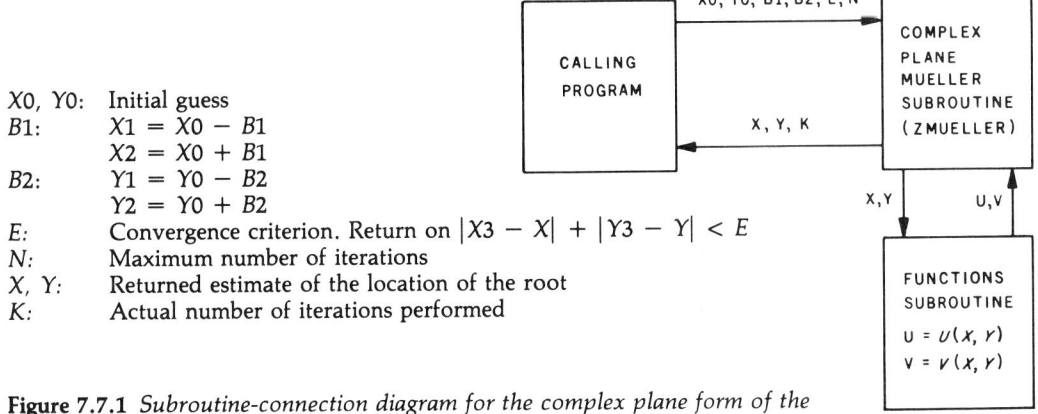

X0, Y0: Initial guess
B1: $X1 = X0 - B1$
$X2 = X0 + B1$
B2: $Y1 = Y0 - B2$
$Y2 = Y0 + B2$
E: Convergence criterion. Return on $|X3 - X| + |Y3 - Y| < E$
N: Maximum number of iterations
X, Y: Returned estimate of the location of the root
K: Actual number of iterations performed

Figure 7.7.1 *Subroutine-connection diagram for the complex plane form of the Mueller subroutine (ZMUELLER).*

Statements/Functions List

$+, -, *, /, <, >$
ABS, ATN, COS, GOSUB, GOTO, IF/THEN, SIN, SQRT

Variables List

A, A1, A2, B, B1, B2, C1, C2, D, D1, D2, E, E1, E2, K, L, L1, L2, N, X0, X1, X2, X3, Y0, Y1, Y2, Y3, U, U1, U2, U3, V, V1, V2, V3

Variables Passed to Subroutine

B1, B2, E, N, X0, Y0

Table 7.7.1 *Functions and variables used in ZMUELLER.*

Listing 7.7.1

```
4425 REM PROGRAM TO DEMONSTRATE THE COMPLEX DOMAIN MUELLER SUBROUTINE
4426 PRINT
4427 PRINT
4428 PRINT "WHAT IS THE INITIAL GUESS:"
4429 PRINT
4430 PRINT "     X0 = ",
4431 INPUT X0
4432 PRINT "WHAT IS THE ASSOCIATED BOUND: ",
4433 INPUT B1
4434 PRINT "     Y0 = ",
4435 INPUT Y0
4436 PRINT "WHAT IS THE ASSOCIATED BOUND: ",
4437 INPUT B2
4438 PRINT
4439 PRINT "INPUT THE CONVERGENCE CRITERION: ",
4440 INPUT E
4441 PRINT "WHAT IS THE LIMIT ON THE NUMBER OF ITERATIONS: ",
4442 INPUT N
4443 GOSUB 47150
4444 PRINT
4445 PRINT
4446 PRINT "THE ESTIMATED ROOT IS:"
4447 PRINT
4448 PRINT "     X = ",X
4449 PRINT "     Y = ",Y
4450 PRINT
4451 PRINT "THE NUMBER OF ITERATIONS WAS ",K
4452 PRINT
4453 PRINT
4454 END
44298 REM ********************
```

```
44299 REM FUNCTIONS SUBROUTINE
44300 U=X*X-Y*Y+1
44301 V=2*X*Y
44302 RETURN
47136 REM ********************
47137 REM MUELLER'S METHOD FOR COMPLEX ROOTS (ZMUELLER)
47138 REM THIS PROGRAM USES THE PARABOLIC FITTING TECHNIQUE
47139 REM ASSOCIATED WITH MUELLER'S METHOD, BUT DOES IT IN
47140 REM THE COMPLEX DOMAIN.
47141 REM THE INPUTS TO THE SUBROUTINE ARE THE INITIAL
47142 REM GUESS, (X0,Y0), THE CONVERGENCE CRITERIA, E,
47143 REM AND THE MAXIMUM NUMBER OF ITERATIONS, N.
47144 REM ALSO REQUIRED ARE BOUNDS ON THE INITIAL GUESS, B1 AND B2.
47145 REM RETURNED IS THE NEW ESTIMATE, (X,Y), AND THE
47146 REM NUMBER OF ITERATIONS PERFORMED, K.
47147 REM IT IS ASSUMED THAT THE FUNCTION F(Z) = U(X,Y)+IV(X,Y)
47148 REM IS AVAILABLE IN THE FUNCTIONS SUBROUTINE.
47149 REM START CALCULATIONS
47150 K=1
47151 X3=X0
47152 Y3=Y0
47153 X1=X3-B1
47154 Y1=Y3-B2
47155 X2=X3+B1
47156 Y2=Y3+B2
47157 D=(X2-X1)*(X2-X1)+(Y2-Y1)*(Y2-Y1)
47158 REM AVOID DIVIDE BY ZERO
47159 IF D=0 THEN D=.0000001
47160 L1=(X3-X2)*(X2-X1)+(Y3-Y2)*(Y2-Y1)
47161 L1=L1/D
47162 L2=(X2-X1)*(Y3-Y2)-(X3-X2)*(Y2-Y1)
47163 L2=L2/D
47164 D1=(X3-X1)*(X2-X1)+(Y3-Y1)*(Y2-Y1)
47165 D1=D1/D
47166 D2=(X2-X1)*(Y3-Y1)-(X3-X1)*(Y2-Y1)
47167 D2=D2/D
47168 REM GET FUNCTION VALUES
47169 X=X1
47170 Y=Y1
47171 GOSUB 44300
47172 U1=U
47173 V1=V
47174 X=X2
47175 Y=Y2
47176 GOSUB 44300
47177 U2=U
47178 V2=V
47179 X=X3
47180 Y=Y3
```

```
47181 GOSUB 44300
47182 U3=U
47183 V3=V
47184 REM CALCULATE MUELLER PARAMETERS
47185 E1=U1*(L1*L1-L2*L2)-2*V1*L1*L2-U2*(D1*D1-D2*D2)
47186 E1=E1+2*V2*D1*D2+U3*(L1+D1)-V3*(L2+D2)
47187 E2=2*L1*L2*U1+V1*(L1*L1-L2*L2)-2*D1*D2*U2-V2*(D1*D1-D2*D2)
47188 E2=E2+U3*(L2+D2)+V3*(L1+D1)
47189 C1=L1*L1*U1-L1*L2*V1-D1*L1*U2+L1*D2*V2+U3*L1
47190 C1=C1-U1*L2*L2-V1*L1*L2+U2*L2*D2+V2*D1*L2-V3*L2
47191 C2=U1*L1*L2+V1*L1*L1-U2*D2*L1-V2*D1*L1+V3*L1
47192 C2=C2+L1*L2*U1-L2*L2*V1-D1*L2*U2+D2*L2*V2+U3*L2
47193 B1=E1*E1-E2*E2-4*(U3*D1*C1-U3*D2*C2-V3*D2*C1-V3*D1*C2)
47194 B2=2*E1*E2-4*(U3*D2*C1+U3*D1*C2+V3*D1*C1-V3*D2*C2)
47195 REM GUARD AGAINST A DIVIDE BY ZERO
47196 IF B1=0 THEN B1=.0000001
47197 A=ATN(B2/B1)
47198 A=A/2
47199 B=SQRT(SQRT(B1*B1+B2*B2))
47200 B1=B*COS(A)
47201 B2=B*SIN(A)
47202 A1=(E1+B1)*(E1+B1)+(E2+B2)*(E2+B2)
47203 A2=(E1-B1)*(E1-B1)+(E2-B2)*(E2-B2)
47204 IF A1>A2 THEN GOTO 47208
47205 A1=E1-B1
47206 A2=E2-B2
47207 GOTO 47210
47208 A1=E1+B1
47209 A2=E2+B2
47210 A=A1*A1+A2*A2
47211 L1=A1*D1*U3-A1*D2*V3+A2*U3*D2+A2*V3*D1
47212 REM GUARD AGAINST DIVIDE BY ZERO
47213 IF A=0 THEN A=.0000001
47214 L1=-2*L1/A
47215 L2=-D1*U3*A2+D2*V3*A2+A1*U3*D2+A1*V3*D1
47216 L2=-2*L2/A
47217 REM CALCULATE NEW ESTIMATE
47218 X=X3+L1*(X3-X2)-L2*(Y3-Y2)
47219 Y=Y3+L2*(X3-X2)+L1*(Y3-Y2)
47220 REM TEST FOR CONVERGENCE
47221 IF ABS(X-X0)+ABS(Y-Y0)<E THEN RETURN
47222 REM TEST FOR NUMBER OF ITERATIONS
47223 IF K>=N THEN RETURN
47224 REM CONTINUE
47225 K=K+1
47226 X0=X
47227 Y0=Y
47228 X1=X2
47229 Y1=Y2
```

Listing 7.7.1 cont.

```
47230 X2=X3
47231 Y2=Y3
47232 X3=X
47233 Y3=Y
47234 GOTO 47157
```

Listing 7.7.1 *The complex plane implementation of Mueller's method (ZMUELLER). A demonstration program is also provided, with sample results appearing in listing 7.7.2.*

Listing 7.7.2

```
RUN


WHAT IS THE INITIAL GUESS:

    X0 = ?0
WHAT IS THE ASSOCIATED BOUND: ?3
    Y0 = ?0
WHAT IS THE ASSOCIATED BOUND: ?3

INPUT THE CONVERGENCE CRITERION: ?.001
WHAT IS THE LIMIT ON THE NUMBER OF ITERATIONS: ?10


THE ESTIMATED ROOT IS:

    X =  2E-15
    Y =  -1

THE NUMBER OF ITERATIONS WAS  2

READY
RUN


WHAT IS THE INITIAL GUESS:

    X0 = ?50
WHAT IS THE ASSOCIATED BOUND: ?300
    Y0 = ?50
WHAT IS THE ASSOCIATED BOUND: ?300

INPUT THE CONVERGENCE CRITERION: ?.0001
WHAT IS THE LIMIT ON THE NUMBER OF ITERATIONS: ?10
```

THE ESTIMATED ROOT IS:

    X = -5.32E-15
    Y = 1

THE NUMBER OF ITERATIONS WAS 2

READY
RUN

WHAT IS THE INITIAL GUESS:

    X0 = ?50
WHAT IS THE ASSOCIATED BOUND: ?.1
    Y0 = ?50
WHAT IS THE ASSOCIATED BOUND: ?.1

INPUT THE CONVERGENCE CRITERION: ?.00001
WHAT IS THE LIMIT ON THE NUMBER OF ITERATIONS: ?10

THE ESTIMATED ROOT IS:

    X = -9.997978E-07
    Y = .99999872

THE NUMBER OF ITERATIONS WAS 2

READY
RUN

WHAT IS THE INITIAL GUESS:

    X0 = ?100
WHAT IS THE ASSOCIATED BOUND: ?.1
    Y0 = ?0
WHAT IS THE ASSOCIATED BOUND: ?.1

INPUT THE CONVERGENCE CRITERION: ?.001
WHAT IS THE LIMIT ON THE NUMBER OF ITERATIONS: ?10

THE ESTIMATED ROOT IS:

Listing 7.7.2 cont.

```
    X =  0
    Y = -1

THE NUMBER OF ITERATIONS WAS  2


READY
```

Listing 7.7.2 *Sample runs using ZMUELLER and its demonstration program. The function examined was $f(z) = z^2 + 1 = (x^2 - y^2 + 1) + i(2xy)$.*

Listing 7.7.3

```
RUN


WHAT IS THE INITIAL GUESS:

    X0 = ?0
WHAT IS THE ASSOCIATED BOUND: ?3
    Y0 = ?0
WHAT IS THE ASSOCIATED BOUND: ?3

INPUT THE CONVERGENCE CRITERION: ?.00001
WHAT IS THE LIMIT ON THE NUMBER OF ITERATIONS: ?10


THE ESTIMATED ROOT IS:

    X = -.80910222
    Y = -.80910222

THE NUMBER OF ITERATIONS WAS  10

READY
RUN


WHAT IS THE INITIAL GUESS:

    X0 = ?0
WHAT IS THE ASSOCIATED BOUND: ?3
    Y0 = ?0
WHAT IS THE ASSOCIATED BOUND: ?3
```

```
INPUT THE CONVERGENCE CRITERION: ?0
WHAT IS THE LIMIT ON THE NUMBER OF ITERATIONS: ?40


THE ESTIMATED ROOT IS:

     X =   -.999381
     Y =   -.999381

THE NUMBER OF ITERATIONS WAS   40


READY
RUN



WHAT IS THE INITIAL GUESS:

     X0 = ?0
WHAT IS THE ASSOCIATED BOUND: ?3
     Y0 = ?0
WHAT IS THE ASSOCIATED BOUND: ?3

INPUT THE CONVERGENCE CRITERION: ?0
WHAT IS THE LIMIT ON THE NUMBER OF ITERATIONS: ?100


THE ESTIMATED ROOT IS:

     X =   -1.0000064
     Y =   -1.0000064

THE NUMBER OF ITERATIONS WAS   100


READY
RUN



WHAT IS THE INITIAL GUESS:

     X0 = ?0
WHAT IS THE ASSOCIATED BOUND: ?3
     Y0 = ?0
WHAT IS THE ASSOCIATED BOUND: ?3
```

Listing 7.7.3 cont.

```
INPUT THE CONVERGENCE CRITERION: ?0
WHAT IS THE LIMIT ON THE NUMBER OF ITERATIONS: ?070


THE ESTIMATED ROOT IS:

    X =   -1.0000064
    Y =   -1.0000064

THE NUMBER OF ITERATIONS WAS    70


READY
```

Listing 7.7.3 *Sample runs of ZMUELLER for the real-valued two-dimensional function* $f(z) = (x + 1)^5$; $\mu = (x + 1)^5$, $\nu = 0$. *Compare with the results shown in section 7.5.*

7.8 Representing Polynomials in the $\mu(x,y) + i\,\nu(x,y)$ Form and Removing Roots

The discussion for the remainder of this chapter will center on the determination of the complex roots of polynomials. In this section, we will consider how polynomials in the complex plane can be put in the form $P(z) = \mu(x,y) + i\,\nu(x,y)$, and why this formulation is often desirable. We will also discuss how previously determined roots can be removed so that the remaining ones can be discovered.

Putting a polynomial into the form $\mu(x,y) + i\,\nu(x,y)$ is useful in that the algorithms discussed earlier in this chapter can be directly called on. Accomplishing this end is easy. Recall from Chapter 1 the complex plane series evaluation subroutine, CMPLXSER. The inputs to this subroutine are the polynomial coefficients, $A(I)$, and the evaluation point, $z = x + iy$. The returned values are z_1 and z_2, the real and imaginary components of the summation, respectively. These are the μ and ν values we want. We can therefore augment the function subroutine called by the root seeking algorithms as shown in figure 7.8.1. This procedure is very simple. The only point of complication is that special parameters, such as N, may be used by both the complex series summation and the root-seeking algorithms. Therefore, the function subroutine must temporarily store such parameters before entering CMPLXSER, and must subsequently recall the values before returning to the root-seeking algorithm.

That takes care of the basic problem of compatibility between the iteration algorithms considered so far in this chapter and polynomial evaluation. Next, we will consider root removal. The procedure to be discussed is also applicable to general functions in the complex plane.

Consider the situation in which a root of the polynomial $P(z)$ has been found using one of the algorithms given in this book. How do we effectively remove that root from $P(z)$ and go on to determine others? Most texts on the subject simply suggest *deflating* $P(z)$ by synthetic division, but they do not point out the potential problems associated with that approach. And those problems can be severe!

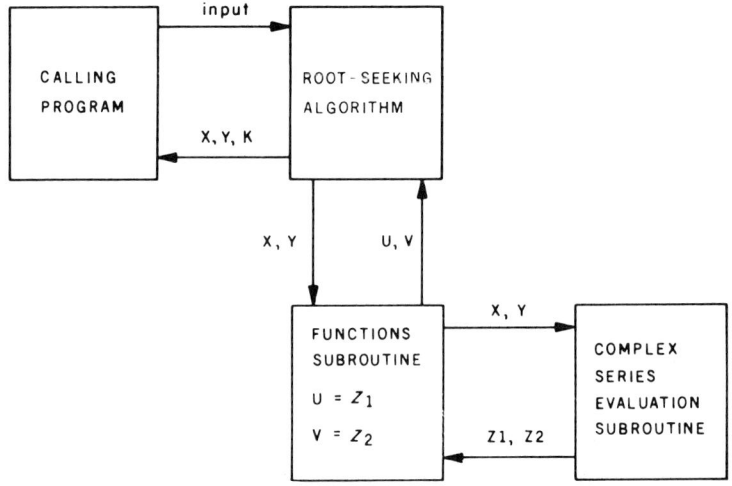

Figure 7.8.1 *Augmented subroutine-connection diagram for using algorithms based on the $f(z) = \mu + iv$ form to treat polynomials in the complex plane. The functions subroutine now acts as an interface that saves and recalls parameters (if necessary—depends on the variables list), and equates U and V to Z_1 and Z_2.*

Deflation is a simple procedure. For example, if the root is real, say x_0, then the synthetic division subroutine (RSYNDIV) given in the preceding chapter can be called on to generate the deflated polynomial $Q(z) = P(z)/(z - x_0)$. If the root is instead complex, say $z_0 = x_0 + iy_0$, then from the discussion in section 7.2, the complex conjugate z_0', is also a root:

$$z_0 = x_0 + iy_0$$
$$z_0' = x_0 - iy_0$$

To deflate $P(z)$ so that only real-valued coefficients are obtained for the resulting polynomial, we can use RSYNDIV and divide by the following quadratic factor:

$$(z - z_0)(z - z_0') = z^2 - 2x_0 z + (x_0^2 + y_0^2)$$

Observe that this factor has only real coefficients.

Once the deflated polynomial $Q(z)$ is obtained, the root-seeking algorithm can again be used to find another root. $Q(z)$ can then be deflated according to that root, and the procedure repeated until all the roots have been found. The procedure is straightforward, but has a flaw—round-off error.

Round-off error can *quickly* accumulate with this procedure and can be visualized as follows. In the first root-seeking step, a root can be found fairly accurately and with little round-off error. However, round-off error is likely to occur in the deflation step in terms of calculating the new coefficients for $Q(z)$. The second root found will therefore be in error. This second root is subsequently used to deflate $Q(z)$, and the error is thus compounded in the next set of coefficients. And the error grows, and grows....

It is therefore advisable to use deflation only once. To treat higher-order polynomials, the iteration algorithms already discussed can be used with new starting points each time, or the procedure discussed in Chapter 6 can be applied (see the next paragraph). In either case, it is desirable to put the polynomial into the form $P(z) = \mu(x,y) + i\,\nu(x,y)$ so that the iterative algorithms discussed in this chapter can be exercised.

Recall from Chapter 6 that for the real-valued function $f(x)$, a root can be made to disappear by simple division:

$$g(x) = \frac{f(x)}{x - a}$$

Note that a is assumed to be a real-valued number. This is a form of deflation. However, if the evaluations of the numerator and the denominator are performed separately (versus algebraically dividing through and then evaluating), the round-off error problem previously discussed can be circumvented. This is because no synthetic division is performed; there are no repeated calculations of coefficients.

The complex plane analogy of the above equation is

$$g(z) = \frac{f(z)}{z - a} \qquad (7.8.1)$$

In this case, the root, a, is assumed to be a complex number: $a = a_x + ia_y$. In the (μ, ν) format, equation (7.8.1) becomes

$$\begin{aligned}g(z) &= \frac{\mu(x,y)}{z - a} + \frac{i\,\nu(x,y)}{z - a} \\ &= \frac{(x - a_x)\,\mu(x,y) + (y - a_y)\,\nu(x,y)}{(x - a_x)^2 + (y - a_y)^2} \\ &\quad + i\,\frac{(x - a_x)\,\nu(x,y) - (y - a_y)\,\mu(x,y)}{(x - a_x)^2 + (y - a_y)^2}\end{aligned}$$

The first term in the last equation can be identified as the U to be returned from the function subroutine. The second term is the V. Therefore, we have

$$U \to \frac{(x - a_x)U + (y - a_y)V}{(x - a_x)^2 + (y - a_y)^2}$$

$$V \to \frac{(x - a_x)V - (y - a_y)U}{(x - a_x)^2 + (y - a_y)^2}$$

(7.8.2)

If two roots are to be removed from a function, the replacement procedure indicated above can be exercised twice, and so on.

This general concept is embodied in the flowchart displayed in figure 7.8.2. In essence, ZMUELLER is repeatedly called to find the remaining roots of a function that has been reduced by the roots already found. The function itself is evaluated in $\mu(x,y)$ and $\nu(x,y)$ form either in the normal function subroutine or by a variant of the complex series evaluation subroutine (CMPLXSER). A BASIC implementation of this flowchart (ALLROOT) appears in listing 7.8.1. Examples of the use of the ALLROOT subroutine are given in listings 7.8.2 through 7.8.5.

In the first example, the function $f(z) = z^2 + 1$ was examined. The two imaginary roots were accurately found using fewer than the maximum number of iterations. In the remaining three examples, the polynomial $P(z) = z(z - 1)(z - 2)(z - 3)(z - 4)$ was probed. Three of the five roots were found on the first try, and the remaining two on the second attempt. For each of the roots actually found, 30 iterations (the maximum chosen) were performed. (This was checked by inserting a PRINT statement into the subroutine.) By increasing the maximum number of iterations to 100, all five roots were discovered in one try, and the results were quite accurate (see listing 7.8.5).

This last example points out one of the important features of ALLROOT. Whereas round-off error *accumulates* when the synthetic division deflation method is used, error does not build in ALLROOT—the algorithm simply fails. Therefore, all the roots obtained (not counting failures that simply return the initial guess) can be taken seriously.

As with ZMUELLER, ALLROOT is a rugged program. When it fails, the failure is obvious, but graceful. It will certainly fail when given a problem that has no solution! When there are several distinct solutions, the higher the maximum number of iterations allowed, the greater the chance of uncovering all the roots in one try. However, high iteration limits, coupled with complex plane polynomial summations, lead to a *very* slow algorithm.

As an exercise, try to use ALLROOT for the following problem. Determine the first five roots of the zeroeth-order Bessel function series, $J_0(x)$. A hint: use the Bessel series coefficients subroutine given in Chapter 1 by changing the target line number of the GOSUB statement in ALLROOT which calls the coefficients subroutine.

FINDING THE COMPLEX ROOTS OF FUNCTIONS 477

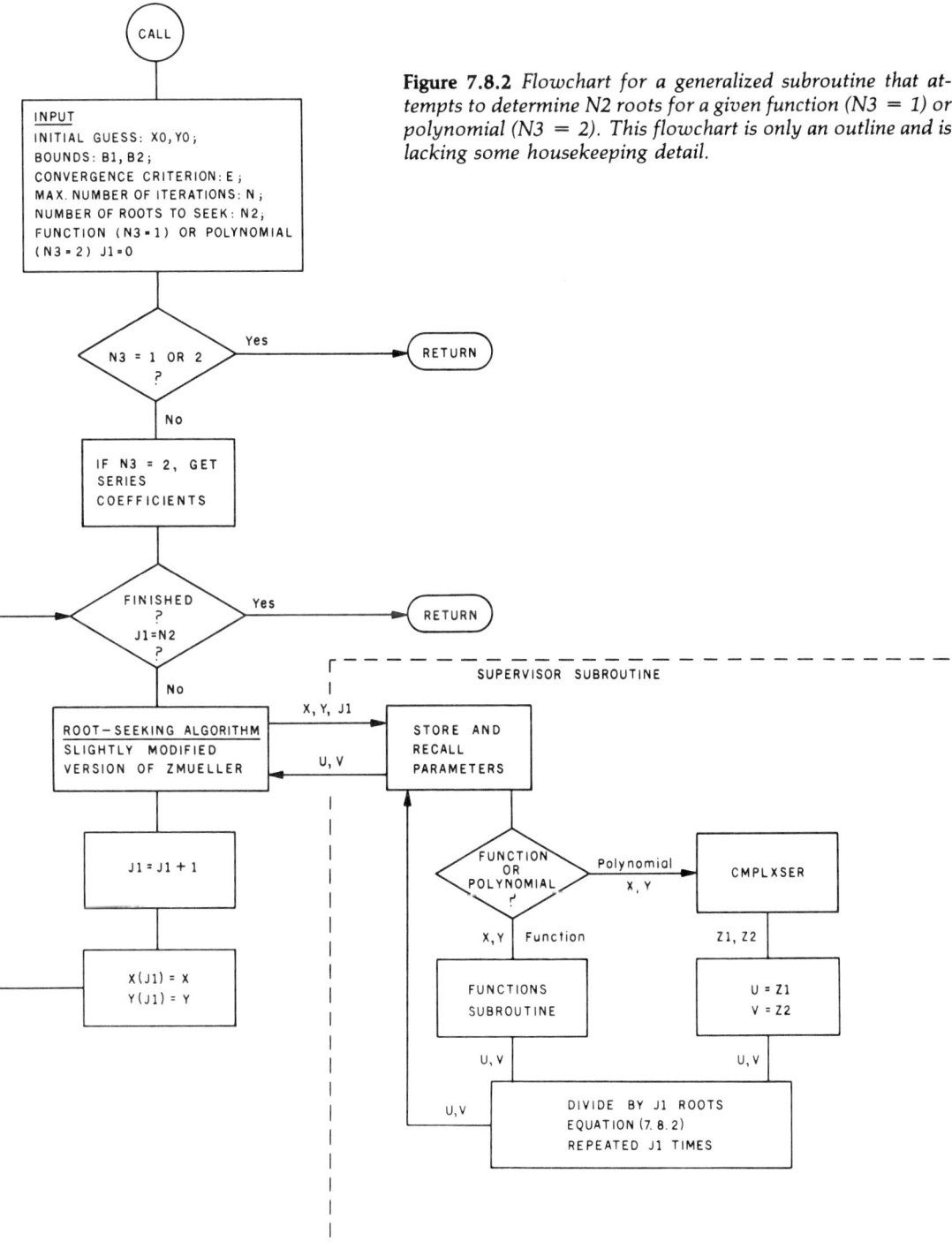

Figure 7.8.2 *Flowchart for a generalized subroutine that attempts to determine N2 roots for a given function (N3 = 1) or polynomial (N3 = 2). This flowchart is only an outline and is lacking some housekeeping detail.*

478 BASIC SCIENTIFIC SUBROUTINES

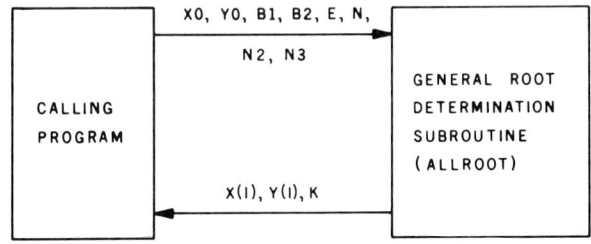

X0, Y0: Initial starting point for *each* root search
B1, B2: Bounds on this inital point
  $X1 = X0 - B1 \qquad Y1 = Y0 - B2$
  $X2 = X0 + B1 \qquad Y2 = Y0 + B2$
E: Convergence criterion
N: Maximum number of iterations per search
N2: Number of roots to be sought
N3: N3 = 1 means the function is in functions subroutine proper
  N3 = 2 means the function is a polynomial
  N3 otherwise is an error
X(I), Y(I): N2 roots found
K: Number of iterations performed in the last search

Figure 7.8.3 *Subroutine-connection diagram for ALLROOT, a general root-determination program. ALLROOT itself is composed of several other subroutines, including a variant of ZMUELLER, CMPLXSER, and several complex-algebra subroutines.*

Statements/Functions List

$+, -, *, /, <, >, \wedge$
ABS, ATN, COS, GOSUB, GOTO, IF/THEN, INT, SIN, SQRT

Variables List

A, A(I), A1, A2, A4, B, B1, B2, C1, C2, D, D1, D2, E, E1, E2, I, J1, J2, K, L1, L2, M, N, N2, N3, N5, U, U1, U2, U3, U5, V, V1, V2, V3, V5, W, X, X(I), X0, X1, X2, X3, X4, Y, Y(I), Y0, Y1, Y2, Y3, Y4, Z1, Z2

Variables Passed to Subroutine

B1, B2, E, N, N2, N3, X0, Y0

Table 7.8.1 *Functions and variables used by ALLROOT, a general root-searching algorithm. This list includes the functions and variables employed by all the other subroutines called in the process of calculation.*

FINDING THE COMPLEX ROOTS OF FUNCTIONS

```
4475 REM PROGRAM TO DEMONSTRATE THE COMPLEX DOMAIN ALLROOT SUBROUTINE
4476 PRINT
4477 PRINT
4478 PRINT "WHAT IS THE INITIAL GUESS:"
4479 PRINT
4480 PRINT "     X0 = ",
4481 INPUT X0
4482 PRINT "WHAT IS THE ASSOCIATED BOUND: ",
4483 INPUT B1
4484 PRINT "     Y0 = ",
4485 INPUT Y0
4486 PRINT "WHAT IS THE ASSOCIATED BOUND: ",
4487 INPUT B2
4488 PRINT
4489 PRINT "INPUT THE CONVERGENCE CRITERION: ",
4490 INPUT E
4491 PRINT "WHAT IS THE LIMIT ON THE NUMBER OF ITERATIONS: ",
4492 INPUT N
4493 PRINT "HOW MANY ROOTS ARE TO BE SOUGHT: ",
4494 INPUT N2
4495 PRINT "IS THE FUNCTION IN THE FUNCTIONS SUBROUTINE (1)"
4496 PRINT "OR IS IT A SERIES (2): ",
4497 INPUT N3
4498 GOSUB 47300
4499 PRINT
4500 PRINT
4501 PRINT "THE ESTIMATED ROOTS ARE:"
4502 FOR I=1 TO N2
4503 PRINT
4504 PRINT "     X = ",X(I)
4505 PRINT "     Y = ",Y(I)
4506 PRINT
4507 NEXT I
4508 PRINT
4509 PRINT "THE LAST NUMBER OF ITERATIONS WAS ",K
4510 PRINT
4511 PRINT
4512 END
40398 REM ********************
40399 REM RECTANGULAR TO POLAR CONVERSION SUBROUTINE (RECT/POL)
40400 U=SQRT(X*X+Y*Y)
40401 REM GUARD AGAINST AMBIGUOUS VECTOR
40402 IF Y=0 THEN Y=(.1)^30
40403 REM GUARD AGAINST DIVIDE BY ZERO
40404 IF X=0 THEN X=(.1)^30
40405 REM SOME BASICS REQUIRE A SIMPLE ARGUMENT
40406 W=Y/X
40407 V=ATN(W)
40408 REM CHECK QUADRANT AND ADJUST
```

Listing 7.8.1

```
40409 IF X<0 THEN V=V+3.1415926535                    Listing 7.8.1 cont.
40410 IF V<0 THEN V=V+6.2831853072
40411 RETURN
40449 REM POLAR TO RECTANGULAR CONVERSION SUBROUTINE (POL/RECT)
40450 X=U*COS(V)
40451 Y=U*SIN(V)
40452 RETURN
41099 REM POLAR POWER SUBROUTINE (ZPOLPOW)
41100 U1=U^N
41101 V1=N*V
41102 V1=V1-6.2831853072*INT(V1/6.2831853072)
41103 RETURN
41198 REM RECTANGULAR COMPLEX NUMBER POWER SUBROUTINE (ZRECTPOW)
41199 REM RECTANGULAR TO POLAR CONVERSION
41200 GOSUB 40400
41201 REM POLAR POWER
41202 GOSUB 41100
41203 REM CHANGE VARIABLE FOR CONVERSION
41204 U=U1
41205 V=V1
41206 REM POLAR TO RECTANGULAR CONVERSION
41207 GOSUB 40450
41208 RETURN
44298 REM ********************
44299 REM FUNCTION SUBROUTINE
44300 U=X*X-Y*Y+1
44301 V=2*X*Y
44302 RETURN
44942 REM ********************
44943 REM COMPLEX SERIES EVALUATION SUBROUTINE (CMPLXSER)
44944 REM THE SERIES COEFFICIENTS ARE A(I), ASSUMED REAL.
44945 REM THE ORDER OF THE POLYNOMIAL IS M.
44946 REM THE SUBROUTINE USES REPEATED CALLS TO THE
44947 REM NTH POWER (Z^N) COMPLEX NUMBER SUBROUTINE.
44948 REM INPUTS TO THE SUBROUTINE ARE X,Y,M, AND THE A(I).
44949 REM OUTPUTS ARE Z1(REAL) AND Z2(IMAGINARY).
44950 Z1=A(0)
44951 Z2=0
44952 REM STORE X AND Y
44953 A1=X
44954 A2=Y
44955 FOR N=1 TO M
44956 REM RECALL ORIGINAL X AND Y
44957 X=A1
44958 Y=A2
44959 REM GO TO Z^N SUBROUTINE
44960 GOSUB 41200
44961 REM FORM PARTIAL SUM
44962 Z1=Z1+A(N)*X
```

Listing 7.8.1 cont.

```
44963 Z2=Z2+A(N)*Y
44964 NEXT N
44965 REM RESTORE X AND Y
44966 X=A1
44967 Y=A2
44968 RETURN
47282 REM *********************
47283 REM GENERAL ROOT DETERMINATION SUBROUTINE (ALLROOT)
47284 REM THE ROUTINE ATTEMPTS TO CALCULATE THE SEVERAL ROOTS OF A
47285 REM GIVEN SERIES OR FUNCTION BY REPEATEDLY USING THE
47286 REM ZMUELLER SUBROUTINE AND REMOVING THE ROOTS ALREADY FOUND
47287 REM BY DIVISION.
47288 REM THE INPUT TO THE SUBROUTINE ARE
47289 REM     X0,Y0 - THE INITIAL GUESS
47290 REM     B1,B2 - THE BOUNDS ON THIS GUESS
47291 REM     E - THE CONVERGENCE CRITERIA
47292 REM     N - THE MAXIMUM NUMBER OF ITERATIONS PER ROOT
47293 REM     N2 - THE NUMBER OF ROOTS BEING SOUGHT
47294 REM     N3 - A FLAG INDICATING A FUNCTION F(Z) (1)
47295 REM          OR A POLYNOMIAL (2)
47296 REM THE PROGRAM RETURNS THE N2 ROOTS FOUND, X(I),Y(I)
47297 REM AND THE LAST NUMBER OF ITERATIONS USED, K.
47298 REM IF K=0 THEN N3 WAS IN ERROR
47299 REM START CALCULATIONS
47300 K=0
47301 IF N3=1 THEN GOTO 47303
47302 IF N3<>2 THEN RETURN
47303 J1=0
47304 REM SAVE THE INITIAL GUESS
47305 X4=X0
47306 Y4=Y0
47307 REM IF N3=2 THEN GET THE SERIES COEFFICIENTS
47308 IF N3=2 THEN GOSUB 47322
47309 REM TEST FOR COMPLETION
47310 IF J1=N2 THEN RETURN
47311 REM GOTO ZMUELLER
47312 GOSUB 47344
47313 J1=J1+1
47314 X(J1)=X
47315 Y(J1)=Y
47316 X0=X4
47317 Y0=Y4
47318 REM TRY ANOTHER PASS
47319 GOTO 47310
47320 REM *********************
47321 REM COEFFICIENTS SUBROUTINE
47322 M=5
47323 A(0)=0
47324 A(1)=24
```

```
47325 A(2)=-50
47326 A(3)=35
47327 A(4)=-10
47328 A(5)=1
47329 RETURN
47330 REM ********************
47331 REM VARIANT ON MUELLER'S METHOD FOR COMPLEX ROOTS
47332 REM THIS PROGRAM USES THE PARABOLIC FITTING TECHNIQUE
47333 REM ASSOCIATED WITH MUELLER'S METHOD, BUT DOES IT IN
47334 REM THE COMPLEX DOMAIN.
47335 REM THE INPUTS TO THE SUBROUTINE ARE THE INITIAL
47336 REM GUESS, (X0,Y0), THE CONVERGENCE CRITERIA, E,
47337 REM AND THE MAXIMUM NUMBER OF ITERATIONS, N.
47338 REM ALSO REQUIRED ARE BOUNDS ON THE INITIAL GUESS, B1 AND B2
47339 REM RETURNED IS THE NEW ESTIMATE, (X,Y), AND THE
47340 REM NUMBER OF ITERATIONS PERFORMED, K.
47341 REM IT IS ASSUMED THAT THE FUNCTION F(Z) = U(X,Y)+IV(X,Y)
47342 REM IS AVAILABLE IN THE FUNCTIONS SUBROUTINE.
47343 REM START CALCULATIONS
47344 K=1
47345 X3=X0
47346 Y3=Y0
47347 X1=X3-B1
47348 Y1=Y3-B2
47349 X2=X3+B1
47350 Y2=Y3+B2
47351 D=(X2-X1)*(X2-X1)+(Y2-Y1)*(Y2-Y1)
47352 REM AVOID DIVIDE BY ZERO
47353 IF D=0 THEN D=.0000001
47354 L1=(X3-X2)*(X2-X1)+(Y3-Y2)*(Y2-Y1)
47355 L1=L1/D
47356 L2=(X2-X1)*(Y3-Y2)-(X3-X2)*(Y2-Y1)
47357 L2=L2/D
47358 D1=(X3-X1)*(X2-X1)+(Y3-Y1)*(Y2-Y1)
47359 D1=D1/D
47360 D2=(X2-X1)*(Y3-Y1)-(X3-X1)*(Y2-Y1)
47361 D2=D2/D
47362 REM GET FUNCTION VALUES
47363 X=X1
47364 Y=Y1
47365 GOSUB 47431
47366 U1=U
47367 V1=V
47368 X=X2
47369 Y=Y2
47370 GOSUB 47431
47371 U2=U
47372 V2=V
47373 X=X3
```

Listing 7.8.1 cont.

```
47374 Y=Y3
47375 GOSUB 47431
47376 U3=U
47377 V3=V
47378 REM CALCULATE MUELLER PARAMETERS
47379 E1=U1*(L1*L1-L2*L2)-2*V1*L1*L2-U2*(D1*D1-D2*D2)
47380 E1=E1+2*V2*D1*D2+U3*(L1+D1)-V3*(L2+D2)
47381 E2=2*L1*L2*U1+V1*(L1*L1-L2*L2)-2*D1*D2*U2-V2*(D1*D1-D2*D2)
47382 E2=E2+U3*(L2+D2)+V3*(L1+D1)
47383 C1=L1*L1*U1-L1*L2*V1-D1*L1*U2+L1*D2*V2+U3*L1
47384 C1=C1-U1*L2*L2-V1*L1*L2+U2*L2*D2+V2*D1*L2-V3*L2
47385 C2=U1*L1*L2+V1*L1*L1-U2*D2*L1-V2*D1*L1+V3*L1
47386 C2=C2+L1*L2*U1-L2*L2*V1-D1*L2*U2+D2*L2*V2+U3*L2
47387 B1=E1*E1-E2*E2-4*(U3*D1*C1-U3*D2*C2-V3*D2*C1-V3*D1*C2)
47388 B2=2*E1*E2-4*(U3*D2*C1+U3*D1*C2+V3*D1*C1-V3*D2*C2)
47389 REM GUARD AGAINST A DIVIDE BY ZERO
47390 IF B1=0 THEN B1=.0000001
47391 A=ATN(B2/B1)
47392 A=A/2
47393 B=SQRT(SQRT(B1*B1+B2*B2))
47394 B1=B*COS(A)
47395 B2=B*SIN(A)
47396 A1=(E1+B1)*(E1+B1)+(E2+B2)*(E2+B2)
47397 A2=(E1-B1)*(E1-B1)+(E2-B2)*(E2-B2)
47398 IF A1>A2 THEN GOTO 47402
47399 A1=E1-B1
47400 A2=E2-B2
47401 GOTO 47404
47402 A1=E1+B1
47403 A2=E2+B2
47404 A=A1*A1+A2*A2
47405 L1=A1*D1*U3-A1*D2*V3+A2*U3*D2+A2*V3*D1
47406 REM GUARD AGAINST DIVIDE BY ZERO
47407 IF A=0 THEN A=.0000001
47408 L1=-2*L1/A
47409 L2=-D1*U3*A2+D2*V3*A2+A1*U3*D2+A1*V3*D1
47410 L2=-2*L2/A
47411 REM CALCULATE NEW ESTIMATE
47412 X=X3+L1*(X3-X2)-L2*(Y3-Y2)
47413 Y=Y3+L2*(X3-X2)+L1*(Y3-Y2)
47414 REM TEST FOR CONVERGENCE
47415 IF ABS(X-X0)+ABS(Y-Y0)<E THEN RETURN
47416 REM TEST FOR NUMBER OF ITERATIONS
47417 IF K>=N THEN RETURN
47418 REM CONTINUE
47419 K=K+1
47420 X0=X
47421 Y0=Y
47422 X1=X2
```

Listing 7.8.1 cont.

```
47423 Y1=Y2
47424 X2=X3
47425 Y2=Y3
47426 X3=X
47427 Y3=Y
47428 GOTO 47351
47429 REM ********************
47430 REM SUPERVISOR SUBROUTINE
47431 N5=N
47432 U5=U1
47433 V5=V1
47434 REM DO WE GO TO THE FUNCTIONS SUBROUTINE OR TO THE SERIES SUBROUTINE?
47435 IF N3=1 THEN GOSUB 44300
47436 IF N3=2 THEN GOSUB 44950
47437 IF N3=1 THEN GOTO 47444
47438 U=Z1
47439 V=Z2
47440 REM RESTORE PARAMETERS
47441 N=N5
47442 U1=U5
47443 V1=V5
47444 IF J1=0 THEN RETURN
47445 REM DIVIDE BY THE J1 ROOTS ALREADY FOUND
47446 FOR J2=1 TO J1
47447 U5=U
47448 U=(X-X(J2))*U+(Y-Y(J2))*V
47449 V=(X-X(J2))*V-(Y-Y(J2))*U5
47450 A4=(X-X(J2))*(X-X(J2))+(Y-Y(J2))*(Y-Y(J2))
47451 REM GUARD AGAINST DIVIDE BY ZERO
47452 IF A4=0 THEN A4=.0000001
47453 V=V/A4
47454 U=U/A4
47455 NEXT J2
47456 REM RETURN TO ZMUELLER
47457 RETURN
```

Listing 7.8.1 *Subroutine (ALLROOT) for sequentially determining the roots of either a general function, $f(z) = \mu + iv$, or a complex polynomial series, $P(z)$. Also shown is a program for demonstrating the operation of ALLROOT (see listings 7.8.2 through 7.8.5).*

Listing 7.8.2

```
          RUN



          WHAT IS THE INITIAL GUESS:
```

```
        X0 = ?4
WHAT IS THE ASSOCIATED BOUND: ?1
        Y0 = ?4
WHAT IS THE ASSOCIATED BOUND: ?1

INPUT THE CONVERGENCE CRITERION: ?.000000001
WHAT IS THE LIMIT ON THE NUMBER OF ITERATIONS: ?30
HOW MANY ROOTS ARE TO BE SOUGHT: ?2
IS THE FUNCTION IN THE FUNCTIONS SUBROUTINE (1)
OR IS IT A SERIES (2): ?1


THE ESTIMATED ROOTS ARE:

        X =   5E-22
        Y =   1


        X =   -2E-22
        Y =   -.99999999


THE LAST NUMBER OF ITERATIONS WAS  6


READY
```

Listing 7.8.2 Sample run of ALLROOT for the complex plane function $f(z) = z^2 + 1$. The functions subroutine is called (N3 = 1), where $\mu = x^2 - y^2 + 1$ and $v = 2xy$. Two roots exist, two were sought, and two were found (very accurately).

Listing 7.8.3

```
RUN


WHAT IS THE INITIAL GUESS:

        X0 = ?-10
WHAT IS THE ASSOCIATED BOUND: ?-1
        Y0 = ?-10
WHAT IS THE ASSOCIATED BOUND: ?-1

INPUT THE CONVERGENCE CRITERION: ?.000000001
WHAT IS THE LIMIT ON THE NUMBER OF ITERATIONS: ?30
HOW MANY ROOTS ARE TO BE SOUGHT: ?5
IS THE FUNCTION IN THE FUNCTIONS SUBROUTINE (1)
OR IS IT A SERIES (2): ?2
```

```
THE ESTIMATED ROOTS ARE:

    X =  .99999748
    Y =  -4.1569869E-06


    X =  7E-22
    Y =  -8E-22


    X =  1.9999998
    Y =  2.7523637E-07


    X =  -10
    Y =  -10


    X =  -10
    Y =  -10


THE LAST NUMBER OF ITERATIONS WAS  1


READY
```

Listing 7.8.3 *Sample run of ALLROOT for the polynomial $P(z) = z(z - 1)(z - 2)(z - 3)(z - 4)$. The initial guess was far from the region containing the roots, but convergence still occurred. However, only three roots were uncovered before a potentially fatal loop was encountered and the search discontinued. The clues regarding the termination are that the initial guess was returned as a root, and that the number of iterations was fewer than the maximum.*

Listing 7.8.4

```
RUN


WHAT IS THE INITIAL GUESS:

    X0 = ?10
WHAT IS THE ASSOCIATED BOUND: ?1
    Y0 = ?10
WHAT IS THE ASSOCIATED BOUND: ?1

INPUT THE CONVERGENCE CRITERION: ?.000000001
WHAT IS THE LIMIT ON THE NUMBER OF ITERATIONS: ?30
HOW MANY ROOTS ARE TO BE SOUGHT: ?5
```

```
IS THE FUNCTION IN THE FUNCTIONS SUBROUTINE (1)
OR IS IT A SERIES (2): ?2

THE ESTIMATED ROOTS ARE:

    X =  3.9999852
    Y = -5.1640863E-06


    X =  2
    Y = -7.7563E-12


    X =  10
    Y =  10


    X =  10
    Y =  10


    X =  10
    Y =  10


THE LAST NUMBER OF ITERATIONS WAS   1

READY
```

Listing 7.8.4 *The previous example is repeated, but using an initial guess which is on the side of the roots opposite to the earlier starting point. The remaining two roots are found.*

Listing 7.8.5

```
RUN


WHAT IS THE INITIAL GUESS:

    X0 = ?10
WHAT IS THE ASSOCIATED BOUND: ?1
    Y0 = ?10
WHAT IS THE ASSOCIATED BOUND: ?1

INPUT THE CONVERGENCE CRITERION: ?0
```

```
WHAT IS THE LIMIT ON THE NUMBER OF ITERATIONS: ?100
HOW MANY ROOTS ARE TO BE SOUGHT: ?5
IS THE FUNCTION IN THE FUNCTIONS SUBROUTINE (1)
OR IS IT A SERIES (2): ?2

THE ESTIMATED ROOTS ARE:

    X =  3.999999
    Y = -6.1143131E-08


    X =  2.0000005
    Y = -8.400083E-23


    X =  .99999963
    Y = -5.3933963E-06


    X =  3.0000019
    Y = -5.9447883E-09


    X = -3.7737457E-24
    Y =  4.2325086E-25


THE LAST NUMBER OF ITERATIONS WAS  100

READY
```

Listing 7.8.5 *A repeat of the previous example, but using 100 iterations per root search sequence. All five roots are found, and with good accuracy.*

7.9 The Quadratic Formula

The determination of the complex roots of a second degree equation is, in principle, a simple matter in that the solution can be written in a closed form. The usual formulae given are

$$r_1 = \frac{-b - \sqrt{b^2 - 4ac}}{2a}$$

$$r_2 = \frac{-b + \sqrt{b^2 - 4ac}}{2a}$$

where the quadratic equation to be solved is

$$y(x) = ax^2 + bx + c$$

These formulae can easily be programmed. However, there is a potential round-off error problem associated with the calculation of r_2, the subtraction operation in the numerator. Hamming (see Ref. 16) offers the following alternative formulae:

$$r_1 = -\text{sign}(b)\left[\frac{|b| + \sqrt{b^2 - 4ac}}{2a}\right]$$

$$r_2 = \frac{c}{ar_1}$$

Hamming's version is also easy to encode, as shown in listing 7.9.1 for the subroutine called QUADRAT. The operation of QUADRAT is demonstrated in listing 7.9.2. The only potential problem in using this program is in dealing with equations having relatively large coefficients, but containing relatively small imaginary components in the roots: round-off error in calculating $b^2 - 4ac$.

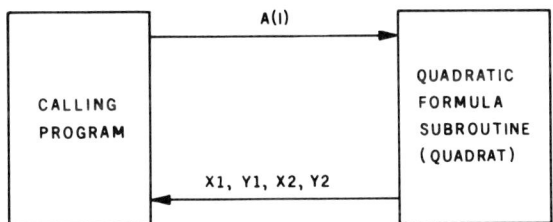

A(I):   Quadratic equation coefficients
X1, Y1, X2, Y2: Calculated roots

Figure 7.9.1 *Subroutine-connection diagram for QUADRAT.*

Statements/Functions List

$+, -, *, /, <, >$
ABS, IF/THEN, SQRT

Variables List

A, A(I), B, C, X1, X2, Y1, Y2

Variables Passed to Subroutine

A(I)

Table 7.9.1 *Functions and variables used by the quadratic formula subroutine (QUADRAT).*

```
4525 REM PROGRAM TO DEMONSTRATE QUADRATIC ROOT SUBROUTINE
4526 PRINT
4527 PRINT "INPUT THE COEFFICIENTS:"
4528 PRINT
4529 PRINT "A(0) = ",
4530 INPUT A(0)
4531 PRINT "A(1) = ",
4532 INPUT A(1)
4533 PRINT "A(2) = ",
4534 INPUT A(2)
4535 GOSUB 47475
4536 PRINT
4537 PRINT
4538 PRINT "RESULTS:"
4539 PRINT
4540 PRINT "     R1 = ",X1,
4541 IF Y1>=0 THEN PRINT " +",
4542 IF Y1<0 THEN PRINT " ",
4543 PRINT Y1," I"
4544 PRINT "     R2 = ",X2,
4545 IF Y2>=0 THEN PRINT " +",
4546 IF Y2<0 THEN PRINT " ",
4547 PRINT Y2," I"
4548 PRINT
4549 PRINT
4550 END
47462 REM ********************
47463 REM QUADRATIC ROOT SUBROUTINE (QUADRAT)
47464 REM THIS PROGRAM CALCULATES THE TWO ROOTS OF
47465 REM A GIVEN SECOND ORDER POLYNOMIAL USING
47466 REM THE QUADRATIC EQUATION EVALUATED IN A
47467 REM MANNER WHICH MINIMIZES ROUND OFF ERROR.
47468 REM THE POLYNOMIAL IS ASSUMED TO BE OF
47469 REM THE FORM
47470 REM      Y = A(2)*X*X +A(1)*X +A(0)
47471 REM THE TWO ROOTS ARE RETURNED AS
47472 REM R1 = X1 + I Y1
47473 REM R2 = X2 + I Y2
47474 REM TEST FOR A(2)=0
47475 IF A(2)<>0 THEN GOTO 47488
47476 REM TEST FOR A(1)=0
47477 IF A(1)<>0 THEN GOTO 47483
47478 X1=0
47479 X2=0
47480 Y1=0
47481 Y2=0
47482 RETURN
47483 X1=-A(0)/A(1)
47484 Y1=0
```

Listing 7.9.1

```
47485 X2=X1
47486 Y2=Y1
47487 RETURN
47488 A=A(1)*A(1)-4*A(2)*A(0)
47489 B=SQRT(ABS(A))
47490 REM ESTABLISH SIGN
47491 IF A(1)=0 THEN C=1
47492 IF A(1)<>0 THEN C=ABS(A(1))/A(1)
47493 REM DETERMINE THE FIRST ROOT
47494 REM CHECK IF ROOT IS COMPLEX
47495 IF A>0 THEN GOTO 47499
47496 X1=-C*ABS(A(1))/(2*A(2))
47497 Y1=-C*B/(2*A(2))
47498 GOTO 47502
47499 X1=-C*(ABS(A(1))+B)/(2*A(2))
47500 Y1=0
47501 REM CALCULATE THE SECOND ROOT
47502 C=X1*X1+Y1*Y1
47503 IF C<>0 THEN GOTO 47506
47504 X2=1000000*1000000*1000000
47505 RETURN
47506 C=A(0)/(C*A(2))
47507 X2=X1*C
47508 Y2=-Y1*C
47509 RETURN
```

Listing 7.9.1 *Modified quadratic formula subroutine (QUADRAT). Also shown is a program for demonstrating QUADRAT (see listing 7.9.2).*

Listing 7.9.2

```
RUN

INPUT THE COEFFICIENTS:

A(0) = ?1
A(1) = ?0
A(2) = ?1


RESULTS:
      R1 =   0  -1 I
      R2 =   0 + 1 I


READY
```

```
RUN

INPUT THE COEFFICIENTS:

A(0) = ?4
A(1) = ?2
A(2) = ?1


RESULTS:

      R1 =   -1  -1.7320508 I
      R2 =   -1 + 1.7320508 I


READY


RUN

INPUT THE COEFFICIENTS:

A(0) = ?1
A(1) = ?1
A(2) = ?1


RESULTS:

      R1 =   -.5  -.8660254 I
      R2 =   -.5 + .8660254 I


READY
```

Listing 7.9.2 Sample runs of QUADRAT, a subroutine for finding the roots of a second-degree polynomial.

7.10 Lin's Method

The discussions in this and the next section will deal with two root-seeking algorithms that capitalize on the specific functional form of polynomials. The first method, that of Lin,[*] applies the knowledge that the complex roots of real-valued coefficient polynomials come in complex conjugate pairs, and that an iterative procedure involving synthetic division can be employed to extract a quadratic factor using only real-number operations. The second technique to be discussed, that of Bairstow,[**] is an extension of the Lin method and contains a Newton iteration step that accelerates convergence.

Lin's method is based on the simple observation that if complex roots occur, they must exist in complex conjugate pairs. This means that any polynomial of degree two or greater must contain a quadratic factor having only real coefficients. We will take as our starting polynomial the following standard form:

$$P(z) = a_0 + a_1 z + a_2 z^2 + \cdots + z^n \quad (7.10.1)$$

Note that it is implicitly assumed that $a_n = 1$. If we arbitrarily pick a quadratic factor $z^2 + Az + B$, then $P(z)$ can be rewritten as

$$P(z) = (z^2 + Az + B)(B_2 + B_3 z + \cdots + z^{n-2}) + B_1 + B_0 \quad (7.10.2)$$

If, by some means, the quadratic factor were chosen so that $B_1 = 0$ and $B_2 = 0$, then the two roots of that factor would also be roots of $P(z)$. Therefore, the goal is to determine A and B so that $B_1 = 0$ and $B_0 = 0$.

By comparing like powers of z in equations (7.10.1) and (7.10.2), we get the following sequence of equations:

$$B_{n-1} = A_{n-1} - A$$
$$B_{n-2} = A_{n-2} - B$$
$$\vdots$$
$$B_{N-J} = A_{N-J} - AB_{N+1-J} - BB_{N+2-J}$$
$$A = \frac{A_1 - BB_3}{B_2}$$
$$B = \frac{A_0}{B_2}$$

---

[*]The discussion in this section is based on that given in *A Practical Guide to Computer Methods*, by T. E. Shoup (Ref. 40).

[**]This is discussed in the next section. For further information, see the reference given above, as well as *Numerical Calculations and Algorithms*, by R. Becket and J. Hurt; *Computer Methods for Science and Engineering*, by R. L. LaFara; and *Introduction to Applied Numerical Analysis*, by R. Hamming.

(This is an application of the uniqueness theorem for polynomials. If two polynomials are equal, then the coefficients of like powers are equal.) If A and B are only guesses, then the last two equations offer a means to partially correct those guesses. This new estimate pair can then be employed to make another pass through the set of equations in order to derive yet another improved estimate.

As usual, the algorithm is not perfect. Convergence is not always ensured, especially when multiple roots are encountered. Also, convergence may be very slow. This is aggravated by the observation that small relative errors in the final estimates of A and B result in much larger relative errors in the finally derived roots.

To demonstrate this last comment, we will consider the quadratic factor $x^2 + 2x + (1 + \epsilon)$, where ϵ represents the error in the coefficient, B. The corresponding roots are $-1 + \sqrt{\epsilon}$ and $-1 - \sqrt{\epsilon}$. Even if the error in the coefficient is small, say 10^{-6}, the error in the root can be large (e.g., 10^{-3}). This observation regarding the error is factored into the algorithm by using the square of the convergence criterion.

The Lin algorithm can be implemented in BASIC as shown in listing 7.10.1. See also figure 7.10.1 and table 7.10.1. A set of interesting examples appears in listing 7.10.2. For these examples, the initial values of A and B were chosen quite arbitrarily: $A = \pi$ and $B = \sqrt{2}$. Other choices could be

1) $A = A_{n-1}$; $B = A_{n-2}$: the iteration tends to seek out the largest roots.

2) $A = A_1$; $B = A_0$: the iteration tends to seek out the smallest roots.

3) $A = 0$; $B = 0$: this appears to be an effective starting point in many cases.

As an exercise, experiment with different starting values.

In the first two examples shown in listing 7.10.2, the polynomial examined was $P(z) = z(z - 1)(z - 2)(z - 3)(z - 4)$. The $z = 0$ root was determined quite accurately, but convergence was slow, as indicated by the error in the $z = 1$ root. The second two examples involved the stress-case polynomial $P(z) = (z + 1)^5$, multiple roots. Convergence occurred, but again, it was very slow.

Lin's algorithm is a reasonably straightforward application of some of the basic properties of polynomials and iterative procedures. Its major drawback is that it is only slowly convergent. Also, it is limited to polynomials of the fourth degree and higher. In the next section, we will consider a more complicated, but a much more rapidly convergent algorithm—Bairstow's method.

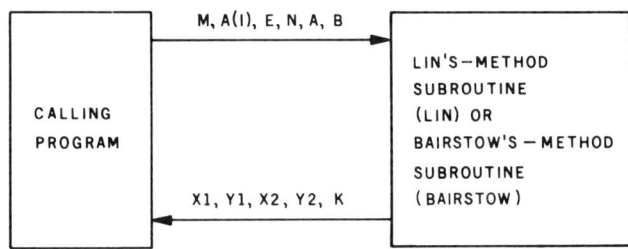

M:	Degree of the polynomial
A(I):	Polynomial coefficients
E:	Convergence criterion
N:	Maximum number of iterations
A, B:	Initial coefficients of the quadratic factor estimate $X^2 + AX + B$
X1, Y1, X2, Y2:	Two roots found
K:	Number of iterations performed

Figure 7.10.1 *Subroutine-connection diagram for both the Lin and Bairstow method subroutines for determining two roots of the polynomial P(z).*

Statements/Functions List

$+, -, *, /, <, >$
ABS, FOR/NEXT, GOTO, IF/THEN, SQRT

Variables List

A, A(I), A1, B, B(I), B1, C, C(I), E, I, J, K, M, N, X1, X2, Y1, Y2

Variables Passed to Subroutine

A, A(I), B, E, M, N

Table 7.10.1 *Functions and variables used by the LIN subroutine.*

Listing 7.10.1

```
4560 REM PROGRAM TO DEMONSTRATE THE LIN SUBROUTINE
4561 PRINT
4562 PRINT
4563 PRINT "INPUT ORDER OF POLYNOMIAL",
4564 INPUT M
4565 PRINT
4566 PRINT "INPUT THE POLYNOMIAL COEFFICIENTS:"
```

```
4567 FOR I=0 TO M
4568 PRINT "     A(",I,") = ",
4569 INPUT A(I)
4570 NEXT I
4571 PRINT
4572 PRINT "WHAT IS THE CONVERGENCE FACTOR: ",
4573 INPUT E
4574 PRINT "WHAT IS THE MAXIMUM NUMBER OF ITERATIONS: ",
4575 INPUT N
4576 A=3.14159
4577 B=SQRT(2)
4578 GOSUB 47525
4579 PRINT
4580 PRINT "THE ROOTS FOUND ARE:"
4581 PRINT
4582 PRINT "     X1 = ",X1
4583 PRINT "     Y1 = ",Y1
4584 PRINT
4585 PRINT "     X2 = ",X2
4586 PRINT "     Y2 = ",Y2
4587 PRINT
4588 PRINT "THE NUMBER OF ITERATIONS WAS ",K
4589 PRINT
4590 PRINT
4591 END
47511 REM ********************
47512 REM POLYNOMIAL COMPLEX ROOTS SUBROUTINE (LIN)
47513 REM USES LIN'S METHOD AS DESCRIBED IN THE REFERENCE
47514 REM A PRACTICAL GUIDE TO COMPUTER METHODS FOR ENGINEERS BY SHOUP.
47515 REM THE INPUT POLYNOMIAL COEFFICIENTS ARE A(0) THROUGH A(M).
47516 REM M IS THE ORDER OF THE POLYNOMIAL.
47517 REM INITIAL GUESSES FOR A AND B ARE REQUIRED.
47518 REM THE RESULTS ARE RETURNED IN X1,Y1 AND X2,Y2.
47519 REM X IS THE REAL PART, AND Y IS THE IMAGINARY.
47520 REM THE MAXIMUM NUMBER OF ITERATIONS IS N.
47521 REM THE NUMBER OF ITERATIONS IS RETURNED IN K
47522 REM THE CONVERGENCE CRITERION IS E.
47523 REM IF NECESSARY, DIMENSION A(I), B(I) AND C(I) IN THE CALLING PROGRAM.
47524 REM NORMALIZE THE A(I) SERIES
47525 FOR I=0 TO M
47526 C(I)=A(I)/A(M)
47527 NEXT I
47528 REM START ITERATION
47529 REM SET INITIAL GUESS FOR THE QUADRATIC COEFFICIENTS
47530 B(0)=0
47531 B(1)=0
47532 B(M-1)=C(M-1)-A
47533 B(M-2)=C(M-2)-A*B(M-1)-B
47534 FOR J=3 TO M
```

Listing 7.10.1 cont.

```
47535 B(M-J)=C(M-J)-A*B(M+1-J)-B*B(M+2-J)
47536 NEXT J
47537 REM GUARD AGAINST DIVIDE BY ZERO
47538 IF B(2)<>0 THEN GOTO 47542
47539 A=A+.0000001
47540 B=B-.0000001
47541 GOTO 47532
47542 A1=(C(1)-B*B(3))/B(2)
47543 B1=C(0)/B(2)
47544 K=K+1
47545 REM TEST FOR THE NUMBER OF ITERATIONS
47546 IF K>=N THEN GOTO 47553
47547 REM TEST FOR CONVERGENCE
47548 IF ABS(A-A1)+ABS(B-B1)<E*E THEN GOTO 47553
47549 A=A1
47550 B=B1
47551 REM RETURN FOR NEXT ITERATION
47552 GOTO 47532
47553 A=A1
47554 B=B1
47555 C=A*A-4*B
47556 REM IS THERE AN IMAGINARY PART
47557 IF C>0 THEN GOTO 47563
47558 Y1=SQRT(-C)
47559 Y2=-Y1
47560 X1=-A
47561 X2=X1
47562 GOTO 47567
47563 Y1=0
47564 Y2=Y1
47565 X1=-A+SQRT(C)
47566 X2=-A-SQRT(C)
47567 X1=X1/2
47568 X2=X2/2
47569 Y1=Y1/2
47570 Y2=Y2/2
47571 RETURN
```

Listing 7.10.1 *Subroutine for applying Lin's root determination method to the polynomial $P(z) = A(0) + A(1)z + \cdots$ (LIN). Also shown is a program for demonstrating the use of LIN. See listing 7.10.2 for sample results.*

Listing 7.10.2

RUN

INPUT ORDER OF POLYNOMIAL?5

```
INPUT THE POLYNOMIAL COEFFICIENTS:               Listing 7.10.2 cont.
    A( 0) = ?0
    A( 1) = ?24
    A( 2) = ?-50
    A( 3) = ?35
    A( 4) = ?-10
    A( 5) = ?1

WHAT IS THE CONVERGENCE FACTOR: ?.00001
WHAT IS THE MAXIMUM NUMBER OF ITERATIONS: ?10

THE ROOTS FOUND ARE:

    X1 =  .96706255
    Y1 =  0

    X2 =  0
    Y2 =  0

THE NUMBER OF ITERATIONS WAS   10


READY
RUN



INPUT ORDER OF POLYNOMIAL?5

INPUT THE POLYNOMIAL COEFFICIENTS:
    A( 0) = ?0
    A( 1) = ?24
    A( 2) = ?-50
    A( 3) = ?35
    A( 4) = ?-10
    A( 5) = ?1

WHAT IS THE CONVERGENCE FACTOR: ?.00000001
WHAT IS THE MAXIMUM NUMBER OF ITERATIONS: ?20

THE ROOTS FOUND ARE:

    X1 =  .9982253
    Y1 =  0

    X2 =  0
    Y2 =  0

THE NUMBER OF ITERATIONS WAS   20
```

Listing 7.10.2 cont.

```
READY
RUN


INPUT ORDER OF POLYNOMIAL?5

INPUT THE POLYNOMIAL COEFFICIENTS:
     A( 0) = ?1
     A( 1) = ?5
     A( 2) = ?10
     A( 3) = ?10
     A( 4) = ?5
     A( 5) = ?1

WHAT IS THE CONVERGENCE FACTOR: ?.000001
WHAT IS THE MAXIMUM NUMBER OF ITERATIONS: ?100

THE ROOTS FOUND ARE:

     X1 =   -.8561115
     Y1 =    .05884386

     X2 =   -.8561115
     Y2 =   -.05884386

THE NUMBER OF ITERATIONS WAS   100


READY
RUN


INPUT ORDER OF POLYNOMIAL?5

INPUT THE POLYNOMIAL COEFFICIENTS:
     A( 0) = ?1
     A( 1) = ?5
     A( 2) = ?10
     A( 3) = ?10
     A( 4) = ?5
     A( 5) = ?1

WHAT IS THE CONVERGENCE FACTOR: ?0
WHAT IS THE MAXIMUM NUMBER OF ITERATIONS: ?1000

THE ROOTS FOUND ARE:
```

```
X1 =   -.93763285
Y1 =   2.5730332E-02

X2 =   -.93763285
Y2 =   -2.5730332E-02
```

THE NUMBER OF ITERATIONS WAS   1000

READY

Listing 7.10.2 Sample runs of LIN. For the first two test cases, the polynomial examined was $P(z) = z(z - 1)(z - 2)(z - 3)(z - 4)$. For the second two runs, the polynomial was $P(z) = (z + 1)^5$.

7.11 Bairstow's Method

Lin's method can be improved upon by applying Newton iteration to the determination of the quadratic factor coefficients A and B [by noting that $B_1 = B_1(A,B)$ and $B_2 = B_2(A,B)$]. This variant is called Bairstow's method. (For an excellent presentation of Bairstow's algorithm, refer to the discussion in LaFara's book, Ref. 37. For further reading regarding an extension of Bairstow's method, see Hamming, Ref. 16.)

Bairstow's algorithm can be implemented as shown in listing 7.11.1. The subroutine input and output formats are the same as those for the LIN subroutine. Sample results are given in listing 7.11.2. The tests that were performed to demonstrate the LIN subroutine are repeated (see listing 7.10.2). For the function $P(z) = z(z - 1)(z - 2)(z - 3)(z - 4)$, the Bairstow subroutine clearly outperformed the Lin algorithm. However, for the multiple root polynomial, $P(z) = (z + 1)^5$, both routines struggled; the Bairstow technique was only marginally better.

If the execution time per iteration is factored in, the Bairstow subroutine is superior to the Lin routine for the first function tested, but inferior for the second. The convergence rate for the Bairstow algorithm is usually better than that of the Lin algorithm. Both clearly have difficulties with multiple roots. Also, both are limited to polynomials of the fourth degree and higher.

Statements/Functions List

+, −, *, /, >
ABS, FOR/NEXT, GOTO, IF/THEN, SQRT

Variables List

A, A1, A(I), B, B1, B(I), C, C(I), D, D2, D(I), E, I, J, K, M, N, X1, X2, Y1, Y2

Variables Passed to Subroutine

A, A(I), B, E, M, N

Table 7.11.1 *Functions and variables used in the BAIRSTOW subroutine.*

Listing 7.11.1

```
4600 REM PROGRAM TO DEMONSTRATE THE BAIRSTOW SUBROUTINE
4601 PRINT
4602 PRINT
4603 PRINT "INPUT ORDER OF POLYNOMIAL",
4604 INPUT M
4605 PRINT
4606 PRINT "INPUT THE POLYNOMIAL COEFFICIENTS:"
4607 FOR I=0 TO M
4608 PRINT "     A(",I,") = ",
4609 INPUT A(I)
4610 NEXT I
4611 PRINT
4612 PRINT "WHAT IS THE CONVERGENCE FACTOR: ",
4613 INPUT E
4614 PRINT "WHAT IS THE MAXIMUM NUMBER OF ITERATIONS: ",
4615 INPUT N
4616 A=3.14159
4617 B=SQRT(2)
4618 GOSUB 47600
4619 PRINT
4620 PRINT "THE ROOTS FOUND ARE:"
4621 PRINT
4622 PRINT "     X1 = ",X1
4623 PRINT "     Y1 = ",Y1
4624 PRINT
4625 PRINT "     X2 = ",X2
4626 PRINT "     Y2 = ",Y2
4627 PRINT
4628 PRINT "THE NUMBER OF ITERATIONS WAS ",K
4629 PRINT
4630 PRINT
4631 END
```

```
47584 REM ********************
47585 REM BAIRSTOW COMPLEX ROOT SUBROUTINE (BAIRSTOW)
47586 REM THIS SUBROUTINE FINDS THE COMPLEX CONJUGATE ROOTS
47587 REM OF A POLYNOMIAL HAVING REAL COEFFICIENTS.
47588 REM SEE COMPUTER METHODS FOR SCIENCE AND ENGINEERING
47589 REM BY R.L. LAFARA.
47590 REM ORDER OF INPUT SERIES IS M >= 4.
47591 REM SERIES COEFFICIENTS ARE A(I).
47592 REM INITIAL GUESSES A AND B ARE REQUIRED.
47593 REM E IS THE CONVERGENCE FACTOR.
47594 REM SUBROUTINE RETURNS X1,Y1 AND X2,Y2.
47595 REM N IS THE MAXIMUM NUMBER OF ITERATIONS.
47596 REM K IS THE NUMBER OF ITERATIONS PERFORMED.
47597 REM IF NECESSARY, DIMENSION A(I),B(I), C(I) AND D(I)
47598 REM IN THE CALLING PROGRAM.
47599 REM USE NORMALIZED SERIES, C(I)
47600 FOR I=0 TO M
47601 C(I)=A(I)/A(M)
47602 NEXT I
47603 REM CHOSE INITIAL ESTIMATES FOR A AND B
47604 K=0
47605 B(M)=1
47606 REM START ITERATION SEQUENCE
47607 B(M-1)=C(M-1)-A
47608 FOR J=2 TO M-1
47609 B(M-J)=C(M-J)-A*B(M+1-J)-B*B(M+2-J)
47610 NEXT J
47611 B(0)=C(0)-B*B(2)
47612 D(M-1)=-1
47613 D(M-2)=-B(M-1)+A
47614 FOR J=3 TO M-1
47615 D(M-J)=-B(M+1-J)-A*D(M+1-J)-B*D(M+2-J)
47616 NEXT J
47617 D(0)=-B*D(2)
47618 D2=-B(2)-B*D(3)
47619 D=D(1)*D2-D(0)*D(2)
47620 A1=-B(1)*D2+B(0)*D(2)
47621 A1=A1/D
47622 B1=-D(1)*B(0)+D(0)*B(1)
47623 B1=B1/D
47624 A=A+A1
47625 B=B+B1
47626 K=K+1
47627 REM TEST FOR THE NUMBER OF ITERATIONS
47628 IF K>=N THEN GOTO 47632
47629 REM TEST FOR CONVERGENCE
47630 IF ABS(A1)+ABS(B1)>E*E THEN GOTO 47607
47631 REM EXTRACT ROOTS FROM QUADRATIC EQUATION
47632 C=A*A-4*B
```

Listing 7.11.1 cont.

```
47633 REM TEST TO SEE IF A COMPLEX ROOT
47634 IF C>0 THEN GOTO 47640
47635 X1=-A
47636 X2=X1
47637 Y1=SQRT(-C)
47638 Y2=-Y1
47639 GOTO 47644
47640 X1=-A+SQRT(C)
47641 X2=-A-SQRT(C)
47642 Y1=0
47643 Y2=Y1
47644 X1=X1/2
47645 X2=X2/2
47646 Y1=Y1/2
47647 Y2=Y2/2
47648 RETURN
```

Listing 7.11.1 *The Bairstow-method subroutine (BAIRSTOW) for finding two of the roots for a polynomial of the fourth degree or greater. Also presented is a program that demonstrates the operation of BAIRSTOW. Sample results are given in listing 7.11.2. The subroutine-connection diagram appears in figure 7.10.1 (previous section).*

Listing 7.11.2

```
RUN

INPUT ORDER OF POLYNOMIAL?5

INPUT THE POLYNOMIAL COEFFICIENTS:
     A( 0) = ?0
     A( 1) = ?24
     A( 2) = ?-50
     A( 3) = ?35
     A( 4) = ?-10
     A( 5) = ?1

WHAT IS THE CONVERGENCE FACTOR: ?.00001
WHAT IS THE MAXIMUM NUMBER OF ITERATIONS: ?10

THE ROOTS FOUND ARE:

     X1 =   1.0000001
     Y1 =   0

     X2 =   0
     Y2 =   0
```

504 BASIC SCIENTIFIC SUBROUTINES

```
THE NUMBER OF ITERATIONS WAS   10                    Listing 7.11.2 cont.

READY
RUN



INPUT ORDER OF POLYNOMIAL?5

INPUT THE POLYNOMIAL COEFFICIENTS:
     A( 0) = ?0
     A( 1) = ?24
     A( 2) = ?-50
     A( 3) = ?35
     A( 4) = ?-10
     A( 5) = ?1

WHAT IS THE CONVERGENCE FACTOR: ?.00000001
WHAT IS THE MAXIMUM NUMBER OF ITERATIONS: ?20

THE ROOTS FOUND ARE:

     X1 =   1.0000001
     Y1 =   0

     X2 =   0
     Y2 =   0

THE NUMBER OF ITERATIONS WAS   11

READY
RUN



INPUT ORDER OF POLYNOMIAL?5

INPUT THE POLYNOMIAL COEFFICIENTS:
     A( 0) = ?1
     A( 1) = ?5
     A( 2) = ?10
     A( 3) = ?10
     A( 4) = ?5
     A( 5) = ?1

WHAT IS THE CONVERGENCE FACTOR: ?.000001
WHAT IS THE MAXIMUM NUMBER OF ITERATIONS: ?100
```

```
THE ROOTS FOUND ARE:

     X1 =   -1.0717389
     Y1 =   3.3130424E-02

     X2 =   -1.0717389
     Y2 =   -3.3130424E-02

THE NUMBER OF ITERATIONS WAS   100


READY
RUN



INPUT ORDER OF POLYNOMIAL?5

INPUT THE POLYNOMIAL COEFFICIENTS:
     A( 0) = ?1
     A( 1) = ?5
     A( 2) = ?10
     A( 3) = ?10
     A( 4) = ?5
     A( 5) = ?1

WHAT IS THE CONVERGENCE FACTOR: ?0
WHAT IS THE MAXIMUM NUMBER OF ITERATIONS: ?1000

THE ROOTS FOUND ARE:

     X1 =   -.8622049
     Y1 =   0

     X2 =   -1.0557327
     Y2 =   0

THE NUMBER OF ITERATIONS WAS   1000


READY
```

Listing 7.11.2 Sample results obtained from the Bairstow subroutine. The first two examples involve the polynomial $P(z) = z(z - 1)(z - 2)(z - 3)(z - 4)$. The second two examples relate to the polynomial $P(z) = (z + 1)^5$. Compare with listing 7.10.2 for Lin's method (previous section).

7.12 Summary—Comparison of Algorithms

The subroutines presented in this chapter represent a very wide range of techniques for finding the roots of functions. Each algorithm has its own particular characteristics that make it more or less suitable than others for a given task.

For example, if the function is a second-degree polynomial, then the best subroutine to use is QUADRAT. If you want to determine all the roots of a function and do not care to do anything but run a program (but you are willing to wait for a slow calculation to finish), then ALLROOT is the answer. If the approximate location of a root is known, and it is isolated, then CZNEWTON is likely to be best. However, if the location of the root is in doubt, and if it is of a multiplicity less than three, then ZMUELLER is a good choice. (ZMUELLER is also good for two-dimensional functions and simultaneous two-dimensional equations.) If the situation is extremely difficult (i.e., if the root is very poorly located and/or is multiple), then ZCIRCLE will find it (but slowly!). Finally, if the problem involves a polynomial of the fourth degree or greater, either LIN or BAIRSTOW can be applied. The routines of this chapter are summarized in table 7.12.1.

Table 7.12.1

Algorithm	Advantages	Disadvantages
Two-dimensional interval search	• Concept simple • Code could be short (though longer than bisection method in one dimension)	• Impractically slow; not presented in text
Circular search (ZCIRCLE)	• Nearly impervious to complications of multiple roots (if M is large enough) • Initial guess need not be good • Accuracy usually *much* better than error criterion	• Slow • Requires several inputs, including a search-radius reduction factor • Uses several concepts, leading to long code
Newton's method in the complex plane (CZNEWTON)	• Short code • Converges quickly if initial guess is close enough to root	• May not converge • Requires derivatives of $\mu(x,y)$
Mueller's method—one dimension (MUELLER)	• Very good convergence stability; more stable than Newton's method • Good convergence rate	• Struggles with multiple roots, but still converges

Algorithm	Advantages	Disadvantages
Mueller's method—two dimensions (MUELLER2)	• See above	• Code long • Convergence stability weakened by successive substitution
Mueller's method—complex plane (ZMUELLER)	• Very powerful; good convergence stability • Extendable to simultaneous functions in two dimensions	• Code long • Execution slow • Slow convergence with multiple roots, but still converges!
Mueller's method for several complex roots (ALLROOT)	• Very general program for both $f = \mu + i\nu$ and polynomial functional forms • Resistant to round-off error • Very good stability	• Long, slow code • Multiple roots cause slow convergence
Quadratic formula (QUADRAT)	• Fast, accurate determination of the roots of second-degree equations	• Very specific
Lin's method (LIN)	• Good stability properties • Simple code; reasonably fast (per iteration)	• Slow convergence, particularly for multiple roots • Specific to polynomials
Bairstow's method (BAIRSTOW)	• Very good stability properties • Faster convergence (usually) than Lin's method	• Slow convergence for multiple roots • Specific to polynomials • More complicated than LIN

Table 7.12.1 *A comparison of algorithms for finding the roots of functions in the complex plane. See also table 7.1.1.*

Chapter 8

Optimization by Steepest Descent

8.1 Introduction

In this chapter, we will briefly examine how the maxima and minima of many functions can be found by using the method of *steepest descent*. The ability to determine the extremes of functions is essential to the process of optimization. For example, given a mathematical model for production costs and sales, the analyst can, in principle, maximize the return on investment by establishing the trade-off between advertising expenses, production run length, and so on. Under this criterion (ROI), the operation would be considered optimized.

Another example is related to the adjustment of the properties of materials. By establishing how the individual chemical and mechanical properties depend on a particular additive, and by placing relative values on those properties (i.e., weighting), an optimum composition for a particular task can be discovered.

There are several key elements in the optimization procedure. First, an *objective function* must be established. This is the mathematical construction to be maximized or minimized. In many cases, the most difficult part of the entire problem is the formulation of this expression. How does one include subjective criteria such as color and safety in an equation that is also meant to involve strength and cost? In any case, after assuming that such a function exists, the next step is to choose a method for solution.

If there is only one parameter to be manipulated, then the problem is reasonably simple. The derivative of the function can be calculated (or numerically estimated), and one of the root-seeking algorithms given in Chapter 6 applied. If the function contains several parameters to be optimized across, however, the problem becomes much more difficult, and is the subject of many texts. In this chapter, only one of the most elementary algorithms will be presented.

The basic problem can be visualized in three dimensions by using figure 8.1.1. The object function is $z(x,y)$. The goal is to find x_m and y_m so that z is a maximum.

510 BASIC SCIENTIFIC SUBROUTINES

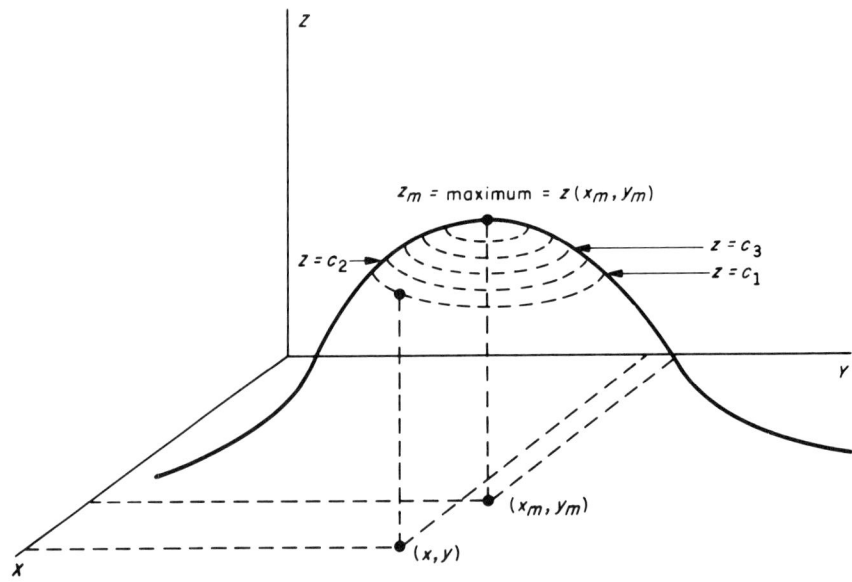

Figure 8.1.1 *Generic representation of the optimization problem. The peak of the function $z(x,y)$ is to be found. Shown above are the contours of equal z [i.e., $z(x,y) = c_i$].*

SUBROUTINE	SUBROUTINE SIZE (bytes) (including support programs)	DEMON-STRATION PROGRAM SIZE (bytes)	EXECUTION SPEED	CONDITIONS/COMMENTS
STEEPDS (steepest-descent optimization)	1768	594	1.4 seconds/iteration	Three dimensions, simple function expression
STEEPDA (steepest-descent optimization using finite difference approximations for the partial derivatives)	2547	594	2 seconds/iteration	Three dimensions, simple function expression

Table 8.1.1 *A summary of the programs presented in Chapter 8.*

The issue involves two simple questions:

1) Given that we are at position (x,y), *which way* to the maximum?
2) Given that we know which way, *how far* should we go?

The first question is usually relatively easy to answer. The answer to the second question is actually what determines the success of the algorithm. We will start with the first question—the direction.

Clearly it would not be productive to change positions in such a manner that the movement was only along a z = constant contour. Rather, the best direction of travel is that in which the contour values change most quickly. If we define the unit vectors in the x and y directions as $\hat{i}$ and $\hat{j}$, respectively, it can be shown in vector notation that the desired direction is

$$\mathbf{g} = \text{GRAD}[z(x,y)] = \frac{\partial z}{\partial x}\hat{i} + \frac{\partial z}{\partial y}\hat{j} \qquad (8.1.1)$$

(See Ref. 38.) If we further define the (x,y) position as $\mathbf{r} = x\hat{i} + y\hat{j}$, then the suggested iteration equation is

$$\mathbf{r}_{i+1} = \mathbf{r}_i + c\mathbf{g}_i \qquad (8.1.2)$$

This vector equation simply states that the new position ($\mathbf{r}_{i+1}$) is obtained by starting at the most recent location ($\mathbf{r}_i$), and moving some distance in the direction of the gradient. The question now is the distance, or the value, of c. That will be considered in the next section.

It is beyond the scope of this chapter to present subroutines for advanced optimization techniques. However, two of the classic approaches should at least be mentioned: the generalized Newton-Raphson method, and the Marquart composite algorithm. As you will see, they are direct extensions of the steepest-descent concept.

The generalized Newton-Raphson iteration scheme can be written in vector notation as

$$\mathbf{r}_{i+1} = \mathbf{r}_i + H\,\text{GRAD}(z) = \mathbf{r}_i + H\mathbf{g} \qquad (8.1.3)$$

In this notation, H is the Hessian matrix $\left[\frac{\partial^2 z}{\partial \mathbf{r}^2}\right]^{-1}$. For two dimensions, this matrix is

$$H^{-1} = \begin{bmatrix} \dfrac{\partial^2 z}{\partial x^2} & \dfrac{\partial^2 z}{\partial x \partial y} \\ \dfrac{\partial^2 z}{\partial y \partial x} & \dfrac{\partial^2 z}{\partial y^2} \end{bmatrix}^{-1} \qquad (8.1.4)$$

Observe that the Hessian matrix automatically answers the question of distance.

The method of steepest descent can be combined with the generalized Newton-

Raphson form to give Marquart's algorithm:

$$\mathbf{r}_{i+1} = \mathbf{r}_i + [cI + H]\,\mathbf{g}_i \qquad (8.1.5)$$

In this equation, I is the identity matrix, which in two dimensions is

$$I = \begin{bmatrix} 1 & 0 \\ 0 & 1 \end{bmatrix}$$

The Marquart procedure combines the directional properties of the steepest-descent procedure (which are useful far from the optimum) with the rapid convergence of Newton-Raphson iteration near the peak. However, the Hessian matrix calculation requires you to provide either analytical forms for the second-order partial derivatives, or finite difference approximations. In either case, the procedure can be both slow and subject to considerable round-off error. We will therefore limit our attention to the more elementary form of steepest descent.

8.2 Steepest Descent with Functional Derivatives

Although the gradient supplies information on the preferred direction of travel, its magnitude is of no help in determining the length of the step. Therefore, it is appropriate to normalize the gradient vector and restate the iteration equation as

$$\mathbf{r}_{i+1} = \mathbf{r}_i + \frac{k\mathbf{g}}{|\mathbf{g}|} \qquad (8.2.1)$$

In two dimensions, the vector equation is equivalent to

$$x_{i+1} = x_i + k\,\frac{\partial z/\partial x}{[(\partial z/\partial x)^2 + (\partial z/\partial y)^2]^{1/2}}$$

$$y_{i+1} = y_i + k\,\frac{\partial z/\partial y}{[(\partial z/\partial x)^2 + (\partial z/\partial y)^2]^{1/2}}$$

We will now concentrate on a means for choosing a value for k at each step in the iteration.

Consider the one-dimensional situation shown in figure 8.2.1. Three sequential evaluations using the same k value were made, and the sequence $y_1 < y_2 < y_3$ is observed. The iteration is clearly heading in the right direction. However, the maximum may be far off, and some form of acceleration should be used in the next step (e.g., a larger value for k). We will choose the following adjustment:

$$\text{If } \left(\frac{y_3 - y_2}{y_2 - y_1}\right) > 0 \quad \text{then} \quad k \to 1.2k$$

However, if $y_3 < y_2$, then the maximum has been passed; k is too large. The reasonable thing to do then is to halve k ($k \rightarrow k/2$), not update the positions, and try again.

The next question is whether or not there are pathological cases that cause a divergence with this algorithm. Figure 8.2.2 shows a situation in which the optimum has been passed, but the algorithm thinks it is still closing in. Therefore, the step size has been *increased* ($1.2x$), although the direction is reversed. The procedure eventually recovers.

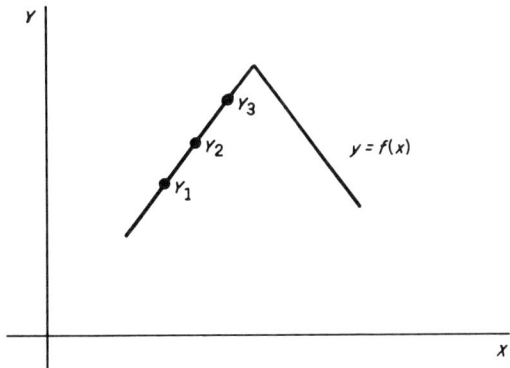

Figure 8.2.1 *One-dimensional example of finding the maximum. In this case, the position of the maximum of $y = f(x)$ is being sought. A sequence of three sequential evaluations using the same k is shown.*

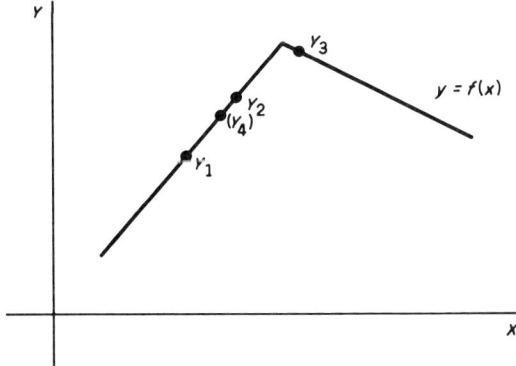

Figure 8.2.2 *A case in which $y_3 > y_2$, but the maximum has been passed. In this situation, k has been increased by a factor of 1.2, and the iteration will attempt to take a larger step back towards (and pass) the maximum on the next iteration (the gradient determines the direction), making the approximation (y_4) momentarily worse. However, this step is not actually taken since $y_4 < y_3$. Instead, the step size is halved and a new estimate tried.*

The choice of the acceleration (1.2) and deceleration (0.5) parameters is not completely arbitrary. The 0.5 deceleration factor is based on the interval-halving algorithm discussed in Chapter 6. It is possible for the estimates to bracket the peak, and by reducing k by 50%, the uncertainty in the location optimum is reduced just as with the bisection method. However, in practical situations this will not occur exclusively, and frequent increases in k (1.2x) can be expected. Therefore, convergence is usually slower than with the bisection algorithm. The 1.2 factor is chosen as a compromise. If it were $\sqrt{2}$ or greater, the algorithm could become unstable. If it were only slightly greater than unity, then the acceleration process would be inhibited. You may wish to experiment with this choice in the subroutines presented in this chapter.

The remaining uncertainty is the initial value for k. This is very much dependent on the specific problem being treated. In general, k should be roughly a fraction of the size of the error in the initial guess.

The above ideas can be incorporated into a BASIC subroutine as shown in listing 8.2.1. Appearing in that listing are the subroutine STEEPDS, a function evaluation routine for $y = \sin x_1 + 2 \cos x_2 - \sin x_3$ and its derivatives, and a demonstration program.

The inputs to STEEPDS are the initial estimates for the parameters, $X(Z)$; the initial step size, K; the error criterion, E; the number of variables, L; and the maximum number of steps, M. A sample run of this program is shown in listing 8.2.2. The object is to optimize y. The maximum value that y can take on is 4. This occurs at $x_1 = \pi/2 + 2m_1\pi$, $x_2 = 2m_2\pi$, and $x_3 = 3\pi/2 + 2m_3\pi$ ($m_1, m_2, m_3 = 0, \pm 1, \pm 2, \ldots$). The iteration homed in on the values corresponding to $m_1 = m_2 = 0$ and $m_3 = -1$. The calculated results and the "true" values are shown below:

Parameter	Calculated Value	True Value	Error
x_1	1.5707964	1.5707963	0.0000001
x_2	−0.00062712	0.0	−0.00062712
x_3	−1.5707963	−1.5707963	0.0000000

The calculated and the true values differ because of round-off error. Even though the procedure is iterative, there are many ways to *numerically* achieve the maximum value for y, which is four. This is the same problem as the one we observed earlier in finding the roots of $y = x^5 + 5x^4 + 10x^3 + 10x^2 + 5x + 1$. There were many values of x near -1 that gave $y = 0$. In this case, the values found give 3.9999996 as the maximum, which is very close to the true value of 4.

There are two ways in which to view this problem. First, if the goal is to maximize profits, then the result that a small range of x_i values lead to the same optimum is not significant; the maximum value is the goal. Second, if the objective is to accurately locate the position of the optimum, then a fundamental difficulty exists. Because the partial derivatives must necessarily be zero at the optimum, small changes in the parameters have an even smaller influence on the maximum value calculated for the objective function. But we are using such changes to locate the optimum! In effect, the procedure is *poorly conditioned* near the maximum.

The low accuracy of the calculated parameters is characteristic of many optimization procedures (e.g., least-squares fitting of experimental data), and is usually not a problem in practical situations.

The main precautions with respect to using STEEPDS involve the choices for the input parameters. Because many functions have several local maxima, starting the iteration far from the largest maximum can result in convergence of one of the local extremes. Also, if the initial value for K is too large, the iteration may show some erratic behavior before closing in on a local maximum. There is often much trial and error associated with optimization. In any case, you should be cognizant of the extraneous optima introduced by round-off error, and of the possibility of local maxima.

STEEPDS naturally seeks the location of the maximum of a function. If the position of the minimum is desired instead, then a small modification is required. If the minimum is negative, then STEEPDS should be used with $-y$ instead of y as the objective function. If the minimum is positive, then the function to optimize is $1/y$. We will examine the latter alteration in the next section.

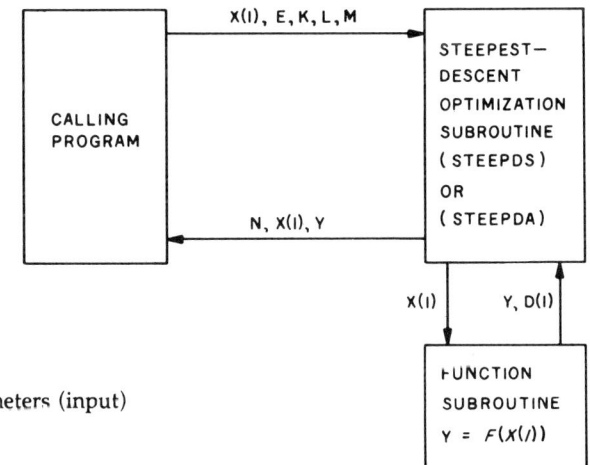

L: Number of parameters
$X(I)$: Initial values for the parameters (input)
E: Convergence criterion
K: Initial value of K
M: Maximum number of iterations
$X(I)$: Calculated position of the optimum (output)
Y: Last Y value calculated
N: Number of complete iterations

Figure 8.2.3 *Subroutine-connection diagram for the steepest-descent optimization subroutines, STEEPDS and STEEPDA. Note that $D(I)$ represents the partial derivatives and is required for STEEPDS only.*

Statements/Functions List

$+, -, *, /, <, >$

ABS, FOR/NEXT, GOSUB, GOTO, IF/THEN, SQRT

Variables List

A, B, E, D, D(I), I, J, K, L, N, X(I), X1(I), Y, Y(I)

Variables Passed to Subroutine

E, K, L, X(I)

Table 8.2.1 *Functions and variables used in steepest-descent subroutines, STEEPDS and STEEPDA. Note that variables A and B appear in STEEPDA only.*

```
4650 REM PROGRAM TO DEMONSTRATE THE USE OF MULTI-DIMENSIONAL
4651 REM STEEPEST DESCENT
4652 PRINT
4653 PRINT
4654 PRINT "HOW MANY DIMENSIONS ARE THERE: ",
4655 INPUT L
4656 DIM X(L),X1(L)
4657 PRINT
4658 PRINT "WHAT IS THE CONVERGENCE CRITERION: ",
4659 INPUT E
4660 PRINT
4661 PRINT "MAXIMUM NUMBER OF ITERATIONS: ",
4662 INPUT M
4663 PRINT
4664 PRINT "WHAT IS THE STARTING CONSTANT, K: ",
4665 INPUT K
4666 PRINT
4667 PRINT "INPUT THE STARTING POINTS:"
4668 PRINT
4669 FOR I=1 TO L
4670 PRINT "X(",I,") = ",
4671 INPUT X(I)
4672 NEXT I
4673 PRINT
4674 PRINT
4675 GOSUB 47700
4676 PRINT "THE RESULTS ARE:"
4677 PRINT
4678 FOR I=1 TO L
4679 PRINT "X(",I,") = ",X(I)
4680 NEXT I
```

Listing 8.2.1

```
4681 PRINT
4682 PRINT "THE NUMBER OF ITERATIONS WAS",N
4683 PRINT
4684 PRINT
4685 END
44298 REM ********************
44299 REM USER FUNCTION SUBROUTINE
44300 Y=SIN(X(1))+2*COS(X(2))-SIN(X(3))
44301 D(1)=COS(X(1))
44302 D(2)=-2*SIN(X(2))
44303 D(3)=-COS(X(3))
44304 RETURN
47684 REM ********************
47685 REM STEEPEST DESCENT OPTIMIZATION SUBROUTINE (STEEPDS)
47686 REM THIS PROGRAM FIND THE LOCAL MAXIMUM OR MINIMUM
47687 REM OF AN L-DIMENSIONAL FUNCTION USING THE METHOD
47688 REM OF STEEPEST DESCENT, OR THE GRADIENT.
47689 REM THE FUNCTION, Y(X(1),X(2)...), IS PLACED IN THE
47690 REM SUBROUTINE AT 44300, ALONG WITH THE L DERIVATIVES
47691 REM OF F, D(I).
47692 REM THE ROUTINE SEEKS USING AN INTERNALLY ADJUSTED
47693 REM MULTIPLIER, K. THE SEARCH IS MADE UNTIL AN ERROR
47694 REM LIMIT, E, IS REACHED.
47695 REM THE USER MUST SUPPLY INITIAL VALUES FOR THE X(I),
47696 REM AS WELL AS K (INITIAL) AND E. THE PROGRAM RETURNS
47697 REM THE LOCALLY OPTIMUM X(I) SET.
47698 REM REMEMBER TO DIMENSION X(I) IN THE CALLING PROGRAM.
47699 REM THE PROGRAM NEEDS THREE VALUES OF Y TO GET STARTED.
47700 N=0
47701 REM START INITIAL PROBE
47702 FOR J=1 TO 3
47703 REM OBTAIN Y AND D(I)
47704 GOSUB 44300
47705 Y(J)=Y
47706 REM UPDATE X(I)
47707 GOSUB 47735
47708 NEXT J
47709 REM WE NOW HAVE A HISTORY TO BASE THE SUBSEQUENT SEARCH ON
47710 REM ACCELERATE SEARCH IF APPROACH IS MONOTONIC
47711 IF (Y(3)-Y(2))/(Y(2)-Y(1))>0 THEN K=K*1.2
47712 REM DECELERATE IF HEADING THE WRONG WAY
47713 IF Y(3)<Y(2) THEN K=K/2
47714 REM UPDATE THE Y(I) IF VALUE HAS DECREASED
47715 IF Y(3)>Y(2) THEN GOTO 47721
47716 REM RESTORE THE X(I)
47717 FOR I=1 TO L
47718 X(I)=X1(I)
47719 NEXT I
47720 GOTO 47724
```

Listing 8.2.1 cont.

```
47721 Y(1)=Y(2)
47722 Y(2)=Y(3)
47723 REM OBTAIN NEW VALUES
47724 GOSUB 44300
47725 Y(3)=Y
47726 REM UPDATE X(I)
47727 GOSUB 47735
47728 REM CHECK FOR CONVERGENCE
47729 N=N+1
47730 IF N>=M THEN RETURN
47731 IF ABS(Y(3)-Y(2))<E THEN RETURN
47732 REM TRY ANOTHER ITERATION
47733 GOTO 47711
47734 REM FIND THE MAGNITUDE OF THE GRADIENT
47735 D=0
47736 FOR I=1 TO L
47737 D=D+D(I)*D(I)
47738 NEXT I
47739 D=SQRT(D)
47740 REM UPDATE THE X(I)
47741 FOR I=1 TO L
47742 REM SAVE OLD VALUES
47743 X1(I)=X(I)
47744 X(I)=X(I)+K*D(I)/D
47745 NEXT I
47746 GOSUB 44300
47747 Y(3)=Y
47748 RETURN
```

Listing 8.2.1 *Subroutine for finding the local maximum of a function by the method of steepest descent (STEEPDS). Also shown is a program to demonstrate STEEPDS for the expressions appearing in the function subroutine (44300). See listing 8.2.2 for a sample run.*

Listing 8.2.2

```
RUN


HOW MANY DIMENSIONS ARE THERE: ?3

WHAT IS THE CONVERGENCE CRITERION: ?.000000001

MAXIMUM NUMBER OF ITERATIONS: ?50

WHAT IS THE STARTING CONSTANT, K: ?1

INPUT THE STARTING POINTS:
```

```
X( 1) = ?1
X( 2) = ?1
X( 3) = ?1
```

1	1.5697684	-.5238603	-2.1882294	3.5471559
2	1.5705963	-.20046106	-1.6828461	3.9536788
3	1.5709445	.49273375	-1.4882161	3.7586782
4	1.5707704	.14613634	-1.5855311	3.9785737
5	1.5708088	-.28531183	-1.563703	3.9191226
6	1.5707896	-.06958775	-1.5746171	3.9951521
7	1.5708022	.18951447	-1.5674982	3.9641862
8	1.5707959	.05996336	-1.5710577	3.9964053
9	1.5707965	-.09555628	-1.5707185	3.990876
10	1.5707962	-.01779646	-1.5708881	3.9996832
11	1.5707966	.07551523	-1.5706474	3.9943002
12	1.5707964	.02885938	-1.5707678	3.9991672
13	1.5707963	.00553146	-1.5708279	3.9999694
14	1.5707964	-.02246203	-1.5707479	3.9994954
15	1.5707964	-.00846528	-1.5707879	3.9999282
16	1.5707963	-.00146691	-1.5708079	3.9999978
17	1.5707965	.0069311	-1.5707747	3.999952
18	1.5707964	.0027321	-1.5707913	3.9999926
19	1.5707963	.00063259	-1.5707996	3.9999996
20	1.5707964	-.00188682	-1.570793	3.9999964
21	1.5707964	-.00062712	-1.5707963	3.9999996

```
THE RESULTS ARE:

X( 1) =   1.5707964
X( 2) =  -6.271106E-04
X( 3) =  -1.5707963

THE NUMBER OF ITERATIONS WAS 21

READY
```

Listing 8.2.2 *Sample run of the program shown in listing 8.2.1. The intermediate results were obtained by inserting a PRINT statement within the subroutine. The first column gives the current iteration number. The second through fourth columns show the corresponding values for x(1), x(2) and x(3), respectively. Displayed in the last column is the currently optimized value of y.*

8.3 Steepest Descent with Approximate Derivatives

The method of steepest descent is based on the obtaining of directional information from the gradient. In the subroutine presented in section 8.2, the gradient was supplied in the form of partial derivatives by the function subroutine. In many cases, however, such derivatives are not available in analytical form, and so numerical approximations must be used. In this section, a steepest-descent subroutine that only requires the function itself will be given; the partial derivatives are estimated using finite difference approximations.

We will take as an example the two-dimensional problem of finding the maximum of $z(x,y)$. The iteration equations are

$$x_i = x_{i+1} + k \frac{\partial z/\partial x}{[(\partial z/\partial x)^2 + (\partial z/\partial y)^2]^{1/2}}$$

$$y_i = y_{i+1} + k \frac{\partial z/\partial y}{[(\partial z/\partial x)^2 + (\partial z/\partial y)^2]^{1/2}}$$

The length of the next step in the procedure is k. If we use the partial derivatives that took the iteration from step $i - 1$ to step i to project ahead by another one-half step, then a new set of derivatives that are evaluated roughly in the middle of the next step can be calculated. These derivatives can be employed to move from position (x_i, y_i) to (x_{i+1}, y_{i+1}). This procedure is outlined mathematically below.

We will define the finite difference derivatives calculated at step i to be

$$D_x(i) = \frac{z(x_i + k\, D_x(i-1)/2,\, y_i) - z(x_i, y_i)}{k\, D_x(i-1)/2} \tag{8.3.1a}$$

$$D_y(i) = \frac{z(x_i,\, y_i + k\, D_y(i-1)/2) - z(x_i, y_i)}{k\, D_y(i-1)/2} \tag{8.3.1b}$$

The new (x,y) position is then

$$x_{i+1} = x_i + k \frac{D_x(i)}{\{[D_x(i)]^2 + [D_y(i)]^2\}^{1/2}} \tag{8.3.2a}$$

$$y_{i+1} = y_i + k \frac{D_y(i)}{\{[D_x(i)]^2 + [D_y(i)]^2\}^{1/2}} \tag{8.3.2b}$$

A generalized form for these equations is incorporated into the steepest-descent subroutine shown in listing 8.3.1 (STEEPDA). Also appearing in listing 8.3.1 is a program to exercise STEEPDA, and a sample run is displayed in listing 8.3.2. The function examined in listing 8.3.2 is the same as that in the example shown in listing 8.2.2; the two listings can be directly compared.

STEEPDA operates in much the same manner as STEEPDS in section 8.2; the input and output variable formats are identical. Because of the numerical approximation to the derivatives, which is initially crude because of the large step size, the iteration bounces around. However, it soon settles down and homes in on a solution. A comparison of the calculated position of the local maximum and the corresponding true values is given below:

Parameter	Calculated Value	True Value	Error
x_1	1.5707419	1.5707963	-0.0000544
x_2	0.00048635	0.0	0.00048635
x_3	4.7121862	4.7123890	-0.0002028

Note that although the maximum value (4) was closely approximated (3.9999998), the calculated location is in considerable error. This was also observed in the previous section, and can be attributed to the partial derivatives being small near the peak.

This same problem occurs with the least-squares curve fitting of experimental data. There is a range of values for the calculated coefficients that gives nearly the same standard deviation, and it is difficult for any numerical procedure to find the set of coefficients exactly. Therefore, you should be warned against placing too much faith in the precision of the coefficients calculated. You must also be aware that the location of a local maximum and not the position of the global optimum may have been found. We will discuss this problem in more detail later.

The steepest-descent algorithm is very flexible in terms of the types of optimizations that can be performed. One particularly important category is the *minimization* of a *positive* objective function. Two examples—the *standard deviation* and the *min-max* fit—are briefly discussed below.

The formulation for the standard deviation problem is simple. The function subroutine is used to calculate the variance between the data (or function) to be fitted and the parametric equation to be used for the fitting. Let this variance be y. Before returning to the main program from the function subroutine, invert y; return $1/y$. The steepest-descent routine will then attempt to maximize $1/y$, and thereby minimize the variance (or standard deviation).

It is the author's experience that up to three least-squares coefficients can usually be reliably determined in this manner. Sometimes the procedure can be extended beyond that. In particularly difficult situations, it is sometimes necessary to build towards the final set of parameters one step at a time. For example, in polynomial least-squares fittings, the following procedure is recommended:

1) Start by fitting the most important coefficient, say x_1, by itself. Use an initial guess somewhere near the expected value, with a step size of $k = 0.1$.
2) Proceed on to estimating x_1 and x_2 by starting with the value calculated above for x_1, and $k = 0.01$.
3) Use x_1 and x_2 from the previous step as initial values, $k = 0.001$ and find (x_1, x_2, x_3).
4) And so on.

This sequence may require larger or smaller values for k than indicated above. It is usually better to creep up on the (local) optimum using small values for k than to bounce around by starting with a large k.

Between each addition of a new parameter to be fitted, a check should be made as to whether or not any significant progress was made in reducing the variance by the last addition of a parameter (e.g., x_3). If not, then either the limit of the procedure was reached, or the iteration homed in on the wrong point. The advantage to using STEEPDA in this mode is that the parameters in the fitting function need not appear in linear combinations as they do in polynomial expressions. Therefore, the technique is capable of dealing with complicated functions.

The optimization can also be with respect to the min-max criterion. In this case, instead of calculating the variance in the function subroutine, the maximum positive and negative differences are recorded. Then, y is set equal to the magnitude of the largest of these two quantities, inverted, and returned to the steepest-descent subroutine. This procedure is identical to least-squares optimization except for the change in the criterion. As an example, consider the min-max polynomial fitting of $\sin(\pi/2)x$ over the interval $-1 \le x \le 1$. Using STEEPDA, the single-parameter min-max fit is

$$\sin \frac{\pi}{2} x \cong 1.22x \qquad \text{for } -1 \le x \le 1$$

(Or, $\sin x \cong 0.78x$. Refer to the figures and discussion in References 1 and 39.) The maximum error observed in this case is 0.22 (at $x = 0$ and $x = 1$). Using $x_1 = \pi/2$ and $x_2 = 1 - \pi/2$ as a new starting point, a two-parameter approximation is obtained:

$$\sin \frac{\pi}{2} x \cong 1.5641x - 0.5725x^3 \qquad \text{for } -1 \le x \le 1$$

The maximum error is reduced to 0.0084. Using the above values and continuing on to three parameters, the fit becomes

$$\sin \frac{\pi}{2} x \cong 1.5732x - 0.6003x^3 + 0.0187x^5 \qquad \text{for } -1 \le x \le 1$$

The associated maximum error in the above fit was 0.009. However, this error is greater than that observed for the two-parameter fit! By starting the iteration from another position, a much lower error maximum of 0.00014 is achieved:

$$\sin \frac{\pi}{2} x \cong 1.5706894x - 0.6432330x^3 + 0.0725556x^5 \qquad \text{for } -1 \le x \le 1$$

The coefficients in this latter approximation are similar to the min-max values provided by Hastings (see Ref. 9):

$$\sin \frac{\pi}{2}x \cong 1.5706268x - 0.6432292x^3 + 0.0727102x^5 \quad \text{for } -1 \leq x \leq 1$$

The min-max error for Hastings' approximation is 0.00011; the two fits are nearly equivalent.

This example illustrates the basic difficulty associated with using the steepest-descent method for min-max curve fitting: the iteration may home in on what is clearly the wrong position. This problem is related to the existence of several solutions to the optimization goal as stated. These solutions can be true analytical maxima, artifacts of the procedure, or artificially created by round-off error.

Figure 8.3.1 shows a situation in which a *near*-min-max fit is obtained. At one step in the procedure, the error maximum at point B is clearly largest. The algorithm reduces this error until some other error maximum dominates. The procedure then seeks to reduce that maximum, but at the expense of the error at B. Soon the sequence switches to minimizing the error at B again. It is thereby quite possible for the iteration to home in on the situation where the errors at points A and B are equal, but greater than the errors at either C or D. This is not the optimum situation, and the resulting coefficients are not those desired. Therefore, all results obtained by this procedure should be checked.

Figures 8.3.2 through 8.3.6 show plots of the error curves for each of the approximations given in this section for $\sin(\pi/2)x$.

In the case of the one-parameter fit, two equal error maxima are apparent: one at $x = 0$ and one at $x = 1$. For the two-parameter fit (see figure 8.3.3), three maxima are apparent. Two of them are equal, but the third is not as large. Apparently the convergence problem discussed above was encountered, and the resulting fit should be classified only as near-min-max. The same difficulty occurred for the three-parameter approximation shown in figure 8.3.4. Although there should be four equal maxima, only two equal maxima were obtained. By starting the iteration from a different position, the maximum error was greatly reduced (compare figures 8.3.4 and 8.3.5), but the convergence problem is still apparent. However, observe that this last approximation bears a close resemblance in both coefficients and error-curve behavior to the true min-max representation (see figure 8.3.6). The fit calculated using STEEPDA is clearly worthy of being called near-min-max in that the maximum error is only a little greater than the optimum value (0.00014 versus 0.00011).

The min-max approximation example given above illustrates the main problem associated with the steepest-descent procedure: it homes in on local maxima. In the case of fitting functions and data by the min-max procedure, it is easy to check the results graphically; we expect to see an equal ripple-error pattern. If the objective function is instead the variance or simply the maximum, then the verification of the results is much more difficult.

```
4700 REM PROGRAM TO DEMONSTRATE THE USE OF MULTI-DIMENSIONAL
4701 REM STEEPEST DESCENT
4702 PRINT
4703 PRINT
4704 PRINT "HOW MANY DIMENSIONS ARE THERE: ",
4705 INPUT L
4706 DIM X(L),X1(L)
4707 PRINT
4708 PRINT "WHAT IS THE CONVERGENCE CRITERION: ",
4709 INPUT E
4710 PRINT
4711 PRINT "MAXIMUM NUMBER OF ITERATIONS: ",
4712 INPUT M
4713 PRINT
4714 PRINT "WHAT IS THE STARTING CONSTANT, K: ",
4715 INPUT K
4716 PRINT
4717 PRINT "INPUT THE STARTING POINTS:"
4718 PRINT
4719 FOR I=1 TO L
4720 PRINT "X(",I,") = ",
4721 INPUT X(I)
4722 NEXT I
4723 PRINT
4724 PRINT
4725 GOSUB 47800
4726 PRINT "THE RESULTS ARE:"
4727 PRINT
4728 FOR I=1 TO L
4729 PRINT "X(",I,") = ",X(I)
4730 NEXT I
4731 PRINT
4732 PRINT "THE NUMBER OF ITERATIONS WAS",N
4733 PRINT
4734 PRINT
4735 END
44298 REM ********************
44299 REM USER FUNCTION SUBROUTINE
44300 Y=SIN(X(1))+2*COS(X(2))-SIN(X(3))
44301 RETURN
47784 REM ********************
47785 REM STEEPEST DESCENT OPTIMIZATION SUBROUTINE (STEEPDA)
47786 REM THIS PROGRAM FIND THE LOCAL MAXIMUM OR MINIMUM
47787 REM OF AN L-DIMENSIONAL FUNCTION USING THE METHOD
47788 REM OF STEEPEST DESCENT, OR THE GRADIENT.
47789 REM THE FUNCTION, Y(X(1),X(2)...), IS PLACED IN THE
47790 REM SUBROUTINE AT 44300. FINITE DIFFERENCES ARE USED TO
47791 REM CALCULATE THE L PARTIAL DERIVATIVES.
47792 REM OF F, D(I).
```

Listing 8.3.1

```
47793 REM THE ROUTINE SEEKS USING AN INTERNALLY ADJUSTED
47794 REM MULTIPLIER, K. THE SEARCH IS MADE UNTIL AN ERROR
47795 REM LIMIT, E, IS REACHED.
47796 REM THE USER MUST SUPPLY INITIAL VALUES FOR THE X(I),
47797 REM AS WELL AS K (INITIAL) AND E. THE PROGRAM RETURNS
47798 REM THE LOCALLY OPTIMUM X(I) SET.
47799 REM REMEMBER TO DIMENSION X(I) IN THE CALLING PROGRAM.
47800 N=0
47801 REM THE PROGRAM NEEDS THREE VALUES OF Y TO GET STARTED.
47802 REM GENERATE STARTING D(I) VALUES.
47803 REM THESE ARE NOT EVEN GOOD GUESSES, AND SLOW THE PROGRAM A LITTLE.
47804 D=1
47805 D(1)=1/SQRT(L)
47806 FOR I=2 TO L
47807 D(I)=D(I-1)
47808 NEXT I
47809 REM START INITIAL PROBE
47810 FOR J=1 TO 3
47811 REM OBTAIN Y
47812 GOSUB 44300
47813 Y(J)=Y
47814 REM OBTAIN APPROXIMATIONS TO THE D(I)
47815 GOSUB 47866
47816 REM UPDATE X(I)
47817 GOSUB 47849
47818 GOSUB 47856
47819 NEXT J
47820 REM WE NOW HAVE A HISTORY TO BASE THE SUBSEQUENT SEARCH ON
47821 REM ACCELERATE SEARCH IF APPROACH IS MONOTONIC
47822 IF (Y(3)-Y(2))/(Y(2)-Y(1))>0 THEN K=K*1.2
47823 REM DECELERATE IF HEADING THE WRONG WAY
47824 IF Y(3)<Y(2) THEN K=K/2
47825 REM UPDATE THE Y(I) IF Y(3)>Y(2)
47826 IF Y(3)>Y(2) THEN GOTO 47832
47827 REM RESTORE THE X(I)
47828 FOR I=1 TO L
47829 X(I)=X1(I)
47830 NEXT I
47831 GOTO 47835
47832 Y(1)=Y(2)
47833 Y(2)=Y(3)
47834 REM OBTAIN NEW VALUES
47835 GOSUB 44300
47836 Y(3)=Y
47837 GOSUB 47866
47838 REM IF D=0 THEN THE PRECISION LIMIT OF THE COMPUTER HAS BEEN REACHED
47839 IF D=0 THEN RETURN
47840 REM UPDATE X(I)
47841 GOSUB 47856
```

Listing 8.3.1 cont.

```
47842 REM CHECK FOR CONVERGENCE
47843 N=N+1
47844 IF N>=M THEN RETURN
47845 IF ABS(Y(3)-Y(2))<E THEN RETURN
47846 REM TRY ANOTHER ITERATION
47847 GOTO 47822
47848 REM FIND THE MAGNITUDE OF THE GRADIENT
47849 D=0
47850 FOR I=1 TO L
47851 D=D+D(I)*D(I)
47852 NEXT I
47853 D=SQRT(D)
47854 RETURN
47855 REM UPDATE THE X(I)
47856 FOR I=1 TO L
47857 REM SAVE OLD VALUES
47858 X1(I)=X(I)
47859 X(I)=X(I)+K*D(I)/D
47860 NEXT I
47861 GOSUB 44300
47862 Y(3)=Y
47863 RETURN
47864 REM FINITE DIFFERENCES SUBROUTINE FOR THE D(I) APPROXIMATION
47865 REM LOOK AHEAD ONE HALF INTERVAL
47866 FOR I=1 TO L
47867 REM SAVE X(I)
47868 A=X(I)
47869 REM FIND INCREMENT
47870 B=D(I)*K/(2*D)
47871 REM MOVE INCREMENT IN X(I)
47872 X(I)=X(I)+B
47873 REM OBTAIN Y
47874 GOSUB 44300
47875 REM GUARD AGAINST DIVIDE BY ZERO NEAR MAXIMUM
47876 IF B=0 THEN B=.00000000001
47877 REM UPDATE D(I)
47878 D(I)=(Y-Y(3))/B
47879 REM GUARD AGAINST LOCKED UP DERIVATIVE
47880 IF D(I)=0 THEN D(I)=.00001
47881 REM RESTORE X(I) AND Y
47882 X(I)=A
47883 Y=Y(3)
47884 NEXT I
47885 REM OBTAIN D
47886 REM GOSUB 47849
47887 RETURN
```

Listing 8.3.1 *Subroutine for applying the steepest-descent algorithm with finite difference approximations for the derivatives of y (STEEPDA). The corresponding subroutine-connection diagram appears in figure 8.2.3, and the lists of functions and variables are given in table 8.2.1. A sample run of the above program is shown in listing 8.3.2.*

```
RUN

HOW MANY DIMENSIONS ARE THERE: ?3

WHAT IS THE CONVERGENCE CRITERION: ?.00000001

MAXIMUM NUMBER OF ITERATIONS: ?50

WHAT IS THE STARTING CONSTANT, K: ?1

INPUT THE STARTING POINTS:

X( 1) = ?1
X( 2) = ?1
X( 3) = ?1


1        .8820197        4.6419139       4.5707379       1.6211749
2       1.613064        -.997305         3.5958548       2.5230424
3       1.6094516       3.25526          5.0181166       -.0342124
4       1.5639965        .0067194        4.5367203       3.9845416
5       1.5942672       -.8331796        4.3890349       3.2929516
6       1.5661157        .1444837        4.6724042       3.9783506
7       1.5663199       -.0101172        4.5922998       3.9926855
8       1.5679244        .0026833        4.6358648       3.9970621
9       1.5691343       -.0039238        4.6679784       3.9989972
10      1.5699714        .0032506        4.6891073       3.9997181
11      1.570407        -.0046102        4.701458        3.9999189
12      1.5707419        .0075987        4.7077878       3.9999316
13      1.5707419       -.0160267        4.7106358       3.9997415
14      1.5707419        .0027092        4.7100002       3.9999898
15      1.5707419        .0011199        4.7110445       3.9999979
16      1.5707419        .0001684        4.7117685       3.9999998
17      1.5707419        .0004863        4.7121862       3.9999998
THE RESULTS ARE:

X( 1) =    1.5707419
X( 2) =    4.8635213E-04
X( 3) =    4.7121862

THE NUMBER OF ITERATIONS WAS 17

READY
```

Listing 8.3.2 *A sample run of STEEPDA for the function* $y(x) = \sin x_1 + 2\cos x_2 - \sin x_3$. *The intermediate results were obtained by inserting a PRINT statement within the subroutine. Compare with listing 8.2.2.*

528 BASIC SCIENTIFIC SUBROUTINES

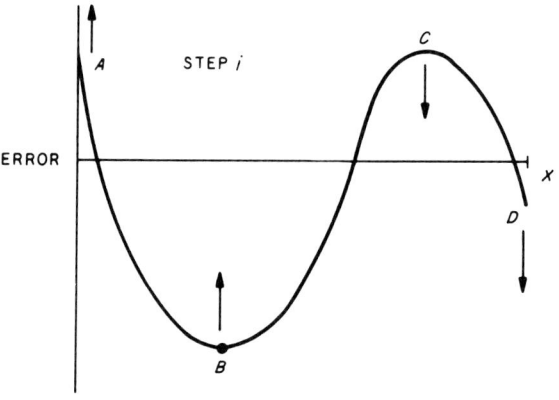

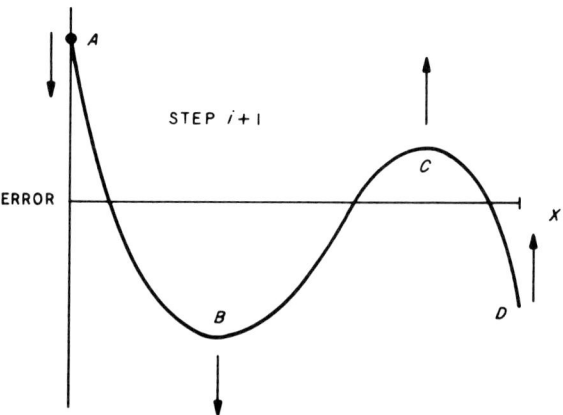

Figure 8.3.1 *Example of a local-optimum configuration in the steepest-descent procedure for determining the min-max coefficients. Initially, point B has the maximum error. One step later, the honor goes to point A. Then it may switch back to B, and so on.*

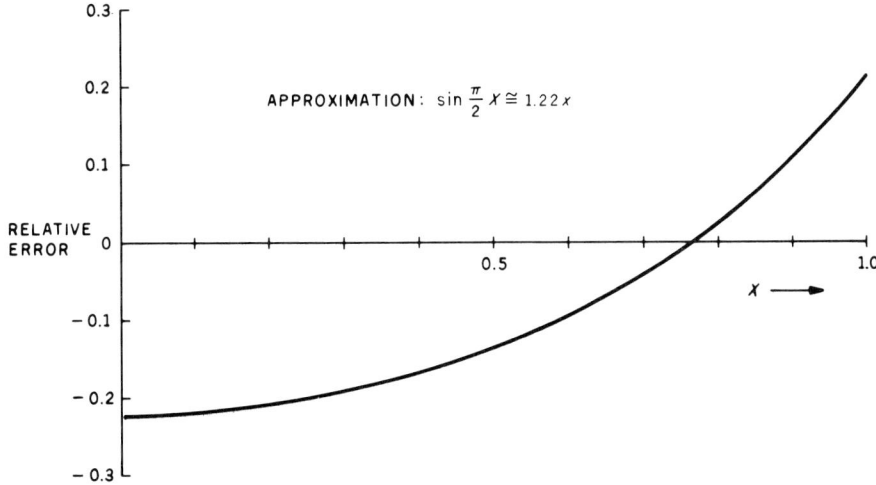

Figure 8.3.2 *Plot of the relative error in the approximation* $\sin(\pi/2)x \cong 1.22x$.

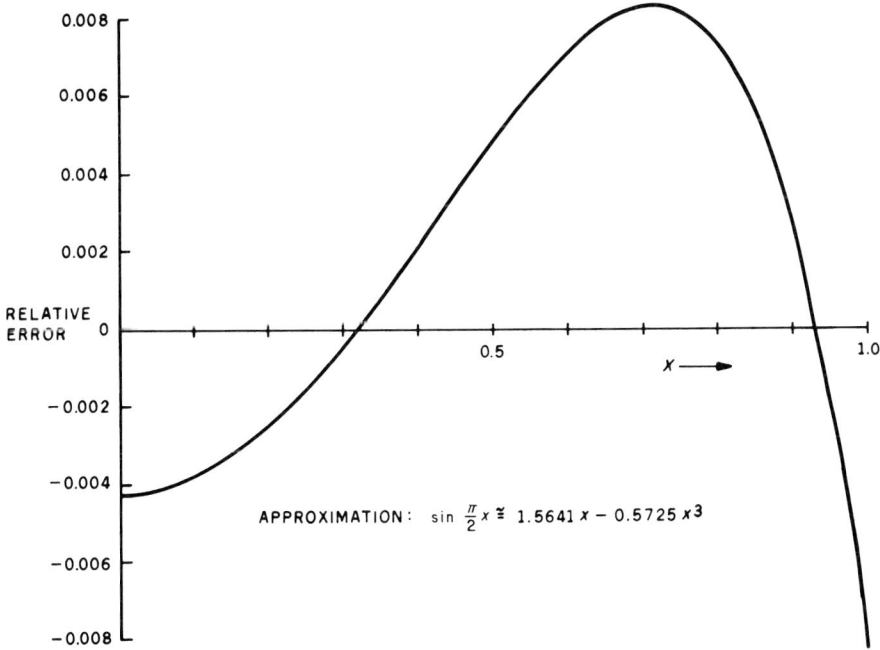

Figure 8.3.3 *Plot of the relative error in the approximation* $\sin(\pi/2)x \cong 1.5641x - 0.5725x^3$.

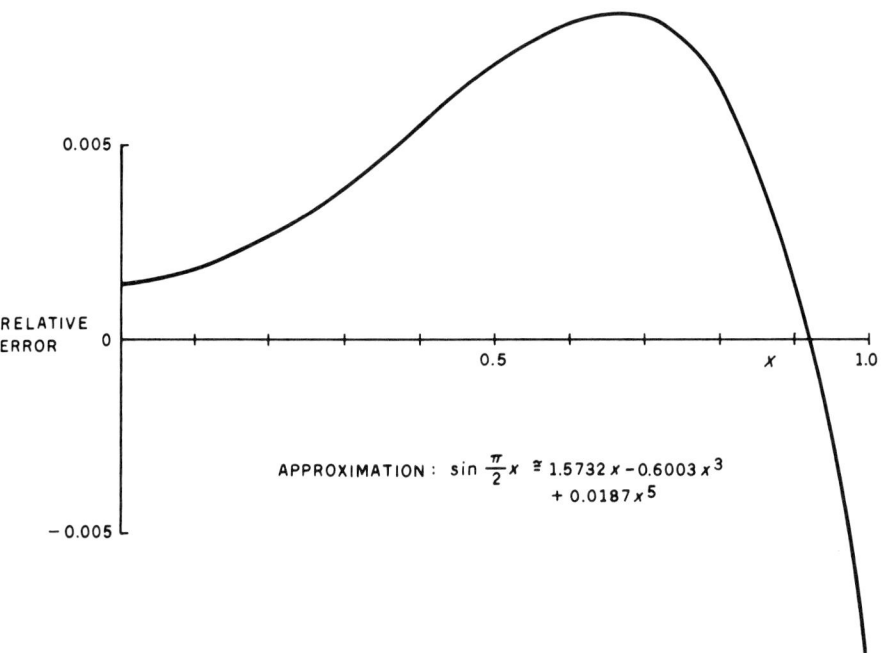

Figure 8.3.4 *Plot of the relative error in the approximation sin $(\pi/2)x \cong 1.5732x - 0.6003x^3 + 0.0187x^5$.*

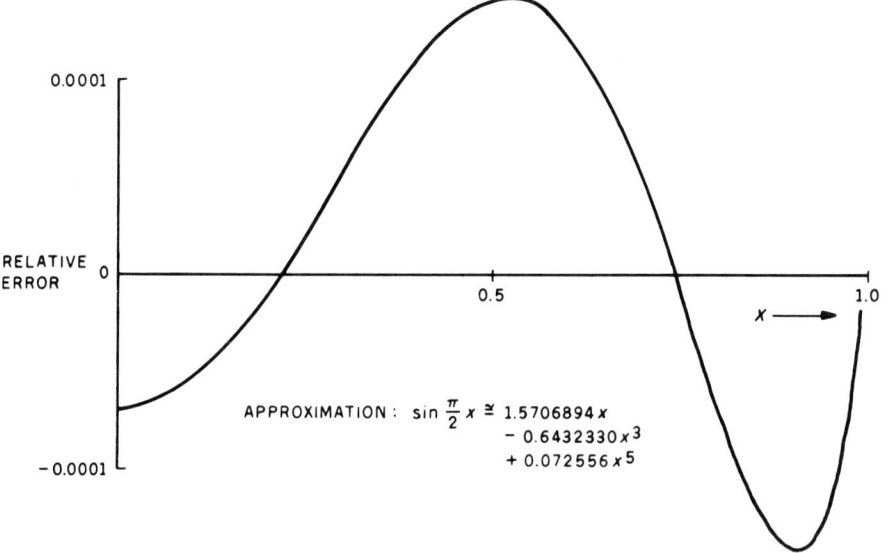

Figure 8.3.5 *Plot of the relative error in the approximation sin $(\pi/2)x \cong 1.5706894x - 0.6432330x^3 + 0.072556x^5$.*

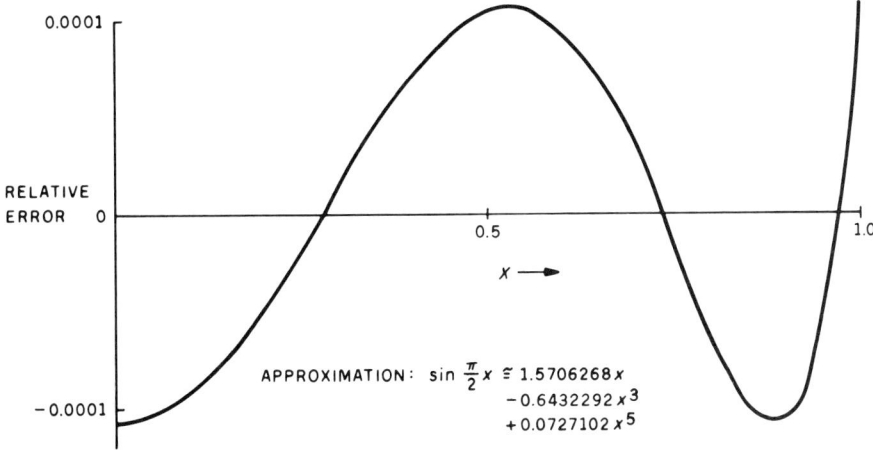

Figure 8.3.6 *Plot of the relative error in the approximation* $\sin(\pi/2)x \cong 1.5706268x - 0.6432292x^3 + 0.0727102x^5$.

8.4 Summary and Conclusions

Quite a few sophisticated (and complicated) algorithms have been developed for the very important subject of optimization. Many of these methods are conceptually based on the method of steepest descent. The discussion in this chapter was limited to simple implementations of that technique. In one of the algorithms presented (STEEPDS), the partial derivatives of the object function are required. In another (STEEPDA), they are numerically approximated. Although the latter subroutine is theoretically less stable because of the numerical approximation, it appears to perform roughly the same in practical situations.

The steepest-descent subroutines given in this chapter bear a strong resemblance in overall concept to the parametric least-squares curve-fitting program (PARAFIT) discussed in Chapter 1. The major difference between the programs is that PARAFIT optimized one parameter at a time, whereas STEEPDS and STEEPDA use the gradient to optimize all of the coefficients simultaneously. The steepest-descent programs are therefore generally more efficient than PARAFIT in their operation.

The use of steepest descent for least-squares and min-max curve fitting was discussed in section 8.3; the conclusion was that the method has merits, but that there are serious reservations in recommending its use for such applications. The major problem is that the desired coefficients are estimated by examining their combined influence on the objective function. Near the optimum, changes in the coefficients have little effect on the value calculated for the objective function, and therefore inferring the values in that manner represents a poorly conditioned situation. Furthermore, the partial derivatives must be well behaved for this procedure to work; such is often *not* the case with min-max optimization. Therefore, for least-squares and min-max polynomial approximations, the subroutines given in Chapters 1 and 2 are recommended. For problems not amenable to calculation by those procedures, you can apply STEEPDS and STEEPDA, but *with caution!*

REFERENCES

1. Ruckdeschel, F. R. *BASIC Scientific Subroutines, Volume I.* Peterborough, NH: BYTE Books, 1980.

2. Ruckdeschel, F. R. "Mits vs. North Star." *Kilobaud Magazine,* 20, August, 1978, p. 44.

3. Feller, W. *An Introduction to Probability Theory and Its Applications.* New York: John Wiley and Sons Inc., 1965.

4. Daniel, C., and F. S. Wood. *Fitting Equations to Data.* New York: Wiley-Interscience, 1971.

5. Beyer, W. H., ed. *Standard Mathematical Tables, 24th ed.* Cleveland OH: CRC Press, 1976.

6. Abramowitz, M., and I. A. Stegun. *Handbook of Mathematical Functions.* New York: Dover Publications, 1964.

7. Hart, J. R. *Computer Approximations.* New York: John Wiley and Sons Inc., 1968.

8. Becket, R., and J. Hurt. *Numerical Calculations and Algorithms.* New York: McGraw-Hill, 1967.

9. Hastings, C. *Approximations for Digital Computers.* Princeton, NJ: Princeton University Press, 1955.

10. Carnahan, B., H. A. Luther, and J. O. Wilkes. *Applied Numerical Methods.* New York: John Wiley and Sons Inc., 1969.

11. Bajpai, A. C., I. M. Calus, and J. A. Fairley. *Statistical Methods for Scientists and Engineers.* New York: John Wiley and Sons Inc., 1978.

12. Jennings, A. *Matrix Computation for Engineers and Scientists.* New York: Wiley-Interscience, 1977.

13. *HP-55 Statistics Programs.* Corvallis, OR: Hewlett-Packard Co., 1975.

14. Crow, E. L., F. A. Davis, and M. W. Maxfield. *Statistics Manual.* New York: Dover Publications, 1960.

15. Fike, C. T. *Computer Evaluation of Mathematical Functions.* Englewood Cliffs, NJ: Prentice-Hall, 1968.

16. Hamming, R. W., and E. A. Feigenbaum. *Introduction to Applied Numerical Analysis.* New York: McGraw-Hill, 1971.

17. Henrici, P. *Computational Analysis with the HP-25 Pocket Calculator.* New York: Wiley-Interscience, 1977.

18. Erdelyi, A. *Asymptotic Expansions.* New York: Dover Publications, 1956.

19. Ball, J. A. *Algorithms for RPN Calculators.* New York: Wiley-Interscience, 1978.

20. Dahlquist, G., and A. Björck. *Numerical Methods.* Englewood Cliffs, NJ: Prentice-Hall, 1969.

21. Beyer, W. H. *Basic Statistical Tables.* Cleveland, OH: Chemical Rubber Co., 1971.

22. Rohlf, F. J., and R. R. Sokal. *Statistical Tables.* San Francisco, CA: W. H. Freeman and Co., 1969.

23. Mack, C. *Essentials of Statistics for Scientists and Technologists.* New York: Plenum Publishing, 1975.

24. *Texas Instruments SR-51 Owners Manual.* Dallas, TX: Texas Instruments, Inc., 1974.

25. Peirce, B. O. *A Short Table of Integrals.* Boston, MA: Ginn and Co., 1957.

26. Carslaw, H. S., and J. C. Jaeger. *Conduction of Heat in Solids.* Oxford, England: Clarendon Press, 1959.

27. Grove, A. S. *Physics and Technology of Semiconductor Devices.* New York: John Wiley and Sons Inc., 1967.

28. Jolley, L. B. W. *Summation of Series.* New York: Dover Publications, 1961.

29. Acton, F. S. *Numerical Methods That Work.* New York: Harper and Row, 1970.

30. Greenhill, A. G. *The Applications of Elliptical Functions.* New York: Dover Publications, 1959.

31. Kaplan, W. *Advanced Calculus.* Reading, MA: Addison-Wesley Publishing, 1952.

32. Gautschi, W. "Computational Aspects of Three Term Recurrence Relations." *SIAM Review,* 9, 1967, pp. 24-82.

33. Volder, J. E. "The CORDIC Trigonometric Computing Technique." *IRE Transactions on Electronic Computers, EC-8*, Sept., 1959.

34. Liffick, Blaise W., ed. *Numbers in Theory and Practice: Programming Techniques Series, Vol. 3.* Peterborough, NH: BYTE Books, 1979.

35. Schmid, H. *Decimal Computation.* New York: John Wiley and Sons Inc., 1974.

36. Egbert, W. E. "Personal Calculator Algorithms III: Inverse Trigonometric Functions." *HP Journal*, 29, No. 3, pp. 22-33.

37. LaFara, R. L. *Computer Methods for Science and Engineering.* Rochelle Park, NJ: Hayden Book Company, 1973.

38. Kreyszig, E. *Advanced Engineering Mathematics.* New York: John Wiley and Sons Inc., 1962.

39. Ruckdeschel, F. R. "Functional Approximations." *BYTE*, November, 1978, pp. 34-46.

40. Shoup, T. E. *A Practical Guide to Computer Methods for Engineers.* Englewood Cliffs, NJ: Prentice-Hall, 1979.

APPENDICES

APPENDIX IA *SOFTWARE INDEX BY NUMBER*

40400   Rectangular-to-Polar Conversion Subroutine (RECT/POL)

40450   Polar-to-Rectangular Conversion Subroutine (POL/RECT)

41100   Polar-Power Subroutine (ZPOLPOW)

41200   Rectangular Complex Number Power Subroutine (RECTPOW)

41900   Matrix Multiplication Subroutine (MATMULT)

41950   Matrix Transpose Subroutine (MATTRANS)

42075   Matrix-Save (B in A) Subroutine (MATSAVBA)

42100   Matrix-Save (C in B) Subroutine (MATSAVCB)

42150   Matrix-Save (A in C) Subroutine (MATSAVAC)

42175   Matrix-Save (C in A) Subroutine (MATSAVCA)

42400   Matrix Inversion Subroutine (MATINV)

43500   Linear Least-Squares Subroutine (LSTSQR1)

43550   Parabolic Least-Squares Subroutine (LSTSQR2)

43600   One-Dimensional Polynomial Coefficient Generation Subroutine (POLYCM)

43650   Multidimensional Least-Squares Polynomial Fitting Subroutine (LEASTSQR)

43750   Standard Deviation Subroutine (SIGMA)

43800	Multidimensional Polynomial Coefficient Generation Subroutine (MLTNLREG)
43950	Forsythe Polynomial Least-Squares Polynomial Fitting Subroutine (LSQRPOLY)
44100	Multidimensional Polynomial Iterated-Regression Subroutine (REGITER)
44250	Parametric Least-Squares Curve-Fit Subroutine (PARAFIT)
44300	Functions Subroutine (user input)
44400	Chi-Square Cumulative Distribution Approximation Subroutine (CHISQA)
44420	Bessel Function Series Subroutine (BESSLSER)
44475	Bessel Function Series Coefficients Subroutine (BESSEL)
44525	Bessel Function Asymptotic-Series Subroutine (BESSEL01)
44580	LN(X!) Asymptotic Series Subroutine (LN(X!))
44600	Chi-Square Probability Density Function Subroutine (CHI-SQR)
44625	Chi-Square Cumulative Distribution Function Subroutine (CHISQ)
44675	Normalized Error Function Subroutine (ASYMERF)
44725	Chebyshev Polynomial Coefficient Evaluation Subroutine (CHEBYSER)
44760	Chebyshev Economization Subroutine (CHEBECON)
44800	Series Reversion Subroutine (REVERSE)
44850	Reciprocal Power Series Subroutine (RECIPRO)
44900	Horner's Shifting Rule Subroutine (HORNER)
44925	Inverse Normal Distribution Subroutine (INVNORM)
44950	Complex Series Evaluation Subroutine (CMPLXSER)
45000	Nth-Root Subroutine (NTHROOT)
45030	General Root Determination Subroutine (GENROOT)

45060	Root Decomposition Subroutine (RTDECOMP)
45125	Tangent Iteration Subroutine (TANITER)
45150	Inverse Tangent Recursion Subroutine (ATANITER)
45200	Inverse Sine Recursion Subroutine (ARCSINIT)
45250	Elliptic Integral Recursion Subroutine (CLIPTIC)
45300	Natural Logarithm Recursion Subroutine (LOGITER)
45350	Integer-Order Bessel Function Recursion Subroutine (INTBESSL)
45400	Legendre Polynomial Coefficient Evaluation Subroutine (LEGNDRE)
45425	Laguerre Polynomial Coefficient Evaluation Subroutine (LAGUERR)
45450	Hermite Polynomial Coefficient Evaluation Subroutine (HERMITE)
45475	Trigonometric CORDIC Subroutine (TRIGCORD)
45550	Inverse Trigonometric CORDIC Subroutine (INVCORD)
45650	Modified CORDIC Exponential Subroutine (EXPCORD)
45725	Modified CORDIC Natural Logarithm Subroutine (LNCORDIC)
45800	Hyperbolic Sine Subroutine (SINH)
45825	Hyperbolic Cosine Subroutine (COSH)
45840	Hyperbolic Tangent Subroutine (TANH)
45875	Inverse Hyperbolic Sine Subroutine (INVSINH)
45900	Inverse Hyperbolic Cosine Subroutine (INVCOSH)
45925	Inverse Hyperbolic Tangent Subroutine (INVTANH)
46000	Lagrange Interpolation Subroutine (LAGRANGE)
46050	Newton Divided-Differences Interpolation Subroutine (NEWTON)

46100   Akima Spline-Fitting Subroutine (AKIMA)

46150   Lagrange Derivative Interpolation Subroutine (D/LAGRNG)

46200   General Integration Subroutine (ITEG)

46350   Bisection Method Subroutine (BISECT)

46375   Newton-Raphson Iteration Subroutine (Z-NEWTON)

46425   Secant Method Subroutine (SECANT)

46475   Modified False-Position Method Subroutine (REGULA)

46525   Aitken Acceleration Subroutine (AITKEN)

46575   Aitken-Steffenson Iteration Subroutine (A/SITER)

46625   Synthetic Division Subroutine (RSYNDIV)

46650   Multiple-Root Determination Subroutine (NEXTROOT)

46700   Complex Root Counting Subroutine (ROOTNUM)

46750   Complex Root Search Subroutine (ZCIRCLE)

46925   Complex Plane Newton-Raphson Iteration Subroutine (CZNEWTON)

46960   Parabolic Root Seeking Subroutine (MUELLER)

47050   Two-Dimensional Parabolic Root Seeking Subroutine (MUELLER2)

47150   Complex Plane Parabolic Root Seeking Subroutine (ZMUELLER)

47300   General Complex Plane Root Determination Subroutine (ALLROOT)

47475   Quadratic Root Determination Subroutine (QUADRAT)

47525   Polynomial Complex Root Determination Subroutine (LIN)

47600   Polynomial Complex Root Determination Subroutine (BAIRSTOW)

47700   Steepest-Descent Optimization Subroutine (STEEPDS)

47800   Steepest-Descent Optimization Subroutine (STEEPDA)

APPENDIX IB *SOFTWARE INDEX BY FUNCTION*

This appendix lists the subroutines presented in Volume II cross-indexed according to function. Program name mnemonics are also given to aid in program identification.

Bessel Functions

    Bessel Function Series Subroutine (BESSLSER): 44420

    Bessel Function Series Coefficients Subroutine (BESSEL): 44475

    Bessel Function Asymptotic-Series Subroutine (BESSEL01): 44525

    Integer-Order Bessel Function Recursion Subroutine (INTBESSL): 45350

Coordinate Conversion

    Rectangular-to-Polar Conversion Subroutine (RECT/POL): 40400

    Polar-to-Rectangular Conversion Subroutine (POL/RECT): 40450

Decimal to Binary Conversion

    Root Decomposition Subroutine (RTDECOMP): 45060

Differentiation

    Lagrange Derivative Interpolation Subroutine (D/LAGRNG): 46150

Economization (Telescoping)

Chebyshev Economization Subroutine (CHEBECON): 44760

Elliptic Integrals

Elliptic Integral Recursion Subroutine (CLIPTIC): 45250

Error Function

Normalized Error Function Subroutine (ASYMERF): 44675

Exponentiation

Polar-Power Subroutine (ZPOLPOW): 41100

Rectangular Complex-Number Power Subroutine (ZRECTPOW): 41200

Nth-Root Subroutine (NTHROOT): 45000

General Root Determination Subroutine (GENROOT): 45030

Modified CORDIC Exponential Subroutine (EXPCORD): 45650

Hyperbolic Functions

Hyperbolic Sine Subroutine (SINH): 45800

Hyperbolic Cosine Subroutine (COSH): 45825

Hyperbolic Tangent Subroutine (TANH): 45840

Integration

General Integration Subroutine (ITEG): 46200

Interpolation

Lagrange Interpolation Subroutine (LAGRANGE): 46000

Newton Divided-Differences Interpolation Subroutine (NEWTON): 46050

Akima Spline-Fitting Subroutine (AKIMA): 46100

Inverse Hyperbolic Functions

Inverse Hyperbolic Sine Subroutine (INVSINH): 45875

Inverse Hyperbolic Cosine Subroutine (INVCOSH): 45900

Inverse Hyperbolic Tangent Subroutine (INVTANH): 45925

Inverse Trigonometric Functions

Inverse Tangent Recursion Subroutine (ATANITER): 45150

Inverse Sine Recursion Subroutine (ARCSINIT): 45200

Inverse Trigonometric CORDIC Subroutine (INVCORD): 45550

Least Squares

One-Dimensional:

Linear Least-Squares Subroutine (LSTSQR1): 43500

Parabolic Least-Squares Subroutine (LSTSQR2): 43550

One-Dimensional Polynomial Coefficient Generation Subroutine (POLYCM): 43600

Forsythe Polynomial Least-Squares Polynomial Fitting Subroutine (LSQRPOLY): 43950

Parametric Least-Squares Curve-Fit Subroutine (PARAFIT): 44250

Multidimensional:

Multidimensional Least-Squares Fitting Subroutine (LEASTSQR): 43650

Standard Deviation Subroutine (SIGMA): 43750

Multidimensional Polynomial Coefficient Generation Subroutine (MULTNLREG): 43800

Multidimensional Polynomial Iterated-Regression Subroutine (REGITER): 44100

Logarithm

LN(X!) Asymptotic Series Subroutine (LN(X!)): 44580

Natural Logarithm Recursion Subroutine (LOGITER): 45300

Modified CORDIC Natural Logarithm Subroutine (LNCORDIC): 45725

Matrix Operations

Matrix Multiplication Subroutine (MATMULT): 41900

Matrix Transpose Subroutine (MATTRANS): 41950

Matrix-Save (B in A) Subroutine (MATSAVBA): 42075

Matrix-Save (C in B) Subroutine (MATSAVCB): 42100

Matrix-Save (A in C) Subroutine (MATSAVAC): 42150

Matrix-Save (C in A) Subroutine (MATSAVCA): 42175

Matrix Inversion Subroutine (MATINV): 42400

Optimization

Parametric Least-Squares Curve-Fit Subroutine (PARAFIT): 44250

Steepest-Descent Optimization Subroutine (STEEPDS): 47700

Steepest-Descent Optimization Subroutine (STEEPDA): 47800

Polynomial Coefficients

Chebyshev Polynomial Coefficient Evaluation Subroutine (CHEBYSER): 44725

Legendre Polynomial Coefficient Evaluation Subroutine (LEGNDRE): 45400

Laguerre Polynomial Coefficient Evaluation Subroutine (LAGUERR): 45425

Hermite Polynomial Coefficient Evaluation Subroutine (HERMITE): 45450

Roots

Real Roots:

    Bisection Method Subroutine (BISECT): 46350

    Newton-Raphson Iteration Subroutine (Z-NEWTON): 46375

    Secant Method Subroutine (SECANT): 46425

    Modified False-Position Method Subroutine (REGULA): 46475

    Aitken Acceleration Subroutine (AITKEN): 46525

    Aitken-Steffenson Iteration Subroutine (A/SITER): 46575

    Multiple-Root Determination Subroutine (NEXTROOT): 46650

    Parabolic Root Seeking Subroutine (MUELLER): 46960

    Two-Dimensional Parabolic Root Seeking Subroutine (MUELLER2): 47050

Complex Roots:

    Complex Root Counting Subroutine (ROOTNUM): 46700

    Complex Root Search Subroutine (ZCIRCLE): 46750

    Complex Plane Newton-Raphson Iteration Subroutine (CZNEWTON): 46925

    Complex Plane Parabolic Root Seeking Subroutine (ZMUELLER): 47150

    General Complex Plane Root Determination Subroutine (ALLROOT): 47300

Polynomial Complex Roots:

    Quadratic Root Determination Subroutine (QUADRAT): 47475

    Polynomial Complex Root Determination Subroutine (LIN): 47525

    Polynomial Complex Root Determination Subroutine (BAIRSTOW): 47600

Series Operations

Series Reversion Subroutine (REVERSE): 44800

Reciprocal Power Series Subroutine (RECIPRO): 44850

Horner's Shifting-Rule Subroutine (HORNER): 44900

Series Summation

Complex Series Evaluation Subroutine (CMPLXSER): 44950

Statistics

Chi-Square Cumulative Distribution Approximation Subroutine (CHISQA): 44400

Chi-Square Probability Density Function Subroutine (CHI-SQR): 44600

Chi-Square Cumulative Distribution Function Subroutine (CHISQ): 44625

Inverse Normal Distribution Subroutine (INVNORM): 44925

Synthetic Division

Synthetic Division Subroutine (RSYNDIV): 44625

Trigonometric Functions

Tangent Iteration Subroutine (TANITER): 45125

Trigonometric CORDIC Subroutine (TRIGCORD): 45475

APPENDIX IIA

Full Listings of North Star BASIC Demonstration and Subroutine Programs

The North Star BASIC demonstration programs given in the text were concatenated in order of appearance and are reproduced in the next several pages. They are followed by the subroutine library.

Total demonstration program set length: 36878 bytes
                                          1891 lines

Total subroutine library length (uncompacted):   85281 bytes
                                                 3684 lines

Demonstration Programs

```
2000 REM LEAST SQUARES DEMONSTRATION PROGRAM
2001 PRINT
2002 PRINT
2003 PRINT "LEAST SQUARES CURVE FIT ROUTINE"
2004 PRINT
2005 PRINT
2006 PRINT "THIS PROGRAM CALCULATES A LINEAR"
2007 PRINT "LEAST SQUARES FIT TO A GIVEN DATA SET. "
2008 PRINT
2009 PRINT "INSTRUCTIONS"
2010 PRINT "--------------"
2011 PRINT
2012 PRINT "THE NUMBER OF DATA COORDINATES PROVIDED "
2013 PRINT "MUST BE GREATER THAN ONE. OTHERWISE, A "
2014 PRINT "DIVIDE BY ZERO ERROR MAY RESULT."
2015 PRINT
2016 PRINT "INPUT THE NUMBER OF DATA POINTS: ",
2017 INPUT N
2018 IF N<2 THEN GOTO 2012
2019 DIM X(N),Y(N)
2020 PRINT
2021 PRINT "THERE ARE TWO INPUT OPTIONS. ONE (1) "
2022 PRINT "INPUTS THE DATA POINTS IN COORDINATE "
2023 PRINT "PAIRS, AND THE OTHER (2) ALLOWS ONE TO "
2024 PRINT "FIRST INPUT THE INDEPENDENT VARIABLE "
```

```
2025 PRINT "VALUES, LATER FOLLOWED BY THE DEPENDENT."
2026 PRINT "WHICH MODE DO YOU DESIRE? (1 OR 2): ",
2027 INPUT Z
2028 PRINT
2029 IF Z=2 THEN GOTO 2032
2030 IF Z=1 THEN GOTO 2042
2031 GOTO 2026
2032 FOR M=0 TO N-1
2033 PRINT M+1,
2034 INPUT X(M)
2035 NEXT M
2036 PRINT
2037 FOR M=0 TO N-1
2038 PRINT M+1,
2039 INPUT Y(M)
2040 NEXT M
2041 GOTO 2047
2042 FOR M=0 TO N-1
2043 PRINT M+1,
2044 INPUT X(M),Y(M)
2045 NEXT M
2046 REM GO TO LINEAR LEAST SQUARES SUBROUTINE
2047 GOSUB 43500
2048 PRINT
2049 PRINT
2050 PRINT "FITTED EQUATION IS: "
2051 PRINT
2052 PRINT "     Y = ",INT(1000000*A)/1000000," ",
2053 IF B>=0 THEN PRINT "+",
2054 PRINT INT(1000000*B)/1000000,"*X"
2055 PRINT
2056 PRINT
2057 PRINT "STANDARD DEVIATION OF FIT: ",
2058 PRINT INT(10000*D)/10000
2059 PRINT
2060 PRINT
2061 END
2100 REM LEAST SQUARES DEMONSTRATION PROGRAM
2101 PRINT
2102 PRINT
2103 PRINT "LEAST SQUARES CURVE FIT ROUTINE"
2104 PRINT
2105 PRINT
2106 PRINT "THIS PROGRAM CALCULATES A PARABOLIC "
2107 PRINT "LEAST SQUARES FIT TO A GIVEN DATA SET. "
2108 PRINT
2109 PRINT "INSTRUCTIONS"
2110 PRINT "------------"
2111 PRINT
2112 PRINT "THE NUMBER OF DATA COORDINATES PROVIDED "
2113 PRINT "MUST BE GREATER THAN TWO. OTHERWISE, A "
2114 PRINT "DIVIDE BY ZERO ERROR MAY RESULT."
2115 PRINT
2116 PRINT "INPUT THE NUMBER OF DATA POINTS: ",
2117 INPUT N
2118 IF N<3 THEN GOTO 2112
2119 DIM X(N),Y(N)
2120 PRINT
2121 PRINT "THERE ARE TWO INPUT OPTIONS. ONE (1) "
```

```
2122 PRINT "INPUTS THE DATA POINTS IN COORDINATE "
2123 PRINT "PAIRS, AND THE OTHER (2) ALLOWS ONE TO "
2124 PRINT "FIRST INPUT THE INDEPENDENT VARIABLE "
2125 PRINT "VALUES, LATER FOLLOWED BY THE DEPENDENT "
2126 PRINT "WHICH MODE DO YOU DESIRE? (1 OR 2): ",
2127 INPUT Z
2128 PRINT
2129 IF Z=2 THEN GOTO 2132
2130 IF Z=1 THEN GOTO 2142
2131 GOTO 2126
2132 FOR M=0 TO N-1
2133 PRINT M+1,
2134 INPUT X(M)
2135 NEXT M
2136 PRINT
2137 FOR M=0 TO N-1
2138 PRINT M+1,
2139 INPUT Y(M)
2140 NEXT M
2141 GOTO 2147
2142 FOR M=0 TO N-1
2143 PRINT M+1,
2144 INPUT X(M),Y(M)
2145 NEXT M
2146 REM GO TO PARABOLIC LEAST SQUARES SUBROUTINE
2147 GOSUB 43550
2148 PRINT
2149 PRINT
2150 PRINT "FITTED EQUATION IS: "
2151 PRINT
2152 PRINT "     Y = ",INT(1000000*A)/1000000," ",
2153 IF B>=0 THEN PRINT "+",
2154 PRINT INT(1000000*B)/1000000,"*X ",
2155 IF C>=0 THEN PRINT "+ ",
2156 PRINT INT(1000000*C)/1000000,"*X*X"
2157 PRINT
2158 PRINT
2159 PRINT "STANDARD DEVIATION OF FIT: ",
2160 PRINT INT(10000*D)/10000
2161 PRINT
2162 PRINT
2163 END
2200 REM PROGRAM TO DEMONSTRATE ONE DIMENSIONAL
2201 REM OPERATION OF THE MULTI-NONLINEAR REGRESSION
2202 REM SUBROUTINE
2203 PRINT "HOW MANY DATA POINTS ARE THERE: ",
2204 INPUT M
2205 PRINT "WHAT IS THE DEGREE OF THE POLYNOMIAL"
2206 PRINT "TO BE FITTED: ",
2207 INPUT N
2208 DIM X(M),Y(M),Z(M,N+1),D(N+1),A(M,M),B(M,2*M),C(M,M)
2209 PRINT
2210 PRINT "INPUT THE DATA IN (X,Y) PAIRS AS PROMPTED:"
2211 PRINT
2212 FOR I=1 TO M
2213 PRINT I,TAB(5),"X , Y = ",
2214 INPUT X(I),Y(I)
2215 NEXT I
2216 PRINT
```

```
2217 PRINT
2218 REM GO TO COEFFICIENTS GENERATION SUBROUTINE
2219 GOSUB 43600
2220 REM GO TO REGRESSION SUBROUTINE
2221 N=N+1
2222 GOSUB 43650
2223 PRINT "THE CALCULATED COEFFICIENTS ARE:"
2224 PRINT
2225 FOR I=1 TO N
2226 PRINT I,TAB(5),INT(1000000*D(I))/1000000
2227 NEXT I
2228 REM GET STANDARD DEVIATION
2229 N=N-1
2230 GOSUB 43750
2231 PRINT
2232 PRINT
2233 PRINT "STANDARD DEVIATION: ",INT(1000000*D)/1000000
2234 PRINT
2235 PRINT
2236 END
2250 REM PROGRAM TO DEMONSTRATE MULTI-DIMENSIONAL
2251 REM OPERATION OF THE MULTI-NONLINEAR REGRESSION
2252 REM SUBROUTINE (MLTNLREG)
2253 PRINT "HOW MANY DATA POINTS ARE THERE: ",
2254 INPUT M
2255 PRINT
2256 PRINT "HOW MANY DIMENSIONS ARE THERE: ",
2257 INPUT L
2258 PRINT
2259 FOR I=1 TO L
2260 PRINT "WHAT IS THE FIT FOR DIMENSION ",I," ",
2261 INPUT M(I)
2262 NEXT I
2263 N=1
2264 FOR I=1 TO L
2265 N=N*(M(I)+1)
2266 NEXT I
2267 DIM X(M,L),Y(M),Z(M,N),D(N),A(M,M),B(M,2*M),C(M,M)
2268 PRINT
2269 PRINT "INPUT THE DATA AS PROMPTED:"
2270 PRINT
2271 FOR I=1 TO M
2272 PRINT "Y(",I,") = ",
2273 INPUT Y(I)
2274 FOR J=1 TO L
2275 PRINT "X(",I,",",J,") = ",
2276 INPUT X(I,J)
2277 NEXT J
2278 PRINT
2279 NEXT I
2280 REM GOTO COEFFICIENTS GENERATION SUBROUTINE
2281 GOSUB 43800
2282 REM GO TO REGRESSION SUBROUTINE
2283 PRINT
2284 GOSUB 43650
2285 PRINT "THE CALCULATED COEFFICIENTS ARE:"
2286 PRINT
2287 FOR I=1 TO N
2288 PRINT I,TAB(5),INT(1000000*D(I))/1000000
```

```
2289 NEXT I
2290 REM GET STANDARD DEVIATION
2291 N=N-1
2292 GOSUB 43800
2293 PRINT
2294 PRINT
2295 PRINT "STANDARD DEVIATION: ",INT(1000000*D)/1000000
2296 PRINT
2297 PRINT
2298 END
2300 REM PROGRAM TO DEMONSTRATE LSQRPOLY
2301 PRINT
2302 PRINT
2303 PRINT "WHAT IS THE ORDER OF THE FIT: ",
2304 INPUT M
2305 PRINT
2306 PRINT "WHAT IS THE ERROR REDUCTION FACTOR: ",
2307 INPUT E
2308 PRINT
2309 PRINT "HOW MANY DATA POINTS ARE THERE: ",
2310 INPUT N
2311 DIM X(N),Y(N),V(N),A(N),B(N),C(N),D(N),C2(N),E(N),F(N)
2312 PRINT
2313 PRINT "INPUT THE DATA POINTS AS PROMPTED: ",
2314 PRINT
2315 FOR I=1 TO N
2316 PRINT I,TAB(5),"X , Y = ",
2317 INPUT X(I),Y(I)
2318 NEXT I
2319 PRINT
2320 PRINT
2321 GOSUB 43950
2322 PRINT "COEFFICIENTS ARE:"
2323 PRINT
2324 FOR I=0 TO L
2325 PRINT I, TAB(5),INT(1000000*C(I))/1000000
2326 NEXT I
2327 PRINT
2328 PRINT
2329 PRINT "STANDARD DEVIATION=",
2330 PRINT INT(100000000*D)/100000000
2331 PRINT
2332 PRINT
2333 END
2350 REM PROGRAM TO DEMONSTRATE MULTIDIMENSIONAL
2351 REM OPERATION OF THE MULTI-NONLINEAR REGRESSION
2352 REM SUBROUTINE WITH ITERATIVE ERROR REDUCTION
2353 PRINT "HOW MANY DATA POINTS ARE THERE: ",
2354 INPUT M
2355 PRINT
2356 PRINT "HOW MANY DIMENSIONS ARE THERE: ",
2357 INPUT L
2358 PRINT
2359 FOR I=1 TO L
2360 PRINT "WHAT IS THE FIT FOR DIMENSION ",I," ",
2361 INPUT M(I)
2362 NEXT I
2363 N=1
2364 FOR I=1 TO L
```

```
2365 N=N*(M(I)+1)
2366 NEXT I
2367 DIM X(M,L),Y(M),Z(M,N),D(N),A(M,M),B(M,2*M),C(M,M),D1(N),Y1(M)
2368 PRINT
2369 PRINT "INPUT THE DATA AS PROMPTED:"
2370 PRINT
2371 FOR I=1 TO M
2372 PRINT "Y(",I,") = ",
2373 INPUT Y(I)
2374 FOR J=1 TO L
2375 PRINT "X(",I,",",J,") = ",
2376 INPUT X(I,J)
2377 NEXT J
2378 PRINT
2379 NEXT I
2380 REM GO TO ITERATION SUPERVISOR
2381 GOSUB 44100
2382 PRINT
2383 PRINT "THE CALCULATED COEFFICIENTS ARE:"
2384 PRINT
2385 FOR I=1 TO N
2386 PRINT I,TAB(5),INT(1000000*D(I))/1000000
2387 NEXT I
2388 PRINT
2389 PRINT
2390 PRINT "STANDARD DEVIATION: ",INT(1000000*D)/1000000
2391 PRINT
2392 PRINT
2393 PRINT"NUMBER OF ITERATIONS: ",
2394 PRINT L1
2395 PRINT
2396 PRINT
2397 END
2400 REM PROGRAM TO DEMONSTRATE THE PARAFIT SUBROUTINE
2401 PRINT
2402 PRINT
2403 N=10
2404 L=3
2405 PRINT "THE INPUT DATA ARE:"
2406 PRINT
2407 FOR I=1 TO N
2408 X(I)=I
2409 Y(I)=2*EXP(-(X(I)-4.5)*(X(I)-4.5)/3)
2410 PRINT "X(",I,") = ",X(I),TAB(15),"Y(",I,") = ",Y(I)
2411 NEXT I
2412 PRINT
2413 PRINT
2414 E=.1
2415 E1=.5
2416 A(1)=10
2417 A(2)=10
2418 A(3)=10
2419 GOSUB 44250
2420 PRINT"THE COEFFICIENTS ARE:"
2421 PRINT A(1)
2422 PRINT A(2)
2423 PRINT A(3)
2424 PRINT
2425 PRINT
```

```
2426 PRINT"THE STANDARD DEVIATION OF THE FIT IS",
2427 PRINT INT(10000000*D)/10000000
2428 PRINT
2429 PRINT
2430 PRINT "THE NUMBER OF ITERATIONS WAS",M
2431 PRINT
2432 PRINT
2433 END
2450 REM PROGRAM TO DEMONSTRATE CHISQA
2451 PRINT
2452 PRINT
2453 PRINT "F(X)",TAB(12),"   X"
2454 PRINT "----",TAB(12),"  ---"
2455 PRINT
2456 M=100
2457 FOR Y=.05 TO 1 STEP .05
2458 GOSUB 44400
2459 PRINT Y,TAB(12),INT(10*X)/10
2460 NEXT Y
2461 PRINT
2462 END
2500 REM PROGRAM TO DEMONSTRATE BESSEL SERIES SUMMATION SUBROUTINE
2501 PRINT
2502 PRINT
2503 PRINT "WHAT IS THE ORDER OF THE BESSEL FUNCTION: ",
2504 INPUT N
2505 PRINT
2506 PRINT "INPUT ARGUMENT",
2507 INPUT X
2508 PRINT
2509 PRINT "INPUT CONVERGENCE CRITERION",
2510 INPUT E
2511 PRINT
2512 PRINT
2513 GOSUB 44425
2514 PRINT "J(",X,") OF ORDER ",N," = ",Y
2515 PRINT
2516 PRINT "NUMBER OF TERMS USED: ",M
2517 PRINT
2518 END
2520 REM PROGRAM TO DEMONSTRATE THE BESSEL COEFFICIENTS SUBROUTINE
2521 PRINT
2522 PRINT
2523 PRINT "WHAT IS THE BESSEL FUNCTION ORDER: ",
2524 INPUT N
2525 PRINT
2526 PRINT "WHAT DEGREE IS DESIRED: ",
2527 INPUT M
2528 DIM A(M+1),B(M+1)
2529 PRINT
2530 PRINT
2531 GOSUB 44475
2532 PRINT"THE COEFFICIENTS ARE:"
2533 PRINT
2534 FOR I=0 TO M
2535 PRINT "A(",I,") = ",A(I)
2536 NEXT I
2537 PRINT
2538 PRINT
```

```
2539 END
2540 NEXT I
2541 RETURN
2550 REM PROGRAM TO DEMONSTRATE THE BESSEL FUNCTION ASYMPTOTIC SERIES
2551 PRINT
2552 PRINT
2553 PRINT "WHAT IS THE DESIRED ERROR BOUND: ",
2554 INPUT E3
2555 PRINT
2556 PRINT
2557 REM BESSEL FUNCTION SUBROUTINE
2558 PRINT "   X              J0(X)               J1(X)              N         E"
2559 PRINT "-------     ----------------    ----------------       -----     --------"
2560 PRINT
2561 FOR X=1 TO 15
2562 GOSUB 44525
2563 PRINT X,TAB(15),INT(100000000*J0)/100000000,TAB(36),INT(100000000*J1)/100000000,
2564 PRINT TAB(55),N,TAB(65),INT(100000000*E)/100000000
2565 NEXT X
2566 PRINT
2567 PRINT
2568 PRINT
2569 END
2575 REM PROGRAM TO DEMONSTRATE LN(X!) SUBROUTINE
2576 PRINT
2577 PRINT
2578 PRINT" X",TAB(8),"LN(X!)",TAB(19),"EXP(LN(X!))"
2579 PRINT "---",TAB(6),"----------",TAB(19),"------------"
2580 PRINT
2581 FOR X=1 TO 10
2582 GOSUB 44580
2583 PRINT X,TAB(5),Y,TAB(18),EXP(Y)
2584 NEXT X
2585 PRINT
2586 PRINT
2587 END
2600 REM PROGRAM TO DEMONSTRATE THE CHI-SQUARE SUBROUTINE
2601 PRINT
2602 PRINT
2603 PRINT "HOW MANY DEGREES OF FREEDOM: ",
2604 INPUT M
2605 PRINT
2606 PRINT "WHAT IS THE RANGE (X1,X2): "
2607 PRINT "X1:",
2608 INPUT X1
2609 PRINT "X2:",
2610 INPUT X2
2611 PRINT
2612 PRINT "WHAT IS THE TABLE STEP SIZE: ",
2613 INPUT X3
2614 PRINT
2615 PRINT
2616 PRINT "   X",TAB(8),"CHI-SQUARE PDF"
2617 PRINT " ---",TAB(8),"--------------"
2618 PRINT
2619 FOR X=X1 TO X2 STEP X3
2620 GOSUB 44600
2621 PRINT X,TAB(8),INT(10000*Y)/10000
2622 NEXT X
```

```
2623 END
2625 REM PROGRAM TO DEMONSTRATE CHISQ
2626 PRINT
2627 PRINT
2628 PRINT "HOW MANY DEGREES OF FREEDOM: ",
2629 INPUT M
2630 PRINT
2631 PRINT "WHAT IS THE RANGE (X1,X2): "
2632 PRINT "X1: ",
2633 INPUT X1
2634 PRINT "X2: ",
2635 INPUT X2
2636 PRINT
2637 PRINT "STEP SIZE: ",
2638 INPUT X3
2639 PRINT
2640 PRINT "SUMMATION TRUNCATION ERROR BOUND: ",
2641 INPUT E
2642 PRINT
2643 PRINT
2644 PRINT "   X",TAB(8),"CHI-SQUARE CDF"
2645 PRINT " ---",TAB(8),"--------------"
2646 FOR X=X1 TO X2 STEP X3
2647 GOSUB 44625
2648 PRINT X,TAB(9),INT(10000*Y)/10000
2649 NEXT X
2650 END
2675 REM PROGRAM TO DEMONSTRATE ASYMERF
2676 PRINT
2677 PRINT
2678 PRINT "INPUT X",
2679 INPUT X
2680 GOSUB 44675
2681 PRINT
2682 PRINT "ERF(X)= ",Y,"   WITH ERROR ESTIMATE= ",INT(100000000*E)/100000000
2683 PRINT "NUMBER OF TERMS EVALUATED WAS",N
2684 PRINT
2685 END
2700 REM PROGRAM TO DEMONSTRATE CHEBYSER SUBROUTINE
2701 PRINT
2702 PRINT
2703 DIM B(10,10)
2704 FOR N-2 TO 10
2705 GOSUB 44725
2706 PRINT "CHEBYSHEV POLYNOMIAL COEFICIENTS"
2707 PRINT "FOR DEGREE",N
2708 PRINT
2709 FOR I=0 TO N
2710 PRINT "A(",I,") = ",B(N,I)
2711 NEXT I
2712 PRINT
2713 PRINT
2714 NEXT N
2715 PRINT
2716 END
2720 REM PROGRAM TO DEMONSTRATE CHEBYSHEV ECONOMIZATION
2721 PRINT
2722 PRINT "WHAT IS THE DEGREE OF"
2723 PRINT "THE INPUT POLYNOMIAL: ",
```

```
2724 INPUT M
2725 PRINT
2726 PRINT "WHAT IS THE DEGREE OF THE"
2727 PRINT "DESIRED ECONOMIZED POLYNOMIAL: ",
2728 INPUT M1
2729 PRINT
2730 PRINT "WHAT IS THE RANGE OF"
2731 PRINT "INPUT POLYNOMIAL: ",
2732 INPUT X0
2733 DIM A(M),B(M,M),C(M)
2734 PRINT
2735 PRINT
2736 PRINT "INPUT THE COEFFICIENTS:"
2737 PRINT
2738 FOR I=0 TO M
2739 PRINT "C(",I,") = ",
2740 INPUT C(I)
2741 NEXT I
2742 PRINT
2743 PRINT
2744 GOSUB 44760
2745 PRINT "THE CHEBYSHEV SERIES COEFFICIENTS ARE:"
2746 PRINT
2747 FOR I=0 TO M
2748 PRINT "A(",I,") = ",A(I)
2749 NEXT I
2750 PRINT
2751 PRINT
2752 PRINT "THE ECONOMIZED POLYNOMIAL"
2753 PRINT "COEFFICIENTS ARE:"
2754 PRINT
2755 FOR I=0 TO M1
2756 PRINT "C(",I,") = ",C(I)
2757 NEXT I
2758 PRINT
2759 PRINT
2760 END
2775 REM PROGRAM TO DEMONSTRATE THE SERIES REVERSION SUBROUTINE
2776 PRINT
2777 PRINT
2778 PRINT "WHAT IS THE DEGREE OF"
2779 PRINT "THE INPUT POLYNOMIAL: ",
2780 INPUT N
2781 PRINT
2782 PRINT "INPUT THE COEFFICIENTS AS PROMPTED:"
2783 PRINT
2784 FOR I=0 TO N
2785 PRINT "A(",I,") = ",
2786 INPUT A(I)
2787 NEXT I
2788 PRINT
2789 PRINT
2790 GOSUB 44800
2791 PRINT "THE REVERSED POLYNOMIAL COEFFICIENTS ARE:"
2792 PRINT
2793 FOR I=0 TO 7
2794 PRINT "B(",I,") = ",B(I)
2795 NEXT I
2796 PRINT
```

```
2797 PRINT
2798 END
2800 REM PROGRAM TO DEMONSTRATE THE SERIES INVERSION SUBROUTINE
2801 PRINT
2802 PRINT
2803 PRINT "WHAT IS THE DEGREE OF THE INPUT POLYNOMIAL: ",
2804 INPUT N
2805 PRINT
2806 PRINT "WHAT IS THE DEGREE OF THE INVERTED POLYNOMIAL: ",
2807 INPUT M
2808 DIM A(M),B(M)
2809 PRINT
2810 PRINT "INPUT THE POLYNOMIAL COEFFICIENTS:"
2811 PRINT
2812 FOR I=0 TO N
2813 PRINT "A(",I,") = ",
2814 INPUT A(I)
2815 NEXT I
2816 GOSUB 44850
2817 PRINT
2818 PRINT
2819 PRINT "THE INVERTED POLYNOMIAL COEFFICIENTS ARE:"
2820 PRINT
2821 FOR I=0 TO M
2822 PRINT "B(",I,") = ",B(I)
2823 NEXT I
2824 PRINT
2825 END
2830 REM TEST PROGRAM FOR HORNER'S RULE
2831 DIM C(4,5)
2832 PRINT
2833 PRINT
2834 PRINT "INPUT THE FIVE COEFFICIENTS:"
2835 PRINT
2836 FOR I=0 TO 4
2837 PRINT "A(",I,") = ",
2838 INPUT A(I)
2839 NEXT I
2840 PRINT
2841 PRINT "WHAT IS THE EXPANSION POINT: ",
2842 INPUT X0
2843 PRINT
2844 GOSUB 44900
2845 PRINT
2846 PRINT "THE SHIFTED COEFFICIENTS ARE:"
2847 PRINT
2848 FOR I=0 TO 4
2849 PRINT "B(",I,") = ",B(I)
2850 NEXT I
2851 PRINT
2852 PRINT
2853 END
2860 REM PROGRAM TO DEMONSTRATE INVERSE NORMAL SUBROUTINE
2861 PRINT
2862 PRINT "P(Z>X)",TAB(11),"X"
2863 PRINT "-------",TAB(10),"---"
2864 PRINT
2865 FOR Y=.5 TO 0 STEP -.02
2866 GOSUB 44925
```

```
2867 PRINT Y,TAB(8),INT(10000*X)/10000
2868 NEXT Y
2869 END
2870 REM PROGRAM TO DEMOSTRATE SINEPROD
2871 E=.000001
2872 PRINT
2873 PRINT
2874 PRINT "   X",TAB(10),"SIN(X) CALC.",TAB(25),"SIN(X) TRUE",TAB(42),"K",
2875 PRINT TAB(52),"ERROR"
2876 PRINT " ---",TAB(10),"-------------",TAB(25),"--------------",TAB(41),"----",
2877 PRINT TAB(52),"-----"
2878 FOR X=0 TO 2 STEP .05
2879 GOSUB 2888
2880 PRINT X,TAB(10),INT(10000000*Y)/10000000,TAB(25),
2881 PRINT INT(10000000*SIN(X))/10000000,TAB(40),K,
2882 PRINT TAB(49),INT((100000000*(Y-SIN(X))))/100000000
2883 NEXT X
2884 PRINT
2885 PRINT
2886 END
2887 REM *********************
2888 REM SINE PRODUCT SERIES SUBROUTINE (SINEPROD)
2889 REM THIS PROGRAM CALCULATES AN APPROXIMATION TO SIN(X)
2890 REM USING REPEATED PRODUCTS.
2891 REM THE INPUTS TO THE PROGRAM ARE THE ARGUMENT, X
2892 REM AND AN ERROR FACTOR, E.
2893 REM THE APPROXIMATION IS RETURNED IN Y.
2894 Y=X
2895 K=1
2896 L=3.14159265358979323846
2897 L=X/L
2898 L=L*L
2899 M=L/(K*K)
2900 Y=Y*(1-M)
2901 IF M<E THEN RETURN
2902 K=K+1
2903 GOTO 2899
3000 REM PROGRAM TO DEMONSTRATE COMPLEX SERIES EVALUATION SUBROUTINE.
3001 REM IT IS ASSUMED THAT THE COEFFICIENTS ARE OBTAINED FROM
3002 REM A SUBROUTINE.
3003 REM GET COEFFICIENTS
3004 GOSUB 3024
3005 REM INPUT COMPLEX NUMBER
3006 PRINT
3007 PRINT
3008 PRINT "INPUT THE COMPLEX NUMBER AS PROMPTED:"
3009 PRINT
3010 PRINT "     REAL PART = ",
3011 INPUT X
3012 PRINT "     COMPLEX PART = ",
3013 INPUT Y
3014 GOSUB 44950
3015 PRINT
3016 PRINT
3017 PRINT "RESULTS ARE:"
3018 PRINT
3019 PRINT "     Z1 = ",Z1
3020 PRINT "     Z2 = ",Z2
3021 PRINT
```

```
3022 END
3023 REM COEFFICIENTS SUBROUTINE
3024 M=5
3025 A(0)=1
3026 A(1)=5
3027 A(2)=10
3028 A(3)=10
3029 A(4)=5
3030 A(5)=1
3031 RETURN
3050 REM PROGRAM TO DEMONSTRATE NTHROOT
3051 PRINT
3052 PRINT
3053 PRINT "WHAT IS THE NUMBER: ",
3054 INPUT Y
3055 PRINT
3056 PRINT "WHAT ROOT IS DESIRED: ",
3057 INPUT N
3058 PRINT
3059 PRINT "WHAT IS THE CONVERGENCE FACTOR: ",
3060 INPUT E
3061 PRINT
3062 REM THE INITIAL VALUE IS X0.
3063 X0=1
3064 PRINT
3065 GOSUB 45000
3066 PRINT "THE",N,"-TH ROOT OF",Y," IS",X
3067 PRINT
3068 PRINT "THE NUMBER OF ITERATIONS WAS",M
3069 PRINT
3070 END
3075 REM PROGRAM TO DEMONSTRATE GENROOT
3076 PRINT
3077 PRINT
3078 PRINT "WHAT IS THE NUMBER: ",
3079 INPUT Y
3080 PRINT
3081 PRINT "WHAT EXPONENT IS DESIRED: ",
3082 INPUT N
3083 PRINT
3084 PRINT "WHAT IS THE CONVERGENCE FACTOR: ",
3085 INPUT E
3086 PRINT
3087 REM NUMBER OF BITS = M
3088 M=32
3089 DIM A(M)
3090 PRINT "THE BINARY REPRESENTATION OF THE FRACTION IS:"
3091 PRINT
3092 REM GET THE BINARY REPRESENTATION
3093 GOSUB 45060
3094 FOR I=1 TO M
3095 PRINT A(I),
3096 NEXT I
3097 PRINT
3098 PRINT
3099 GOSUB 45030
3100 PRINT "THE",N,"-TH POWER OF",Y," IS",X
3101 PRINT
3102 PRINT
```

```
3103 END
3125 REM PROGRAM TO DEMONSTRATE TANITER SUBROUTINE
3126 PRINT
3127 PRINT
3128 PRINT "WHAT IS THE ARGUMENT: ",
3129 INPUT X
3130 PRINT
3131 PRINT "WHAT IS THE CONVERGENCE FACTOR: ",
3132 INPUT E
3133 PRINT
3134 PRINT
3135 GOSUB 45125
3136 PRINT "THE TANGENT OF",X," IS",Y
3137 PRINT
3138 PRINT "THE ARCTANGENT OF THE TANGENT IS",ATN(Y)
3139 PRINT
3140 END
3150 REM PROGRAM TO DEMONSTRATE ATANITER
3151 PRINT
3152 PRINT
3153 PRINT "WHAT IS THE ARGUMENT: ",
3154 INPUT X
3155 PRINT
3156 PRINT "WHAT IS THE CONVERGENCE FACTOR: ",
3157 INPUT E
3158 PRINT
3159 GOSUB 45150
3160 PRINT
3161 PRINT "THE ARCTANGENT OF",X," EQUALS",Y
3162 PRINT
3163 PRINT "THE NUMBER OF ITERATIONS WAS",M
3164 PRINT
3165 PRINT
3166 END
3175 REM PROGRAM TO DEMONSTRATE ARCSINE RECURSION
3176 PRINT
3177 PRINT
3178 E=.00000001
3179 PRINT "  X",TAB(10),"ARCSIN(X)",TAB(25),"STEPS",TAB(37),"ERROR"
3180 PRINT "  ---",TAB(10),"---------",TAB(25),"-----",TAB(35),"------------"
3181 PRINT
3182 FOR X=0 TO 1 STEP .05
3183 GOSUB 45200
3184 PRINT X,TAB(9),INT(10000000*Y)/10000000,TAB(25),
3185 PRINT M,TAB(34),INT(10000000*(SIN(Y)-X)+.5)/10000000
3186 NEXT X
3187 PRINT
3188 PRINT
3189 END
3200 REM PROGRAM TO DEMONSTRATE EVALUATING ELLIPTIC INTEGRALS OF
3201 REM THE FIRST AND SECOND KINDS (COMPLETE)
3202 PRINT
3203 PRINT
3204 DIM A(200),B(200)
3205 E=.0000001
3206 PRINT "  K",TAB(14),"K(K)",TAB(29),"E(K)",TAB(41),"STEPS"
3207 PRINT "  ---",TAB(10),"-------------",TAB(25),"-------------",TAB(41),"-------"
3208 PRINT
3209 FOR K=0 TO 1 STEP .05
```

```
3210 GOSUB 45250
3211 PRINT K,TAB(10),E1,TAB(25),E2,TAB(42),N
3212 NEXT K
3213 END
3225 REM PROGRAM TO DEMONSTRATE LOG BY RECURSION
3226 PRINT
3227 PRINT
3228 E=.000001
3229 PRINT "   X",TAB(10),"LOG(X)",TAB(23),"STEPS",TAB(35),"ERROR"
3230 PRINT "  ---",TAB(7),"------------",TAB(23),"-----",TAB(32),"-------------"
3231 PRINT
3232 FOR X=.05 TO .95 STEP .05
3233 GOSUB 45300
3234 PRINT X,TAB(7),Y,TAB(24),M,TAB(32),INT(100000000*(Y-LOG(X)))/100000000
3235 NEXT X
3236 PRINT
3237 PRINT
3238 END
3250 REM PROGRAM TO DEMONSTRATE INTBESSL SUBROUTINE
3251 PRINT
3252 PRINT
3253 M=20
3254 PRINT "   X",TAB(10),"J0(X)",TAB(25),"J1(X)",TAB(40),"J2(X)",
3255 PRINT TAB(55),"J3(X)",TAB(70),"J4(X)"
3256 PRINT "  ---",TAB(7),"-----------",TAB(22),"-----------",TAB(37),
3257 PRINT "-----------",TAB(52),"-----------",TAB(67),"-----------"
3258 PRINT
3259 FOR X=.1 TO 3 STEP .1
3260 GOSUB 45350
3261 I=100000000
3262 PRINT X,TAB(7),INT(I*Y(0))/I,TAB(22),INT(I*Y(1))/I,TAB(37),INT(I*Y(2))/I,
3263 PRINT TAB(52),INT(I*Y(3))/I,TAB(67),INT(I*Y(4))/I
3264 NEXT X
3265 PRINT
3266 PRINT
3267 END
3275 REM PROGRAM TO DEMONSTRATE THE LEGENDRE COEFFICIENTS SUBROUTINE
3276 PRINT
3277 PRINT
3278 DIM A(10),B(10,10)
3279 FOR N=2 TO 10
3280 PRINT
3281 PRINT "LEGENDRE POLYNOMIAL COEFFICIENTS"
3282 PRINT "FOR ORDER ",N
3283 PRINT
3284 GOSUB 45400
3285 FOR K=0 TO N
3286 PRINT "A(",K,") = ",A(K)
3287 NEXT K
3288 PRINT
3289 NEXT N
3290 PRINT
3291 PRINT
3292 END
3300 REM PROGRAM TO DEMONSTRATE THE LAGUERRE COEFFICIENTS SUBROUTINE
3301 PRINT
3302 PRINT
3303 DIM A(10),B(10,10)
3304 FOR N=2 TO 10
```

```
3305 PRINT
3306 PRINT "LAGUERRE POLYNOMIAL COEFFICIENTS"
3307 PRINT "FOR ORDER ",N
3308 PRINT
3309 GOSUB 45425
3310 FOR K=0 TO N
3311 PRINT "A(",K,") = ",A(K)
3312 NEXT K
3313 PRINT
3314 NEXT N
3315 PRINT
3316 PRINT
3317 END
3320 REM PROGRAM TO DEMONSTRATE THE HERMITE COEFFICIENTS SUBROUTINE
3321 PRINT
3322 PRINT
3323 DIM A(10),B(10,10)
3324 FOR N=2 TO 10
3325 PRINT
3326 PRINT "HERMITE POLYNOMIAL COEFFICIENTS"
3327 PRINT "FOR ORDER ",N
3328 PRINT
3329 GOSUB 45450
3330 FOR K=0 TO N
3331 PRINT "A(",K,") = ",A(K)
3332 NEXT K
3333 PRINT
3334 NEXT N
3335 PRINT
3336 PRINT
3337 END
3350 REM PROGRAM TO DEMONSTRATE USE OF THE CORDIC
3351 REM TRIGONOMETRIC APPROXIMATION SUBROUTINE.
3352 PRINT
3353 PRINT
3354 REM PROGRAM PRINTS OUT 21 VALUES OF SINE AND
3355 REM COSINE AND COMPARES THE RESULT WITH VALUES
3356 REM CALCULATED INTERNALLY.
3357 DIM A(12),W(12)
3358 B=3.141592653589793238
3359 PRINT "   ANGLE (RADIANS)",TAB(24),"CALCULATED SINE",TAB(40),
3360 PRINT "        DIFFERENCE",TAB(67),"CALCULATED COSINE",TAB(84),"   DIFFERENCE"
3361 PRINT "   ------------------",TAB(24),"---------------",TAB(40),
3362 PRINT "        ----------",TAB(67),"--------------------",TAB(84),"   ----------"
3363 PRINT
3364 B=.025*B
3365 M=1E14
3366 FOR J=0 TO 20
3367 A=J*B
3368 GOSUB 45475
3369 PRINT INT(M*A)/M,TAB(23),INT(M*U)/M,TAB(44),
3370 PRINT INT((U-SIN(A))*M)/M,TAB(66),INT(M*V)/M,TAB(88),INT((V-COS(A))*M)/M
3371 NEXT J
3372 PRINT
3373 PRINT
3374 END
3425 REM PROGRAM TO DEMONSTRATE THE USE OF THE INVERSE
3426 REM TRIGONOMETRIC CORDIC SUBROUTINE
3427 PRINT
```

```
3428 PRINT
3429 PRINT "DO YOU WANT THE INVERSE SINE (S)"
3430 DIM A(12),W(12)
3431 PRINT "                          COSINE (C)"
3432 PRINT "                     OR TANGENT (T): ",
3433 INPUT A$
3434 PRINT
3435 PRINT "WHAT IS THE VALUE: ",
3436 INPUT Z1
3437 U=0
3438 V=0
3439 W=0
3440 IF A$="S" THEN U=Z1
3441 IF A$="C" THEN V=Z1
3442 IF A$="T" THEN W=Z1
3443 REM GET RESULT
3444 GOSUB 45550
3445 PRINT
3446 PRINT
3447 IF A$="S" THEN PRINT "ARCSIN(",
3448 IF A$="C" THEN PRINT "ARCOS(",
3449 IF A$="T" THEN PRINT "ARCTAN(",
3450 PRINT Z1,") = ",A," RADIANS"
3451 PRINT
3452 PRINT
3453 END
3460 REM PROGRAM TO DEMONSTRATE CORDIC EXPONENTIAL SUBROUTINE (EXPCORD)
3461 PRINT
3462 PRINT
3463 PRINT" X", TAB(20),"     EXP(X)",TAB(50),"RELATIVE DIFFERENCE"
3464 PRINT" -", TAB(20),"     ------",TAB(50),"-------------------"
3465 PRINT
3466 M=1E14
3467 FOR X=-1 TO 4 STEP .1
3468 GOSUB 45650
3469 PRINTX,TAB(20),Y,TAB(50),INT(M*(Y-EXP(X))/Y)/M
3470 NEXT X
3471 PRINT
3472 PRINT
3473 END
3475 REM PROGRAM TO DEMONSTRATE THE MODIFIED CORDIC NATURAL LOGARITHM SUBROUTINE
3476 PRINT
3477 PRINT
3478 PRINT" X", TAB(20),"     LN(X)",TAB(50),"RELATIVE DIFFERENCE"
3479 PRINT" -", TAB(20),"     ------",TAB(50),"-------------------"
3480 PRINT
3481 DIM A(15),W(15)
3482 M=1E14
3483 FOR X=.1 TO 3 STEP .1
3484 GOSUB 45725
3485 PRINTX,TAB(20),INT(M*Y)/M,TAB(49),INT(M*(Y-LOG(X))/(Y+.000001))/M
3486 NEXT X
3487 PRINT
3488 PRINT
3489 END
3500 REM PROGRAM TO DEMONSTRATE THE HYPERBOLIC SUBROUTINES
3501 PRINT
3502 PRINT
3503 PRINT "   X ",TAB(15),"   SINH(X)",TAB(42),"   COSH(X)",TAB(71),"   TANH(X)"
```

```
3504 PRINT"---",TAB(15),"--------------",TAB(42),"------------------",TAB(71),"---------------
3505 PRINT
3506 FOR X=-5 TO 5 STEP .2
3507 PRINT X,
3508 REM GET SINH(X)
3509 GOSUB 45800
3510 PRINT TAB(12),Y,
3511 REM GET COSH(X)
3512 GOSUB 45825
3513 PRINT TAB(40),Y,
3514 REM GET TANH(X)
3515 GOSUB 45840
3516 PRINT TAB(68),Y
3517 NEXT X
3518 PRINT
3519 PRINT
3520 END
3525 REM PROGRAM TO DEMONSTRATE THE INVERSE HYPERBOLIC FUNCTION SUBROUTINES
3526 PRINT
3527 PRINT
3528 DIM A(15),W(15)
3529 PRINT "   X ",TAB(12),"ARCSINH(X)",TAB(32),"ARCCOSH(X)",TAB(52),"ARCTANH(X)
3530 PRINT "  ---",TAB(12),"----------",TAB(32),"----------",TAB(52),"----------
3531 PRINT
3532 FOR X=-3 TO 3 STEP .2
3533 PRINT " ",X,
3534 REM GET ARCSINH(X)
3535 GOSUB 45875
3536 PRINT TAB(11),Y,
3537 REM GET ARCOSH(X)
3538 GOSUB 45900
3539 PRINT TAB(31),Y,
3540 REM GET ARCTANH(X)
3541 GOSUB 45925
3542 PRINT TAB(51),Y
3543 NEXT X
3544 PRINT
3545 PRINT
3546 END
3550 REM DEMONSTRATION PROGRAM FOR LAGRANGE INTERPOLATION OF SIN(X)
3551 PRINT
3552 PRINT
3553 DIM L(9),Y(15),X(15)
3554 V=14
3555 REM INPUT TABLE
3556 FOR I=1 TO V
3557 READ X(I),Y(I)
3558 NEXT I
3559 REM SINE TABLE VALUES FROM
3560 REM HANDBOOK OF MATHEMATICAL FUNCTIONS
3561 REM BY ABRAMOWITZ, M., AND STEGUN, I.A.
3562 REM NBS, JUNE 1964
3563 DATA 0,0,.125,.12467473
3564 DATA .217,.21530095,.299,.29456472
3565 DATA .376,.36720285,.450,.43496553
3566 DATA .520,.49688014,.589,.55552980
3567 DATA .656,.60995199,.721,.66013615
3568 DATA .7853981634,0.7071067812
3569 DATA .849,.75062005,.911,.79011709
```

```
3570 DATA .972,.82601466
3571 REM INPUT INTERPOLATION POINT
3572 PRINT "INPUT X: ",
3573 INPUT X
3574 PRINT "INPUT THE ORDER OF THE INTERPOLATION: ",
3575 INPUT N
3576 REM GOTO INTERPOLATION SUBROUTINE
3577 GOSUB 46000
3578 PRINT "SIN(X)= : ",Y
3579 PRINT "ERROR CHECK: ",N
3580 PRINT
3581 GOTO 3572
3582 END
3583 IF J=K THEN GOTO 3585
3584 L(K)=L(K)*(X-X(J+I))/(X(I+K)-X(J+I))
3585 NEXT J
3586 Y=Y+L(K)*Y(I+K)
3587 NEXT K
3588 RETURN
3600 REM PROGRAM TO DEMONSTRATE THE NEWTON INTERPOLATION
3601 PRINT
3602 PRINT
3603 REM SUBROUTINE FOR SIN(X)
3604 V=14
3605 DIM X(V+1),Y(V+1),Y1(V),Y2(V-1),Y3(V-2)
3606 REM INPUT TABLE
3607 FOR I=1 TO V
3608 READ X(I),Y(I)
3609 NEXT I
3610 REM SIN TABLE VALUES FROM
3611 REM HANDBOOK OF MATHEMATICAL FUNCTIONS
3612 REM BY ABRAMOWITZ, M., AND STEGUN, I.A.
3613 REM NBS, JUNE 1964
3614 DATA 0,0,.125,.12467473,.217,.21530095
3615 DATA .299,.29456472,.376,.36720285
3616 DATA .450,.43496553,.520,.49688014
3617 DATA .589,.55552980,.656,.60995199
3618 DATA .721,.66013615,.7853981634,0.7071067812
3619 DATA .849,.75072005
3620 DATA .911,.79011709,.972,.82601466
3621 REM INPUT INTERPOLATION POINT
3622 PRINT "INPUT X :",
3623 INPUT X
3624 REM GO TO INTERPOLATION SUBROUTINE
3625 GOSUB 46050
3626 PRINT "SIN(",X,") = ",Y
3627 PRINT"ERROR ESTIMATE (W/O ROUNDOFF): ",E
3628 PRINT "ERROR CHECK: ",N
3629 PRINT
3630 PRINT
3631 GOTO 3622
3632 END
3650 REM TABLE SPACING PROGRAM FOR SIN(X) IN THE FIRST OCTANT
3651 PRINT
3652 PRINT
3653 REM PROGRAM GENERATES TABLE LOCATIONS WHICH APPROXIMATELY
3654 REM MINIMIZE THE MAXIMUM ERROR FOR A GIVEN NUMBER OF TABLE POINTS.
3655 REM A CUBIC FIT IS ASSUMED.
3656 REM G=CORRECTION TO GUESS.
```

```
3657 REM L IS USED TO ESTABLISH THE RANGE OF INTERPOLATION, IN THIS CASE PI/4.
3658 REM FOR ANOTHER FUNCTION, CHANGE NEXT TWO LINES.
3659 L1=3.141595/4
3660 L=SIN(L1)
3661 REM Q=PRINT FLAG
3662 Q=0
3663 REM ESTABLISH THE INITIAL GUESS FOR QUARTER INTERVAL SPACING
3664 REM NOTE THAT THE FINAL NUMBER OF INTERPOLATION INTERVALS IN THE
3665 REM INTERPOLATION RANGE IS N
3666 N=20
3667 H=L1/(4*N)
3668 PRINT "CONVERGENCE FACTOR:"
3669 REM S STORES THE LENGTH OF THE FIRST INTERVAL
3670 S=H
3671 REM PRINT HEADING
3672 IF Q=0 THEN GOTO 3681
3673 PRINT
3674 PRINT
3675 PRINT "INTERVAL",TAB(12),"POSITION",TAB(25),"INTERVAL SIZE"
3676 PRINT "--------",TAB(12),"--------",TAB(25),"----------------"
3677 REM A IS A MEASURE OF THE ERROR WHICH IS TO BE CONSTANT
3678 REM IT IS USED TO DETERMINE THE SPACING FOR ALL SUBSEQUENT INTERVALS
3679 REM THE FUNCTION IN THE NEXT LINE IS THE FOURTH DERIVATIVE (ABSOLUTE VALUE)
3680 REM OF THE FUNCTION TO BE EVALUATED
3681 A=H*(SIN(H/2))^(1/4)
3682 REM START INTERVAL SECTIONING
3683 X=0
3684 FOR I=1 TO N+7
3685 REM IF CONVERGENCE HAS BEEN ACHIEVED, PRINT RESULTS
3686 IF Q=0 THEN GOTO 3689
3687 PRINT I,TAB(10),INT(100000000*X)/100000000,TAB(26),INT(100000000*H)/1000000
3688 REM MOVE ON TO THE NEXT INTERVAL
3689 X=X+H
3690 REM CALCULATE THE NEW SPACING FOR THAT INTERVAL
3691 H=A*(1/ABS(SIN(X)))^(1/4)
3692 REM AT THIS POINT A CORRECTION IS APPLIED TO MAKE THE SUM
3693 REM OF THE INTERVALS (N OF THEM) EQUAL L1, THE RANGE)
3694 REM NOTE THAT THE FUNCTION ITSELF IS USED ON THE NEXT LINE,
3695 REM NOT ITS DERIVATIVE
3696 IF I=N THEN G=L/ABS(SIN(X))
3697 NEXT I
3698 IF Q=1 THEN GOTO 3705
3699 PRINT G
3700 REM CHECK FOR CONVERGENCE
3701 IF ABS(G-1)<.01/N THEN Q=1
3702 REM ADJUST SPACING OF THE FIRST INTERVAL AND TRY AGAIN
3703 H=G*S
3704 GOTO 3670
3705 PRINT
3706 PRINT
3707 END
3725 REM PROGRAM TO DEMONSTRATE THE AKIMA SPLINE FITTING SUBROUTINE
3726 PRINT
3727 PRINT
3728 REM V=NUMBER OF TABLE VALUES
3729 V=14
3730 DIM X(V+1),Y(V+1),M(V+4),Z(V+1)
3731 REM INPUT TABLE
3732 FOR I=1 TO V
```

```
3733 READ X(I),Y(I)
3734 NEXT I
3735 REM SIN TABLE VALUES FROM
3736 REM HANDBOOK OF MATHEMATICAL FUNCTIONS
3737 REM BY ABRAMOWITZ, M., AND STEGUN, I.A.
3738 REM NBS, JUNE 1964
3739 DATA 0,0,.125,.12467473,.217,.21530095
3740 DATA .299,.29456472,.376,.36720285
3741 DATA .450,.43496553,.520,.49688014
3742 DATA .589,.55552980,.656,.60995199
3743 DATA .721,.66013615,.7853981634,0.7071067812
3744 DATA .849,.75072005
3745 DATA .911,.79011709,.972,.82601466
3746 PRINT
3747 PRINT " X ",TAB(10),"SIN(X) HANDBOOK",TAB(30),"AKIMA INTERPOLATION",
3748 PRINT TAB(54),"OBSERVED ERROR"
3749 PRINT "---",TAB(10),"---------------",TAB(30),"--------------------",
3750 PRINT TAB(54),"--------------"
3751 PRINT
3752 FOR X=0 TO .75 STEP .05
3753 GOSUB 46100
3754 PRINT X,TAB(10),INT(10000000*SIN(X))/10000000,TAB(32),INT(10000000*Y)/10000000,
3755 PRINT TAB(56),INT(1000000*(SIN(X)-Y))/1000000
3756 NEXT X
3757 PRINT
3758 PRINT
3759 END
3775 REM DEMONSTRATION PROGRAM FOR LAGRANGE DERIVATIVE INTERPOLATION OF SIN(X)
3776 PRINT
3777 PRINT
3778 V=15
3779 N=3
3780 DIM L(9),M(9,9),X(V),Y(V)
3781 REM INPUT TABLE
3782 FOR I=1 TO V
3783 READ X(I),Y(I)
3784 NEXT I
3785 REM SINE TABLE VALUES FROM
3786 REM HANDBOOK OF MATHEMATICAL FUNCTIONS
3787 REM BY ABRAMOWITZ, M., AND STEGUN, I.A.
3788 REM NBS, JUNE 1964
3789 DATA -.125,-.12467473
3790 DATA 0,0,.125,.12467473
3791 DATA .217,.21530095,.299,.29456472
3792 DATA .376,.36720285,.450,.43496553
3793 DATA .520,.49688014,.589,.55552980
3794 DATA .656,.60995199,.721,.66013615
3795 DATA .7853981634,0.7071067812
3796 DATA .849,.75062005,.911,.79011709
3797 DATA .972,.82601466
3798 PRINT
3799 PRINT " X ",TAB(10),"COS(X) HANDBOOK",TAB(30),"LAGRANGE DERIVATIVE",
3800 PRINT TAB(54),"OBSERVED ERROR"
3801 PRINT "---",TAB(10),"---------------",TAB(30),"--------------------",
3802 PRINT TAB(54),"--------------"
3803 PRINT
3804 FOR X=0 TO .75 STEP .05
3805 GOSUB 46150
3806 PRINT X,TAB(10),COS(X),TAB(32),Y,TAB(56),INT(100000000*(COS(X)-Y))/100000000
```

```
3807 NEXT X
3808 END
3825 REM PROGRAM TO DEMONSTRATE THE GENERAL INTEGRATION SUBROUTINE
3826 REM EXAMPLE IS THE INTERGRAL OF SIN(X) FROM X1 TO X2
3827 REM V=NUMBER OF TABLE VALUES
3828 V=14
3829 DIM X(V),Y(V),Z(V),M(V+3)
3830 REM INPUT TABLE
3831 FOR I=1 TO V
3832 READ X(I),Y(I)
3833 NEXT I
3834 REM SIN TABLE VALUES FROM
3835 REM HANDBOOK OF MATHEMATICAL FUNCTIONS
3836 REM BY ABRAMOWITZ, M., AND STEGUN, I.A.
3837 REM NBS, JUNE 1964
3838 DATA 0,0,.125,.12467473,.217,.21530095
3839 DATA .299,.29456472,.376,.36720285
3840 DATA .450,.43496553,.520,.49688014
3841 DATA .589,.55552980,.656,.60995199
3842 DATA .721,.66013615,.7853981634,0.7071067812
3843 DATA .849,.75072005
3844 DATA .911,.79011709,.972,.82601466
3845 PRINT "INPUT START POINT, END POINT: ",
3846 INPUT X1,X2
3847 GOSUB 46200
3848 PRINT "INTEGRAL FROM ",X1," TO ",X2," EQUALS ",Z
3849 PRINT "ERROR CHECK: ",Z1
3850 PRINT
3851 PRINT
3852 GOTO 3845
3875 REM PROGRAM TO CALCULATE THE ERROR FUNCTION TABLE POSITIONS
3876 PRINT
3877 PRINT
3878 REM STARTING POSITION IS X=0
3879 X=0
3880 D=0
3881 I=0
3882 REM INITIAL INTERVAL SIZE IS H0=.025
3883 H0=.025
3884 REM TERMINATION POSITION IS E=10
3885 E=20
3886 PRINT
3887 PRINT
3888 PRINT "POSITION NUMBER",TAB(21),"POSITION",TAB(35),"INTERVAL SIZE"
3889 PRINT "----------------",TAB(21),"--------",TAB(35),"-------------"
3890 PRINT
3891 REM START ITERATION
3892 IF INT(I/4)<>I/4 THEN GOTO 3895
3893 PRINT TAB(6),I/4+1,TAB(22),INT(1000*X)/1000,TAB(36),INT(1000*D)/1000
3894 D=0
3895 H=EXP(-X*X/4)*SQRT(SQRT((3-6*X*X+4*X*X*X*X)/3))
3896 H=H0/H
3897 I=I+1
3898 X=X+H
3899 D=D+H
3900 IF X<E THEN GOTO 3892
3901 PRINT
3902 PRINT
3903 END
```

```
3925 REM ROOT TESTING PROGRAM (ROOTTEST)
3926 PRINT
3927 PRINT
3928 PRINT "THIS PROGRAM WILL HELP YOU"
3929 PRINT "DETERMINE WHERE TO LOOK FOR"
3930 PRINT "ROOTS OF A POLYNOMIAL."
3931 PRINT
3932 PRINT "WHAT IS THE DEGREE OF THE POLYNOMIAL: ",
3933 INPUT N
3934 DIM A(N)
3935 PRINT
3936 PRINT "INPUT THE POLYNOMIAL COEFFICIENTS"
3937 PRINT "AS PROMPTED:"
3938 PRINT
3939 FOR I=0 TO N
3940 PRINT "A(",I,") = ",
3941 INPUT A(I)
3942 NEXT I
3943 REM NORMALIZE SO A(N)=1
3944 FOR I=0 TO N
3945 A(I)=A(I)/A(N)
3946 NEXT I
3947 PRINT
3948 PRINT
3949 PRINT "THERE ARE",N," ROOTS."
3950 PRINT
3951 PRINT
3952 REM FIND THE MAXIMUM VALUE OF ROOT
3953 A=A(N-1)*A(N-1)-2*A(N-2)
3954 IF A>0 THEN GOTO 3958
3955 PRINT "THERE ARE AT LEAST TWO COMPLEX"
3956 PRINT "ROOTS. THE ANALYSIS ENDS."
3957 GOTO 3998
3958 A=SQRT(A)
3959 PRINT
3960 PRINT "THE MAGNITUDE OF THE LARGEST ROOT"
3961 PRINT "IS NOT GREATER THAN",A
3962 PRINT
3963 PRINT
3964 A=-1
3965 IF INT(N/2)=N/2 THEN A=-A
3966 A=A(0)/A
3967 REM B WILL FLAG A NEGATIVE ROOT
3968 B=0
3969 IF A>0 THEN GOTO 3972
3970 PRINT "THERE IS AT LEAST ONE NEGATIVE REAL ROOT."
3971 B=1
3972 PRINT
3973 REM TEST FOR DESCARTES RULE NUMBER 1
3974 C=0
3975 FOR I=1 TO N
3976 IF A(I-1)*A(I)<0 THEN C=C+1
3977 NEXT I
3978 IF C=1 THEN PRINT "THERE IS AT MOST ONE POSITIVE ",
3979 IF C>1 THEN PRINT "THERE ARE AT MOST",C," POSITIVE ",
3980 IF C=1 THEN PRINT "REAL ROOT."
3981 IF C>1 THEN PRINT "REAL ROOTS."
3982 PRINT
3983 REM TEST FOR DESCARTES RULE NUMBER 2
```

```
3984 D=0
3985 FOR I=1 TO N
3986 IF A(I-1)*A(I)>0 THEN D=D+1
3987 NEXT I
3988 IF D=1 THEN PRINT "THERE IS AT MOST ONE NEGATIVE ",
3989 IF D>1 THEN PRINT "THERE ARE AT MOST",D," NEGATIVE ",
3990 IF D=1 THEN PRINT "REAL ROOT."
3991 IF D>1 THEN PRINT "REAL ROOTS."
3992 PRINT
3993 IF INT(C/2)=C/2 THEN GOTO 3995
3994 PRINT "THERE IS AT LEAST ONE POSITIVE REAL ROOT."
3995 IF INT(D/2)=D/2 THEN GOTO 3998
3996 IF B=1 THEN GOTO 3998
3997 PRINT "THERE IS AT LEAST ONE NEGATIVE REAL ROOT."
3998 PRINT
3999 PRINT
4000 END
4025 REM PROGRAM TO DEMONSTRATE THE BISECTION SUBROUTINE
4026 PRINT
4027 PRINT
4028 PRINT "WHAT IS THE INITIAL RANGE (X0,X1):"
4029 PRINT
4030 PRINT "X0 = ",
4031 INPUT X0
4032 PRINT "X1 = ",
4033 INPUT X1
4034 PRINT
4035 PRINT "WHAT IS THE CONVERGENCE CRITERION: ",
4036 INPUT E
4037 REM GO TO BISECTION METHOD SUBROUTINE
4038 GOSUB 46350
4039 PRINT
4040 PRINT
4041 PRINT "THE CALCULATED ZERO IS X =",X
4042 PRINT
4043 PRINT "THE ASSOCIATED Y VALUE IS Y = ",Y
4044 PRINT
4045 PRINT "THE NUMBER OF STEPS WAS",M
4046 PRINT
4047 PRINT
4048 END
4050 REM PROGRAM TO DEMONSTRATE NEWTON'S METHOD
4051 PRINT
4052 PRINT
4053 PRINT "INPUT THE INITIAL GUESS: ",
4054 INPUT X0
4055 PRINT
4056 PRINT "INPUT THE CONVERGENCE FACTOR: ",
4057 INPUT E
4058 PRINT
4059 PRINT "MAXIMUM NUMBER OF ITERATIONS: ",
4060 INPUT M
4061 PRINT
4062 GOSUB 46375
4063 PRINT "THE CALCULATED ROOT IS X =",X
4064 PRINT
4065 PRINT "NUMBER OF ITERATIONS: ",N
4066 PRINT
4067 PRINT
```

```
4068 END
4075 REM PROGRAM TO DEMONSTRATE THE SECANT SUBROUTINE
4076 PRINT
4077 PRINT
4078 PRINT "INPUT THE TWO INITIAL GUESSES:"
4079 PRINT
4080 PRINT "X0 = ",
4081 INPUT X0
4082 PRINT "X1 = ",
4083 INPUT X1
4084 PRINT
4085 PRINT "INPUT THE CONVERGENCE FACTOR: ",
4086 INPUT E
4087 PRINT
4088 PRINT "MAXIMUM NUMBER OF ITERATIONS: ",
4089 INPUT M
4090 PRINT
4091 GOSUB 46425
4092 PRINT "THE CALCULATED ROOT IS X =",X
4093 PRINT
4094 PRINT "NUMBER OF ITERATIONS: ",N
4095 PRINT
4096 PRINT
4097 END
4100 REM PROGRAM TO DEMONSTRATE THE MODIFIED REGULA FALSI SUBROUTINE
4101 PRINT
4102 PRINT
4103 PRINT "INPUT THE TWO BRACKETTING GUESSES:"
4104 PRINT
4105 PRINT "X0 = ",
4106 INPUT X0
4107 PRINT "X1 = ",
4108 INPUT X1
4109 PRINT
4110 PRINT "INPUT THE CONVERGENCE FACTOR: ",
4111 INPUT E
4112 PRINT
4113 PRINT "MAXIMUM NUMBER OF ITERATIONS: ",
4114 INPUT M
4115 PRINT
4116 GOSUB 46475
4117 PRINT
4118 PRINT "THE CALCULATED ROOT IS X =",X
4119 PRINT
4120 PRINT "NUMBER OF ITERATIONS: ",N
4121 PRINT
4122 PRINT
4123 END
4125 REM PROGRAM TO DEMONSTRATE THE AITKEN ACCELERATION SUBROUTINE
4126 PRINT
4127 PRINT
4128 PRINT "INPUT THE INITIAL GUESS: ",
4129 INPUT X0
4130 PRINT
4131 PRINT "INPUT THE CONVERGENCE CRITERION: ",
4132 INPUT E
4133 PRINT
4134 PRINT "INPUT THE CONVERGENCE FACTOR: ",
4135 INPUT C
```

```
4136 PRINT
4137 PRINT "MAXIMUM NUMBER OF ITERATIONS: ",
4138 INPUT M
4139 PRINT
4140 GOSUB 46525
4141 PRINT "THE CALCULATED ROOT IS X =",X
4142 PRINT
4143 PRINT "NUMBER OF ITERATIONS: ",N
4144 PRINT
4145 PRINT
4146 END
4150 REM PROGRAM TO DEMONSTRATE AITKEN STEFFENSON ITERATION
4151 PRINT
4152 PRINT
4153 PRINT "INPUT THE INITIAL GUESS: ",
4154 INPUT X0
4155 PRINT
4156 PRINT "INPUT THE CONVERGENCE CRITERION: ",
4157 INPUT E
4158 PRINT
4159 PRINT "INPUT THE CONVERGENCE FACTOR: ",
4160 INPUT C
4161 PRINT
4162 PRINT "MAXIMUM NUMBER OF ITERATIONS: ",
4163 INPUT M
4164 PRINT
4165 GOSUB 46575
4166 PRINT "THE CALCULATED ROOT IS X =",X
4167 PRINT
4168 PRINT "NUMBER OF ITERATIONS: ",N
4169 PRINT
4170 PRINT
4171 END
4175 REM PROGRAM TO DEMONSTRATE SYNTHETIC DIVISION (RSYNDIV)
4176 PRINT
4177 PRINT "WHAT IS THE DEGREE OF THE POLYNOMIAL"
4178 PRINT "TO BE DIVIDED INTO: ",
4179 INPUT N1
4180 DIM C(N1)
4181 PRINT
4182 PRINT "INPUT THE POLYNOMIAL COEFFICIENTS AS PROMPTED:"
4183 PRINT
4184 FOR I=0 TO N1
4185 PRINT "C(",I,") = ",
4186 INPUT C(I)
4187 NEXT I
4188 PRINT
4189 PRINT "WHAT IS THE ORDER OF THE POLYNOMIAL "
4190 PRINT "BE DIVIDED BY: ",
4191 INPUT N2
4192 DIM A(N1-N2),B(N2)
4193 PRINT
4194 PRINT "INPUT THE POLYNOMIAL COEFFICIENTS AS PROMPTED:"
4195 PRINT
4196 FOR I=0 TO N2
4197 PRINT "B(",I,") = ",
4198 INPUT B(I)
4199 NEXT I
4200 GOSUB 46625
```

```
4201 PRINT
4202 PRINT
4203 PRINT "THE COEFFICIENTS OF THE RESULTING POLYNOMIAL ARE:"
4204 PRINT
4205 FOR I=0 TO N1-N2
4206 PRINT "A(",I,") = ",A(I)
4207 NEXT I
4208 PRINT
4209 PRINT
4210 END
4225 REM PROGRAM TO DEMONSTRATE NEXTROOT SUBROUTINE
4226 PRINT
4227 PRINT
4228 PRINT "HOW MANY ROOTS HAVE BEEN DETERMINED: ",
4229 INPUT L
4230 PRINT
4231 PRINT
4232 PRINT "INPUT THE ROOTS AS PROMPTED:"
4233 PRINT
4234 FOR I=1 TO L
4235 PRINT "A(",I,") = ",
4236 INPUT A(I)
4237 NEXT I
4238 PRINT
4239 PRINT "WHAT IS THE INITIAL GUESS: ",
4240 INPUT X0
4241 PRINT
4242 PRINT "WHAT IS THE CONVERGENCE CRITERION: ",
4243 INPUT E
4244 PRINT
4245 PRINT "MAXIMUM NUMBER OF ITERATIONS: ",
4246 INPUT M
4247 PRINT
4248 GOSUB 46650
4249 PRINT
4250 PRINT "THE CALCULATED ROOT IS X =",X
4251 PRINT
4252 PRINT "NUMBER OF ITERATIONS: ",N
4253 PRINT
4254 PRINT
4255 END
4275 REM PROGRAM TO DEMONSTRATE THE COMPLEX ROOT COUNTING SUBROUTINE
4276 PRINT
4277 PRINT
4278 PRINT "WHERE IS THE CENTER OF THE SEARCH CIRCLE (X0,Y0):"
4279 PRINT
4280 PRINT "     X0 = ",
4281 INPUT X0
4282 PRINT "     Y0 = ",
4283 INPUT Y0
4284 PRINT
4285 PRINT "WHAT IS THE RADIUS OF THIS CIRCLE: ",
4286 INPUT W
4287 PRINT "HOW MANY EVALUATION POINTS PER QUADRANT: ",
4288 INPUT M
4289 DIM N(4*M)
4290 GOSUB 46700
4291 PRINT
4292 PRINT
```

```
4293 PRINT "NUMBER OF COMPLETE CYCLES FOUND:",
4294 PRINT N
4295 PRINT "RESIDUAL: ",
4296 PRINT A
4297 PRINT
4298 PRINT
4299 END
4300 REM PROGRAM TO DEMONSTRATE THE ZERO SEARCHING ALGORITHM
4301 PRINT
4302 PRINT
4303 PRINT "WHAT IS THE INITIAL GUESS FOR X AND Y: "
4304 PRINT "     X0 = ",
4305 INPUT X0
4306 PRINT "     Y0 = ",
4307 INPUT Y0
4308 PRINT
4309 PRINT "WHAT IS THE RADIUS OF THE FIRST SEARCH CIRCLE: ",
4310 INPUT W
4311 PRINT "BY WHAT FRACTION IS THIS CIRCLE TO BE REDUCED ON EACH ITERATION: ",
4312 INPUT E
4313 PRINT "HOW MANY POINTS ARE TO BE SEARCHED PER QUADRANT: ",
4314 INPUT M
4315 PRINT "HOW MANY ITERATIONS ARE TO BE EMPLOYED: ",
4316 INPUT N
4317 DIM X(4*M),Y(4*M),U(4*M),V(4*M)
4318 GOSUB 46760
4319 PRINT
4320 PRINT
4321 PRINT "THE APPROXIMATE SOLUTION IS Z = ",X,
4322 IF Y>=0 THEN PRINT " +",
4323 IF Y<0 THEN PRINT " ",
4324 PRINT Y," I"
4325 PRINT
4326 PRINT "NUMBER OF ITERATIONS: ",K
4327 PRINT
4328 END
4350 REM PROGRAM TO DEMONSTRATE THE NEWTON ROOT
4351 REM PROGRAM IN THE COMPLEX DOMAIN
4352 PRINT
4353 PRINT
4354 PRINT "WHAT IS THE INITIAL GUESS:"
4355 PRINT "     X0 = ",
4356 INPUT X0
4357 PRINT "     Y0 = ",
4358 INPUT Y0
4359 PRINT "WHAT IS THE CONVERGENCE CRITERION: ",
4360 INPUT E
4361 PRINT "WHAT IS THE LIMIT ON THE NUMBER OF ITERATIONS: ",
4362 INPUT N
4363 GOSUB 46925
4364 PRINT
4365 PRINT
4366 PRINT "THE ROOT ESTIMATE IS:"
4367 PRINT "     X0 = ",X
4368 PRINT "     Y0 = ",Y
4369 PRINT
4370 PRINT "THE NUMBER OF ITERATIONS PERFORMED WAS: ",
4371 PRINT K
4372 PRINT
```

```
4373 PRINT
4374 END
4375 REM PROGRAM TO DEMONSTRATE MUELLER'S METHOD
4376 PRINT
4377 PRINT
4378 PRINT "WHAT IS THE INITIAL GUESS: ",
4379 INPUT X0
4380 PRINT "WHAT IS THE BOUND ON THIS GUESS: ",
4381 INPUT D
4382 PRINT "WHAT IS THE ERROR CRITERION: ",
4383 INPUT E
4384 PRINT "HOW MANY INTERATIONS (MAXIMUM): ",
4385 INPUT N
4386 PRINT
4387 GOSUB 46960
4388 PRINT
4389 PRINT "THE ESTIMATED ROOT IS: ",X
4390 PRINT "THE NUMBER OF ITERATIONS PERFORMED WAS: ",K
4391 PRINT
4392 END
4400 REM PROGRAM TO DEMONSTRATE A TWO DIMENSIONAL VERSION OF MUELLER'S METHOD
4401 PRINT
4402 PRINT
4403 PRINT "WHAT ARE THE INITIAL GUESSES AND THEIR BOUNDS:"
4404 PRINT "     X0 = ",
4405 INPUT X0
4406 PRINT "     BOUND ON X0 = ",
4407 INPUT B1
4408 PRINT "     Y0 = ",
4409 INPUT Y0
4410 PRINT "     BOUND ON Y0 = ",
4411 INPUT B2
4412 PRINT "WHAT IS THE ERROR CRITERION: ",
4413 INPUT E
4414 PRINT "HOW MANY INTERATIONS (MAXIMUM): ",
4415 INPUT N
4416 PRINT
4417 GOSUB 47050
4418 PRINT
4419 PRINT "THE ESTIMATED ROOT IS (X,Y) = (",X,",",Y,")"
4420 PRINT "THE NUMBER OF ITERATIONS PERFORMED WAS: ",K
4421 PRINT
4422 END
4425 REM PROGRAM TO DEMONSTRATE THE COMPLEX DOMAIN MUELLER SUBROUTINE
4426 PRINT
4427 PRINT
4428 PRINT "WHAT IS THE INITIAL GUESS:"
4429 PRINT
4430 PRINT "     X0 = ",
4431 INPUT X0
4432 PRINT "WHAT IS THE ASSOCIATED BOUND: ",
4433 INPUT B1
4434 PRINT "     Y0 = ",
4435 INPUT Y0
4436 PRINT "WHAT IS THE ASSOCIATED BOUND: ",
4437 INPUT B2
4438 PRINT
4439 PRINT "INPUT THE CONVERGENCE CRITERION: ",
4440 INPUT E
```

```
4441 PRINT "WHAT IS THE LIMIT ON THE NUMBER OF ITERATIONS: ",
4442 INPUT N
4443 GOSUB 47150
4444 PRINT
4445 PRINT
4446 PRINT "THE ESTIMATED ROOT IS:"
4447 PRINT
4448 PRINT "     X = ",X
4449 PRINT "     Y = ",Y
4450 PRINT
4451 PRINT "THE NUMBER OF ITERATIONS WAS ",K
4452 PRINT
4453 PRINT
4454 END
4475 REM PROGRAM TO DEMONSTRATE THE COMPLEX DOMAIN ALLROOT SUBROUTINE
4476 PRINT
4477 PRINT
4478 PRINT "WHAT IS THE INITIAL GUESS:"
4479 PRINT
4480 PRINT "     X0 = ",
4481 INPUT X0
4482 PRINT "WHAT IS THE ASSOCIATED BOUND: ",
4483 INPUT B1
4484 PRINT "     Y0 = ",
4485 INPUT Y0
4486 PRINT "WHAT IS THE ASSOCIATED BOUND: ",
4487 INPUT B2
4488 PRINT
4489 PRINT "INPUT THE CONVERGENCE CRITERION: ",
4490 INPUT E
4491 PRINT "WHAT IS THE LIMIT ON THE NUMBER OF ITERATIONS: ",
4492 INPUT N
4493 PRINT "HOW MANY ROOTS ARE TO BE SOUGHT: ",
4494 INPUT N2
4495 PRINT "IS THE FUNCTION IN THE FUNCTIONS SUBROUTINE (1)"
4496 PRINT "OR IS IT A SERIES (2): ",
4497 INPUT N3
4498 GOSUB 47300
4499 PRINT
4500 PRINT
4501 PRINT "THE ESTIMATED ROOTS ARE:"
4502 FOR I=1 TO N2
4503 PRINT
4504 PRINT "     X = ",X(I)
4505 PRINT "     Y = ",Y(I)
4506 PRINT
4507 NEXT I
4508 PRINT
4509 PRINT "THE LAST NUMBER OF ITERATIONS WAS ",K
4510 PRINT
4511 PRINT
4512 END
4525 REM PROGRAM TO DEMONSTRATE QUADRATIC ROOT SUBROUTINE
4526 PRINT
4527 PRINT "INPUT THE COEFFICIENTS:"
4528 PRINT
4529 PRINT "A(0) = ",
4530 INPUT A(0)
4531 PRINT "A(1) = ",
```

```
4532 INPUT A(1)
4533 PRINT "A(2) = ",
4534 INPUT A(2)
4535 GOSUB 47475
4536 PRINT
4537 PRINT
4538 PRINT "RESULTS:"
4539 PRINT
4540 PRINT "     R1 = ",X1,
4541 IF Y1>=0 THEN PRINT " +",
4542 IF Y1<0 THEN PRINT " ",
4543 PRINT Y1," I"
4544 PRINT "     R2 = ",X2,
4545 IF Y2>=0 THEN PRINT " +",
4546 IF Y2<0 THEN PRINT " ",
4547 PRINT Y2," I"
4548 PRINT
4549 PRINT
4550 END
4560 REM PROGRAM TO DEMONSTRATE THE LIN SUBROUTINE
4561 PRINT
4562 PRINT
4563 PRINT "INPUT ORDER OF POLYNOMIAL",
4564 INPUT M
4565 PRINT
4566 PRINT "INPUT THE POLYNOMIAL COEFFICIENTS:"
4567 FOR I=0 TO M
4568 PRINT "    A(",I,") = ",
4569 INPUT A(I)
4570 NEXT I
4571 PRINT
4572 PRINT "WHAT IS THE CONVERGENCE FACTOR: ",
4573 INPUT E
4574 PRINT "WHAT IS THE MAXIMUM NUMBER OF ITERATIONS: ",
4575 INPUT N
4576 A=3.14159
4577 B=SQRT(2)
4578 GOSUB 47525
4579 PRINT
4580 PRINT "THE ROOTS FOUND ARE:"
4581 PRINT
4582 PRINT "    X1 = ",X1
4583 PRINT "    Y1 = ",Y1
4584 PRINT
4585 PRINT "    X2 = ",X2
4586 PRINT "    Y2 = ",Y2
4587 PRINT
4588 PRINT "THE NUMBER OF ITERATIONS WAS ",K
4589 PRINT
4590 PRINT
4591 END
4600 REM PROGRAM TO DEMONSTRATE THE BAIRSTOW SUBROUTINE
4601 PRINT
4602 PRINT
4603 PRINT "INPUT ORDER OF POLYNOMIAL",
4604 INPUT M
4605 PRINT
4606 PRINT "INPUT THE POLYNOMIAL COEFFICIENTS:"
4607 FOR I=0 TO M
```

```
4608 PRINT "     A(",I,") = ",
4609 INPUT A(I)
4610 NEXT I
4611 PRINT
4612 PRINT "WHAT IS THE CONVERGENCE FACTOR: ",
4613 INPUT E
4614 PRINT "WHAT IS THE MAXIMUM NUMBER OF ITERATIONS: ",
4615 INPUT N
4616 A=3.14159
4617 B=SQRT(2)
4618 GOSUB 47600
4619 PRINT
4620 PRINT "THE ROOTS FOUND ARE:"
4621 PRINT
4622 PRINT "     X1 = ",X1
4623 PRINT "     Y1 = ",Y1
4624 PRINT
4625 PRINT "     X2 = ",X2
4626 PRINT "     Y2 = ",Y2
4627 PRINT
4628 PRINT "THE NUMBER OF ITERATIONS WAS ",K
4629 PRINT
4630 PRINT
4631 END
4650 REM PROGRAM TO DEMONSTRATE THE USE OF MULTI-DIMENSIONAL
4651 REM STEEPEST DESCENT
4652 PRINT
4653 PRINT
4654 PRINT "HOW MANY DIMENSIONS ARE THERE: ",
4655 INPUT L
4656 DIM X(L),X1(L)
4657 PRINT
4658 PRINT "WHAT IS THE CONVERGENCE CRITERION: ",
4659 INPUT E
4660 PRINT
4661 PRINT "MAXIMUM NUMBER OF ITERATIONS: ",
4662 INPUT M
4663 PRINT
4664 PRINT "WHAT IS THE STARTING CONSTANT, K: ",
4665 INPUT K
4666 PRINT
4667 PRINT "INPUT THE STARTING POINTS:"
4668 PRINT
4669 FOR I=1 TO L
4670 PRINT "X(",I,") = ",
4671 INPUT X(I)
4672 NEXT I
4673 PRINT
4674 PRINT
4675 GOSUB 47700
4676 PRINT "THE RESULTS ARE:"
4677 PRINT
4678 FOR I=1 TO L
4679 PRINT "X(",I,") = ",X(I)
4680 NEXT I
4681 PRINT
4682 PRINT "THE NUMBER OF ITERATIONS WAS",N
4683 PRINT
4684 PRINT
```

```
4685 END
4700 REM PROGRAM TO DEMONSTRATE THE USE OF MULTI-DIMENSIONAL
4701 REM STEEPEST DESCENT
4702 PRINT
4703 PRINT
4704 PRINT "HOW MANY DIMENSIONS ARE THERE: ",
4705 INPUT L
4706 DIM X(L),X1(L)
4707 PRINT
4708 PRINT "WHAT IS THE CONVERGENCE CRITERION: ",
4709 INPUT E
4710 PRINT
4711 PRINT "MAXIMUM NUMBER OF ITERATIONS: ",
4712 INPUT M
4713 PRINT
4714 PRINT "WHAT IS THE STARTING CONSTANT, K: ",
4715 INPUT K
4716 PRINT
4717 PRINT "INPUT THE STARTING POINTS:"
4718 PRINT
4719 FOR I=1 TO L
4720 PRINT "X(",I,") = ",
4721 INPUT X(I)
4722 NEXT I
4723 PRINT
4724 PRINT
4725 GOSUB 47800
4726 PRINT "THE RESULTS ARE:"
4727 PRINT
4728 FOR I=1 TO L
4729 PRINT "X(",I,") = ",X(I)
4730 NEXT I
4731 PRINT
4732 PRINT "THE NUMBER OF ITERATIONS WAS",N
4733 PRINT
4734 PRINT
4735 END
```

Subroutine Programs

```
40398 REM *********************
40399 REM RECTANGULAR TO POLAR CONVERSION SUBROUTINE (RECT/POL)
40400 U=SQRT(X*X+Y*Y)
40401 REM GUARD AGAINST AMBIGUOUS VECTOR
40402 IF Y=0 THEN Y=(.1)^30
40403 REM GUARD AGAINST DIVIDE BY ZERO
40404 IF X=0 THEN X=(.1)^30
40405 REM SOME BASICS REQUIRE A SIMPLE ARGUMENT
40406 W=Y/X
40407 V=ATN(W)
40408 REM CHECK QUADRANT AND ADJUST
40409 IF X<0 THEN V=V+3.1415926535
40410 IF V<0 THEN V=V+6.2831853072
40411 RETURN
40449 REM POLAR TO RECTANGULAR CONVERSION SUBROUTINE (POL/RECT)
40450 X=U*COS(V)
40451 Y=U*SIN(V)
```

```
40452 RETURN
41099 REM POLAR POWER SUBROUTINE (ZPOLPOW)
41100 U1=U^N
41101 V1=N*V
41102 V1=V1-6.2831853072*INT(V1/6.2831853072)
41103 RETURN
41198 REM RECTANGULAR COMPLEX NUMBER POWER SUBROUTINE (ZRECTPOW)
41199 REM RECTANGULAR TO POLAR CONVERSION
41200 GOSUB 40400
41201 REM POLAR POWER
41202 GOSUB 41100
41203 REM CHANGE VARIABLE FOR CONVERSION
41204 U=U1
41205 V=V1
41206 REM POLAR TO RECTANGULAR CONVERSION
41207 GOSUB 40450
41208 RETURN
41897 REM ********************
41898 REM MATRIX MULTIPLICATION SUBROUTINE (MATMULT)
41899 REM C=A X B    A IS M1 BY N1    B IS M2 BY N2    C IS M1 BY N2
41900 FOR I=1 TO M1
41901 FOR J=1 TO N2
41902 C(I,J)=0
41903 FOR K=1 TO N1
41904 C(I,J)=C(I,J)+A(I,K)*B(K,J)
41905 NEXT K
41906 NEXT J
41907 NEXT I
41908 RETURN
41947 REM ********************
41948 REM MATRIX TRANSPOSE SUBROUTINE (MATTRANS)
41949 REM B=TRANSPOSE(A)
41950 FOR I=1 TO N
41951 FOR J=1 TO M
41952 B(I,J)=A(J,I)
41953 NEXT J
41954 NEXT I
41955 RETURN
42072 REM ********************
42073 REM MATRIX SAVE (B IN A) SUBROUTINE (MATSAVBA)
42074 REM N1,N2 AND N3 ARE INPUT INDICES
42075 IF N1*N2*N3=0 THEN GOTO 42085
42076 REM CHECK DIMENSION
42077 FOR I1=1 TO N1
42078 FOR I2=1 TO N2
42079 FOR I3=1 TO N3
42080 A(I1,I2,I3)=B(I1,I2,I3)
42081 NEXT I3
42082 NEXT I2
42083 NEXT I1
42084 RETURN
42085 IF N1*N2=0 THEN GOTO 42092
42086 FOR I1=1 TO N1
42087 FOR I2=1 TO N2
42088 A(I1,I2)=B(I1,I2)
42089 NEXT I2
42090 NEXT I1
42091 RETURN
42092 IF N1=0 THEN RETURN
```

```
42093 FOR I1=1 TO N1
42094 A(I1)=B(I1)
42095 NEXT I1
42096 RETURN
42097 REM *********************
42098 REM MATRIX SAVE (C IN B) SUBROUTINE (MATSAVCB)
42099 REM N1,N2 AND N3 ARE INPUT INDICES
42100 IF N1*N2*N3=0 THEN GOTO 42110
42101 REM CHECK DIMENSION
42102 FOR I1=1 TO N1
42103 FOR I2=1 TO N2
42104 FOR I3=1 TO N3
42105 B(I1,I2,I3)=C(I1,I2,I3)
42106 NEXT I3
42107 NEXT I2
42108 NEXT I1
42109 RETURN
42110 IF N1*N2=0 THEN GOTO 42117
42111 FOR I1=1 TO N1
42112 FOR I2=1 TO N2
42113 B(I1,I2)=C(I1,I2)
42114 NEXT I2
42115 NEXT I1
42116 RETURN
42117 IF N1=0 THEN RETURN
42118 FOR I1=1 TO N1
42119 B(I1)=C(I1)
42120 NEXT I1
42121 RETURN
42147 REM ********************
42148 REM MATRIX SAVE (A IN C) SUBROUTINE (MATSAVAC)
42149 REM N1,N2 AND N3 ARE INPUT INDICES
42150 IF N1*N2*N3=0 THEN GOTO 42160
42151 REM CHECK DIMENSION
42152 FOR I1=1 TO N1
42153 FOR I2=1 TO N2
42154 FOR I3=1 TO N3
42155 C(I1,I2,I3)=A(I1,I2,I3)
42156 NEXT I3
42157 NEXT I2
42158 NEXT I1
42159 RETURN
42160 IF N1*N2=0 THEN GOTO 42167
42161 FOR I1=1 TO N1
42162 FOR I2=1 TO N2
42163 C(I1,I2)=A(I1,I2)
42164 NEXT I2
42165 NEXT I1
42166 RETURN
42167 IF N1=0 THEN RETURN
42168 FOR I1=1 TO N1
42169 C(I1)=A(I1)
42170 NEXT I1
42171 RETURN
42172 REM ********************
42173 REM MATRIX SAVE (C IN A) SUBROUTINE (MATSAVCA)
42174 REM N1,N2 AND N3 ARE INPUT INDICES
42175 IF N1*N2*N3=0 THEN GOTO 42185
42176 REM CHECK DIMENSION
```

```
42177 FOR I1=1 TO N1
42178 FOR I2=1 TO N2
42179 FOR I3=1 TO N3
42180 A(I1,I2,I3)=C(I1,I2,I3)
42181 NEXT I3
42182 NEXT I2
42183 NEXT I1
42184 RETURN
42185 IF N1*N2=0 THEN GOTO 42192
42186 FOR I1=1 TO N1
42187 FOR I2=1 TO N2
42188 A(I1,I2)=C(I1,I2)
42189 NEXT I2
42190 NEXT I1
42191 RETURN
42192 IF N1=0 THEN RETURN
42193 FOR I1=1 TO N1
42194 A(I1)=C(I1)
42195 NEXT I1
42196 RETURN
42394 REM ********************
42395 REM MATRIX INVERSION SUBROUTINE (MATINV)
42396 REM GAUSS-JORDAN ELIMINATION
42397 REM MATRIX A IS INPUT, MATRIX B IS OUTPUT
42398 REM DIM A=N X N      TEMPORARY DIM B=N X 2N
42399 REM FIRST CREATE MATRIX WITH A ON THE LEFT AND I ON THE RIGHT
42400 FOR I=1 TO N
42401 FOR J=1 TO N
42402 B(I,J+N)=0
42403 B(I,J)=A(I,J)
42404 NEXT J
42405 B(I,I+N)=1
42406 NEXT I
42407 REM PERFORM ROW ORIENTED OPERATIONS TO CONVERT THE LEFT HAND
42408 REM SIDE OF B TO THE IDENTITY MATRIX. THE INVERSE OF A WILL
42409 REM THEN BE ON THE RIGHT.
42410 FOR K=1 TO N
42411 IF K=N THEN GOTO 42424
42412 M=K
42413 REM FIND MAXIMUM ELEMENT
42414 FOR I=K+1 TO N
42415 IF ABS(B(I,K))>ABS(B(M,K)) THEN M=I
42416 NEXT I
42417 IF M=K THEN GOTO 42424
42418 FOR J=K TO 2*N
42419 B=B(K,J)
42420 B(K,J)=B(M,J)
42421 B(M,J)=B
42422 NEXT J
42423 REM DIVIDE ROW K
42424 FOR J=K+1 TO 2*N
42425 B(K,J)=B(K,J)/B(K,K)
42426 NEXT J
42427 IF K=1 THEN GOTO 42434
42428 FOR I=1 TO K-1
42429 FOR J=K+1 TO 2*N
42430 B(I,J)=B(I,J)-B(I,K)*B(K,J)
42431 NEXT J
42432 NEXT I
```

```
42433 IF K=N THEN GOTO 42441
42434 FOR I=K+1 TO N
42435 FOR J=K+1 TO 2*N
42436 B(I,J)=B(I,J)-B(I,K)*B(K,J)
42437 NEXT J
42438 NEXT I
42439 NEXT K
42440 REM RETRIEVE INVERSE FROM THE RIGHT SIDE OF B
42441 FOR I=1 TO N
42442 FOR J=1 TO N
42443 B(I,J)=B(I,J+N)
42444 NEXT J
42445 NEXT I
42446 RETURN
43491 REM *********************
43492 REM LINEAR LEAST SQUARES SUBROUTINE (LSTSQR1)
43493 REM THE INPUT DATA SET IS (X(M),Y(M)).
43494 REM THE NUMBER OF DATA POINTS IS N.
43495 REM THE NUMBER OF DIFFERENT POINTS MUST BE GREATER THAN ONE.
43496 REM X(M) AND Y(M) MUST BE DIMENSIONED IN THE CALLING PROGRAM.
43497 REM THE SUBROUTINE ALSO CALCULATES THE UNBIASED ESTIMATE
43498 REM OF THE STANDARD DEVIATION, D.
43499 REM THE RETURNED PARAMETERS ARE A,B AND D.
43500 A1=0
43501 A2=0
43502 B0=0
43503 B1=0
43504 FOR M=0 TO N-1
43505 A1=A1+X(M)
43506 A2=A2+X(M)*X(M)
43507 B0=B0+Y(M)
43508 B1=B1+Y(M)*X(M)
43509 NEXT M
43510 A1=A1/N
43511 A2=A2/N
43512 B0=B0/N
43513 B1=B1/N
43514 D=A1*A1-A2
43515 A=A1*B1-A2*B0
43516 A=A/D
43517 B=A1*B0-B1
43518 B=B/D
43519 REM *********************
43520 REM EVALUATION OF STANDARD DEVIATION (UNBIASED ESTIMATE)
43521 D=0
43522 FOR M=0 TO N-1
43523 D1=Y(M)-A-B*X(M)
43524 D=D+D1*D1
43525 NEXT M
43526 D=SQRT(D/(N-2))
43527 RETURN
43541 REM *********************
43542 REM PARABOLIC LEAST SQUARES SUBROUTINE (LSTSQR2)
43543 REM THE INPUT DATA SET IS (X(M),Y(M)).
43544 REM THE NUMBER OF DATA POINTS IS N.
43545 REM THE NUMBER OF DIFFERENT POINTS MUST BE GREATER THAN THREE.
43546 REM X(M) AND Y(M) MUST BE DIMENSIONED IN THE CALLING PROGRAM.
43547 REM THE SUBROUTINE ALSO CALCULATES THE UNBIASED ESTIMATE
43548 REM OF THE STANDARD DEVIATION, D.
```

```
43549 REM THE RETURNED PARAMETERS ARE A,B,C AND D.
43550 A0=1
43551 A1=0
43552 A2=0
43553 A3=0
43554 A4=0
43555 B0=0
43556 B1=0
43557 B2=0
43558 FOR M=0 TO N-1
43559 A1=A1+X(M)
43560 A2=A2+X(M)*X(M)
43561 A3=A3+X(M)*X(M)*X(M)
43562 A4=A4+X(M)*X(M)*X(M)*X(M)
43563 B0=B0+Y(M)
43564 B1=B1+Y(M)*X(M)
43565 B2=B2+Y(M)*X(M)*X(M)
43566 NEXT M
43567 A1=A1/N
43568 A2=A2/N
43569 A3=A3/N
43570 A4=A4/N
43571 B0=B0/N
43572 B1=B1/N
43573 B2=B2/N
43574 D=A0*(A2*A4-A3*A3)-A1*(A1*A4-A3*A2)+A2*(A1*A3-A2*A2)
43575 A=B0*(A2*A4-A3*A3)+B1*(A3*A2-A1*A4)+B2*(A1*A3-A2*A2)
43576 A=A/D
43577 B=B0*(A3*A2-A1*A4)+B1*(A0*A4-A2*A2)+B2*(A2*A1-A0*A3)
43578 B=B/D
43579 C=B0*(A1*A3-A2*A2)+B1*(A1*A2-A0*A3)+B2*(A0*A2-A1*A1)
43580 C=C/D
43581 REM ********************
43582 REM EVALUATION OF STANDARD DEVIATION (UNBIASED ESTIMATE)
43583 D=0
43584 FOR M=0 TO N-1
43585 D1=Y(M)-A-B*X(M)-C*X(M)*X(M)
43586 D=D+D1*D1
43587 NEXT M
43588 D=SQRT(D/(N-3))
43589 RETURN
43590 REM ********************
43591 REM COEFFICIENT MATRIX GENERATION SUBROUTINE FOR THE
43592 REM ONE DIMENSIONAL POLYNOMIAL REGRESSION (POLYCM).
43593 REM THE INPUT DATA SET CONSISTS OF N PAIRS OF
43594 REM (X(I),Y(I)) VALUES.
43595 REM THE REGRESSION ORDER IS N.
43596 REM THE MATRIX RETURNED, Z, IS M ROWS BY N+1 COLUMNS.
43597 REM DIMENSION THIS MATRIX IN THE CALLING PROGRAM.
43598 REM RECALL THAT THE REGRESSION ROUTINE INPUT WILL USE N+1,,
43599 REM YOU MUST SET N TO N+1 BEFORE ENTERING IT.
43600 FOR I=1 TO M
43601 B=1
43602 FOR J=1 TO N+1
43603 Z(I,J)=B
43604 B=B*X(I)
43605 NEXT J
43606 NEXT I
43607 REM Z IS M ROWS BY N+1 COLUMNS
```

```
43608 RETURN
43628 REM ***********************
43629 REM LEAST SQUARES FITTING SUBROUTINE (LEASTSQR)
43630 REM GENERAL SUBROUTINE FOR MULTIDIMENSIONAL,
43631 REM NONLINEAR REGRESSION.
43632 REM THE EQUATION FITTED HAS THE FORM
43633 REM Y=D(1)X1 + D(2)X2 +  ... + D(N)XN
43634 REM CHANGE IN NOTATION IS DUE TO A DIMENSION CONFLICT.
43635 REM THE COEFFICIENTS ARE RETURNED BY THE PROGRAM IN D(I).
43636 REM THE XI CAN BE SIMPLE POWERS OF X, OR FUNCTIONS.
43637 REM NOTE THAT THE XI ARE ASSUMED TO BE INDEPENDENT.
43638 REM THE MEASURED RESPONSES ARE Y(I)- THERE ARE M OF THEM.
43639 REM Y IS M ROW COLUMN VECTOR, AND Z(I,J) IS A
43640 REM M ROW BY N COLUMN MATRIX.
43641 REM M>=N
43642 REM THE SUBROUTINE INPUTS ARE Y(I), Z(I,J), M AND N.
43643 REM THE WORKING MATRICES WITHIN THE PROGRAM ARE A(I,J),
43644 REM B(I,J) AND C(I,J).
43645 REM THE SUBROUTINE CALLS SEVERAL OTHER MATRIX OPERATION
43646 REM ROUTINES TO PERFORM THE CALCULATION.
43647 REM DIMENSION A,B,C,Y AND Z IN THE CALLING PROGRAM.
43648 REM START PROCEDURE.
43649 REM STORE M AND N
43650 M4=M
43651 N4=N
43652 REM MOVE Z(I,J) TO A(I,J)
43653 FOR I=1 TO M
43654 FOR J=1 TO N
43655 A(I,J)=Z(I,J)
43656 NEXT J
43657 NEXT I
43658 REM A IS M BY N
43659 REM FIND TRANSPOSE OF A AND PUT RESULT IN B
43660 GOSUB 41950
43661 REM B IS N BY M
43662 REM MOVE A TO C, B TO A, AND C TO B
43663 REM WILL HAVE Z(TRANSPOSE) IN A, Z IN B
43664 N1=M
43665 N2=N
43666 N3=0
43667 GOSUB 42150
43668 N1=N
43669 N2=M
43670 GOSUB 42075
43671 N1=M
43672 N2=N
43673 GOSUB 42100
43674 REM MULTIPLY A AND B
43675 REM A IS N BY M
43676 REM B IS M BY N
43677 M1=N
43678 N1=M
43679 M2=M
43680 GOSUB 41900
43681 REM RESULT IS IN C, AN N BY N MATRIX.. MOVE TO A AND FIND INVERSE
43682 N1=N
43683 GOSUB 42175
43684 REM A IS N BY N
43685 GOSUB 42400
```

```
43686 REM RESTORE M
43687 M=M4
43688 REM INVERSE IS IN B. MOVE TO A, PUT Z(TRANSPOSE) IN B, AND MULTIPLY
43689 REM B IS N BY N
43690 GOSUB 42075
43691 FOR I=1 TO M
43692 FOR J=1 TO N
43693 B(J,I)=Z(I,J)
43694 NEXT J
43695 NEXT I
43696 REM B IS N BY M    A IS N BY N
43697 M2=N
43698 N2=M
43699 GOSUB 41900
43700 REM PRODUCT IS N BY M
43701 REM PRODUCT IS IN C. MOVE TO A. LOAD Y IN B, AND MULTIPLY
43702 N1=N
43703 N2=M
43704 GOSUB 42175
43705 REM B IS A COLUMN VECTOR (M BY 1)
43706 FOR I=1 TO M
43707 B(I,1)=Y(I)
43708 NEXT I
43709 N2=1\N1=M
43710 M2=M
43711 REM A IS N BY M    B IS M BY 1
43712 GOSUB 41900
43713 REM PRODUCT IS N BY 1
43714 REM REGRESSION COEFFICIENTS ARE IN C(I,1)
43715 REM MOVE THEM TO D(I)
43716 FOR I=1 TO N
43717 D(I)=C(I,1)
43718 NEXT I
43719 RETURN
43742 REM *********************
43743 REM STANDARD DEVIATION SUBROUTINE (SIGMA).
43744 REM THIS SUBROUTINE CALCULATES THE STANDARD
43745 REM DEVIATION FOR A POLYNOMIAL FIT.
43746 REM THE INPUTS ARE THE NUMBER OF DATA POINTS, M,
43747 REM THE DEGREE OF THE FIT, N,
43748 REM THE POLYNOMIAL COEFFICIENTS, D(I),
43749 REM THE ORIGINAL DATA SET, X(I), Y(I).
43750 D=0
43751 FOR I=1 TO M
43752 Y=0
43753 B=1
43754 FOR J=1 TO N+1
43755 Y=Y+D(J)*B
43756 B=B*X(I)
43757 NEXT J
43758 D=D+(Y-Y(I))*(Y-Y(I))
43759 NEXT I
43760 IF M-N-1>0 THEN GOTO 43763
43761 D=0
43762 RETURN
43763 D=D/(M-N-1)
43764 D=SQRT(D)
43765 RETURN
43783 REM *********************
```

```
43784 REM COEFFICIENT MATRIX GENERATION SUBROUTINE FOR MULTIPLE
43785 REM NONLINEAR REGRESSION (MLTNLREG).
43786 REM ALSO CALCULATES THE STANDARD DEVIATION, D, EVEN
43787 REM THOUGH THERE IS SOME REDUNDANT COMPUTING.
43788 REM THE MAXIMUM NUMBER OF DIMENSIONS IS 9.
43789 REM THE INPUT DATA SET CONSISTS OF M DATA SETS OF THE FORM
43790 REM     Y(I), X(I,1), X(I,2),......., X(I,L)
43791 REM THE NUMBER OF DIMENSIONS IS L.
43792 REM THE ORDER OF THE FIT TO EACH DIMENSION IS M(J).
43793 REM THE RESULT IS AN (M1+1)*(M2+1)...*(ML+1)+1
43794 REM COLUMN BY M ROW MATRIX, Z.
43795 REM THIS MATRIX IS ARRANGED AS FOLLOWS ( EXAMPLE- L=2, M(1)=2, M(2)=2)
43796 REM   1   X1   X1*X1   X2   X2*X1   X2*X1*X1   X2*X2   X2*X2*X1   X2*X2*X1*X1
43797 REM THIS MATRIX SHOULD BE DIMENSIONED IN THE CALLING PROGRAM
43798 REM AS SHOULD ALSO THE X(I,J) MATRIX OF DATA VALUES.
43799 REM CALCULATE THE TOTAL NUMBER OF DIMENSIONS
43800 N=1
43801 FOR I=1 TO L
43802 N=N*(M(I)+1)
43803 NEXT I
43804 D=0
43805 FOR I=1 TO M
43806 REM BRANCH ACCORDING TO DIMENSION
43807 REM RETURN IF DIMENSION IS GREATER THAN 9
43808 IF L>0 THEN GOTO 43811
43809 L=0
43810 RETURN
43811 IF L<=9 THEN GOTO 43814
43812 L=0
43813 RETURN
43814 J=0
43815 REM MINIMAL BASIC VERSION. REPLACE FOLLOWING WITH ON/GOTO
43816 IF L=1 THEN GOSUB 43840
43817 IF L=2 THEN GOSUB 43849
43818 IF L=3 THEN GOSUB 43857
43819 IF L=4 THEN GOSUB 43865
43820 IF L=5 THEN GOSUB 43873
43821 IF L=6 THEN GOSUB 43881
43822 IF L=7 THEN GOSUB 43889
43823 IF L=8 THEN GOSUB 43897
43824 IF L=9 THEN GOSUB 43905
43825 REM ARRAY GENERATED FOR ROW I
43826 Y=0
43827 FOR K=1 TO N
43828 Y=Y+D(K)*Z(I,K)
43829 NEXT K
43830 D=D+(Y(I)-Y)*(Y(I)-Y)
43831 NEXT I
43832 REM CALCULATE STANDARD DEVIATION
43833 IF M-N>0 THEN GOTO 43836
43834 D=0
43835 RETURN
43836 D=D/(M-N)
43837 D=SQRT(D)
43838 RETURN
43839 RETURN
43840 B=1
43841 C=B
43842 FOR I1=0 TO M(1)
```

```
43843 J=J+1
43844 Z(I,J)=B
43845 B=B*X(I,1)
43846 NEXT I1
43847 B=C
43848 RETURN
43849 B=1
43850 C=B
43851 FOR I2=0 TO M(2)
43852 GOSUB 43841
43853 B=B*X(I,2)
43854 NEXT I2
43855 B=C
43856 RETURN
43857 B=1
43858 C=B
43859 FOR I3=0 TO M(3)
43860 GOSUB 43850
43861 B=B*X(I,3)
43862 NEXT I3
43863 B=C
43864 RETURN
43865 B=1
43866 C=B
43867 FOR I4=0 TO M(4)
43868 GOSUB 43858
43869 B=B*X(I,4)
43870 NEXT I4
43871 B=C
43872 RETURN
43873 B=1
43874 C=B
43875 FOR I5=0 TO M(5)
43876 GOSUB 43866
43877 B=B*X(I,5)
43878 NEXT I5
43879 B=C
43880 RETURN
43881 B=1
43882 C=B
43883 FOR I6=0 TO M(6)
43884 GOSUB 43874
43885 B=B*X(I,6)
43886 NEXT I6
43887 B=C
43888 RETURN
43889 B=1
43890 C=B
43891 FOR I7=0 TO M(7)
43892 GOSUB 43882
43893 B=B*X(I,7)
43894 NEXT I7
43895 B=C
43896 RETURN
43897 B=1
43898 C=B
43899 FOR I8=0 TO M(8)
43900 GOSUB 43890
43901 B=B*X(I,8)
```

```
43902 NEXT I8
43903 B=C
43904 RETURN
43905 B=1
43906 FOR I9=0 TO M(9)
43907 GOSUB 43898
43908 B=B*X(I,9)
43909 NEXT I9
43910 RETURN
43932 REM ********************
43933 REM LEAST SQUARES POLYNOMIAL FITTING SUBROUTINE (LSQRPOLY).
43934 REM THIS PROGRAM LEAST SQUARES FITS A POLYNOMIAL TO INPUT DATA.
43935 REM FORSYTHE ORTHOGONAL POLYNOMIALS ARE USED IN THE FITTING.
43936 REM THE NUMBER OF DATA POINTS IS N.
43937 REM THE DATA IS INPUT TO THE SUBROUTINE IN X(I),Y(I) PAIRS.
43938 REM THE COEFFICIENTS ARE RETURNED IN C(I).
43939 REM THE SMOOTHED DATA IS RETURNED IN V(I).
43940 REM THE ORDER OF THE FIT IS SPECIFIED BY M.
43941 REM THE STANDARD DEVIATION OF THE FIT IS RETURNED IN D.
43942 REM THERE ARE TWO OPTIONS AVAILABLE BY USE OF THE PARAMETER E.
43943 REM IF E=0 THE FIT IS TO ORDER M.
43944 REM IF E>0 THE ORDER OF FIT INCREASES TOWARDS M, BUT
43945 REM WILL STOP IF THE RELATIVE STANDARD DEVIATION DOES NOT
43946 REM DECREASE BY MORE THAN E BETWEEN SUCCESSIVE FITS.
43947 REM THE ORDER OF THE FIT THEN OBTAINED IS L.
43948 REM THE ARRAYS X,Y,V,A,B,C,C2,D,E AND F MUST BE DIMENSIONED.
43949 REM A(I) AND B(I) ARE SIMPLY WORK ARRAYS
43950 N1=M+1
43951 V1=10000000
43952 REM INITIALIZE THE ARRAYS
43953 FOR I=1 TO N1
43954 A(I)=0
43955 B(I)=0
43956 F(I)=0
43957 NEXT I
43958 FOR I=1 TO N
43959 V(I)=0
43960 D(I)=0
43961 NEXT I
43962 D1=SQRT(N)
43963 W=D1
43964 FOR I=1 TO N
43965 E(I)=1/W
43966 NEXT I
43967 F1=D1
43968 A1=0
43969 FOR I=1 TO N
43970 A1=A1+X(I)*E(I)*E(I)
43971 NEXT I
43972 C1=0
43973 FOR I=1 TO N
43974 C1=C1+Y(I)*E(I)
43975 NEXT I
43976 B(1)=1/F1
43977 F(1)=B(1)*C1
43978 FOR I=1 TO N
43979 V(I)=V(I)+E(I)*C1
43980 NEXT I
43981 M=1
```

```
43982 REM SAVE LATEST RESULTS
43983 FOR I=1 TO L
43984 C2(I)=C(I)
43985 NEXT I
43986 L2=L
43987 V2=V
43988 F2=F1
43989 A2=A1
43990 F1=0
43991 FOR I=1 TO N
43992 B1=E(I)
43993 E(I)=(X(I)-A2)*E(I)-F2*D(I)
43994 D(I)=B1
43995 F1=F1+E(I)*E(I)
43996 NEXT I
43997 F1=SQRT(F1)
43998 FOR I=1 TO N
43999 E(I)=E(I)/F1
44000 NEXT I
44001 A1=0
44002 FOR I=1 TO N
44003 A1=A1+X(I)*E(I)*E(I)
44004 NEXT I
44005 C1=0
44006 FOR I=1 TO N
44007 C1=C1+E(I)*Y(I)
44008 NEXT I
44009 M=M+1
44010 I=0
44011 L=M-I
44012 B2=B(L)
44013 D1=0
44014 IF L>1 THEN D1=B(L-1)
44015 D1=D1-A2*B(L)-F2*A(L)
44016 B(L)=D1/F1
44017 A(L)=B2
44018 I=I+1
44019 IF I<>M THEN GOTO 44011
44020 FOR I=1 TO N
44021 V(I)=V(I)+E(I)*C1
44022 NEXT I
44023 FOR I=1 TO N1
44024 F(I)=F(I)+B(I)*C1
44025 C(I)=F(I)
44026 NEXT I
44027 V=0
44028 FOR I=1 TO N
44029 V=V+(V(I)-Y(I))*(V(I)-Y(I))
44030 NEXT I
44031 REM NOTE THE DIVISION IS BY THE NUMBER OF DEGREES OF FREEDOM
44032 V=SQRT (V/(N-L-1))
44033 L=M
44034 IF E=0 THEN GOTO 44040
44035 REM TEST FOR MIMIMAL IMPROVEMENT
44036 IF ABS(V1-V)/V<E THEN GOTO 44053
44037 REM IF ERROR IS LARGER, QUIT
44038 IF E*V>E*V1 THEN GOTO 44053
44039 V1=V
44040 IF M=N1 THEN GOTO 44043
```

```
44041 GOTO 43983
44042 REM SHIFT THE C(I) DOWN SO C(O) IS THE CONSTANT TERM
44043 FOR I=1 TO L
44044 C(I-1)=C(I)
44045 NEXT I
44046 C(L)=0
44047 REM L IS THE ORDER OF THE POLYNOMIAL FITTED
44048 L=L-1
44049 D=V
44050 RETURN
44051 REM SEQUENCE HAS BEEN ABORTED
44052 REM RECOVER LAST VALUES
44053 L=L2
44054 V=V2
44055 FOR I=1 TO L
44056 C(I)=C2(I)
44057 NEXT I
44058 GOTO 44043
44082 REM ********************
44083 REM MULTI-DIMENSIONAL POLYNOMIAL REGRESSION
44084 REM ITERATION SUBROUTINE (REGITER).
44085 REM THIS PROGRAM SUPERVISES THE CALLING OF SEVERAL
44086 REM OTHER SUBROUTINES IN ORDER TO ITERATIVELY
44087 REM FIT LEAST SQUARES POLYNOMIALS IN MORE THAN
44088 REM ONE DIMENSION.
44089 REM THE PROGRAM REPEATEDLY CALCULATES IMPROVED COEFFICIENTS
44090 REM UNTIL THE STANDARD DEVIATION IS NO LONGER REDUCED.
44091 REM THE INPUTS TO THE SUBROUTINE ARE THE NUMBER OF
44092 REM DIMENSIONS, L, THE DEGREE OF FIT FOR EACH
44093 REM DIMENSION, M(I), AND THE INPUT DATA, X(I,J) AND Y(I).
44094 REM THE COEFFICIENTS ARE RETURNED IN D(I), WITH THE
44095 REM STANDARD DEVIATION IN D.
44096 REM ALSO RETURNED IS THE NUMBER OF ITERATIONS TRIED, L1.
44097 REM THE ORIGINAL Y(I) VALUES ARE SAVED IN Y1(I).
44098 REM THE CURRENT COEFFICIENTS ARE STORED IN D1(I).
44099 REM THE PREVIOUS STANDARD DEVIATION IS SAVED IN D1.
44100 L1=0
44101 REM SAVE Y(I)
44102 FOR I=1 TO M
44103 Y1(I)=Y(I)
44104 NEXT I
44105 REM ZERO D1(I)
44106 FOR I=1 TO N
44107 D1(I)=0
44108 NEXT I
44109 REM SET THE INITIAL STANDARD DEVIATION HIGH
44110 D1=10000000
44111 REM GO TO COEFFICIENTS SUBROUTINE
44112 GOSUB 43800
44113 REM GO TO REGRESSION SUBROUTINE
44114 GOSUB 43650
44115 REM GET STANDARD DEVIATION
44116 GOSUB 43800
44117 REM IF STANDARD DEVIATION IS DECREASING, CONTINUE
44118 IF D1>D THEN GOTO 44131
44119 REM TERMINATE ITERATION
44120 FOR I=1 TO N
44121 D(I)=D1(I)
44122 NEXT I
```

```
44123 REM RESTORE Y(I)
44124 FOR I=1 TO M
44125 Y(I)=Y1(I)
44126 NEXT I
44127 REM GET THE FINAL STANDARD DEVIATION
44128 GOSUB 43800
44129 RETURN
44130 REM SAVE THE STANDARD DEVIATION
44131 D1=D
44132 L1=L1+1
44133 REM AUGMENT COEFFICIENT MATRIX
44134 FOR I=1 TO N
44135 D(I)=D1(I)+D(I)
44136 D1(I)=D(I)
44137 NEXT I
44138 REM RESTORE Y(I)
44139 FOR I=1 TO M
44140 Y(I)=Y1(I)
44141 NEXT I
44142 REM REDUCE Y(I) ACCORDING TO THE D(I)
44143 GOSUB 44147
44144 REM WE NOW HAVE A SET OF ERROR VALUES
44145 GOTO 44112
44146 REM *********
44147 FOR I=1 TO M
44148 J=0
44149 REM MINIMAL BASIC VERSION. REPLACE FOLLOWING WITH ON/GOTO
44150 IF L=1 THEN GOSUB 44167
44151 IF L=2 THEN GOSUB 44176
44152 IF L=3 THEN GOSUB 44184
44153 IF L=4 THEN GOSUB 44192
44154 IF L=5 THEN GOSUB 44200
44155 IF L=6 THEN GOSUB 44208
44156 IF L=7 THEN GOSUB 44216
44157 IF L=8 THEN GOSUB 44224
44158 IF L=9 THEN GOSUB 44232
44159 REM ARRAY GENERATED FOR ROW I
44160 Y=0
44161 FOR K=1 TO N
44162 Y=Y+D(K)*Z(I,K)
44163 NEXT K
44164 Y(I)=Y(I)-Y
44165 NEXT I
44166 RETURN
44167 B=1
44168 C=B
44169 FOR I1=0 TO M(1)
44170 J=J+1
44171 Z(I,J)=B
44172 B=B*X(I,1)
44173 NEXT I1
44174 B=C
44175 RETURN
44176 B=1
44177 C=B
44178 FOR I2=0 TO M(2)
44179 GOSUB 44168
44180 B=B*X(I,2)
44181 NEXT I2
```

```
44182 B=C
44183 RETURN
44184 B=1
44185 C=B
44186 FOR I3=0 TO M(3)
44187 GOSUB 44177
44188 B=B*X(I,3)
44189 NEXT I3
44190 B=C
44191 RETURN
44192 B=1
44193 C=B
44194 FOR I4=0 TO M(4)
44195 GOSUB 44185
44196 B=B*X(I,4)
44197 NEXT I4
44198 B=C
44199 RETURN
44200 B=1
44201 C=B
44202 FOR I5=0 TO M(5)
44203 GOSUB 44193
44204 B=B*X(I,5)
44205 NEXT I5
44206 B=C
44207 RETURN
44208 B=1
44209 C=B
44210 FOR I6=0 TO M(6)
44211 GOSUB 44201
44212 B=B*X(I,6)
44213 NEXT I6
44214 B=C
44215 RETURN
44216 B=1
44217 C=B
44218 FOR I7=0 TO M(7)
44219 GOSUB 44209
44220 B=B*X(I,7)
44221 NEXT I7
44222 B=C
44223 RETURN
44224 B=1
44225 C=B
44226 FOR I8=0 TO M(8)
44227 GOSUB 44217
44228 B=B*X(I,8)
44229 NEXT I8
44230 B=C
44231 RETURN
44232 B=1
44233 FOR I9=0 TO M(9)
44234 GOSUB 44225
44235 B=B*X(I,9)
44236 NEXT I9
44237 RETURN
44238 REM ********************
44239 REM PARAMETRIC LEAST SQUARES CURVE FIT SUBROUTINE (PARAFIT).
44240 REM THIS PROGRAM LEAST SQUARES FITS A FUNCTION TO A SET OF
```

```
44241 REM DATA VALUES BY SUCCESSIVELY (PARAMETER BY PARAMETER) REDUCING THE VARIANCE
44242 REM CONVERGENCE DEPENDS ON THE INITIAL VALUES.- CONVERGENCE IS NOT ASSURE
44243 REM N PAIRS OF DATA VALUES, (X(I),Y(I)), ARE GIVEN.
44244 REM THERE ARE L PARAMETERS, A(J), TO BE OPTIMIZED ACROSS.
44245 REM REQUIRED ARE INITIAL VALUES FOR THE PARAMETER A(L) AND E.
44246 REM ANOTHER IMPORTANT PARAMETER WHICH AFFECTS STABILITY IS E1,
44247 REM WHICH IS INITIALLY CONVERTED TO E1(L) FOR THE FIRST INTERVALS.
44248 REM THE PARAMETERS ARE MULTIPLIED BY (1-E1(I)) ON EACH PASS.
44249 REM DIMENSION X(I),Y(I),A(I) AND E1(I) IN THE CALLING PROGRAM.
44250 FOR I=1 TO L
44251 E1(I)=E1
44252 NEXT I
44253 M=0
44254 REM SET UP TEST RESIDUAL
44255 L1=1000000
44256 REM MAKE SWEEP THROUGH ALL PARAMETERS
44257 FOR I=1 TO L
44258 A0=A(I)
44259 REM GET VALUE OF RESIDUAL
44260 A(I)=A0
44261 GOSUB 44286
44262 REM STORE RESULT IN M0
44263 M0=L2
44264 REM REPEAT FOR M1
44265 A(I)=A0*(1-E1(I))
44266 GOSUB 44286
44267 M1=L2
44268 REM CHANGE INTERVAL SIZE IF CALLED FOR
44269 REM IF VARIANCE WAS INCREASED, HALVE E1(I)
44270 IF M1>M0 THEN E1(I)=-E1(I)/2
44271 REM IF VARIANCE WAS REDUCED, INCREASE STEP SIZE BY INCREASING E1(I)
44272 IF M1<M0 THEN E1(I)=1.2*E1(I)
44273 REM IF VARIANCE HAS INCREASED, TRY TO REDUCE IT
44274 IF M1>M0 THEN A(I)=A0
44275 IF M1>M0 THEN GOTO 44261
44276 NEXT I
44277 REM END OF A COMPLETE PASS
44278 REM TEST FOR CONVERGENCE
44279 M=M+1
44280 IF L2=0 THEN RETURN
44281 IF ABS((L1-L2)/L2)<E THEN RETURN
44282 REM IF THIS POINT IS REACHED, ANOTHER PASS IS CALLED FOR
44283 L1=L2
44284 GOTO 44257
44285 REM RESIDUAL GENERATION SUBROUTINE
44286 L2=0
44287 FOR J=1 TO N
44288 X=X(J)
44289 REM OBTAIN FUNCTION
44290 GOSUB 44300
44291 L2=L2+(Y(J)-Y)*(Y(J)-Y)
44292 NEXT J
44293 D=SQRT(L2/(N-L))
44294 RETURN
44298 REM ********************
44299 REM FUNCTIONS SUBROUTINE
44300 Y=A(1)*EXP(-(X-A(2))*(X-A(2))/A(3))
44301 RETURN
44393 REM ********************
```

```
44394 REM CHI-SQUARE CUMMULATIVE DISTRIBUTION APPROXIMATION (CHISQA).
44395 REM REFERENCE- STATISTICS MANUAL
44396 REM CROW, MAXFIELD AND DAVIS (DOVER, 1960).
44397 REM THE INPUT VALUE IS Y, THE PROBABILITY.
44398 REM THE OUTPUT VALUE IS THE CORRESPONDING
44399 REM CHI-SQUARE STATISTIC.
44400 X=Y
44401 REM GUARD AGAINST 0 DISCONTINUITY
44402 IF X=0 THEN X=EXP(-100)
44403 IF X>.5 THEN GOTO 44408
44404 X=-LOG(X)
44405 REM REGRESSED TABLE CORRECTION
44406 Z=-.803+1.312*X-.2118*X*X+.016*X*X*X
44407 GOTO 44413
44408 X=1-X
44409 REM GUARD AGAINST 0 DISCONTINUITY
44410 IF X=0 THEN GOTO 44415
44411 X=-LOG(X)
44412 Z=.803-1.312*X+.2118*X*X-.016*X*X*X
44413 X=2/(9*M)
44414 X=1-X+Z*SQRT(X)
44415 X=M*X*X*X
44416 RETURN
44419 REM ********************
44420 REM BESSEL FUNCTION SERIES SUBROUTINE (BESSLSER)
44421 REM THE ORDER IS N, THE ARGUMENT X.
44422 REM THE RETURNED VALUE IS IN Y.
44423 REM THE NUMBER OF TERMS USED IS RETURNED IN M.
44424 REM E IS THE CONVERGENCE CRITERION
44425 A=1
44426 IF N<=1 THEN GOTO 44431
44427 REM CALCULATE N!
44428 FOR I=1 TO N
44429 A=A*I
44430 NEXT I
44431 A=1/A
44432 IF N=0 THEN GOTO 44437
44433 REM CALCULATE MULTIPLYING TERM
44434 FOR I=1 TO N
44435 A=A*X/2
44436 NEXT I
44437 B0=1
44438 B2=1
44439 M=0
44440 REM ASSEMBLE SERIES SUM
44441 M=M+1
44442 B1=-(X*X*B0)/(M*(M+N)*4)
44443 B2=B2+B1
44444 B0=B1
44445 REM TEST FOR CONVERGENCE
44446 IF ABS(B1)>E THEN GOTO 44441
44447 REM FORM FINAL ANSWER
44448 Y=A*B2
44449 RETURN
44468 REM ********************
44469 REM BESSEL FUNCTION SERIES COEFFICIENT EVALUATION SUBROUTINE (BESSEL)
44470 REM M+1 IS THE NUMBER OF COEFFICIENTS DESIRED.
44471 REM N IS THE ORDER OF THE BESSEL FUNCTION.
44472 REM THE COEFFICIENTS ARE RETURNED IN A(I).
```

```
44473 REM DIMENSION A(I) AND B(I) IN THE CALLING PROGRAM.
44474 REM A1,B1 AND B(I) ARE DUMMY VARIABLES.
44475 A1=1
44476 B1=1
44477 FOR I=1 TO N
44478 B(I-1)=0
44479 B1=B1*I
44480 A1=A1/2
44481 NEXT I
44482 B1=A1/B1
44483 A1=1
44484 FOR I=0 TO M STEP 2
44485 A(I)=A1*B1
44486 A(I+1)=0
44487 A1=-A1/((I+2)*(N+N+I+2))
44488 NEXT I
44489 A1=A1/2
44490 FOR I=0 TO M
44491 B(I+N)=A(I)
44492 NEXT I
44493 FOR I=0 TO N+M
44494 A(I)=B(I)
44495 NEXT I
44496 RETURN
44515 REM ********************
44516 REM BESSEL FUNCTION ASYMPTOTIC SERIES SUBROUTINE (BESSEL01)
44517 REM THIS PROGRAM CALCULATES THE ZEROTH AND FIRST ORDER BESSEL
44518 REM FUNCTIONS USING AN ASYMPTOTIC SERIES EXPANSION.
44519 REM THE REQUIRED INPUT ARE X AND A CONVERGENCE FACTOR E.
44520 REM RETURNED ARE THE TWO BESSEL FUNCTIONS, J0(X) AND J1(X)
44521 REM REFERENCE-  ALGORITHMS FOR RPN CALCULATORS
44522 REM     BY BALL, J.A.,   WILEY AND SONS,
44523 REM CALCULATE P AND Q POLYNOMIALS
44524 REM P0(X)=M1  P1(X)=M2   Q0(X)=N1   Q1(X)=N2
44525 A=1
44526 A1=1
44527 A2=1
44528 B=1
44529 C=1
44530 E1=1000000
44531 M=-1
44532 X1=1/(8*X)
44533 X1=X1*X1
44534 M1=1
44535 M2=1
44536 N1=-1/(8*X)
44537 N2=-3*N1
44538 N=0
44539 M=M+2
44540 A=A*M*M
44541 M=M+2
44542 A=A*M*M
44543 C=C*X1
44544 A1=A1*A2
44545 A2=A2+1
44546 A1=A1*A2
44547 A2=A2+1
44548 E2=A*C/A1
44549 E4=1+(M+2)/M+(M+2)*(M+2)/(A2*8*X)+(M+2)*(M+4)/(A2*8*X)
```

```
44550 E4=E4*E2
44551 REM TEST FOR DIVERGENCE
44552 IF ABS(E4)>E1 THEN GOTO 44562
44553 E1=ABS(E2)
44554 M1=M1-E2
44555 M2=M2+E2*(M+2)/M
44556 N1=N1+E2*(M+2)*(M+2)/(A2*8*X)
44557 N2=N2-E2*(M+2)*(M+4)/(A2*8*X)
44558 N=N+1
44559 REM TEST FOR CONVERGENCE CRITERION
44560 IF E1<E3 THEN GOTO 44562
44561 GOTO 44539
44562 A=3.1415926536
44563 E=E2
44564 B=SQRT(2/(A*X))
44565 J0=B*(M1*COS(X-A/4)-N1*SIN(X-A/4))
44566 J1=B*(M2*COS(X-3*A/4)-N2*SIN(X-3*A/4))
44567 RETURN
44572 REM *********************
44573 REM SERIES APPROXIMATION SUBROUTINE FOR LN(X!)   (LN(X!))
44574 REM ACCURACY BETTER THAN 6 PLACES FOR X>=3.
44575 REM ACCURACY BETTER THAN 12 PLACES FOR X>10.
44576 REM ADVANTAGE IS THAT VERY LARGE VALUES OF THE ARGUMENT CAN BE USED
44577 REM WITHOUT FEAR OF OVERFLOW.
44578 REM REFERENCE-  CRC MATH TABLES.
44579 REM X IS THE INPUT, Y IS THE OUTPUT
44580 X1=1/(X*X)
44581 Y=(X+.5)*LOG(X)-X*(1-X1/12+X1*X1/360-X1*X1*X1/1260+X1*X1*X1*X1/1680)
44582 Y=Y+0.918938533205
44583 RETURN
44592 REM *********************
44593 REM CHI-SQUARE FUNCTION SUBROUTINE (CHI-SQR)
44594 REM THIS PROGRAM TAKES A GIVEN DEGREE OF FREEDOM, M
44595 REM AND VALUE, X, AND CALCULATES THE CHI-SQUARE
44596 REM DENSITY DISTRIBUTION FUNCTION VALUE, Y.
44597 REM REFERENCE- TEXAS INSTRUMENTS SR-51 OWNERS MANUAL (1974).
44598 REM SUBROUTINE LN(X!) IS ALSO CALLED.
44599 REM SAVE X
44600 M1=X
44601 REM PERFORM CALCULATION
44602 X=M/2-1
44603 REM GOTO LN(X!) SUBROUTINE
44604 GOSUB 44580
44605 X=M1
44606 C=-X/2+(M/2-1)*LOG(X)-(M/2)*LOG(2)-Y
44607 Y=EXP(C)
44608 RETURN
44615 REM *********************
44616 REM CHI-SQUARE CUMMULATIVE DISTRIBUTION (CHISQ)
44617 REM THE PROGRAM IS FAIRLY ACCURATE AND CALLS UPON THE
44618 REM CHI-SQUARE PROBABILITY DENSITY FUNCTION SUBROUTINE (CHI-SQR).
44619 REM THE INPUT PARAMETER IS M, THE NUMBER OF DEGREES OF FREEDOM.
44620 REM ALSO REQUIRED IS THE ORDINATE VALUE. THE PROGRAM RETURNS Y,
44621 REM THE CUMMULATIVE DISTRIBUTION INTEGRAL FROM 0 TO X.
44622 REM REFERENCE- HEWLETT-PACKARD STATISTICS PROGRAMS, 1974.
44623 REM THIS PROGRAM ALSO REQUIRES AN ACCURACY PARAMETER, E, TO
44624 REM DETERMINE THE LEVEL OF SUMMATION.
44625 Y1=1
44626 X2=X
```

```
44627 M2=M+2
44628 X2=X2/M2
44629 Y1=Y1+X2
44630 IF X2<E THEN GOTO 44637
44631 M2=M2+2
44632 REM THIS FORM IS USED TO AVOID OVERFLOW
44633 X2=X2*X/M2
44634 REM LOOP TO CONTINUE SUM
44635 GOTO 44629
44636 REM OBTAIN Y, THE PROBABILITY DENSITY FUNCTION
44637 GOSUB 44600
44638 Y=Y1*Y*2*X/M
44639 RETURN
44661 REM ********************
44662 REM ASYMPTOTIC SERIES EXPANSION OF THE INTEGRAL OF
44663 REM   2 EXP(-X*X)/SQRT(PI)  - THE NORMALIZED ERROR FUNCTION (ASYMERF)
44664 REM THIS PROGRAM DETEMINES THE VALUES OF THE ABOVE
44665 REM INTEGRAND USING AN ASYMPTOTIC SERIES WHICH IS
44666 REM EVALUATED TO THE LEVEL OF MAXIMUM ACCURACY.
44667 REM THE INTEGRAL IS FROM 0 TO X.
44668 REM THE INPUT PARAMETER IS X>0. THE RESULTS ARE
44669 REM RETURNED IN Y AND Y1, WITH THE ERROR MEASURE IN E.
44670 REM THE PROGRAM ALSO RETURNS THE NUMBER OF TERMS USED.
44671 REM NOTE- THE ERROR IS ROUGHLY EQUAL TO
44672 REM FIRST TERM NEGLECTED IN THE SERIES SUMMATION.
44673 REM REFERENCE-  A SHORT TABLE OF INTEGRALS BY B.O. PEIRCE
44674 REM    GINN AND COMPANY   1957
44675 N=1
44676 Y=1
44677 C2=1/(2*X*X)
44678 Y=Y-C2
44679 N=N+2
44680 C1=C2
44681 C2=-C1*N/(2*X*X)
44682 REM TEST FOR DIVERGENCE
44683 REM THE BREAK POINT IS ROUGHLY N=X*X
44684 IF ABS(C2)>ABS(C1) THEN GOTO 44687
44685 REM CONTINUE SUMMATION
44686 GOTO 44678
44687 N=(N+1)/2
44688 E=EXP(-X*X)/(X*1.772453850905516)
44689 Y1=Y*E
44690 Y=1-Y1
44691 E=E*C2
44692 RETURN
44717 REM ********************
44718 REM CHEBYCHEV SERIES COEFFICIENT EVALUATION SUBROUTINE (CHEBYSER)
44719 REM THE ORDER OF THE POLYNOMIAL IS N.
44720 REM THE COEFFICIENTS ARE RETURNED IN THE
44721 REM ARRAY B(I,J). I IS THE DEGREE OF THE POLYNOMIAL,
44722 REM J IS THE COEFFICIENT ORDER.
44723 REM DIMENSION B(I,J) IN THE CALLING PROGRAM.
44724 REM ESTABLISH T0 AND T1 COEFFICIENTS
44725 B(0,0)=1
44726 B(1,0)=0
44727 B(1,1)=1
44728 REM RETURN IF ORDER IS LESS THAN TWO
44729 IF N<2 THEN RETURN
44730 FOR I=2 TO N
```

```
44731 FOR J=1 TO I
44732 REM BASIC RECURSION RELATION
44733 B(I,J)=B(I-1,J-1)+B(I-1,J-1)-B(I-2,J)
44734 NEXT J
44735 B(I,0)=-B(I-2,0)
44736 NEXT I
44737 RETURN
44746 REM *********************
44747 REM CHEBYSHEV ECONOMIZATION SUBROUTINE (CHEBECON)
44748 REM ROUTINE TAKES THE INPUT POLYNOMIAL COEFFICIENTS, C(I),
44749 REM AND RETURNS THE CHEBYSCHEV SERIES COEFFICIENTS, A(I).
44750 REM THE DEGREE OF THE SERIES PASSED TO THE ROUTINE IS M.
44751 REM THE DEGREE OF THE SERIES RETURNED IS M1.
44752 REM THE MAXIMUM RANGE OF X IS X0- X0 IS USED FOR SCALING.
44753 REM THE CHEBYSCHEV SERIES COEFFICIENT (B(I,J) SUBROUTINE IS
44754 REM CALLED- I IS THE ORDER OF THE CHEBYSCHEV POLYNOMIAL.
44755 REM NOTE THAT THE INPUT SERIES COEFFICIENTS ARE NULLED DURING THE PROCESS,
44756 REM AND THEN SET EQUAL TO THE ECONOMIZED SERIES COEFFICIENTS.
44757 REM THE CHEBYSHEV SERIES IS VALID ONLY OVER THE RANGE ABS(X/X0)<=1.
44758 REM DIMENSION A(I),B(I,J),C(I) IN THE CALLING PROGRAM.
44759 REM START BY SCALING THE INPUT COEFFICIENTS ACCORDING TO C(I)
44760 B=X0
44761 FOR I=1 TO M
44762 C(I)=C(I)*B
44763 B=B*X0
44764 NEXT I
44765 REM GET CHEBYSCHEV SERIES COEFFICIENTS.
44766 REM POLYNOMIAL SERIES IS REDUCED FROM THE HIGHEST ORDER DOWN
44767 FOR N=M TO 0 STEP -1
44768 GOSUB 44725
44769 A(N)=C(N)/B(N,N)
44770 FOR L=0 TO N
44771 REM CHEBYSCHEV SERIES OF ORDER L IS SUBTRACTED OUT OF THE POLYNOMIAL
44772 C(L)=C(L)-A(N)*B(N,L)
44773 NEXT L
44774 NEXT N
44775 REM PERFORM TRUNCATION
44776 FOR I=0 TO M1
44777 FOR J=0 TO I
44778 C(J)=C(J)+A(I)*B(I,J)
44779 NEXT J
44780 NEXT I
44781 REM CONVERT BACK TO THE INTERVAL X0
44782 B=1/X0
44783 FOR I=1 TO M1
44784 C(I)=C(I)*B
44785 B=B/X0
44786 NEXT I
44787 RETURN
44789 REM *********************
44790 REM SERIES REVERSION SUBROUTINE (REVERSE)
44791 REM THIS PROGRAM TAKES A POLYNOMIAL, Y=A(0) + A(1)X + ..
44792 REM AND RETURNS A POLYNOMIAL X = B(0) + B(1)Y + ...
44793 REM REFERENCE   CRC STANDARD MATHEMATICAL TABLES
44794 REM              24TH EDITION
44795 REM THE INPUT SERIES COEFFICIENTS ARE A(0),A(1), ETC.
44796 REM A(1) MUST BE NONZERO.
44797 REM THE OUTPUT SERIES COEFFICIENTS ARE B(0),B(1),....,B(7).
44798 REM THE DEGREE OF REVERSION IS LIMITED TO SEVEN.
```

```
44799 REM A1,A2,.... ARE DUMMY VARIABLES.
44800 A1=A(1)
44801 B(1)=1/A1
44802 A=1/A1
44803 B=A*A
44804 A=A*B
44805 B(2)=-A2/A
44806 A3=A(3)
44807 A=A*B
44808 B(3)=A*(2*A2*A2-A1*A3)
44809 A4=A(4)
44810 A=A*B
44811 B(4)=A*(5*A1*A2*A3-A1*A1*A4-5*A2*A2*A2)
44812 A5=A(5)
44813 A=A*B
44814 B(5)=6*A1*A1*A2*A4+3*A1*A1*A3*A3+14*A2*A2*A2*A2
44815 B(5)=B(5)-A1*A1*A1*A5-21*A1*A2*A2*A3
44816 B(5)=A*B(5)
44817 A6=A(6)
44818 A=A*B
44819 B(6)=7*A1*A1*A1*A2*A5+7*A1*A1*A1*A3*A4+84*A1*A2*A2*A2*A3
44820 B(6)=B(6)-A1*A1*A1*A1*A6-28*A1*A1*A2*A2*A4
44821 B(6)=B(6)-28*A1*A1*A2*A3*A3-42*A2*A2*A2*A2*A2
44822 B(6)=A*B(6)
44823 A7=A(7)
44824 A=A*B
44825 B(7)=8*A1*A1*A1*A1*A2*A6+8*A1*A1*A1*A1*A3*A5
44826 B(7)=B(7)+4*A1*A1*A1*A1*A4*A4+120*A1*A1*A2*A2*A2*A4
44827 B(7)=B(7)+180*A1*A1*A2*A2*A3*A3+132*A2*A2*A2*A2*A2*A2
44828 B(7)=B(7)-A1*A1*A1*A1*A1*A7-36*A1*A1*A1*A2*A2*A5
44829 B(7)=B(7)-72*A1*A1*A1*A2*A3*A4-12*A1*A1*A1*A3*A3*A3
44830 B(7)=B(7)-330*A1*A2*A2*A2*A2*A3
44831 B(7)=A*B(7)
44832 B(0)=0
44833 A=A(0)
44834 FOR I=1 TO 7
44835 B(0)=B(0)-B(I)*A
44836 A=A*A(0)
44837 NEXT I
44838 RETURN
44841 REM **********************
44842 REM RECIPROCAL POWER SERIES SUBROUTINE (RECIPRO)
44843 REM REFERENCE- COMPUTATIONAL ANALYSIS BY HENRICI.
44844 REM THE INPUT SERIES COEFFICIENTS ARE A(I).
44845 REM THE OUTPUT SERIES COEFFICIENTS ARE B(I).
44846 REM THE DEGREE OF THE INPUT POLYNOMIAL IS N..
44847 REM THE DEGREE OF THE INVERTED POLYNOMIAL IS M.
44848 REM DIMENSION A(I) AND B(I) IN THE CALLING PROGRAM
44849 REM THE PROGRAM WILL TAKE CARE OF THE NORMALIZATION USING L
44850 L=A(0)
44851 FOR I=0 TO N
44852 A(I)=A(I)/L
44853 B(I)=0
44854 NEXT I
44855 REM CLEAR ARRAYS
44856 FOR I=N+1 TO M
44857 A(I)=0
44858 B(I)=0
44859 NEXT I
```

```
44860 REM CALCULATE THE B(I) COEFFICIENTS
44861 B(0)=1
44862 FOR I=1 TO M
44863 J=1
44864 B(I)=B(I)-A(J)*B(I-J)
44865 J=J+1
44866 IF J<=I THEN GOTO 44864
44867 NEXT I
44868 REM UN-NORMALIZE THE A(I) AND B(I)
44869 FOR I=0 TO M
44870 A(I)=A(I)*L
44871 B(I)=B(I)/L
44872 NEXT I
44873 RETURN
44892 REM ********************
44893 REM HORNER'S SHIFTING RULE SUBROUTINE (HORNER)
44894 REM THIS SUBROUTINE TAKES A GIVEN QUARTIC
44895 REM POLYNOMIAL AND CONVERTS IT TO A TAYLOR EXPANSION.
44896 REM THE INPUT SERIES COEFFICIENTS ARE A(I).
44897 REM THE EXPANSION POINT IS X0.
44898 REM THE SHIFTED COEFFICIENTS ARE RETURNED IN B(I).
44899 REM C(4,5) MUST BE DIMENSIONED IN THE CALLING PROGRAM.
44900 FOR J=0 TO 4
44901 C(J,0)=A(4-J)
44902 NEXT J
44903 FOR I=0 TO 4
44904 C(0,I+1)=C(0,I)
44905 J=1
44906 IF J>4-I THEN GOTO 44910
44907 C(J,I+1)=X0*C(J-1,I+1)+C(J,I)
44908 J=J+1
44909 GOTO 44906
44910 NEXT I
44911 FOR I=0 TO 4
44912 B(4-I)=C(I,4-I+1)
44913 NEXT I
44914 RETURN
44915 REM ********************
44916 REM INVERSE NORMAL DISTRIBUTION SUBROUTINE (INVNORM)
44917 REM THIS PROGRAM CALCULATES AN APPROXIMATION
44918 REM TO THE INTEGRAL OF THE NORMAL DISTRIBUTION
44919 REM FUNCTION FROM X TO INFINITY (THE TAIL).
44920 REM A RATIONAL POLYNOMIAL IS USED.
44921 REM THE INPUT IS Y, WITH THE RESULT RETURNED IN X.
44922 REM THE ACCURACY IS BETTER THAN 0.0005 IN THE RANGE 0<Y<=.5
44923 REM REFERENCE- ABRAMOWITZ AND STEGUN
44924 REM DEFINE COEFFICIENTS
44925 C0=2.515517
44926 C1=0.802853
44927 C2=0.010328
44928 D1=1.432788
44929 D2=0.189269
44930 D3=0.001308
44931 IF Y=0 THEN X=10000000000000
44932 IF Y=0 THEN RETURN
44933 Z=SQRT(-LOG(Y*Y))
44934 X=1+D1*Z+D2*Z*Z+D3*Z*Z*Z
44935 X=(C0+C1*Z+C2*Z*Z)/X
44936 X=Z-X
```

```
44937 RETURN
44942 REM *********************
44943 REM COMPLEX SERIES EVALUATION SUBROUTINE (CMPLXSER)
44944 REM THE SERIES COEFFICIENTS ARE A(I), ASSUMED REAL.
44945 REM THE ORDER OF THE POLYNOMIAL IS M.
44946 REM THE SUBROUTINE USES REPEATED CALLS TO THE
44947 REM NTH POWER (Z^N) COMPLEX NUMBER SUBROUTINE.
44948 REM INPUTS TO THE SUBROUTINE ARE X,Y,M, AND THE A(I).
44949 REM OUTPUTS ARE Z1(REAL) AND Z2(IMAGINARY).
44950 Z1=A(0)
44951 Z2=0
44952 REM STORE X AND Y
44953 A1=X
44954 A2=Y
44955 FOR N=1 TO M
44956 REM RECALL ORIGINAL X AND Y
44957 X=A1
44958 Y=A2
44959 REM GO TO Z^N SUBROUTINE
44960 GOSUB 41200
44961 REM FORM PARTIAL SUM
44962 Z1=Z1+A(N)*X
44963 Z2=Z2+A(N)*Y
44964 NEXT N
44965 REM RESTORE X AND Y
44966 X=A1
44967 Y=A2
44968 RETURN
44991 REM *********************
44992 REM NTH ROOT SUBROUTINE (NTHROOT)
44993 REM USES NEWTON-RAPHSON ITERATION.
44994 REM REFERENCE- HART, COMPUTER APPROXIMATIONS.
44995 REM EXPONENT IS 1/N, INPUT PARAMETER IS N
44996 REM NOTE THAT N MUST BE AN INTEGER.
44997 REM ARGUMENT IS Y, DESIRED ACCURACY IS E.
44998 REM RETURNED VALUE IS X.
44999 REM INITIAL VALUE IS X0.
45000 IF N<=1 THEN RETURN
45001 IF Y<0 THEN RETURN
45002 IF INT(N)<>N THEN RETURN
45003 IF E<=0 THEN RETURN
45004 IF Y>0 THEN GOTO 45007
45005 X=0
45006 RETURN
45007 M=0
45008 REM FIND N-1 POWER OF X0
45009 X2=1
45010 FOR I=1 TO N-1
45011 X2=X2*X0
45012 NEXT I
45013 REM ITERATE
45014 X1=((N-1)*X0+Y/X2)/N
45015 X=X1
45016 M=M+1
45017 IF ABS((X0-X1)/X1)<E THEN RETURN
45018 X0=X1
45019 GOTO 45009
45020 REM *********************
45021 REM GENERAL ROOT DETERMINATION SUBROUTINE (GENROOT)
```

```
45022 REM HIGH ACCURACY ITERATION INVOLVING THE SQUARE ROOT.
45023 REM ROUTINE DECOMPOSES EXPONENT INTO A BINARY REPRESENTATION
45024 REM AND THEN APPLIES NEWTON-RAPHSON ITERATION.
45025 REM Y IS THE INPUT, N IS THE EXPONENT, AND X THE RETURNED ROOT.
45026 REM E IS THE DESIRED ACCURACY OF THE ITERATION.
45027 REM M IS THE DESIRED NUMBER OF BITS IN THE REPRESENTATION OF N.
45028 REM SAVE Y FOR RETURNING FROM SUBROUTINE.
45029 REM SAVE N
45030 N3=N
45031 IF Y<0THEN RETURN
45032 IF E<=0 THEN RETURN
45033 IF Y>0 THEN GOTO 45036
45034 X=0
45035 RETURN
45036 X3=Y
45037 REM IF THE EXPONENT IS NEGATIVE, INVERT PROBLEM
45038 IF N>=0 THEN GOTO 45042
45039 N=-N
45040 Y=1/Y
45041 REM BREAK N DOWN INTO POWERS OF 1/2
45042 GOSUB 45060
45043 REM FIND MULTIPLIERS
45044 PRINT
45045 GOSUB 45074
45046 Y=X3
45047 N=N3
45048 RETURN
45053 REM ********************
45054 REM ROOT DECOMPOSITION SUBROUTINE (RTDECOMP)
45055 REM DECOMPOSE ROOT N INTO A BINARY REPRESENTATION.
45056 REM M IS THE NUMBER OF BINARY DIGITS.
45057 REM N IS THE INPUT DECIMAL NUMBER.
45058 REM N1 IS THE INTEGER PART OF N.
45059 REM A(I) IS THE BINARY REPRESENTATION OF THE REMAINING FRACTION.
45060 N1=INT(N)
45061 N2=N-N1
45062 REM N2 IS BETWEEN 0 AND 1
45063 REM DECOMPOSE N2 INTO FRACTIONS
45064 A=.5
45065 FOR I=1 TO M
45066 A(I)=0
45067 IF A<N2 THEN A(I)=1
45068 IF A<N2 THEN N2=N2-A
45069 A=A/2
45070 NEXT I
45071 RETURN
45072 REM ********************
45073 REM FIND MULTIPLYING FACTORS
45074 FOR I=1 TO M
45075 REM FIND SQUARE ROOT OF Y
45076 GOSUB 45109
45077 REM REPLACE Y WITH ITS SQUARE ROOT
45078 Y=X1
45079 IF A(I)=1 THEN GOTO 45083
45080 A(I)=1
45081 GOTO 45084
45082 REM A(I) IS SET EQUAL TO THE LATEST SQUARE ROOT OF Y
45083 A(I)=Y
45084 NEXT I
```

```
45085 REM ASSEMBLE RESULTS
45086 REM RETRIEVE Y
45087 Y=X3
45088 REM RETRIEVE N
45089 N=N3
45090 REM TAKE CARE OF N1 MULTIPLICATIONS
45091 X2=1
45092 FOR I=1 TO N1
45093 X2=X2*Y
45094 NEXT I
45095 REM TAKE CARE OF ROOT PORTION
45096 FOR I=1 TO M
45097 X2=X2*A(I)
45098 NEXT I
45099 REM THE FINAL ROOT IS X
45100 X=X2
45101 RETURN
45102 REM ********************
45103 REM SQUARE ROOT SUBROUTINE (SQROOT)
45104 REM USES NEWTON-RAPHSON ITERATION.
45105 REM CALLED HERON'S RULE.
45106 REM REFERENCE- HART, COMPUTER APPROXIMATIONS.
45107 REM ARGUMENT IS Y, RETURNED VALUE IS X1.
45108 REM DESIRED ACCURACY IS E.
45109 X0=1
45110 X1=(X0+Y/X0)/2
45111 IF ABS((X1-X0)/X1)<E THEN RETURN
45112 X0=X1
45113 GOTO 45110
45117 REM ********************
45118 REM TANGENT ITERATION SUBROUTINE (TANITER)
45119 REM USES THE INVERSE TANGENT.
45120 REM BASED ON NEWTON-RAPHSON ITERATION.
45121 REM X IS THE ARGUMENT, Y IS THE RESULT.
45122 REM THE DESIRED ACCURACY IS E.
45123 REM NOTE, THE ALLOWABLE RANGE OF THE ARGUMENT IS -PI/2 TO PI/2.
45124 REM INITIAL GUESS IS X0=1
45125 X0=1
45126 REM CHECK FOR DIVIDE BY ZERO
45127 IF X<>0 THEN GOTO 45131
45128 Y=0
45129 RETURN
45130 REM CHECK FOR OUT OF BOUNDS
45131 IF ABS(X)>=3.1415926535/2 THEN RETURN
45132 IF E<=0 THEN RETURN
45133 REM CAN CALL ARCTANGENT SUBROUTINE HERE.
45134 X1=X0+(X-ATN(X0))*(1+X0*X0)
45135 REM TEST FOR ACCURACY
45136 IF ABS((X0-X1)/X1)<E THEN GOTO 45139
45137 X0=X1
45138 GOTO 45134
45139 Y=X1
45140 RETURN
45142 REM ********************
45143 REM INVERSE TANGENT RECURSION SUBROUTINE (ATANITER)
45144 REM USES GAUSS ITERATION
45145 REM REFERENCE- ACTON, NUMERICAL METHODS THAT WORK
45146 REM ARGUMENT IS X, RESULT IS Y
45147 REM DESIRED ACCURACY IS E
```

```
45148 REM HERON'S RULE (ITERATION) FOR THE SQUARE ROOT IS ALSO USED
45149 REM A0,A1,B0,B1 ARE DUMMY VARIABLES
45150 IF E<0 THEN RETURN
45151 M=0
45152 Y=1+X*X
45153 REM FIND SQUARE ROOT OF 1/Y
45154 GOSUB 45177
45155 X2=1/X1
45156 A0=X2
45157 B0=1
45158 A1=(A0+B0)/2
45159 Y=A1*B0
45160 REM FIND SQUARE ROOT
45161 GOSUB 45177
45162 B1=X1
45163 REM CHECK ACCURACY
45164 M=M+1
45165 IF ABS((A1-B1)/B1)<E THEN GOTO 45170
45166 A0=A1
45167 B0=B1
45168 GOTO 45158
45169 REM COMPUTE FINAL RESULT
45170 Y=A1*B1
45171 REM OBTAIN SQUARE ROOT
45172 GOSUB 45177
45173 Y=X*X2/X1
45174 RETURN
45175 REM ********************
45176 REM SQUARE ROOT SUBROUTINE
45177 X0=1
45178 X1=(X0+Y/X0)/2
45179 IF ABS((X0-X1)/X1)<E THEN RETURN
45180 X0=X1
45181 GOTO 45178
45194 REM ********************
45195 REM ARCSIN(X) RECURSION SUBROUTINE
45196 REM INPUT IS X (-1<X<1)
45197 REM OUTPUT IS Y=ARCSIN(X)
45198 REM CONVERGENCE CRITERIA IS E
45199 REM REFERENCE- COMPUTATIONAL ANALYSIS BY HENRICI
45200 M=0
45201 REM GUARD AGAINST FAILURE
45202 IF E<=0 THEN RETURN
45203 IF X<>0 THEN GOTO 45207
45204 Y=0
45205 RETURN
45206 REM CHECK RANGE
45207 IF ABS(X)>1 THEN RETURN
45208 U0=X*SQRT(1-X*X)
45209 U1=X
45210 U2=U1*SQRT(2*U1/(U1+U0))
45211 Y=U2
45212 M=M+1
45213 IF ABS(U2-U1)<E THEN RETURN
45214 U0=U1
45215 U1=U2
45216 GOTO 45210
45236 REM ********************
45237 REM COMPLETE ELLIPTIC INTEGRAL OF THE FIRST
```

```
45238 REM AND SECOND KIND (CLIPTIC)
45239 REM THE INPUT PARAMETER IS K, WHICH SHOULD
45240 REM BE BETWEEN 0 AND 1.
45241 REM TECHNIQUE USES GAUSS' FORMULA FOR THE
45242 REM ARITHMOGEOMETRICAL MEAN.
45243 REM REFERENCE- BALL, ALGORITHMS FOR RPN CALCULATORS.
45244 REM E IS A MEASURE OF THE CONVERGENCE ACCURACY.
45245 REM DEPENDING ON E, A(I) AND B(I) MAY HAVE TO BE DIMENSIONED
45246 REM IN THE CALLING PROGRAM.
45247 REM THE RETURNED VALUES ARE E1, THE ELLIPTIC
45248 REM INTEGRAL OF THE FIRST KIND, AND E2,
45249 REM THE INTEGRAL OF THE SECOND KIND.
45250 A(0)=1+K
45251 B(0)=1-K
45252 N=0
45253 IF K<0 THEN RETURN
45254 IF K>1 THEN RETURN
45255 IF E<=0 THEN RETURN
45256 IF K<1 THEN GOTO 45261
45257 E2=1
45258 E1=1000000000
45259 E1=E1*E1*E1*E1
45260 RETURN
45261 N=N+1
45262 REM GENERATE IMPROVED VALUES
45263 A(N)=(A(N-1)+B(N-1))/2
45264 B(N)=SQRT(A(N-1)*B(N-1))
45265 IF ABS(A(N)-B(N))>E THEN GOTO 45261
45266 E1=1.5707963268/A(N)
45267 E2=2
45268 M=1
45269 FOR I=1 TO N
45270 E2=E2-M*(A(I)*A(I)-B(I)*B(I))
45271 M=M*2
45272 NEXT I
45273 E2=E2*E1/2
45274 RETURN
45294 REM ********************
45295 REM LN(X) RECURSION SUBROUTINE
45296 REM INPUT IS X (0<X<1)
45297 REM OUTPUT IS Y=LN(X)
45298 REM CONVERGENCE CRITERIA IS E
45299 REM REFERENCE-  COMPUTATIONAL ANALYSIS BY HENRICI
45300 M=0
45301 REM GUARD AGAINST FAILURE
45302 IF E<=0 THEN RETURN
45303 IF X<1 THEN GOTO 45307
45304 Y=0
45305 RETURN
45306 REM CHECK RANGE
45307 IF X<=0 THEN RETURN
45308 U0=(X*X-1/(X*X))/4
45309 U1=(X-1/X)/2
45310 U2=U1*SQRT(2*U1/(U1+U0))
45311 Y=U2
45312 M=M+1
45313 IF ABS(U2-U1)<E THEN RETURN
45314 U0=U1
45315 U1=U2
```

```
45316 GOTO 45310
45341 REM ********************
45342 REM INTEGER ORDER BESSEL FUNCTION SUBROUTINE (INTBESSL)
45343 REM CALCULATES BESSEL FUNCTIONS OF ORDER 0 THROUGH 4
45344 REM FOR X>0.
45345 REM MILLER'S METHOD USED, SEE HENRICI
45346 REM ARGUMENT IS X
45347 REM NUMBER OF STEPS =M
45348 REM RETURNED RESULTS ARE Y(I)
45349 REM TEST FOR RANGE
45350 IF X<=0 THEN RETURN
45351 IF M<=0 THEN RETURN
45352 Y(0)=1
45353 Y(1)=0
45354 C=0
45355 N=M
45356 REM UPDATE RESULTS
45357 FOR I=4 TO 1 STEP -1
45358 Y(I)=Y(I-1)
45359 NEXT I
45360 REM APPLY RECURSION RELATION
45361 Y(0)=2*N*Y(1)/X-Y(2)
45362 N=N-1
45363 IF N=0 THEN GOTO 45367
45364 IF INT(N/2)<>N/2 THEN GOTO 45357
45365 C=C+2*Y(0)
45366 GOTO 45357
45367 C=C+Y(0)
45368 REM SCALE THE RESULTS
45369 FOR I=0 TO 4
45370 Y(I)=Y(I)/C
45371 NEXT I
45372 RETURN
45393 REM ********************
45394 REM LEGENDRE SERIES COEFFICIENT EVALUATION SUBROUTINE (LEGNDRE)
45395 REM BY MEANS OF RECURSION RELATION
45396 REM THE ORDER OF THE POLYNOMIAL IS N
45397 REM THE COEFFICIENTS ARE RETURNED IN A(I)
45398 REM DIMENSION A(I) AND B(I,J) IN THE CALLING PROGRAM
45399 REM ESTABLISH P0 AND P1 COEFFICIENTS
45400 B(0,0)-1
45401 B(1,0)=0
45402 B(1,1)=1
45403 REM RETURN IF ORDER IS LESS THAN TWO
45404 IF N<2 THEN RETURN
45405 FOR I=2 TO N
45406 B(I,0)=-(I-1)*B(I-2,0)/I
45407 FOR J=1 TO I
45408 REM BASIC RECURSION RELATION
45409 B(I,J)=(I+I-1)*B(I-1,J-1)-(I-1)*B(I-2,J)
45410 B(I,J)=B(I,J)/I
45411 NEXT J
45412 NEXT I
45413 FOR I=0 TO N
45414 A(I)=B(N,I)
45415 NEXT I
45416 RETURN
45418 REM ********************
45419 REM LAGUERRE POLYNOMIAL COEFFICIENT EVALUATION SUBROUTINE (LAGUERR)
```

```
45420 REM BY MEANS OF RECURSION RELATION
45421 REM THE ORDER OF THE POLYNOMIAL IS N
45422 REM THE COEFFICIENTS ARE RETURNED IN A(I)
45423 REM DIMENSION A(I) AND B(I,J) IN THE CALLING PROGRAM
45424 REM ESTABLISH L0 AND L1 COEFFICIENTS
45425 B(0,0)=1
45426 B(1,0)=1
45427 B(1,1)=-1
45428 REM RETURN IF ORDER IS LESS THAN TWO
45429 IF N<2 THEN RETURN
45430 FOR I=2 TO N
45431 B(I,0)=(2*I-1)*B(I-1,0)-(I-1)*(I-1)*B(I-2,0)
45432 FOR J=1 TO I
45433 REM BASIC RECURSION RELATION
45434 B(I,J)=(2*I-1)*B(I-1,J)-B(I-1,J-1)-(I-1)*(I-1)*B(I-2,J)
45435 NEXT J
45436 NEXT I
45437 FOR I=0 TO N
45438 A(I)=B(N,I)
45439 NEXT I
45440 RETURN
45443 REM ********************
45444 REM HERMITE POLYNOMIAL COEFFICIENT EVALUATION SUBROUTINE (HERMITE)
45445 REM BY MEANS OF RECURSION RELATION
45446 REM THE ORDER OF THE POLYNOMIAL IS N
45447 REM THE COEFFICIENTS ARE RETURNED IN A(I)
45448 REM DIMENSION A(I) AND B(I,J) IN THE CALLING PROGRAM
45449 REM ESTABLISH H0 AND H1 COEFFICIENTS
45450 B(0,0)=1
45451 B(1,0)=0
45452 B(1,1)=2
45453 REM RETURN IF ORDER IS LESS THAN TWO
45454 IF N<2 THEN RETURN
45455 FOR I=2 TO N
45456 B(I,0)=-2*(I-1)*B(I-2,0)
45457 FOR J=1 TO I
45458 REM BASIC RECURSION RELATION
45459 B(I,J)=2*B(I-1,J-1)-2*(I-1)*B(I-2,J)
45460 NEXT J
45461 NEXT I
45462 FOR I=0 TO N
45463 A(I)=B(N,I)
45464 NEXT I
45465 RETURN
45467 REM ********************
45468 REM TRIGONOMETRIC CORDIC SUBROUTINE (TRIGCORD)
45469 REM THIS SUBROUTINE CALCULATES THE SINE AND COSINE
45470 REM OF AN ANGLE USING THE CORDIC ROTATION METHOD.
45471 REM THE INPUT ANGLE IS A.
45472 REM THE SINE IS RETURNED IN U, AND THE COSINE IN V.
45473 REM REMEMBER TO DIMENSION W(I) AND A(I) IN THE CALLING PROGRAM
45474 REM IF THE ANGLE IS ZERO, SET FUNCTIONS AND RETURN.
45475 IF A<>0 THEN GOTO 45480
45476 U=0
45477 V=1
45478 RETURN
45479 REM GET THE TANGENT COEFFICIENTS
45480 GOSUB 45518
45481 REM REM DETERMINE THE WEIGHTS, W(I)1060
```

```
45482 GOSUB 45506
45483 U0=P
45484 V0=0
45485 REM PERFORM THE ROTATIONS UP TO THE RESIDUAL
45486 FOR I=1 TO N
45487 REM UPDATE U0 AND V0
45488 U1=U0-W(I)*A(I)*V0
45489 V1=V0+W(I)*A(I)*U0
45490 U0=U1
45491 V0=V1
45492 NEXT I
45493 REM PERFORM THE RESIDUAL ROTATION USING Z
45494 REM USE U0 AND V0 AS DUMMY VARIABLES
45495 U0=1-Z*Z/2
45496 V0=Z*(1+Z*Z/3)
45497 REM U AND V ARE THE FINAL RESULTS
45498 V=U0*(U1-V0*V1)
45499 U=U0*(V1+V0*U1)
45500 REM U=SIN(A)
45501 REM V=COS(A)
45502 RETURN
45503 REM TRIG CORDIC WEIGHTS SUBROUTINE
45504 REM THE WEIGHTS ARE W(I)= PLUS OR MINUS 1
45505 REM THE INPUT ANGLE IS A
45506 Z=A
45507 FOR I=1 TO N
45508 W(I)=-1
45509 IF Z>0 THEN W(I)=1
45510 Z=Z-W(I)*A0
45511 A0=A0/2
45512 NEXT I
45513 REM Z IS THE RESIDUAL ANGLE
45514 REM ********************
45515 RETURN
45516 REM TRIG CORDIC COEFFICIENT SUBROUTINE
45517 REM FOR CASE N=12
45518 N=12
45519 REM THE TANGENTS ARE GIVEN IN A(I)
45520 A(1)=1
45521 A(2)=.414213562373095
45522 A(3)=.198912367379658
45523 A(4)=.098491403357164425
45524 A(5)=.049126849769467725
45525 A(6)=.02454862210892544
45526 A(7)=.01227246237956627
45527 A(8)=.006136000157623401
45528 A(9)=.003067971201422665
45529 A(10)=.001533981991088666
45530 A(11)=.0007669905443430926
45531 A(12)=.0003834952157714441
45532 REM P REPRESENTS P(N)
45533 P=.6366197879720413
45534 REM A0 IS PI/4
45535 A0=.7853981633974484
45536 RETURN
45541 REM ********************
45542 REM INVERSE TRIGONOMETRIC CORDIC SUBROUTINE (INVCORD).
45543 REM THIS SUBROUTINE CALCULATES THE ANGLE CORRESPONDING
45544 REM TO A GIVEN SINE, COSINE OR TANGENT USING THE
```

```
45545 REM CORDIC ROTATION METHOD.
45546 REM THE INPUT IS U=SIN(A), V=COS(A), OR W=TAN(A).
45547 REM THE RETURNED VALUE IS A.
45548 REM REMEMBER TO DIMENSION W(I) AND A(I) IN THE CALLING PROGRAM.
45549 REM TRANSLATE THE U,V,W INPUTS
45550 IF U=0 THEN GOTO 45555
45551 REM INVERSE SINE IS WANTED
45552 IF ABS(U)>=.0001 THEN V=SQRT(1-U*U)
45553 IF ABS(U)<.0001 THEN V=1-U*U/2-U*U*U*U/8
45554 GOTO 45571
45555 IF V=0 THEN GOTO 45560
45556 REM INVERSE COSINE IS WANTED
45557 IF ABS(V)>=.0001 THEN U=SQRT(1-V*V)
45558 IF ABS(V)<.0001 THEN U=1-V*V/2-V*V*V*V/8
45559 GOTO 45571
45560 IF W=0 THEN GOTO 45568
45561 REM INVERSE TANGENT IS WANTED
45562 IF ABS(W)<=10000 THEN U=1/SQRT(1+1/(W*W))
45563 IF ABS(W)>10000 THEN U=1-1/(2*W*W)+3/(8*W*W*W*W)
45564 IF ABS(U)>=.0001 THEN V=SQRT(1-U*U)
45565 IF ABS(U)<.0001 THEN V=1-U*U/2-U*U*U*U/8
45566 U=U*ABS(W)/W
45567 GOTO 45571
45568 A=0
45569 RETURN
45570 REM GET COEFFICIENTS
45571 GOSUB 45616
45572 REM TEST FOR SPECIAL VALUES
45573 IF ABS(U)<1 THEN GOTO 45578
45574 IF ABS(V)>0 THEN GOTO 45578
45575 REM SPECIAL CASE FOUND
45576 A=2*A0*ABS(U)/U
45577 RETURN
45578 IF ABS(V)<1 THEN GOTO 45583
45579 IF ABS(U)>0 THEN GOTO 45583
45580 A=0
45581 RETURN
45582 REM SWITCH U WITH V AND INITIALIZE
45583 A=U
45584 U=V
45585 V=A
45586 U0=P
45587 U1=U0
45588 V0=0
45589 V1=V0
45590 REMERFORM THE ROTATIONS UP TO THE RESIDUAL
45591 FOR I=1 TO N
45592 REM IS ROTATION TO BE PLUS OR MINUS?
45593 W(I)=-1
45594 IF V0<V THEN W(I)=1
45595 REM UPDATE U0 AND V0
45596 U1=U0-W(I)*A(I)*V0
45597 V1=V0+W(I)*A(I)*U0
45598 U0=U1
45599 V0=V1
45600 NEXT I
45601 REM THE SET OF W(I) WEIGHTS HAVE NOW BEEN DETERMINED
45602 REM PERFORM THE RESIDUAL ANGLE APPROXIMATION
45603 Z=V*U1-U*V1
```

```
45604 Z=Z+Z*Z*Z/6
45605 REM ASSEMBLE RESULTS
45606 FOR I=1 TO N
45607 Z=Z+W(I)*A0
45608 A0=A0/2
45609 NEXT I
45610 REM RESULT IS IN Z
45611 A=Z
45612 RETURN
45613 REM ********************
45614 REM TRIG CORDIC COEFFICIENT SUBROUTINE
45615 REM FOR CASE N=12
45616 N=12
45617 REM THE TANGENTS ARE GIVEN IN A(I)
45618 A(1)=1
45619 A(2)=.414213562373095
45620 A(3)=.198912367379658
45621 A(4)=.09849140335716425
45622 A(5)=.04912684976946725
45623 A(6)=.02454862210892544
45624 A(7)=.01227246237956627
45625 A(8)=.006136000157623401
45626 A(9)=.003067971201422665
45627 A(10)=.001533981991088666
45628 A(11)=.0007669905443430926
45629 A(12)=.000383495215771441
45630 REM P REPRESENTS P(N)
45631 P=.6366197879720413
45632 REM A0 IS PI/4
45633 A0=.7853981633974484
45634 RETURN
45644 REM ********************
45645 REM MODIFIED CORDIC EXPONENTIAL SUBROUTINE (EXPCORD).
45646 REM THIS PROGRAM TAKES AN INPUT VALUE AND RETURNS Y=EXP(X).
45647 REM X MAY BE ANY POSITIVE OR NEGATIVE VALUE.
45648 REM REMEMBER TO DIMENSION A(I) AND W(I) IN THE CALLING PROGRAM.
45649 REM GET COEFFICIENTS
45650 GOSUB 45684
45651 REM REDUCE THE RANGE OF X
45652 K=INT(X)
45653 X=X-K
45654 REM DETERMINE THE WEIGHTING COEFFICIENTS, W(I)
45655 GOSUB 45673
45656 REM CACLULATE PRODUCTS
45657 Y=1
45658 FOR I=1 TO N
45659 IF W(I)>0 THEN Y=Y*A(I)
45660 NEXT I
45661 REM PERFORM RESIDUAL MULTIPLICATION
45662 Y=Y*(1+Z*(1+Z/2*(1+Z/3*(1+Z/4))))
45663 REM ACCOUNT FOR FACTOR EXP(K)
45664 IF K<0 THEN E=1/E
45665 IF ABS(K)<1 THEN GOTO 45671
45666 FOR I=1 TO ABS(K)
45667 Y=Y*E
45668 NEXT I
45669 REM RESTORE X
45670 X=X+K
45671 RETURN
```

```
45672 REM WEIGHT DETERMINATION SUBROUTINE
45673 A=.5
45674 Z=X
45675 FOR I=1 TO N
45676 W(I)=0
45677 IF Z>A THEN W(I)=1
45678 Z=Z-W(I)*A
45679 A=A/2
45680 NEXT I
45681 RETURN
45682 REM ********************
45683 REM EXPONENTIAL COEFFICIENTS SUBROUTINE
45684 N=9
45685 E=2.718281828459045
45686 A(1)=1.648721270700128
45687 A(2)=1.284025416687742
45688 A(3)=1.133148453066826
45689 A(4)=1.064494458917859
45690 A(5)=1.031743407499103
45691 A(6)=1.015747708586686
45692 A(7)=1.007843097206448
45693 A(8)=1.003913889338348
45694 A(9)=1.001955033591003
45695 RETURN
45719 REM ********************
45720 REM MODIFIED CORDIC NATURAL LOGARITHM SUBROUTINE (LNCORDIC).
45721 REM THIS PROGRAM TAKES AN INPUT VALUE AND RETURNS Y=LN(X).
45722 REM X MAY BE ANY POSITIVE VALUE.
45723 REM REMEMBER TO DIMENSION A(I) AND W(I) IN THE CALLING PROGRAM.
45724 REM GET COEFFICIENTS
45725 GOSUB 45770
45726 REM IF X<=0 THEN AN ERROR EXISTS, RETURN
45727 IF X<=0 THEN RETURN
45728 K=0
45729 REM SAVE X
45730 X1=X
45731 REM REDUCE THE RANGE OF X
45732 IF X<E THEN GOTO 45739
45733 REM DIVIDE OUT A POWER OF E
45734 K=K+1
45735 X=X/E
45736 GOTO 45732
45737 REM TEST IF X>=1. IF SO GO TO NEXT STEP
45738 REM OTHERWISE, BRING X TO >1
45739 IF X>=1 THEN GOTO 45744
45740 K=K-1
45741 X=X*E
45742 GOTO 45739
45743 REM DETERMINE THE WEIGHTING COEFFICIENTS, W(I)
45744 GOSUB 45761
45745 REM CALCULATE RESIDUAL FACTOR BASED ON Z
45746 REM WANT LN(Z), WHERE Z IS NEAR UNITY
45747 Z=Z-1
45748 Z=Z*(1-(Z/2)*(1+(Z/3)*(1-Z/4)))
45749 REM ASSEMBLE RESULTS
45750 A=1/2
45751 FOR I=1 TO N
45752 Z=Z+W(I)*A
45753 A=A/2
```

```
45754 NEXT I
45755 REM Z IS NOW THE MANTISSA, K THE CHARACTERISTIC
45756 Y=K+Z
45757 REM RESTORE X
45758 X=X1
45759 RETURN
45760 REM WEIGHT DETERMINATION SUBROUTINE
45761 Z=X
45762 FOR I=1 TO N
45763 W(I)=0
45764 IF Z>A(I) THEN W(I)=1
45765 IF W(I)=1 THEN Z=Z/A(I)
45766 NEXT I
45767 RETURN
45768 REM ********************
45769 REM EXPONENTIAL COEFFICIENTS SUBROUTINE
45770 N=15
45771 E=2.718281828459045
45772 A(1)=1.648721270700128
45773 A(2)=1.284025416687742
45774 A(3)=1.133148453066826
45775 A(4)=1.064494458917859
45776 A(5)=1.031743407499103
45777 A(6)=1.015747708586686
45778 A(7)=1.007843097206448
45779 A(8)=1.003913889338348
45780 A(9)=1.001955033591003
45781 A(10)=1.000977039492417
45782 A(11)=1.000488400478694
45783 A(12)=1.000244170429748
45784 A(13)=1.000122077763384
45785 A(14)=1.000061037018933
45786 A(15)=1.000030518043791
45787 RETURN
45790 REM ********************
45791 REM HYPERBOLIC SINE SUBROUTINE (SINH).
45792 REM THIS PROGRAM USES THE DEFINITION OF THE
45793 REM HYPERBOLIC SINE AND THE MODIFIED CORDIC
45794 REM EXPONENTIAL SUBROUTINE TO APPROXIMATE
45795 REM ARCSINH(X) OVER THE ENTIRE RANGE OF REAL X.
45796 REM THE INPUT TO THE SUBROUTINE IS X.
45797 REM THE RETURNED VALUE IS Y=ARCSINH(X).
45798 REM START CALCULATION
45799 REM IS X SMALL ENOUGH TO CAUSE ROUND OFF ERROR?
45800 IF ABS(X)<.35 THEN GOTO 45808
45801 REM CALCULATE SINH(X) USING EXPONENTIAL DEFINITION
45802 REM GET EXP(X)
45803 GOSUB 45650
45804 REM CALCULATE SINH(X)
45805 Y=(Y-(1/Y))/2
45806 RETURN
45807 REM SERIES APPROXIMATION
45808 Z=1
45809 Y=1
45810 FOR I=1 TO 8
45811 Z=Z*X*X/((2*I)*(2*I+1))
45812 Y=Y+Z
45813 NEXT I
45814 Y=X*Y
```

```
45815 RETURN
45816 REM *********************
45817 REM HYPERBOLIC COSINE SUBROUTINE (COSH).
45818 REM THIS PROGRAM USES THE DEFINITION OF THE
45819 REM HYPERBOLIC COSINE AND THE MODIFIED CORDIC
45820 REM EXPONENTIAL SUBROUTINE TO APPROXIMATE
45821 REM ARCOSH(X) OVER THE ENTIRE RANGE OF REAL X.
45822 REM THE RETURNED VALUE IS Y=ARCOSH(X).
45823 REM START CALCULATION
45824 REM GET EXP(X)
45825 GOSUB 45650
45826 Y=(Y+(1/Y))/2
45827 RETURN
45828 REM *********************
45829 REM HYPERBOLIC TANGENT SUBROUTINE (TANH).
45830 REM THIS PROGRAM USES THE DEFINITION
45831 REM    TAN(X)=SINH(X)/COSH(X)
45832 REM TO CALCULATE THE HYPERBOLIC TANGENT.
45833 REM THE INPUT IS X.
45834 REM THE OUTPUT IS Y=TANH(X).
45835 REM START CALCULATION
45836 REM GET SINH(X)
45840 GOSUB 45800
45841 V=Y
45842 REM GET COSH(X)
45843 GOSUB 45825
45844 Y=V/Y
45845 RETURN
45866 REM *********************
45867 REM ARCSINH(X) SUBROUTINE (INVSINH).
45868 REM THIS ROUTINE CALCULATES THE INVERSE
45869 REM HYPERBOLIC SINE USING THE MODIFIED
45870 REM CORDIC NATURAL LOGARITHM SUBROUTINE.
45871 REM THE INPUT IS X.
45872 REM THE OUTPUT IS Y=ARCSINH(X).
45873 REM START CALCULATION
45874 REM TEST FOR ZERO ARGUMENT
45875 IF X<>0 THEN GOTO 45879
45876 Y=0
45877 RETURN
45878 REM SAVE X
45879 X2=X
45880 X=ABS(X)
45881 X=X+SQRT(X*X+1)
45882 REM GET LOGARITHM
45883 GOSUB 45725
45884 REM INSERT SIGN
45885 Y=(X2/ABS(X2))*Y
45886 REM RESTORE X
45887 X=X2
45888 RETURN
45891 REM *********************
45892 REM ARCCOSH(X) SUBROUTINE (INVCOSH).
45893 REM THIS ROUTINE CALCULATES THE INVERSE
45894 REM HYPERBOLIC COSINE USING THE MODIFIED
45895 REM CORDIC NATURAL LOGARITHM SUBROUTINE.
45896 REM THE INPUT IS X.
45897 REM THE OUTPUT IS Y=ARCOSH(X).
45898 REM BEGIN CALCULATION
```

```
45899 REM TEST FOR ARGUMENT LESS THAN OR EQUAL TO UNITY
45900 IF X>1 THEN GOTO 45904
45901 Y=0
45902 RETURN
45903 REM SAVE X
45904 X2=X
45905 X=ABS(X)
45906 X=X+SQRT(X-1)*SQRT(X+1)
45907 REM GET LOGARITHM
45908 GOSUB 45725
45909 REM RESTORE X
45910 X=X2
45911 RETURN
45916 REM ARCTANH(X) SUBROUTINE (INVTANH)
45917 REM *********************
45918 REM THIS PROGRAM CALCULATES THE INVERSE
45919 REM HYPERBOLIC TANGENT USING THE MODIFIED
45920 REM CORDIC NATURAL LOGARITHM SUBROUTINE.
45921 REM THE INPUT IS X.
45922 REM THE OUTPUT IS Y=ARCTANH(X).
45923 REM START CALCULATION
45924 REM TEST FOR X>= +/- 1
45925 IF ABS(X)<1 THEN GOTO 45929
45926 Y=(X/ABS(X))*1000000*1000000*1000000
45927 RETURN
45928 REM TEST FOR ZERO ARGUMENT
45929 IF X<>0 THEN GOTO 45933
45930 Y=0
45931 RETURN
45932 REM SAVE X
45933 X2=X
45934 X=(1+X)/(1-X)
45935 REM GET LOGARITHM
45936 GOSUB 45725
45937 REM RESTORE X
45938 X=X2
45939 RETURN
45988 REM *********************
45989 REM LAGRANGE INTERPOLATION SUBROUTINE (LAGRANGE)
45990 REM N IS THE LEVEL OF THE INTERPOLATION (EG., N=2 IS QUADRATIC).
45991 REM V IS THE TOTAL NUMBER OF TABLE VALUES.
45992 REM (X(I),Y(I)) ARE THE COORDINATE TABLE VALUES, Y(I) BEING THE
45993 REM DEPENDENT VARIABLE. THE X(I) MAY BE ARBITRARILY SPACED.
45994 REM X IS THE INTERPOLATION POINT WHICH IS ASSUMED TO BE IN THE
45995 REM INTERVAL WITH AT LEAST ONE TABLE VALUE TO THE LEFT, AND N TO THE RIGHT.
45996 REM IF THIS IS VIOLATED, N WILL BE SET TO ZERO.
45997 REM IT IS ASSUMED THAT THE TABLE VALUES ARE IN ASCENDING X(I) ORDER.
45998 REM X(I), Y(I) AND L(I) MUST BE DIMENSIONED IN THE CALLING PROGRAM.
45999 REM CHECK TO SEE IF INTERPOLATION POINT IS IN APPROPRIATE RANGE
46000 IF X<X(1) THEN GOTO 46003
46001 IF X<=X(V-N) THEN GOTO 46006
46002 REM AN ERROR HAS BEEN ENCOUNTERED
46003 N=0
46004 RETURN
46005 REM FIND THE RELEVANT TABLE INTERVAL
46006 I=0
46007 I=I+1
46008 IF X>X(I) THEN GOTO 46007
46009 I=I-1
```

```
46010 REM BEGIN INTERPOLATION
46011 FOR J=0 TO N
46012 L(J)=1
46013 NEXT J
46014 Y=0
46015 FOR K=0 TO N
46016 FOR J=0 TO N
46017 IF J=K THEN GOTO 46019
46018 L(K)=L(K)*(X-X(J+I))/(X(I+K)-X(J+I))
46019 NEXT J
46020 Y=Y+L(K)*Y(I+K)
46021 NEXT K
46022 RETURN
46039 REM ********************
46040 REM NEWTON DIVIDED DIFFERENCES INTERPOLATION SUBROUTINE (NEWTON)
46041 REM CALCULATES CUBIC INTERPOLATIONS FOR A GIVEN TABLE.
46042 REM (X(I),Y(I)) ARE THE V COORDINATE PAIRS OF DATA.
46043 REM X(I) IS THE INDEPENDENT VARIABLE, Y(I) THE DEPENDENT.
46044 REM THE INTERPOLATION POINT IS X. IT IS ASSUMED THAT THERE
46045 REM IS AT LEAST ONE DATA POINT TO THE LEFT, AND THREE TO THE RIGHT.
46046 REM IF THIS IS VIOLATED, THEN N IS SET TO 0.
46047 REM E IS THE ERROR ESTIMATE.
46048 REM X,Y,Y1,Y2 AND Y3 ARE ASSUMED DIMENSIONED IN THE CALLING PROGRAM
46049 REM CHECK TO SEE IF X IS IN THE INTERVAL
46050 N=1
46051 IF X>=X(1) THEN GOTO 46054
46052 N=0
46053 RETURN
46054 IF X<=X(V-3) THEN GOTO 46058
46055 N=0
46056 RETURN
46057 REM GENERATE DIVIDED DIFFERENCES
46058 FOR I=1 TO V-1
46059 Y1(I)=(Y(I+1)-Y(I))/(X(I+1)-X(I))
46060 NEXT I
46061 FOR I=1 TO V-2
46062 Y2(I)=(Y1(I+1)-Y1(I))/(X(I+1)-X(I))
46063 NEXT I
46064 FOR I=1 TO V-3
46065 Y3(I)=(Y2(I+1)-Y2(I))/(X(I+1)-X(I))
46066 NEXT I
46067 REM FIND RELEVANT TABLE INTERVAL
46068 I=0
46069 I=I+1
46070 IF X>X(I) THEN GOTO 46069
46071 I=I-1
46072 REM BEGIN INTERPOLATION
46073 A=X-X(I)
46074 B=A*(X-X(I+1))
46075 C=B*(X-X(I+2))
46076 Y=Y(I)+A*Y1(I)+B*Y2(I)+C*Y3(I)
46077 REM CALCULATE NEXT TERM IN THE EXPANSION FOR AN ERROR ESTIMATE
46078 E=C*(X-X(I+3))*Y/24
46079 RETURN
46090 REM *********************
46091 REM AKIMA SPLINE FITTING SUBROUTINE (AKIMA)
46092 REM THE INPUT TABLE IS (X(I),Y(I)), WHERE Y(I)
46093 REM IS THE DEPENDENT VARIABLE.
46094 REM THE INTERPOLATION POINT IS X, WHICH IS ASSUMED
```

```
46095 REM TO BE IN THE RANGE OF THE TABLE WITH AT LEAST
46096 REM ONE TABLE POINT TO THE LEFT, AND THREE TO THE RIGHT.
46097 REM Y IS RETURNED AS THE INTERPOLATED VALUE.
46098 REM N IS RETURNED AS AN ERROR CHECK (N=0 IMPLIES ERROR).
46099 REM DIMENSION M,X,Y AND Z IN THE CALLING PROGRAM
46100 N=1
46101 REM CHECK TO SEE IF X IS IN THE TABLE RANGE
46102 IF X>=X(1) THEN GOTO 46105
46103 N=0
46104 RETURN
46105 IF X<=X(V-3) THEN GOTO 46108
46106 N=0
46107 RETURN
46108 X(0)=2*X(1)-X(2)
46109 REM CALCULATE AKIMA COEFFICIENTS
46110 FOR I=1 TO V-1
46111 REM SHIFT I TO I+2
46112 M(I+2)=(Y(I+1)-Y(I))/(X(I+1)-X(I))
46113 NEXT I
46114 M(V+2)=2*M(V+1)-M(V)
46115 M(V+3)=2*M(V+2)-M(V+1)
46116 M(2)=2*M(3)-M(4)
46117 M(1)=2*M(2)-M(3)
46118 FOR I=1 TO V
46119 A=ABS(M(I+3)-M(I+2))
46120 B=ABS(M(I+1)-M(I))
46121 IF A+B<>0 THEN GOTO 46124
46122 Z(I)=(M(I+2)+M(I+1))/2
46123 GOTO 46125
46124 Z(I)=(A*M(I+1)+B*M(I+2))/(A+B)
46125 NEXT I
46126 REM FIND RELEVANT TABLE INTERVAL
46127 I=0
46128 I=I+1
46129 IF X>=X(I) THEN GOTO 46128
46130 I=I-1
46131 REM BEGIN INTERPOLATION
46132 B=X(I+1)-X(I)
46133 A=X-X(I)
46134 Y=Y(I)+Z(I)*A+(3*M(I+2)-2*Z(I)-Z(I+1))*A*A/B
46135 Y=Y+(Z(I)+Z(I+1)-2*M(I+2))*A*A*A/(B*B)
46136 RETURN
46138 REM **********************
46139 REM LAGRANGE DERIVATIVE INTERPOLATION SUBROUTINE (D/LAGRNG)
46140 REM N IS THE LEVEL OF THE INTERPOLATION (EG., N=2 IS QUADRATIC).
46141 REM V IS THE TOTAL NUMBER OF TABLE VALUES.
46142 REM (X(I),Y(I)) ARE THE COORDINATE TABLE VALUES, Y(I) BEING THE
46143 REM DEPENDENT VARIABLE. THE X(I) MAY BE ARBITRARILY SPACED.
46144 REM X IS THE INTERPOLATION POINT WHICH IS ASSUMED TO BE IN THE
46145 REM INTERVAL WITH AT LEAST ONE TABLE VALUE TO THE LEFT, AND N TO THE RIGHT.
46146 REM Y IS RETURNED AS THE DESIRED DERIVATIVE.
46147 REM N IS RETURNED AS THE ERROR CHECK (N=0 IMPLIES ERROR).
46148 REM DIMENSION L(I),M(I,J),X(I) AND Y(I) IN THE CALLING PROGRAM
46149 REM CHECK TO SEE IF X IS IN INTERVAL
46150 IF X>X(1) THEN GOTO 46153
46151 N=0
46152 RETURN
46153 IF X<=X(V-N) THEN GOTO 46157
46154 N=0
```

```
46155 RETURN
46156 REM FIND THE RELEVANT TABLE INTERVAL
46157 I=0
46158 I=I+1
46159 IF X>X(I) THEN GOTO 46158
46160 I=I-1
46161 REM BEGIN INTERPOLATION
46162 FOR J=0 TO N
46163 L(J)=0
46164 FOR K=0 TO N
46165 M(J,K)=1
46166 NEXT K
46167 NEXT J
46168 Y=0
46169 FOR K=0 TO N
46170 FOR J=0 TO N
46171 IF J=K THEN GOTO 46179
46172 FOR L=0 TO N
46173 IF L=K THEN GOTO 46178
46174 IF L<>J THEN GOTO 46177
46175 M(L,K)=M(L,K)/(X(I+K)-X(I+J))
46176 GOTO 46178
46177 M(L,K)=M(L,K)*(X-X(J+I))/(X(I+K)-X(I+J))
46178 NEXT L
46179 NEXT J
46180 FOR L=0 TO N
46181 IF L=K THEN GOTO 46183
46182 L(K)=L(K)+M(L,K)
46183 NEXT L
46184 Y=Y+L(K)*Y(I+K)
46185 NEXT K
46186 RETURN
46189 REM ********************
46190 REM GENERAL INTEGRATION SUBROUTINE (ITEG)
46191 REM INTERPOLATION BY AKIMA (OR OTHER).
46192 REM INTEGRATION BY ENHANCED TRAPAZOIDAL RULE.
46193 REM REM WITH RICHARDSON EXTRAPOLATION TO GIVE CUBIC ACCURACY.
46194 REM CAN BE USED UNDER VERY GENERAL CONDITIONS.
46195 REM THE INTEGRATION RANGE IS (X1,X2).
46196 REM IT IS ASSUMED THAT X1<X2, AND THAT THERE IS AT LEAST ONE TABLE
46197 REM VALUE TO THE LEFT OF X1, AND THREE TO THE RIGHT OF X2.
46198 REM THE RESULT IS RETURNED IN Z
46199 REM AN ERROR CHECK IS RETURNED IN Z1- Z1=0 IMPLIES ERROR.
46200 Z=0
46201 Z1=0
46202 REM CHECK TO SEE IF END POINTS ARE IN ALLOWABLE RANGE
46203 IF X1<X(1) THEN RETURN
46204 IF X2>X(V-3) THEN RETURN
46205 REM IF X1>X2 THEN SWITCH AND SET FLAG
46206 IF X1<X2 THEN GOTO 46211
46207 X3=X1
46208 X1=X2
46209 X2=X3
46210 Z1=1
46211 IF X2=X1 THEN RETURN
46212 REM START TRAPAZOIDAL INTEGRATIONS
46213 REM FIRST INTEGRATION TO GET I1
46214 GOSUB 46232
46215 REM SECOND ROUND TO GET I2
```

```
46216 GOSUB 46267
46217 REM RICHARDSON EXTRAPOLATION
46218 Z=4*I2/3-I1/3
46219 REM CHECK TO SEE IF THE END POINTS HAVE BEEN REVERSED
46220 IF Z1=0 THEN GOTO 46225
46221 Z=-Z
46222 X2=X1
46223 X1=X3
46224 REM RESET ERROR FLAG
46225 Z1=1
46226 RETURN
46227 REM Z IS THE INTEGRAL DESIRED
46228 RETURN
46229 REM ********************
46230 REM ROUTINE FOR THE FIRST TRAPAZOIDAL INTEGRATION, I1
46231 REM N1 KEEPS TRACK OF THE NUMBER OF INTERVALS
46232 I1=0
46233 N1=0
46234 X=X1
46235 REM FIND THE BEGINNING OF THE INTERVAL
46236 REM GO TO BRANCH WHICH CALLS THE INTERPOLATION SUBROUTINE
46237 REM FIND THE INTERVAL, I, AND THE LEFT END POINT, Y
46238 GOSUB 46299
46239 REM IS THERE AT LEAST ONE TABLE INTERVAL?
46240 IF X2>X(I+1) THEN GOTO 46250
46241 REM IF NOT, INTEGRAL IS SIMPLE
46242 N1=N1+1
46243 D=Y
46244 X=X2
46245 REM FIND END POINT Y VALUE
46246 GOSUB 46299
46247 I1=(Y+D)*(X2-X1)/2
46248 RETURN
46249 REM AT LEAST ONE TABLE INTERVAL MUST BE SUMMED OVER
46250 J1=I
46251 I1=I1+(Y+Y(I+1))*(X(I+1)-X)/2
46252 REM ANY MORE INTERVALS? IF NOT, FINISH INTEGRAL WITH END POINT
46253 IF X2<X(J1+3) THEN GOTO 46259
46254 REM OTHERWISE, KEEP SUMMING
46255 N1=N1+1
46256 I1=I1+(Y(J1+1)+Y(J1+3))*(X(J1+3)-X(J1+1))/2
46257 J1=J1+2
46258 GOTO 46253
46259 X=X2
46260 REM FIND LAST Y VALUE
46261 GOSUB 46299
46262 I1=I1+(Y+Y(J1+1))*(X2-X(J1+1))/2
46263 N1=N1+1
46264 RETURN
46265 REM ********************
46266 INTEGRATION FOR I2
46267 I2=0
46268 X=X1
46269 GOSUB 46299
46270 D=Y
46271 IF X2>X(I+1) THEN GOTO 46280
46272 X=X1+(X2-X1)/2
46273 GOSUB 46299
46274 I2=I2+(D+Y)*(X2-X1)/4
```

```
46275 D=Y
46276 X=X2
46277 GOSUB 46299
46278 I2=I2+(D+Y)*(X2-X1)/4
46279 RETURN
46280 X=X1+(X(I+1)-X1)/2
46281 J1=I
46282 GOSUB 46299
46283 I2=I2+(Y+D)*(X-X1)/2
46284 I2=I2+(Y+Y(J1+1))*(X(J1+1)-X)/2
46285 IF X2<X(J1+2) THEN GOTO 46289
46286 I2=I2+(Y(J1+1)+Y(J1+2))*(X(J1+2)-X(J1+1))/2
46287 J1=J1+1
46288 GOTO 46285
46289 X=X2-(X2-X(J1+1))/2
46290 GOSUB 46299
46291 D=Y
46292 I2=I2+(Y(J1+1)+D)*(X2-X)/2
46293 X=X2
46294 GOSUB 46299
46295 I2=I2+(D+Y)*(X2-X(J1+1))/4
46296 RETURN
46297 REM BRANCH TO AN INTERPOLATION SUBROUTINE FROM HERE
46298 REM GO TO AKIMA SPLINE INTERPOLATION SUBROUTINE
46299 GOSUB 46100
46300 REM JUST RETURNED FROM INTERPOLATION SUBROUTINE
46301 REM RETURN TO PROGRAM
46302 RETURN
46333 REM *********************
46334 REM BISECTION METHOD SUBROUTINE (BISECT)
46335 REM THIS PROGRAM ITERATIVELY SEEKS THE ZERO
46336 REM OF A FUNCTION USING THE METHOD OF INTERVAL
46337 REM HALVING UNTIL THE INTERVAL IS LESS THAN
46338 REM E IN WIDTH.
46339 REM IT IS ASSUMED THAT THE FUNCTION Y=Y(X)
46340 REM IS AVAILABLE FROM THE FUNCTION SUBROUTINE
46341 REM LOCATED AT 44300.
46342 REM THIS SUBROUTINE REQUIRES AS INPUT THE INITIAL
46343 REM RANGE VALUES (X0 AND X1), AS WELL AS THE
46344 REM CONVERGENCE CRITERION, E.
46345 REM THE ZERO MUST BE WITHIN THE RANGE SPECIFIED
46346 REM OR AN ERRONEOUS VALUE WILL BE RETURNED IN X.
46347 REM THIS SUBROUTINE RETURNS THE ESTIMATE OF THE ROOT
46348 REM IN X, AND THE CORRESPONDING Y VALUE.
46349 REM ALSO RETURNED IS THE NUMBER OF STEPS (M).
46350 M=0
46351 X=X0
46352 GOSUB 44300
46353 Y0=Y
46354 X=X1
46355 GOSUB 44300
46356 X=(X0+X1)/2
46357 GOSUB 44300
46358 M=M+1
46359 IF Y*Y0=0 THEN RETURN
46360 IF Y*Y0<0 THEN X1=X
46361 IF Y*Y0>0 THEN X0=X
46362 IF ABS(X1-X0)>E THEN GOTO 46351
46363 RETURN
```

```
46364 REM ********************
46365 REM NEWTON'S METHOD SUBROUTINE (Z-NEWTON)
46366 REM THIS PROGRAM CALCULATES THE ZEROS OF A
46367 REM FUNCTION BY NEWTON'S METHOD.
46368 REM THE ROUTINE REQUIRES AN INITIAL GUESS, X0,
46369 REM AND A CONVERGENCE FACTOR, E.
46370 REM ALSO REQUIRED IS A LIMIT ON THE NUMBER
46371 REM OF ITERATIONS, M. THE NUMBER USED IS
46372 REM RETURNED IN N.
46373 REM IT IS ASSUMED THAT THE FUNCTION AND ITS
46374 REM DERIVATIVE ARE IN THE SUBROUTINE AT 44300.
46375 N=0
46376 REM GET Y AND Y1
46377 X=X0
46378 GOSUB 44300
46379 REM UPDATE ESTIMATE
46380 X0=X0-Y/Y1
46381 N=N+1
46382 IF N>=M THEN RETURN
46383 IF ABS(Y/Y1)>E THEN GOTO 46377
46384 X=X0
46385 RETURN
46413 REM ********************
46414 REM SECANT METHOD SUBROUTINE (SECANT)
46415 REM THIS SUBROUTINE CALCULATES THE ZEROES OF A
46416 REM FUNCTION USING THE SECANT METHOD.
46417 REM TWO INITIAL GUESSES ARE REQUIRED, X0 AND X1.
46418 REM THE CONVERGENCE CRITERION IS E.
46419 REM THE MAXIMUM NUMBER OF ITERATIONS IS M.
46420 REM THE NUMBER OF ITERATIONS PERFORMED IS
46421 REM RETURNED IN N.
46422 REM THE RESULT IS RETURNED IN X.
46423 REM IT IS ASSUMED THAT THE FUNCTION, Y(X),
46424 REM IS IN THE SUBROUTINE AT 44300.
46425 N=0
46426 REM START ITERATION
46427 X=X0
46428 GOSUB 44300
46429 Y0=Y
46430 X=X1
46431 REM GET NEXT POINT
46432 GOSUB 44300
46433 Y1=Y
46434 REM CALCULATE NEW ESTIMATE
46435 REM IF Y1=Y0 THEN THERE WILL BE AN OVERFLOW
46436 REM GUARD AGAINST THIS ARTIFICIALLY
46437 IF Y1=Y0 THEN Y1=Y1+.001
46438 X=(X0*Y1-X1*Y0)/(Y1-Y0)
46439 N=N+1
46440 REM TEST FOR CONVERGENCE
46441 IF N>=M THEN RETURN
46442 IF ABS(X1-X0)<E THEN RETURN
46443 REM UPDATE POSITIONS
46444 X0=X1
46445 X1=X
46446 GOTO 46427
46463 REM ********************
46464 REM MODIFIED FALSE POSITION SUBROUTINE (REGULA)
46465 REM SUBROUTINE USES HAMMING'S MODIFICATION TO
```

```
46466 REM SPEED CONVERGENCE.
46467 REM IT IS ASSUMED THAT THE FUNCTION Y(X) IS
46468 REM IN THE SUBROUTINE AT 44300.
46469 REM THE TWO INTITIAL GUESSES ARE X0 AND X1.
46470 REM THESE TWO GUESSES MUST BRACKET THE ZERO.
46471 REM THE CONVERGENCE CRITERION IS E.
46472 REM THE MAXIMUM NUMBER OF GUESSES IS M.
46473 REM THE RESULT IS RETURNED IN X.
46474 REM THE NUMBER OF ITERATIONS IS RETURNED IN N.
46475 N=0
46476 REM ME X0<X1
46477 IF X0<X1 THEN GOTO 46481
46478 X=X0
46479 X0=X1
46480 X1=X
46481 X=X0
46482 REM GET Y0 AND Y1
46483 GOSUB 44300
46484 Y0=Y
46485 X=X1
46486 REM INITIAL GUESSES FOR A AND B ARE REQUIRED
46487 GOSUB 44300
46488 Y1=Y
46489 REM CALCULATE A NEW ESTIMATE, X
46490 X=(X0*Y1-X1*Y0)/(Y1-Y0)
46491 REM TEST FOR CONVERGENCE
46492 N=N+1
46493 IF N>=M THEN RETURN
46494 IF ABS(X1-X)<E THEN RETURN
46495 REM GET A NEW Y(X) VALUE
46496 GOSUB 44300
46497 REM APPLY HAMMING'S MODIFICATION
46498 IF Y1*Y=0 THEN RETURN
46499 IF Y0*Y>0 THEN GOTO 46504
46500 X1=X
46501 Y1=Y
46502 Y0=Y0/2
46503 GOTO 46490
46504 X0=X
46505 Y0=Y
46506 Y1=Y1/2
46507 GOTO 46490
46509 REM ********************
46510 REM AITKEN ACCELERATION SUBROUTINE (AITKEN)
46511 REM THIS ROUTINE CALCULATES THE ZEROS OF A FUNCTION
46512 REM BY ITERATION, AND EMPLOYS AITKEN ACCELERATION TO
46513 REM SPEED UP CONVERGENCE.
46514 REM REM THE SUBROUTINE REQUIRES AN INITIAL GUESS, X0,
46515 REM AND TWO CONVERGENCE FACTORS, C AND E.
46516 REM E RELATES TO THE ACCURACY OF THE ESTIMATE, AND C
46517 REM IS USED TO AID THE CONVERGENCE.
46518 REM ALSO REQUIRED IS AN ITERATION LIMIT, M.
46519 REM C=-1 IS A NORMAL VALUE. IF DIVERGENCE OCCURS,
46520 REM SMALLER AND/OR POSITIVE VALUES SHOULD BE TRIED.
46521 REM THE RESULT IS RETURNED IN X.
46522 REM THE NUMBER OF ITERATIONS IS RETURNED IN N.
46523 REM IT IS ASSUMED THAT THE FUNCTION Y(X) IS IN
46524 REM THE SUBROUTINE AT 44300.
46525 N=0
```

```
46526 X=X0
46527 REM GET Y
46528 GOSUB 44300
46529 Y=X+C*Y
46530 REM ARE THERE ENOUGH POINTS FOR ACCELERATION?
46531 IF N>0 THEN GOTO 46536
46532 X1=Y
46533 X=X1
46534 N=N+1
46535 GOTO 46528
46536 X2=Y
46537 N=N+1
46538 REM GUARD AGAINST A ZERO DENOMINATOR
46539 IF X2-2*X1+X0=0 THEN X0=X0+.001
46540 REM PERFORM ACCELERATION
46541 K=(X2-X1)*(X2-X1)/(X2-2*X1+X0)
46542 X2=X2-K
46543 REM TEST FOR CONVERGENCE
46544 IF N>=M THEN RETURN
46545 IF ABS(K)<E THEN RETURN
46546 X0=X1
46547 X1=X2
46548 X=X1
46549 GOTO 46528
46559 REM ********************
46560 REM AITKEN-STEFFENSON ITERATION SUBROUTINE (A/SITER)
46561 REM THIS ROUTINE CALCULATES THE ZEROS OF A FUNCTION
46562 REM BY ITERATION, AND EMPLOYS AITKEN ACCELERATION TO
46563 REM SPEED UP CONVERGENCE.
46564 REM THE SUBROUTINE REQUIRES AN INITIAL GUESS, X0,
46565 REM AND TWO CONVERGENCE FACTORS, C AND E.
46566 REM E RELATES TO THE ACCURACY OF THE ESTIMATE, AND C
46567 REM IS USED TO AID THE CONVERGENCE.
46568 REM ALSO REQUIRED IS A LIMIT TO THE NUMBER OF ITERATIONS, M.
46569 REM C=-1 IS A NORMAL VALUE. IF DIVERGENCE OCCURS,
46570 REM SMALLER AND/OR POSITIVE VALUES SHOULD BE TRIED.
46571 REM THE RESULT IS RETURNED IN X.
46572 REM THE NUMBER OF ITERATIONS IS RETURNED IN N.
46573 REM IT IS ASSUMED THAT THE FUNCTION Y(X) IS IN
46574 REM THE SUBROUTINE AT 44300.
46575 N=0
46576 M1=0
46577 X=X0
46578 REM GET Y
46579 GOSUB 44300
46580 Y=X+C*Y
46581 REM ARE THERE ENOUGH POINTS FOR ACCELERATION?
46582 IF M1>0 THEN GOTO 46588
46583 N=N+1
46584 M1=M1+1
46585 X=X1
46586 X1=Y
46587 GOTO 46579
46588 X2=Y
46589 REM PERFORM ACCELERATION
46590 REM GUARD AGAINST ZERO DENOMINATOR
46591 K=(X2-2*X1+X0)
46592 IF K=0 THEN K=.001
46593 K=(X1-X0)*(X1-X0)/K
```

```
46594 X0=X0-K
46595 REM TEST FOR CONVERGENCE
46596 IF N>=M THEN RETURN
46597 IF ABS(X-X0)<E THEN RETURN
46598 REM REPEAT PROCESS
46599 GOTO 46576
46617 REM ********************
46618 REM SYNTHETIC DIVISION SUBROUTINE (RSYNDIV)
46619 REM ASSUMES REAL POLYNOMIAL COEFFICIENTS.
46620 REM FORM CALCULATED IS A(X)=C(X)/B(X).
46621 REM THE INPUT POLYNOMIAL COEFFICIENTS ARE
46622 REM C(I) AND B(I), THE RESULT IS A(I).
46623 REM C(X) IS OF ORDER N1, B(X) IS OF ORDER N2.
46624 REM RESULT IS OF ORDER N1-N2 (AT MOST).
46625 FOR I=N1 TO N2 STEP -1
46626 A(I-N2)=C(I)/B(N2)
46627 IF I=N2 THEN GOTO 46631
46628 FOR J=0 TO N2
46629 C(I-J)=C(I-J)-A(I-N2)*B(N2-J)
46630 NEXT J
46631 NEXT I
46632 RETURN
46641 REM ********************
46642 REM SUBROUTINE FOR DETERMINING ADDITIONAL ROOTS OF
46643 REM A FUNCTION GIVEN A SET OF ALREADY ESTABLISHED ROOTS (NEXTROOT)
46644 REM USE IS RESTRICTED TO REAL ROOTS.
46645 REM METHOD APPLIED IS NEWTON-RAPHSON ITERATION.
46646 REM THE L ESTABLISHED ROOTS ARE A(I).
46647 REM THE FUNCTION Y AND ITS DERIVATIVE ARE PLACED IN SUBROUTINE 44300.
46648 REM THE INITIAL GUESS IS X0.
46649 REM THE ACCURACY CRITERIA IS E.
46650 N=0
46651 REM GIVEN X0, FIND F/F'.
46652 X=X0
46653 GOSUB 44300
46654 B=Y1/Y
46655 FOR I=1 TO L
46656 B=B-1/(X0-A(I))
46657 NEXT I
46658 REM NEWTON-RAPHSON ITERATION
46659 X1=X0-1/B
46660 N=N+1
46661 REM TEST FOR CONVERGENCE
46662 IF N>=M THEN GOTO 46666
46663 IF ABS(X1-X0)<E THEN GOTO 46666
46664 X0=X1
46665 GOTO 46652
46666 X=X1
46667 RETURN
46679 REM ********************
46680 REM COMPLEX ROOT COUNTING SUBROUTINE (ROOTNUM)
46681 REM THIS ROUTINE CALCULATES THE NUMBER OF COMPLEX
46682 REM ROOTS WITHIN A CIRCLE OF RADIUS W CENTERED
46683 REM ON (X0,Y0) BY COUNTING (U,V) TRANSISTIONS
46684 REM AROUND THE CIRCUMFERENCE.
46685 REM THE INPUT PARAMETERS ARE
46686 REM     W - THE RADIUS OF THE CIRCLE
46687 REM     (X0,Y0) - THE CENTER OF THE CIRCLE
46688 REM     M - THE NUMBER OF EVALUATION POINTS PER QUADRANT
```

```
46689 REM THE ROUTINE RETURNS THE NUMBER OF ROOTS FOUND, N
46690 REM AND THE NUMBER A, WHERE A<>0 INDICATES A FAILURE
46691 REM IN THE ALGORITHM.
46692 REM IT IS ASSUMED THAT THE FUNCTION IS COMPLEX IN THE
46693 REM DOMAIN BEING SEARCHED (U AND V BOTH HAVE TRANSITIONS).
46694 REM IT IS ALSO ASSUMED THAT THE SECTOR SPACING IS CLOSE
46695 REM ENOUGH TO CATCH ALL TRANSISTIONS.
46696 REM NOTE THAT N(I) MUST BE DIMENSIONED TO 4M IN THE
46697 REM CALLING PROGRAM.
46698 REM OBSERVE THAT U(X,Y) AND V(X,Y) ARE EXPECTED TO BE
46699 REM FOUND IN THE FUNCTIONS SUBROUTINE.
46700 A=3.14159/(2*M)
46701 REM START CALCULATION BY ESTABLISHING THE N(I) ARRAY
46702 FOR I=1 TO 4*M
46703 X=W*COS(A*(I-1))+X0
46704 Y=W*SIN(A*(I-1))+Y0
46705 GOSUB 44300
46706 IF U>=0 THEN IF V>=0 THEN N(I)=1
46707 IF U<0 THEN IF V>=0 THEN N(I)=2
46708 IF U<0 THEN IF V<0 THEN N(I)=3
46709 IF U>=0 THEN IF V<0 THEN N(I)=4
46710 NEXT I
46711 REM COUNT COMPLETE CYCLES COUNTERCLOCKWISE
46712 N=N(1)
46713 A=0
46714 FOR I=2 TO 4*M
46715 IF N=N(I) THEN GOTO 46721
46716 IF N<>4 THEN IF N=N(I)+1 THEN A=A-1
46717 IF N=1 THEN IF N(I)=4 THEN A=A-1
46718 IF N=4 THEN IF N(I)=1 THEN A=A+1
46719 IF N+1=N(I) THEN A=A+1
46720 N=N(I)
46721 NEXT I
46722 REM COMPLETE CIRCLE
46723 IF N<>4 THEN IF N=N(1)+1 THEN A=A-1
46724 IF N=4 THEN IF N(1)=1 THEN A=A+1
46725 IF N=1 THEN IF N(1)=4 THEN A=A-1
46726 IF N+1=N(1) THEN A=A+1
46727 A=ABS(A)
46728 N=INT(A/4)
46729 A=A-4*INT(A/4)
46730 RETURN
46737 REM *********************
46738 REM COMPLEX ROOT SEARCH SUBROUTINE (ZCIRCLE)
46739 REM THIS PROGRAM SEARCHES FOR THE COMPLEX ROOTS
46740 REM OF AN ANALYTICAL FUNCTION BY ENCIRCLING THE
46741 REM ZERO AND ESTIMATING WHERE IT IS. THE CIRCLE
46742 REM IS SUBSEQUENTLY TIGHTENED BY A FACTOR E, AND
46743 REM A NEW ESTIMATE MADE.
46744 REM THE INPUTS TO THE SUBROUTINE ARE
46745 REM   (X0,Y0) - THE INITIAL GUESSES
46746 REM   W - THE INITIAL RADIUS OF THE SEARCH CIRCLE
46747 REM   E - THE FACTOR BY WHICH THE CIRCLE IS REDUCED
46748 REM   N - THE NUMBER OF ITERATIONS
46749 REM   M - THE NUMBER OF EVALUATION POINTS PER QUADRANT
46750 REM THE RESULTS IS RETURNED IN Z=X+IY (X,Y).
46751 REM ALSO, THE NUMBER OF ITERATIONS PERFORMED, OR
46752 REM IN PROGRESS, IS RETURNED IN K.
46753 REM X(I),Y(I),U(I) AND V(I) MUST BE DIMENSIONED
```

```
46754 REM IN THE CALLING PROGRAM TO 4M.
46755 REM IT IS ASSUMED THAT THE FUNCTION IS DECOMPOSED
46756 REM INTO ITS REAL AND IMAGINARY PARTS, U(X,Y) AND
46757 REM V(X,Y), AND THAT THESE ARE ACCESSIBLE BY A CALL
46758 REM TO THE FUNCTION SUBROUTINE WHICH RETURNS U AND V.
46759 REM START CALCULATION BY FINDING THE EVALUATION POINTS
46760 M=M*4
46761 K=1
46762 A=6.283185/M
46763 FOR I=1 TO M
46764 X(I)=W*COS(A*(I-1))+X0
46765 Y(I)=W*SIN(A*(I-1))+Y0
46766 NEXT I
46767 REM DETERMINE THE CORRESPONDING U(I) AND(I)
46768 FOR I=1 TO M
46769 X=X(I)
46770 Y=Y(I)
46771 GOSUB 44300
46772 U(I)=U
46773 V(I)=V
46774 NEXT I
46775 REM FIND THE POSITION AT WHICH U CHANGES SIGN IN THE
46776 REM COUNTERCLOCKWISE DIRECTION
46777 I=1
46778 U=U(I)
46779 GOSUB 46885
46780 IF U*U(I)<0 THEN GOTO 46785
46781 REM GUARD AGAINST INFINITE LOOP
46782 IF I=1 THEN GOTO 46881
46783 GOTO 46779
46784 REM TRANSITION FOUND
46785 M1=I
46786 REM SEARCH FOR THE OTHER TRANSITION, STARTING
46787 REM ON THE OTHER SIDE OF THE CIRCLE
46788 I=M1+M/2
46789 IF I>M THEN I=I-M
46790 J=I
46791 U=U(I)
46792 REM FLIP DIRECTIONS ALTERNATELY
46793 GOSUB 46885
46794 IF U*U(I)<0 THEN GOTO 46801
46795 IF U*U(J)<0 THEN GOTO 46804
46796 REM TEST FOR INFINITE LOOP
46797 IF I=M1+M/2 THEN GOTO 46881
46798 IF J=M1+M/2 THEN GOTO 46881
46799 GOTO 46793
46800 REM TRANSITION FOUND
46801 M3=I
46802 GOTO 46808
46803 REM TRANSITION FOUND
46804 IF J=M THEN J=0
46805 M3=J+1
46806 REM M1 AND M3 HAVE BEEN DETERMINED. NOW FOR M2 AND M4.
46807 REM NOW FOR THE V TRANSITIONS
46808 I=M1+M/4
46809 IF I>M THEN I=I-M
46810 J=I
46811 V=V(I)
46812 GOSUB 46885
```

```
46813 IF V*V(I)<0 THEN GOTO 46820
46814 IF V*V(J)<0 THEN GOTO 46822
46815 REM AGAIN, GUARD AGAINST THE INFINITE LOOP
46816 IF I=M1+M/4 THEN GOTO 46881
46817 IF J=M1+M/4 THEN GOTO 46881
46818 GOTO 46812
46819 REM M2 HAS BEEN FOUND
46820 M2=I
46821 GOTO 46825
46822 IF J=M THEN J=0
46823 M2=J+1
46824 REM M2 HAS BEEN FOUND. NOW FOR M4
46825 I=M2+M/2
46826 IF I>M THEN I=I-M
46827 J=I
46828 V=V(I)
46829 GOSUB 46885
46830 IF U*V(I)<0 THEN GOTO 46836
46831 IF V*V(J)<0 THEN GOTO 46838
46832 REM GUARD AGAINST THE INFINITE LOOP AGAIN
46833 IF I=M2+M/2 THEN GOTO 46881
46834 IF J=M2+M/2 THEN GOTO 46881
46835 GOTO 46829
46836 M4=I
46837 GOTO 46842
46838 IF J=M THEN J=0
46839 M4=J+1
46840 REM ALL THE INTERSECTIONS HAVE BEEN DETERMINED
46841 REM INTERPOLATE TO FIND THE FOUR (X,Y) COORDINATES
46842 I=M1
46843 GOSUB 46891
46844 X1=X
46845 Y1=Y
46846 I=M2
46847 GOSUB 46891
46848 X2=X
46849 Y2=Y
46850 I=M3
46851 GOSUB 46891
46852 X3=X
46853 Y3=Y
46854 I=M4
46855 GOSUB 46891
46856 X4=X
46857 Y4=Y
46858 REM CALCULATE THE INTERSECTION OF THE LINES
46859 REM GUARD AGAINST A DIVIDE BY ZERO
46860 IF X1<>X3 THEN GOTO 46864
46861 X=X1
46862 Y=(Y1+Y3)/2
46863 GOTO 46868
46864 M1=(Y3-Y1)/(X3-X1)
46865 IF X2<>X4 THEN GOTO 46868
46866 M2=100000000
46867 GOTO 46869
46868 M2=(Y2-Y4)/(X2-X4)
46869 B1=Y1-M1*X1
46870 B2=Y2-M2*X2
46871 X=-(B1-B2)/(M1-M2)
```

```
46872 Y=(M1*B2+M2*B1)/(M1+M2)
46873 REM IS ANOTHER ITERATION IN ORDER?
46874 IF K=N THEN RETURN
46875 X0=X
46876 Y0=Y
46877 K=K+1
46878 W=W*E
46879 GOTO 46763
46880 REM INFINITE LOOP ENCOUNTERED, RETURN
46881 X=0
46882 Y=0
46883 RETURN
46884 REM AUXILLIARY SUBROUTINE
46885 I=I+1
46886 J=J-1
46887 IF I>M THEN I=I-M
46888 IF J<1 THEN J=J+M
46889 RETURN
46890 REM AUXILLIARY SUBROUTINE FOR INTERPOLATION
46891 J=I-1
46892 IF J<1 THEN J=J+M
46893 REM REGULA FALSI INTERPOLATION FOR THE ZERO
46894 X=(X(I)*U(J)+X(J)*U(I))/(U(I)+U(J))
46895 Y=(Y(I)*V(J)+Y(J)*V(I))/(V(I)+V(J))
46896 RETURN
46912 REM *********************
46913 REM COMPLEX ROOT SEEKING USING NEWTON'S METHOD (CZNEWTON)
46914 REM THIS ROUTINE USES THE COMPLEX DOMAIN FORM OF
46915 REM NEWTON'S METHOD FOR ITERATIVELY SEARCHING FOR ROOTS.
46916 REM IT IS ASSUMED THAT THE FUNCTION AND ITS FIRST PARTIAL
46917 REM DERIVATIVES ARE AVAILABLE FROM THE FUNCTIONS SUBROUTINE
46918 REM IN THE FORM  F(Z) = U(X,Y) + I V(X,Y).
46919 REM THE REQUIRED DERIVATIVES ARE DU/DX AND DU/DY.
46920 REM THE INPUTS TO THE SUBROUTINE ARE THE INITIAL GUESS, X0, Y0,
46921 REM THE CONVERGENCE CRITERIA, E, AND THE MAXIMUM NUMBER OF
46922 REM ITERATIONS TO BE PERFORMED, N.
46923 REM THE RESULTING APPROXIMATION TO THE ROOT IS RETURNED IN
46924 REM (X,Y), AND THE NUMBER OF ITERATIONS IN K.
46925 K=0
46926 K=K+1
46927 REM GET U, V AND THE DERIVATIVES
46928 X=X0
46929 Y=Y0
46930 GOSUB 44300
46931 A=U1*U1+U2*U2
46932 X=X0+(V*U2-U*U1)/A
46933 Y=Y0-(V*U1+U*U2)/A
46934 REM CHECK FOR CONVERGENCE IN EUCLIDEAN SPACE
46935 IF (X0-X)*(X0-X)+(Y0-Y)*(Y0-Y)<=E*E THEN RETURN
46936 IF K>=N THEN RETURN
46937 X0=X
46938 Y0=Y
46939 GOTO 46926
46944 REM *********************
46945 REM PARABOLIC ROOT SEEKING SUBROUTINE (MUELLER)
46946 REM THIS PROGRAM ITERATIVELY SEEKS THE ROOT OF A
46947 REM FUNCTION BY FITTING A PARABOLA TO THREE POINTS
46948 REM AND CALCULATING THE NEAREST ROOT AS DESCRIBED IN
46949 REM BECKET AND HURT, NUMERICAL CALCULATIONS AND ALGORITHMS.
```

```
46950 REM THE SUBROUTINE INPUTS ARE
46951 REM     X0 - THE INITIAL GUESS
46952 REM     D - A BOUND ON THE ERROR IN THIS GUESS
46953 REM     E - THE CONVERGENCE CRITERIA
46954 REM     N - THE MAXIMUM NUMBER OF ITERATIONS
46955 REM THE PROGRAM RETURNS THE VALUE OF THE ROOT FOUND, X,
46956 REM AND THE NUMBER OF ITERATIONS PERFORMED, K.
46957 REM IT IS ASSUMED THAT THE FUNCTION Y(X) IS AVAILABLE
46958 REM IN THE FUNCTIONS SUBROUTINE.
46959 REM SET UP THE THREE EVALUATION POINTS
46960 K=1
46961 X3=X0
46962 X1=X3-D
46963 X2=X3+D
46964 REM CALCULATE MUELLER PARAMETERS
46965 REM GUARD AGAINST DIVIDE BY ZERO
46966 IF X2-X1=0 THEN X2=X2*(1.0000001)
46967 IF X2-X1=0 THEN X2=X2+.0000001
46968 L1=(X3-X2)/(X2-X1)
46969 D1=(X3-X1)/(X2-X1)
46970 IF K>1 THEN GOTO 46978
46971 REM GET VALUES OF FUNCTION
46972 X=X1
46973 GOSUB 44300
46974 E1=Y
46975 X=X2
46976 GOSUB 44300
46977 E2=Y
46978 X=X3
46979 GOSUB 44300
46980 E3=Y
46981 A1=L1*L1*E1-D1*D1*E2+(L1+D1)*E3
46982 C1=L1*(L1*E1-D1*E2+E3)
46983 B=A1*A1-4*D1*C1*E3
46984 REM TEST FOR COMPLEX ROOT, MEANING THE PARABOLA IS INVERTED
46985 IF B<0 THEN B=0
46986 REM CHOOSE CLOSEST ROOT
46987 IF A1<0 THEN A1=A1-SQRT(B)
46988 IF A1>0 THEN A1=A1+SQRT(B)
46989 REM GUARD AGAINST A DIVIDE BY ZERO
46990 IF ABS(A1)+ABS(B)=0 THEN A1=4*D1*E3
46991 REM CALCULATE RELATIVE DISTANCE OF NEXT GUESS
46992 REM GUARD AGAINST DIVIDE BY ZERO
46993 IF A1=0 THEN A1=.0000001
46994 L=-2*D1*E3/A1
46995 REM CALCULATE NEXT ESTIMATE
46996 X=X3+L*(X3-X2)
46997 REM TEST FOR CONVERGENCE
46998 IF ABS(X-X3)<E THEN RETURN
46999 REM TEST FOR NUMBER OF ITERATIONS
47000 IF K>=N THEN RETURN
47001 REM OTHERWISE, MAKE ANOTHER PASS
47002 K=K+1
47003 REM SAVE SOME CALCULATIONS:
47004 X1=X2
47005 X2=X3
47006 X3=X
47007 E1=E2
47008 E2=E3
```

```
47009 GOTO 46967
47033 REM *********************
47034 REM PARABOLIC ROOT SEEKING SUBROUTINE (MUELLER2)
47035 REM FOR A TWO DIMENSIONAL FUNCTION, W(X,Y).
47036 REM THIS PROGRAM ITERATIVELY SEEKS THE ROOT OF A
47037 REM FUNCTION BY FITTING A PARABOLA TO THREE POINTS
47038 REM AND CALCULATING THE NEAREST ROOT AS DESCRIBED IN
47039 REM BECKET AND HURT, NUMERICAL CALCULATIONS AND ALGORITHMS.
47040 REM THE SUBROUTINE INPUTS ARE
47041 REM      X0,Y0 - THE INITIAL GUESS
47042 REM      B1,B2 - A BOUND ON THE ERROR IN THIS GUESS
47043 REM      E - THE CONVERGENCE CRITERIA
47044 REM      N - THE MAXIMUM NUMBER OF ITERATIONS
47045 REM THE PROGRAM RETURNS THE VALUE OF THE ROOT FOUND, (X,Y),
47046 REM AND THE NUMBER OF ITERATIONS PERFORMED, K.
47047 REM IT IS ASSUMED THAT THE FUNCTION U(X,Y) IS AVAILABLE
47048 REM IN THE FUNCTIONS SUBROUTINE.
47049 REM SET UP THE THREE EVALUATION POINTS
47050 K=1
47051 X3=X0
47052 X1=X3-B1
47053 X2=X3+B1
47054 REM CALCULATE MUELLER PARAMETERS
47055 REM GUARD AGAINST DIVIDE BY ZERO
47056 IF X2-X1=0 THEN X2=X2*(1.0000001)
47057 IF X2-X1=0 THEN X2=X2+.0000001
47058 L1=(X3-X2)/(X2-X1)
47059 D1=(X3-X1)/(X2-X1)
47060 REM GET VALUES OF FUNCTION
47061 Y=Y0
47062 X=X1
47063 GOSUB 44300
47064 E1=W
47065 X=X2
47066 GOSUB 44300
47067 E2=W
47068 X=X3
47069 GOSUB 44300
47070 E3=W
47071 GOSUB  47111
47072 REM CALCULATE NEW X ESTIMATE
47073 B1=L*(X3-X2)
47074 X=X3+B1
47075 REM TEST FOR CONVERGENCE
47076 IF ABS(B1)+ABS(B2)<E THEN RETURN
47077 X0=X
47078 REM REPEAT FOR THE Y DIRECTION
47079 Y3=Y0
47080 Y1=Y3-B2
47081 Y2=Y3+B2
47082 REM CALCULATE MUELLER PARAMETERS
47083 REM GUARD AGAINST A DIVIDE BY ZERO
47084 IF Y2-Y1=0 THEN Y2=Y2*(1.0000001)
47085 IF Y2-Y1=0 THEN Y2=Y2+.0000001
47086 L1=(Y3-Y2)/(Y2-Y1)
47087 D1=(Y3-Y1)/(Y2-Y1)
47088 REM GET VALUES OF FUNCTION
47089 Y=Y1
47090 GOSUB 44300
```

```
47091 E1=W
47092 Y=Y2
47093 GOSUB 44300
47094 E2=W
47095 Y=Y3
47096 GOSUB 44300
47097 E3=W
47098 GOSUB 47111
47099 REM CALCULATE NEW Y ESTIMATE
47100 B2=L*(Y3-Y2)
47101 Y=Y3+B2
47102 REM TEST FOR CONVERGENCE
47103 IF ABS(B1)+ABS(B2)<E THEN RETURN
47104 REM TEST FOR NUMBER OF ITERATIONS
47105 IF K>=N THEN RETURN
47106 Y0=Y
47107 K=K+1
47108 REM START ANOTHER PASS
47109 GOTO 47051
47110 REM UTILITY SUBROUTINE
47111 A1=L1*L1*E1-D1*D1*E2+(L1+D1)*E3
47112 C1=L1*(L1*E1-D1*E2+E3)
47113 B=A1*A1-4*D1*C1*E3
47114 REM TEST FOR COMPLEX ROOT, MEANING THE PARABOLA IS INVERTED
47115 IF B<0 THEN B=0
47116 REM CHOOSE CLOSEST ROOT
47117 IF A1<0 THEN A1=A1-SQRT(B)
47118 IF A1>0 THEN A1=A1+SQRT(B)
47119 REM GUARD AGAINST A DIVIDE BY ZERO
47120 IF ABS(A1)+ABS(B)=0 THEN A1=4*D1*E3
47121 REM CALCULATE RELATIVE DISTANCE OF NEXT GUESS
47122 REM GUARD AGAINST DIVIDE BY ZERO
47123 IF A1=0 THEN A1=.0000001
47124 L=-2*D1*E3/A1
47125 RETURN
47136 REM ********************
47137 REM MUELLER'S METHOD FOR COMPLEX ROOTS (ZMUELLER)
47138 REM THIS PROGRAM USES THE PARABOLIC FITTING TECHNIQUE
47139 REM ASSOCIATED WITH MUELLER'S METHOD, BUT DOES IT IN
47140 REM THE COMPLEX DOMAIN.
47141 REM THE INPUTS TO THE SUBROUTINE ARE THE INITIAL
47142 REM GUESS, (X0,Y0), THE CONVERGENCE CRITERIA, E,
47143 REM AND THE MAXIMUM NUMBER OF ITERATIONS, N.
47144 REM ALSO REQUIRED ARE BOUNDS ON THE INITIAL GUESS, B1 AND B2.
47145 REM RETURNED IS THE NEW ESTIMATE, (X,Y), AND THE
47146 REM NUMBER OF ITERATIONS PERFORMED, K.
47147 REM IT IS ASSUMED THAT THE FUNCTION F(Z) = U(X,Y)+IV(X,Y)
47148 REM IS AVAILABLE IN THE FUNCTIONS SUBROUTINE.
47149 REM START CALCULATIONS
47150 K=1
47151 X3=X0
47152 Y3=Y0
47153 X1=X3-B1
47154 Y1=Y3-B2
47155 X2=X3+B1
47156 Y2=Y3+B2
47157 D=(X2-X1)*(X2-X1)+(Y2-Y1)*(Y2-Y1)
47158 REM AVOID DIVIDE BY ZERO
47159 IF D=0 THEN D=.0000001
```

```
47160 L1=(X3-X2)*(X2-X1)+(Y3-Y2)*(Y2-Y1)
47161 L1=L1/D
47162 L2=(X2-X1)*(Y3-Y2)-(X3-X2)*(Y2-Y1)
47163 L2=L2/D
47164 D1=(X3-X1)*(X2-X1)+(Y3-Y1)*(Y2-Y1)
47165 D1=D1/D
47166 D2=(X2-X1)*(Y3-Y1)-(X3-X1)*(Y2-Y1)
47167 D2=D2/D
47168 REM GET FUNCTION VALUES
47169 X=X1
47170 Y=Y1
47171 GOSUB 44300
47172 U1=U
47173 V1=V
47174 X=X2
47175 Y=Y2
47176 GOSUB 44300
47177 U2=U
47178 V2=V
47179 X=X3
47180 Y=Y3
47181 GOSUB 44300
47182 U3=U
47183 V3=V
47184 REM CALCULATE MUELLER PARAMETERS
47185 E1=U1*(L1*L1-L2*L2)-2*V1*L1*L2-U2*(D1*D1-D2*D2)
47186 E1=E1+2*V2*D1*D2+U3*(L1+D1)-V3*(L2+D2)
47187 E2=2*L1*L2*U1+V1*(L1*L1-L2*L2)-2*D1*D2*U2-V2*(D1*D1-D2*D2)
47188 E2=E2+U3*(L2+D2)+V3*(L1+D1)
47189 C1=L1*L1*U1-L1*L2*V1-D1*L1*U2+L1*D2*V2+U3*L1
47190 C1=C1-U1*L2*L2-V1*L1*L2+U2*L2*D2+V2*D1*L2-V3*L2
47191 C2=U1*L1*L2+V1*L1*L1-U2*D2*L1-V2*D1*L1+V3*L1
47192 C2=C2+L1*L2*U1-L2*L2*V1-D1*L2*U2+D2*L2*V2+U3*L2
47193 B1=E1*E1-E2*E2-4*(U3*D1*C1-U3*D2*C2-V3*D2*C1-V3*D1*C2)
47194 B2=2*E1*E2-4*(U3*D2*C1+U3*D1*C2+V3*D1*C1-V3*D2*C2)
47195 REM GUARD AGAINST A DIVIDE BY ZERO
47196 IF B1=0 THEN B1=.0000001
47197 A=ATN(B2/B1)
47198 A=A/2
47199 B=SQRT(SQRT(B1*B1+B2*B2))
47200 B1=B*COS(A)
47201 B2=B*SIN(A)
47202 A1=(E1+B1)*(E1+B1)+(E2+B2)*(E2+B2)
47203 A2=(E1-B1)*(E1-B1)+(E2-B2)*(E2-B2)
47204 IF A1>A2 THEN GOTO 47208
47205 A1=E1-B1
47206 A2=E2-B2
47207 GOTO 47210
47208 A1=E1+B1
47209 A2=E2+B2
47210 A=A1*A1+A2*A2
47211 L1=A1*D1*U3-A1*D2*V3+A2*U3*D2+A2*V3*D1
47212 REM GUARD AGAINST DIVIDE BY ZERO
47213 IF A=0 THEN A=.0000001
47214 L1=-2*L1/A
47215 L2=-D1*U3*A2+D2*V3*A2+A1*U3*D2+A1*V3*D1
47216 L2=-2*L2/A
47217 REM CALCULATE NEW ESTIMATE
47218 X=X3+L1*(X3-X2)-L2*(Y3-Y2)
```

```
47219 Y=Y3+L2*(X3-X2)+L1*(Y3-Y2)
47220 REM TEST FOR CONVERGENCE
47221 IF ABS(X-X0)+ABS(Y-Y0)<E THEN RETURN
47222 REM TEST FOR NUMBER OF ITERATIONS
47223 IF K>=N THEN RETURN
47224 REM CONTINUE
47225 K=K+1
47226 X0=X
47227 Y0=Y
47228 X1=X2
47229 Y1=Y2
47230 X2=X3
47231 Y2=Y3
47232 X3=X
47233 Y3=Y
47234 GOTO 47157
47282 REM ********************
47283 REM GENERAL ROOT DETERMINATION SUBROUTINE (ALLROOT)
47284 REM THE ROUTINE ATTEMPTS TO CALCULATE THE SEVERAL ROOTS OF A
47285 REM GIVEN SERIES OR FUNCTION BY REPEATEDLY USING THE
47286 REM ZMUELLER SUBROUTINE AND REMOVING THE ROOTS ALREADY FOUND
47287 REM BY DIVISION.
47288 REM THE INPUT TO THE SUBROUTINE ARE
47289 REM      X0,Y0 - THE INITIAL GUESS
47290 REM      B1,B2 - THE BOUNDS ON THIS GUESS
47291 REM      E - THE CONVERGENCE CRITERIA
47292 REM      N - THE MAXIMUM NUMBER OF ITERATIONS PER ROOT
47293 REM      N2 - THE NUMBER OF ROOTS BEING SOUGHT
47294 REM      N3 - A FLAG INDICATING A FUNCTION F(Z) (1)
47295 REM           OR A POLYNOMIAL (2)
47296 REM THE PROGRAM RETURNS THE N2 ROOTS FOUND, X(I),Y(I)
47297 REM AND THE LAST NUMBER OF ITERATIONS USED, K.
47298 REM IF K=0 THEN N3 WAS IN ERROR
47299 REM START CALCULATIONS
47300 K=0
47301 IF N3=1 THEN GOTO 47303
47302 IF N3<>2 THEN RETURN
47303 J1=0
47304 REM SAVE THE INITIAL GUESS
47305 X4=X0
47306 Y4=Y0
47307 REM IF N3=2 THEN GET THE SERIES COEFFICIENTS
47308 IF N3=2 THEN GOSUB 47322
47309 REM TEST FOR COMPLETION
47310 IF J1=N2 THEN RETURN
47311 REM GOTO ZMUELLER
47312 GOSUB 47344
47313 J1=J1+1
47314 X(J1)=X
47315 Y(J1)=Y
47316 X0=X4
47317 Y0=Y4
47318 REM TRY ANOTHER PASS
47319 GOTO 47310
47320 REM ********************
47321 REM COEFFICIENTS SUBROUTINE
47322 M=5
47323 A(0)=0
47324 A(1)=24
```

```
47325 A(2)=-50
47326 A(3)=35
47327 A(4)=-10
47328 A(5)=1
47329 RETURN
47330 REM *********************
47331 REM VARIANT ON MUELLER'S METHOD FOR COMPLEX ROOTS
47332 REM THIS PROGRAM USES THE PARABOLIC FITTING TECHNIQUE
47333 REM ASSOCIATED WITH MUELLER'S METHOD, BUT DOES IT IN
47334 REM THE COMPLEX DOMAIN.
47335 REM THE INPUTS TO THE SUBROUTINE ARE THE INITIAL
47336 REM GUESS, (X0,Y0), THE CONVERGENCE CRITERIA, E,
47337 REM AND THE MAXIMUM NUMBER OF ITERATIONS, N.
47338 REM ALSO REQUIRED ARE BOUNDS ON THE INITIAL GUESS, B1 AND B2.
47339 REM RETURNED IS THE NEW ESTIMATE, (X,Y), AND THE
47340 REM NUMBER OF ITERATIONS PERFORMED, K.
47341 REM IT IS ASSUMED THAT THE FUNCTION F(Z) = U(X,Y)+IV(X,Y)
47342 REM IS AVAILABLE IN THE FUNCTIONS SUBROUTINE.
47343 REM START CALCULATIONS
47344 K=1
47345 X3=X0
47346 Y3=Y0
47347 X1=X3-B1
47348 Y1=Y3-B2
47349 X2=X3+B1
47350 Y2=Y3+B2
47351 D=(X2-X1)*(X2-X1)+(Y2-Y1)*(Y2-Y1)
47352 REM AVOID DIVIDE BY ZERO
47353 IF D=0 THEN D=.0000001
47354 L1=(X3-X2)*(X2-X1)+(Y3-Y2)*(Y2-Y1)
47355 L1=L1/D
47356 L2=(X2-X1)*(Y3-Y2)-(X3-X2)*(Y2-Y1)
47357 L2=L2/D
47358 D1=(X3-X1)*(X2-X1)+(Y3-Y1)*(Y2-Y1)
47359 D1=D1/D
47360 D2=(X2-X1)*(Y3-Y1)-(X3-X1)*(Y2-Y1)
47361 D2=D2/D
47362 REM GET FUNCTION VALUES
47363 X=X1
47364 Y=Y1
47365 GOSUB 47431
47366 U1=U
47367 V1=V
47368 X=X2
47369 Y=Y2
47370 GOSUB 47431
47371 U2=U
47372 V2=V
47373 X=X3
47374 Y=Y3
47375 GOSUB 47431
47376 U3=U
47377 V3=V
47378 REM CALCULATE MUELLER PARAMETERS
47379 E1=U1*(L1*L1-L2*L2)-2*V1*L1*L2-U2*(D1*D1-D2*D2)
47380 E1=E1+2*V2*D1*D2+U3*(L1+D1)-V3*(L2+D2)
47381 E2=2*L1*L2*U1+V1*(L1*L1-L2*L2)-2*D1*D2*U2-V2*(D1*D1-D2*D2)
47382 E2=E2+U3*(L2+D2)+V3*(L1+D1)
47383 C1=L1*L1*U1-L1*L2*V1-D1*L1*U2+L1*D2*V2+U3*L1
```

```
47384 C1=C1-U1*L2*L2-V1*L1*L2+U2*L2*D2+V2*D1*L2-V3*L2
47385 C2=U1*L1*L2+V1*L1*L1-U2*D2*L1-V2*D1*L1+V3*L1
47386 C2=C2+L1*L2*U1-L2*L2*V1-D1*L2*U2+D2*L2*V2+U3*L2
47387 B1=E1*E1-E2*E2-4*(U3*D1*C1-U3*D2*C2-V3*D2*C1-V3*D1*C2)
47388 B2=2*E1*E2-4*(U3*D2*C1+U3*D1*C2+V3*D1*C1-V3*D2*C2)
47389 REM GUARD AGAINST A DIVIDE BY ZERO
47390 IF B1=0 THEN B1=.0000001
47391 A=ATN(B2/B1)
47392 A=A/2
47393 B=SQRT(SQRT(B1*B1+B2*B2))
47394 B1=B*COS(A)
47395 B2=B*SIN(A)
47396 A1=(E1+B1)*(E1+B1)+(E2+B2)*(E2+B2)
47397 A2=(E1-B1)*(E1-B1)+(E2-B2)*(E2-B2)
47398 IF A1>A2 THEN GOTO 47402
47399 A1=E1-B1
47400 A2=E2-B2
47401 GOTO 47404
47402 A1=E1+B1
47403 A2=E2+B2
47404 A=A1*A1+A2*A2
47405 L1=A1*D1*U3-A1*D2*V3+A2*U3*D2+A2*V3*D1
47406 REM GUARD AGAINST DIVIDE BY ZERO
47407 IF A=0 THEN A=.0000001
47408 L1=-2*L1/A
47409 L2=-D1*U3*A2+D2*V3*A2+A1*U3*D2+A1*V3*D1
47410 L2=-2*L2/A
47411 REM CALCULATE NEW ESTIMATE
47412 X=X3+L1*(X3-X2)-L2*(Y3-Y2)
47413 Y=Y3+L2*(X3-X2)+L1*(Y3-Y2)
47414 REM TEST FOR CONVERGENCE
47415 IF ABS(X-X0)+ABS(Y-Y0)<E THEN RETURN
47416 REM TEST FOR NUMBER OF ITERATIONS
47417 IF K>=N THEN RETURN
47418 REM CONTINUE
47419 K=K+1
47420 X0=X
47421 Y0=Y
47422 X1=X2
47423 Y1=Y2
47424 X2=X3
47425 Y2=Y3
47426 X3=X
47427 Y3=Y
47428 GOTO 47351
47429 REM ********************
47430 REM SUPERVISOR SUBROUTINE
47431 N5=N
47432 U5=U1
47433 V5=V1
47434 REM DO WE GO TO THE FUNCTIONS SUBROUTINE OR TO THE SERIES SUBROUTINE?
47435 IF N3=1 THEN GOSUB 44300
47436 IF N3=2 THEN GOSUB 44950
47437 IF N3=1 THEN GOTO 47444
47438 U=Z1
47439 V=Z2
47440 REM RESTORE PARAMETERS
47441 N=N5
47442 U1=U5
```

```
47443 V1=V5
47444 IF J1=0 THEN RETURN
47445 REM DIVIDE BY THE J1 ROOTS ALREADY FOUND
47446 FOR J2=1 TO J1
47447 U5=U
47448 U=(X-X(J2))*U+(Y-Y(J2))*V
47449 V=(X-X(J2))*V-(Y-Y(J2))*U5
47450 A4=(X-X(J2))*(X-X(J2))+(Y-Y(J2))*(Y-Y(J2))
47451 REM GUARD AGAINST DIVIDE BY ZERO
47452 IF A4=0 THEN A4=.0000001
47453 V=V/A4
47454 U=U/A4
47455 NEXT J2
47456 REM RETURN TO ZMUELLER
47457 RETURN
47462 REM ********************
47463 REM QUADRATIC ROOT SUBROUTINE (QUADRAT)
47464 REM THIS PROGRAM CALCULATES THE TWO ROOTS OF
47465 REM A GIVEN SECOND ORDER POLYNOMIAL USING
47466 REM THE QUADRATIC EQUATION EVALUATED IN A
47467 REM MANNER WHICH MINIMIZES ROUND OFF ERROR.
47468 REM THE POLYNOMIAL IS ASSUMED TO BE OF
47469 REM THE FORM
47470 REM      Y = A(2)*X*X +A(1)*X +A(0)
47471 REM THE TWO ROOTS ARE RETURNED AS
47472 REM R1 = X1 + I Y1
47473 REM R2 = X2 + I Y2
47474 REM TEST FOR A(2)=0
47475 IF A(2)<>0 THEN GOTO 47488
47476 REM TEST FOR A(1)=0
47477 IF A(1)<>0 THEN GOTO 47483
47478 X1=0
47479 X2=0
47480 Y1=0
47481 Y2=0
47482 RETURN
47483 X1=-A(0)/A(1)
47484 Y1=0
47485 X2=X1
47486 Y2=Y1
47487 RETURN
47488 A=A(1)*A(1)-4*A(2)*A(0)
47489 B=SQRT(ABS(A))
47490 REM ESTABLISH SIGN
47491 IF A(1)=0 THEN C=1
47492 IF A(1)<>0 THEN C=ABS(A(1))/A(1)
47493 REM DETERMINE THE FIRST ROOT
47494 REM CHECK IF ROOT IS COMPLEX
47495 IF A>0 THEN GOTO 47499
47496 X1=-C*ABS(A(1))/(2*A(2))
47497 Y1=-C*B/(2*A(2))
47498 GOTO 47502
47499 X1=-C*(ABS(A(1))+B)/(2*A(2))
47500 Y1=0
47501 REM CALCULATE THE SECOND ROOT
47502 C=X1*X1+Y1*Y1
47503 IF C<>0 THEN GOTO 47506
47504 X2=1000000*1000000*1000000
47505 RETURN
```

```
47506 C=A(0)/(C*A(2))
47507 X2=X1*C
47508 Y2=-Y1*C
47509 RETURN
47511 REM *********************
47512 REM POLYNOMIAL COMPLEX ROOTS SUBROUTINE (LIN)
47513 REM USES LIN'S METHOD AS DESCRIBED IN THE REFERENCE
47514 REM A PRACTICAL GUIDE TO COMPUTER METHODS FOR ENGINEERS BY SHOUP.
47515 REM THE INPUT POLYNOMIAL COEFFICIENTS ARE A(0) THROUGH A(M).
47516 REM M IS THE ORDER OF THE POLYNOMIAL.
47517 REM INITIAL GUESSES FOR A AND B ARE REQUIRED.
47518 REM THE RESULTS ARE RETURNED IN X1,Y1 AND X2,Y2.
47519 REM X IS THE REAL PART, AND Y IS THE IMAGINARY.
47520 REM THE MAXIMUM NUMBER OF ITERATIONS IS N.
47521 REM THE NUMBER OF ITERATIONS IS RETURNED IN K
47522 REM THE CONVERGENCE CRITERION IS E.
47523 REM IF NECESSARY, DIMENSION A(I), B(I) AND C(I) IN THE CALLING PROGRAM.
47524 REM NORMALIZE THE A(I) SERIES
47525 FOR I=0 TO M
47526 C(I)=A(I)/A(M)
47527 NEXT I
47528 REM START ITERATION
47529 REM SET INITIAL GUESS FOR THE QUADRATIC COEFFICIENTS
47530 B(0)=0
47531 B(1)=0
47532 B(M-1)=C(M-1)-A
47533 B(M-2)=C(M-2)-A*B(M-1)-B
47534 FOR J=3 TO M
47535 B(M-J)=C(M-J)-A*B(M+1-J)-B*B(M+2-J)
47536 NEXT J
47537 REM GUARD AGAINST DIVIDE BY ZERO
47538 IF B(2)<>0 THEN GOTO 47542
47539 A=A+.0000001
47540 B=B-.0000001
47541 GOTO 47532
47542 A1=(C(1)-B*B(3))/B(2)
47543 B1=C(0)/B(2)
47544 K=K+1
47545 REM TEST FOR THE NUMBER OF ITERATIONS
47546 IF K>=N THEN GOTO 47553
47547 REM TEST FOR CONVERGENCE
47548 IF ABS(A-A1)+ABS(B-B1)<E*E THEN GOTO 47553
47549 A=A1
47550 B=B1
47551 REM RETURN FOR NEXT ITERATION
47552 GOTO 47532
47553 A=A1
47554 B=B1
47555 C=A*A-4*B
47556 REM IS THERE AN IMAGINARY PART
47557 IF C>0 THEN GOTO 47563
47558 Y1=SQRT(-C)
47559 Y2=-Y1
47560 X1=-A
47561 X2=X1
47562 GOTO 47567
47563 Y1=0
47564 Y2=Y1
47565 X1=-A+SQRT(C)
```

```
47566 X2=-A-SQRT(C)
47567 X1=X1/2
47568 X2=X2/2
47569 Y1=Y1/2
47570 Y2=Y2/2
47571 RETURN
47584 REM *********************
47585 REM BAIRSTOW COMPLEX ROOT SUBROUTINE (BAIRSTOW)
47586 REM THIS SUBROUTINE FINDS THE COMPLEX CONJUGATE ROOTS
47587 REM OF A POLYNOMIAL HAVING REAL COEFFICIENTS.
47588 REM SEE COMPUTER METHODS FOR SCIENCE AND ENGINEERING
47589 REM BY R.L. LAFARA.
47590 REM ORDER OF INPUT SERIES IS M >= 4.
47591 REM SERIES COEFFICIENTS ARE A(I).
47592 REM INITIAL GUESSES A AND B ARE REQUIRED.
47593 REM E IS THE CONVERGENCE FACTOR.
47594 REM SUBROUTINE RETURNS X1,Y1 AND X2,Y2.
47595 REM N IS THE MAXIMUM NUMBER OF ITERATIONS.
47596 REM K IS THE NUMBER OF ITERATIONS PERFORMED.
47597 REM IF NECESSARY, DIMENSION A(I),B(I), C(I) AND D(I)
47598 REM IN THE CALLING PROGRAM.
47599 REM USE NORMALIZED SERIES, C(I)
47600 FOR I=0 TO M
47601 C(I)=A(I)/A(M)
47602 NEXT I
47603 REM CHOSE INITIAL ESTIMATES FOR A AND B
47604 K=0
47605 B(M)=1
47606 REM START ITERATION SEQUENCE
47607 B(M-1)=C(M-1)-A
47608 FOR J=2 TO M-1
47609 B(M-J)=C(M-J)-A*B(M+1-J)-B*B(M+2-J)
47610 NEXT J
47611 B(0)=C(0)-B*B(2)
47612 D(M-1)=-1
47613 D(M-2)=-B(M-1)+A
47614 FOR J=3 TO M-1
47615 D(M-J)=-B(M+1-J)-A*D(M+1-J)-B*D(M+2-J)
47616 NEXT J
47617 D(0)=-B*D(2)
47618 D2=-B(2)-B*D(3)
47619 D=D(1)*D2-D(0)*D(2)
47620 A1=-B(1)*D2+B(0)*D(2)
47621 A1=A1/D
47622 B1=-D(1)*B(0)+D(0)*B(1)
47623 B1=B1/D
47624 A=A+A1
47625 B=B+B1
47626 K=K+1
47627 REM TEST FOR THE NUMBER OF ITERATIONS
47628 IF K>=N THEN GOTO 47632
47629 REM TEST FOR CONVERGENCE
47630 IF ABS(A1)+ABS(B1)>E*E THEN GOTO 47607
47631 REM EXTRACT ROOTS FROM QUADRATIC EQUATION
47632 C=A*A-4*B
47633 REM TEST TO SEE IF A COMPLEX ROOT
47634 IF C>0 THEN GOTO 47640
47635 X1=-A
47636 X2=X1
```

```
47637 Y1=SQRT(-C)
47638 Y2=-Y1
47639 GOTO 47644
47640 X1=-A+SQRT(C)
47641 X2=-A-SQRT(C)
47642 Y1=0
47643 Y2=Y1
47644 X1=X1/2
47645 X2=X2/2
47646 Y1=Y1/2
47647 Y2=Y2/2
47648 RETURN
47684 REM ********************
47685 REM STEEPEST DESCENT OPTIMIZATION SUBROUTINE (STEEPDS)
47686 REM THIS PROGRAM FIND THE LOCAL MAXIMUM OR MINIMUM
47687 REM OF AN L-DIMENSIONAL FUNCTION USING THE METHOD
47688 REM OF STEEPEST DESCENT, OR THE GRADIENT.
47689 REM THE FUNCTION, Y(X(1),X(2)...), IS PLACED IN THE
47690 REM SUBROUTINE AT 44300, ALONG WITH THE L DERIVATIVES
47691 REM OF F, D(I).
47692 REM THE ROUTINE SEEKS USING AN INTERNALLY ADJUSTED
47693 REM MULTIPLIER, K. THE SEARCH IS MADE UNTIL AN ERROR
47694 REM LIMIT, E, IS REACHED.
47695 REM THE USER MUST SUPPLY INITIAL VALUES FOR THE X(I),
47696 REM AS WELL AS K (INITIAL) AND E. THE PROGRAM RETURNS
47697 REM THE LOCALLY OPTIMUM X(I) SET.
47698 REM REMEMBER TO DIMENSION X(I) IN THE CALLING PROGRAM.
47699 REM THE PROGRAM NEEDS THREE VALUES OF Y TO GET STARTED.
47700 N=0
47701 REM START INITIAL PROBE
47702 FOR J=1 TO 3
47703 REM OBTAIN Y AND D(I)
47704 GOSUB 44300
47705 Y(J)=Y
47706 REM UPDATE X(I)
47707 GOSUB 47735
47708 NEXT J
47709 REM WE NOW HAVE A HISTORY TO BASE THE SUBSEQUENT SEARCH ON
47710 REM ACCELERATE SEARCH IF APPROACH IS MONOTONIC
47711 IF (Y(3)-Y(2))/(Y(2)-Y(1))>0 THEN K=K*1.2
47712 REM DECELERATE IF HEADING THE WRONG WAY
47713 IF Y(3)<Y(2) THEN K=K/2
47714 REM UPDATE THE Y(I) IF VALUE HAS DECREASED
47715 IF Y(3)>Y(2) THEN GOTO 47721
47716 REM RESTORE THE X(I)
47717 FOR I=1 TO L
47718 X(I)=X1(I)
47719 NEXT I
47720 GOTO 47724
47721 Y(1)=Y(2)
47722 Y(2)=Y(3)
47723 REM OBTAIN NEW VALUES
47724 GOSUB 44300
47725 Y(3)=Y
47726 REM UPDATE X(I)
47727 GOSUB 47735
47728 REM CHECK FOR CONVERGENCE
47729 N=N+1
47730 IF N>=M THEN RETURN
```

```
47731 IF ABS(Y(3)-Y(2))<E THEN RETURN
47732 REM TRY ANOTHER ITERATION
47733 GOTO 47711
47734 REM FIND THE MAGNITUDE OF THE GRADIENT
47735 D=0
47736 FOR I=1 TO L
47737 D=D+D(I)*D(I)
47738 NEXT I
47739 D=SQRT(D)
47740 REM UPDATE THE X(I)
47741 FOR I=1 TO L
47742 REM SAVE OLD VALUES
47743 X1(I)=X(I)
47744 X(I)=X(I)+K*D(I)/D
47745 NEXT I
47746 GOSUB 44300
47747 Y(3)=Y
47748 RETURN
47784 REM ********************
47785 REM STEEPEST DESCENT OPTIMIZATION SUBROUTINE (STEEPDA)
47786 REM THIS PROGRAM FIND THE LOCAL MAXIMUM OR MINIMUM
47787 REM OF AN L-DIMENSIONAL FUNCTION USING THE METHOD
47788 REM OF STEEPEST DESCENT, OR THE GRADIENT.
47789 REM THE FUNCTION, Y(X(1),X(2)...), IS PLACED IN THE
47790 REM SUBROUTINE AT 44300. FINITE DIFFERENCES ARE USED TO
47791 REM CALCULATE THE L PARTIAL DERIVATIVES.
47792 REM OF F, D(I).
47793 REM THE ROUTINE SEEKS USING AN INTERNALLY ADJUSTED
47794 REM MULTIPLIER, K. THE SEARCH IS MADE UNTIL AN ERROR
47795 REM LIMIT, E, IS REACHED.
47796 REM THE USER MUST SUPPLY INITIAL VALUES FOR THE X(I),
47797 REM AS WELL AS K (INITIAL) AND E. THE PROGRAM RETURNS
47798 REM THE LOCALLY OPTIMUM X(I) SET.
47799 REM REMEMBER TO DIMENSION X(I) IN THE CALLING PROGRAM.
47800 N=0
47801 REM THE PROGRAM NEEDS THREE VALUES OF Y TO GET STARTED.
47802 REM GENERATE STARTING D(I) VALUES.
47803 REM THESE ARE NOT EVEN GOOD GUESSES, AND SLOW THE PROGRAM A LITTLE.
47804 D=1
47805 D(1)=1/SQRT(L)
47806 FOR I=2 TO L
47807 D(I)=D(I-1)
47808 NEXT I
47809 REM START INITIAL PROBE
47810 FOR J=1 TO 3
47811 REM OBTAIN Y
47812 GOSUB 44300
47813 Y(J)=Y
47814 REM OBTAIN APPROXIMATIONS TO THE D(I)
47815 GOSUB 47866
47816 REM UPDATE X(I)
47817 GOSUB 47849
47818 GOSUB 47856
47819 NEXT J
47820 REM WE NOW HAVE A HISTORY TO BASE THE SUBSEQUENT SEARCH ON
47821 REM ACCELERATE SEARCH IF APPROACH IS MONOTONIC
47822 IF (Y(3)-Y(2))/(Y(2)-Y(1))>0 THEN K=K*1.2
47823 REM DECELERATE IF HEADING THE WRONG WAY
47824 IF Y(3)<Y(2) THEN K=K/2
```

```
47825 REM UPDATE THE Y(I) IF Y(3)>Y(2)
47826 IF Y(3)>Y(2) THEN GOTO 47832
47827 REM RESTORE THE X(I)
47828 FOR I=1 TO L
47829 X(I)=X1(I)
47830 NEXT I
47831 GOTO 47835
47832 Y(1)=Y(2)
47833 Y(2)=Y(3)
47834 REM OBTAIN NEW VALUES
47835 GOSUB 44300
47836 Y(3)=Y
47837 GOSUB 47866
47838 REM IF D=0 THEN THE PRECISION LIMIT OF THE COMPUTER HAS BEEN REACHED
47839 IF D=0 THEN RETURN
47840 REM UPDATE X(I)
47841 GOSUB 47856
47842 REM CHECK FOR CONVERGENCE
47843 N=N+1
47844 IF N>=M THEN RETURN
47845 IF ABS(Y(3)-Y(2))<E THEN RETURN
47846 REM TRY ANOTHER ITERATION
47847 GOTO 47822
47848 REM FIND THE MAGNITUDE OF THE GRADIENT
47849 D=0
47850 FOR I=1 TO L
47851 D=D+D(I)*D(I)
47852 NEXT I
47853 D=SQRT(D)
47854 RETURN
47855 REM UPDATE THE X(I)
47856 FOR I=1 TO L
47857 REM SAVE OLD VALUES
47858 X1(I)=X(I)
47859 X(I)=X(I)+K*D(I)/D
47860 NEXT I
47861 GOSUB 44300
47862 Y(3)=Y
47863 RETURN
47864 REM FINITE DIFFERENCES SUBROUTINE FOR THE D(I) APPROXIMATION
47865 REM LOOK AHEAD ONE HALF INTERVAL
47866 FOR I=1 TO L
47867 REM SAVE X(I)
47868 A=X(I)
47869 REM FIND INCREMENT
47870 B=D(I)*K/(2*D)
47871 REM MOVE INCREMENT IN X(I)
47872 X(I)=X(I)+B
47873 REM OBTAIN Y
47874 GOSUB 44300
47875 REM GUARD AGAINST DIVIDE BY ZERO NEAR MAXIMUM
47876 IF B=0 THEN B=.00000000001
47877 REM UPDATE D(I)
47878 D(I)=(Y-Y(3))/B
47879 REM GUARD AGAINST LOCKED UP DERIVATIVE
47880 IF D(I)=0 THEN D(I)=.00001
47881 REM RESTORE X(I) AND Y
47882 X(I)=A
47883 Y=Y(3)
```

```
47884 NEXT I
47885 REM OBTAIN D
47886 REM GOSUB 47849
47887 RETURN
```

APPENDIX IIB *Compacted North Star BASIC Subroutine Listing*

The North Star BASIC subroutine library shown in Appendix IIA was compacted by removing the unnecessary REMark statements and spaces.

Compacted subroutine library length: 29460 bytes
                                       2351 lines

General Information

North Star BASIC contains several important functions that are used in the subroutines in this book. Most of these functions are supplied in Chapter VI of Volume I as subroutines, or can be derived from those programs.

Sine and Cosine

The SINE and COSINE subroutines given in Chapter VI of Volume I can be used to replace the function SIN( ) and COS( ). These functions are used in the following statements:

SIN( ):	40451	44565	44566	46704	46765	47201	47395
COS( ):	40450	44565	44566	46703	46764	47200	47394

Arctangent

The arctangent (inverse tangent) function is not included in North Star BASIC Version 6, Release 3 and earlier versions. Subroutine ARCTAN (see Volume I) can be used to replace the ATN( ) function that appears in the following statements:

40407	45134	47197	47391

```
40400U=SQRT(X*X+Y*Y)
40402IFY=0THENY=(.1)^30
40404IFX=0THENX=(.1)^30
40406W=Y/X
40407V=ATN(W)
```

```
40409 IF X<0 THEN V=V+3.1415926535
40410 IF V<0 THEN V=V+6.2831853072
40411 RETURN
40450 X=U*COS(V)
40451 Y=U*SIN(V)
40452 RETURN
41100 U1=U^N
41101 V1=N*V
41102 V1=V1-6.2831853072*INT(V1/6.2831853072)
41103 RETURN
41200 GOSUB 40400
41202 GOSUB 41100
41204 U=U1
41205 V=V1
41207 GOSUB 40450
41208 RETURN
41900 FOR I=1 TO M1
41901 FOR J=1 TO N2
41902 C(I,J)=0
41903 FOR K=1 TO N1
41904 C(I,J)=C(I,J)+A(I,K)*B(K,J)
41905 NEXT K
41906 NEXT J
41907 NEXT I
41908 RETURN
41950 FOR I=1 TO N
41951 FOR J=1 TO M
41952 B(I,J)=A(J,I)
41953 NEXT J
41954 NEXT I
41955 RETURN
42075 IF N1*N2*N3=0 THEN GOTO 42085
42077 FOR I1=1 TO N1
42078 FOR I2=1 TO N2
42079 FOR I3=1 TO N3
42080 A(I1,I2,I3)=B(I1,I2,I3)
42081 NEXT I3
42082 NEXT I2
42083 NEXT I1
42084 RETURN
42085 IF N1*N2=0 THEN GOTO 42092
42086 FOR I1=1 TO N1
42087 FOR I2=1 TO N2
42088 A(I1,I2)=B(I1,I2)
42089 NEXT I2
42090 NEXT I1
42091 RETURN
42092 IF N1=0 THEN RETURN
42093 FOR I1=1 TO N1
42094 A(I1)=B(I1)
42095 NEXT I1
42096 RETURN
42100 IF N1*N2*N3=0 THEN GOTO 42110
42102 FOR I1=1 TO N1
42103 FOR I2=1 TO N2
42104 FOR I3=1 TO N3
42105 B(I1,I2,I3)=C(I1,I2,I3)
42106 NEXT I3
42107 NEXT I2
```

```
42108NEXTI1
42109RETURN
42110IFN1*N2=0THENGOTO42117
42111FORI1=1TON1
42112FORI2=1TON2
42113B(I1,I2)=C(I1,I2)
42114NEXTI2
42115NEXTI1
42116RETURN
42117IFN1=0THENRETURN
42118FORI1=1TON1
42119B(I1)=C(I1)
42120NEXTI1
42121RETURN
42150IFN1*N2*N3=0THENGOTO42160
42152FORI1=1TON1
42153FORI2=1TON2
42154FORI3=1TON3
42155C(I1,I2,I3)=A(I1,I2,I3)
42156NEXTI3
42157NEXTI2
42158NEXTI1
42159RETURN
42160IFN1*N2=0THENGOTO42167
42161FORI1=1TON1
42162FORI2=1TON2
42163C(I1,I2)=A(I1,I2)
42164NEXTI2
42165NEXTI1
42166RETURN
42167IFN1=0THENRETURN
42168FORI1=1TON1
42169C(I1)=A(I1)
42170NEXTI1
42171RETURN
42175IFN1*N2*N3=0THENGOTO42185
42177FORI1=1TON1
42178FORI2=1TON2
42179FORI3=1TON3
42180A(I1,I2,I3)=C(I1,I2,I3)
42181NEXTI3
42182NEXTI2
42183NEXTI1
42184RETURN
42185IFN1*N2=0THENGOTO42192
42186FORI1=1TON1
42187FORI2=1TON2
42188A(I1,I2)=C(I1,I2)
42189NEXTI2
42190NEXTI1
42191RETURN
42192IFN1=0THENRETURN
42193FORI1=1TON1
42194A(I1)=C(I1)
42195NEXTI1
42196RETURN
42400FORI=1TON
42401FORJ=1TON
42402B(I,J+N)=0
```

```
42403 B(I,J)=A(I,J)
42404 NEXTJ
42405 B(I,I+N)=1
42406 NEXTI
42410 FORK=1TON
42411 IFK=NTHENGOTO42424
42412 M=K
42414 FORI=K+1TON
42415 IFABS(B(I,K))>ABS(B(M,K))THENM=I
42416 NEXTI
42417 IFM=KTHENGOTO42424
42418 FORJ=KTO2*N
42419 B=B(K,J)
42420 B(K,J)=B(M,J)
42421 B(M,J)=B
42422 NEXTJ
42424 FORJ=K+1TO2*N
42425 B(K,J)=B(K,J)/B(K,K)
42426 NEXTJ
42427 IFK=1THENGOTO42434
42428 FORI=1TOK-1
42429 FORJ=K+1TO2*N
42430 B(I,J)=B(I,J)-B(I,K)*B(K,J)
42431 NEXTJ
42432 NEXTI
42433 IFK=NTHENGOTO42441
42434 FORI=K+1TON
42435 FORJ=K+1TO2*N
42436 B(I,J)=B(I,J)-B(I,K)*B(K,J)
42437 NEXTJ
42438 NEXTI
42439 NEXTK
42441 FORI=1TON
42442 FORJ=1TON
42443 B(I,J)=B(I,J+N)
42444 NEXTJ
42445 NEXTI
42446 RETURN
43500 A1=0
43501 A2=0
43502 B0=0
43503 B1=0
43504 FORM=0TON-1
43505 A1=A1+X(M)
43506 A2=A2+X(M)*X(M)
43507 B0=B0+Y(M)
43508 B1=B1+Y(M)*X(M)
43509 NEXTM
43510 A1=A1/N
43511 A2=A2/N
43512 B0=B0/N
43513 B1=B1/N
43514 D=A1*A1-A2
43515 A=A1*B1-A2*B0
43516 A=A/D
43517 B=A1*B0-B1
43518 B=B/D
43521 D=0
43522 FORM=0TON-1
```

```
43523D1=Y(M)-A-B*X(M)
43524D=D+D1*D1
43525NEXTM
43526D=SQRT(D/(N-2))
43527RETURN
43550A0=1
43551A1=0
43552A2=0
43553A3=0
43554A4=0
43555B0=0
43556B1=0
43557B2=0
43558FORM=0TON-1
43559A1=A1+X(M)
43560A2=A2+X(M)*X(M)
43561A3=A3+X(M)*X(M)*X(M)
43562A4=A4+X(M)*X(M)*X(M)*X(M)
43563B0=B0+Y(M)
43564B1=B1+Y(M)*X(M)
43565B2=B2+Y(M)*X(M)*X(M)
43566NEXTM
43567A1=A1/N
43568A2=A2/N
43569A3=A3/N
43570A4=A4/N
43571B0=B0/N
43572B1=B1/N
43573B2=B2/N
43574D=A0*(A2*A4-A3*A3)-A1*(A1*A4-A3*A2)+A2*(A1*A3-A2*A2)
43575A=B0*(A2*A4-A3*A3)+B1*(A3*A2-A1*A4)+B2*(A1*A3-A2*A2)
43576A=A/D
43577B=B0*(A3*A2-A1*A4)+B1*(A0*A4-A2*A2)+B2*(A2*A1-A0*A3)
43578B=B/D
43579C=B0*(A1*A3-A2*A2)+B1*(A1*A2-A0*A3)+B2*(A0*A2-A1*A1)
43580C=C/D
43583D=0
43584FORM=0TON-1
43585D1=Y(M)-A-B*X(M)-C*X(M)*X(M)
43586D=D+D1*D1
43587NEXTM
43588D=SQRT(D/(N-3))
43589RETURN
43600FORI=1TOM
43601B=1
43602FORJ=1TON+1
43603Z(I,J)=B
43604B=B*X(I)
43605NEXTJ
43606NEXTI
43608RETURN
43650M4=M
43651N4=N
43653FORI=1TOM
43654FORJ=1TON
43655A(I,J)=Z(I,J)
43656NEXTJ
43657NEXTI
43660GOSUB41950
```

```
43664N1=M
43665N2=N
43666N3=0
43667GOSUB42150
43668N1=N
43669N2=M
43670GOSUB42075
43671N1=M
43672N2=N
43673GOSUB42100
43677M1=N
43678N1=M
43679M2=M
43680GOSUB41900
43682N1=N
43683GOSUB42175
43685GOSUB42400
43687M=M4
43690GOSUB42075
43691FORI=1TOM
43692FORJ=1TON
43693B(J,I)=Z(I,J)
43694NEXTJ
43695NEXTI
43697M2=N
43698N2=M
43699GOSUB41900
43702N1=N
43703N2=M
43704GOSUB42175
43706FORI=1TOM
43707B(I,1)=Y(I)
43708NEXTI
43709N2=1\N1=M
43710M2=M
43712GOSUB41900
43716FORI=1TON
43717D(I)=C(I,1)
43718NEXTI
43719RETURN
43750D=0
43751FORI=1TOM
43752Y=0
43753B=1
43754FORJ=1TON+1
43755Y=Y+D(J)*B
43756B=B*X(I)
43757NEXTJ
43758D=D+(Y-Y(I))*(Y-Y(I))
43759NEXTI
43760IFM-N-1>0THENGOTO43763
43761D=0
43762RETURN
43763D=D/(M-N-1)
43764D=SQRT(D)
43765RETURN
43800N=1
43801FORI=1TOL
43802N=N*(M(I)+1)
```

```
43803NEXTI
43804D=0
43805FORI=1TOM
43808IFL>0THENGOTO43811
43809L=0
43810RETURN
43811IFL<=9THENGOTO43814
43812L=0
43813RETURN
43814J=0
43816IFL=1THENGOSUB43840
43817IFL=2THENGOSUB43849
43818IFL=3THENGOSUB43857
43819IFL=4THENGOSUB43865
43820IFL=5THENGOSUB43873
43821IFL=6THENGOSUB43881
43822IFL=7THENGOSUB43889
43823IFL=8THENGOSUB43897
43824IFL=9THENGOSUB43905
43826Y=0
43827FORK=1TON
43828Y=Y+D(K)*Z(I,K)
43829NEXTK
43830D=D+(Y(I)-Y)*(Y(I)-Y)
43831NEXTI
43833IFM-N>0THENGOTO43836
43834D=0
43835RETURN
43836D=D/(M-N)
43837D=SQRT(D)
43838RETURN
43839RETURN
43840B=1
43841C=B
43842FORI1=0TOM(1)
43843J=J+1
43844Z(I,J)=B
43845B=B*X(I,1)
43846NEXTI1
43847B=C
43848RETURN
43849B=1
43850C=B
43851FORI2=0TOM(2)
43852GOSUB43841
43853B=B*X(I,2)
43854NEXTI2
43855B=C
43856RETURN
43857B=1
43858C=B
43859FORI3=0TOM(3)
43860GOSUB43850
43861B=B*X(I,3)
43862NEXTI3
43863B=C
43864RETURN
43865B=1
43866C=B
```

```
43867FORI4=0TOM(4)
43868GOSUB43858
43869B=B*X(I,4)
43870NEXTI4
43871B=C
43872RETURN
43873B=1
43874C=B
43875FORI5=0TOM(5)
43876GOSUB43866
43877B=B*X(I,5)
43878NEXTI5
43879B=C
43880RETURN
43881B=1
43882C=B
43883FORI6=0TOM(6)
43884GOSUB43874
43885B=B*X(I,6)
43886NEXTI6
43887B=C
43888RETURN
43889B=1
43890C=B
43891FORI7=0TOM(7)
43892GOSUB43882
43893B=B*X(I,7)
43894NEXTI7
43895B=C
43896RETURN
43897B=1
43898C=B
43899FORI8=0TOM(8)
43900GOSUB43890
43901B=B*X(I,8)
43902NEXTI8
43903B=C
43904RETURN
43905B=1
43906FORI9=0TOM(9)
43907GOSUB43898
43908B=B*X(I,9)
43909NEXTI9
43910RETURN
43950N1=M+1
43951V1=10000000
43953FORI=1TON1
43954A(I)=0
43955B(I)=0
43956F(I)=0
43957NEXTI
43958FORI=1TON
43959V(I)=0
43960D(I)=0
43961NEXTI
43962D1=SQRT(N)
43963W=D1
43964FORI=1TON
43965E(I)=1/W
```

```
43966NEXTI
43967F1=D1
43968A1=0
43969FORI=1TON
43970A1=A1+X(I)*E(I)*E(I)
43971NEXTI
43972C1=0
43973FORI=1TON
43974C1=C1+Y(I)*E(I)
43975NEXTI
43976B(1)=1/F1
43977F(1)=B(1)*C1
43978FORI=1TON
43979V(I)=V(I)+E(I)*C1
43980NEXTI
43981M=1
43983FORI=1TOL
43984C2(I)=C(I)
43985NEXTI
43986L2=L
43987V2=V
43988F2=F1
43989A2=A1
43990F1=0
43991FORI=1TON
43992B1=E(I)
43993E(I)=(X(I)-A2)*E(I)-F2*D(I)
43994D(I)=B1
43995F1=F1+E(I)*E(I)
43996NEXTI
43997F1=SQRT(F1)
43998FORI=1TON
43999E(I)=E(I)/F1
44000NEXTI
44001A1=0
44002FORI=1TON
44003A1=A1+X(I)*E(I)*E(I)
44004NEXTI
44005C1=0
44006FORI=1TON
44007C1=C1+E(I)*Y(I)
44008NEXTI
44009M=M+1
44010I=0
44011L=M-I
44012B2=B(L)
44013D1=0
44014IFL>1THEND1=B(L-1)
44015D1=D1-A2*B(L)-F2*A(L)
44016B(L)=D1/F1
44017A(L)=B2
44018I=I+1
44019IFI<>MTHENGOTO44011
44020FORI=1TON
44021V(I)=V(I)+E(I)*C1
44022NEXTI
44023FORI=1TON1
44024F(I)=F(I)+B(I)*C1
44025C(I)=F(I)
```

```
44026NEXTI
44027V=0
44028FORI=1TON
44029V=V+(V(I)-Y(I))*(V(I)-Y(I))
44030NEXTI
44032V=SQRT(V/(N-L-1))
44033L=M
44034IFE=0THENGOTO44040
44036IFABS(V1-V)/V<ETHENGOTO44053
44038IFE*V>E*V1THENGOTO44053
44039V1=V
44040IFM=N1THENGOTO44043
44041GOTO43983
44043FORI=1TOL
44044C(I-1)=C(I)
44045NEXTI
44046C(L)=0
44048L=L-1
44049D=V
44050RETURN
44053L=L2
44054V=V2
44055FORI=1TOL
44056C(L)=C2(L)
44057NEXTI
44058GOTO44043
44100L1=0
44102FORI=1TOM
44103Y1(I)=Y(I)
44104NEXTI
44106FORI=1TON
44107D1(I)=0
44108NEXTI
44110D1=10000000
44112GOSUB43800
44114GOSUB43650
44116GOSUB43800
44118IFD1>DTHENGOTO44131
44120FORI=1TON
44121D(I)=D1(I)
44122NEXTI
44124FORI=1TOM
44125Y(I)=Y1(I)
44126NEXTI
44128GOSUB43800
44129RETURN
44131D1=D
44132L1=L1+1
44134FORI=1TON
44135D(I)=D1(I)+D(I)
44136D1(I)=D(I)
44137NEXTI
44139FORI=1TOM
44140Y(I)=Y1(I)
44141NEXTI
44143GOSUB44147
44145GOTO44112
44147FORI=1TOM
44148J=0
```

```
44150IFL=1THENGOSUB44167
44151IFL=2THENGOSUB44176
44152IFL=3THENGOSUB44184
44153IFL=4THENGOSUB44192
44154IFL=5THENGOSUB44200
44155IFL=6THENGOSUB44208
44156IFL=7THENGOSUB44216
44157IFL=8THENGOSUB44224
44158IFL=9THENGOSUB44232
44160Y=0
44161FORK=1TON
44162Y=Y+D(K)*Z(I,K)
44163NEXTK
44164Y(I)=Y(I)-Y
44165NEXTI
44166RETURN
44167B=1
44168C=B
44169FORI1=0TOM(1)
44170J=J+1
44171Z(I,J)=B
44172B=B*X(I,1)
44173NEXTI1
44174B=C
44175RETURN
44176B=1
44177C=B
44178FORI2=0TOM(2)
44179GOSUB44168
44180B=B*X(I,2)
44181NEXTI2
44182B=C
44183RETURN
44184B=1
44185C=B
44186FORI3=0TOM(3)
44187GOSUB44177
44188B=B*X(I,3)
44189NEXTI3
44190B=C
44191RETURN
44192B=1
44193C=B
44194FORI4=0TOM(4)
44195GOSUB44185
44196B=B*X(I,4)
44197NEXTI4
44198B=C
44199RETURN
44200B=1
44201C=B
44202FORI5=0TOM(5)
44203GOSUB44193
44204B=B*X(I,5)
44205NEXTI5
44206B=C
44207RETURN
44208B=1
44209C=B
```

```
44210FORI6=0TOM(6)
44211GOSUB44201
44212B=B*X(I,6)
44213NEXTI6
44214B=C
44215RETURN
44216B=1
44217C=B
44218FORI7=0TOM(7)
44219GOSUB44209
44220B=B*X(I,7)
44221NEXTI7
44222B=C
44223RETURN
44224B=1
44225C=B
44226FORI8=0TOM(8)
44227GOSUB44217
44228B=B*X(I,8)
44229NEXTI8
44230B=C
44231RETURN
44232B=1
44233FORI9=0TOM(9)
44234GOSUB44225
44235B=B*X(I,9)
44236NEXTI9
44237RETURN
44250FORI=1TOL
44251E1(I)=E1
44252NEXTI
44253M=0
44255L1=1000000
44257FORI=1TOL
44258A0=A(I)
44260A(I)=A0
44261GOSUB44286
44263M0=L2
44265A(I)=A0*(1-E1(I))
44266GOSUB44286
44267M1=L2
44270IFM1>M0THENE1(I)=-E1(I)/2
44272IFM1<M0THENE1(I)=1.2*E1(I)
44274IFM1>M0THENA(I)=A0
44275IFM1>M0THENGOTO44261
44276NEXTI
44279M=M+1
44280IFL2=0THENRETURN
44281IFABS((L1-L2)/L2)<ETHENRETURN
44283L1=L2
44284GOTO44257
44286L2=0
44287FORJ=1TON
44288X=X(J)
44290GOSUB44300
44291L2=L2+(Y(J)-Y)*(Y(J)-Y)
44292NEXTJ
44293D=SQRT(L2/(N-L))
44294RETURN
```

```
44300Y=A(1)*EXP(-(X-A(2))*(X-A(2))/A(3))
44301RETURN
44400X=Y
44402IFX=0THENX=EXP(-100)
44403IFX>.5THENGOTO44408
44404X=-LOG(X)
44406Z=-.803+1.312*X-.2118*X*X+.016*X*X*X
44407GOTO44413
44408X=1-X
44410IFX=0THENGOTO44415
44411X=-LOG(X)
44412Z=.803-1.312*X+.2118*X*X-.016*X*X*X
44413X=2/(9*M)
44414X=1-X+Z*SQRT(X)
44415X=M*X*X*X
44416RETURN
44425A=1
44428FORI=1TON
44429A=A*I
44430NEXTI
44431A=1/A
44432IFN=0THENGOTO44437
44434FORI=1TON
44435A=A*X/2
44436NEXTI
44437B0=1
44438B2=1
44439M=0
44441M=M+1
44442B1=-(X*X*B0)/(M*(M+N)*4)
44443B2=B2+B1
44444B0=B1
44446IFABS(B1)>ETHENGOTO44441
44448Y=A*B2
44449RETURN
44475A1=1
44476B1=1
44477FORI=1TON
44478B(I-1)=0
44479B1=B1*I
44480A1=A1/2
44481NEXTI
44482B1=A1/B1
44483A1=1
44484FORI=0TOMSTEP2
44485A(I)=A1*B1
44486A(I+1)=0
44487A1=-A1/((I+2)*(N+N+I+2))
44488NEXTI
44489A1=A1/2
44490FORI=0TOM
44491B(I+N)=A(I)
44492NEXTI
44493FORI=0TON+M
44494A(I)=B(I)
44495NEXTI
44496RETURN
44525A=1
44526A1=1
```

```
44527 A2=1
44528 B=1
44529 C=1
44530 E1=1000000
44531 M=-1
44532 X1=1/(8*X)
44533 X1=X1*X1
44534 M1=1
44535 M2=1
44536 N1=-1/(8*X)
44537 N2=-3*N1
44538 N=0
44539 M=M+2
44540 A=A*M*M
44541 M=M+2
44542 A=A*M*M
44543 C=C*X1
44544 A1=A1*A2
44545 A2=A2+1
44546 A1=A1*A2
44547 A2=A2+1
44548 E2=A*C/A1
44549 E4=1+(M+2)/M+(M+2)*(M+2)/(A2*8*X)+(M+2)*(M+4)/(A2*8*X)
44550 E4=E4*E2
44552 IFABS(E4)>E1THENGOTO44562
44553 E1=ABS(E2)
44554 M1=M1-E2
44555 M2=M2+E2*(M+2)/M
44556 N1=N1+E2*(M+2)*(M+2)/(A2*8*X)
44557 N2=N2-E2*(M+2)*(M+4)/(A2*8*X)
44558 N=N+1
44560 IFE1<E3THENGOTO44562
44561 GOTO44539
44562 A=3.1415926536
44563 E=E2
44564 B=SQRT(2/(A*X))
44565 J0=B*(M1*COS(X-A/4)-N1*SIN(X-A/4))
44566 J1=B*(M2*COS(X-3*A/4)-N2*SIN(X-3*A/4))
44567 RETURN
44580 X1=1/(X*X)
44581 Y=(X+.5)*LOG(X)-X*(1-X1/12+X1*X1/360-X1*X1*X1/1260+X1*X1*X1*X1/1680)
44582 Y=Y+0.918938533205
44583 RETURN
44600 M1=X
44602 X=M/2-1
44604 GOSUB44580
44605 X=M1
44606 C=-X/2+(M/2-1)*LOG(X)-(M/2)*LOG(2)-Y
44607 Y=EXP(C)
44608 RETURN
44625 Y1=1
44626 X2=X
44627 M2=M+2
44628 X2=X2/M2
44629 Y1=Y1+X2
44630 IFX2<ETHENGOTO44637
44631 M2=M2+2
44633 X2=X2*X/M2
44635 GOTO44629
```

```
44637GOSUB44600
44638Y=Y1*Y*2*X/M
44639RETURN
44675N=1
44676Y=1
44677C2=1/(2*X*X)
44678Y=Y-C2
44679N=N+2
44680C1=C2
44681C2=-C1*N/(2*X*X)
44686GOTO44678
44687N=(N+1)/2
44688E=EXP(-X*X)/(X*1.772453850905516)
44689Y1=Y*E
44690Y=1-Y1
44691E=E*C2
44692RETURN
44725B(0,0)=1
44726B(1,0)=0
44727B(1,1)=1
44729IFN<2THENRETURN
44730FORI=2TON
44731FORJ=1TOI
44733B(I,J)=B(I-1,J-1)+B(I-1,J-1)-B(I-2,J)
44734NEXTJ
44735B(I,0)=-B(I-2,0)
44736NEXTI
44737RETURN
44760B=X0
44761FORI=1TOM
44762C(I)=C(I)*B
44763B=B*X0
44764NEXTI
44767FORN=MTO0STEP-1
44768GOSUB44725
44769A(N)=C(N)/B(N,N)
44770FORL=0TON
44772C(L)=C(L)-A(N)*B(N,L)
44773NEXTL
44774NEXTN
44776FORI=0TOM1
44777FORJ=0TOI
44778C(J)=C(J)+A(I)*B(I,J)
44779NEXTJ
44780NEXTI
44782B=1/X0
44783FORI=1TOM1
44784C(I)=C(I)*B
44785B=B/X0
44786NEXTI
44787RETURN
44800A1=A(1)
44801B(1)=1/A1
44802A=1/A1
44803B=A*A
44804A=A*B
44805B(2)=-A2/A
44806A3=A(3)
44807A=A*B
```

```
44808 B(3)=A*(2*A2*A2-A1*A3)
44809 A4=A(4)
44810 A=A*B
44811 B(4)=A*(5*A1*A2*A3-A1*A1*A4-5*A2*A2*A2)
44812 A5=A(5)
44813 A=A*B
44814 B(5)=6*A1*A1*A2*A4+3*A1*A1*A3*A3+14*A2*A2*A2*A2
44815 B(5)=B(5)-A1*A1*A1*A5-21*A1*A2*A2*A3
44816 B(5)=A*B(5)
44817 A6=A(6)
44818 A=A*B
44819 B(6)=7*A1*A1*A1*A2*A5+7*A1*A1*A1*A3*A4+84*A1*A2*A2*A2*A3
44820 B(6)=B(6)-A1*A1*A1*A1*A6-28*A1*A1*A2*A2*A4
44821 B(6)=B(6)-28*A1*A1*A2*A3*A3-42*A2*A2*A2*A2*A2
44822 B(6)=A*B(6)
44823 A7=A(7)
44824 A=A*B
44825 B(7)=8*A1*A1*A1*A1*A2*A6+8*A1*A1*A1*A1*A3*A5
44826 B(7)=B(7)+4*A1*A1*A1*A1*A4*A4+120*A1*A1*A2*A2*A2*A4
44827 B(7)=B(7)+180*A1*A1*A2*A2*A3*A3+132*A1*A2*A2*A2*A2*A2
44828 B(7)=B(7)-A1*A1*A1*A1*A1*A7-36*A1*A1*A1*A2*A2*A5
44829 B(7)=B(7)-72*A1*A1*A1*A2*A3*A4-12*A1*A1*A1*A3*A3*A3
44830 B(7)=B(7)-330*A1*A2*A2*A2*A2*A3
44831 B(7)=A*B(7)
44832 B(0)=0
44833 A=A(0)
44834 FOR I=1 TO 7
44835 B(0)=B(0)-B(I)*A
44836 A=A*A(0)
44837 NEXT I
44838 RETURN
44850 L=A(0)
44851 FOR I=0 TO N
44852 A(I)=A(I)/L
44853 B(I)=0
44854 NEXT I
44856 FOR I=N+1 TO M
44857 A(I)=0
44858 B(I)=0
44859 NEXT I
44861 B(0)=1
44862 FOR I=1 TO M
44863 J=1
44864 B(I)=B(I)-A(J)*B(I-J)
44865 J=J+1
44866 IF J<=I THEN GOTO 44864
44867 NEXT I
44869 FOR I=0 TO M
44870 A(I)=A(I)*L
44871 B(I)=B(I)/L
44872 NEXT I
44873 RETURN
44900 FOR J=0 TO 4
44901 C(J,0)=A(4-J)
44902 NEXT J
44903 FOR I=0 TO 4
44904 C(0,I+1)=C(0,I)
44905 J=1
44906 IF J>4-I THEN GOTO 44910
```

```
44907C(J,I+1)=X0*C(J-1,I+1)+C(J,I)
44908J=J+1
44909GOTO44906
44910NEXTI
44911FORI=0TO4
44912B(4-I)=C(I,4-I+1)
44913NEXTI
44914RETURN
44925C0=2.515517
44926C1=0.802853
44927C2=0.010328
44928D1=1.432788
44929D2=0.189269
44930D3=0.001308
44931IFY=0THENX=10000000000000
44932IFY=0THENRETURN
44933Z=SQRT(-LOG(Y*Y))
44934X=1+D1*Z+D2*Z*Z+D3*Z*Z*Z
44935X=(C0+C1*Z+C2*Z*Z)/X
44936X=Z-X
44937RETURN
44950Z1=A(0)
44951Z2=0
44953A1=X
44954A2=Y
44955FORN=1TOM
44957X=A1
44958Y=A2
44960GOSUB41200
44962Z1=Z1+A(N)*X
44963Z2=Z2+A(N)*Y
44964NEXTN
44966X=A1
44967Y=A2
44968RETURN
45000IFN<=1THENRETURN
45001IFY<0THENRETURN
45002IFINT(N)<>NTHENRETURN
45003IFE<=0THENRETURN
45004IFY>0THENGOTO45007
45005X=0
45006RETURN
45007M=0
45009X2=1
45010FORI=1TON-1
45011X2=X2*X0
45012NEXTI
45014X1=((N-1)*X0+Y/X2)/N
45015X=X1
45016M=M+1
45017IFABS((X0-X1)/X1)<ETHENRETURN
45018X0=X1
45019GOTO45009
45030N3=N
45031IFY<0THENRETURN
45032IFE<=0THENRETURN
45033IFY>0THENGOTO45036
45034X=0
45035RETURN
```

```
45036X3=Y
45038IFN>=0THENGOTO45042
45039N=-N
45040Y=1/Y
45042GOSUB45060
45044PRINT
45045GOSUB45074
45046Y=X3
45047N=N3
45048RETURN
45060N1=INT(N)
45061N2=N-N1
45064A=.5
45065FORI=1TOM
45066A(I)=0
45067IFA<N2THENA(I)=1
45068IFA<N2THENN2=N2-A
45069A=A/2
45070NEXTI
45071RETURN
45074FORI=1TOM
45076GOSUB45109
45078Y=X1
45079IFA(I)=1THENGOTO45083
45080A(I)=1
45081GOTO45084
45083A(I)=Y
45084NEXTI
45087Y=X3
45089N=N3
45091X2=1
45092FORI=1TON1
45093X2=X2*Y
45094NEXTI
45096FORI=1TOM
45097X2=X2*A(I)
45098NEXTI
45100X=X2
45101RETURN
45109X0=1
45110X1=(X0+Y/X0)/2
45111IFABS((X1-X0)/X1)<ETHENRETURN
45112X0=X1
45113GOTO45110
45125X0=1
45127IFX<>0THENGOTO45131
45128Y=0
45129RETURN
45131IFABS(X)>=3.1415926535/2THENRETURN
45132IFE<=0THENRETURN
45134X1=X0+(X-ATN(X0))*(1+X0*X0)
45136IFABS((X0-X1)/X1)<ETHENGOTO45139
45137X0=X1
45138GOTO45134
45139Y=X1
45140RETURN
45150IFE<0THENRETURN
45151M=0
45152Y=1+X*X
```

```
45154 GOSUB45177
45155 X2=1/X1
45156 A0=X2
45157 B0=1
45158 A1=(A0+B0)/2
45159 Y=A1*B0
45161 GOSUB45177
45162 B1=X1
45164 M=M+1
45165 IFABS((A1-B1)/B1)<ETHENGOTO45170
45166 A0=A1
45167 B0=B1
45168 GOTO45158
45170 Y=A1*B1
45172 GOSUB45177
45173 Y=X*X2/X1
45174 RETURN
45177 X0=1
45178 X1=(X0+Y/X0)/2
45179 IFABS((X0-X1)/X1)<ETHENRETURN
45180 X0=X1
45181 GOTO45178
45200 M=0
45202 IFE<=0THENRETURN
45203 IFX<>0THENGOTO45207
45204 Y=0
45205 RETURN
45207 IFABS(X)>1THENRETURN
45208 U0=X*SQRT(1-X*X)
45209 U1=X
45210 U2=U1*SQRT(2*U1/(U1+U0))
45211 Y=U2
45212 M=M+1
45213 IFABS(U2-U1)<ETHENRETURN
45214 U0=U1
45215 U1=U2
45216 GOTO45210
45250 A(0)=1+K
45251 B(0)=1-K
45252 N=0
45253 IFK<0THENRETURN
45254 IFK>1THENRETURN
45255 IFE<=0THENRETURN
45256 IFK<1THENGOTO45261
45257 E2=1
45258 E1=1000000000
45259 E1=E1*E1*E1*E1
45260 RETURN
45261 N=N+1
45263 A(N)=(A(N-1)+B(N-1))/2
45264 B(N)=SQRT(A(N-1)*B(N-1))
45265 IFABS(A(N)-B(N))>ETHENGOTO45261
45266 E1=1.5707963268/A(N)
45267 E2=2
45268 M=1
45269 FORI=1TON
45270 E2=E2-M*(A(I)*A(I)-B(I)*B(I))
45271 M=M*2
45272 NEXTI
```

```
45273 E2=E2*E1/2
45274 RETURN
45300 M=0
45302 IFE<=0THENRETURN
45303 IFX<1THENGOTO45307
45304 Y=0
45305 RETURN
45307 IFX<=0THENRETURN
45308 U0=(X*X-1/(X*X))/4
45309 U1=(X-1/X)/2
45310 U2=U1*SQRT(2*U1/(U1+U0))
45311 Y=U2
45312 M=M+1
45313 IFABS(U2-U1)<ETHENRETURN
45314 U0=U1
45315 U1=U2
45316 GOTO45310
45350 IFX<=0THENRETURN
45351 IFM<=0THENRETURN
45352 Y(0)=1
45353 Y(1)=0
45354 C=0
45355 N=M
45357 FORI=4TO1STEP-1
45358 Y(I)=Y(I-1)
45359 NEXTI
45361 Y(0)=2*N*Y(1)/X-Y(2)
45362 N=N-1
45363 IFN=0THENGOTO45367
45364 IFINT(N/2)<>N/2THENGOTO45357
45365 C=C+2*Y(0)
45366 GOTO45356
45367 C=C+Y(0)
45369 FORI=0TO4
45370 Y(I)=Y(I)/C
45371 NEXTI
45372 RETURN
45400 B(0,0)=1
45401 B(1,0)=0
45402 B(1,1)=1
45404 IFN<2THENRETURN
45405 FORI=2TON
45406 B(I,0)=-(I-1)*B(I-2,0)/I
45407 FORJ=1TOI
45409 B(I,J)=(I+I-1)*B(I-1,J-1)-(I-1)*B(I-2,J)
45410 B(I,J)=B(I,J)/I
45411 NEXTJ
45412 NEXTI
45413 FORI=0TON
45414 A(I)=B(N,I)
45415 NEXTI
45416 RETURN
45425 B(0,0)=1
45426 B(1,0)=1
45427 B(1,1)=-1
45429 IFN<2THENRETURN
45430 FORI=2TON
45431 B(I,0)=(2*I-1)*B(I-1,0)-(I-1)*(I-1)*B(I-2,0
45432 FORJ=1TOI
```

```
45434 B(I,J)=(2*I-1)*B(I-1,J)-B(I-1,J-1)-(I-1)*(I
45435 NEXTJ
45436 NEXTI
45437 FORI=0TON
45438 A(I)=B(N,I)
45439 NEXTI
45440 RETURN
45450 B(0,0)=1
45451 B(1,0)=0
45452 B(1,1)=2
45454 IFN<2THENRETURN
45455 FORI=2TON
45456 B(I,0)=-2*(I-1)*B(I-2,0)
45457 FORJ=1TOI
45459 B(I,J)=2*B(I-1,J-1)-2*(I-1)*B(I-2,J)
45460 NEXTJ
45461 NEXTI
45462 FORI=0TON
45463 A(I)=B(N,I)
45464 NEXTI
45465 RETURN
45475 IFA<>0THENGOTO45480
45476 U=0
45477 V=1
45478 RETURN
45480 GOSUB45518
45482 GOSUB45506
45483 U0=P
45484 V0=0
45486 FORI=1TON
45488 U1=U0-W(I)*A(I)*V0
45489 V1=V0+W(I)*A(I)*U0
45490 U0=U1
45491 V0=V1
45492 NEXTI
45495 U0=1-Z*Z/2
45496 V0=Z*(1+Z*Z/3)
45498 V=U0*(U1-V0*V1)
45499 U=U0*(V1+V0*U1)
45502 RETURN
45506 Z=A
45507 FORI=1TON
45508 W(I)=-1
45509 IFZ>0THENW(I)=1
45510 Z=Z-W(I)*A0
45511 A0=A0/2
45512 NEXTI
45515 RETURN
45518 N=12
45520 A(1)=1
45521 A(2)=.414213562373095
45522 A(3)=.198912367379658
45523 A(4)=.09849140335716425
45524 A(5)=.04912684976946725
45525 A(6)=.02454862210892544
45526 A(7)=.01227246237956627
45527 A(8)=.006136000157623401
45528 A(9)=.003067971201422665
45529 A(10)=.001533981991088666
```

```
45530A(11)=.0007669905443430926
45531A(12)=.000383495215771441
45533P=.6366197879720413
45535A0=.7853981633974484
45536RETURN
45550IFU=0THENGOTO45555
45552IFABS(U)>=.0001THENV=SQRT(1-U*U)
45553IFABS(U)<.0001THENV=1-U*U/2-U*U*U*U/8
45554GOTO45571
45555IFV=0THENGOTO45560
45557IFABS(V)>=.0001THENU=SQRT(1-V*V)
45558IFABS(V)<.0001THENU=1-V*V/2-V*V*V*V/8
45559GOTO45571
45560IFW=0THENGOTO45568
45562IFABS(W)<=10000THENU=1/SQRT(1+1/(W*W))
45563IFABS(W)>10000THENU=1-1/(2*W*W)+3/(8*W*W*W*W)
45564IFABS(U)>=.0001THENV=SQRT(1-U*U)
45565IFABS(U)<.0001THENV=1-U*U/2-U*U*U*U/8
45566U=U*ABS(W)/W
45567GOTO45571
45568A=0
45569RETURN
45571GOSUB45616
45573IFABS(U)<1THENGOTO45578
45574IFABS(V)>0THENGOTO45578
45576A=2*A0*ABS(U)/U
45577RETURN
45578IFABS(V)<1THENGOTO45583
45579IFABS(U)>0THENGOTO45583
45580A=0
45581RETURN
45583A=U
45584U=V
45585V=A
45586U0=P
45587U1=U0
45588V0=0
45589V1=V0
45591FORI=1TON
45593W(I)=-1
45594IFV0<VTHENW(I)=1
45596U1=U0-W(I)*A(I)*V0
45597V1=V0+W(I)*A(I)*U0
45598U0=U1
45599V0=V1
45600NEXTI
45603Z=V*U1-U*V1
45604Z=Z+Z*Z*Z/6
45606FORI=1TON
45607Z=Z+W(I)*A0
45608A0=A0/2
45609NEXTI
45611A=Z
45612RETURN
45616N=12
45618A(1)=1
45619A(2)=.414213562373095
45620A(3)=.198912367379658
45621A(4)=.09849140335716425
```

```
45622A(5)=.04912684976946725
45623A(6)=.02454862210892544
45624A(7)=.01227246237956627
45625A(8)=.006136000157623401
45626A(9)=.003067971201422665
45627A(10)=.001533981991088666
45628A(11)=.0007669905443430926
45629A(12)=.000383495215771441
45631P=.6366197879720413
45633A0=.7853981633974484
45634RETURN
45650GOSUB45684
45652K=INT(X)
45653X=X-K
45655GOSUB45673
45657Y=1
45658FORI=1TON
45659IFW(I)>0THENY=Y*A(I)
45660NEXTI
45662Y=Y*(1+Z*(1+Z/2*(1+Z/3*(1+Z/4))))
45664IFK<0THENE=1/E
45665IFABS(K)<1THENGOTO45671
45666FORI=1TOABS(K)
45667Y=Y*E
45668NEXTI
45670X=X+K
45671RETURN
45673A=.5
45674Z=X
45675FORI=1TON
45676W(I)=0
45677IFZ>ATHENW(I)=1
45678Z=Z-W(I)*A
45679A=A/2
45680NEXTI
45681RETURN
45684N=9
45685E=2.718281828459045
45686A(1)=1.648721270700128
45687A(2)=1.284025416687742
45688A(3)=1.133148453066826
45689A(4)=1.064494458917859
45690A(5)=1.031743407499103
45691A(6)=1.015747708586686
45692A(7)=1.007843097206448
45693A(8)=1.003913889338348
45694A(9)=1.001955033591003
45695RETURN
45725GOSUB45770
45727IFX<=0THENRETURN
45728K=0
45730X1=X
45732IFX<ETHENGOTO45739
45734K=K+1
45735X=X/E
45736GOTO45732
45739IFX>=1THENGOTO45744
45740K=K-1
45741X=X*E
```

```
45742GOTO45739
45744GOSUB45761
45747Z=Z-1
45748Z=Z*(1-(Z/2)*(1+(Z/3)*(1-Z/4)))
45750A=1/2
45751FORI=1TON
45752Z=Z+W(I)*A
45753A=A/2
45754NEXTI
45756Y=K+Z
45758X=X1
45759RETURN
45761Z=X
45762FORI=1TON
45763W(I)=0
45764IFZ>A(I)THENW(I)=1
45765IFW(I)=1THENZ=Z/A(I)
45766NEXTI
45767RETURN
45770N=15
45771E=2.718281828459045
45772A(1)=1.648721270700128
45773A(2)=1.284025416687742
45774A(3)=1.133148453066826
45775A(4)=1.064494458917859
45776A(5)=1.031743407499103
45777A(6)=1.015747708586686
45778A(7)=1.007843097206448
45779A(8)=1.003913889338348
45780A(9)=1.001955033591003
45781A(10)=1.000977039492417
45782A(11)=1.000488400478694
45783A(12)=1.000244170429748
45784A(13)=1.000122077763384
45785A(14)=1.000061037018933
45786A(15)=1.000030518043791
45787RETURN
45800IFABS(X)<.35THENGOTO45808
45803GOSUB45650
45805Y=(Y-(1/Y))/2
45806RETURN
45808Z=1
45809Y=1
45810FORI=1TO8
45811Z=Z*X*X/((2*I)*(2*I+1))
45812Y=Y+Z
45813NEXTI
45814Y=X*Y
45815RETURN
45825GOSUB45650
45826Y=(Y+(1/Y))/2
45827RETURN
45840GOSUB45800
45841V=Y
45843GOSUB45825
45844Y=V/Y
45845RETURN
45875IFX<>0THENGOTO45879
45876Y=0
```

```
45877RETURN
45879X2=X
45880X=ABS(X)
45881X=X+SQRT(X*X+1)
45883GOSUB45725
45885Y=(X2/ABS(X2))*Y
45887X=X2
45888RETURN
45900IFX>1THENGOTO45904
45901Y=0
45902RETURN
45904X2=X
45905X=ABS(X)
45906X=X+SQRT(X-1)*SQRT(X+1)
45908GOSUB45725
45910X=X2
45911RETURN
45925IFABS(X)<1THENGOTO45929
45926Y=(X/ABS(X))*1000000*1000000*1000000
45927RETURN
45929IFX<>0THENGOTO45933
45930Y=0
45931RETURN
45933X2=X
45934X=(1+X)/(1-X)
45936GOSUB45725
45938X=X2
45939RETURN
46000IFX<X(1)THENGOTO46003
46001IFX<=X(V-N)THENGOTO46006
46003N=0
46004RETURN
46006I=0
46007I=I+1
46008IFX>X(I)THENGOTO46007
46009I=I-1
46011FORJ=0TON
46012L(J)=1
46013NEXTJ
46014Y=0
46015FORK=0TON
46016FORJ=0TON
46017IFJ=KTHENGOTO46019
46018L(K)=L(K)*(X-X(J+I))/(X(I+K)-X(J+I))
46019NEXTJ
46020Y=Y+L(K)*Y(I+K)
46021NEXTK
46022RETURN
46050N=1
46051IFX>=X(1)THENGOTO46054
46052N=0
46053RETURN
46054IFX<=X(V-3)THENGOTO46058
46055N=0
46056RETURN
46058FORI=1TOV-1
46059Y1(I)=(Y(I+1)-Y(I))/(X(I+1)-X(I))
46060NEXTI
46061FORI=1TOV-2
```

```
46062 Y2(I)=(Y1(I+1)-Y1(I))/(X(I+1)-X(I))
46063 NEXTI
46064 FORI=1TOV-3
46065 Y3(I)=(Y2(I+1)-Y2(I))/(X(I+1)-X(I))
46066 NEXTI
46068 I=0
46069 I=I+1
46070 IFX>X(I)THENGOTO46069
46071 I=I-1
46073 A=X-X(I)
46074 B=A*(X-X(I+1))
46075 C=B*(X-X(I+2))
46076 Y=Y(I)+A*Y1(I)+B*Y2(I)+C*Y3(I)
46078 E=C*(X-X(I+3))*Y/24
46079 RETURN
46100 N=1
46102 IFX>=X(1)THENGOTO46105
46103 N=0
46104 RETURN
46105 IFX<=X(V-3)THENGOTO46108
46106 N=0
46107 RETURN
46108 X(0)=2*X(1)-X(2)
46110 FORI=1TOV-1
46112 M(I+2)=(Y(I+1)-Y(I))/(X(I+1)-X(I))
46113 NEXTI
46114 M(V+2)=2*M(V+1)-M(V)
46115 M(V+3)=2*M(V+2)-M(V+1)
46116 M(2)=2*M(3)-M(4)
46117 M(1)=2*M(2)-M(3)
46118 FORI=1TOV
46119 A=ABS(M(I+3)-M(I+2))
46120 B=ABS(M(I+1)-M(I))
46121 IFA+B<>0THENGOTO46124
46122 Z(I)=(M(I+2)+M(I+1))/2
46123 GOTO46125
46124 Z(I)=(A*M(I+1)+B*M(I+2))/(A+B)
46125 NEXTI
46127 I=0
46128 I=I+1
46129 IFX>=X(I)THENGOTO46128
46130 I=I-1
46132 B=X(I+1)-X(I)
46133 A=X-X(I)
46134 Y=Y(I)+Z(I)*A+(3*M(I+2)-2*Z(I)-Z(I+1))*A*A/B
46135 Y=Y+(Z(I)+Z(I+1)-2*M(I+2))*A*A*A/(B*B)
46136 RETURN
46150 IFX>X(1)THENGOTO46153
46151 N=0
46152 RETURN
46153 IFX<=X(V-N)THENGOTO46157
46154 N=0
46155 RETURN
46157 I=0
46158 I=I+1
46159 IFX>X(I)THENGOTO46158
46160 I=I-1
46162 FORJ=0TON
46163 L(J)=0
```

```
46164FORK=0TON
46165M(J,K)=1
46166NEXTK
46167NEXTJ
46168Y=0
46169FORK=0TON
46170FORJ=0TON
46171IFJ=KTHENGOTO46179
46172FORL=0TON
46173IFL=KTHENGOTO46178
46174IFL<>JTHENGOTO46177
46175M(L,K)=M(L,K)/(X(I+K)-X(I+J))
46176GOTO46178
46177M(L,K)=M(L,K)*(X-X(J+I))/(X(I+K)-X(I+J))
46178NEXTL
46179NEXTJ
46180FORL=0TON
46181IFL=KTHENGOTO46183
46182L(K)=L(K)+M(L,K)
46183NEXTL
46184Y=Y+L(K)*Y(I+K)
46185NEXTK
46186RETURN
46200Z=0
46201Z1=0
46203IFX1<X(1)THENRETURN
46204IFX2>X(V-3)THENRETURN
46206IFX1<X2THENGOTO46211
46207X3=X1
46208X1=X2
46209X2=X3
46210Z1=1
46211IFX2=X1THENRETURN
46214GOSUB46232
46216GOSUB46267
46218Z=4*I2/3-I1/3
46220IFZ1=0THENGOTO46225
46221Z=-Z
46222X2=X1
46223X1=X3
46225Z1=1
46226RETURN
46228RETURN
46232I1=0
46233N1=0
46234X=X1
46238GOSUB46299
46240IFX2>X(I+1)THENGOTO46250
46242N1=N1+1
46243D=Y
46244X=X2
46246GOSUB46299
46247I1=(Y+D)*(X2-X1)/2
46248RETURN
46250J1=I
46251I1=I1+(Y+Y(I+1))*(X(I+1)-X)/2
46253IFX2<X(J1+3)THENGOTO46259
46255N1=N1+1
46256I1=I1+(Y(J1+1)+Y(J1+3))*(X(J1+3)-X(J1+1))/2
```

```
46257J1=J1+2
46258GOTO46253
46259X=X2
46261GOSUB46299
46262I1=I1+(Y+Y(J1+1))*(X2-X(J1+1))/2
46263N1=N1+1
46264RETURN
46266INTEGRATIONFORI2
46267I2=0
46268X=X1
46269GOSUB46299
46270D=Y
46271IFX2>X(I+1)THENGOTO46280
46272X=X1+(X2-X1)/2
46273GOSUB46299
46274I2=I2+(D+Y)*(X2-X1)/4
46275D=Y
46276X=X2
46277GOSUB46299
46278I2=I2+(D+Y)*(X2-X1)/4
46279RETURN
46280X=X1+(X(I+1)-X1)/2
46281J1=I
46282GOSUB46299
46283I2=I2+(Y+D)*(X-X1)/2
46284I2=I2+(Y+Y(J1+1))*(X(J1+1)-X)/2
46285IFX2<X(J1+2)THENGOTO46289
46286I2=I2+(Y(J1+1)+Y(J1+2))*(X(J1+2)-X(J1+1))/2
46287J1=J1+1
46288GOTO46285
46289X=X2-(X2-X(J1+1))/2
46290GOSUB46299
46291D=Y
46292I2=I2+(Y(J1+1)+D)*(X2-X)/2
46293X=X2
46294GOSUB46299
46295I2=I2+(D+Y)*(X2-X(J1+1))/4
46296RETURN
46299GOSUB46100
46302RETURN
46350M=0
46351X=X0
46352GOSUB44300
46353Y0=Y
46354X=X1
46355GOSUB44300
46356X=(X0+X1)/2
46357GOSUB44300
46358M=M+1
46359IFY*Y0=0THENRETURN
46360IFY*Y0<0THENX1=X
46361IFY*Y0>0THENX0=X
46362IFABS(X1-X0)>ETHENGOTO46351
46363RETURN
46375N=0
46377X=X0
46378GOSUB44300
46380X0=X0-Y/Y1
46381N=N+1
```

```
46382IFN>=MTHENRETURN
46383IFABS(Y/Y1)>ETHENGOTO46377
46384X=X0
46385RETURN
46425N=0
46427X=X0
46428GOSUB44300
46429Y0=Y
46430X=X1
46432GOSUB44300
46433Y1=Y
46437IFY1=Y0THENY1=Y1+.001
46438X=(X0*Y1-X1*Y0)/(Y1-Y0)
46439N=N+1
46441IFN>=MTHENRETURN
46442IFABS(X1-X0)<ETHENRETURN
46444X0=X1
46445X1=X
46446GOTO46427
46475N=0
46477IFX0<X1THENGOTO46481
46478X=X0
46479X0=X1
46480X1=X
46481X=X0
46483GOSUB44300
46484Y0=Y
46485X=X1
46487GOSUB44300
46488Y1=Y
46490X=(X0*Y1-X1*Y0)/(Y1-Y0)
46492N=N+1
46493IFN>=MTHENRETURN
46494IFABS(X1-X)<ETHENRETURN
46496GOSUB44300
46498IFY1*Y=0THENRETURN
46499IFY0*Y>0THENGOTO46504
46500X1=X
46501Y1=Y
46502Y0=Y0/2
46503GOTO46490
46504X0=X
46505Y0=Y
46506Y1=Y1/2
46507GOTO46490
46525N=0
46526X=X0
46528GOSUB44300
46529Y=X+C*Y
46531IFN>0THENGOTO46536
46532X1=Y
46533X=X1
46534N=N+1
46535GOTO46528
46536X2=Y
46537N=N+1
46539IFX2-2*X1+X0=0THENX0=X0+.001
46541K=(X2-X1)*(X2-X1)/(X2-2*X1+X0)
46542X2=X2-K
```

```
46544IFN>=MTHENRETURN
46545IFABS(K)<ETHENRETURN
46546X0=X1
46547X1=X2
46548X=X1
46549GOTO46528
46575N=0
46576M1=0
46577X=X0
46579GOSUB44300
46580Y=X+C*Y
46582IFM1>0THENGOTO46588
46583N=N+1
46584M1=M1+1
46585X=X1
46586X1=Y
46587GOTO46579
46588X2=Y
46591K=(X2-2*X1+X0)
46592IFK=0THENK=.001
46593K=(X1-X0)*(X1-X0)/K
46594X0=X0-K
46596IFN>=MTHENRETURN
46597IFABS(X-X0)<ETHENRETURN
46599GOTO46576
46625FORI=N1TON2STEP-1
46626A(I-N2)=C(I)/B(N2)
46627IFI=N2THENGOTO46631
46628FORJ=0TON2
46629C(I-J)=C(I-J)-A(I-N2)*B(N2-J)
46630NEXTJ
46631NEXTI
46632RETURN
46650N=0
46652X=X0
46653GOSUB44300
46654B=Y1/Y
46655FORI=1TOL
46656B=B-1/(X0-A(I))
46657NEXTI
46659X1=X0-1/B
46660N=N+1
46662IFN>=MTHENGOTO46666
46663IFABS(X1-X0)<ETHENGOTO46666
46664X0=X1
46665GOTO46652
46666X=X1
46667RETURN
46700A=3.14159/(2*M)
46702FORI=1TO4*M
46703X=W*COS(A*(I-1))+X0
46704Y=W*SIN(A*(I-1))+Y0
46705GOSUB44300
46706IFU>=0THENIFV>=0THENN(I)=1
46707IFU<0THENIFV>=0THENN(I)=2
46708IFU<0THENIFV<0THENN(I)=3
46709IFU>=0THENIFV<0THENN(I)=4
46710NEXTI
46712N=N(1)
```

```
46713A=0
46714FORI=2TO4*M
46715IFN=N(I)THENGOTO46721
46716IFN<>4THENIFN=N(I)+1THENA=A-1
46717IFN=1THENIFN(I)=4THENA=A-1
46718IFN=4THENIFN(I)=1THENA=A+1
46719IFN+1=N(I)THENA=A+1
46720N=N(I)
46721NEXTI
46723IFN<>4THENIFN=N(1)+1THENA=A-1
46724IFN=4THENIFN(1)=1THENA=A+1
46725IFN=1THENIFN(1)=4THENA=A-1
46726IFN+1=N(1)THENA=A+1
46727A=ABS(A)
46728N=INT(A/4)
46729A=A-4*INT(A/4)
46730RETURN
46760M=M*4
46761K=1
46762A=6.283185/M
46763FORI=1TOM
46764X(I)=W*COS(A*(I-1))+X0
46765Y(I)=W*SIN(A*(I-1))+Y0
46766NEXTI
46768FORI=1TOM
46769X=X(I)
46770Y=Y(I)
46771GOSUB44300
46772U(I)=U
46773V(I)=V
46774NEXTI
46777I=1
46778U=U(I)
46779GOSUB46885
46780IFU*U(I)<0THENGOTO46785
46782IFI=1THENGOTO46881
46783GOTO46779
46785M1=I
46788I=M1+M/2
46789IFI>MTHENI=I-M
46790J=I
46791U=U(I)
46793GOSUB46885
46794IFU*U(I)<0THENGOTO46801
46795IFU*U(J)<0THENGOTO46804
46797IFI=M1+M/2THENGOTO46881
46798IFJ=M1+M/2THENGOTO46881
46799GOTO46793
46801M3=I
46802GOTO46808
46804IFJ=MTHENJ=0
46805M3=J+1
46808I=M1+M/4
46809IFI>MTHENI=I-M
46810J=I
46811V=V(I)
46812GOSUB46885
46813IFV*V(I)<0THENGOTO46820
46814IFV*V(J)<0THENGOTO46822
```

```
46816IFI=M1+M/4THENGOTO46881
46817IFJ=M1+M/4THENGOTO46881
46818GOTO46812
46820M2=I
46821GOTO46825
46822IFJ=MTHENJ=0
46823M2=J+1
46825I=M2+M/2
46826IFI>MTHENI=I-M
46827J=I
46828V=V(I)
46829GOSUB46885
46830IFU*V(I)<0THENGOTO46836
46831IFV*V(J)<0THENGOTO46838
46833IFI=M2+M/2THENGOTO46881
46834IFJ=M2+M/2THENGOTO46881
46835GOTO46829
46836M4=I
46837GOTO46842
46838IFJ=MTHENJ=0
46839M4=J+1
46842I=M1
46843GOSUB46891
46844X1=X
46845Y1=Y
46846I=M2
46847GOSUB46891
46848X2=X
46849Y2=Y
46850I=M3
46851GOSUB46891
46852X3=X
46853Y3=Y
46854I=M4
46855GOSUB46891
46856X4=X
46857Y4=Y
46860IFX1<>X3THENGOTO46864
46861X=X1
46862Y=(Y1+Y3)/2
46863GOTO46868
46864M1=(Y3-Y1)/(X3-X1)
46865IFX2<>X4THENGOTO46868
46866M2=100000000
46867GOTO46869
46868M2=(Y2-Y4)/(X2-X4)
46869B1=Y1-M1*X1
46870B2=Y2-M2*X2
46871X=-(B1-B2)/(M1-M2)
46872Y=(M1*B2+M2*B1)/(M1+M2)
46874IFK=NTHENRETURN
46875X0=X
46876Y0=Y
46877K=K+1
46878W=W*E
46879GOTO46763
46881X=0
46882Y=0
46883RETURN
```

```
46885I=I+1
46886J=J-1
46887IFI>MTHENI=I-M
46888IFJ<1THENJ=J+M
46889RETURN
46891J=I-1
46892IFJ<1THENJ=J+M
46894X=(X(I)*U(J)+X(J)*U(I))/(U(I)+U(J))
46895Y=(Y(I)*V(J)+Y(J)*V(I))/(V(I)+V(J))
46896RETURN
46925K=0
46926K=K+1
46928X=X0
46929Y=Y0
46930GOSUB44300
46931A=U1*U1+U2*U2
46932X=X0+(V*U2-U*U1)/A
46933Y=Y0-(V*U1+U*U2)/A
46935IF(X0-X)*(X0-X)+(Y0-Y)*(Y0-Y)<=E*ETHENRETURN
46936IFK>=NTHENRETURN
46937X0=X
46938Y0=Y
46939GOTO46926
46960K=1
46961X3=X0
46962X1=X3-D
46963X2=X3+D
46966IFX2-X1=0THENX2=X2*(1.0000001)
46967IFX2-X1=0THENX2=X2+.0000001
46968L1=(X3-X2)/(X2-X1)
46969D1=(X3-X1)/(X2-X1)
46970IFK>1THENGOTO46978
46972X=X1
46973GOSUB44300
46974E1=Y
46975X=X2
46976GOSUB44300
46977E2=Y
46978X=X3
46979GOSUB44300
46980E3=Y
46981A1=L1*L1*E1-D1*D1*E2+(L1+D1)*E3
46982C1=L1*(L1*E1-D1*E2+E3)
46983B=A1*A1-4*D1*C1*E3
46985IFB<0THENB=0
46987IFA1<0THENA1=A1-SQRT(B)
46988IFA1>0THENA1=A1+SQRT(B)
46990IFABS(A1)+ABS(B)=0THENA1=4*D1*E3
46993IFA1=0THENA1=.0000001
46994L=-2*D1*E3/A1
46996X=X3+L*(X3-X2)
46998IFABS(X-X3)<ETHENRETURN
47000IFK>=NTHENRETURN
47002K=K+1
47004X1=X2
47005X2=X3
47006X3=X
47007E1=E2
47008E2=E3
```

```
47009 GOTO46967
47050 K=1
47051 X3=X0
47052 X1=X3-B1
47053 X2=X3+B1
47056 IFX2-X1=0THENX2=X2*(1.0000001)
47057 IFX2-X1=0THENX2=X2+.0000001
47058 L1=(X3-X2)/(X2-X1)
47059 D1=(X3-X1)/(X2-X1)
47061 Y=Y0
47062 X=X1
47063 GOSUB44300
47064 E1=W
47065 X=X2
47066 GOSUB44300
47067 E2=W
47068 X=X3
47069 GOSUB44300
47070 E3=W
47071 GOSUB47111
47073 B1=L*(X3-X2)
47074 X=X3+B1
47076 IFABS(B1)+ABS(B2)<ETHENRETURN
47077 X0=X
47079 Y3=Y0
47080 Y1=Y3-B2
47081 Y2=Y3+B2
47084 IFY2-Y1=0THENY2=Y2*(1.0000001)
47085 IFY2-Y1=0THENY2=Y2+.0000001
47086 L1=(Y3-Y2)/(Y2-Y1)
47087 D1=(Y3-Y1)/(Y2-Y1)
47089 Y=Y1
47090 GOSUB44300
47091 E1=W
47092 Y=Y2
47093 GOSUB44300
47094 E2=W
47095 Y=Y3
47096 GOSUB44300
47097 E3=W
47098 GOSUB47111
47100 B2=L*(Y3-Y2)
47101 Y=Y3+B2
47103 IFABS(B1)+ABS(B2)<ETHENRETURN
47105 IFK>=NTHENRETURN
47106 Y0=Y
47107 K=K+1
47109 GOTO47051
47111 A1=L1*L1*E1-D1*D1*E2+(L1+D1)*E3
47112 C1=L1*(L1*E1-D1*E2+E3)
47113 B=A1*A1-4*D1*C1*E3
47115 IFB<0THENB=0
47117 IFA1<0THENA1=A1-SQR(B)
47118 IFA1>0THENA1=A1+SQR(B)
47120 IFABS(A1)+ABS(B)=0THENA1=4*D1*E3
47123 IFA1=0THENA1=.0000001
47124 L=-2*D1*E3/A1
47125 RETURN
47150 K=1
```

```
47151X3=X0
47152Y3=Y0
47153X1=X3-B1
47154Y1=Y3-B2
47155X2=X3+B1
47156Y2=Y3+B2
47157D=(X2-X1)*(X2-X1)+(Y2-Y1)*(Y2-Y1)
47159IFD=0THEND=.0000001
47160L1=(X3-X2)*(X2-X1)+(Y3-Y2)*(Y2-Y1)
47161L1=L1/D
47162L2=(X2-X1)*(Y3-Y2)-(X3-X2)*(Y2-Y1)
47163L2=L2/D
47164D1=(X3-X1)*(X2-X1)+(Y3-Y1)*(Y2-Y1)
47165D1=D1/D
47166D2=(X2-X1)*(Y3-Y1)-(X3-X1)*(Y2-Y1)
47167D2=D2/D
47169X=X1
47170Y=Y1
47171GOSUB44300
47172U1=U
47173V1=V
47174X=X2
47175Y=Y2
47176GOSUB44300
47177U2=U
47178V2=V
47179X=X3
47180Y=Y3
47181GOSUB44300
47182U3=U
47183V3=V
47185E1=U1*(L1*L1-L2*L2)-2*V1*L1*L2-U2*(D1*D1-D2*D2)
47186E1=E1+2*V2*D1*D2+U3*(L1+D1)-V3*(L2+D2)
47187E2=2*L1*L2*U1+V1*(L1*L1-L2*L2)-2*D1*D2*U2-V2*(D1*D1-D2*D2)
47188E2=E2+U3*(L2+D2)+V3*(L1+D1)
47189C1=L1*L1*U1-L1*L2*V1-D1*L1*U2+L1*D2*V2+U3*L1
47190C1=C1-U1*L2*L2-V1*L1*L2+U2*L2*D2+V2*D1*L2-V3*L2
47191C2=U1*L1*L2+V1*L1*L1-U2*D2*L1-V2*D1*L1+V3*L1
47192C2=C2+L1*L2*U1-L2*L2*V1-D1*L2*U2+D2*L2*V2+U3*L2
47193B1=E1*E1-E2*E2-4*(U3*D1*C1-U3*D2*C2-V3*D2*C1-V3*D1*C2)
47194B2=2*E1*E2-4*(U3*D2*C1+U3*D1*C2+V3*D1*C1-V3*D2*C2)
47196IFB1=0THENB1=.0000001
47197A=ATN(B2/B1)
47198A=A/2
47199B=SQRT(SQRT(B1*B1+B2*B2))
47200B1=B*COS(A)
47201B2=B*SIN(A)
47202A1=(E1+B1)*(E1+B1)+(E2+B2)*(E2+B2)
47203A2=(E1-B1)*(E1-B1)+(E2-B2)*(E2-B2)
47204IFA1>A2THENGOTO47208
47205A1=E1-B1
47206A2=E2-B2
47207GOTO47210
47208A1=E1+B1
47209A2=E2+B2
47210A=A1*A1+A2*A2
47211L1=A1*D1*U3-A1*D2*V3+A2*U3*D2+A2*V3*D1
47213IFA=0THENA=.0000001
47214L1=-2*L1/A
```

```
47215 L2=-D1*U3*A2+D2*V3*A2+A1*U3*D2+A1*V3*D1
47216 L2=-2*L2/A
47218 X=X3+L1*(X3-X2)-L2*(Y3-Y2)
47219 Y=Y3+L2*(X3-X2)+L1*(Y3-Y2)
47221 IFABS(X-X0)+ABS(Y-Y0)<ETHENRETURN
47223 IFK>=NTHENRETURN
47225 K=K+1
47226 X0=X
47227 Y0=Y
47228 X1=X2
47229 Y1=Y2
47230 X2=X3
47231 Y2=Y3
47232 X3=X
47233 Y3=Y
47234 GOTO47157
47300 K=0
47301 IFN3=1THENGOTO47303
47302 IFN3<>2THENRETURN
47303 J1=0
47305 X4=X0
47306 Y4=Y0
47308 IFN3=2THENGOSUB47322
47310 IFJ1=N2THENRETURN
47312 GOSUB47344
47313 J1=J1+1
47314 X(J1)=X
47315 Y(J1)=Y
47316 X0=X4
47317 Y0=Y4
47319 GOTO47310
47322 M=5
47323 A(0)=0
47324 A(1)=24
47325 A(2)=-50
47326 A(3)=35
47327 A(4)=-10
47328 A(5)=1
47329 RETURN
47344 K=1
47345 X3=X0
47346 Y3=Y0
47347 X1=X3-B1
47348 Y1=Y3-B2
47349 X2=X3+B1
47350 Y2=Y3+B2
47351 D=(X2-X1)*(X2-X1)+(Y2-Y1)*(Y2-Y1)
47353 IFD=0THEND=.0000001
47354 L1=(X3-X2)*(X2-X1)+(Y3-Y2)*(Y2-Y1)
47355 L1=L1/D
47356 L2=(X2-X1)*(Y3-Y2)-(X3-X2)*(Y2-Y1)
47357 L2=L2/D
47358 D1=(X3-X1)*(X2-X1)+(Y3-Y1)*(Y2-Y1)
47359 D1=D1/D
47360 D2=(X2-X1)*(Y3-Y1)-(X3-X1)*(Y2-Y1)
47361 D2=D2/D
47363 X=X1
47364 Y=Y1
47365 GOSUB47431
```

```
47366U1=U
47367V1=V
47368X=X2
47369Y=Y2
47370GOSUB47431
47371U2=U
47372V2=V
47373X=X3
47374Y=Y3
47375GOSUB47431
47376U3=U
47377V3=V
47379E1=U1*(L1*L1-L2*L2)-2*V1*L1*L2-U2*(D1*D1-D2*D2)
47380E1=E1+2*V2*D1*D2+U3*(L1+D1)-V3*(L2+D2)
47381E2=2*L1*L2*U1+V1*(L1*L1-L2*L2)-2*D1*D2*U2-V2*(D1*D1-D2*D2)
47382E2=E2+U3*(L2+D2)+V3*(L1+D1)
47383C1=L1*L1*U1-L1*L2*V1-D1*L1*U2+L1*D2*V2+U3*L1
47384C1=C1-U1*L2*L2-V1*L1*L2+U2*L2*D2+V2*D1*L2-V3*L2
47385C2=U1*L1*L2+V1*L1*L1-U2*D2*L1-V2*D1*L1+V3*L1
47386C2=C2+L1*L2*U1-L2*L2*V1-D1*L2*U2+D2*L2*V2+U3*L2
47387B1=E1*E1-E2*E2-4*(U3*D1*C1-U3*D2*C2-V3*D2*C1-V3*D1*C2)
47388B2=2*E1*E2-4*(U3*D2*C1+U3*D1*C2+V3*D1*C1-V3*D2*C2)
47390IFB1=0THENB1=.0000001
47391A=ATN(B2/B1)
47392A=A/2
47393B=SQRT(SQRT(B1*B1+B2*B2))
47394B1=B*COS(A)
47395B2=B*SIN(A)
47396A1=(E1+B1)*(E1+B1)+(E2+B2)*(E2+B2)
47397A2=(E1-B1)*(E1-B1)+(E2-B2)*(E2-B2)
47398IFA1>A2THENGOTO47402
47399A1=E1-B1
47400A2=E2-B2
47401GOTO47404
47402A1=E1+B1
47403A2=E2+B2
47404A=A1*A1+A2*A2
47405L1=A1*D1*U3-A1*D2*V3+A2*U3*D2+A2*V3*D1
47407IFA=0THENA=.0000001
47408L1=-2*L1/A
47409L2=-D1*U3*A2+D2*V3*A2+A1*U3*D2+A1*V3*D1
47410L2=-2*L2/A
47412X=X3+L1*(X3-X2)-L2*(Y3-Y2)
47413Y=Y3+L2*(X3-X2)+L1*(Y3-Y2)
47415IFABS(X-X0)+ABS(Y-Y0)<ETHENRETURN
47417IFK>=NTHENRETURN
47419K=K+1
47420X0=X
47421Y0=Y
47422X1=X2
47423Y1=Y2
47424X2=X3
47425Y2=Y3
47426X3=X
47427Y3=Y
47428GOTO47351
47431N5=N
47432U5=U1
47433V5=V1
```

```
47435 IF N3=1 THEN GOSUB 44300
47436 IF N3=2 THEN GOSUB 44950
47437 IF N3=1 THEN GOTO 47444
47438 U=Z1
47439 V=Z2
47441 N=N5
47442 U1=U5
47443 V1=V5
47444 IF J1=0 THEN RETURN
47446 FOR J2=1 TO J1
47447 U5=U
47448 U=(X-X(J2))*U+(Y-Y(J2))*V
47449 V=(X-X(J2))*V-(Y-Y(J2))*U5
47450 A4=(X-X(J2))*(X-X(J2))+(Y-Y(J2))*(Y-Y(J2))
47452 IF A4=0 THEN A4=.0000001
47453 V=V/A4
47454 U=U/A4
47455 NEXT J2
47457 RETURN
47475 IF A(2)<>0 THEN GOTO 47488
47477 IF A(1)<>0 THEN GOTO 47483
47478 X1=0
47479 X2=0
47480 Y1=0
47481 Y2=0
47482 RETURN
47483 X1=-A(0)/A(1)
47484 Y1=0
47485 X2=X1
47486 Y2=Y1
47487 RETURN
47488 A=A(1)*A(1)-4*A(2)*A(0)
47489 B=SQRT(ABS(A))
47491 IF A(1)=0 THEN C=1
47492 IF A(1)<>0 THEN C=ABS(A(1))/A(1)
47495 IF A>0 THEN GOTO 47499
47496 X1=-C*ABS(A(1))/(2*A(2))
47497 Y1=-C*B/(2*A(2))
47498 GOTO 47502
47499 X1=-C*(ABS(A(1))+B)/(2*A(2))
47500 Y1=0
47502 C=X1*X1+Y1*Y1
47503 IF C<>0 THEN GOTO 47506
47504 X2=1000000*1000000*1000000
47505 RETURN
47506 C=A(0)/(C*A(2))
47507 X2=X1*C
47508 Y2=-Y1*C
47509 RETURN
47525 FOR I=0 TO M
47526 C(I)=A(I)/A(M)
47527 NEXT I
47530 B(0)=0
47531 B(1)=0
47532 B(M-1)=C(M-1)-A
47533 B(M-2)=C(M-2)-A*B(M-1)-B
47534 FOR J=3 TO M
47535 B(M-J)=C(M-J)-A*B(M+1-J)-B*B(M+2-J)
47536 NEXT J
```

```
47538IFB(2)<>0THENGOTO47542
47539A=A+.0000001
47540B=B-.0000001
47541GOTO47532
47542A1=(C(1)-B*B(3))/B(2)
47543B1=C(0)/B(2)
47544K=K+1
47546IFK>=NTHENGOTO47553
47548IFABS(A-A1)+ABS(B-B1)<E*ETHENGOTO47553
47549A=A1
47550B=B1
47552GOTO47532
47553A=A1
47554B=B1
47555C=A*A-4*B
47557IFC>0THENGOTO47563
47558Y1=SQRT(-C)
47559Y2=-Y1
47560X1=-A
47561X2=X1
47562GOTO47567
47563Y1=0
47564Y2=Y1
47565X1=-A+SQRT(C)
47566X2=-A-SQRT(C)
47567X1=X1/2
47568X2=X2/2
47569Y1=Y1/2
47570Y2=Y2/2
47571RETURN
47600FORI=0TOM
47601C(I)=A(I)/A(M)
47602NEXTI
47604K=0
47605B(M)=1
47607B(M-1)=C(M-1)-A
47608FORJ=2TOM-1
47609B(M-J)=C(M-J)-A*B(M+1-J)-B*B(M+2-J)
47610NEXTJ
47611B(0)=C(0)-B*B(2)
47612D(M-1)=-1
47613D(M-2)=-B(M-1)+A
47614FORJ=3TOM-1
47615D(M-J)=-B(M+1-J)-A*D(M+1-J)-B*D(M+2-J)
47616NEXTJ
47617D(0)=-B*D(2)
47618D2=-B(2)-B*D(3)
47619D=D(1)*D2-D(0)*D(2)
47620A1=-B(1)*D2+B(0)*D(2)
47621A1=A1/D
47622B1=-D(1)*B(0)+D(0)*B(1)
47623B1=B1/D
47624A=A+A1
47625B=B+B1
47626K=K+1
47628IFK>=NTHENGOTO47632
47630IFABS(A1)+ABS(B1)>E*ETHENGOTO47607
47632C=A*A-4*B
47634IFC>0THENGOTO47640
```

```
47635 X1=-A
47636 X2=X1
47637 Y1=SQRT(-C)
47638 Y2=-Y1
47639 GOTO47644
47640 X1=-A+SQRT(C)
47641 X2=-A-SQRT(C)
47642 Y1=0
47643 Y2=Y1
47644 X1=X1/2
47645 X2=X2/2
47646 Y1=Y1/2
47647 Y2=Y2/2
47648 RETURN
47700 N=0
47702 FORJ=1TO3
47704 GOSUB44300
47705 Y(J)=Y
47707 GOSUB47735
47708 NEXTJ
47711 IF(Y(3)-Y(2))/(Y(2)-Y(1))>0THENK=K*1.2
47713 IFY(3)<Y(2)THENK=K/2
47715 IFY(3)>Y(2)THENGOTO47721
47717 FORI=1TOL
47718 X(I)=X1(I)
47719 NEXTI
47720 GOTO47724
47721 Y(1)=Y(2)
47722 Y(2)=Y(3)
47724 GOSUB44300
47725 Y(3)=Y
47727 GOSUB47735
47729 N=N+1
47730 IFN>=MTHENRETURN
47731 IFABS(Y(3)-Y(2))<ETHENRETURN
47733 GOTO47711
47735 D=0
47736 FORI=1TOL
47737 D=D+D(I)*D(I)
47738 NEXTI
47739 D=SQRT(D)
47741 FORI=1TOL
47743 X1(I)=X(I)
47744 X(I)=X(I)+K*D(I)/D
47745 NEXTI
47746 GOSUB44300
47747 Y(3)=Y
47748 RETURN
47800 N=0
47804 D=1
47805 D(1)=1/SQRT(L)
47806 FORI=2TOL
47807 D(I)=D(I-1)
47808 NEXTI
47810 FORJ=1TO3
47812 GOSUB44300
47813 Y(J)=Y
47815 GOSUB47866
47817 GOSUB47849
```

```
47818GOSUB47856
47819NEXTJ
47822IF(Y(3)-Y(2))/(Y(2)-Y(1))>0THENK=K*1.2
47824IFY(3)<Y(2)THENK=K/2
47826IFY(3)>Y(2)THENGOTO47832
47828FORI=1TOL
47829X(I)=X1(I)
47830NEXTI
47831GOTO47835
47832Y(1)=Y(2)
47833Y(2)=Y(3)
47835GOSUB44300
47836Y(3)=Y
47837GOSUB47866
47839IFD=0THENRETURN
47841GOSUB47856
47843N=N+1
47844IFN>=MTHENRETURN
47845IFABS(Y(3)-Y(2))<ETHENRETURN
47847GOTO47822
47849D=0
47850FORI=1TOL
47851D=D+D(I)*D(I)
47852NEXTI
47853D=SQRT(D)
47854RETURN
47856FORI=1TOL
47858X1(I)=X(I)
47859X(I)=X(I)+K*D(I)/D
47860NEXTI
47861GOSUB44300
47862Y(3)=Y
47863RETURN
47866FORI=1TOL
47868A=X(I)
47870B=D(I)*K/(2*D)
47872X(I)=X(I)+B
47874GOSUB44300
47876IFB=0THENB=.00000000001
47878D(I)=(Y-Y(3))/B
47880IFD(I)=0THEND(I)=.00001
47882X(I)=A
47883Y=Y(3)
47884NEXTI
47887RETURN
```

APPENDIX III Conversion to Other BASIC Dialects and Microsoft BASIC Program Listings

Presented in this appendix are aids for converting the subroutines listed in Appendix IIA to other dialects of BASIC, in particular, Microsoft. As was discussed in the Introduction, the language subset employed in the subroutines requires little revision to allow those programs to be used with most BASIC interpreters. Most of the changes are apparent from a consideration of the alterations needed for compatibility with the very popular Microsoft dialect. The following items are specific to North Star-to-Microsoft translation. Only the subroutines (line numbers 40000 and up) are treated.

Also included in this appendix is a program listing of a nearly universal Microsoft BASIC version. Although there are differences between the Microsoft BASIC dialects used in the Altair, TRS-80, PET, Apple, and other computers, they are all essentially close enough so that this listing should be applicable with few changes.

The North Star "," and Microsoft ";"

In North Star BASIC, the use of a comma in a print statement permits free-format listing with no carriage returns until the end of the line is reached. "Free format" means that the items listed are printed immediately next to one another on the line, with a separating space in the case of numeric fields. The use of a comma in Microsoft BASIC leads to a tabular listing. Each numerical field, for example, starts in a particular column. The Microsoft equivalent of the North Star comma is the semicolon.

The Random Number Generator, RND

The difference between the North Star and Microsoft RND functions is important and is discussed in Chapter V of Volume I. In North Star BASIC, using $RND(X)$ with $0 < X < 1$ sets a new seed, and with $X = 0$, the random number sequence in progress is simply continued. With Microsoft BASIC, by contrast, the use of $X < 0$ usually sets a new seed, and the sequence in progress is continued with $X > 0$. $X = 0$ generally results in the return of the last number generated. To translate between the two dialects of BASIC, RND(0) in North Star BASIC should be changed to RND(0.999). The particular value chosen for the argument, 0.999, works with all known versions of Microsoft BASIC. (The RND function does not appear in the subroutines given in Volume II, but it is extensively used in the programs presented in Volume I.)

FOR/NEXT Loops

The North Star BASIC interpreter compares the loop argument with the limit *before* the loop is executed. The Microsoft interpreter performs the comparison *after* the loop is executed. Thus, all FOR/NEXT loops are processed *at least once* in Microsoft BASIC, even if the argument is beyond the limit. For example, the loop starting with "FOR I = 7 to 5" will be skipped in North Star BASIC, and executed once in Microsoft. To avoid this seemingly illogical behavior in Microsoft BASIC, test the loop variable with an "IF" statement before the "FOR" statement. This test has been included in several places in the Microsoft listing given in this appendix. For example, see lines 40208 and 40211.

The Square-Root Function

The square-root function in North Star BASIC is SQRT( ), whereas in Microsoft BASIC it is SQR( ). The translation is trivial, but important. See the following lines:

40400	43526	43588	43764	43837
43962	43997	44032	44293	44414
44564	44933	45208	45210	45264
45310	45552	45557	45562	45564
45881	45906	46987	46988	47117
47118	47199	47393	47489	47558
47565	47566	47637	47640	47641
47739	47805	47853		

The TAB Function

The North Star and Microsoft dialects of BASIC are, on the whole, consistent in their use of the TAB function. However, the Applesoft dialect contains an idiosyncrasy: TAB(0) causes a carriage shift to column 255. Thus, TAB(0) should be avoided. Atari BASIC has no TAB.

General Information

North Star and Microsoft BASIC contain several important functions that are used in the subroutines in this book, but that may not be available in other interpreters. However, most of these functions are supplied in Chapter VI of Volume I as subroutines, or can be derived from those programs.

Sine and Cosine

The SINE and COSINE subroutines given in Chapter VI of Volume I can be used to replace the functions SIN( ) and COS( ). These functions are used in the following statements:

SIN( ):	40451	44565	44566	46704	46765	47201	47395
COS( ):	40450	44565	44566	46703	46764	47200	47394

Arctangent

The arctangent (inverse tangent) function is not included in many dialects of BASIC, such as North Star Version 6, Release 3 and earlier versions. Subroutine ARCTAN (see Volume I) can be used to replace the ATN( ) function that appears in the following statements:

40407   45134   47197   47391

Other Functions

Other functions used by the subroutines that are available in both North Star and Microsoft BASIC, but that may not be present in smaller interpreters, are the natural logarithm [LOG( )], the exponent [EXP( )], and the power (^). The first two can be replaced by the LN( ) and EXP( ) subroutines given in Chapter VI of Volume I. The third function can be generated by observing that $Y \wedge X = EXP[X*LOG(Y)]$. The statement numbers at which these functions appear are shown below:

LOG( ):	44404	44411	44581	44606	44933
EXP( ):	44402	44607	44688		
^ :	40402	40404	41100		

CP/M Microsoft BASIC Program Listings

The following is a concatenated listing of the CP/M Microsoft BASIC versions of the demonstration programs and subroutines presented in the text. This particular dialect is very closely compatible with all Microsoft extended BASIC interpreters, including the TRS-80 (Level II), Apple II (Applesoft), PET/CBM, and Altair. Small incompatibilities are the double-precision indicators "#" and "D". The former is automatically added by the MBASIC version to the end of fixed-point format double-precision numbers such as 1.234567891011#. Similarly, "D" replaces "E" in double-precision exponential notation numbers such as 0.1234567891011D-10. If your BASIC does not support this notation, remove "#" and change "D" to "E".

The size of the concatenated Microsoft BASIC demonstration programs is 38110 bytes. The associated subroutine library size is 88590 bytes. Note that both of these numbers are significantly larger than the corresponding North Star BASIC values given in Appendix IIA.

Demonstration Programs

```
2000 REM LEAST SQUARES DEMONSTRATION PROGRAM
2001 PRINT
2002 PRINT
2003 PRINT "LEAST SQUARES CURVE FIT ROUTINE"
2004 PRINT
2005 PRINT
2006 PRINT "THIS PROGRAM CALCULATES A LINEAR"
2007 PRINT "LEAST SQUARES FIT TO A GIVEN DATA SET. "
2008 PRINT
2009 PRINT "INSTRUCTIONS"
2010 PRINT "------------"
2011 PRINT
2012 PRINT "THE NUMBER OF DATA COORDINATES PROVIDED "
2013 PRINT "MUST BE GREATER THAN ONE. OTHERWISE, A "
2014 PRINT "DIVIDE BY ZERO ERROR MAY RESULT."
2015 PRINT
2016 PRINT "INPUT THE NUMBER OF DATA POINTS: ";
2017 INPUT N
2018 IF N<2 THEN GOTO 2012
2019 DIM X(N),Y(N)
2020 PRINT
2021 PRINT "THERE ARE TWO INPUT OPTIONS. ONE (1) "
2022 PRINT "INPUTS THE DATA POINTS IN COORDINATE "
2023 PRINT "PAIRS, AND THE OTHER (2) ALLOWS ONE TO "
2024 PRINT "FIRST INPUT THE INDEPENDENT VARIABLE "
2025 PRINT "VALUES, LATER FOLLOWED BY THE DEPENDENT."
2026 PRINT "WHICH MODE DO YOU DESIRE? (1 OR 2): ";
2027 INPUT Z
2028 PRINT
2029 IF Z=2 THEN GOTO 2032
2030 IF Z=1 THEN GOTO 2042
2031 GOTO 2026
2032 FOR M=0 TO N-1
2033 PRINT M+1;
2034 INPUT X(M)
2035 NEXT M
2036 PRINT
```

```
2037 FOR M=0 TO N-1
2038 PRINT M+1;
2039 INPUT Y(M)
2040 NEXT M
2041 GOTO 2047
2042 FOR M=0 TO N-1
2043 PRINT M+1;
2044 INPUT X(M),Y(M)
2045 NEXT M
2046 REM GO TO LINEAR LEAST SQUARES SUBROUTINE
2047 GOSUB 43500
2048 PRINT
2049 PRINT
2050 PRINT "FITTED EQUATION IS: "
2051 PRINT
2052 PRINT "      Y = ";INT(1E+06*A)/1E+06;" ";
2053 IF B>=0 THEN PRINT "+";
2054 PRINT INT(1E+06*B)/1E+06;"*X"
2055 PRINT
2056 PRINT
2057 PRINT "STANDARD DEVIATION OF FIT: ";
2058 PRINT INT(10000*D)/10000
2059 PRINT
2060 PRINT
2061 END
2100 REM LEAST SQUARES DEMONSTRATION PROGRAM
2101 PRINT
2102 PRINT
2103 PRINT "LEAST SQUARES CURVE FIT ROUTINE"
2104 PRINT
2105 PRINT
2106 PRINT "THIS PROGRAM CALCULATES A PARABOLIC "
2107 PRINT "LEAST SQUARES FIT TO A GIVEN DATA SET. "
2108 PRINT
2109 PRINT "INSTRUCTIONS"
2110 PRINT "-------------"
2111 PRINT
2112 PRINT "THE NUMBER OF DATA COORDINATES PROVIDED "
2113 PRINT "MUST BE GREATER THAN TWO. OTHERWISE, A "
2114 PRINT "DIVIDE BY ZERO ERROR MAY RESULT."
2115 PRINT
2116 PRINT "INPUT THE NUMBER OF DATA POINTS: ";
2117 INPUT N
2118 IF N<3 THEN GOTO 2112
2119 DIM X(N),Y(N)
2120 PRINT
2121 PRINT "THERE ARE TWO INPUT OPTIONS. ONE (1) "
2122 PRINT "INPUTS THE DATA POINTS IN COORDINATE "
2123 PRINT "PAIRS, AND THE OTHER (2) ALLOWS ONE TO "
2124 PRINT "FIRST INPUT THE INDEPENDENT VARIABLE "
2125 PRINT "VALUES, LATER FOLLOWED BY THE DEPENDENT "
2126 PRINT "WHICH MODE DO YOU DESIRE? (1 OR 2): ";
2127 INPUT Z
2128 PRINT
2129 IF Z=2 THEN GOTO 2132
2130 IF Z=1 THEN GOTO 2142
2131 GOTO 2126
2132 FOR M=0 TO N-1
2133 PRINT M+1;
```

```
2134 INPUT X(M)
2135 NEXT M
2136 PRINT
2137 FOR M=0 TO N-1
2138 PRINT M+1;
2139 INPUT Y(M)
2140 NEXT M
2141 GOTO 2147
2142 FOR M=0 TO N-1
2143 PRINT M+1;
2144 INPUT X(M),Y(M)
2145 NEXT M
2146 REM GO TO PARABOLIC LEAST SQUARES SUBROUTINE
2147 GOSUB 43550
2148 PRINT
2149 PRINT
2150 PRINT "FITTED EQUATION IS: "
2151 PRINT
2152 PRINT "    Y = ";INT(1E+06*A)/1E+06;" ";
2153 IF B>=0 THEN PRINT "+";
2154 PRINT INT(1E+06*B)/1E+06;"*X ";
2155 IF C>=0 THEN PRINT "+ ";
2156 PRINT INT(1E+06*C)/1E+06;"*X*X"
2157 PRINT
2158 PRINT
2159 PRINT "STANDARD DEVIATION OF FIT: ";
2160 PRINT INT(10000*D)/10000
2161 PRINT
2162 PRINT
2163 END
2200 REM PROGRAM TO DEMONSTRATE ONE DIMENSIONAL
2201 REM OPERATION OF THE MULTI-NONLINEAR REGRESSION
2202 REM SUBROUTINE
2203 PRINT "HOW MANY DATA POINTS ARE THERE: ";
2204 INPUT M
2205 PRINT "WHAT IS THE DEGREE OF THE POLYNOMIAL"
2206 PRINT "TO BE FITTED: ";
2207 INPUT N
2208 DIM X(M),Y(M),Z(M,N+1),D(N+1),A(M,M),B(M,2*M),C(M,M)
2209 PRINT
2210 PRINT "INPUT THE DATA IN (X,Y) PAIRS AS PROMPTED:"
2211 PRINT
2212 FOR I=1 TO M
2213 PRINT I;TAB(5);"X , Y = ";
2214 INPUT X(I),Y(I)
2215 NEXT I
2216 PRINT
2217 PRINT
2218 REM GO TO COEFFICIENTS GENERATION SUBROUTINE
2219 GOSUB 43600
2220 REM GO TO REGRESSION SUBROUTINE
2221 N=N+1
2222 GOSUB 43650
2223 PRINT "THE CALCULATED COEFFICIENTS ARE:"
2224 PRINT
2225 FOR I=1 TO N
2226 PRINT I;TAB(5);INT(1E+06*D(I))/1E+06
2227 NEXT I
2228 REM GET STANDARD DEVIATION
```

```
2229 N=N-1
2230 GOSUB 43750
2231 PRINT
2232 PRINT
2233 PRINT "STANDARD DEVIATION: ";INT(1E+06*D)/1E+06
2234 PRINT
2235 PRINT
2236 END
2250 REM PROGRAM TO DEMONSTRATE MULTI-DIMENSIONAL
2251 REM OPERATION OF THE MULTI-NONLINEAR REGRESSION
2252 REM SUBROUTINE (MLTNLREG)
2253 PRINT "HOW MANY DATA POINTS ARE THERE: ";
2254 INPUT M
2255 PRINT
2256 PRINT "HOW MANY DIMENSIONS ARE THERE: ";
2257 INPUT L
2258 PRINT
2259 FOR I=1 TO L
2260 PRINT "WHAT IS THE FIT FOR DIMENSION ";I;" ";
2261 INPUT M(I)
2262 NEXT I
2263 N=1
2264 FOR I=1 TO L
2265 N=N*(M(I)+1)
2266 NEXT I
2267 DIM X(M,L),Y(M),Z(M,N),D(N),A(M,M),B(M,2*M),C(M,M)
2268 PRINT
2269 PRINT "INPUT THE DATA AS PROMPTED:"
2270 PRINT
2271 FOR I=1 TO M
2272 PRINT "Y(";I;") = ";
2273 INPUT Y(I)
2274 FOR J=1 TO L
2275 PRINT "X(";I;",";J;") = ";
2276 INPUT X(I,J)
2277 NEXT J
2278 PRINT
2279 NEXT I
2280 REM GOTO COEFFICIENTS GENERATION SUBROUTINE
2281 GOSUB 43800
2282 REM GO TO REGRESSION SUBROUTINE
2283 PRINT
2284 GOSUB 43650
2285 PRINT "THE CALCULATED COEFFICIENTS ARE:"
2286 PRINT
2287 FOR I=1 TO N
2288 PRINT I;TAB(5);INT(1E+06*D(I))/1E+06
2289 NEXT I
2290 REM GET STANDARD DEVIATION
2291 N=N-1
2292 GOSUB 43800
2293 PRINT
2294 PRINT
2295 PRINT "STANDARD DEVIATION: ";INT(1E+06*D)/1E+06
2296 PRINT
2297 PRINT
2298 END
2300 REM PROGRAM TO DEMONSTRATE LSQRPOLY
2301 PRINT
```

```
2302 PRINT
2303 PRINT "WHAT IS THE ORDER OF THE FIT: ";
2304 INPUT M
2305 PRINT
2306 PRINT "WHAT IS THE ERROR REDUCTION FACTOR: ";
2307 INPUT E
2308 PRINT
2309 PRINT "HOW MANY DATA POINTS ARE THERE: ";
2310 INPUT N
2311 DIM X(N),Y(N),V(N),A(N),B(N),C(N),D(N),C2(N),E(N),F(N)
2312 PRINT
2313 PRINT "INPUT THE DATA POINTS AS PROMPTED: ";
2314 PRINT
2315 FOR I=1 TO N
2316 PRINT I;TAB(5);"X , Y = ";
2317 INPUT X(I),Y(I)
2318 NEXT I
2319 PRINT
2320 PRINT
2321 GOSUB 43950
2322 PRINT "COEFFICIENTS ARE:"
2323 PRINT
2324 FOR I=0 TO L
2325 PRINT I; TAB(5);INT(1E+06*C(I))/1E+06
2326 NEXT I
2327 PRINT
2328 PRINT
2329 PRINT "STANDARD DEVIATION=";
2330 PRINT INT(100000000#*D)/100000000#
2331 PRINT
2332 PRINT
2333 END
2350 REM PROGRAM TO DEMONSTRATE MULTIDIMENSIONAL
2351 REM OPERATION OF THE MULTI-NONLINEAR REGRESSION
2352 REM SUBROUTINE WITH ITERATIVE ERROR REDUCTION
2353 PRINT "HOW MANY DATA POINTS ARE THERE: ";
2354 INPUT M
2355 PRINT
2356 PRINT "HOW MANY DIMENSIONS ARE THERE: ";
2357 INPUT L
2358 PRINT
2359 FOR I=1 TO L
2360 PRINT "WHAT IS THE FIT FOR DIMENSION ";I;" ";
2361 INPUT M(I)
2362 NEXT I
2363 N=1
2364 FOR I=1 TO L
2365 N=N*(M(I)+1)
2366 NEXT I
2367 DIM X(M,L),Y(M),Z(M,N),D(N),A(M,M),B(M,2*M),C(M,M),D1(N),Y1(M)
2368 PRINT
2369 PRINT "INPUT THE DATA AS PROMPTED:"
2370 PRINT
2371 FOR I=1 TO M
2372 PRINT "Y(";I;") = ";
2373 INPUT Y(I)
2374 FOR J=1 TO L
2375 PRINT "X(";I;",";J;") = ";
2376 INPUT X(I,J)
```

```
2377 NEXT J
2378 PRINT
2379 NEXT I
2380 REM GO TO ITERATION SUPERVISOR
2381 GOSUB 44100
2382 PRINT
2383 PRINT "THE CALCULATED COEFFICIENTS ARE:"
2384 PRINT
2385 FOR I=1 TO N
2386 PRINT I;TAB(5);INT(1E+06*D(I))/1E+06
2387 NEXT I
2388 PRINT
2389 PRINT
2390 PRINT "STANDARD DEVIATION: ";INT(1E+06*D)/1E+06
2391 PRINT
2392 PRINT
2393 PRINT"NUMBER OF ITERATIONS: ";
2394 PRINT L1
2395 PRINT
2396 PRINT
2397 END
2400 REM PROGRAM TO DEMONSTRATE THE PARAFIT SUBROUTINE
2401 PRINT
2402 PRINT
2403 N=10
2404 L=3
2405 PRINT "THE INPUT DATA ARE:"
2406 PRINT
2407 FOR I=1 TO N
2408 X(I)=I
2409 Y(I)=2*EXP(-(X(I)-4.5)*(X(I)-4.5)/3)
2410 PRINT "X(";I;") = ";X(I);TAB(15);"Y(";I;") = ";Y(I)
2411 NEXT I
2412 PRINT
2413 PRINT
2414 E=.1
2415 E1=.5
2416 A(1)=10
2417 A(2)=10
2418 A(3)=10
2419 GOSUB 44250
2420 PRINT"THE COEFFICIENTS ARE:"
2421 PRINT A(1)
2422 PRINT A(2)
2423 PRINT A(3)
2424 PRINT
2425 PRINT
2426 PRINT"THE STANDARD DEVIATION OF THE FIT IS";
2427 PRINT INT(10000000#*D)/10000000#
2428 PRINT
2429 PRINT
2430 PRINT "THE NUMBER OF ITERATIONS WAS";M
2431 PRINT
2432 PRINT
2433 END
2450 REM PROGRAM TO DEMONSTRATE CHISQA
2451 PRINT
2452 PRINT
2453 PRINT "P(X)";TAB(12);"    X"
```

```
2454 PRINT "----";TAB(12);"  ---"
2455 PRINT
2456 M=100
2457 FOR Y=.05 TO 1 STEP .05
2458 GOSUB 44400
2459 PRINT Y;TAB(12);INT(10*X)/10
2460 NEXT Y
2461 PRINT
2462 END
2500 REM PROGRAM TO DEMONSTRATE BESSEL SERIES SUMMATION SUBROUTINE
2501 PRINT
2502 PRINT
2503 PRINT "WHAT IS THE ORDER OF THE BESSEL FUNCTION: ";
2504 INPUT N
2505 PRINT
2506 PRINT "INPUT ARGUMENT";
2507 INPUT X
2508 PRINT
2509 PRINT "INPUT CONVERGENCE CRITERION";
2510 INPUT E
2511 PRINT
2512 PRINT
2513 GOSUB 44425
2514 PRINT "J(";X;") OF ORDER ";N;" = ";Y
2515 PRINT
2516 PRINT "NUMBER OF TERMS USED: ";M
2517 PRINT
2518 END
2520 REM PROGRAM TO DEMONSTRATE THE BESSEL COEFFICIENTS SUBROUTINE
2521 PRINT
2522 PRINT
2523 PRINT "WHAT IS THE BESSEL FUNCTION ORDER: ";
2524 INPUT N
2525 PRINT
2526 PRINT "WHAT DEGREE IS DESIRED: ";
2527 INPUT M
2528 DIM A(M+1),B(M+1)
2529 PRINT
2530 PRINT
2531 GOSUB 44475
2532 PRINT"THE COEFFICIENTS ARE:"
2533 PRINT
2534 FOR I=0 TO M
2535 PRINT "A(";I;") = ";A(I)
2536 NEXT I
2537 PRINT
2538 PRINT
2539 END
2540 NEXT I
2541 RETURN
2550 REM PROGRAM TO DEMONSTRATE THE BESSEL FUNCTION ASYMPTOTIC SERIES
2551 PRINT
2552 PRINT
2553 PRINT "WHAT IS THE DESIRED ERROR BOUND: ";
2554 INPUT E3
2555 PRINT
2556 PRINT
2557 REM BESSEL FUNCTION SUBROUTINE
2558 PRINT "   X            J0(X)            J1(X)              N              E"
```

```
2559 PRINT "-------       ---------------      -------------    -----     -----------"
2560 PRINT
2561 FOR X=1 TO 15
2562 GOSUB 44525
2563 PRINT X;TAB(15);INT(100000000#*J0)/100000000#;TAB(36);INT(100000000#*J1)/100000000#;
2564 PRINT TAB(55);N;TAB(65);INT(100000000#*E)/100000000#
2565 NEXT X
2566 PRINT
2567 PRINT
2568 PRINT
2569 END
2575 REM PROGRAM TO DEMOSTRATE LN(X!) SUBROUTINE
2576 PRINT
2577 PRINT
2578 PRINT" X";TAB(8);"LN(X!)";TAB(19);"EXP(LN(X!))"
2579 PRINT "---";TAB(6);"----------";TAB(19);"-----------"
2580 PRINT
2581 FOR X=1 TO 10
2582 GOSUB 44580
2583 PRINT X;TAB(5);Y;TAB(18);EXP(Y)
2584 NEXT X
2585 PRINT
2586 PRINT
2587 END
2600 REM PROGRAM TO DEMONSTRATE THE CHI-SQUARE SUBROUTINE
2601 PRINT
2602 PRINT
2603 PRINT "HOW MANY DEGREES OF FREEDOM: ";
2604 INPUT M
2605 PRINT
2606 PRINT "WHAT IS THE RANGE (X1,X2): "
2607 PRINT "X1:";
2608 INPUT X1
2609 PRINT "X2:";
2610 INPUT X2
2611 PRINT
2612 PRINT "WHAT IS THE TABLE STEP SIZE: ";
2613 INPUT X3
2614 PRINT
2615 PRINT
2616 PRINT "   X";TAB(8);"CHI-SQUARE PDF"
2617 PRINT "  ---";TAB(8);"--------------"
2618 PRINT
2619 FOR X=X1 TO X2 STEP X3
2620 GOSUB 44600
2621 PRINT X;TAB(8);INT(10000*Y)/10000
2622 NEXT X
2623 END
2625 REM PROGRAM TO DEMONSTRATE CHISQ
2626 PRINT
2627 PRINT
2628 PRINT "HOW MANY DEGREES OF FREEDOM: ";
2629 INPUT M
2630 PRINT
2631 PRINT "WHAT IS THE RANGE (X1,X2): "
2632 PRINT "X1: ";
2633 INPUT X1
2634 PRINT "X2: ";
2635 INPUT X2
```

```
2636 PRINT
2637 PRINT "STEP SIZE: ";
2638 INPUT X3
2639 PRINT
2640 PRINT "SUMMATION TRUNCATION ERROR BOUND: ";
2641 INPUT E
2642 PRINT
2643 PRINT
2644 PRINT "   X";TAB(8);"CHI-SQUARE CDF"
2645 PRINT "  ---";TAB(8);"---------------"
2646 FOR X=X1 TO X2 STEP X3
2647 GOSUB 44625
2648 PRINT X;TAB(9);INT(10000*Y)/10000
2649 NEXT X
2650 END
2675 REM PROGRAM TO DEMONSTRATE ASYMERF
2676 PRINT
2677 PRINT
2678 PRINT "INPUT X";
2679 INPUT X
2680 GOSUB 44675
2681 PRINT
2682 PRINT "ERF(X)= ";Y;"   WITH ERROR ESTIMATE= ";INT(100000000#*E)/100000000#
2683 PRINT "NUMBER OF TERMS EVALUATED WAS";N
2684 PRINT
2685 END
2700 REM PROGRAM TO DEMONSTRATE CHEBYSER SUBROUTINE
2701 PRINT
2702 PRINT
2703 DIM B(10,10)
2704 FOR N=2 TO 10
2705 GOSUB 44725
2706 PRINT "CHEBYSHEV POLYNOMIAL COEFICIENTS"
2707 PRINT "FOR DEGREE";N
2708 PRINT
2709 FOR I=0 TO N
2710 PRINT "A(";I;") = ";B(N,I)
2711 NEXT I
2712 PRINT
2713 PRINT
2714 NEXT N
2715 PRINT
2716 END
2720 REM PROGRAM TO DEMONSTRATE CHEBYSHEV ECONOMIZATION
2721 PRINT
2722 PRINT "WHAT IS THE DEGREE OF"
2723 PRINT "THE INPUT POLYNOMIAL: ";
2724 INPUT M
2725 PRINT
2726 PRINT "WHAT IS THE DEGREE OF THE"
2727 PRINT "DESIRED ECONOMIZED POLYNOMIAL: ";
2728 INPUT M1
2729 PRINT
2730 PRINT "WHAT IS THE RANGE OF"
2731 PRINT "INPUT POLYNOMIAL: ";
2732 INPUT X0
2733 DIM A(M),B(M,M),C(M)
2734 PRINT
2735 PRINT
```

```
2736 PRINT "INPUT THE COEFFICIENTS:"
2737 PRINT
2738 FOR I=0 TO M
2739 PRINT "C(";I;") = ";
2740 INPUT C(I)
2741 NEXT I
2742 PRINT
2743 PRINT
2744 GOSUB 44760
2745 PRINT "THE CHEBYSHEV SERIES COEFFICIENTS ARE:"
2746 PRINT
2747 FOR I=0 TO M
2748 PRINT "A(";I;") = ";A(I)
2749 NEXT I
2750 PRINT
2751 PRINT
2752 PRINT "THE ECONOMIZED POLYNOMIAL"
2753 PRINT "COEFFICIENTS ARE:"
2754 PRINT
2755 FOR I=0 TO M1
2756 PRINT "C(";I;") = ";C(I)
2757 NEXT I
2758 PRINT
2759 PRINT
2760 END
2775 REM PROGRAM TO DEMONSTRATE THE SERIES REVERSION SUBROUTINE
2776 PRINT
2777 PRINT
2778 PRINT "WHAT IS THE DEGREE OF"
2779 PRINT "THE INPUT POLYNOMIAL: ";
2780 INPUT N
2781 PRINT
2782 PRINT "INPUT THE COEFFICIENTS AS PROMPTED:"
2783 PRINT
2784 FOR I=0 TO N
2785 PRINT "A(";I;") = ";
2786 INPUT A(I)
2787 NEXT I
2788 PRINT
2789 PRINT
2790 GOSUB 44800
2791 PRINT "THE REVERSED POLYNOMIAL COEFFICIENTS ARE:"
2792 PRINT
2793 FOR I=0 TO 7
2794 PRINT "B(";I;") = ";B(I)
2795 NEXT I
2796 PRINT
2797 PRINT
2798 END
2800 REM PROGRAM TO DEMONSTRATE THE SERIES INVERSION SUBROUTINE
2801 PRINT
2802 PRINT
2803 PRINT "WHAT IS THE DEGREE OF THE INPUT POLYNOMIAL: ";
2804 INPUT N
2805 PRINT
2806 PRINT "WHAT IS THE DEGREE OF THE INVERTED POLYNOMIAL: ";
2807 INPUT M
2808 DIM A(M),B(M)
2809 PRINT
```

```
2810 PRINT "INPUT THE POLYNOMIAL COEFFICIENTS:"
2811 PRINT
2812 FOR I=0 TO N
2813 PRINT "A(";I;") = ";
2814 INPUT A(I)
2815 NEXT I
2816 GOSUB 44850
2817 PRINT
2818 PRINT
2819 PRINT "THE INVERTED POLYNOMIAL COEFFICIENTS ARE:"
2820 PRINT
2821 FOR I=0 TO M
2822 PRINT "B(";I;") = ";B(I)
2823 NEXT I
2824 PRINT
2825 END
2830 REM TEST PROGRAM FOR HORNER'S RULE
2831 DIM C(4,5)
2832 PRINT
2833 PRINT
2834 PRINT "INPUT THE FIVE COEFFICIENTS:"
2835 PRINT
2836 FOR I=0 TO 4
2837 PRINT "A(";I;") = ";
2838 INPUT A(I)
2839 NEXT I
2840 PRINT
2841 PRINT "WHAT IS THE EXPANSION POINT: ";
2842 INPUT X0
2843 PRINT
2844 GOSUB 44900
2845 PRINT
2846 PRINT "THE SHIFTED COEFFICIENTS ARE:"
2847 PRINT
2848 FOR I=0 TO 4
2849 PRINT "B(";I;") = ";B(I)
2850 NEXT I
2851 PRINT
2852 PRINT
2853 END
2860 REM PROGRAM TO DEMONSTRATE INVERSE NORMAL SUBROUTINE
2861 PRINT
2862 PRINT "P(Z>X)";TAB(11);"X"
2863 PRINT "------";TAB(10);"---"
2864 PRINT
2865 FOR Y=.5 TO 0 STEP -.02
2866 GOSUB 44925
2867 PRINT Y;TAB(8);INT(10000*X)/10000
2868 NEXT Y
2869 END
2870 REM PROGRAM TO DEMOSTRATE SINEPROD
2871 E=1E-06
2872 PRINT
2873 PRINT
2874 PRINT "   X";TAB(10);"SIN(X) CALC.";TAB(25);"SIN(X) TRUE";TAB(42);"K";
2875 PRINT TAB(52);"ERROR"
2876 PRINT "  ---";TAB(10);"------------";TAB(25);"-----------";TAB(41);"---";
2877 PRINT TAB(52);"-----"
2878 FOR X=0 TO 2 STEP .05
```

```
2879 GOSUB 2888
2880 PRINT X;TAB(10);INT(10000000#*Y)/10000000#;TAB(25);
2881 PRINT INT(10000000#*SIN(X))/10000000#;TAB(40);K;
2882 PRINT TAB(49);INT((100000000#*(Y-SIN(X))))/100000000#
2883 NEXT X
2884 PRINT
2885 PRINT
2886 END
2887 REM ********************
2888 REM SINE PRODUCT SERIES SUBROUTINE (SINEPROD)
2889 REM THIS PROGRAM CALCULATES AN APPROXIMATION TO SIN(X)
2890 REM USING REPEATED PRODUCTS.
2891 REM THE INPUTS TO THE PROGRAM ARE THE ARGUMENT; X
2892 REM AND AN ERROR FACTOR; E.
2893 REM THE APPROXIMATION IS RETURNED IN Y.
2894 Y=X
2895 K=1
2896 L=3.1415926535889793#
2897 L=X/L
2898 L=L*L
2899 M=L/(K*K)
2900 Y=Y*(1-M)
2901 IF M<E THEN RETURN
2902 K=K+1
2903 GOTO 2899
3000 REM PROGRAM TO DEMONSTRATE COMPLEX SERIES EVALUATION SUBROUTINE.
3001 REM IT IS ASSUMED THAT THE COEFFICIENTS ARE OBTAINED FROM
3002 REM A SUBROUTINE.
3003 REM GET COEFFICIENTS
3004 GOSUB 3024
3005 REM INPUT COMPLEX NUMBER
3006 PRINT
3007 PRINT
3008 PRINT "INPUT THE COMPLEX NUMBER AS PROMPTED:"
3009 PRINT
3010 PRINT "     REAL PART = ";
3011 INPUT X
3012 PRINT "     COMPLEX PART = ";
3013 INPUT Y
3014 GOSUB 44950
3015 PRINT
3016 PRINT
3017 PRINT "RESULTS ARE:"
3018 PRINT
3019 PRINT "     Z1 = ";Z1
3020 PRINT "     Z2 = ";Z2
3021 PRINT
3022 END
3023 REM COEFFICIENTS SUBROUTINE
3024 M=5
3025 A(0)=1
3026 A(1)=5
3027 A(2)=10
3028 A(3)=10
3029 A(4)=5
3030 A(5)=1
3031 RETURN
3050 REM PROGRAM TO DEMONSTRATE NTHROOT
3051 PRINT
```

```
3052 PRINT
3053 PRINT "WHAT IS THE NUMBER: ";
3054 INPUT Y
3055 PRINT
3056 PRINT "WHAT ROOT IS DESIRED: ";
3057 INPUT N
3058 PRINT
3059 PRINT "WHAT IS THE CONVERGENCE FACTOR: ";
3060 INPUT E
3061 PRINT
3062 REM THE INITIAL VALUE IS X0.
3063 X0=1
3064 PRINT
3065 GOSUB 45000
3066 PRINT "THE";N;"-TH ROOT OF";Y;" IS";X
3067 PRINT
3068 PRINT "THE NUMBER OF ITERATIONS WAS";M
3069 PRINT
3070 END
3075 REM PROGRAM TO DEMONSTRATE GENROOT
3076 PRINT
3077 PRINT
3078 PRINT "WHAT IS THE NUMBER: ";
3079 INPUT Y
3080 PRINT
3081 PRINT "WHAT EXPONENT IS DESIRED: ";
3082 INPUT N
3083 PRINT
3084 PRINT "WHAT IS THE CONVERGENCE FACTOR: ";
3085 INPUT E
3086 PRINT
3087 REM NUMBER OF BITS = M
3088 M=32
3089 DIM A(M)
3090 PRINT "THE BINARY REPRESENTATION OF THE FRACTION IS:"
3091 PRINT
3092 REM GET THE BINARY REPRESENTATION
3093 GOSUB 45060
3094 FOR I=1 TO M
3095 PRINT A(I);
3096 NEXT I
3097 PRINT
3098 PRINT
3099 GOSUB 45030
3100 PRINT "THE";N;"-TH POWER OF";Y;" IS";X
3101 PRINT
3102 PRINT
3103 END
3125 REM PROGRAM TO DEMONSTRATE TANITER SUBROUTINE
3126 PRINT
3127 PRINT
3128 PRINT "WHAT IS THE ARGUMENT: ";
3129 INPUT X
3130 PRINT
3131 PRINT "WHAT IS THE CONVERGENCE FACTOR: ";
3132 INPUT E
3133 PRINT
3134 PRINT
3135 GOSUB 45125
```

```
3136 PRINT "THE TANGENT OF";X;" IS";Y
3137 PRINT
3138 PRINT "THE ARCTANGENT OF THE TANGENT IS";ATN(Y)
3139 PRINT
3140 END
3150 REM PROGRAM TO DEMONSTRATE ATANITER
3151 PRINT
3152 PRINT
3153 PRINT "WHAT IS THE ARGUMENT: ";
3154 INPUT X
3155 PRINT
3156 PRINT "WHAT IS THE CONVERGENCE FACTOR: ";
3157 INPUT E
3158 PRINT
3159 GOSUB 45150
3160 PRINT
3161 PRINT "THE ARCTANGENT OF";X;" EQUALS";Y
3162 PRINT
3163 PRINT "THE NUMBER OF ITERATIONS WAS";M
3164 PRINT
3165 PRINT
3166 END
3175 REM PROGRAM TO DEMONSTRATE ARCSINE RECURSION
3176 PRINT
3177 PRINT
3178 E=1E-08
3179 PRINT "   X";TAB(10);"ARCSIN(X)";TAB(25);"STEPS";TAB(37);"ERROR"
3180 PRINT "  ---";TAB(10);"---------";TAB(25);"-----";TAB(35);"---------"
3181 PRINT
3182 FOR X=0 TO 1 STEP .05
3183 GOSUB 45200
3184 PRINT X;TAB(9);INT(10000000#*Y)/10000000#;TAB(25);
3185 PRINT M;TAB(34);INT(10000000#*(SIN(Y)-X)+.5)/10000000#
3186 NEXT X
3187 PRINT
3188 PRINT
3189 END
3200 REM PROGRAM TO DEMONSTRATE EVALUATING ELLIPTIC INTEGRALS OF
3201 REM THE FIRST AND SECOND KINDS (COMPLETE)
3202 PRINT
3203 PRINT
3204 DIM A(200),B(200)
3205 E=1E-07
3206 PRINT "   K";TAB(14);"K(K)";TAB(29);"E(K)";TAB(41);"STEPS"
3207 PRINT "  ---";TAB(10);"------------";TAB(25);"------------";TAB(41);"-----"
3208 PRINT
3209 FOR K=0 TO 1 STEP .05
3210 GOSUB 45250
3211 PRINT K;TAB(10);E1;TAB(25);E2;TAB(42);N
3212 NEXT K
3213 END
3225 REM PROGRAM TO DEMONSTRATE LOG BY RECURSION
3226 PRINT
3227 PRINT
3228 E=1E-06
3229 PRINT "   X";TAB(10);"LOG(X)";TAB(23);"STEPS";TAB(35);"ERROR"
3230 PRINT "  ---";TAB(7);"------------";TAB(23);"-----";TAB(32);"------------"
3231 PRINT
3232 FOR X=.05 TO .95 STEP .05
```

```
3233 GOSUB 45300
3234 PRINT X;TAB(7);Y;TAB(24);M;TAB(32);INT(100000000#*(Y-LOG(X)))/100000000#
3235 NEXT X
3236 PRINT
3237 PRINT
3238 END
3250 REM PROGRAM TO DEMONSTRATE INTBESSL SUBROUTINE
3251 PRINT
3252 PRINT
3253 M=15
3254 PRINT "  X";TAB(10);"J0(X)";TAB(25);"J1(X)";TAB(40);"J2(X)";
3255 PRINT TAB(55);"J3(X)";TAB(70);"J4(X)"
3256 PRINT " ---";TAB(7);"-----------";TAB(22);"-----------";TAB(37);
3257 PRINT "-----------";TAB(52);"-----------";TAB(67);"-----------"
3258 PRINT
3259 FOR X=.1 TO 3 STEP .1
3260 GOSUB 45350
3261 I=100000000#
3262 PRINT X;TAB(7);INT(I*Y(0))/I;TAB(22);INT(I*Y(1))/I;TAB(37);INT(I*Y(2))/I;
3263 PRINT TAB(52);INT(I*Y(3))/I;TAB(67);INT(I*Y(4))/I
3264 NEXT X
3265 PRINT
3266 PRINT
3267 END
3275 REM PROGRAM TO DEMONSTRATE THE LEGENDRE COEFFICIENTS SUBROUTINE
3276 PRINT
3277 PRINT
3278 DIM A(10),B(10,10)
3279 FOR N=2 TO 10
3280 PRINT
3281 PRINT "LEGENDRE POLYNOMIAL COEFFICIENTS"
3282 PRINT "FOR ORDER ";N
3283 PRINT
3284 GOSUB 45400
3285 FOR K=0 TO N
3286 PRINT "A(";K;") = ";A(K)
3287 NEXT K
3288 PRINT
3289 NEXT N
3290 PRINT
3291 PRINT
3292 END
3300 REM PROGRAM TO DEMONSTRATE THE LAGUERRE COEFFICIENTS SUBROUTINE
3301 PRINT
3302 PRINT
3303 DIM A(10),B(10,10)
3304 FOR N=2 TO 10
3305 PRINT
3306 PRINT "LAGUERRE POLYNOMIAL COEFFICIENTS"
3307 PRINT "FOR ORDER ";N
3308 PRINT
3309 GOSUB 45425
3310 FOR K=0 TO N
3311 PRINT "A(";K;") = ";A(K)
3312 NEXT K
3313 PRINT
3314 NEXT N
3315 PRINT
3316 PRINT
```

```
3317 END
3320 REM PROGRAM TO DEMONSTRATE THE HERMITE COEFFICIENTS SUBROUTINE
3321 PRINT
3322 PRINT
3323 DIM A(10),B(10,10)
3324 FOR N=2 TO 10
3325 PRINT
3326 PRINT "HERMITE POLYNOMIAL COEFFICIENTS"
3327 PRINT "FOR ORDER ";N
3328 PRINT
3329 GOSUB 45450
3330 FOR K=0 TO N
3331 PRINT "A(";K;") = ";A(K)
3332 NEXT K
3333 PRINT
3334 NEXT N
3335 PRINT
3336 PRINT
3337 END
3350 REM PROGRAM TO DEMONSTRATE USE OF THE CORDIC
3351 REM TRIGONOMETRIC APPROXIMATION SUBROUTINE.
3352 PRINT
3353 PRINT
3354 REM PROGRAM PRINTS OUT 21 VALUES OF SINE AND
3355 REM COSINE AND COMPARES THE RESULT WITH VALUES
3356 REM CALCULATED INTERNALLY.
3357 DIM A(12),W(12)
3358 B=3.141592653589793#
3359 PRINT "    ANGLE (RADIANS)";TAB(24);"CALCULATED SINE";TAB(40);
3360 PRINT "        DIFFERENCE";TAB(67);"CALCULATED COSINE";TAB(84);"    DIFFERENCE"
3361 PRINT "    ----------------";TAB(24);"-----------------";TAB(40);
3362 PRINT "        ----------";TAB(67);"-----------------";TAB(84);"    ----------"
3363 PRINT
3364 B=.025*B
3365 M=1E+14
3366 FOR J=0 TO 20
3367 A=J*B
3368 GOSUB 45475
3369 PRINT INT(M*A)/M;TAB(23);INT(M*U)/M;TAB(44);
3370 PRINT INT((U-SIN(A))*M)/M;TAB(66);INT(M*V)/M;TAB(88);INT((V-COS(A))*M)/M
3371 NEXT J
3372 PRINT
3373 PRINT
3374 END
3375 REM CORDIC TRIGONOMETRIC COEFFICIENT DETERMINATION PROGRAM
3376 REM WRITTEN IN DOUBLE PRECISION MICROSOFT BASIC
3377 DIM C#(51),T#(51),G#(51)
3378 C#(0)=.7071067811865476#
3379 T#(0)=1#
3380 G#(0)=1#-C#(0)
3381 P#=1#
3382 FOR I=0 TO 50
3383 P#=P#*C#(I)
3384 PRINT I, C#(I),T#(I),P#
3385 Y#=(1#+C#(I))/2#
3386 GOSUB 3395
3387 C#(I+1)=X#
3388 G#(I+1)=G#(I)/(2*(1+C#(I+1)))
3389 Y#=G#(I)/(1#+C#(I))
```

```
3390 GOSUB 3395
3391 T#(I+1)=X#
3392 NEXT I
3393 END
3394 REM SQUARE ROOT DETERMINATION SUBROUTINE
3395 X#=1
3396 X1#=X#
3397 X#=(X#+Y#/X#)/2
3398 IF ABS((X#-X1#)/X#)<1E-15 THEN RETURN
3399 GOTO 3396
3425 REM PROGRAM TO DEMONSTRATE THE USE OF THE INVERSE
3426 REM TRIGONOMETRIC CORDIC SUBROUTINE
3427 PRINT
3428 PRINT
3429 PRINT "DO YOU WANT THE INVERSE SINE (S)"
3430 DIM A(12),W(12)
3431 PRINT "                        COSINE (C)"
3432 PRINT "                     OR TANGENT (T): ";
3433 INPUT A$
3434 PRINT
3435 PRINT "WHAT IS THE VALUE: ";
3436 INPUT Z1
3437 U=0
3438 V=0
3439 W=0
3440 IF A$="S" THEN U=Z1
3441 IF A$="C" THEN V=Z1
3442 IF A$="T" THEN W=Z1
3443 REM GET RESULT
3444 GOSUB 45550
3445 PRINT
3446 PRINT
3447 IF A$="S" THEN PRINT "ARCSIN(";
3448 IF A$="C" THEN PRINT "ARCOS(";
3449 IF A$="T" THEN PRINT "ARCTAN(";
3450 PRINT Z1;") = ";A;" RADIANS"
3451 PRINT
3452 PRINT
3453 END
3460 REM PROGRAM TO DEMONSTRATE CORDIC EXPONENTIAL SUBROUTINE (EXPCORD)
3461 PRINT
3462 PRINT
3463 PRINT"   X"; TAB(20);"     EXP(X)";TAB(50);"RELATIVE DIFFERENCE"
3464 PRINT"   -"; TAB(20);"     ------";TAB(50);"-------------------"
3465 PRINT
3466 M=1E+14
3467 FOR X=-1 TO 4 STEP .1
3468 GOSUB 45650
3469 PRINTX;TAB(20);Y;TAB(50);INT(M*(Y-EXP(X))/Y)/M
3470 NEXT X
3471 PRINT
3472 PRINT
3473 END
3475 REM PROGRAM TO DEMONSTRATE THE MODIFIED CORDIC NATURAL LOGARITHM SUBROUTIN
3476 PRINT
3477 PRINT
3478 PRINT"   X"; TAB(20);"     LN(X)";TAB(50);"RELATIVE DIFFERENCE"
3479 PRINT"   -"; TAB(20);"     -----";TAB(50);"-------------------"
3480 PRINT
```

```
3481 DIM A(15),W(15)
3482 M=1E+14
3483 FOR X=.1 TO 3 STEP .1
3484 GOSUB 45725
3485 PRINTX;TAB(20);INT(M*Y)/M;TAB(49);INT(M*(Y-LOG(X))/(Y+1E-06))/M
3486 NEXT X
3487 PRINT
3488 PRINT
3489 END
3500 REM PROGRAM TO DEMONSTRATE THE HYPERBOLIC SUBROUTINES
3501 PRINT
3502 PRINT
3503 PRINT "   X ";TAB(15);"   SINH(X)";TAB(42);"   COSH(X)";TAB(71);"   TANH(X)"
3504 PRINT"---";TAB(15);"-------------";TAB(42);"-------------";TAB(71);"-------------"
3505 PRINT
3506 FOR X=-5 TO 5 STEP .2
3507 PRINT X;
3508 REM GET SINH(X)
3509 GOSUB 45800
3510 PRINT TAB(12);Y;
3511 REM GET COSH(X)
3512 GOSUB 45825
3513 PRINT TAB(40);Y;
3514 REM GET TANH(X)
3515 GOSUB 45840
3516 PRINT TAB(68);Y
3517 NEXT X
3518 PRINT
3519 PRINT
3520 END
3525 REM PROGRAM TO DEMONSTRATE THE INVERSE HYPERBOLIC FUNCTION SUBROUTINES
3526 PRINT
3527 PRINT
3528 DIM A(15),W(15)
3529 PRINT "   X ";TAB(12);"ARCSINH(X)";TAB(32);"ARCCOSH(X)";TAB(52);"ARCTANH(X)"
3530 PRINT "   ---";TAB(12);"----------";TAB(32);"----------";TAB(52);"----------"
3531 PRINT
3532 FOR X=-3 TO 3 STEP .2
3533 PRINT " ";X;
3534 REM GET ARCSINH(X)
3535 GOSUB 45875
3536 PRINT TAB(11);Y;
3537 REM GET ARCOSH(X)
3538 GOSUB 45900
3539 PRINT TAB(31);Y;
3540 REM GET ARCTANH(X)
3541 GOSUB 45925
3542 PRINT TAB(51);Y
3543 NEXT X
3544 PRINT
3545 PRINT
3546 END
3550 REM DEMONSTRATION PROGRAM FOR LAGRANGE INTERPOLATION OF SIN(X)
3551 PRINT
3552 PRINT
3553 DIM L(9),Y(15),X(15)
3554 V=14
3555 REM INPUT TABLE
3556 FOR I=1 TO V
```

```
3557 READ X(I),Y(I)
3558 NEXT I
3559 REM SINE TABLE VALUES FROM
3560 REM HANDBOOK OF MATHEMATICAL FUNCTIONS
3561 REM BY ABRAMOWITZ, M., AND STEGUN, I.A.
3562 REM NBS; JUNE 1964
3563 DATA 0,0,.125,.12467473
3564 DATA .217,.21530095,.299,.29456472
3565 DATA .376,.36720285,.450,.43496553
3566 DATA .520,.49688014,.589,.55552980
3567 DATA .656,.60995199,.721,.66013615
3568 DATA .7853981634,0.7071067812
3569 DATA .849,.75062005,.911,.79011709
3570 DATA .972,.82601466
3571 REM INPUT INTERPOLATION POINT
3572 PRINT "INPUT X: ";
3573 INPUT X
3574 PRINT "INPUT THE ORDER OF THE INTERPOLATION: ";
3575 INPUT N
3576 REM GOTO INTERPOLATION SUBROUTINE
3577 GOSUB 46000
3578 PRINT "SIN(X)= : ";Y
3579 PRINT "ERROR CHECK: ";N
3580 PRINT
3581 GOTO 3572
3582 END
3583 IF J=K THEN GOTO 3585
3584 L(K)=L(K)*(X-X(J+I))/(X(I+K)-X(J+I))
3585 NEXT J
3586 Y=Y+L(K)*Y(I+K)
3587 NEXT K
3588 RETURN
3600 REM PROGRAM TO DEMONSTRATE THE NEWTON INTERPOLATION
3601 PRINT
3602 PRINT
3603 REM SUBROUTINE FOR SIN(X)
3604 V=14
3605 DIM X(V+1),Y(V+1),Y1(V),Y2(V-1),Y3(V-2)
3606 REM INPUT TABLE
3607 FOR I=1 TO V
3608 READ X(I),Y(I)
3609 NEXT I
3610 REM SIN TABLE VALUES FROM
3611 REM HANDBOOK OF MATHEMATICAL FUNCTIONS
3612 REM BY ABRAMOWITZ, M., AND STEGUN, I.A.
3613 REM NBS; JUNE 1964
3614 DATA 0,0,.125,.12467473,.217,.21530095
3615 DATA .299,.29456472,.376,.36720285
3616 DATA .450,.43496553,.520,.49688014
3617 DATA .589,.55552980,.656,.60995199
3618 DATA .721,.66013615,.7853981634,0.7071067812
3619 DATA .849,.75072005
3620 DATA .911,.79011709,.972,.82601466
3621 REM INPUT INTERPOLATION POINT
3622 PRINT "INPUT X :";
3623 INPUT X
3624 REM GO TO INTERPOLATION SUBROUTINE
3625 GOSUB 46050
3626 PRINT "SIN(";X;") = ";Y
```

```
3627 PRINT"ERROR ESTIMATE (W/O ROUNDOFF): ";E
3628 PRINT "ERROR CHECK: ";N
3629 PRINT
3630 PRINT
3631 GOTO 3622
3632 END
3650 REM TABLE SPACING PROGRAM FOR SIN(X) IN THE FIRST OCTANT
3651 PRINT
3652 PRINT
3653 REM PROGRAM GENERATES TABLE LOCATIONS WHICH APPROXIMATELY
3654 REM MINIMIZE THE MAXIMUM ERROR FOR A GIVEN NUMBER OF TABLE POINTS.
3655 REM A CUBIC FIT IS ASSUMED.
3656 REM G=CORRECTION TO GUESS.
3657 REM L IS USED TO ESTABLISH THE RANGE OF INTERPOLATION; IN THIS CASE PI/4.
3658 REM FOR ANOTHER FUNCTION; CHANGE NEXT TWO LINES.
3659 L1=3.1416/4
3660 L=SIN(L1)
3661 REM Q=PRINT FLAG
3662 Q=0
3663 REM ESTABLISH THE INITIAL GUESS FOR QUARTER INTERVAL SPACING
3664 REM NOTE THAT THE FINAL NUMBER OF INTERPOLATION INTERVALS IN THE
3665 REM INTERPOLATION RANGE IS N
3666 N=20
3667 H=L1/(4*N)
3668 PRINT "CONVERGENCE FACTOR:"
3669 REM S STORES THE LENGTH OF THE FIRST INTERVAL
3670 S=H
3671 REM PRINT HEADING
3672 IF Q=0 THEN GOTO 3681
3673 PRINT
3674 PRINT
3675 PRINT "INTERVAL";TAB(12);"POSITION";TAB(25);"INTERVAL SIZE"
3676 PRINT "--------";TAB(12);"--------";TAB(25);"--------------"
3677 REM A IS A MEASURE OF THE ERROR WHICH IS TO BE CONSTANT
3678 REM IT IS USED TO DETERMINE THE SPACING FOR ALL SUBSEQUENT INTERVALS
3679 REM THE FUNCTION IN THE NEXT LINE IS THE FOURTH DERIVATIVE (ABSOLUTE VALUE)
3680 REM OF THE FUNCTION TO BE EVALUATED
3681 A=H*(SIN(H/2))^(1/4)
3682 REM START INTERVAL SECTIONING
3683 X=0
3684 FOR I=1 TO N+7
3685 REM IF CONVERGENCE HAS BEEN ACHIEVED; PRINT RESULTS
3686 IF Q=0 THEN GOTO 3689
3687 PRINT I;TAB(10);INT(100000000#*X)/100000000#;TAB(26);INT(100000000#*H)/100000000#
3688 REM MOVE ON TO THE NEXT INTERVAL
3689 X=X+H
3690 REM CALCULATE THE NEW SPACING FOR THAT INTERVAL
3691 H=A*(1/ABS(SIN(X)))^(1/4)
3692 REM AT THIS POINT A CORRECTION IS APPLIED TO MAKE THE SUM
3693 REM OF THE INTERVALS (N OF THEM) EQUAL L1; THE RANGE)
3694 REM NOTE THAT THE FUNCTION ITSELF IS USED ON THE NEXT LINE;
3695 REM NOT ITS DERIVATIVE
3696 IF I=N THEN G=L/ABS(SIN(X))
3697 NEXT I
3698 IF Q=1 THEN GOTO 3705
3699 PRINT G
3700 REM CHECK FOR CONVERGENCE
3701 IF ABS(G-1)<.01/N THEN Q=1
3702 REM ADJUST SPACING OF THE FIRST INTERVAL AND TRY AGAIN
```

```
3703 H=G*S
3704 GOTO 3670
3705 PRINT
3706 PRINT
3707 END
3725 REM PROGRAM TO DEMONSTRATE THE AKIMA SPLINE FITTING SUBROUTINE
3726 PRINT
3727 PRINT
3728 REM V=NUMBER OF TABLE VALUES
3729 V=14
3730 DIM X(V+1),Y(V+1),M(V+4),Z(V+1)
3731 REM INPUT TABLE
3732 FOR I=1 TO V
3733 READ X(I),Y(I)
3734 NEXT I
3735 REM SIN TABLE VALUES FROM
3736 REM HANDBOOK OF MATHEMATICAL FUNCTIONS
3737 REM BY ABRAMOWITZ, M., AND STEGUN, I.A.
3738 REM NBS; JUNE 1964
3739 DATA 0,0,.125,.12467473,.217,.21530095
3740 DATA .299,.29456472,.376,.36720285
3741 DATA .450,.43496553,.520,.49688014
3742 DATA .589,.55552980,.656,.60995199
3743 DATA .721,.66013615,.7853981634,0.7071067812
3744 DATA .849,.75072005
3745 DATA .911,.79011709,.972,.82601466
3746 PRINT
3747 PRINT " X ";TAB(10);"SIN(X) HANDBOOK";TAB(30);"AKIMA INTERPOLATION";
3748 PRINT TAB(54);"OBSERVED ERROR"
3749 PRINT "---";TAB(10);"-----------------";TAB(30);"---------------------";
3750 PRINT TAB(54);"--------------"
3751 PRINT
3752 FOR X=0 TO .75 STEP .05
3753 GOSUB 46100
3754 PRINT X;TAB(10);INT(10000000#*SIN(X))/10000000#;TAB(32);INT(10000000#*Y)/1000000
3755 PRINT TAB(56);INT(1E+06*(SIN(X)-Y))/1E+06
3756 NEXT X
3757 PRINT
3758 PRINT
3759 END
3775 REM DEMONSTRATION PROGRAM FOR LAGRANGE DERIVATIVE INTERPOLATION OF SIN(X)
3776 PRINT
3777 PRINT
3778 V=15
3779 N=3
3780 DIM L(9),M(9,9),X(V),Y(V)
3781 REM INPUT TABLE
3782 FOR I=1 TO V
3783 READ X(I),Y(I)
3784 NEXT I
3785 REM SINE TABLE VALUES FROM
3786 REM HANDBOOK OF MATHEMATICAL FUNCTIONS
3787 REM BY ABRAMOWITZ, M., AND STEGUN, I.A.
3788 REM NBS; JUNE 1964
3789 DATA -.125,-.12467473
3790 DATA 0,0,.125,.12467473
3791 DATA .217,.21530095,.299,.29456472
3792 DATA .376,.36720285,.450,.43496553
3793 DATA .520,.49688014,.589,.55552980
```

```
3794 DATA .656,.60995199,.721,.66013615
3795 DATA .7853981634,0.7071067812
3796 DATA .849,.75062005,.911,.79011709
3797 DATA .972,.82601466
3798 PRINT
3799 PRINT " X ";TAB(10);"COS(X) HANDBOOK";TAB(30);"LAGRANGE DERIVATIVE";
3800 PRINT TAB(54);"OBSERVED ERROR"
3801 PRINT "---";TAB(10);"---------------";TAB(30);"------------------";
3802 PRINT TAB(54);"--------------"
3803 PRINT
3804 FOR X=0 TO .75 STEP .05
3805 GOSUB 46150
3806 PRINT X;TAB(10);COS(X);TAB(32);Y;TAB(56);INT(100000000#*(COS(X)-Y))/100000000#
3807 NEXT X
3808 END
3825 REM PROGRAM TO DEMONSTRATE THE GENERAL INTEGRATION SUBROUTINE
3826 REM EXAMPLE IS THE INTERGRAL OF SIN(X) FROM X1 TO X2
3827 REM V=NUMBER OF TABLE VALUES
3828 V=14
3829 DIM X(V),Y(V),Z(V),M(V+3)
3830 REM INPUT TABLE
3831 FOR I=1 TO V
3832 READ X(I),Y(I)
3833 NEXT I
3834 REM SIN TABLE VALUES FROM
3835 REM HANDBOOK OF MATHEMATICAL FUNCTIONS
3836 REM BY ABRAMOWITZ, M., AND STEGUN, I.A.
3837 REM NBS, JUNE 1964
3838 DATA 0,0,.125,.12467473,.217,.21530095
3839 DATA .299,.29456472,.376,.36720285
3840 DATA .450,.43496553,.520,.49688014
3841 DATA .589,.55552980,.656,.60995199
3842 DATA .721,.66013615,.7853981634,0.7071067812
3843 DATA .849,.75072005
3844 DATA .911,.79011709,.972,.82601466
3845 PRINT "INPUT START POINT, END POINT: ";
3846 INPUT X1,X2
3847 GOSUB 46200
3848 PRINT "INTEGRAL FROM ";X1;" TO ";X2;" EQUALS ";Z
3849 PRINT "ERROR CHECK: ";Z1
3850 PRINT
3851 PRINT
3852 GOTO 3845
3875 REM PROGRAM TO CALCULATE THE ERROR FUNCTION TABLE POSITIONS
3876 PRINT
3877 PRINT
3878 REM STARTING POSITION IS X=0
3879 X=0
3880 D=0
3881 I=0
3882 REM INITIAL INTERVAL SIZE IS H0=.025
3883 H0=.025
3884 REM TERMINATION POSITION IS E=10
3885 E=20
3886 PRINT
3887 PRINT
3888 PRINT "POSITION NUMBER";TAB(21);"POSITION";TAB(35);"INTERVAL SIZE"
3889 PRINT "---------------";TAB(21);"--------";TAB(35);"-------------"
3890 PRINT
```

```
3891 REM START ITERATION
3892 IF INT(I/4)<>I/4 THEN GOTO 3895
3893 PRINT TAB(6);I/4+1;TAB(22);INT(1000*X)/1000;TAB(36);INT(1000*D)/1000
3894 D=0
3895 H=EXP(-X*X/4)*SQR(SQR((3-6*X*X+4*X*X*X*X)/3))
3896 H=H0/H
3897 I=I+1
3898 X=X+H
3899 D=D+H
3900 IF X<E THEN GOTO 3892
3901 PRINT
3902 PRINT
3903 END
3925 REM ROOT TESTING PROGRAM (ROOTTEST)
3926 PRINT
3927 PRINT
3928 PRINT "THIS PROGRAM WILL HELP YOU"
3929 PRINT "DETERMINE WHERE TO LOOK FOR"
3930 PRINT "ROOTS OF A POLYNOMIAL."
3931 PRINT
3932 PRINT "WHAT IS THE DEGREE OF THE POLYNOMIAL: ";
3933 INPUT N
3934 DIM A(N)
3935 PRINT
3936 PRINT "INPUT THE POLYNOMIAL COEFFICIENTS"
3937 PRINT "AS PROMPTED:"
3938 PRINT
3939 FOR I=0 TO N
3940 PRINT "A(";I;") = ";
3941 INPUT A(I)
3942 NEXT I
3943 REM NORMALIZE SO A(N)=1
3944 FOR I=0 TO N
3945 A(I)=A(I)/A(N)
3946 NEXT I
3947 PRINT
3948 PRINT
3949 PRINT "THERE ARE";N;" ROOTS."
3950 PRINT
3951 PRINT
3952 REM FIND THE MAXIMUM VALUE OF ROOT
3953 A=A(N-1)*A(N-1)-2*A(N-2)
3954 IF A>0 THEN GOTO 3958
3955 PRINT "THERE ARE AT LEAST TWO COMPLEX"
3956 PRINT "ROOTS. THE ANALYSIS ENDS."
3957 GOTO 3998
3958 A=SQR(A)
3959 PRINT
3960 PRINT "THE MAGNITUDE OF THE LARGEST ROOT"
3961 PRINT "IS NOT GREATER THAN";A
3962 PRINT
3963 PRINT
3964 A=-1
3965 IF INT(N/2)=N/2 THEN A=-A
3966 A=A(0)/A
3967 REM B WILL FLAG A NEGATIVE ROOT
3968 B=0
3969 IF A>0 THEN GOTO 3972
3970 PRINT "THERE IS AT LEAST ONE NEGATIVE REAL ROOT."
```

```
3971 B=1
3972 PRINT
3973 REM TEST FOR DESCARTES RULE NUMBER 1
3974 C=0
3975 FOR I=1 TO N
3976 IF A(I-1)*A(I)<0 THEN C=C+1
3977 NEXT I
3978 IF C=1 THEN PRINT "THERE IS AT MOST ONE POSITIVE ";
3979 IF C>1 THEN PRINT "THERE ARE AT MOST";C;" POSITIVE ";
3980 IF C=1 THEN PRINT "REAL ROOT."
3981 IF C>1 THEN PRINT "REAL ROOTS."
3982 PRINT
3983 REM TEST FOR DESCARTES RULE NUMBER 2
3984 D=0
3985 FOR I=1 TO N
3986 IF A(I-1)*A(I)>0 THEN D=D+1
3987 NEXT I
3988 IF D=1 THEN PRINT "THERE IS AT MOST ONE NEGATIVE ";
3989 IF D>1 THEN PRINT "THERE ARE AT MOST";D;" NEGATIVE ";
3990 IF D=1 THEN PRINT "REAL ROOT."
3991 IF D>1 THEN PRINT "REAL ROOTS."
3992 PRINT
3993 IF INT(C/2)=C/2 THEN GOTO 3995
3994 PRINT "THERE IS AT LEAST ONE POSITIVE REAL ROOT."
3995 IF INT(D/2)=D/2 THEN GOTO 3998
3996 IF B=1 THEN GOTO 3998
3997 PRINT "THERE IS AT LEAST ONE NEGATIVE REAL ROOT."
3998 PRINT
3999 PRINT
4000 END
4025 REM PROGRAM TO DEMONSTRATE THE BISECTION SUBROUTINE
4026 PRINT
4027 PRINT
4028 PRINT "WHAT IS THE INITIAL RANGE (X0,X1):"
4029 PRINT
4030 PRINT "X0 = ";
4031 INPUT X0
4032 PRINT "X1 = ";
4033 INPUT X1
4034 PRINT
4035 PRINT "WHAT IS THE CONVERGENCE CRITERION: ";
4036 INPUT E
4037 REM GO TO BISECTION METHOD SUBROUTINE
4038 GOSUB 46350
4039 PRINT
4040 PRINT
4041 PRINT "THE CALCULATED ZERO IS X =";X
4042 PRINT
4043 PRINT "THE ASSOCIATED Y VALUE IS Y = ";Y
4044 PRINT
4045 PRINT "THE NUMBER OF STEPS WAS";M
4046 PRINT
4047 PRINT
4048 END
4050 REM PROGRAM TO DEMONSTRATE NEWTON'S METHOD
4051 PRINT
4052 PRINT
4053 PRINT "INPUT THE INITIAL GUESS: ";
4054 INPUT X0
```

```
4055 PRINT
4056 PRINT "INPUT THE CONVERGENCE FACTOR: ";
4057 INPUT E
4058 PRINT
4059 PRINT "MAXIMUM NUMBER OF ITERATIONS: ";
4060 INPUT M
4061 PRINT
4062 GOSUB 46375
4063 PRINT "THE CALCULATED ROOT IS X =";X
4064 PRINT
4065 PRINT "NUMBER OF ITERATIONS: ";N
4066 PRINT
4067 PRINT
4068 END
4075 REM PROGRAM TO DEMONSTRATE THE SECANT SUBROUTINE
4076 PRINT
4077 PRINT
4078 PRINT "INPUT THE TWO INITIAL GUESSES:"
4079 PRINT
4080 PRINT "X0 = ";
4081 INPUT X0
4082 PRINT "X1 = ";
4083 INPUT X1
4084 PRINT
4085 PRINT "INPUT THE CONVERGENCE FACTOR: ";
4086 INPUT E
4087 PRINT
4088 PRINT "MAXIMUM NUMBER OF ITERATIONS: ";
4089 INPUT M
4090 PRINT
4091 GOSUB 46425
4092 PRINT "THE CALCULATED ROOT IS X =";X
4093 PRINT
4094 PRINT "NUMBER OF ITERATIONS: ";N
4095 PRINT
4096 PRINT
4097 END
4100 REM PROGRAM TO DEMONSTRATE THE MODIFIED REGULA FALSI SUBROUTINE
4101 PRINT
4102 PRINT
4103 PRINT "INPUT THE TWO BRACKETTING GUESSES:"
4104 PRINT
4105 PRINT "X0 = ";
4106 INPUT X0
4107 PRINT "X1 = ";
4108 INPUT X1
4109 PRINT
4110 PRINT "INPUT THE CONVERGENCE FACTOR: ";
4111 INPUT E
4112 PRINT
4113 PRINT "MAXIMUM NUMBER OF ITERATIONS: ";
4114 INPUT M
4115 PRINT
4116 GOSUB 46475
4117 PRINT
4118 PRINT "THE CALCULATED ROOT IS X =";X
4119 PRINT
4120 PRINT "NUMBER OF ITERATIONS: ";N
4121 PRINT
```

```
4122 PRINT
4123 END
4125 REM PROGRAM TO DEMONSTRATE THE AITKEN ACCELERATION SUBROUTINE
4126 PRINT
4127 PRINT
4128 PRINT "INPUT THE INITIAL GUESS: ";
4129 INPUT X0
4130 PRINT
4131 PRINT "INPUT THE CONVERGENCE CRITERION: ";
4132 INPUT E
4133 PRINT
4134 PRINT "INPUT THE CONVERGENCE FACTOR: ";
4135 INPUT C
4136 PRINT
4137 PRINT "MAXIMUM NUMBER OF ITERATIONS: ";
4138 INPUT M
4139 PRINT
4140 GOSUB 46525
4141 PRINT "THE CALCULATED ROOT IS X =";X
4142 PRINT
4143 PRINT "NUMBER OF ITERATIONS: ";N
4144 PRINT
4145 PRINT
4146 END
4150 REM PROGRAM TO DEMONSTRATE AITKEN STEFFENSON ITERATION
4151 PRINT
4152 PRINT
4153 PRINT "INPUT THE INITIAL GUESS: ";
4154 INPUT X0
4155 PRINT
4156 PRINT "INPUT THE CONVERGENCE CRITERION: ";
4157 INPUT E
4158 PRINT
4159 PRINT "INPUT THE CONVERGENCE FACTOR: ";
4160 INPUT C
4161 PRINT
4162 PRINT "MAXIMUM NUMBER OF ITERATIONS: ";
4163 INPUT M
4164 PRINT
4165 GOSUB 46575
4166 PRINT "THE CALCULATED ROOT IS X =";X
4167 PRINT
4168 PRINT "NUMBER OF ITERATIONS: ";N
4169 PRINT
4170 PRINT
4171 END
4175 REM PROGRAM TO DEMONSTRATE SYNTHETIC DIVISION (RSYNDIV)
4176 PRINT
4177 PRINT "WHAT IS THE DEGREE OF THE POLYNOMIAL"
4178 PRINT "TO BE DIVIDED INTO: ";
4179 INPUT N1
4180 DIM C(N1)
4181 PRINT
4182 PRINT "INPUT THE POLYNOMIAL COEFFICIENTS AS PROMPTED:"
4183 PRINT
4184 FOR I=0 TO N1
4185 PRINT "C(";I;") = ";
4186 INPUT C(I)
4187 NEXT I
```

```
4188 PRINT
4189 PRINT "WHAT IS THE ORDER OF THE POLYNOMIAL "
4190 PRINT "BE DIVIDED BY: ";
4191 INPUT N2
4192 DIM A(N1-N2),B(N2)
4193 PRINT
4194 PRINT "INPUT THE POLYNOMIAL COEFFICIENTS AS PROMPTED:"
4195 PRINT
4196 FOR I=0 TO N2
4197 PRINT "B(";I;") = ";
4198 INPUT B(I)
4199 NEXT I
4200 GOSUB 46625
4201 PRINT
4202 PRINT
4203 PRINT "THE COEFFICIENTS OF THE RESULTING POLYNOMIAL ARE:"
4204 PRINT
4205 FOR I=0 TO N1-N2
4206 PRINT "A(";I;") = ";A(I)
4207 NEXT I
4208 PRINT
4209 PRINT
4210 END
4225 REM PROGRAM TO DEMONSTRATE NEXTROOT SUBROUTINE
4226 PRINT
4227 PRINT
4228 PRINT "HOW MANY ROOTS HAVE BEEN DETERMINED: ";
4229 INPUT L
4230 PRINT
4231 PRINT
4232 PRINT "INPUT THE ROOTS AS PROMPTED:"
4233 PRINT
4234 FOR I=1 TO L
4235 PRINT "A(";I;") = ";
4236 INPUT A(I)
4237 NEXT I
4238 PRINT
4239 PRINT "WHAT IS THE INITIAL GUESS: ";
4240 INPUT X0
4241 PRINT
4242 PRINT "WHAT IS THE CONVERGENCE CRITERION: ";
4243 INPUT E
4244 PRINT
4245 PRINT "MAXIMUM NUMBER OF ITERATIONS: ";
4246 INPUT M
4247 PRINT
4248 GOSUB 46650
4249 PRINT
4250 PRINT "THE CALCULATED ROOT IS X =";X
4251 PRINT
4252 PRINT "NUMBER OF ITERATIONS: ";N
4253 PRINT
4254 PRINT
4255 END
4275 REM PROGRAM TO DEMONSTRATE THE COMPLEX ROOT COUNTING SUBROUTINE
4276 PRINT
4277 PRINT
4278 PRINT "WHERE IS THE CENTER OF THE SEARCH CIRCLE (X0,Y0):"
4279 PRINT
```

```
4280 PRINT "      X0 = ";
4281 INPUT X0
4282 PRINT "      Y0 = ";
4283 INPUT Y0
4284 PRINT
4285 PRINT "WHAT IS THE RADIUS OF THIS CIRCLE: ";
4286 INPUT W
4287 PRINT "HOW MANY EVALUATION POINTS PER QUADRANT: ";
4288 INPUT M
4289 DIM N(4*M)
4290 GOSUB 46700
4291 PRINT
4292 PRINT
4293 PRINT "NUMBER OF COMPLETE CYCLES FOUND:";
4294 PRINT N
4295 PRINT "RESIDUAL: ";
4296 PRINT A
4297 PRINT
4298 PRINT
4299 END
4300 REM PROGRAM TO DEMONSTRATE THE ZERO SEARCHING ALGORITHM
4301 PRINT
4302 PRINT
4303 PRINT "WHAT IS THE INITIAL GUESS FOR X AND Y: "
4304 PRINT "      X0 = ";
4305 INPUT X0
4306 PRINT "      Y0 = ";
4307 INPUT Y0
4308 PRINT
4309 PRINT "WHAT IS THE RADIUS OF THE FIRST SEARCH CIRCLE: ";
4310 INPUT W
4311 PRINT "BY WHAT FRACTION IS THIS CIRCLE TO BE REDUCED ON EACH ITERATION: ";
4312 INPUT E
4313 PRINT "HOW MANY POINTS ARE TO BE SEARCHED PER QUADRANT: ";
4314 INPUT M
4315 PRINT "HOW MANY ITERATIONS ARE TO BE EMPLOYED: ";
4316 INPUT N
4317 DIM X(4*M),Y(4*M),U(4*M),V(4*M)
4318 GOSUB 46760
4319 PRINT
4320 PRINT
4321 PRINT "THE APPROXIMATE SOLUTION IS Z = ";X;
4322 IF Y>=0 THEN PRINT " +";
4323 IF Y<0 THEN PRINT " ";
4324 PRINT Y;" I"
4325 PRINT
4326 PRINT "NUMBER OF ITERATIONS: ";K
4327 PRINT
4328 END
4350 REM PROGRAM TO DEMONSTRATE THE NEWTON ROOT
4351 REM PROGRAM IN THE COMPLEX DOMAIN
4352 PRINT
4353 PRINT
4354 PRINT "WHAT IS THE INITIAL GUESS:"
4355 PRINT "      X0 = ";
4356 INPUT X0
4357 PRINT "      Y0 = ";
4358 INPUT Y0
4359 PRINT "WHAT IS THE CONVERGENCE CRITERION: ";
```

```
4360 INPUT E
4361 PRINT "WHAT IS THE LIMIT ON THE NUMBER OF ITERATIONS: ";
4362 INPUT N
4363 GOSUB 46925
4364 PRINT
4365 PRINT
4366 PRINT "THE ROOT ESTIMATE IS:"
4367 PRINT "     X0 = ";X
4368 PRINT "     Y0 = ";Y
4369 PRINT
4370 PRINT "THE NUMBER OF ITERATIONS PERFORMED WAS: ";
4371 PRINT K
4372 PRINT
4373 PRINT
4374 END
4375 REM PROGRAM TO DEMONSTRATE MUELLER'S METHOD
4376 PRINT
4377 PRINT
4378 PRINT "WHAT IS THE INITIAL GUESS: ";
4379 INPUT X0
4380 PRINT "WHAT IS THE BOUND ON THIS GUESS: ";
4381 INPUT D
4382 PRINT "WHAT IS THE ERROR CRITERION: ";
4383 INPUT E
4384 PRINT "HOW MANY INTERATIONS (MAXIMUM): ";
4385 INPUT N
4386 PRINT
4387 GOSUB 46960
4388 PRINT
4389 PRINT "THE ESTIMATED ROOT IS: ";X
4390 PRINT "THE NUMBER OF ITERATIONS PERFORMED WAS: ";K
4391 PRINT
4392 END
4400 REM PROGRAM TO DEMONSTRATE A TWO DIMENSIONAL VERSION OF MUELLER'S METHOD
4401 PRINT
4402 PRINT
4403 PRINT "WHAT ARE THE INITIAL GUESSES AND THEIR BOUNDS:"
4404 PRINT "     X0 = ";
4405 INPUT X0
4406 PRINT "     BOUND ON X0 = ";
4407 INPUT B1
4408 PRINT "     Y0 = ";
4409 INPUT Y0
4410 PRINT "     BOUND ON Y0 = ";
4411 INPUT B2
4412 PRINT "WHAT IS THE ERROR CRITERION: ";
4413 INPUT E
4414 PRINT "HOW MANY INTERATIONS (MAXIMUM): ";
4415 INPUT N
4416 PRINT
4417 GOSUB 47050
4418 PRINT
4419 PRINT "THE ESTIMATED ROOT IS (X,Y) = (";X;",";Y;")"
4420 PRINT "THE NUMBER OF ITERATIONS PERFORMED WAS: ";K
4421 PRINT
4422 END
4425 REM PROGRAM TO DEMONSTRATE THE COMPLEX DOMAIN MUELLER SUBROUTINE
4426 PRINT
4427 PRINT
```

```
4428 PRINT "WHAT IS THE INITIAL GUESS:"
4429 PRINT
4430 PRINT "     X0 = ";
4431 INPUT X0
4432 PRINT "WHAT IS THE ASSOCIATED BOUND: ";
4433 INPUT B1
4434 PRINT "     Y0 = ";
4435 INPUT Y0
4436 PRINT "WHAT IS THE ASSOCIATED BOUND: ";
4437 INPUT B2
4438 PRINT
4439 PRINT "INPUT THE CONVERGENCE CRITERION: ";
4440 INPUT E
4441 PRINT "WHAT IS THE LIMIT ON THE NUMBER OF ITERATIONS: ";
4442 INPUT N
4443 GOSUB 47150
4444 PRINT
4445 PRINT
4446 PRINT "THE ESTIMATED ROOT IS:"
4447 PRINT
4448 PRINT "     X = ";X
4449 PRINT "     Y = ";Y
4450 PRINT
4451 PRINT "THE NUMBER OF ITERATIONS WAS ";K
4452 PRINT
4453 PRINT
4454 END
4475 REM PROGRAM TO DEMONSTRATE THE COMPLEX DOMAIN ALLROOT SUBROUTINE
4476 PRINT
4477 PRINT
4478 PRINT "WHAT IS THE INITIAL GUESS:"
4479 PRINT
4480 PRINT "     X0 = ";
4481 INPUT X0
4482 PRINT "WHAT IS THE ASSOCIATED BOUND: ";
4483 INPUT B1
4484 PRINT "     Y0 = ";
4485 INPUT Y0
4486 PRINT "WHAT IS THE ASSOCIATED BOUND: ";
4487 INPUT B2
4488 PRINT
4489 PRINT "INPUT THE CONVERGENCE CRITERION: ";
4490 INPUT E
4491 PRINT "WHAT IS THE LIMIT ON THE NUMBER OF ITERATIONS: ";
4492 INPUT N
4493 PRINT "HOW MANY ROOTS ARE TO BE SOUGHT: ";
4494 INPUT N2
4495 PRINT "IS THE FUNCTION IN THE FUNCTIONS SUBROUTINE (1)"
4496 PRINT "OR IS IT A SERIES (2): ";
4497 INPUT N3
4498 GOSUB 47300
4499 PRINT
4500 PRINT
4501 PRINT "THE ESTIMATED ROOTS ARE:"
4502 FOR I=1 TO N2
4503 PRINT
4504 PRINT "     X = ";X(I)
4505 PRINT "     Y = ";Y(I)
4506 PRINT
```

```
4507 NEXT I
4508 PRINT
4509 PRINT "THE LAST NUMBER OF ITERATIONS WAS ";K
4510 PRINT
4511 PRINT
4512 END
4525 REM PROGRAM TO DEMONSTRATE QUADRATIC ROOT SUBROUTINE
4526 PRINT
4527 PRINT "INPUT THE COEFFICIENTS:"
4528 PRINT
4529 PRINT "A(0) = ";
4530 INPUT A(0)
4531 PRINT "A(1) = ";
4532 INPUT A(1)
4533 PRINT "A(2) = ";
4534 INPUT A(2)
4535 GOSUB 47475
4536 PRINT
4537 PRINT
4538 PRINT "RESULTS:"
4539 PRINT
4540 PRINT "     R1 = ";X1;
4541 IF Y1>=0 THEN PRINT " +";
4542 IF Y1<0 THEN PRINT " ";
4543 PRINT Y1;" I"
4544 PRINT "     R2 = ";X2;
4545 IF Y2>=0 THEN PRINT " +";
4546 IF Y2<0 THEN PRINT " ";
4547 PRINT Y2;" I"
4548 PRINT
4549 PRINT
4550 END
4560 REM PROGRAM TO DEMONSTRATE THE LIN SUBROUTINE
4561 PRINT
4562 PRINT
4563 PRINT "INPUT ORDER OF POLYNOMIAL";
4564 INPUT M
4565 PRINT
4566 PRINT "INPUT THE POLYNOMIAL COEFFICIENTS:"
4567 FOR I=0 TO M
4568 PRINT "     A(";I;") = ";
4569 INPUT A(I)
4570 NEXT I
4571 PRINT
4572 PRINT "WHAT IS THE CONVERGENCE FACTOR: ";
4573 INPUT E
4574 PRINT "WHAT IS THE MAXIMUM NUMBER OF ITERATIONS: ";
4575 INPUT N
4576 A=3.14159
4577 B=SQR(2)
4578 GOSUB 47525
4579 PRINT
4580 PRINT "THE ROOTS FOUND ARE:"
4581 PRINT
4582 PRINT "     X1 = ";X1
4583 PRINT "     Y1 = ";Y1
4584 PRINT
4585 PRINT "     X2 = ";X2
4586 PRINT "     Y2 = ";Y2
```

```
4587 PRINT
4588 PRINT "THE NUMBER OF ITERATIONS WAS ";K
4589 PRINT
4590 PRINT
4591 END
4600 REM PROGRAM TO DEMONSTRATE THE BAIRSTOW SUBROUTINE
4601 PRINT
4602 PRINT
4603 PRINT "INPUT ORDER OF POLYNOMIAL";
4604 INPUT M
4605 PRINT
4606 PRINT "INPUT THE POLYNOMIAL COEFFICIENTS:"
4607 FOR I=0 TO M
4608 PRINT "     A(";I;") = ";
4609 INPUT A(I)
4610 NEXT I
4611 PRINT
4612 PRINT "WHAT IS THE CONVERGENCE FACTOR: ";
4613 INPUT E
4614 PRINT "WHAT IS THE MAXIMUM NUMBER OF ITERATIONS: ";
4615 INPUT N
4616 A=3.14159
4617 B=SQR(2)
4618 GOSUB 47600
4619 PRINT
4620 PRINT "THE ROOTS FOUND ARE:"
4621 PRINT
4622 PRINT "     X1 = ";X1
4623 PRINT "     Y1 = ";Y1
4624 PRINT
4625 PRINT "     X2 = ";X2
4626 PRINT "     Y2 = ";Y2
4627 PRINT
4628 PRINT "THE NUMBER OF ITERATIONS WAS ";K
4629 PRINT
4630 PRINT
4631 END
4650 REM PROGRAM TO DEMONSTRATE THE USE OF MULTI-DIMENSIONAL
4651 REM STEEPEST DESCENT
4652 PRINT
4653 PRINT
4654 PRINT "HOW MANY DIMENSIONS ARE THERE: ";
4655 INPUT L
4656 DIM X(L),X1(L)
4657 PRINT
4658 PRINT "WHAT IS THE CONVERGENCE CRITERION: ";
4659 INPUT E
4660 PRINT
4661 PRINT "MAXIMUM NUMBER OF ITERATIONS: ";
4662 INPUT M
4663 PRINT
4664 PRINT "WHAT IS THE STARTING CONSTANT, K: ";
4665 INPUT K
4666 PRINT
4667 PRINT "INPUT THE STARTING POINTS:"
4668 PRINT
4669 FOR I=1 TO L
4670 PRINT "X(";I;") = ";
4671 INPUT X(I)
```

```
4672 NEXT I
4673 PRINT
4674 PRINT
4675 GOSUB 47700
4676 PRINT "THE RESULTS ARE:"
4677 PRINT
4678 FOR I=1 TO L
4679 PRINT "X(";I;") = ";X(I)
4680 NEXT I
4681 PRINT
4682 PRINT "THE NUMBER OF ITERATIONS WAS";N
4683 PRINT
4684 PRINT
4685 END
4700 REM PROGRAM TO DEMONSTRATE THE USE OF MULTI-DIMENSIONAL
4701 REM STEEPEST DESCENT
4702 PRINT
4703 PRINT
4704 PRINT "HOW MANY DIMENSIONS ARE THERE: ";
4705 INPUT L
4706 DIM X(L),X1(L)
4707 PRINT
4708 PRINT "WHAT IS THE CONVERGENCE CRITERION: ";
4709 INPUT E
4710 PRINT
4711 PRINT "MAXIMUM NUMBER OF ITERATIONS: ";
4712 INPUT M
4713 PRINT
4714 PRINT "WHAT IS THE STARTING CONSTANT, K: ";
4715 INPUT K
4716 PRINT
4717 PRINT "INPUT THE STARTING POINTS:"
4718 PRINT
4719 FOR I=1 TO L
4720 PRINT "X(";I;") = ";
4721 INPUT X(I)
4722 NEXT I
4723 PRINT
4724 PRINT
4725 GOSUB 47800
4726 PRINT "THE RESULTS ARE:"
4727 PRINT
4728 FOR I=1 TO L
4729 PRINT "X(";I;") = ";X(I)
4730 NEXT I
4731 PRINT
4732 PRINT "THE NUMBER OF ITERATIONS WAS";N
4733 PRINT
4734 PRINT
4735 END
```

Subroutine Programs

```
40398 REM ********************
40399 REM RECTANGULAR TO POLAR CONVERSION SUBROUTINE (RECT/POL)
40400 U=SQR(X*X+Y*Y)
40401 REM GUARD AGAINST AMBIGUOUS VECTOR
```

```
40402 IF Y=0 THEN Y=(.1)^30
40403 REM GUARD AGAINST DIVIDE BY ZERO
40404 IF X=0 THEN X=(.1)^30
40405 REM SOME BASICS REQUIRE A SIMPLE ARGUMENT
40406 W=Y/X
40407 V=ATN(W)
40408 REM CHECK QUADRANT AND ADJUST
40409 IF X<0 THEN V=V+3.1415926535#
40410 IF V<0 THEN V=V+6.2831853072#
40411 RETURN
40449 REM POLAR TO RECTANGULAR CONVERSION SUBROUTINE (POL/RECT)
40450 X=U*COS(V)
40451 Y=U*SIN(V)
40452 RETURN
41099 REM POLAR POWER SUBROUTINE (ZPOLPOW)
41100 U1=U^N
41101 V1=N*V
41102 V1=V1-6.2831853072#*INT(V1/6.2831853072#)
41103 RETURN
41198 REM RECTANGULAR COMPLEX NUMBER POWER SUBROUTINE (ZRECTPOW)
41199 REM RECTANGULAR TO POLAR CONVERSION
41200 GOSUB 40400
41201 REM POLAR POWER
41202 GOSUB 41100
41203 REM CHANGE VARIABLE FOR CONVERSION
41204 U=U1
41205 V=V1
41206 REM POLAR TO RECTANGULAR CONVERSION
41207 GOSUB 40450
41208 RETURN
41897 REM ********************
41898 REM MATRIX MULTIPLICATION SUBROUTINE (MATMULT)
41899 REM C=A X B   A IS M1 BY N1   B IS M2 BY N2   C IS M1 BY N2
41900 FOR I=1 TO M1
41901 FOR J=1 TO N2
41902 C(I,J)=0
41903 FOR K=1 TO N1
41904 C(I,J)=C(I,J)+A(I,K)*B(K,J)
41905 NEXT K
41906 NEXT J
41907 NEXT I
41908 RETURN
41947 REM ********************
41948 REM MATRIX TRANSPOSE SUBROUTINE (MATTRANS)
41949 REM B=TRANSPOSE(A)
41950 FOR I=1 TO N
41951 FOR J=1 TO M
41952 B(I,J)=A(J,I)
41953 NEXT J
41954 NEXT I
41955 RETURN
42072 REM ********************
42073 REM MATRIX SAVE (B IN A) SUBROUTINE (MATSAVBA)
42074 REM N1,N2 AND N3 ARE INPUT INDICES
42075 IF N1*N2*N3=0 THEN GOTO 42085
42076 REM CHECK DIMENSION
42077 FOR I1=1 TO N1
42078 FOR I2=1 TO N2
42079 FOR I3=1 TO N3
```

```
42080 A(I1,I2,I3)=B(I1,I2,I3)
42081 NEXT I3
42082 NEXT I2
42083 NEXT I1
42084 RETURN
42085 IF N1*N2=0 THEN GOTO 42092
42086 FOR I1=1 TO N1
42087 FOR I2=1 TO N2
42088 A(I1,I2)=B(I1,I2)
42089 NEXT I2
42090 NEXT I1
42091 RETURN
42092 IF N1=0 THEN RETURN
42093 FOR I1=1 TO N1
42094 A(I1)=B(I1)
42095 NEXT I1
42096 RETURN
42097 REM ********************
42098 REM MATRIX SAVE (C IN B) SUBROUTINE (MATSAVCB)
42099 REM N1,N2 AND N3 ARE INPUT INDICES
42100 IF N1*N2*N3=0 THEN GOTO 42110
42101 REM CHECK DIMENSION
42102 FOR I1=1 TO N1
42103 FOR I2=1 TO N2
42104 FOR I3=1 TO N3
42105 B(I1,I2,I3)=C(I1,I2,I3)
42106 NEXT I3
42107 NEXT I2
42108 NEXT I1
42109 RETURN
42110 IF N1*N2=0 THEN GOTO 42117
42111 FOR I1=1 TO N1
42112 FOR I2=1 TO N2
42113 B(I1,I2)=C(I1,I2)
42114 NEXT I2
42115 NEXT I1
42116 RETURN
42117 IF N1=0 THEN RETURN
42118 FOR I1=1 TO N1
42119 B(I1)=C(I1)
42120 NEXT I1
42121 RETURN
42147 REM ********************
42148 REM MATRIX SAVE (A IN C) SUBROUTINE (MATSAVAC)
42149 REM N1,N2 AND N3 ARE INPUT INDICES
42150 IF N1*N2*N3=0 THEN GOTO 42160
42151 REM CHECK DIMENSION
42152 FOR I1=1 TO N1
42153 FOR I2=1 TO N2
42154 FOR I3=1 TO N3
42155 C(I1,I2,I3)=A(I1,I2,I3)
42156 NEXT I3
42157 NEXT I2
42158 NEXT I1
42159 RETURN
42160 IF N1*N2=0 THEN GOTO 42167
42161 FOR I1=1 TO N1
42162 FOR I2=1 TO N2
42163 C(I1,I2)=A(I1,I2)
```

```
42164 NEXT I2
42165 NEXT I1
42166 RETURN
42167 IF N1=0 THEN RETURN
42168 FOR I1=1 TO N1
42169 C(I1)=A(I1)
42170 NEXT I1
42171 RETURN
42172 REM ********************
42173 REM MATRIX SAVE (C IN A) SUBROUTINE (MATSAVCA)
42174 REM N1,N2 AND N3 ARE INPUT INDICES
42175 IF N1*N2*N3=0 THEN GOTO 42185
42176 REM CHECK DIMENSION
42177 FOR I1=1 TO N1
42178 FOR I2=1 TO N2
42179 FOR I3=1 TO N3
42180 A(I1,I2,I3)=C(I1,I2,I3)
42181 NEXT I3
42182 NEXT I2
42183 NEXT I1
42184 RETURN
42185 IF N1*N2=0 THEN GOTO 42192
42186 FOR I1=1 TO N1
42187 FOR I2=1 TO N2
42188 A(I1,I2)=C(I1,I2)
42189 NEXT I2
42190 NEXT I1
42191 RETURN
42192 IF N1=0 THEN RETURN
42193 FOR I1=1 TO N1
42194 A(I1)=C(I1)
42195 NEXT I1
42196 RETURN
42394 REM ********************
42395 REM MATRIX INVERSION SUBROUTINE (MATINV)
42396 REM GAUSS-JORDAN ELIMINATION
42397 REM MATRIX A IS INPUT, MATRIX B IS OUTPUT
42398 REM DIM A=N X N    TEMPORARY DIM B=N X 2N
42399 REM FIRST CREATE MATRIX WITH A ON THE LEFT AND I ON THE RIGHT
42400 FOR I=1 TO N
42401 FOR J=1 TO N
42402 B(I,J+N)=0
42403 B(I,J)=A(I,J)
42404 NEXT J
42405 B(I,I+N)=1
42406 NEXT I
42407 REM PERFORM ROW ORIENTED OPERATIONS TO CONVERT THE LEFT HAND
42408 REM SIDE OF B TO THE IDENTITY MATRIX. THE INVERSE OF A WILL
42409 REM THEN BE ON THE RIGHT.
42410 FOR K=1 TO N
42411 IF K=N THEN GOTO 42424
42412 M=K
42413 REM FIND MAXIMUM ELEMENT
42414 FOR I=K+1 TO N
42415 IF ABS(B(I,K))>ABS(B(M,K)) THEN M=I
42416 NEXT I
42417 IF M=K THEN GOTO 42424
42418 FOR J=K TO 2*N
42419 B=B(K,J)
```

```
42420 B(K,J)=B(M,J)
42421 B(M,J)=B
42422 NEXT J
42423 REM DIVIDE ROW K
42424 FOR J=K+1 TO 2*N
42425 B(K,J)=B(K,J)/B(K,K)
42426 NEXT J
42427 IF K=1 THEN GOTO 42434
42428 FOR I=1 TO K-1
42429 FOR J=K+1 TO 2*N
42430 B(I,J)=B(I,J)-B(I,K)*B(K,J)
42431 NEXT J
42432 NEXT I
42433 IF K=N THEN GOTO 42441
42434 FOR I=K+1 TO N
42435 FOR J=K+1 TO 2*N
42436 B(I,J)=B(I,J)-B(I,K)*B(K,J)
42437 NEXT J
42438 NEXT I
42439 NEXT K
42440 REM RETRIEVE INVERSE FROM THE RIGHT SIDE OF B
42441 FOR I=1 TO N
42442 FOR J=1 TO N
42443 B(I,J)=B(I,J+N)
42444 NEXT J
42445 NEXT I
42446 RETURN
43491 REM ********************
43492 REM LINEAR LEAST SQUARES SUBROUTINE (LSTSQR1)
43493 REM THE INPUT DATA SET IS (X(M),Y(M)).
43494 REM THE NUMBER OF DATA POINTS IS N.
43495 REM THE NUMBER OF DIFFERENT POINTS MUST BE GREATER THAN ONE.
43496 REM X(M) AND Y(M) MUST BE DIMENSIONED IN THE CALLING PROGRAM.
43497 REM THE SUBROUTINE ALSO CALCULATES THE UNBIASED ESTIMATE
43498 REM OF THE STANDARD DEVIATION, D.
43499 REM THE RETURNED PARAMETERS ARE A,B AND D.
43500 A1=0
43501 A2=0
43502 B0=0
43503 B1=0
43504 FOR M=0 TO N-1
43505 A1=A1+X(M)
43506 A2=A2+X(M)*X(M)
43507 B0=B0+Y(M)
43508 B1=B1+Y(M)*X(M)
43509 NEXT M
43510 A1=A1/N
43511 A2=A2/N
43512 B0=B0/N
43513 B1=B1/N
43514 D=A1*A1-A2
43515 A=A1*B1-A2*B0
43516 A=A/D
43517 B=A1*B0-B1
43518 B=B/D
43519 REM ********************
43520 REM EVALUATION OF STANDARD DEVIATION (UNBIASED ESTIMATE)
43521 D=0
43522 FOR M=0 TO N-1
```

```
43523 D1=Y(M)-A-B*X(M)
43524 D=D+D1*D1
43525 NEXT M
43526 D=SQR(D/(N-2))
43527 RETURN
43541 REM *********************
43542 REM PARABOLIC LEAST SQUARES SUBROUTINE (LSTSQR2)
43543 REM THE INPUT DATA SET IS (X(M),Y(M)).
43544 REM THE NUMBER OF DATA POINTS IS N.
43545 REM THE NUMBER OF DIFFERENT POINTS MUST BE GREATER THAN THREE.
43546 REM X(M) AND Y(M) MUST BE DIMENSIONED IN THE CALLING PROGRAM.
43547 REM THE SUBROUTINE ALSO CALCULATES THE UNBIASED ESTIMATE
43548 REM OF THE STANDARD DEVIATION, D.
43549 REM THE RETURNED PARAMETERS ARE A,B,C AND D.
43550 A0=1
43551 A1=0
43552 A2=0
43553 A3=0
43554 A4=0
43555 B0=0
43556 B1=0
43557 B2=0
43558 FOR M=0 TO N-1
43559 A1=A1+X(M)
43560 A2=A2+X(M)*X(M)
43561 A3=A3+X(M)*X(M)*X(M)
43562 A4=A4+X(M)*X(M)*X(M)*X(M)
43563 B0=B0+Y(M)
43564 B1=B1+Y(M)*X(M)
43565 B2=B2+Y(M)*X(M)*X(M)
43566 NEXT M
43567 A1=A1/N
43568 A2=A2/N
43569 A3=A3/N
43570 A4=A4/N
43571 B0=B0/N
43572 B1=B1/N
43573 B2=B2/N
43574 D=A0*(A2*A4-A3*A3)-A1*(A1*A4-A3*A2)+A2*(A1*A3-A2*A2)
43575 A=B0*(A2*A4-A3*A3)+B1*(A3*A2-A1*A4)+B2*(A1*A3-A2*A2)
43576 A=A/D
43577 B=B0*(A3*A2-A1*A4)+B1*(A0*A4-A2*A2)+B2*(A2*A1-A0*A3)
43578 B=B/D
43579 C=B0*(A1*A3-A2*A2)+B1*(A1*A2-A0*A3)+B2*(A0*A2-A1*A1)
43580 C=C/D
43581 REM *********************
43582 REM EVALUATION OF STANDARD DEVIATION (UNBIASED ESTIMATE)
43583 D=0
43584 FOR M=0 TO N-1
43585 D1=Y(M)-A-B*X(M)-C*X(M)*X(M)
43586 D=D+D1*D1
43587 NEXT M
43588 D=SQR(D/(N-3))
43589 RETURN
43590 REM *********************
43591 REM COEFFICIENT MATRIX GENERATION SUBROUTINE FOR THE
43592 REM ONE DIMENSIONAL POLYNOMIAL REGRESSION (POLYCM).
43593 REM THE INPUT DATA SET CONSISTS OF N PAIRS OF
43594 REM (X(I),Y(I)) VALUES.
```

```
43595 REM THE REGRESSION ORDER IS N.
43596 REM THE MATRIX RETURNED, Z, IS M ROWS BY N+1 COLUMNS.
43597 REM DIMENSION THIS MATRIX IN THE CALLING PROGRAM.
43598 REM RECALL THAT THE REGRESSION ROUTINE INPUT WILL USE N+1,.
43599 REM YOU MUST SET N TO N+1 BEFORE ENTERING IT.
43600 FOR I=1 TO M
43601 B=1
43602 FOR J=1 TO N+1
43603 Z(I,J)=B
43604 B=B*X(I)
43605 NEXT J
43606 NEXT I
43607 REM Z IS M ROWS BY N+1 COLUMNS
43608 RETURN
43628 REM *********************
43629 REM LEAST SQUARES FITTING SUBROUTINE (LEASTSQR)
43630 REM GENERAL SUBROUTINE FOR MULTIDIMENSIONAL,
43631 REM NONLINEAR REGRESSION.
43632 REM THE EQUATION FITTED HAS THE FORM
43633 REM Y=D(1)X1 + D(2)X2 +  ... + D(N)XN
43634 REM CHANGE IN NOTATION IS DUE TO A DIMENSION CONFLICT.
43635 REM THE COEFFICIENTS ARE RETURNED BY THE PROGRAM IN D(I).
43636 REM THE XI CAN BE SIMPLE POWERS OF X, OR FUNCTIONS.
43637 REM NOTE THAT THE XI ARE ASSUMED TO BE INDEPENDENT.
43638 REM THE MEASURED RESPONSES ARE Y(I)- THERE ARE M OF THEM.
43639 REM Y IS M ROW COLUMN VECTOR, AND Z(I,J) IS A
43640 REM M ROW BY N COLUMN MATRIX.
43641 REM M>=N
43642 REM THE SUBROUTINE INPUTS ARE Y(I), Z(I,J), M AND N.
43643 REM THE WORKING MATRICES WITHIN THE PROGRAM ARE A(I,J),
43644 REM B(I,J) AND C(I,J).
43645 REM THE SUBROUTINE CALLS SEVERAL OTHER MATRIX OPERATION
43646 REM ROUTINES TO PERFORM THE CALCULATION.
43647 REM DIMENSION A,B,C,Y AND Z IN THE CALLING PROGRAM.
43648 REM START PROCEDURE.
43649 REM STORE M AND N
43650 M4=M
43651 N4=N
43652 REM MOVE Z(I,J) TO A(I,J)
43653 FOR I=1 TO M
43654 FOR J=1 TO N
43655 A(I,J)=Z(I,J)
43656 NEXT J
43657 NEXT I
43658 REM A IS M BY N
43659 REM FIND TRANSPOSE OF A AND PUT RESULT IN B
43660 GOSUB 41950
43661 REM B IS N BY M
43662 REM MOVE A TO C, B TO A, AND C TO B
43663 REM WILL HAVE Z(TRANSPOSE) IN A, Z IN B
43664 N1=M
43665 N2=N
43666 N3=0
43667 GOSUB 42150
43668 N1=N
43669 N2=M
43670 GOSUB 42075
43671 N1=M
43672 N2=N
```

```
43673 GOSUB 42100
43674 REM MULTIPLY A AND B
43675 REM A IS N BY M
43676 REM B IS M BY N
43677 M1=N
43678 N1=N
43679 M2=M
43680 GOSUB 41900
43681 REM RESULT IS IN C, AN N BY N MATRIX.. MOVE TO A AND FIND INVERSE
43682 N1=N
43683 GOSUB 42175
43684 REM A IS N BY N
43685 GOSUB 42400
43686 REM RESTORE M
43687 M=M4
43688 REM INVERSE IS IN B. MOVE TO A, PUT Z(TRANSPOSE) IN B, AND MULTIPLY
43689 REM B IS N BY N
43690 GOSUB 42075
43691 FOR I=1 TO M
43692 FOR J=1 TO N
43693 B(J,I)=Z(I,J)
43694 NEXT J
43695 NEXT I
43696 REM B IS N BY M   A IS N BY N
43697 M2=N
43698 N2=M
43699 GOSUB 41900
43700 REM PRODUCT IS N BY M
43701 REM PRODUCT IS IN C. MOVE TO A. LOAD Y IN B, AND MULTIPLY
43702 N1=N
43703 N2=M
43704 GOSUB 42175
43705 REM B IS A COLUMN VECTOR (M BY 1)
43706 FOR I=1 TO M
43707 B(I,1)=Y(I)
43708 NEXT I
43709 N2=1:N1=M
43710 M2=M
43711 REM A IS N BY M   B IS M BY 1
43712 GOSUB 41900
43713 REM PRODUCT IS N BY 1
43714 REM REGRESSION COEFFICIENTS ARE IN C(I,1)
43715 REM MOVE THEM TO D(I)
43716 FOR I=1 TO N
43717 D(I)=C(I,1)
43718 NEXT I
43719 RETURN
43742 REM ********************
43743 REM STANDARD DEVIATION SUBROUTINE (SIGMA).
43744 REM THIS SUBROUTINE CALCULATES THE STANDARD
43745 REM DEVIATION FOR A POLYNOMIAL FIT.
43746 REM THE INPUTS ARE THE NUMBER OF DATA POINTS, M,
43747 REM THE DEGREE OF THE FIT, N,
43748 REM THE POLYNOMIAL COEFFICIENTS, D(I),
43749 REM THE ORIGINAL DATA SET, X(I), Y(I).
43750 D=0
43751 FOR I=1 TO M
43752 Y=0
43753 B=1
```

```
43754 FOR J=1 TO N+1
43755 Y=Y+D(J)*B
43756 B=B*X(I)
43757 NEXT J
43758 D=D+(Y-Y(I))*(Y-Y(I))
43759 NEXT I
43760 IF M-N-1>0 THEN GOTO 43763
43761 D=0
43762 RETURN
43763 D=D/(M-N-1)
43764 D=SQR(D)
43765 RETURN
43783 REM ********************
43784 REM COEFFICIENT MATRIX GENERATION SUBROUTINE FOR MULTIPLE
43785 REM NONLINEAR REGRESSION (MLTNLREG).
43786 REM ALSO CALCULATES THE STANDARD DEVIATION, D, EVEN
43787 REM THOUGH THERE IS SOME REDUNDANT COMPUTING.
43788 REM THE MAXIMUM NUMBER OF DIMENSIONS IS 9.
43789 REM THE INPUT DATA SET CONSISTS OF M DATA SETS OF THE FORM
43790 REM     Y(I), X(I,1), X(I,2),......, X(I,L)
43791 REM THE NUMBER OF DIMENSIONS IS L.
43792 REM THE ORDER OF THE FIT TO EACH DIMENSION IS M(J).
43793 REM THE RESULT IS AN (M1+1)*(M2+1)...*(ML+1)+1
43794 REM COLUMN BY M ROW MATRIX, Z.
43795 REM THIS MATRIX IS ARRANGED AS FOLLOWS ( EXAMPLE- L=2, M(1)=2, M(2)=2)
43796 REM   1   X1   X1*X1   X2   X2*X1   X2*X1*X1   X2*X2   X2*X2*X1   X2*X2*X1*X1
43797 REM THIS MATRIX SHOULD BE DIMENSIONED IN THE CALLING PROGRAM
43798 REM AS SHOULD ALSO THE X(A,J) MATRIX OF DATA VALUES.
43799 REM CALCULATE THE TOTAL NUMBER OF DIMENSIONS
43800 N=1
43801 FOR I=1 TO L
43802 N=N*(M(I)+1)
43803 NEXT I
43804 D=0
43805 FOR I=1 TO M
43806 REM BRANCH ACCORDING TO DIMENSION
43807 REM RETURN IF DIMENSION IS GREATER THAN 9
43808 IF L>0 THEN GOTO 43811
43809 L=0
43810 RETURN
43811 IF L<=9 THEN GOTO 43814
43812 L=0
43813 RETURN
43814 J=0
43815 REM MINIMAL BASIC VERSION. REPLACE FOLLOWING WITH ON/GOTO
43816 IF L=1 THEN GOSUB 43840
43817 IF L=2 THEN GOSUB 43849
43818 IF L=3 THEN GOSUB 43857
43819 IF L=4 THEN GOSUB 43865
43820 IF L=5 THEN GOSUB 43873
43821 IF L=6 THEN GOSUB 43881
43822 IF L=7 THEN GOSUB 43889
43823 IF L=8 THEN GOSUB 43897
43824 IF L=9 THEN GOSUB 43905
43825 REM ARRAY GENERATED FOR ROW I
43826 Y=0
43827 FOR K=1 TO N
43828 Y=Y+D(K)*Z(I,K)
43829 NEXT K
```

```
43830 D=D+(Y(I)-Y)*(Y(I)-Y)
43831 NEXT I
43832 REM CALCULATE STANDARD DEVIATION
43833 IF M-N>0 THEN GOTO 43836
43834 D=0
43835 RETURN
43836 D=D/(M-N)
43837 D=SQR(D)
43838 RETURN
43839 RETURN
43840 B=1
43841 C=B
43842 FOR I1=0 TO M(1)
43843 J=J+1
43844 Z(I,J)=B
43845 B=B*X(I,1)
43846 NEXT I1
43847 B=C
43848 RETURN
43849 B=1
43850 C=B
43851 FOR I2=0 TO M(2)
43852 GOSUB 43841
43853 B=B*X(I,2)
43854 NEXT I2
43855 B=C
43856 RETURN
43857 B=1
43858 C=B
43859 FOR I3=0 TO M(3)
43860 GOSUB 43850
43861 B=B*X(I,3)
43862 NEXT I3
43863 B=C
43864 RETURN
43865 B=1
43866 C=B
43867 FOR I4=0 TO M(4)
43868 GOSUB 43858
43869 B=B*X(I,4)
43870 NEXT I4
43871 B=C
43872 RETURN
43873 B=1
43874 C=B
43875 FOR I5=0 TO M(5)
43876 GOSUB 43866
43877 B=B*X(I,5)
43878 NEXT I5
43879 B=C
43880 RETURN
43881 B=1
43882 C=B
43883 FOR I6=0 TO M(6)
43884 GOSUB 43874
43885 B=B*X(I,6)
43886 NEXT I6
43887 B=C
43888 RETURN
```

```
43889 B=1
43890 C=B
43891 FOR I7=0 TO M(7)
43892 GOSUB 43882
43893 B=B*X(I,7)
43894 NEXT I7
43895 B=C
43896 RETURN
43897 B=1
43898 C=B
43899 FOR I8=0 TO M(8)
43900 GOSUB 43890
43901 B=B*X(I,8)
43902 NEXT I8
43903 B=C
43904 RETURN
43905 B=1
43906 FOR I9=0 TO M(9)
43907 GOSUB 43898
43908 B=B*X(I,9)
43909 NEXT I9
43910 RETURN
43932 REM ********************
43933 REM LEAST SQUARES POLYNOMIAL FITTING SUBROUTINE (LSQRPOLY).
43934 REM THIS PROGRAM LEAST SQUARES FITS A POLYNOMIAL TO INPUT DATA.
43935 REM FORSYTHE ORTHOGONAL POLYNOMIALS ARE USED IN THE FITTING.
43936 REM THE NUMBER OF DATA POINTS IS N.
43937 REM THE DATA IS INPUT TO THE SUBROUTINE IN X(I),Y(I) PAIRS.
43938 REM THE COEFFICIENTS ARE RETURNED IN C(I).
43939 REM THE SMOOTHED DATA IS RETURNED IN V(I).
43940 REM THE ORDER OF THE FIT IS SPECIFIED BY M.
43941 REM THE STANDARD DEVIATION OF THE FIT IS RETURNED IN D.
43942 REM THERE ARE TWO OPTIONS AVAILABLE BY USE OF THE PARAMETER E.
43943 REM IF E=0 THE FIT IS TO ORDER M.
43944 REM IF E>0 THE ORDER OF FIT INCREASES TOWARDS M, BUT
43945 REM WILL STOP IF THE RELATIVE STANDARD DEVIATION DOES NOT
43946 REM DECREASE BY MORE THAN E BETWEEN SUCCESSIVE FITS.
43947 REM THE ORDER OF THE FIT THEN OBTAINED IS L.
43948 REM THE ARRAYS X,Y,V,A,B,C,C2,D,E AND F MUST BE DIMENSIONED.
43949 REM A(I) AND B(I) ARE SIMPLY WORK ARRAYS
43950 N1=M+1
43951 V1=10000000#
43952 REM INITIALIZE THE ARRAYS
43953 FOR I=1 TO N1
43954 A(I)=0
43955 B(I)=0
43956 F(I)=0
43957 NEXT I
43958 FOR I=1 TO N
43959 V(I)=0
43960 D(I)=0
43961 NEXT I
43962 D1=SQR(N)
43963 W=D1
43964 FOR I=1 TO N
43965 E(I)=1/W
43966 NEXT I
43967 F1=D1
43968 A1=0
```

```
43969 FOR I=1 TO N
43970 A1=A1+X(I)*E(I)*E(I)
43971 NEXT I
43972 C1=0
43973 FOR I=1 TO N
43974 C1=C1+Y(I)*E(I)
43975 NEXT I
43976 B(1)=1/F1
43977 F(1)=B(1)*C1
43978 FOR I=1 TO N
43979 V(I)=V(I)+E(I)*C1
43980 NEXT I
43981 M=1
43982 REM SAVE LATEST RESULTS
43983 FOR I=1 TO L
43984 C2(I)=C(I)
43985 NEXT I
43986 L2=L
43987 V2=V
43988 F2=F1
43989 A2=A1
43990 F1=0
43991 FOR I=1 TO N
43992 B1=E(I)
43993 E(I)=(X(I)-A2)*E(I)-F2*D(I)
43994 D(I)=B1
43995 F1=F1+E(I)*E(I)
43996 NEXT I
43997 F1=SQR(F1)
43998 FOR I=1 TO N
43999 E(I)=E(I)/F1
44000 NEXT I
44001 A1=0
44002 FOR I=1 TO N
44003 A1=A1+X(I)*E(I)*E(I)
44004 NEXT I
44005 C1=0
44006 FOR I=1 TO N
44007 C1=C1+E(I)*Y(I)
44008 NEXT I
44009 M=M+1
44010 I=0
44011 L=M-I
44012 B2=B(L)
44013 D1=0
44014 IF L>1 THEN D1=B(L-1)
44015 D1=D1-A2*B(L)-F2*A(L)
44016 B(L)=D1/F1
44017 A(L)=B2
44018 I=I+1
44019 IF I<>M THEN GOTO 44011
44020 FOR I=1 TO N
44021 V(I)=V(I)+E(I)*C1
44022 NEXT I
44023 FOR I=1 TO N1
44024 F(I)=F(I)+B(I)*C1
44025 C(I)=F(I)
44026 NEXT I
44027 V=0
```

```
44028 FOR I=1 TO N
44029 V=V+(V(I)-Y(I))*(V(I)-Y(I))
44030 NEXT I
44031 REM NOTE THE DIVISION IS BY THE NUMBER OF DEGREES OF FREEDOM
44032 V=SQR (V/(N-L-1))
44033 L=M
44034 IF E=0 THEN GOTO 44040
44035 REM TEST FOR MIMIMAL IMPROVEMENT
44036 IF ABS(V1-V)/V<E THEN GOTO 44053
44037 REM IF ERROR IS LARGER, QUIT
44038 IF E*V>E*V1 THEN GOTO 44053
44039 V1=V
44040 IF M=N1 THEN GOTO 44043
44041 GOTO 43983
44042 REM SHIFT THE C(I) DOWN SO C(0) IS THE CONSTANT TERM
44043 FOR I=1 TO L
44044 C(I-1)=C(I)
44045 NEXT I
44046 C(L)=0
44047 REM L IS THE ORDER OF THE POLYNOMIAL FITTED
44048 L=L-1
44049 D=V
44050 RETURN
44051 REM SEQUENCE HAS BEEN ABORTED
44052 REM RECOVER LAST VALUES
44053 L=L2
44054 V=V2
44055 FOR I=1 TO L
44056 C(L)=C2(L)
44057 NEXT I
44058 GOTO 44043
44082 REM **********************
44083 REM MULTI-DIMENSIONAL POLYNOMIAL REGRESSION
44084 REM ITERATION SUBROUTINE (REGITER).
44085 REM THIS PROGRAM SUPERVISES THE CALLING OF SEVERAL
44086 REM OTHER SUBROUTINES IN ORDER TO ITERATIVELY
44087 REM FIT LEAST SQUARES POLYNOMIALS IN MORE THAN
44088 REM ONE DIMENSION.
44089 REM THE PROGRAM REPEATEDLY CALCULATES IMPROVED COEFFICIENTS
44090 REM UNTIL THE STANDARD DEVIATION IS NO LONGER REDUCED.
44091 REM THE INPUTS TO THE SUBROUTINE ARE THE NUMBER OF
44092 REM DIMENSIONS, L, THE DEGREE OF FIT FOR EACH
44093 REM DIMENSION, M(I), AND THE INPUT DATA, X(I,J) AND Y(I).
44094 REM THE COEFFICIENTS ARE RETURNED IN D(I), WITH THE
44095 REM STANDARD DEVIATION IN D.
44096 REM ALSO RETURNED IS THE NUMBER OF ITERATIONS TRIED, L1.
44097 REM THE ORIGINAL Y(I) VALUES ARE SAVED IN Y1(I).
44098 REM THE CURRENT COEFFICIENTS ARE STORED IN D1(I).
44099 REM THE PREVIOUS STANDARD DEVIATION IS SAVED IN D1.
44100 L1=0
44101 REM SAVE Y(I)
44102 FOR I=1 TO M
44103 Y1(I)=Y(I)
44104 NEXT I
44105 REM ZERO D1(I)
44106 FOR I=1 TO N
44107 D1(I)=0
44108 NEXT I
44109 REM SET THE INITIAL STANDARD DEVIATION HIGH
```

```
44110 D1=10000000#
44111 REM GO TO COEFFICIENTS SUBROUTINE
44112 GOSUB 43800
44113 REM GO TO REGRESSION SUBROUTINE
44114 GOSUB 43650
44115 REM GET STANDARD DEVIATION
44116 GOSUB 43800
44117 REM IF STANDARD DEVIATION IS DECREASING, CONTINUE
44118 IF D1>D THEN GOTO 44131
44119 REM TERMINATE ITERATION
44120 FOR I=1 TO N
44121 D(I)=D1(I)
44122 NEXT I
44123 REM RESTORE Y(I)
44124 FOR I=1 TO M
44125 Y(I)=Y1(I)
44126 NEXT I
44127 REM GET THE FINAL STANDARD DEVIATION
44128 GOSUB 43800
44129 RETURN
44130 REM SAVE THE STANDARD DEVIATION
44131 D1=D
44132 L1=L1+1
44133 REM AUGMENT COEFFICIENT MATRIX
44134 FOR I=1 TO N
44135 D(I)=D1(I)+D(I)
44136 D1(I)=D(I)
44137 NEXT I
44138 REM RESTORE Y(I)
44139 FOR I=1 TO M
44140 Y(I)=Y1(I)
44141 NEXT I
44142 REM REDUCE Y(I) ACCORDING TO THE D(I)
44143 GOSUB 44147
44144 REM WE NOW HAVE A SET OF ERROR VALUES
44145 GOTO 44112
44146 REM *********
44147 FOR I=1 TO M
44148 J=0
44149 REM MINIMAL BASIC VERSION. REPLACE FOLLOWING WITH ON/GOTO
44150 IF L=1 THEN GOSUB 44167
44151 IF L=2 THEN GOSUB 44176
44152 IF L=3 THEN GOSUB 44184
44153 IF L=4 THEN GOSUB 44192
44154 IF L=5 THEN GOSUB 44200
44155 IF L=6 THEN GOSUB 44208
44156 IF L=7 THEN GOSUB 44216
44157 IF L=8 THEN GOSUB 44224
44158 IF L=9 THEN GOSUB 44232
44159 REM ARRAY GENERATED FOR ROW I
44160 Y=0
44161 FOR K=1 TO N
44162 Y=Y+D(K)*Z(I,K)
44163 NEXT K
44164 Y(I)=Y(I)-Y
44165 NEXT I
44166 RETURN
44167 B=1
44168 C=B
```

```
44169 FOR I1=0 TO M(1)
44170 J=J+1
44171 Z(I,J)=B
44172 B=B*X(I,1)
44173 NEXT I1
44174 B=C
44175 RETURN
44176 B=1
44177 C=B
44178 FOR I2=0 TO M(2)
44179 GOSUB 44168
44180 B=B*X(I,2)
44181 NEXT I2
44182 B=C
44183 RETURN
44184 B=1
44185 C=B
44186 FOR I3=0 TO M(3)
44187 GOSUB 44177
44188 B=B*X(I,3)
44189 NEXT I3
44190 B=C
44191 RETURN
44192 B=1
44193 C=B
44194 FOR I4=0 TO M(4)
44195 GOSUB 44185
44196 B=B*X(I,4)
44197 NEXT I4
44198 B=C
44199 RETURN
44200 B=1
44201 C=B
44202 FOR I5=0 TO M(5)
44203 GOSUB 44193
44204 B=B*X(I,5)
44205 NEXT I5
44206 B=C
44207 RETURN
44208 B=1
44209 C=B
44210 FOR I6=0 TO M(6)
44211 GOSUB 44201
44212 B=B*X(I,6)
44213 NEXT I6
44214 B=C
44215 RETURN
44216 B=1
44217 C=B
44218 FOR I7=0 TO M(7)
44219 GOSUB 44209
44220 B=B*X(I,7)
44221 NEXT I7
44222 B=C
44223 RETURN
44224 B=1
44225 C=B
44226 FOR I8=0 TO M(8)
44227 GOSUB 44217
```

```
44228 B=B*X(I,8)
44229 NEXT I8
44230 B=C
44231 RETURN
44232 B=1
44233 FOR I9=0 TO M(9)
44234 GOSUB 44225
44235 B=B*X(I,9)
44236 NEXT I9
44237 RETURN
44238 REM ********************
44239 REM PARAMETRIC LEAST SQUARES CURVE FIT SUBROUTINE (PARAFIT).
44240 REM THIS PROGRAM LEAST SQUARES FITS A FUNCTION TO A SET OF
44241 REM DATA VALUES BY SUCCESSIVELY (PARAMETER BY PARAMETER) REDUCING THE VARIANCE.
44242 REM CONVERGENCE DEPENDS ON THE INITIAL VALUES.- CONVERGENCE IS NOT ASSURED.
44243 REM N PAIRS OF DATA VALUES, (X(I),Y(I)), ARE GIVEN.
44244 REM THERE ARE L PARAMETERS, A(J), TO BE OPTIMIZED ACROSS.
44245 REM REQUIRED ARE INITIAL VALUES FOR THE PARAMETER A(L) AND E.
44246 REM ANOTHER IMPORTANT PARAMETER WHICH AFFECTS STABILITY IS E1,
44247 REM WHICH IS INITIALLY CONVERTED TO E1(L) FOR THE FIRST INTERVALS.
44248 REM THE PARAMETERS ARE MULTIPLIED BY (1-E1(I)) ON EACH PASS.
44249 REM DIMENSION X(I),Y(I),A(I) AND E1(I) IN THE CALLING PROGRAM.
44250 FOR I=1 TO L
44251 E1(I)=E1
44252 NEXT I
44253 M=0
44254 REM SET UP TEST RESIDUAL
44255 L1=1E+06
44256 REM MAKE SWEEP THROUGH ALL PARAMETERS
44257 FOR I=1 TO L
44258 A0=A(I)
44259 REM GET VALUE OF RESIDUAL
44260 A(I)=A0
44261 GOSUB 44286
44262 REM STORE RESULT IN M0
44263 M0=L2
44264 REM REPEAT FOR M1
44265 A(I)=A0*(1-E1(I))
44266 GOSUB 44286
44267 M1=L2
44268 REM CHANGE INTERVAL SIZE IF CALLED FOR
44269 REM IF VARIANCE WAS INCREASED, HALVE E1(I)
44270 IF M1>M0 THEN E1(I)=-E1(I)/2
44271 REM IF VARIANCE WAS REDUCED, INCREASE STEP SIZE BY INCREASING E1(I)
44272 IF M1<M0 THEN E1(I)=1.2*E1(I)
44273 REM IF VARIANCE HAS INCREASED, TRY TO REDUCE IT
44274 IF M1>M0 THEN A(I)=A0
44275 IF M1>M0 THEN GOTO 44261
44276 NEXT I
44277 REM END OF A COMPLETE PASS
44278 REM TEST FOR CONVERGENCE
44279 M=M+1
44280 IF L2=0 THEN RETURN
44281 IF ABS((L1-L2)/L2)<E THEN RETURN
44282 REM IF THIS POINT IS REACHED, ANOTHER PASS IS CALLED FOR
44283 L1=L2
44284 GOTO 44257
44285 REM RESIDUAL GENERATION SUBROUTINE
44286 L2=0
```

```
44287 FOR J=1 TO N
44288 X=X(J)
44289 REM OBTAIN FUNCTION
44290 GOSUB 44300
44291 L2=L2+(Y(J)-Y)*(Y(J)-Y)
44292 NEXT J
44293 D=SQR(L2/(N-L))
44294 RETURN
44298 REM *********************
44299 REM FUNCTIONS SUBROUTINE
44300 Y=A(1)*EXP(-(X-A(2))*(X-A(2))/A(3))
44301 RETURN
44393 REM CHI-SQUARE CUMMULATIVE DISTRIBUTION APPROXIMATION (CHISQA).
44394 REM *********************
44395 REM REFERENCE- STATISTICS MANUAL
44396 REM CROW, MAXFIELD AND DAVIS (DOVER, 1960).
44397 REM THE INPUT VALUE IS Y, THE PROBABILITY.
44398 REM THE OUTPUT VALUE IS THE CORRESPONDING
44399 REM CHI-SQUARE STATISTIC.
44400 X=Y
44401 REM GUARD AGAINST 0 DISCONTINUITY
44402 IF X=0 THEN X=EXP(-100)
44403 IF X>.5 THEN GOTO 44408
44404 X=-LOG(X)
44405 REM REGRESSED TABLE CORRECTION
44406 Z=-.803+1.312*X-.2118*X*X+.016*X*X*X
44407 GOTO 44413
44408 X=1-X
44409 REM GUARD AGAINST 0 DISCONTINUITY
44410 IF X=0 THEN GOTO 44415
44411 X=-LOG(X)
44412 Z=.803-1.312*X+.2118*X*X-.016*X*X*X
44413 X=2/(9*M)
44414 X=1-X+Z*SQR(X)
44415 X=M*X*X*X
44416 RETURN
44419 REM *********************
44420 REM BESSEL FUNCTION SERIES SUBROUTINE (BESSLSER)
44421 REM THE ORDER IS N, THE ARGUMENT X.
44422 REM THE RETURNED VALUE IS IN Y.
44423 REM THE NUMBER OF TERMS USED IS RETURNED IN M.
44424 REM E IS THE CONVERGENCE CRITERION
44425 A=1
44426 IF N<=1 THEN GOTO 44431
44427 REM CALCULATE N!
44428 FOR I=1 TO N
44429 A=A*I
44430 NEXT I
44431 A=1/A
44432 IF N=0 THEN GOTO 44437
44433 REM CALCULATE MULTIPLYING TERM
44434 FOR I=1 TO N
44435 A=A*X/2
44436 NEXT I
44437 B0=1
44438 B2=1
44439 M=0
44440 REM ASSEMBLE SERIES SUM
44441 M=M+1
```

```
44442 B1=-(X*X*B0)/(M*(M+N)*4)
44443 B2=B2+B1
44444 B0=B1
44445 REM TEST FOR CONVERGENCE
44446 IF ABS(B1)>E THEN GOTO 44441
44447 REM FORM FINAL ANSWER
44448 Y=A*B2
44449 RETURN
44468 REM ********************
44469 REM BESSEL FUNCTION SERIES COEFFICIENT EVALUATION SUBROUTINE (BESSEL)
44470 REM M+1 IS THE NUMBER OF COEFFICIENTS DESIRED.
44471 REM N IS THE ORDER OF THE BESSEL FUNCTION.
44472 REM THE COEFFICIENTS ARE RETURNED IN A(I).
44473 REM DIMENSION A(I) AND B(I) IN THE CALLING PROGRAM.
44474 REM A1,B1 AND B(I) ARE DUMMY VARIABLES.
44475 A1=1
44476 B1=1:IF N=0 THEN GOTO 44482
44477 FOR I=1 TO N
44478 B(I-1)=0
44479 B1=B1*I
44480 A1=A1/2
44481 NEXT I
44482 B1=A1/B1
44483 A1=1
44484 FOR I=0 TO M STEP 2
44485 A(I)=A1*B1
44486 A(I+1)=0
44487 A1=-A1/((I+2)*(N+N+I+2))
44488 NEXT I
44489 A1=A1/2
44490 FOR I=0 TO M
44491 B(I+N)=A(I)
44492 NEXT I
44493 FOR I=0 TO N+M
44494 A(I)=B(I)
44495 NEXT I
44496 RETURN
44515 REM ********************
44516 REM BESSEL FUNCTION ASYMPTOTIC SERIES SUBROUTINE (BESSEL01)
44517 REM THIS PROGRAM CALCULATES THE ZEROTH AND FIRST ORDER BESSEL
44518 REM FUNCTIONS USING AN ASYMPTOTIC SERIES EXPANSION.
44519 REM THE REQUIRED INPUT ARE X AND A CONVERGENCE FACTOR E.
44520 REM RETURNED ARE THE TWO BESSEL FUNCTIONS, J0(X) AND J1(X)
44521 REM REFERENCE-  ALGORITHMS FOR RPN CALCULATORS
44522 REM      BY BALL, J.A.,   WILEY AND SONS,
44523 REM CALCULATE P AND Q POLYNOMIALS
44524 REM P0(X)=M1   P1(X)=M2   Q0(X)=N1   Q1(X)=N2
44525 A=1
44526 A1=1
44527 A2=1
44528 B=1
44529 C=1
44530 E1=1E+06
44531 M=-1
44532 X1=1/(8*X)
44533 X1=X1*X1
44534 M1=1
44535 M2=1
44536 N1=-1/(8*X)
```

```
44537 N2=-3*N1
44538 N=0
44539 M=M+2
44540 A=A*M*M
44541 M=M+2
44542 A=A*M*M
44543 C=C*X1
44544 A1=A1*A2
44545 A2=A2+1
44546 A1=A1*A2
44547 A2=A2+1
44548 E2=A*C/A1
44549 E4=1+(M+2)/M+(M+2)*(M+2)/(A2*8*X)+(M+2)*(M+4)/(A2*8*X)
44550 E4=E4*E2
44551 REM TEST FOR DIVERGENCE
44552 IF ABS(E4)>E1 THEN GOTO 44562
44553 E1=ABS(E2)
44554 M1=M1-E2
44555 M2=M2+E2*(M+2)/M
44556 N1=N1+E2*(M+2)*(M+2)/(A2*8*X)
44557 N2=N2-E2*(M+2)*(M+4)/(A2*8*X)
44558 N=N+1
44559 REM TEST FOR CONVERGENCE CRITERION
44560 IF E1<E3 THEN GOTO 44562
44561 GOTO 44539
44562 A=3.1415926536#
44563 E=E2
44564 B=SQR(2/(A*X))
44565 J0=B*(M1*COS(X-A/4)-N1*SIN(X-A/4))
44566 J1=B*(M2*COS(X-3*A/4)-N2*SIN(X-3*A/4))
44567 RETURN
44572 REM *********************
44573 REM SERIES APPROXIMATION SUBROUTINE FOR LN(X!)   (LN(X!))
44574 REM ACCURACY BETTER THAN 6 PLACES FOR X>=3.
44575 REM ACCURACY BETTER THAN 12 PLACES FOR X>10.
44576 REM ADVANTAGE IS THAT VERY LARGE VALUES OF THE ARGUMENT CAN BE USED
44577 REM WITHOUT FEAR OF OVERFLOW.
44578 REM REFERENCE-  CRC MATH TABLES.
44579 REM X IS THE INPUT, Y IS THE OUTPUT
44580 X1=1/(X*X)
44581 Y=(X+.5)*LOG(X)-X*(1-X1/12+X1*X1/360-X1*X1*X1/1260+X1*X1*X1*X1/1680)
44582 Y=Y+.918938533205#
44583 RETURN
44592 REM *********************
44593 REM CHI-SQUARE FUNCTION SUBROUTINE (CHI-SQR)
44594 REM THIS PROGRAM TAKES A GIVEN DEGREE OF FREEDOM, M
44595 REM AND VALUE, X, AND CALCULATES THE CHI-SQUARE
44596 REM DENSITY DISTRIBUTION FUNCTION VALUE, Y.
44597 REM REFERENCE- TEXAS INSTRUMENTS SR-51 OWNERS MANUAL (1974).
44598 REM SUBROUTINE LN(X!) IS ALSO CALLED.
44599 REM SAVE X
44600 M1=X
44601 REM PERFORM CALCULATION
44602 X=M/2-1
44603 REM GOTO LN(X!) SUBROUTINE
44604 GOSUB 44580
44605 X=M1
44606 C=-X/2+(M/2-1)*LOG(X)-(M/2)*LOG(2)-Y
44607 Y=EXP(C)
```

```
44608 RETURN
44615 REM *********************
44616 REM CHI-SQUARE CUMMULATIVE DISTRIBUTION (CHISQ)
44617 REM THE PROGRAM IS FAIRLY ACCURATE AND CALLS UPON THE
44618 REM CHI-SQUARE PROBABILITY DENSITY FUNCTION SUBROUTINE (CHI-SQR).
44619 REM THE INPUT PARAMETER IS M, THE NUMBER OF DEGREES OF FREEDOM.
44620 REM ALSO REQUIRED IS THE ORDINATE VALUE. THE PROGRAM RETURNS Y,
44621 REM THE CUMMULATIVE DISTRIBUTION INTEGRAL FROM 0 TO X.
44622 REM REFERENCE- HEWLETT-PACKARD STATISTICS PROGRAMS, 1974.
44623 REM THIS PROGRAM ALSO REQUIRES AN ACCURACY PARAMETER, E, TO
44624 REM DETERMINE THE LEVEL OF SUMMATION.
44625 Y1=1
44626 X2=X
44627 M2=M+2
44628 X2=X2/M2
44629 Y1=Y1+X2
44630 IF X2<E THEN GOTO 44637
44631 M2=M2+2
44632 REM THIS FORM IS USED TO AVOID OVERFLOW
44633 X2=X2*X/M2
44634 REM LOOP TO CONTINUE SUM
44635 GOTO 44629
44636 REM OBTAIN Y, THE PROBABILITY DENSITY FUNCTION
44637 GOSUB 44600
44638 Y=Y1*Y*2*X/M
44639 RETURN
44661 REM *********************
44662 REM ASYMPTOTIC SERIES EXPANSION OF THE INTEGRAL OF
44663 REM   2 EXP(-X*X)/SQRT(PI)  - THE NORMALIZED ERROR FUNCTION (ASYMERF)
44664 REM THIS PROGRAM DETEMINES THE VALUES OF THE ABOVE
44665 REM INTEGRAND USING AN ASYMPTOTIC SERIES WHICH IS
44666 REM EVALUATED TO THE LEVEL OF MAXIMUM ACCURACY.
44667 REM THE INTEGRAL IS FROM 0 TO X.
44668 REM THE INPUT PARAMETER IS X>0. THE RESULTS ARE
44669 REM RETURNED IN Y AND Y1, WITH THE ERROR MEASURE IN E.
44670 REM THE PROGRAM ALSO RETURNS THE NUMBER OF TERMS USED.
44671 REM NOTE- THE ERROR IS ROUGHLY EQUAL TO
44672 REM FIRST TERM NEGLECTED IN THE SERIES SUMMATION.
44673 REM REFERENCE-  A SHORT TABLE OF INTEGRALS BY B.O. PEIRCE
44674 REM     GINN AND COMPANY    1957
44675 N=1
44676 Y=1
44677 C2=1/(2*X*X)
44678 Y=Y-C2
44679 N=N+2
44680 C1=C2
44681 C2=-C1*N/(2*X*X)
44682 REM TEST FOR DIVERGENCE
44683 REM THE BREAK POINT IS ROUGHLY N=X*X
44684 IF ABS(C2)>ABS(C1) THEN GOTO 44687
44685 REM CONTINUE SUMMATION
44686 GOTO 44678
44687 N=(N+1)/2
44688 E=EXP(-X*X)/(X*1.772453850905516#)
44689 Y1=Y*E
44690 Y=1-Y1
44691 E=E*C2
44692 RETURN
44717 REM *********************
```

```
44718 REM CHEBYCHEV SERIES COEFFICIENT EVALUATION SUBROUTINE (CHEBYSER)
44719 REM THE ORDER OF THE POLYNOMIAL IS N.
44720 REM THE COEFFICIENTS ARE RETURNED IN THE
44721 REM ARRAY B(I,J). I IS THE DEGREE OF THE POLYNOMIAL,
44722 REM J IS THE COEFFICIENT ORDER.
44723 REM DIMENSION B(I,J) IN THE CALLING PROGRAM.
44724 REM ESTABLISH T0 AND T1 COEFFICIENTS
44725 B(0,0)=1
44726 B(1,0)=0
44727 B(1,1)=1
44728 REM RETURN IF ORDER IS LESS THAN TWO
44729 IF N<2 THEN RETURN
44730 FOR I=2 TO N
44731 FOR J=1 TO I
44732 REM BASIC RECURSION RELATION
44733 B(I,J)=B(I-1,J-1)+B(I-1,J-1)-B(I-2,J)
44734 NEXT J
44735 B(I,0)=-B(I-2,0)
44736 NEXT I
44737 RETURN
44746 REM ********************
44747 REM CHEBYSHEV ECONOMIZATION SUBROUTINE (CHEBECON)
44748 REM ROUTINE TAKES THE INPUT POLYNOMIAL COEFFICIENTS, C(I),
44749 REM AND RETURNS THE CHEBYSCHEV SERIES COEFFICIENTS, A(I).
44750 REM THE DEGREE OF THE SERIES PASSED TO THE ROUTINE IS M.
44751 REM THE DEGREE OF THE SERIES RETURNED IS M1.
44752 REM THE MAXIMUM RANGE OF X IS X0- X0 IS USED FOR SCALING.
44753 REM THE CHEBYSCHEV SERIES COEFFICIENT (B(I,J) SUBROUTINE IS
44754 REM CALLED- I IS THE ORDER OF THE CHEBYSCHEV POLYNOMIAL.
44755 REM NOTE THAT THE INPUT SERIES COEFFICIENTS ARE NULLED DURING THE PROCESS
44756 REM AND THEN SET EQUAL TO THE ECONOMIZED SERIES COEFFICIENTS.
44757 REM THE CHEBYSHEV SERIES IS VALID ONLY OVER THE RANGE ABS(X/X0)<=1.
44758 REM DIMENSION A(I),B(I,J),C(I) IN THE CALLING PROGRAM.
44759 REM START BY SCALING THE INPUT COEFFICIENTS ACCORDING TO C(I)
44760 B=X0
44761 FOR I=1 TO M
44762 C(I)=C(I)*B
44763 B=B*X0
44764 NEXT I
44765 REM GET CHEBYSCHEV SERIES COEFFICIENTS.
44766 REM POLYNOMIAL SERIES IS REDUCED FROM THE HIGHEST ORDER DOWN
44767 FOR N=M TO 0 STEP -1
44768 GOSUB 44725
44769 A(N)=C(N)/B(N,N)
44770 FOR L=0 TO N
44771 REM CHEBYSCHEV SERIES OF ORDER L IS SUBTRACTED OUT OF THE POLYNOMIAL
44772 C(L)=C(L)-A(N)*B(N,L)
44773 NEXT L
44774 NEXT N
44775 REM PERFORM TRUNCATION
44776 FOR I=0 TO M1
44777 FOR J=0 TO I
44778 C(J)=C(J)+A(I)*B(I,J)
44779 NEXT J
44780 NEXT I
44781 REM CONVERT BACK TO THE INTERVAL X0
44782 B=1/X0
44783 FOR I=1 TO M1
44784 C(I)=C(I)*B
```

```
44785 B=B/X0
44786 NEXT I
44787 RETURN
44789 REM ********************
44790 REM SERIES REVERSION SUBROUTINE (REVERSE)
44791 REM THIS PROGRAM TAKES A POLYNOMIAL, Y=A(0) + A(1)X + ..
44792 REM AND RETURNS A POLYNOMIAL X = B(0) + B(1)Y + ...
44793 REM REFERENCE   CRC STANDARD MATHEMATICAL TABLES
44794 REM                24TH EDITION
44795 REM THE INPUT SERIES COEFFICIENTS ARE A(0),A(1), ETC.
44796 REM A(1) MUST BE NONZERO.
44797 REM THE OUTPUT SERIES COEFFICIENTS ARE B(0),B(1),....,B(7).
44798 REM THE DEGREE OF REVERSION IS LIMITED TO SEVEN.
44799 REM A1,A2,.... ARE DUMMY VARIABLES.
44800 A1=A(1)
44801 B(1)=1/A1
44802 A=1/A1
44803 B=A*A
44804 A=A*B
44805 B(2)=-A2/A
44806 A3=A(3)
44807 A=A*B
44808 B(3)=A*(2*A2*A2-A1*A3)
44809 A4=A(4)
44810 A=A*B
44811 B(4)=A*(5*A1*A2*A3-A1*A1*A4-5*A2*A2*A2)
44812 A5=A(5)
44813 A=A*B
44814 B(5)=6*A1*A1*A2*A4+3*A1*A1*A3*A3+14*A2*A2*A2*A2
44815 B(5)=B(5)-A1*A1*A1*A5-21*A1*A2*A2*A3
44816 B(5)=A*B(5)
44817 A6=A(6)
44818 A=A*B
44819 B(6)=7*A1*A1*A1*A2*A5+7*A1*A1*A1*A3*A4+84*A1*A2*A2*A2*A3
44820 B(6)=B(6)-A1*A1*A1*A1*A6-28*A1*A1*A2*A2*A4
44821 B(6)=B(6)-28*A1*A1*A2*A3*A3-42*A2*A2*A2*A2*A2
44822 B(6)=A*B(6)
44823 A7=A(7)
44824 A=A*B
44825 B(7)=8*A1*A1*A1*A1*A2*A6+8*A1*A1*A1*A1*A3*A5
44826 B(7)=B(7)+4*A1*A1*A1*A1*A4*A4+120*A1*A1*A2*A2*A2*A4
44827 B(7)=B(7)+180*A1*A1*A2*A2*A3*A3+132*A2*A2*A2*A2*A2*A2
44828 B(7)=B(7)-A1*A1*A1*A1*A1*A7-36*A1*A1*A1*A1*A2*A2*A5
44829 B(7)=B(7)-72*A1*A1*A1*A2*A3*A4-12*A1*A1*A1*A3*A3*A3
44830 B(7)=B(7)-330*A1*A2*A2*A2*A2*A3
44831 B(7)=A*B(7)
44832 B(0)=0
44833 A=A(0)
44834 FOR I=1 TO 7
44835 B(0)=B(0)-B(I)*A
44836 A=A*A(0)
44837 NEXT I
44838 RETURN
44841 REM ********************
44842 REM RECIPROCAL POWER SERIES SUBROUTINE (RECIPRO)
44843 REM REFERENCE- COMPUTATIONAL ANALYSIS BY HENRICI.
44844 REM THE INPUT SERIES COEFFICIENTS ARE A(I).
44845 REM THE OUTPUT SERIES COEFFICIENTS ARE B(I).
44846 REM THE DEGREE OF THE INPUT POLYNOMIAL IS N..
```

```
44847 REM THE DEGREE OF THE INVERTED POLYNOMIAL IS M.
44848 REM DIMENSION A(I) AND B(I) IN THE CALLING PROGRAM
44849 REM THE PROGRAM WILL TAKE CARE OF THE NORMALIZATION USING L
44850 L=A(0)
44851 FOR I=0 TO N
44852 A(I)=A(I)/L
44853 B(I)=0
44854 NEXT I
44855 REM CLEAR ARRAYS
44856 FOR I=N+1 TO M
44857 A(I)=0
44858 B(I)=0
44859 NEXT I
44860 REM CALCULATE THE B(I) COEFFICIENTS
44861 B(0)=1
44862 FOR I=1 TO M
44863 J=1
44864 B(I)=B(I)-A(J)*B(I-J)
44865 J=J+1
44866 IF J<=I THEN GOTO 44864
44867 NEXT I
44868 REM UN-NORMALIZE THE A(I) AND B(I)
44869 FOR I=0 TO M
44870 A(I)=A(I)*L
44871 B(I)=B(I)/L
44872 NEXT I
44873 RETURN
44892 REM *********************
44893 REM HORNER'S SHIFTING RULE SUBROUTINE (HORNER)
44894 REM THIS SUBROUTINE TAKES A GIVEN QUARTIC
44895 REM POLYNOMIAL AND CONVERTS IT TO A TAYLOR EXPANSION.
44896 REM THE INPUT SERIES COEFFICIENTS ARE A(I).
44897 REM THE EXPANSION POINT IS X0.
44898 REM THE SHIFTED COEFFICIENTS ARE RETURNED IN B(I).
44899 REM C(4,5) MUST BE DIMENSIONED IN THE CALLING PROGRAM.
44900 FOR J=0 TO 4
44901 C(J,0)=A(4-J)
44902 NEXT J
44903 FOR I=0 TO 4
44904 C(0,I+1)=C(0,I)
44905 J=1
44906 IF J>4-I THEN GOTO 44910
44907 C(J,I+1)=X0*C(J-1,I+1)+C(J,I)
44908 J=J+1
44909 GOTO 44906
44910 NEXT I
44911 FOR I=0 TO 4
44912 B(4-I)=C(I,4-I+1)
44913 NEXT I
44914 RETURN
44915 REM *********************
44916 REM INVERSE NORMAL DISTRIBUTION SUBROUTINE (INVNORM)
44917 REM THIS PROGRAM CALCULATES AN APPROXIMATION
44918 REM TO THE INTEGRAL OF THE NORMAL DISTRIBUTION
44919 REM FUNCTION FROM X TO INFINITY (THE TAIL).
44920 REM A RATIONAL POLYNOMIAL IS USED.
44921 REM THE INPUT IS Y, WITH THE RESULT RETURNED IN X.
44922 REM THE ACCURACY IS BETTER THAN 0.0005 IN THE RANGE 0<Y<=.5
44923 REM REFERENCE- ABRAMOWITZ AND STEGUN
```

```
44924 REM DEFINE COEFFICIENTS
44925 C0=2.51552
44926 C1=.802853
44927 C2=.010328
44928 D1=1.43279
44929 D2=.189269
44930 D3=1.308E-03
44931 IF Y=0 THEN X=10000000000000#
44932 IF Y=0 THEN RETURN
44933 Z=SQR(-LOG(Y*Y))
44934 X=1+D1*Z+D2*Z*Z+D3*Z*Z*Z
44935 X=(C0+C1*Z+C2*Z*Z)/X
44936 X=Z-X
44937 RETURN
44942 REM *********************
44943 REM COMPLEX SERIES EVALUATION SUBROUTINE (CMPLXSER)
44944 REM THE SERIES COEFFICIENTS ARE A(I), ASSUMED REAL.
44945 REM THE ORDER OF THE POLYNOMIAL IS M.
44946 REM THE SUBROUTINE USES REPEATED CALLS TO THE
44947 REM NTH POWER (Z^N) COMPLEX NUMBER SUBROUTINE.
44948 REM INPUTS TO THE SUBROUTINE ARE X,Y,M, AND THE A(I).
44949 REM OUTPUTS ARE Z1(REAL) AND Z2(IMAGINARY).
44950 Z1=A(0)
44951 Z2=0
44952 REM STORE X AND Y
44953 A1=X
44954 A2=Y
44955 FOR N=1 TO M
44956 REM RECALL ORIGINAL X AND Y
44957 X=A1
44958 Y=A2
44959 REM GO TO Z^N SUBROUTINE
44960 GOSUB 41200
44961 REM FORM PARTIAL SUM
44962 Z1=Z1+A(N)*X
44963 Z2=Z2+A(N)*Y
44964 NEXT N
44965 REM RESTORE X AND Y
44966 X=A1
44967 Y=A2
44968 RETURN
44991 REM *********************
44992 REM NTH ROOT SUBROUTINE (NTHROOT)
44993 REM USES NEWTON-RAPHSON ITERATION.
44994 REM REFERENCE- HART, COMPUTER APPROXIMATIONS.
44995 REM EXPONENT IS 1/N, INPUT PARAMETER IS N
44996 REM NOTE THAT N MUST BE AN INTEGER.
44997 REM ARGUMENT IS Y, DESIRED ACCURACY IS E.
44998 REM RETURNED VALUE IS X.
44999 REM INITIAL VALUE IS X0.
45000 IF N<=1 THEN RETURN
45001 IF Y<0 THEN RETURN
45002 IF INT(N)<>N THEN RETURN
45003 IF E<=0 THEN RETURN
45004 IF Y>0 THEN GOTO 45007
45005 X=0
45006 RETURN
45007 M=0
45008 REM FIND N-1 POWER OF X0
```

```
45009 X2=1
45010 FOR I=1 TO N-1
45011 X2=X2*X0
45012 NEXT I
45013 REM ITERATE
45014 X1=((N-1)*X0+Y/X2)/N
45015 X=X1
45016 M=M+1
45017 IF ABS((X0-X1)/X1)<E THEN RETURN
45018 X0=X1
45019 GOTO 45009
45020 REM ********************
45021 REM GENERAL ROOT DETERMINATION SUBROUTINE (GENROOT)
45022 REM HIGH ACCURACY ITERATION INVOLVING THE SQUARE ROOT.
45023 REM ROUTINE DECOMPOSES EXPONENT INTO A BINARY REPRESENTATION
45024 REM AND THEN APPLIES NEWTON-RAPHSON ITERATION.
45025 REM Y IS THE INPUT, N IS THE EXPONENT, AND X THE RETURNED ROOT.
45026 REM E IS THE DESIRED ACCURACY OF THE ITERATION.
45027 REM M IS THE DESIRED NUMBER OF BITS IN THE REPRESENTATION OF N.
45028 REM SAVE Y FOR RETURNING FROM SUBROUTINE.
45029 REM SAVE N
45030 N3=N
45031 IF Y<0THEN RETURN
45032 IF E<=0 THEN RETURN
45033 IF Y>0 THEN GOTO 45036
45034 X=0
45035 RETURN
45036 X3=Y
45037 REM IF THE EXPONENT IS NEGATIVE, INVERT PROBLEM
45038 IF N>=0 THEN GOTO 45042
45039 N=-N
45040 Y=1/Y
45041 REM BREAK N DOWN INTO POWERS OF 1/2
45042 GOSUB 45060
45043 REM FIND MULTIPLIERS
45044 PRINT
45045 GOSUB 45074
45046 Y=X3
45047 N=N3
45048 RETURN
45053 REM ********************
45054 REM ROOT DECOMPOSITION SUBROUTINE (RTDECOMP)
45055 REM DECOMPOSE ROOT N INTO A BINARY REPRESENTATION.
45056 REM M IS THE NUMBER OF BINARY DIGITS.
45057 REM N IS THE INPUT DECIMAL NUMBER.
45058 REM N1 IS THE INTEGER PART OF N.
45059 REM A(I) IS THE BINARY REPRESENTATION OF THE REMAINING FRACTION.
45060 N1=INT(N)
45061 N2=N-N1
45062 REM N2 IS BETWEEN 0 AND 1
45063 REM DECOMPOSE N2 INTO FRACTIONS
45064 A=.5
45065 FOR I=1 TO M
45066 A(I)=0
45067 IF A<N2 THEN A(I)=1
45068 IF A<N2 THEN N2=N2-A
45069 A=A/2
45070 NEXT I
45071 RETURN
```

```
45072 REM ********************
45073 REM FIND MULTIPLYING FACTORS
45074 FOR I=1 TO M
45075 REM FIND SQUARE ROOT OF Y
45076 GOSUB 45109
45077 REM REPLACE Y WITH ITS SQUARE ROOT
45078 Y=X1
45079 IF A(I)=1 THEN GOTO 45083
45080 A(I)=1
45081 GOTO 45084
45082 REM A(I) IS SET EQUAL TO THE LATEST SQUARE ROOT OF Y
45083 A(I)=Y
45084 NEXT I
45085 REM ASSEMBLE RESULTS
45086 REM RETRIEVE Y
45087 Y=X3
45088 REM RETRIEVE N
45089 N=N3
45090 REM TAKE CARE OF N1 MULTIPLICATIONS
45091 X2=1:IF N1=0 THEN GOTO 45096
45092 FOR I=1 TO N1
45093 X2=X2*Y
45094 NEXT I
45095 REM TAKE CARE OF ROOT PORTION
45096 FOR I=1 TO M
45097 X2=X2*A(I)
45098 NEXT I
45099 REM THE FINAL ROOT IS X
45100 X=X2
45101 RETURN
45102 REM ********************
45103 REM SQUARE ROOT SUBROUTINE (SQROOT)
45104 REM USES NEWTON-RAPHSON ITERATION.
45105 REM CALLED HERON'S RULE.
45106 REM REFERENCE- HART, COMPUTER APPROXIMATIONS.
45107 REM ARGUMENT IS Y, RETURNED VALUE IS X1.
45108 REM DESIRED ACCURACY IS E.
45109 X0-1
45110 X1=(X0+Y/X0)/2
45111 IF ABS((X1-X0)/X1)<E THEN RETURN
45112 X0=X1
45113 GOTO 45110
45117 REM ********************
45118 REM TANGENT ITERATION SUBROUTINE (TANITER)
45119 REM USES THE INVERSE TANGENT.
45120 REM BASED ON NEWTON-RAPHSON ITERATION.
45121 REM X IS THE ARGUMENT, Y IS THE RESULT.
45122 REM THE DESIRED ACCURACY IS E.
45123 REM NOTE, THE ALLOWABLE RANGE OF THE ARGUMENT IS -PI/2 TO PI/2.
45124 REM INITIAL GUESS IS X0=1
45125 X0=1
45126 REM CHECK FOR DIVIDE BY ZERO
45127 IF X<>0 THEN GOTO 45131
45128 Y=0
45129 RETURN
45130 REM CHECK FOR OUT OF BOUNDS
45131 IF ABS(X)>=3.1415926535#/2 THEN RETURN
45132 IF E<=0 THEN RETURN
45133 REM CAN CALL ARCTANGENT SUBROUTINE HERE.
```

```
45134 X1=X0+(X-ATN(X0))*(1+X0*X0)
45135 REM TEST FOR ACCURACY
45136 IF ABS((X0-X1)/X1)<E THEN GOTO 45139
45137 X0=X1
45138 GOTO 45134
45139 Y=X1
45140 RETURN
45142 REM ********************
45143 REM INVERSE TANGENT RECURSION SUBROUTINE (ATANITER)
45144 REM USES GAUSS ITERATION
45145 REM REFERENCE- ACTON, NUMERICAL METHODS THAT WORK
45146 REM ARGUMENT IS X, RESULT IS Y
45147 REM DESIRED ACCURACY IS E
45148 REM HERON'S RULE (ITERATION) FOR THE SQUARE ROOT IS ALSO USED
45149 REM A0,A1,B0,B1 ARE DUMMY VARIABLES
45150 IF E<0 THEN RETURN
45151 M=0
45152 Y=1+X*X
45153 REM FIND SQUARE ROOT OF 1/Y
45154 GOSUB 45177
45155 X2=1/X1
45156 A0=X2
45157 B0=1
45158 A1=(A0+B0)/2
45159 Y=A1*B0
45160 REM FIND SQUARE ROOT
45161 GOSUB 45177
45162 B1=X1
45163 REM CHECK ACCURACY
45164 M=M+1
45165 IF ABS((A1-B1)/B1)<E THEN GOTO 45170
45166 A0=A1
45167 B0=B1
45168 GOTO 45158
45169 REM COMPUTE FINAL RESULT
45170 Y=A1*B1
45171 REM OBTAIN SQUARE ROOT
45172 GOSUB 45177
45173 Y=X*X2/X1
45174 RETURN
45175 REM ********************
45176 REM SQUARE ROOT SUBROUTINE
45177 X0=1
45178 X1=(X0+Y/X0)/2
45179 IF ABS((X0-X1)/X1)<E THEN RETURN
45180 X0=X1
45181 GOTO 45178
45194 REM ********************
45195 REM ARCSIN(X) RECURSION SUBROUTINE
45196 REM INPUT IS X (-1<X<1)
45197 REM OUTPUT IS Y=ARCSIN(X)
45198 REM CONVERGENCE CRITERIA IS E
45199 REM REFERENCE- COMPUTATIONAL ANALYSIS BY HENRICI
45200 M=0
45201 REM GUARD AGAINST FAILURE
45202 IF E<=0 THEN RETURN
45203 IF X<>0 THEN GOTO 45207
45204 Y=0
45205 RETURN
```

```
45206 REM CHECK RANGE
45207 IF ABS(X)>1 THEN RETURN
45208 U0=X*SQR(1-X*X)
45209 U1=X
45210 U2=U1*SQR(2*U1/(U1+U0))
45211 Y=U2
45212 M=M+1
45213 IF ABS(U2-U1)<E THEN RETURN
45214 U0=U1
45215 U1=U2
45216 GOTO 45210
45236 REM ********************
45237 REM COMPLETE ELLIPTIC INTEGRAL OF THE FIRST
45238 REM AND SECOND KIND (CLIPTIC)
45239 REM THE INPUT PARAMETER IS K, WHICH SHOULD
45240 REM BE BETWEEN 0 AND 1.
45241 REM TECHNIQUE USES GAUSS' FORMULA FOR THE
45242 REM ARITHMOGEOMETRICAL MEAN.
45243 REM REFERENCE- BALL, ALGORITHMS FOR RPN CALCULATORS.
45244 REM E IS A MEASURE OF THE CONVERGENCE ACCURACY.
45245 REM DEPENDING ON E, A(I) AND B(I) MAY HAVE TO BE DIMENSIONED
45246 REM IN THE CALLING PROGRAM.
45247 REM THE RETURNED VALUES ARE E1, THE ELLIPTIC
45248 REM INTEGRAL OF THE FIRST KIND, AND E2,
45249 REM THE INTEGRAL OF THE SECOND KIND.
45250 A(0)=1+K
45251 B(0)=1-K
45252 N=0
45253 IF K<0 THEN RETURN
45254 IF K>1 THEN RETURN
45255 IF E<=0 THEN RETURN
45256 IF K<1 THEN GOTO 45261
45257 E2=1
45258 E1=1000000000#
45259 E1=E1*E1*E1*E1
45260 RETURN
45261 N=N+1
45262 REM GENERATE IMPROVED VALUES
45263 A(N)=(A(N-1)+B(N-1))/2
45264 B(N)=SQR(A(N-1)*B(N-1))
45265 IF ABS(A(N)-B(N))>E THEN GOTO 45261
45266 E1=1.5707963268#/A(N)
45267 E2=2
45268 M=1
45269 FOR I=1 TO N
45270 E2=E2-M*(A(I)*A(I)-B(I)*B(I))
45271 M=M*2
45272 NEXT I
45273 E2=E2*E1/2
45274 RETURN
45294 REM ********************
45295 REM LN(X) RECURSION SUBROUTINE
45296 REM INPUT IS X (0<X<1)
45297 REM OUTPUT IS Y=LN(X)
45298 REM CONVERGENCE CRITERIA IS E
45299 REM REFERENCE-  COMPUTATIONAL ANALYSIS BY HENRICI
45300 M=0
45301 REM GUARD AGAINST FAILURE
45302 IF E<=0 THEN RETURN
```

```
45303 IF X<1 THEN GOTO 45307
45304 Y=0
45305 RETURN
45306 REM CHECK RANGE
45307 IF X<=0 THEN RETURN
45308 U0=(X*X-1/(X*X))/4
45309 U1=(X-1/X)/2
45310 U2=U1*SQR(2*U1/(U1+U0))
45311 Y=U2
45312 M=M+1
45313 IF ABS(U2-U1)<E THEN RETURN
45314 U0=U1
45315 U1=U2
45316 GOTO 45310
45341 REM ********************
45342 REM INTEGER ORDER BESSEL FUNCTION SUBROUTINE (INTBESSL)
45343 REM CALCULATES BESSEL FUNCTIONS OF ORDER 0 THROUGH 4
45344 REM FOR X>0.
45345 REM MILLER'S METHOD USED, SEE HENRICI
45346 REM ARGUMENT IS X
45347 REM NUMBER OF STEPS =M
45348 REM RETURNED RESULTS ARE Y(I)
45349 REM TEST FOR RANGE
45350 IF X<=0 THEN RETURN
45351 IF M<=0 THEN RETURN
45352 Y(0)=1
45353 Y(1)=0
45354 C=0
45355 N=M
45356 REM UPDATE RESULTS
45357 FOR I=4 TO 1 STEP -1
45358 Y(I)=Y(I-1)
45359 NEXT I
45360 REM APPLY RECURSION RELATION
45361 Y(0)=2*N*Y(1)/X-Y(2)
45362 N=N-1
45363 IF N=0 THEN GOTO 45367
45364 IF INT(N/2)<>N/2 THEN GOTO 45357
45365 C=C+2*Y(0)
45366 GOTO 45357
45367 C=C+Y(0)
45368 REM SCALE THE RESULTS
45369 FOR I=0 TO 4
45370 Y(I)=Y(I)/C
45371 NEXT I
45372 RETURN
45393 REM ********************
45394 REM LEGENDRE SERIES COEFFICIENT EVALUATION SUBROUTINE (LEGNDRE)
45395 REM BY MEANS OF RECURSION RELATION
45396 REM THE ORDER OF THE POLYNOMIAL IS N
45397 REM THE COEFFICIENTS ARE RETURNED IN A(I)
45398 REM DIMENSION A(I) AND B(I,J) IN THE CALLING PROGRAM
45399 REM ESTABLISH P0 AND P1 COEFFICIENTS
45400 B(0,0)=1
45401 B(1,0)=0
45402 B(1,1)=1
45403 REM RETURN IF ORDER IS LESS THAN TWO
45404 IF N<2 THEN RETURN
45405 FOR I=2 TO N
```

```
45406 B(I,0)=-(I-1)*B(I-2,0)/I
45407 FOR J=1 TO I
45408 REM BASIC RECURSION RELATION
45409 B(I,J)=(I+I-1)*B(I-1,J-1)-(I-1)*B(I-2,J)
45410 B(I,J)=B(I,J)/I
45411 NEXT J
45412 NEXT I
45413 FOR I=0 TO N
45414 A(I)=B(N,I)
45415 NEXT I
45416 RETURN
45418 REM ********************
45419 REM LAGUERRE POLYNOMIAL COEFFICIENT EVALUATION SUBROUTINE (LAGUERR)
45420 REM BY MEANS OF RECURSION RELATION
45421 REM THE ORDER OF THE POLYNOMIAL IS N
45422 REM THE COEFFICIENTS ARE RETURNED IN A(I)
45423 REM DIMENSION A(I) AND B(I,J) IN THE CALLING PROGRAM
45424 REM ESTABLISH L0 AND L1 COEFFICIENTS
45425 B(0,0)=1
45426 B(1,0)=1
45427 B(1,1)=-1
45428 REM RETURN IF ORDER IS LESS THAN TWO
45429 IF N<2 THEN RETURN
45430 FOR I=2 TO N
45431 B(I,0)=(2*I-1)*B(I-1,0)-(I-1)*(I-1)*B(I-2,0)
45432 FOR J=1 TO I
45433 REM BASIC RECURSION RELATION
45434 B(I,J)=(2*I-1)*B(I-1,J)-B(I-1,J-1)-(I-1)*(I-1)*B(I-2,J)
45435 NEXT J
45436 NEXT I
45437 FOR I=0 TO N
45438 A(I)=B(N,I)
45439 NEXT I
45440 RETURN
45443 REM ********************
45444 REM HERMITE POLYNOMIAL COEFFICIENT EVALUATION SUBROUTINE (HERMITE)
45445 REM BY MEANS OF RECURSION RELATION
45446 REM THE ORDER OF THE POLYNOMIAL IS N
45447 REM THE COEFFICIENTS ARE RETURNED IN A(I)
45448 REM DIMENSION A(I) AND B(I,J) IN THE CALLING PROGRAM
45449 REM ESTABLISH H0 AND H1 COEFFICIENTS
45450 B(0,0)=1
45451 B(1,0)=0
45452 B(1,1)=2
45453 REM RETURN IF ORDER IS LESS THAN TWO
45454 IF N<2 THEN RETURN
45455 FOR I=2 TO N
45456 B(I,0)=-2*(I-1)*B(I-2,0)
45457 FOR J=1 TO I
45458 REM BASIC RECURSION RELATION
45459 B(I,J)=2*B(I-1,J-1)-2*(I-1)*B(I-2,J)
45460 NEXT J
45461 NEXT I
45462 FOR I=0 TO N
45463 A(I)=B(N,I)
45464 NEXT I
45465 RETURN
45467 REM ********************
45468 REM TRIGONOMETRIC CORDIC SUBROUTINE (TRIGCORD)
```

```
45469 REM THIS SUBROUTINE CALCULATES THE SINE AND COSINE
45470 REM OF AN ANGLE USING THE CORDIC ROTATION METHOD.
45471 REM THE INPUT ANGLE IS A.
45472 REM THE SINE IS RETURNED IN U, AND THE COSINE IN V.
45473 REM REMEMBER TO DIMENSION W(I) AND A(I) IN THE CALLING PROGRAM
45474 REM IF THE ANGLE IS ZERO, SET FUNCTIONS AND RETURN.
45475 IF A<>0 THEN GOTO 45480
45476 U=0
45477 V=1
45478 RETURN
45479 REM GET THE TANGENT COEFFICIENTS
45480 GOSUB 45518
45481 REM REM DETERMINE THE WEIGHTS, W(I)1060
45482 GOSUB 45506
45483 U0=P
45484 V0=0
45485 REM PERFORM THE ROTATIONS UP TO THE RESIDUAL
45486 FOR I=1 TO N
45487 REM UPDATE U0 AND V0
45488 U1=U0-W(I)*A(I)*V0
45489 V1=V0+W(I)*A(I)*U0
45490 U0=U1
45491 V0=V1
45492 NEXT I
45493 REM PERFORM THE RESIDUAL ROTATION USING Z
45494 REM USE U0 AND V0 AS DUMMY VARIABLES
45495 U0=1-Z*Z/2
45496 V0=Z*(1+Z*Z/3)
45497 REM U AND V ARE THE FINAL RESULTS
45498 V=U0*(U1-V0*V1)
45499 U=U0*(V1+V0*U1)
45500 REM U=SIN(A)
45501 REM V=COS(A)
45502 RETURN
45503 REM TRIG CORDIC WEIGHTS SUBROUTINE
45504 REM THE WEIGHTS ARE W(I)= PLUS OR MINUS 1
45505 REM THE INPUT ANGLE IS A
45506 Z=A
45507 FOR I=1 TO N
45508 W(I)=-1
45509 IF Z>0 THEN W(I)=1
45510 Z=Z-W(I)*A0
45511 A0=A0/2
45512 NEXT I
45513 REM Z IS THE RESIDUAL ANGLE
45514 REM ********************
45515 RETURN
45516 REM TRIG CORDIC COEFFICIENT SUBROUTINE
45517 REM FOR CASE N=12
45518 N=12
45519 REM THE TANGENTS ARE GIVEN IN A(I)
45520 A(1)=1
45521 A(2)=.414213562373095#
45522 A(3)=.198912367379658#
45523 A(4)=.098491403357164425#
45524 A(5)=.04912684976946725#
45525 A(6)=.02454862210892544#
45526 A(7)=.01227246237956627#
45527 A(8)=6.136000157623401D-03
```

```
45528 A(9)=3.0679712014226650-03
45529 A(10)=1.533981991088666D-03
45530 A(11)=7.669905443430926D-04
45531 A(12)=3.83495215771441D-04
45532 REM P REPRESENTS P(N)
45533 P=.63661977879720413#
45534 REM A0 IS PI/4
45535 A0=.7853981633974484#
45536 RETURN
45541 REM ********************
45542 REM INVERSE TRIGONOMETRIC CORDIC SUBROUTINE (INVCORD).
45543 REM THIS SUBROUTINE CALCULATES THE ANGLE CORRESPONDING
45544 REM TO A GIVEN SINE, COSINE OR TANGENT USING THE
45545 REM CORDIC ROTATION METHOD.
45546 REM THE INPUT IS U=SIN(A), V=COS(A), OR W=TAN(A).
45547 REM THE RETURNED VALUE IS A.
45548 REM REMEMBER TO DIMENSION W(I) AND A(I) IN THE CALLING PROGRAM.
45549 REM TRANSLATE THE U,V,W INPUTS
45550 IF U=0 THEN GOTO 45555
45551 REM INVERSE SINE IS WANTED
45552 IF ABS(U)>=1E-04 THEN V=SQR(1-U*U)
45553 IF ABS(U)<1E-04 THEN V=1-U*U/2-U*U*U*U/8
45554 GOTO 45571
45555 IF V=0 THEN GOTO 45560
45556 REM INVERSE COSINE IS WANTED
45557 IF ABS(V)>=1E-04 THEN U=SQR(1-V*V)
45558 IF ABS(V)<1E-04 THEN U=1-V*V/2-V*V*V*V/8
45559 GOTO 45571
45560 IF W=0 THEN GOTO 45568
45561 REM INVERSE TANGENT IS WANTED
45562 IF ABS(W)<=10000 THEN U=1/SQR(1+1/(W*W))
45563 IF ABS(W)>10000 THEN U=1-1/(2*W*W)+3/(8*W*W*W*W)
45564 IF ABS(U)>=1E-04 THEN V=SQR(1-U*U)
45565 IF ABS(U)<1E-04 THEN V=1-U*U/2-U*U*U*U/8
45566 U=U*ABS(W)/W
45567 GOTO 45571
45568 A=0
45569 RETURN
45570 REM GET COEFFICIENTS
45571 GOSUB 45616
45572 REM TEST FOR SPECIAL VALUES
45573 IF ABS(U)<1 THEN GOTO 45578
45574 IF ABS(V)>0 THEN GOTO 45578
45575 REM SPECIAL CASE FOUND
45576 A=2*A0*ABS(U)/U
45577 RETURN
45578 IF ABS(V)<1 THEN GOTO 45583
45579 IF ABS(U)>0 THEN GOTO 45583
45580 A=0
45581 RETURN
45582 REM SWITCH U WITH V AND INITIALIZE
45583 A=U
45584 U=V
45585 V=A
45586 U0=P
45587 U1=U0
45588 V0=0
45589 V1=V0
45590 REMERFORM THE ROTATIONS UP TO THE RESIDUAL
```

```
45591 FOR I=1 TO N
45592 REM IS ROTATION TO BE PLUS OR MINUS?
45593 W(I)=-1
45594 IF V0<V THEN W(I)=1
45595 REM UPDATE U0 AND V0
45596 U1=U0-W(I)*A(I)*V0
45597 V1=V0+W(I)*A(I)*U0
45598 U0=U1
45599 V0=V1
45600 NEXT I
45601 REM THE SET OF W(I) WEIGHTS HAVE NOW BEEN DETERMINED
45602 REM PERFORM THE RESIDUAL ANGLE APPROXIMATION
45603 Z=V*U1-U*V1
45604 Z=Z+Z*Z*Z/6
45605 REM ASSEMBLE RESULTS
45606 FOR I=1 TO N
45607 Z=Z+W(I)*A0
45608 A0=A0/2
45609 NEXT I
45610 REM RESULT IS IN Z
45611 A=Z
45612 RETURN
45613 REM ********************
45614 REM TRIG CORDIC COEFFICIENT SUBROUTINE
45615 REM FOR CASE N=12
45616 N=12
45617 REM THE TANGENTS ARE GIVEN IN A(I)
45618 A(1)=1
45619 A(2)=.414213562373095#
45620 A(3)=.198912367379658#
45621 A(4)=.09849140335716425#
45622 A(5)=.04912684976946725#
45623 A(6)=.02454862210892544#
45624 A(7)=.01227246237956627#
45625 A(8)=6.136000157623401D-03
45626 A(9)=3.067971201422665D-03
45627 A(10)=1.533981991088666D-03
45628 A(11)=7.669905443430926D-04
45629 A(12)=3.834952155771441D-04
45630 REM P REPRESENTS P(N)
45631 P=.6366197879720413#
45632 REM A0 IS PI/4
45633 A0=.7853981633974484#
45634 RETURN
45644 REM ********************
45645 REM MODIFIED CORDIC EXPONENTIAL SUBROUTINE (EXPCORD).
45646 REM THIS PROGRAM TAKES AN INPUT VALUE AND RETURNS Y=EXP(X).
45647 REM X MAY BE ANY POSITIVE OR NEGATIVE VALUE.
45648 REM REMEMBER TO DIMENSION A(I) AND W(I) IN THE CALLING PROGRAM.
45649 REM GET COEFFICIENTS
45650 GOSUB 45684
45651 REM REDUCE THE RANGE OF X
45652 K=INT(X)
45653 X=X-K
45654 REM DETERMINE THE WEIGHTING COEFFICIENTS, W(I)
45655 GOSUB 45673
45656 REM CACLULATE PRODUCTS
45657 Y=1
45658 FOR I=1 TO N
```

```
45659 IF W(I)>0 THEN Y=Y*A(I)
45660 NEXT I
45661 REM PERFORM RESIDUAL MULTIPLICATION
45662 Y=Y*(1+Z*(1+Z/2*(1+Z/3*(1+Z/4))))
45663 REM ACCOUNT FOR FACTOR EXP(K)
45664 IF K<0 THEN E=1/E
45665 IF ABS(K)<1 THEN GOTO 45671
45666 FOR I=1 TO ABS(K)
45667 Y=Y*E
45668 NEXT I
45669 REM RESTORE X
45670 X=X+K
45671 RETURN
45672 REM WEIGHT DETERMINATION SUBROUTINE
45673 A=.5
45674 Z=X
45675 FOR I=1 TO N
45676 W(I)=0
45677 IF Z>A THEN W(I)=1
45678 Z=Z-W(I)*A
45679 A=A/2
45680 NEXT I
45681 RETURN
45682 REM ********************
45683 REM EXPONENTIAL COEFFICIENTS SUBROUTINE
45684 N=9
45685 E=2.718281828459045#
45686 A(1)=1.648721270700128#
45687 A(2)=1.284025416687742#
45688 A(3)=1.133148453066826#
45689 A(4)=1.064494458917859#
45690 A(5)=1.031743407499103#
45691 A(6)=1.015747708586686#
45692 A(7)=1.007843097206448#
45693 A(8)=1.003913889338348#
45694 A(9)=1.001955033591003#
45695 RETURN
45719 REM ********************
45720 REM MODIFIED CORDIC NATURAL LOGARITHM SUBROUTINE (LNCORDIC).
45721 REM THIS PROGRAM TAKES AN INPUT VALUE AND RETURNS Y-LN(X).
45722 REM X MAY BE ANY POSITIVE VALUE.
45723 REM REMEMBER TO DIMENSION A(I) AND W(I) IN THE CALLING PROGRAM.
45724 REM GET COEFFICIENTS
45725 GOSUB 45770
45726 REM IF X<=0 THEN AN ERROR EXISTS, RETURN
45727 IF X<=0 THEN RETURN
45728 K=0
45729 REM SAVE X
45730 X1=X
45731 REM REDUCE THE RANGE OF X
45732 IF X<E THEN GOTO 45739
45733 REM DIVIDE OUT A POWER OF E
45734 K=K+1
45735 X=X/E
45736 GOTO 45732
45737 REM TEST IF X>=1. IF SO GO TO NEXT STEP
45738 REM OTHERWISE, BRING X TO >1
45739 IF X>=1 THEN GOTO 45744
45740 K=K-1
```

```
45741 X=X*E
45742 GOTO 45739
45743 REM DETERMINE THE WEIGHTING COEFFICIENTS, W(I)
45744 GOSUB 45761
45745 REM CALCULATE RESIDUAL FACTOR BASED ON Z
45746 REM WANT LN(Z), WHERE Z IS NEAR UNITY
45747 Z=Z-1
45748 Z=Z*(1-(Z/2)*(1+(Z/3)*(1-Z/4)))
45749 REM ASSEMBLE RESULTS
45750 A=1/2
45751 FOR I=1 TO N
45752 Z=Z+W(I)*A
45753 A=A/2
45754 NEXT I
45755 REM Z IS NOW THE MANTISSA, K THE CHARACTERISTIC
45756 Y=K+Z
45757 REM RESTORE X
45758 X=X1
45759 RETURN
45760 REM WEIGHT DETERMINATION SUBROUTINE
45761 Z=X
45762 FOR I=1 TO N
45763 W(I)=0
45764 IF Z>A(I) THEN W(I)=1
45765 IF W(I)=1 THEN Z=Z/A(I)
45766 NEXT I
45767 RETURN
45768 REM *********************
45769 REM EXPONENTIAL COEFFICIENTS SUBROUTINE
45770 N=15
45771 E=2.718281828459045#
45772 A(1)=1.648721270700128#
45773 A(2)=1.284025416687742#
45774 A(3)=1.133148453066826#
45775 A(4)=1.064494458917859#
45776 A(5)=1.031743407499103#
45777 A(6)=1.015747708586686#
45778 A(7)=1.007843097206448#
45779 A(8)=1.003913889338348#
45780 A(9)=1.001955033591003#
45781 A(10)=1.000977039492417#
45782 A(11)=1.000488400478694#
45783 A(12)=1.000244170429748#
45784 A(13)=1.000122077763384#
45785 A(14)=1.000061037018933#
45786 A(15)=1.000030518043791#
45787 RETURN
45790 REM *********************
45791 REM HYPERBOLIC SINE SUBROUTINE (SINH).
45792 REM THIS PROGRAM USES THE DEFINITION OF THE
45793 REM HYPERBOLIC SINE AND THE MODIFIED CORDIC
45794 REM EXPONENTIAL SUBROUTINE TO APPROXIMATE
45795 REM ARCSINH(X) OVER THE ENTIRE RANGE OF REAL X.
45796 REM THE INPUT TO THE SUBROUTINE IS X.
45797 REM THE RETURNED VALUE IS Y=ARCSINH(X).
45798 REM START CALCULATION
45799 REM IS X SMALL ENOUGH TO CAUSE ROUND OFF ERROR?
45800 IF ABS(X)<.35 THEN GOTO 45808
45801 REM CALCULATE SINH(X) USING EXPONENTIAL DEFINITION
```

```
45802 REM GET EXP(X)
45803 GOSUB 45650
45804 REM CALCULATE SINH(X)
45805 Y=(Y-(1/Y))/2
45806 RETURN
45807 REM SERIES APPROXIMATION
45808 Z=1
45809 Y=1
45810 FOR I=1 TO 8
45811 Z=Z*X*X/((2*I)*(2*I+1))
45812 Y=Y+Z
45813 NEXT I
45814 Y=X*Y
45815 RETURN
45816 REM ********************
45817 REM HYPERBOLIC COSINE SUBROUTINE (COSH).
45818 REM THIS PROGRAM USES THE DEFINITION OF THE
45819 REM HYPERBOLIC COSINE AND THE MODIFIED CORDIC
45820 REM EXPONENTIAL SUBROUTINE TO APPROXIMATE
45821 REM ARCOSH(X) OVER THE ENTIRE RANGE OF REAL X.
45822 REM THE RETURNED VALUE IS Y=ARCOSH(X).
45823 REM START CALCULATION
45824 REM GET EXP(X)
45825 GOSUB 45650
45826 Y=(Y+(1/Y))/2
45827 RETURN
45828 REM ********************
45829 REM HYPERBOLIC TANGENT SUBROUTINE (TANH).
45830 REM THIS PROGRAM USES THE DEFINITION
45831 REM    TAN(X)=SINH(X)/COSH(X)
45832 REM TO CALCULATE THE HYPERBOLIC TANGENT.
45833 REM THE INPUT IS X.
45834 REM THE OUTPUT IS Y=TANH(X).
45835 REM START CALCULATION
45836 REM GET SINH(X)
45840 GOSUB 45800
45841 V=Y
45842 REM GET COSH(X)
45843 GOSUB 45825
45844 Y=V/Y
45845 RETURN
45866 REM ********************
45867 REM ARCSINH(X) SUBROUTINE (INVSINH).
45868 REM THIS ROUTINE CALCULATES THE INVERSE
45869 REM HYPERBOLIC SINE USING THE MODIFIED
45870 REM CORDIC NATURAL LOGARITHM SUBROUTINE.
45871 REM THE INPUT IS X.
45872 REM THE OUTPUT IS Y=ARCSINH(X).
45873 REM START CALCULATION
45874 REM TEST FOR ZERO ARGUMENT
45875 IF X<>0 THEN GOTO 45879
45876 Y=0
45877 RETURN
45878 REM SAVE X
45879 X2=X
45880 X=ABS(X)
45881 X=X+SQR(X*X+1)
45882 REM GET LOGARITHM
45883 GOSUB 45725
```

```
45884 REM INSERT SIGN
45885 Y=(X2/ABS(X2))*Y
45886 REM RESTORE X
45887 X=X2
45888 RETURN
45891 REM ********************
45892 REM ARCCOSH(X) SUBROUTINE (INVCOSH).
45893 REM THIS ROUTINE CALCULATES THE INVERSE
45894 REM HYPERBOLIC COSINE USING THE MODIFIED
45895 REM CORDIC NATURAL LOGARITHM SUBROUTINE.
45896 REM THE INPUT IS X.
45897 REM THE OUTPUT IS Y=ARCOSH(X).
45898 REM BEGIN CALCULATION
45899 REM TEST FOR ARGUMENT LESS THAN OR EQUAL TO UNITY
45900 IF X>1 THEN GOTO 45904
45901 Y=0
45902 RETURN
45903 REM SAVE X
45904 X2=X
45905 X=ABS(X)
45906 X=X+SQR(X-1)*SQR(X+1)
45907 REM GET LOGARITHM
45908 GOSUB 45725
45909 REM RESTORE X
45910 X=X2
45911 RETURN
45916 REM ARCTANH(X) SUBROUTINE (INVTANH)
45917 REM ********************
45918 REM THIS PROGRAM CALCULATES THE INVERSE
45919 REM HYPERBOLIC TANGENT USING THE MODIFIED
45920 REM CORDIC NATURAL LOGARITHM SUBROUTINE.
45921 REM THE INPUT IS X.
45922 REM THE OUTPUT IS Y=ARCTANH(X).
45923 REM START CALCULATION
45924 REM TEST FOR X>= +/- 1
45925 IF ABS(X)<1 THEN GOTO 45929
45926 Y=(X/ABS(X))*1E+06*1E+06*1E+06
45927 RETURN
45928 REM TEST FOR ZERO ARGUMENT
45929 IF X<>0 THEN GOTO 45933
45930 Y=0
45931 RETURN
45932 REM SAVE X
45933 X2=X
45934 X=(1+X)/(1-X)
45935 REM GET LOGARITHM
45936 GOSUB 45725
45937 REM RESTORE X
45938 X=X2
45939 RETURN
45988 REM ********************
45989 REM LAGRANGE INTERPOLATION SUBROUTINE (LAGRANGE)
45990 REM N IS THE LEVEL OF THE INTERPOLATION (EG., N=2 IS QUADRATIC).
45991 REM V IS THE TOTAL NUMBER OF TABLE VALUES.
45992 REM (X(I),Y(I)) ARE THE COORDINATE TABLE VALUES, Y(I) BEING THE
45993 REM DEPENDENT VARIABLE. THE X(I) MAY BE ARBITRARILY SPACED.
45994 REM X IS THE INTERPOLATION POINT WHICH IS ASSUMED TO BE IN THE
45995 REM INTERVAL WITH AT LEAST ONE TABLE VALUE TO THE LEFT, AND N TO THE RIGH
45996 REM IF THIS IS VIOLATED, N WILL BE SET TO ZERO.
```

```
45997 REM IT IS ASSUMED THAT THE TABLE VALUES ARE IN ASCENDING X(I) ORDER.
45998 REM X(I), Y(I) AND L(I) MUST BE DIMENSIONED IN THE CALLING PROGRAM.
45999 REM CHECK TO SEE IF INTERPOLATION POINT IS IN APPROPRIATE RANGE
46000 IF X<X(1) THEN GOTO 46003
46001 IF X<=X(V-N) THEN GOTO 46006
46002 REM AN ERROR HAS BEEN ENCOUNTERED
46003 N=0
46004 RETURN
46005 REM FIND THE RELEVANT TABLE INTERVAL
46006 I=0
46007 I=I+1
46008 IF X>X(I) THEN GOTO 46007
46009 I=I-1
46010 REM BEGIN INTERPOLATION
46011 FOR J=0 TO N
46012 L(J)=1
46013 NEXT J
46014 Y=0
46015 FOR K=0 TO N
46016 FOR J=0 TO N
46017 IF J=K THEN GOTO 46019
46018 L(K)=L(K)*(X-X(J+I))/(X(I+K)-X(J+I))
46019 NEXT J
46020 Y=Y+L(K)*Y(I+K)
46021 NEXT K
46022 RETURN
46039 REM ********************
46040 REM NEWTON DIVIDED DIFFERENCES INTERPOLATION SUBROUTINE (NEWTON)
46041 REM CALCULATES CUBIC INTERPOLATIONS FOR A GIVEN TABLE.
46042 REM (X(I),Y(I)) ARE THE V COORDINATE PAIRS OF DATA.
46043 REM X(I) IS THE INDEPENDENT VARIABLE, Y(I) THE DEPENDENT.
46044 REM THE INTERPOLATION POINT IS X. IT IS ASSUMED THAT THERE
46045 REM IS AT LEAST ONE DATA POINT TO THE LEFT, AND THREE TO THE RIGHT.
46046 REM IF THIS IS VIOLATED, THEN N IS SET TO 0.
46047 REM E IS THE ERROR ESTIMATE.
46048 REM X,Y,Y1,Y2 AND Y3 ARE ASSUMED DIMENSIONED IN THE CALLING PROGRAM
46049 REM CHECK TO SEE IF X IS IN THE INTERVAL
46050 N=1
46051 IF X>=X(1) THEN GOTO 46054
46052 N=0
46053 RETURN
46054 IF X<=X(V-3) THEN GOTO 46058
46055 N=0
46056 RETURN
46057 REM GENERATE DIVIDED DIFFERENCES
46058 FOR I=1 TO V-1
46059 Y1(I)=(Y(I+1)-Y(I))/(X(I+1)-X(I))
46060 NEXT I
46061 FOR I=1 TO V-2
46062 Y2(I)=(Y1(I+1)-Y1(I))/(X(I+1)-X(I))
46063 NEXT I
46064 FOR I=1 TO V-3
46065 Y3(I)=(Y2(I+1)-Y2(I))/(X(I+1)-X(I))
46066 NEXT I
46067 REM FIND RELEVANT TABLE INTERVAL
46068 I=0
46069 I=I+1
46070 IF X>X(I) THEN GOTO 46069
46071 I=I-1
```

```
46072 REM BEGIN INTERPOLATION
46073 A=X-X(I)
46074 B=A*(X-X(I+1))
46075 C=B*(X-X(I+2))
46076 Y=Y(I)+A*Y1(I)+B*Y2(I)+C*Y3(I)
46077 REM CALCULATE NEXT TERM IN THE EXPANSION FOR AN ERROR ESTIMATE
46078 E=C*(X-X(I+3))*Y/24
46079 RETURN
46090 REM *********************
46091 REM AKIMA SPLINE FITTING SUBROUTINE (AKIMA)
46092 REM THE INPUT TABLE IS (X(I),Y(I)), WHERE Y(I)
46093 REM IS THE DEPENDENT VARIABLE.
46094 REM THE INTERPOLATION POINT IS X, WHICH IS ASSUMED
46095 REM TO BE IN THE RANGE OF THE TABLE WITH AT LEAST
46096 REM ONE TABLE POINT TO THE LEFT, AND THREE TO THE RIGHT.
46097 REM Y IS RETURNED AS THE INTERPOLATED VALUE.
46098 REM N IS RETURNED AS AN ERROR CHECK (N=0 IMPLIES ERROR).
46099 REM DIMENSION M,X,Y AND Z IN THE CALLING PROGRAM
46100 N=1
46101 REM CHECK TO SEE IF X IS IN THE TABLE RANGE
46102 IF X>=X(1) THEN GOTO 46105
46103 N=0
46104 RETURN
46105 IF X<=X(V-3) THEN GOTO 46108
46106 N=0
46107 RETURN
46108 X(0)=2*X(1)-X(2)
46109 REM CALCULATE AKIMA COEFFICIENTS
46110 FOR I=1 TO V-1
46111 REM SHIFT I TO I+2
46112 M(I+2)=(Y(I+1)-Y(I))/(X(I+1)-X(I))
46113 NEXT I
46114 M(V+2)=2*M(V+1)-M(V)
46115 M(V+3)=2*M(V+2)-M(V+1)
46116 M(2)=2*M(3)-M(4)
46117 M(1)=2*M(2)-M(3)
46118 FOR I=1 TO V
46119 A=ABS(M(I+3)-M(I+2))
46120 B=ABS(M(I+1)-M(I))
46121 IF A+B<>0 THEN GOTO 46124
46122 Z(I)=(M(I+2)+M(I+1))/2
46123 GOTO 46125
46124 Z(I)=(A*M(I+1)+B*M(I+2))/(A+B)
46125 NEXT I
46126 REM FIND RELEVANT TABLE INTERVAL
46127 I=0
46128 I=I+1
46129 IF X>=X(I) THEN GOTO 46128
46130 I=I-1
46131 REM BEGIN INTERPOLATION
46132 B=X(I+1)-X(I)
46133 A=X-X(I)
46134 Y=Y(I)+Z(I)*A+(3*M(I+2)-2*Z(I)-Z(I+1))*A*A/B
46135 Y=Y+(Z(I)+Z(I+1)-2*M(I+2))*A*A*A/(B*B)
46136 RETURN
46138 REM *********************
46139 REM LAGRANGE DERIVATIVE INTERPOLATION SUBROUTINE (D/LAGRNG)
46140 REM N IS THE LEVEL OF THE INTERPOLATION (EG., N=2 IS QUADRATIC).
46141 REM V IS THE TOTAL NUMBER OF TABLE VALUES.
```

```
46142 REM (X(I),Y(I)) ARE THE COORDINATE TABLE VALUES, Y(I) BEING THE
46143 REM DEPENDENT VARIABLE. THE X(I) MAY BE ARBITRARILY SPACED.
46144 REM X IS THE INTERPOLATION POINT WHICH IS ASSUMED TO BE IN THE
46145 REM INTERVAL WITH AT LEAST ONE TABLE VALUE TO THE LEFT, AND N TO THE RIGHT.
46146 REM Y IS RETURNED AS THE DESIRED DERIVATIVE.
46147 REM N IS RETURNED AS THE ERROR CHECK (N=0 IMPLIES ERROR).
46148 REM DIMENSION L(I),M(I,J),X(I) AND Y(I) IN THE CALLING PROGRAM
46149 REM CHECK TO SEE IF X IS IN INTERVAL
46150 IF X>X(1) THEN GOTO 46153
46151 N=0
46152 RETURN
46153 IF X<=X(V-N) THEN GOTO 46157
46154 N=0
46155 RETURN
46156 REM FIND THE RELEVANT TABLE INTERVAL
46157 I=0
46158 I=I+1
46159 IF X>X(I) THEN GOTO 46158
46160 I=I-1
46161 REM BEGIN INTERPOLATION
46162 FOR J=0 TO N
46163 L(J)=0
46164 FOR K=0 TO N
46165 M(J,K)=1
46166 NEXT K
46167 NEXT J
46168 Y=0
46169 FOR K=0 TO N
46170 FOR J=0 TO N
46171 IF J=K THEN GOTO 46179
46172 FOR L=0 TO N
46173 IF L=K THEN GOTO 46178
46174 IF L<>J THEN GOTO 46177
46175 M(L,K)=M(L,K)/(X(I+K)-X(I+J))
46176 GOTO 46178
46177 M(L,K)=M(L,K)*(X-X(J+I))/(X(I+K)-X(I+J))
46178 NEXT L
46179 NEXT J
46180 FOR L=0 TO N
46181 IF L=K THEN GOTO 46183
46182 L(K)=L(K)+M(L,K)
46183 NEXT L
46184 Y=Y+L(K)*Y(I+K)
46185 NEXT K
46186 RETURN
46189 REM *********************
46190 REM GENERAL INTEGRATION SUBROUTINE (ITEG)
46191 REM INTERPOLATION BY AKIMA (OR OTHER).
46192 REM INTEGRATION BY ENHANCED TRAPAZOIDAL RULE.
46193 REM REM WITH RICHARDSON EXTRAPOLATION TO GIVE CUBIC ACCURACY.
46194 REM CAN BE USED UNDER VERY GENERAL CONDITIONS.
46195 REM THE INTEGRATION RANGE IS (X1,X2).
46196 REM IT IS ASSUMED THAT X1<X2, AND THAT THERE IS AT LEAST ONE TABLE
46197 REM VALUE TO THE LEFT OF X1, AND THREE TO THE RIGHT OF X2.
46198 REM THE RESULT IS RETURNED IN Z
46199 REM AN ERROR CHECK IS RETURNED IN Z1- Z1=0 IMPLIES ERROR.
46200 Z=0
46201 Z1=0
46202 REM CHECK TO SEE IF END POINTS ARE IN ALLOWABLE RANGE
```

```
46203 IF X1<X(1) THEN RETURN
46204 IF X2>X(V-3) THEN RETURN
46205 REM IF X1>X2 THEN SWITCH AND SET FLAG
46206 IF X1<X2 THEN GOTO 46211
46207 X3=X1
46208 X1=X2
46209 X2=X3
46210 Z1=1
46211 IF X2=X1 THEN RETURN
46212 REM START TRAPAZOIDAL INTEGRATIONS
46213 REM FIRST INTEGRATION TO GET I1
46214 GOSUB 46232
46215 REM SECOND ROUND TO GET I2
46216 GOSUB 46267
46217 REM RICHARDSON EXTRAPOLATION
46218 Z=4*I2/3-I1/3
46219 REM CHECK TO SEE IF THE END POINTS HAVE BEEN REVERSED
46220 IF Z1=0 THEN GOTO 46225
46221 Z=-Z
46222 X2=X1
46223 X1=X3
46224 REM RESET ERROR FLAG
46225 Z1=1
46226 RETURN
46227 REM Z IS THE INTEGRAL DESIRED
46228 RETURN
46229 REM ********************
46230 REM ROUTINE FOR THE FIRST TRAPAZOIDAL INTEGRATION, I1
46231 REM N1 KEEPS TRACK OF THE NUMBER OF INTERVALS
46232 I1=0
46233 N1=0
46234 X=X1
46235 REM FIND THE BEGINNING OF THE INTERVAL
46236 REM GO TO BRANCH WHICH CALLS THE INTERPOLATION SUBROUTINE
46237 REM FIND THE INTERVAL, I, AND THE LEFT END POINT, Y
46238 GOSUB 46299
46239 REM IS THERE AT LEAST ONE TABLE INTERVAL?
46240 IF X2>X(I+1) THEN GOTO 46250
46241 REM IF NOT, INTEGRAL IS SIMPLE
46242 N1=N1+1
46243 D=Y
46244 X=X2
46245 REM FIND END POINT Y VALUE
46246 GOSUB 46299
46247 I1=(Y+D)*(X2-X1)/2
46248 RETURN
46249 REM AT LEAST ONE TABLE INTERVAL MUST BE SUMMED OVER
46250 J1=I
46251 I1=I1+(Y+Y(I+1))*(X(I+1)-X)/2
46252 REM ANY MORE INTERVALS? IF NOT, FINISH INTEGRAL WITH END POINT
46253 IF X2<X(J1+3) THEN GOTO 46259
46254 REM OTHERWISE, KEEP SUMMING
46255 N1=N1+1
46256 I1=I1+(Y(J1+1)+Y(J1+3))*(X(J1+3)-X(J1+1))/2
46257 J1=J1+2
46258 GOTO 46253
46259 X=X2
46260 REM FIND LAST Y VALUE
46261 GOSUB 46299
```

```
46262 I1=I1+(Y+Y(J1+1))*(X2-X(J1+1))/2
46263 N1=N1+1
46264 RETURN
46265 REM ********************
46266 INTEGRATION FOR I2
46267 I2=0
46268 X=X1
46269 GOSUB 46299
46270 D=Y
46271 IF X2>X(I+1) THEN GOTO 46280
46272 X=X1+(X2-X1)/2
46273 GOSUB 46299
46274 I2=I2+(D+Y)*(X2-X1)/4
46275 D=Y
46276 X=X2
46277 GOSUB 46299
46278 I2=I2+(D+Y)*(X2-X1)/4
46279 RETURN
46280 X=X1+(X(I+1)-X1)/2
46281 J1=I
46282 GOSUB 46299
46283 I2=I2+(Y+D)*(X-X1)/2
46284 I2=I2+(Y+Y(J1+1))*(X(J1+1)-X)/2
46285 IF X2<X(J1+2) THEN GOTO 46289
46286 I2=I2+(Y(J1+1)+Y(J1+2))*(X(J1+2)-X(J1+1))/2
46287 J1=J1+1
46288 GOTO 46285
46289 X=X2-(X2-X(J1+1))/2
46290 GOSUB 46299
46291 D=Y
46292 I2=I2+(Y(J1+1)+D)*(X2-X)/2
46293 X=X2
46294 GOSUB 46299
46295 I2=I2+(D+Y)*(X2-X(J1+1))/4
46296 RETURN
46297 REM BRANCH TO AN INTERPOLATION SUBROUTINE FROM HERE
46298 REM GO TO AKIMA SPLINE INTERPOLATION SUBROUTINE
46299 GOSUB 46100
46300 REM JUST RETURNED FROM INTERPOLATION SUBROUTINE
46301 REM RETURN TO PROGRAM
46302 RETURN
46333 REM ********************
46334 REM BISECTION METHOD SUBROUTINE (BISECT)
46335 REM THIS PROGRAM ITERATIVELY SEEKS THE ZERO
46336 REM OF A FUNCTION USING THE METHOD OF INTERVAL
46337 REM HALVING UNTIL THE INTERVAL IS LESS THAN
46338 REM E IN WIDTH.
46339 REM IT IS ASSUMED THAT THE FUNCTION Y=Y(X)
46340 REM IS AVAILABLE FROM THE FUNCTION SUBROUTINE
46341 REM LOCATED AT 44300.
46342 REM THIS SUBROUTINE REQUIRES AS INPUT THE INITIAL
46343 REM RANGE VALUES (X0 AND X1), AS WELL AS THE
46344 REM CONVERGENCE CRITERION, E.
46345 REM THE ZERO MUST BE WITHIN THE RANGE SPECIFIED
46346 REM OR AN ERRONEOUS VALUE WILL BE RETURNED IN X.
46347 REM THIS SUBROUTINE RETURNS THE ESTIMATE OF THE ROOT
46348 REM IN X, AND THE CORRESPONDING Y VALUE.
46349 REM ALSO RETURNED IS THE NUMBER OF STEPS (M).
46350 M=0
```

```
46351 X=X0
46352 GOSUB 44300
46353 Y0=Y
46354 X=X1
46355 GOSUB 44300
46356 X=(X0+X1)/2
46357 GOSUB 44300
46358 M=M+1
46359 IF Y*Y0=0 THEN RETURN
46360 IF Y*Y0<0 THEN X1=X
46361 IF Y*Y0>0 THEN X0=X
46362 IF ABS(X1-X0)>E THEN GOTO 46351
46363 RETURN
46364 REM ********************
46365 REM NEWTON'S METHOD SUBROUTINE (Z-NEWTON)
46366 REM THIS PROGRAM CALCULATES THE ZEROS OF A
46367 REM FUNCTION BY NEWTON'S METHOD.
46368 REM THE ROUTINE REQUIRES AN INITIAL GUESS, X0,
46369 REM AND A CONVERGENCE FACTOR, E.
46370 REM ALSO REQUIRED IS A LIMIT ON THE NUMBER
46371 REM OF ITERATIONS, M. THE NUMBER USED IS
46372 REM RETURNED IN N.
46373 REM IT IS ASSUMED THAT THE FUNCTION AND ITS
46374 REM DERIVATIVE ARE IN THE SUBROUTINE AT 44300.
46375 N=0
46376 REM GET Y AND Y1
46377 X=X0
46378 GOSUB 44300
46379 REM UPDATE ESTIMATE
46380 X0=X0-Y/Y1
46381 N=N+1
46382 IF N>=M THEN RETURN
46383 IF ABS(Y/Y1)>E THEN GOTO 46377
46384 X=X0
46385 RETURN
46413 REM ********************
46414 REM SECANT METHOD SUBROUTINE (SECANT)
46415 REM THIS SUBROUTINE CALCULATES THE ZEROES OF A
46416 REM FUNCTION USING THE SECANT METHOD.
46417 REM TWO INITIAL GUESSES ARE REQUIRED, X0 AND X1.
46418 REM THE CONVERGENCE CRITERION IS E.
46419 REM THE MAXIMUM NUMBER OF ITERATIONS IS M.
46420 REM THE NUMBER OF ITERATIONS PERFORMED IS
46421 REM RETURNED IN N.
46422 REM THE RESULT IS RETURNED IN X.
46423 REM IT IS ASSUMED THAT THE FUNCTION, Y(X),
46424 REM IS IN THE SUBROUTINE AT 44300.
46425 N=0
46426 REM START ITERATION
46427 X=X0
46428 GOSUB 44300
46429 Y0=Y
46430 X=X1
46431 REM GET NEXT POINT
46432 GOSUB 44300
46433 Y1=Y
46434 REM CALCULATE NEW ESTIMATE
46435 REM IF Y1=Y0 THEN THERE WILL BE AN OVERFLOW
46436 REM GUARD AGAINST THIS ARTIFICIALLY
```

```
46437 IF Y1=Y0 THEN Y1=Y1+1E-03
46438 X=(X0*Y1-X1*Y0)/(Y1-Y0)
46439 N=N+1
46440 REM TEST FOR CONVERGENCE
46441 IF N>=M THEN RETURN
46442 IF ABS(X1-X0)<E THEN RETURN
46443 REM UPDATE POSITIONS
46444 X0=X1
46445 X1=X
46446 GOTO 46427
46463 REM ********************
46464 REM MODIFIED FALSE POSITION SUBROUTINE (REGULA)
46465 REM SUBROUTINE USES HAMMING'S MODIFICATION TO
46466 REM SPEED CONVERGENCE.
46467 REM IT IS ASSUMED THAT THE FUNCTION Y(X) IS
46468 REM IN THE SUBROUTINE AT 44300.
46469 REM THE TWO INITIAL GUESSES ARE X0 AND X1.
46470 REM THESE TWO GUESSES MUST BRACKET THE ZERO.
46471 REM THE CONVERGENCE CRITERION IS E.
46472 REM THE MAXIMUM NUMBER OF GUESSES IS M.
46473 REM THE RESULT IS RETURNED IN X.
46474 REM THE NUMBER OF ITERATIONS IS RETURNED IN N.
46475 N=0
46476 REM ME X0<X1
46477 IF X0<X1 THEN GOTO 46481
46478 X=X0
46479 X0=X1
46480 X1=X
46481 X=X0
46482 REM GET Y0 AND Y1
46483 GOSUB 44300
46484 Y0=Y
46485 X=X1
46486 REM INITIAL GUESSES FOR A AND B ARE REQUIRED
46487 GOSUB 44300
46488 Y1=Y
46489 REM CALCULATE A NEW ESTIMATE, X
46490 X=(X0*Y1-X1*Y0)/(Y1-Y0)
46491 REM TEST FOR CONVERGENCE
46492 N=N+1
46493 IF N>=M THEN RETURN
46494 IF ABS(X1-X)<E THEN RETURN
46495 REM GET A NEW Y(X) VALUE
46496 GOSUB 44300
46497 REM APPLY HAMMING'S MODIFICATION
46498 IF Y1*Y=0 THEN RETURN
46499 IF Y0*Y>0 THEN GOTO 46504
46500 X1=X
46501 Y1=Y
46502 Y0=Y0/2
46503 GOTO 46490
46504 X0=X
46505 Y0=Y
46506 Y1=Y1/2
46507 GOTO 46490
46509 REM ********************
46510 REM AITKEN ACCELERATION SUBROUTINE (AITKEN)
46511 REM THIS ROUTINE CALCULATES THE ZEROS OF A FUNCTION
46512 REM BY ITERATION, AND EMPLOYS AITKEN ACCELERATION TO
```

```
46513 REM SPEED UP CONVERGENCE.
46514 REM REM THE SUBROUTINE REQUIRES AN INITIAL GUESS, X0,
46515 REM AND TWO CONVERGENCE FACTORS, C AND E.
46516 REM E RELATES TO THE ACCURACY OF THE ESTIMATE, AND C
46517 REM IS USED TO AID THE CONVERGENCE.
46518 REM ALSO REQUIRED IS AN ITERATION LIMIT, M.
46519 REM C=-1 IS A NORMAL VALUE. IF DIVERGENCE OCCURS,
46520 REM SMALLER AND/OR POSITIVE VALUES SHOULD BE TRIED.
46521 REM THE RESULT IS RETURNED IN X.
46522 REM THE NUMBER OF ITERATIONS IS RETURNED IN N.
46523 REM IT IS ASSUMED THAT THE FUNCTION Y(X) IS IN
46524 REM THE SUBROUTINE AT 44300.
46525 N=0
46526 X=X0
46527 REM GET Y
46528 GOSUB 44300
46529 Y=X+C*Y
46530 REM ARE THERE ENOUGH POINTS FOR ACCELERATION?
46531 IF N>0 THEN GOTO 46536
46532 X1=Y
46533 X=X1
46534 N=N+1
46535 GOTO 46528
46536 X2=Y
46537 N=N+1
46538 REM GUARD AGAINST A ZERO DENOMINATOR
46539 IF X2-2*X1+X0=0 THEN X0=X0+1E-03
46540 REM PERFORM ACCELERATION
46541 K=(X2-X1)*(X2-X1)/(X2-2*X1+X0)
46542 X2=X2-K
46543 REM TEST FOR CONVERGENCE
46544 IF N>=M THEN RETURN
46545 IF ABS(K)<E THEN RETURN
46546 X0=X1
46547 X1=X2
46548 X=X1
46549 GOTO 46528
46559 REM ********************
46560 REM AITKEN-STEFFENSON ITERATION SUBROUTINE (A/SITER)
46561 REM THIS ROUTINE CALCULATES THE ZEROS OF A FUNCTION
46562 REM BY ITERATION, AND EMPLOYS AITKEN ACCELERATION TO
46563 REM SPEED UP CONVERGENCE.
46564 REM REM THE SUBROUTINE REQUIRES AN INITIAL GUESS, X0,
46565 REM AND TWO CONVERGENCE FACTORS, C AND E.
46566 REM E RELATES TO THE ACCURACY OF THE ESTIMATE, AND C
46567 REM IS USED TO AID THE CONVERGENCE.
46568 REM ALSO REQUIRED IS A LIMIT TO THE NUMBER OF ITERATIONS, M.
46569 REM C=-1 IS A NORMAL VALUE. IF DIVERGENCE OCCURS,
46570 REM SMALLER AND/OR POSITIVE VALUES SHOULD BE TRIED.
46571 REM THE RESULT IS RETURNED IN X.
46572 REM THE NUMBER OF ITERATIONS IS RETURNED IN N.
46573 REM IT IS ASSUMED THAT THE FUNCTION Y(X) IS IN
46574 REM THE SUBROUTINE AT 44300.
46575 N=0
46576 M1=0
46577 X=X0
46578 REM GET Y
46579 GOSUB 44300
46580 Y=X+C*Y
```

```
46581 REM ARE THERE ENOUGH POINTS FOR ACCELERATION?
46582 IF M1>0 THEN GOTO 46588
46583 N=N+1
46584 M1=M1+1
46585 X=X1
46586 X1=Y
46587 GOTO 46579
46588 X2=Y
46589 REM PERFORM ACCELERATION
46591 K=(X2-2*X1+X0)
46592 IF K=0 THEN K=1E-03
46593 K=(X1-X0)*(X1-X0)/K
46594 X0=X0-K
46595 REM TEST FOR CONVERGENCE
46596 IF N>=M THEN RETURN
46597 IF ABS(X-X0)<E THEN RETURN
46598 REM REPEAT PROCESS
46599 GOTO 46576
46617 REM ********************
46618 REM SYNTHETIC DIVISION SUBROUTINE (RSYNDIV)
46619 REM ASSUMES REAL POLYNOMIAL COEFFICIENTS.
46620 REM FORM CALCULATED IS A(X)=C(X)/B(X).
46621 REM THE INPUT POLYNOMIAL COEFFICIENTS ARE
46622 REM C(I) AND B(I), THE RESULT IS A(I).
46623 REM C(X) IS OF ORDER N1, B(X) IS OF ORDER N2.
46624 REM RESULT IS OF ORDER N1-N2 (AT MOST).
46625 FOR I=N1 TO N2 STEP -1
46626 A(I-N2)=C(I)/B(N2)
46627 IF I=N2 THEN GOTO 46631
46628 FOR J=0 TO N2
46629 C(I-J)=C(I-J)-A(I-N2)*B(N2-J)
46630 NEXT J
46631 NEXT I
46632 RETURN
46641 REM ********************
46642 REM SUBROUTINE FOR DETERMINING ADDITIONAL ROOTS OF
46643 REM A FUNCTION GIVEN A SET OF ALREADY ESTABLISHED ROOTS (NEXTROOT)
46644 REM USE IS RESTRICTED TO REAL ROOTS.
46645 REM METHOD APPLIED IS NEWTON-RAPHSON ITERATION.
46646 REM THE L ESTABLISHED ROOTS ARE A(I).
46647 REM THE FUNCTION Y AND ITS DERIVATIVE ARE PLACED IN SUBROUTINE 44300.
46648 REM THE INITIAL GUESS IS X0.
46649 REM THE ACCURACY CRITERIA IS E.
46650 N=0
46651 REM GIVEN X0, FIND F/F'.
46652 X=X0
46653 GOSUB 44300
46654 B=Y1/Y
46655 FOR I=1 TO L
46656 B=B-1/(X0-A(I))
46657 NEXT I
46658 REM NEWTON-RAPHSON ITERATION
46659 X1=X0-1/B
46660 N=N+1
46661 REM TEST FOR CONVERGENCE
46662 IF N>=M THEN GOTO 46666
46663 IF ABS(X1-X0)<E THEN GOTO 46666
46664 X0=X1
46665 GOTO 46652
```

```
46666 X=X1
46667 RETURN
46679 REM *********************
46680 REM COMPLEX ROOT COUNTING SUBROUTINE (ROOTNUM)
46681 REM THIS ROUTINE CALCULATES THE NUMBER OF COMPLEX
46682 REM ROOTS WITHIN A CIRCLE OF RADIUS W CENTERED
46683 REM ON (X0,Y0) BY COUNTING (U,V) TRANSISTIONS
46684 REM AROUND THE CIRCUMFERENCE.
46685 REM THE INPUT PARAMETERS ARE
46686 REM     W - THE RADIUS OF THE CIRCLE
46687 REM     (X0,Y0) - THE CENTER OF THE CIRCLE
46688 REM     M - THE NUMBER OF EVALUATION POINTS PER QUADRANT
46689 REM THE ROUTINE RETURNS THE NUMBER OF ROOTS FOUND, N
46690 REM AND THE NUMBER A, WHERE A<>0 INDICATES A FAILURE
46691 REM IN THE ALGORITHM.
46692 REM IT IS ASSUMED THAT THE FUNCTION IS COMPLEX IN THE
46693 REM DOMAIN BEING SEARCHED (U AND V BOTH HAVE TRANSITIONS).
46694 REM IT IS ALSO ASSUMED THAT THE SECTOR SPACING IS CLOSE
46695 REM ENOUGH TO CATCH ALL TRANSITIONS.
46696 REM NOTE THAT N(I) MUST BE DIMENSIONED TO 4M IN THE
46697 REM CALLING PROGRAM.
46698 REM OBSERVE THAT U(X,Y) AND V(X,Y) ARE EXPECTED TO BE
46699 REM FOUND IN THE FUNCTIONS SUBROUTINE.
46700 A=3.14159/(2*M)
46701 REM START CALCULATION BY ESTABLISHING THE N(I) ARRAY
46702 FOR I=1 TO 4*M
46703 X=W*COS(A*(I-1))+X0
46704 Y=W*SIN(A*(I-1))+Y0
46705 GOSUB 44300
46706 IF U>=0 THEN IF V>=0 THEN N(I)=1
46707 IF U<0 THEN IF V>=0 THEN N(I)=2
46708 IF U<0 THEN IF V<0 THEN N(I)=3
46709 IF U>=0 THEN IF V<0 THEN N(I)=4
46710 NEXT I
46711 REM COUNT COMPLETE CYCLES COUNTERCLOCKWISE
46712 N=N(1)
46713 A=0
46714 FOR I=2 TO 4*M
46715 IF N=N(I) THEN GOTO 46721
46716 IF N<>4 THEN IF N=N(I)+1 THEN A=A-1
46717 IF N=1 THEN IF N(I)=4 THEN A=A-1
46718 IF N=4 THEN IF N(I)=1 THEN A=A+1
46719 IF N+1=N(I) THEN A=A+1
46720 N=N(I)
46721 NEXT I
46722 REM COMPLETE CIRCLE
46723 IF N<>4 THEN IF N=N(1)+1 THEN A=A-1
46724 IF N=4 THEN IF N(1)=1 THEN A=A+1
46725 IF N=1 THEN IF N(1)=4 THEN A=A-1
46726 IF N+1=N(1) THEN A=A+1
46727 A=ABS(A)
46728 N=INT(A/4)
46729 A=A-4*INT(A/4)
46730 RETURN
46737 REM ********************
46738 REM COMPLEX ROOT SEARCH SUBROUTINE (ZCIRCLE)
46739 REM THIS PROGRAM SEARCHES FOR THE COMPLEX ROOTS
46740 REM OF AN ANALYTICAL FUNCTION BY ENCIRCLING THE
46741 REM ZERO AND ESTIMATING WHERE IT IS. THE CIRCLE
```

```
46742 REM IS SUBSEQUENTLY TIGHTENED BY A FACTOR E, AND
46743 REM A NEW ESTIMATE MADE.
46744 REM THE INPUTS TO THE SUBROUTINE ARE
46745 REM   (X0,Y0) - THE INITIAL GUESSES
46746 REM    W - THE INITIAL RADIUS OF THE SEARCH CIRCLE
46747 REM    E - THE FACTOR BY WHICH THE CIRCLE IS REDUCED
46748 REM    N - THE NUMBER OF ITERATIONS
46749 REM    M - THE NUMBER OF EVALUATION POINTS PER QUADRANT
46750 REM THE RESULTS IS RETURNED IN Z=X+IY (X,Y).
46751 REM ALSO, THE NUMBER OF ITERATIONS PERFORMED, OR
46752 REM IN PROGRESS, IS RETURNED IN K.
46753 REM X(I),Y(I),U(I) AND V(I) MUST BE DIMENSIONED
46754 REM IN THE CALLING PROGRAM TO 4M.
46755 REM IT IS ASSUMED THAT THE FUNCTION IS DECOMPOSED
46756 REM INTO ITS REAL AND IMAGINARY PARTS, U(X,Y) AND
46757 REM V(X,Y), AND THAT THESE ARE ACCESSIBLE BY A CALL
46758 REM TO THE FUNCTION SUBROUTINE WHICH RETURNS U AND V.
46759 REM START CALCULATION BY FINDING THE EVALUATION POINTS
46760 M=M*4
46761 K=1
46762 A=6.28319/M
46763 FOR I=1 TO M
46764 X(I)=W*COS(A*(I-1))+X0
46765 Y(I)=W*SIN(A*(I-1))+Y0
46766 NEXT I
46767 REM DETERMINE THE CORRESPONDING U(I) AND(I)
46768 FOR I=1 TO M
46769 X=X(I)
46770 Y=Y(I)
46771 GOSUB 44300
46772 U(I)=U
46773 V(I)=V
46774 NEXT I
46775 REM FIND THE POSITION AT WHICH U CHANGES SIGN IN THE
46776 REM COUNTERCLOCKWISE DIRECTION
46777 I=1
46778 U=U(I)
46779 GOSUB 46885
46780 IF U*U(I)<0 THEN GOTO 46785
46781 REM GUARD AGAINST INFINITE LOOP
46782 IF I=1 THEN GOTO 46881
46783 GOTO 46779
46784 REM TRANSITION FOUND
46785 M1=I
46786 REM SEARCH FOR THE OTHER TRANSITION, STARTING
46787 REM ON THE OTHER SIDE OF THE CIRCLE
46788 I=M1+M/2
46789 IF I>M THEN I=I-M
46790 J=I
46791 U=U(I)
46792 REM FLIP DIRECTIONS ALTERNATELY
46793 GOSUB 46885
46794 IF U*U(I)<0 THEN GOTO 46801
46795 IF U*U(J)<0 THEN GOTO 46804
46796 REM TEST FOR INFINITE LOOP
46797 IF I=M1+M/2 THEN GOTO 46881
46798 IF J=M1+M/2 THEN GOTO 46881
46799 GOTO 46793
46800 REM TRANSITION FOUND
```

```
46801 M3=I
46802 GOTO 46808
46803 REM TRANSITION FOUND
46804 IF J=M THEN J=0
46805 M3=J+1
46806 REM M1 AND M3 HAVE BEEN DETERMINED. NOW FOR M2 AND M4.
46807 REM NOW FOR THE V TRANSITIONS
46808 I=M1+M/4
46809 IF I>M THEN I=I-M
46810 J=I
46811 V=V(I)
46812 GOSUB 46885
46813 IF V*V(I)<0 THEN GOTO 46820
46814 IF V*V(J)<0 THEN GOTO 46822
46815 REM AGAIN, GUARD AGAINST THE INFINITE LOOP
46816 IF I=M1+M/4 THEN GOTO 46881
46817 IF J=M1+M/4 THEN GOTO 46881
46818 GOTO 46812
46819 REM M2 HAS BEEN FOUND
46820 M2=I
46821 GOTO 46825
46822 IF J=M THEN J=0
46823 M2=J+1
46824 REM M2 HAS BEEN FOUND. NOW FOR M4
46825 I=M2+M/2
46826 IF I>M THEN I=I-M
46827 J=I
46828 V=V(I)
46829 GOSUB 46885
46830 IF U*V(I)<0 THEN GOTO 46836
46831 IF V*V(J)<0 THEN GOTO 46838
46832 REM GUARD AGAINST THE INFINITE LOOP AGAIN
46833 IF I=M2+M/2 THEN GOTO 46881
46834 IF J=M2+M/2 THEN GOTO 46881
46835 GOTO 46829
46836 M4=I
46837 GOTO 46842
46838 IF J=M THEN J=0
46839 M4=J+1
46840 REM ALL THE INTERSECTIONS HAVE BEEN DETERMINED
46841 REM INTERPOLATE TO FIND THE FOUR (X,Y) COORDINATES
46842 I=M1
46843 GOSUB 46891
46844 X1=X
46845 Y1=Y
46846 I=M2
46847 GOSUB 46891
46848 X2=X
46849 Y2=Y
46850 I=M3
46851 GOSUB 46891
46852 X3=X
46853 Y3=Y
46854 I=M4
46855 GOSUB 46891
46856 X4=X
46857 Y4=Y
46858 REM CALCULATE THE INTERSECTION OF THE LINES
46859 REM GUARD AGAINST A DIVIDE BY ZERO
```

```
46860 IF X1<>X3 THEN GOTO 46864
46861 X=X1
46862 Y=(Y1+Y3)/2
46863 GOTO 46868
46864 M1=(Y3-Y1)/(X3-X1)
46865 IF X2<>X4 THEN GOTO 46868
46866 M2=100000000#
46867 GOTO 46869
46868 M2=(Y2-Y4)/(X2-X4)
46869 B1=Y1-M1*X1
46870 B2=Y2-M2*X2
46871 X=-(B1-B2)/(M1-M2)
46872 Y=(M1*B2+M2*B1)/(M1+M2)
46873 REM IS ANOTHER ITERATION IN ORDER?
46874 IF K=N THEN RETURN
46875 X0=X
46876 Y0=Y
46877 K=K+1
46878 W=W*E
46879 GOTO 46763
46880 REM INFINITE LOOP ENCOUNTERED, RETURN
46881 X=0
46882 Y=0
46883 RETURN
46884 REM AUXILLIARY SUBROUTINE
46885 I=I+1
46886 J=J-1
46887 IF I>M THEN I=I-M
46888 IF J<1 THEN J=J+M
46889 RETURN
46890 REM AUXILLIARY SUBROUTINE FOR INTERPOLATION
46891 J=I-1
46892 IF J<1 THEN J=J+M
46893 REM REGULA FALSI INTERPOLATION FOR THE ZERO
46894 X=(X(I)*U(J)+X(J)*U(I))/(U(I)+U(J))
46895 Y=(Y(I)*V(J)+Y(J)*V(I))/(V(I)+V(J))
46896 RETURN
46912 REM ********************
46913 REM COMPLEX ROOT SEEKING USING NEWTON'S METHOD (CZNEWTON)
46914 REM THIS ROUTINE USES THE COMPLEX DOMAIN FORM OF
46915 REM NEWTON'S METHOD FOR ITERATIVELY SEARCHING FOR ROOTS.
46916 REM IT IS ASSUMED THAT THE FUNCTION AND ITS FIRST PARTIAL
46917 REM DERIVATIVES ARE AVAILABLE FROM THE FUNCTIONS SUBROUTINE
46918 REM IN THE FORM  F(Z) = U(X,Y) + I V(X,Y).
46919 REM THE REQUIRED DERIVATIVES ARE DU/DX AND DU/DY.
46920 REM THE INPUTS TO THE SUBROUTINE ARE THE INITIAL GUESS, X0, Y0,
46921 REM THE CONVERGENCE CRITERIA, E, AND THE MAXIMUM NUMBER OF
46922 REM ITERATIONS TO BE PERFORMED, N.
46923 REM THE RESULTING APPROXIMATION TO THE ROOT IS RETURNED IN
46924 REM (X,Y), AND THE NUMBER OF ITERATIONS IN K.
46925 K=0
46926 K=K+1
46927 REM GET U, V AND THE DERIVATIVES
46928 X=X0
46929 Y=Y0
46930 GOSUB 44300
46931 A=U1*U1+U2*U2
46932 X=X0+(V*U2-U*U1)/A
46933 Y=Y0-(V*U1+U*U2)/A
```

```
46934 REM CHECK FOR CONVERGENCE IN EUCLIDEAN SPACE
46935 IF (X0-X)*(X0-X)+(Y0-Y)*(Y0-Y)<=E*E THEN RETURN
46936 IF K>=N THEN RETURN
46937 X0=X
46938 Y0=Y
46939 GOTO 46926
46944 REM ********************
46945 REM PARABOLIC ROOT SEEKING SUBROUTINE (MUELLER)
46946 REM THIS PROGRAM ITERATIVELY SEEKS THE ROOT OF A
46947 REM FUNCTION BY FITTING A PARABOLA TO THREE POINTS
46948 REM AND CALCULATING THE NEAREST ROOT AS DESCRIBED IN
46949 REM BECKET AND HURT, NUMERICAL CALCULATIONS AND ALGORITHMS.
46950 REM THE SUBROUTINE INPUTS ARE
46951 REM     X0 - THE INITIAL GUESS
46952 REM     D - A BOUND ON THE ERROR IN THIS GUESS
46953 REM     E - THE CONVERGENCE CRITERIA
46954 REM     N - THE MAXIMUM NUMBER OF ITERATIONS
46955 REM THE PROGRAM RETURNS THE VALUE OF THE ROOT FOUND, X,
46956 REM AND THE NUMBER OF ITERATIONS PERFORMED, K.
46957 REM IT IS ASSUMED THAT THE FUNCTION Y(X) IS AVAILABLE
46958 REM IN THE FUNCTIONS SUBROUTINE.
46959 REM SET UP THE THREE EVALUATION POINTS
46960 K=1
46961 X3=X0
46962 X1=X3-D
46963 X2=X3+D
46964 REM CALCULATE MUELLER PARAMETERS
46965 REM GUARD AGAINST DIVIDE BY ZERO
46966 IF X2-X1=0 THEN X2=X2*(1.0000001#)
46967 IF X2-X1=0 THEN X2=X2+1E-07
46968 L1=(X3-X2)/(X2-X1)
46969 D1=(X3-X1)/(X2-X1)
46970 IF K>1 THEN GOTO 46978
46971 REM GET VALUES OF FUNCTION
46972 X=X1
46973 GOSUB 44300
46974 E1=Y
46975 X=X2
46976 GOSUB 44300
46977 E2=Y
46978 X=X3
46979 GOSUB 44300
46980 E3=Y
46981 A1=L1*L1*E1-D1*D1*E2+(L1+D1)*E3
46982 C1=L1*(L1*E1-D1*E2+E3)
46983 B=A1*A1-4*D1*C1*E3
46984 REM TEST FOR COMPLEX ROOT, MEANING THE PARABOLA IS INVERTED
46985 IF B<0 THEN B=0
46986 REM CHOOSE CLOSEST ROOT
46987 IF A1<0 THEN A1=A1-SQR(B)
46988 IF A1>0 THEN A1=A1+SQR(B)
46989 REM GUARD AGAINST A DIVIDE BY ZERO
46990 IF ABS(A1)+ABS(B)=0 THEN A1=4*D1*E3
46991 REM CALCULATE RELATIVE DISTANCE OF NEXT GUESS
46992 REM GUARD AGAINST DIVIDE BY ZERO
46993 IF A1=0 THEN A1=1E-07
46994 L=-2*D1*E3/A1
46995 REM CALCULATE NEXT ESTIMATE
46996 X=X3+L*(X3-X2)
```

```
46997 REM TEST FOR CONVERGENCE
46998 IF ABS(X-X3)<E THEN RETURN
46999 REM TEST FOR NUMBER OF ITERATIONS
47000 IF K>=N THEN RETURN
47001 REM OTHERWISE, MAKE ANOTHER PASS
47002 K=K+1
47003 REM SAVE SOME CALCULATIONS:
47004 X1=X2
47005 X2=X3
47006 X3=X
47007 E1=E2
47008 E2=E3
47009 GOTO 46967
47033 REM *********************
47034 REM PARABOLIC ROOT SEEKING SUBROUTINE (MUELLER2)
47035 REM FOR A TWO DIMENSIONAL FUNCTION, W(X,Y).
47036 REM THIS PROGRAM ITERATIVELY SEEKS THE ROOT OF A
47037 REM FUNCTION BY FITTING A PARABOLA TO THREE POINTS
47038 REM AND CALCULATING THE NEAREST ROOT AS DESCRIBED IN
47039 REM BECKET AND HURT, NUMERICAL CALCULATIONS AND ALGORITHMS.
47040 REM THE SUBROUTINE INPUTS ARE
47041 REM     X0,Y0 - THE INITIAL GUESS
47042 REM     B1,B2 - A BOUND ON THE ERROR IN THIS GUESS
47043 REM     E - THE CONVERGENCE CRITERIA
47044 REM     N - THE MAXIMUM NUMBER OF ITERATIONS
47045 REM THE PROGRAM RETURNS THE VALUE OF THE ROOT FOUND, (X,Y),
47046 REM AND THE NUMBER OF ITERATIONS PERFORMED, K.
47047 REM IT IS ASSUMED THAT THE FUNCTION U(X,Y) IS AVAILABLE
47048 REM IN THE FUNCTIONS SUBROUTINE.
47049 REM SET UP THE THREE EVALUATION POINTS
47050 K=1
47051 X3=X0
47052 X1=X3-B1
47053 X2=X3+B1
47054 REM CALCULATE MUELLER PARAMETERS
47055 REM GUARD AGAINST DIVIDE BY ZERO
47056 IF X2-X1=0 THEN X2=X2*(1.0000001#)
47057 IF X2-X1=0 THEN X2=X2+1E-07
47058 L1=(X3-X2)/(X2-X1)
47059 D1=(X3-X1)/(X2-X1)
47060 REM GET VALUES OF FUNCTION
47061 Y=Y0
47062 X=X1
47063 GOSUB 44300
47064 E1=W
47065 X=X2
47066 GOSUB 44300
47067 E2=W
47068 X=X3
47069 GOSUB 44300
47070 E3=W
47071 GOSUB  47111
47072 REM CALCULATE NEW X ESTIMATE
47073 B1=L*(X3-X2)
47074 X=X3+B1
47075 REM TEST FOR CONVERGENCE
47076 IF ABS(B1)+ABS(B2)<E THEN RETURN
47077 X0=X
47078 REM REPEAT FOR THE Y DIRECTION
```

```
47079 Y3=Y0
47080 Y1=Y3-B2
47081 Y2=Y3+B2
47082 REM CALCULATE MUELLER PARAMETERS
47083 REM GUARD AGAINST A DIVIDE BY ZERO
47084 IF Y2-Y1=0 THEN Y2=Y2*(1.0000001#)
47085 IF Y2-Y1=0 THEN Y2=Y2+1E-07
47086 L1=(Y3-Y2)/(Y2-Y1)
47087 D1=(Y3-Y1)/(Y2-Y1)
47088 REM GET VALUES OF FUNCTION
47089 Y=Y1
47090 GOSUB 44300
47091 E1=W
47092 Y=Y2
47093 GOSUB 44300
47094 E2=W
47095 Y=Y3
47096 GOSUB 44300
47097 E3=W
47098 GOSUB 47111
47099 REM CALCULATE NEW Y ESTIMATE
47100 B2=L*(Y3-Y2)
47101 Y=Y3+B2
47102 REM TEST FOR CONVERGENCE
47103 IF ABS(B1)+ABS(B2)<E THEN RETURN
47104 REM TEST FOR NUMBER OF ITERATIONS
47105 IF K>=N THEN RETURN
47106 Y0=Y
47107 K=K+1
47108 REM START ANOTHER PASS
47109 GOTO 47051
47110 REM UTILITY SUBROUTINE
47111 A1=L1*L1*E1-D1*D1*E2+(L1+D1)*E3
47112 C1=L1*(L1*E1-D1*E2+E3)
47113 B=A1*A1-4*D1*C1*E3
47114 REM TEST FOR COMPLEX ROOT, MEANING THE PARABOLA IS INVERTED
47115 IF B<0 THEN B=0
47116 REM CHOOSE CLOSEST ROOT
47117 IF A1<0 THEN A1=A1-SQR(B)
47118 IF A1>0 THEN A1=A1+SQR(B)
47119 REM GUARD AGAINST A DIVIDE BY ZERO
47120 IF ABS(A1)+ABS(B)=0 THEN A1=4*D1*E3
47121 REM CALCULATE RELATIVE DISTANCE OF NEXT GUESS
47122 REM GUARD AGAINST DIVIDE BY ZERO
47123 IF A1=0 THEN A1=1E-07
47124 L=-2*D1*E3/A1
47125 RETURN
47136 REM ********************
47137 REM MUELLER'S METHOD FOR COMPLEX ROOTS (ZMUELLER)
47138 REM THIS PROGRAM USES THE PARABOLIC FITTING TECHNIQUE
47139 REM ASSOCIATED WITH MUELLER'S METHOD, BUT DOES IT IN
47140 REM THE COMPLEX DOMAIN.
47141 REM THE INPUTS TO THE SUBROUTINE ARE THE INITIAL
47142 REM GUESS, (X0,Y0), THE CONVERGENCE CRITERIA, E,
47143 REM AND THE MAXIMUM NUMBER OF ITERATIONS, N.
47144 REM ALSO REQUIRED ARE BOUNDS ON THE INITIAL GUESS, B1 AND B2.
47145 REM RETURNED IS THE NEW ESTIMATE, (X,Y), AND THE
47146 REM NUMBER OF ITERATIONS PERFORMED, K.
47147 REM IT IS ASSUMED THAT THE FUNCTION F(Z) = U(X,Y)+IV(X,Y)
```

```
47148 REM IS AVAILABLE IN THE FUNCTIONS SUBROUTINE.
47149 REM START CALCULATIONS
47150 K=1
47151 X3=X0
47152 Y3=Y0
47153 X1=X3-B1
47154 Y1=Y3-B2
47155 X2=X3+B1
47156 Y2=Y3+B2
47157 D=(X2-X1)*(X2-X1)+(Y2-Y1)*(Y2-Y1)
47158 REM AVOID DIVIDE BY ZERO
47159 IF D=0 THEN D=1E-07
47160 L1=(X3-X2)*(X2-X1)+(Y3-Y2)*(Y2-Y1)
47161 L1=L1/D
47162 L2=(X2-X1)*(Y3-Y2)-(X3-X2)*(Y2-Y1)
47163 L2=L2/D
47164 D1=(X3-X1)*(X2-X1)+(Y3-Y1)*(Y2-Y1)
47165 D1=D1/D
47166 D2=(X2-X1)*(Y3-Y1)-(X3-X1)*(Y2-Y1)
47167 D2=D2/D
47168 REM GET FUNCTION VALUES
47169 X=X1
47170 Y=Y1
47171 GOSUB 44300
47172 U1=U
47173 V1=V
47174 X=X2
47175 Y=Y2
47176 GOSUB 44300
47177 U2=U
47178 V2=V
47179 X=X3
47180 Y=Y3
47181 GOSUB 44300
47182 U3=U
47183 V3=V
47184 REM CALCULATE MUELLER PARAMETERS
47185 E1=U1*(L1*L1-L2*L2)-2*V1*L1*L2-U2*(D1*D1-D2*D2)
47186 E1=E1+2*V2*D1*D2+U3*(L1+D1)-V3*(L2+D2)
47187 E2=2*L1*L2*U1+V1*(L1*L1-L2*L2)-2*D1*D2*U2-V2*(D1*D1-D2*D2)
47188 E2=E2+U3*(L2+D2)+V3*(L1+D1)
47189 C1=L1*L1*U1-L1*L2*V1-D1*L1*U2+L1*D2*V2+U3*L1
47190 C1=C1-U1*L2*L2-V1*L1*L2+U2*L2*D2+V2*D1*L2-V3*L2
47191 C2=U1*L1*L2+V1*L1*L1-U2*D2*L1-V2*D1*L1+V3*L1
47192 C2=C2+L1*L2*U1-L2*L2*V1-D1*L2*U2+D2*L2*V2+U3*L2
47193 B1=E1*E1-E2*E2-4*(U3*D1*C1-U3*D2*C2-V3*D2*C1-V3*D1*C2)
47194 B2=2*E1*E2-4*(U3*D2*C1+U3*D1*C2+V3*D1*C1-V3*D2*C2)
47195 REM GUARD AGAINST A DIVIDE BY ZERO
47196 IF B1=0 THEN B1=1E-07
47197 A=ATN(B2/B1)
47198 A=A/2
47199 B=SQR(SQR(B1*B1+B2*B2))
47200 B1=B*COS(A)
47201 B2=B*SIN(A)
47202 A1=(E1+B1)*(E1+B1)+(E2+B2)*(E2+B2)
47203 A2=(E1-B1)*(E1-B1)+(E2-B2)*(E2-B2)
47204 IF A1>A2 THEN GOTO 47208
47205 A1=E1-B1
47206 A2=E2-B2
```

```
47207 GOTO 47210
47208 A1=E1+B1
47209 A2=E2+B2
47210 A=A1*A1+A2*A2
47211 L1=A1*D1*U3-A1*D2*V3+A2*U3*D2+A2*V3*D1
47212 REM GUARD AGAINST DIVIDE BY ZERO
47213 IF A=0 THEN A=1E-07
47214 L1=-2*L1/A
47215 L2=-D1*U3*A2+D2*V3*A2+A1*U3*D2+A1*V3*D1
47216 L2=-2*L2/A
47217 REM CALCULATE NEW ESTIMATE
47218 X=X3+L1*(X3-X2)-L2*(Y3-Y2)
47219 Y=Y3+L2*(X3-X2)+L1*(Y3-Y2)
47220 REM TEST FOR CONVERGENCE
47221 IF ABS(X-X0)+ABS(Y-Y0)<E THEN RETURN
47222 REM TEST FOR NUMBER OF ITERATIONS
47223 IF K>=N THEN RETURN
47224 REM CONTINUE
47225 K=K+1
47226 X0=X
47227 Y0=Y
47228 X1=X2
47229 Y1=Y2
47230 X2=X3
47231 Y2=Y3
47232 X3=X
47233 Y3=Y
47234 GOTO 47157
47282 REM ********************
47283 REM GENERAL ROOT DETERMINATION SUBROUTINE (ALLROOT)
47284 REM THE ROUTINE ATTEMPTS TO CALCULATE THE SEVERAL ROOTS OF A
47285 REM GIVEN SERIES OR FUNCTION BY REPEATEDLY USING THE
47286 REM ZMUELLER SUBROUTINE AND REMOVING THE ROOTS ALREADY FOUND
47287 REM BY DIVISION.
47288 REM THE INPUT TO THE SUBROUTINE ARE
47289 REM      X0,Y0 - THE INITIAL GUESS
47290 REM      B1,B2 - THE BOUNDS ON THIS GUESS
47291 REM      E - THE CONVERGENCE CRITERIA
47292 REM      N - THE MAXIMUM NUMBER OF ITERATIONS PER ROOT
47293 REM      N2 - THE NUMBER OF ROOTS BEING SOUGHT
47294 REM      N3 - A FLAG INDICATING A FUNCTION F(Z) (1)
47295 REM           OR A POLYNOMIAL (2)
47296 REM THE PROGRAM RETURNS THE N2 ROOTS FOUND, X(I),Y(I)
47297 REM AND THE LAST NUMBER OF ITERATIONS USED, K.
47298 REM IF K=0 THEN N3 WAS IN ERROR
47299 REM START CALCULATIONS
47300 K=0
47301 IF N3=1 THEN GOTO 47303
47302 IF N3<>2 THEN RETURN
47303 J1=0
47304 REM SAVE THE INITIAL GUESS
47305 X4=X0
47306 Y4=Y0
47307 REM IF N3=2 THEN GET THE SERIES COEFFICIENTS
47308 IF N3=2 THEN GOSUB 47322
47309 REM TEST FOR COMPLETION
47310 IF J1=N2 THEN RETURN
47311 REM GOTO ZMUELLER
47312 GOSUB 47344
```

```
47313 J1=J1+1
47314 X(J1)=X
47315 Y(J1)=Y
47316 X0=X4
47317 Y0=Y4
47318 REM TRY ANOTHER PASS
47319 GOTO 47310
47320 REM ********************
47321 REM COEFFICIENTS SUBROUTINE
47322 M=5
47323 A(0)=0
47324 A(1)=24
47325 A(2)=-50
47326 A(3)=35
47327 A(4)=-10
47328 A(5)=1
47329 RETURN
47330 REM ********************
47331 REM VARIANT ON MUELLER'S METHOD FOR COMPLEX ROOTS
47332 REM THIS PROGRAM USES THE PARABOLIC FITTING TECHNIQUE
47333 REM ASSOCIATED WITH MUELLER'S METHOD, BUT DOES IT IN
47334 REM THE COMPLEX DOMAIN.
47335 REM THE INPUTS TO THE SUBROUTINE ARE THE INITIAL
47336 REM GUESS, (X0,Y0), THE CONVERGENCE CRITERIA, E,
47337 REM AND THE MAXIMUM NUMBER OF ITERATIONS, N.
47338 REM ALSO REQUIRED ARE BOUNDS ON THE INITIAL GUESS, B1 AND B2.
47339 REM RETURNED IS THE NEW ESTIMATE, (X,Y), AND THE
47340 REM NUMBER OF ITERATIONS PERFORMED, K.
47341 REM IT IS ASSUMED THAT THE FUNCTION F(Z) = U(X,Y)+IV(X,Y)
47342 REM IS AVAILABLE IN THE FUNCTIONS SUBROUTINE.
47343 REM START CALCULATIONS
47344 K=1
47345 X3=X0
47346 Y3=Y0
47347 X1=X3-B1
47348 Y1=Y3-B2
47349 X2=X3+B1
47350 Y2=Y3+B2
47351 D=(X2-X1)*(X2-X1)+(Y2-Y1)*(Y2-Y1)
47352 REM AVOID DIVIDE BY ZERO
47353 IF D=0 THEN D=1E-07
47354 L1=(X3-X2)*(X2-X1)+(Y3-Y2)*(Y2-Y1)
47355 L1=L1/D
47356 L2=(X2-X1)*(Y3-Y2)-(X3-X2)*(Y2-Y1)
47357 L2=L2/D
47358 D1=(X3-X1)*(X2-X1)+(Y3-Y1)*(Y2-Y1)
47359 D1=D1/D
47360 D2=(X2-X1)*(Y3-Y1)-(X3-X1)*(Y2-Y1)
47361 D2=D2/D
47362 REM GET FUNCTION VALUES
47363 X=X1
47364 Y=Y1
47365 GOSUB 47431
47366 U1=U
47367 V1=V
47368 X=X2
47369 Y=Y2
47370 GOSUB 47431
47371 U2=U
```

```
47372 V2=V
47373 X=X3
47374 Y=Y3
47375 GOSUB 47431
47376 U3=U
47377 V3=V
47378 REM CALCULATE MUELLER PARAMETERS
47379 E1=U1*(L1*L1-L2*L2)-2*V1*L1*L2-U2*(D1*D1-D2*D2)
47380 E1=E1+2*V2*D1*D2+U3*(L1+D1)-V3*(L2+D2)
47381 E2=2*L1*L2*U1+V1*(L1*L1-L2*L2)-2*D1*D2*U2-V2*(D1*D1-D2*D2)
47382 E2=E2+U3*(L2+D2)+V3*(L1+D1)
47383 C1=L1*L1*U1-L1*L2*V1-D1*L1*U2+L1*D2*V2+U3*L1
47384 C1=C1-U1*L2*L2-V1*L1*L2+U2*L2*D2+V2*D1*L2-V3*L2
47385 C2=U1*L1*L2+V1*L1*L1-U2*D2*L1-V2*D1*L1+V3*L1
47386 C2=C2+L1*L2*U1-L2*L2*V1-D1*L2*U2+D2*L2*V2+U3*L2
47387 B1=E1*E1-E2*E2-4*(U3*D1*C1-U3*D2*C2-V3*D2*C1-V3*D1*C2)
47388 B2=2*E1*E2-4*(U3*D2*C1+U3*D1*C2+V3*D1*C1-V3*D2*C2)
47389 REM GUARD AGAINST A DIVIDE BY ZERO
47390 IF B1=0 THEN B1=1E-07
47391 A=ATN(B2/B1)
47392 A=A/2
47393 B=SQR(SQR(B1*B1+B2*B2))
47394 B1=B*COS(A)
47395 B2=B*SIN(A)
47396 A1=(E1+B1)*(E1+B1)+(E2+B2)*(E2+B2)
47397 A2=(E1-B1)*(E1-B1)+(E2-B2)*(E2-B2)
47398 IF A1>A2 THEN GOTO 47402
47399 A1=E1-B1
47400 A2=E2-B2
47401 GOTO 47404
47402 A1=E1+B1
47403 A2=E2+B2
47404 A=A1*A1+A2*A2
47405 L1=A1*D1*U3-A1*D2*V3+A2*U3*D2+A2*V3*D1
47406 REM GUARD AGAINST DIVIDE BY ZERO
47407 IF A=0 THEN A=1E-07
47408 L1=-2*L1/A
47409 L2=-D1*U3*A2+D2*V3*A2+A1*U3*D2+A1*V3*D1
47410 L2=-2*L2/A
47411 REM CALCULATE NEW ESTIMATE
47412 X=X3+L1*(X3-X2)-L2*(Y3-Y2)
47413 Y=Y3+L2*(X3-X2)+L1*(Y3-Y2)
47414 REM TEST FOR CONVERGENCE
47415 IF ABS(X-X0)+ABS(Y-Y0)<E THEN RETURN
47416 REM TEST FOR NUMBER OF ITERATIONS
47417 IF K>=N THEN RETURN
47418 REM CONTINUE
47419 K=K+1
47420 X0=X
47421 Y0=Y
47422 X1=X2
47423 Y1=Y2
47424 X2=X3
47425 Y2=Y3
47426 X3=X
47427 Y3=Y
47428 GOTO 47351
47429 REM *******************
47430 REM SUPERVISOR SUBROUTINE
```

```
47431 N5=N
47432 U5=U1
47433 V5=V1
47434 REM DO WE GO TO THE FUNCTIONS SUBROUTINE OR TO THE SERIES SUBROUTINE?
47435 IF N3=1 THEN GOSUB 44300
47436 IF N3=2 THEN GOSUB 44950
47437 IF N3=1 THEN GOTO 47444
47438 U=Z1
47439 V=Z2
47440 REM RESTORE PARAMETERS
47441 N=N5
47442 U1=U5
47443 V1=V5
47444 IF J1=0 THEN RETURN
47445 REM DIVIDE BY THE J1 ROOTS ALREADY FOUND
47446 FOR J2=1 TO J1
47447 U5=U
47448 U=(X-X(J2))*U+(Y-Y(J2))*V
47449 V=(X-X(J2))*V-(Y-Y(J2))*U5
47450 A4=(X-X(J2))*(X-X(J2))+(Y-Y(J2))*(Y-Y(J2))
47451 REM GUARD AGAINST DIVIDE BY ZERO
47452 IF A4=0 THEN A4=1E-07
47453 V=V/A4
47454 U=U/A4
47455 NEXT J2
47456 REM RETURN TO ZMUELLER
47457 RETURN
47462 REM ********************
47463 REM QUADRATIC ROOT SUBROUTINE (QUADRAT)
47464 REM THIS PROGRAM CALCULATES THE TWO ROOTS OF
47465 REM A GIVEN SECOND ORDER POLYNOMIAL USING
47466 REM THE QUADRATIC EQUATION EVALUATED IN A
47467 REM MANNER WHICH MINIMIZES ROUND OFF ERROR.
47468 REM THE POLYNOMIAL IS ASSUMED TO BE OF
47469 REM THE FORM
47470 REM       Y = A(2)*X*X +A(1)*X +A(0)
47471 REM THE TWO ROOTS ARE RETURNED AS
47472 REM R1 = X1 + I Y1
47473 REM R2 = X2 + I Y2
47474 REM TEST FOR A(2)=0
47475 IF A(2)<>0 THEN GOTO 47488
47476 REM TEST FOR A(1)=0
47477 IF A(1)<>0 THEN GOTO 47483
47478 X1=0
47479 X2=0
47480 Y1=0
47481 Y2=0
47482 RETURN
47483 X1=-A(0)/A(1)
47484 Y1=0
47485 X2=X1
47486 Y2=Y1
47487 RETURN
47488 A=A(1)*A(1)-4*A(2)*A(0)
47489 B=SQR(ABS(A))
47490 REM ESTABLISH SIGN
47491 IF A(1)=0 THEN C=1
47492 IF A(1)<>0 THEN C=ABS(A(1))/A(1)
47493 REM DETERMINE THE FIRST ROOT
```

```
47494 REM CHECK IF ROOT IS COMPLEX
47495 IF A>0 THEN GOTO 47499
47496 X1=-C*ABS(A(1))/(2*A(2))
47497 Y1=-C*B/(2*A(2))
47498 GOTO 47502
47499 X1=-C*(ABS(A(1))+B)/(2*A(2))
47500 Y1=0
47501 REM CALCULATE THE SECOND ROOT
47502 C=X1*X1+Y1*Y1
47503 IF C<>0 THEN GOTO 47506
47504 X2=1E+06*1E+06*1E+06
47505 RETURN
47506 C=A(0)/(C*A(2))
47507 X2=X1*C
47508 Y2=-Y1*C
47509 RETURN
47511 REM ********************
47512 REM POLYNOMIAL COMPLEX ROOTS SUBROUTINE (LIN)
47513 REM USES LIN'S METHOD AS DESCRIBED IN THE REFERENCE
47514 REM A PRACTICAL GUIDE TO COMPUTER METHODS FOR ENGINEERS BY SHOUP.
47515 REM THE INPUT POLYNOMIAL COEFFICIENTS ARE A(0) THROUGH A(M).
47516 REM M IS THE ORDER OF THE POLYNOMIAL.
47517 REM INITIAL GUESSES FOR A AND B ARE REQUIRED.
47518 REM THE RESULTS ARE RETURNED IN X1,Y1 AND X2,Y2.
47519 REM X IS THE REAL PART, AND Y IS THE IMAGINARY.
47520 REM THE MAXIMUM NUMBER OF ITERATIONS IS N.
47521 REM THE NUMBER OF ITERATIONS IS RETURNED IN K
47522 REM THE CONVERGENCE CRITERION IS E.
47523 REM IF NECESSARY, DIMENSION A(I), B(I) AND C(I) IN THE CALLING PROGRAM.
47524 REM NORMALIZE THE A(I) SERIES
47525 FOR I=0 TO M
47526 C(I)=A(I)/A(M)
47527 NEXT I
47528 REM START ITERATION
47529 REM SET INITIAL GUESS FOR THE QUADRATIC COEFFICIENTS
47530 B(0)=0
47531 B(1)=0
47532 B(M-1)=C(M-1)-A
47533 B(M-2)=C(M-2)-A*B(M-1)-B
47534 FOR J=3 TO M
47535 B(M-J)=C(M-J)-A*B(M+1-J)-B*B(M+2-J)
47536 NEXT J
47537 REM GUARD AGAINST DIVIDE BY ZERO
47538 IF B(2)<>0 THEN GOTO 47542
47539 A=A+1E-07
47540 B=B-1E-07
47541 GOTO 47532
47542 A1=(C(1)-B*B(3))/B(2)
47543 B1=C(0)/B(2)
47544 K=K+1
47545 REM TEST FOR THE NUMBER OF ITERATIONS
47546 IF K>=N THEN GOTO 47553
47547 REM TEST FOR CONVERGENCE
47548 IF ABS(A-A1)+ABS(B-B1)<E*E THEN GOTO 47553
47549 A=A1
47550 B=B1
47551 REM RETURN FOR NEXT ITERATION
47552 GOTO 47532
47553 A=A1
```

```
47554 B=B1
47555 C=A*A-4*B
47556 REM IS THERE AN IMAGINARY PART
47557 IF C>0 THEN GOTO 47563
47558 Y1=SQR(-C)
47559 Y2=-Y1
47560 X1=-A
47561 X2=X1
47562 GOTO 47567
47563 Y1=0
47564 Y2=Y1
47565 X1=-A+SQR(C)
47566 X2=-A-SQR(C)
47567 X1=X0/2
47568 X2=X2/2
47569 Y1=Y1/2
47570 Y2=Y2/2
47571 RETURN
47584 REM ********************
47585 REM BAIRSTOW COMPLEX ROOT SUBROUTINE (BAIRSTOW)
47586 REM THIS SUBROUTINE FINDS THE COMPLEX CONJUGATE ROOTS
47587 REM OF A POLYNOMIAL HAVING REAL COEFFICIENTS.
47588 REM SEE COMPUTER METHODS FOR SCIENCE AND ENGINEERING
47589 REM BY R.L. LAFARA.
47590 REM ORDER OF INPUT SERIES IS M >= 4.
47591 REM SERIES COEFFICIENTS ARE A(I).
47592 REM INITIAL GUESSES A AND B ARE REQUIRED.
47593 REM E IS THE CONVERGENCE FACTOR.
47594 REM SUBROUTINE RETURNS X1,Y1 AND X2,Y2.
47595 REM N IS THE MAXIMUM NUMBER OF ITERATIONS.
47596 REM K IS THE NUMBER OF ITERATIONS PERFORMED.
47597 REM IF NECESSARY, DIMENSION A(I),B(I), C(I) AND D(I)
47598 REM IN THE CALLING PROGRAM.
47599 REM USE NORMALIZED SERIES, C(I)
47600 FOR I=0 TO M
47601 C(I)=A(I)/A(M)
47602 NEXT I
47603 REM CHOSE INITIAL ESTIMATES FOR A AND B
47604 K=0
47605 B(M)=1
47606 REM START ITERATION SEQUENCE
47607 B(M-1)=C(M-1)-A
47608 FOR J=2 TO M-1
47609 B(M-J)=C(M-J)-A*B(M+1-J)-B*B(M+2-J)
47610 NEXT J
47611 B(0)=C(0)-B*B(2)
47612 D(M-1)=-1
47613 D(M-2)=-B(M-1)+A
47614 FOR J=3 TO M-1
47615 D(M-J)=-B(M+1-J)-A*D(M+1-J)-B*D(M+2-J)
47616 NEXT J
47617 D(0)=-B*D(2)
47618 D2=-B(2)-B*D(3)
47619 D=D(1)*D2-D(0)*D(2)
47620 A1=-B(1)*D2+B(0)*D(2)
47621 A1=A1/D
47622 B1=-D(1)*B(0)+D(0)*B(1)
47623 B1=B1/D
47624 A=A+A1
```

```
47625 B=B+B1
47626 K=K+1
47627 REM TEST FOR THE NUMBER OF ITERATIONS
47628 IF K>=N THEN GOTO 47632
47629 REM TEST FOR CONVERGENCE
47630 IF ABS(A1)+ABS(B1)>E*E THEN GOTO 47607
47631 REM EXTRACT ROOTS FROM QUADRATIC EQUATION
47632 C=A*A-4*B
47633 REM TEST TO SEE IF A COMPLEX ROOT
47634 IF C>0 THEN GOTO 47640
47635 X1=-A
47636 X2=X1
47637 Y1=SQR(-C)
47638 Y2=-Y1
47639 GOTO 47644
47640 X1=-A+SQR(C)
47641 X2=-A-SQR(C)
47642 Y1=0
47643 Y2=Y1
47644 X1=X1/2
47645 X2=X2/2
47646 Y1=Y1/2
47647 Y2=Y2/2
47648 RETURN
47684 REM ********************
47685 REM STEEPEST DESCENT OPTIMIZATION SUBROUTINE (STEEPDS)
47686 REM THIS PROGRAM FIND THE LOCAL MAXIMUM OR MINIMUM
47687 REM OF AN L-DIMENSIONAL FUNCTION USING THE METHOD
47688 REM OF STEEPEST DESCENT, OR THE GRADIENT.
47689 REM THE FUNCTION, Y(X(1),X(2)...), IS PLACED IN THE
47690 REM SUBROUTINE AT 44300, ALONG WITH THE L DERIVATIVES
47691 REM OF F, D(I).
47692 REM THE ROUTINE SEEKS USING AN INTERNALLY ADJUSTED
47693 REM MULTIPLIER, K. THE SEARCH IS MADE UNTIL AN ERROR
47694 REM LIMIT, E, IS REACHED.
47695 REM THE USER MUST SUPPLY INITIAL VALUES FOR THE X(I),
47696 REM AS WELL AS K (INITIAL) AND E. THE PROGRAM RETURNS
47697 REM THE LOCALLY OPTIMUM X(I) SET.
47698 REM REMEMBER TO DIMENSION X(I) IN THE CALLING PROGRAM.
47699 REM THE PROGRAM NEEDS THREE VALUES OF Y TO GET STARTED.
47700 N=0
47701 REM START INITIAL PROBE
47702 FOR J=1 TO 3
47703 REM OBTAIN Y AND D(I)
47704 GOSUB 44300
47705 Y(J)=Y
47706 REM UPDATE X(I)
47707 GOSUB 47735
47708 NEXT J
47709 REM WE NOW HAVE A HISTORY TO BASE THE SUBSEQUENT SEARCH ON
47710 REM ACCELERATE SEARCH IF APPROACH IS MONOTONIC
47711 IF (Y(3)-Y(2))/(Y(2)-Y(1))>0 THEN K=K*1.2
47712 REM DECELERATE IF HEADING THE WRONG WAY
47713 IF Y(3)<Y(2) THEN K=K/2
47714 REM UPDATE THE Y(I) IF VALUE HAS DECREASED
47715 IF Y(3)>Y(2) THEN GOTO 47721
47716 REM RESTORE THE X(I)
47717 FOR I=1 TO L
47718 X(I)=X1(I)
```

```
47719 NEXT I
47720 GOTO 47724
47721 Y(1)=Y(2)
47722 Y(2)=Y(3)
47723 REM OBTAIN NEW VALUES
47724 GOSUB 44300
47725 Y(3)=Y
47726 REM UPDATE X(I)
47727 GOSUB 47735
47728 REM CHECK FOR CONVERGENCE
47729 N=N+1
47730 IF N>=M THEN RETURN
47731 IF ABS(Y(3)-Y(2))<E THEN RETURN
47732 REM TRY ANOTHER ITERATION
47733 GOTO 47711
47734 REM FIND THE MAGNITUDE OF THE GRADIENT
47735 D=0
47736 FOR I=1 TO L
47737 D=D+D(I)*D(I)
47738 NEXT I
47739 D=SQR(D)
47740 REM UPDATE THE X(I)
47741 FOR I=1 TO L
47742 REM SAVE OLD VALUES
47743 X1(I)=X(I)
47744 X(I)=X(I)+K*D(I)/D
47745 NEXT I
47746 GOSUB 44300
47747 Y(3)=Y
47748 RETURN
47784 REM ********************
47785 REM STEEPEST DESCENT OPTIMIZATION SUBROUTINE (STEEPDA)
47786 REM THIS PROGRAM FIND THE LOCAL MAXIMUM OR MINIMUM
47787 REM OF AN L-DIMENSIONAL FUNCTION USING THE METHOD
47788 REM OF STEEPEST DESCENT, OR THE GRADIENT.
47789 REM THE FUNCTION, Y(X(1),X(2)...), IS PLACED IN THE
47790 REM SUBROUTINE AT 44300. FINITE DIFFERENCES ARE USED TO
47791 REM CALCULATE THE L PARTIAL DERIVATIVES.
47792 REM OF F, D(I).
47793 REM THE ROUTINE SEEKS USING AN INTERNALLY ADJUSTED
47794 REM MULTIPLIER, K. THE SEARCH IS MADE UNTIL AN ERROR
47795 REM LIMIT, E, IS REACHED.
47796 REM THE USER MUST SUPPLY INITIAL VALUES FOR THE X(I),
47797 REM AS WELL AS K (INITIAL) AND E. THE PROGRAM RETURNS
47798 REM THE LOCALLY OPTIMUM X(I) SET.
47799 REM REMEMBER TO DIMENSION X(I) IN THE CALLING PROGRAM.
47800 N=0
47801 REM THE PROGRAM NEEDS THREE VALUES OF Y TO GET STARTED.
47802 REM GENERATE STARTING D(I) VALUES.
47803 REM THESE ARE NOT EVEN GOOD GUESSES, AND SLOW THE PROGRAM A LITTLE.
47804 D=1
47805 D(1)=1/SQR(L)
47806 FOR I=2 TO L
47807 D(I)=D(I-1)
47808 NEXT I
47809 REM START INITIAL PROBE
47810 FOR J=1 TO 3
47811 REM OBTAIN Y
47812 GOSUB 44300
```

```
47813 Y(J)=Y
47814 REM OBTAIN APPROXIMATIONS TO THE D(I)
47815 GOSUB 47866
47816 REM UPDATE X(I)
47817 GOSUB 47849
47818 GOSUB 47856
47819 NEXT J
47820 REM WE NOW HAVE A HISTORY TO BASE THE SUBSEQUENT SEARCH ON
47821 REM ACCELERATE SEARCH IF APPROACH IS MONOTONIC
47822 IF (Y(3)-Y(2))/(Y(2)-Y(1))>0 THEN K=K*1.2
47823 REM DECELERATE IF HEADING THE WRONG WAY
47824 IF Y(3)<Y(2) THEN K=K/2
47825 REM UPDATE THE Y(I) IF Y(3)>Y(2)
47826 IF Y(3)>Y(2) THEN GOTO 47832
47827 REM RESTORE THE X(I)
47828 FOR I=1 TO L
47829 X(I)=X1(I)
47830 NEXT I
47831 GOTO 47835
47832 Y(1)=Y(2)
47833 Y(2)=Y(3)
47834 REM OBTAIN NEW VALUES
47835 GOSUB 44300
47836 Y(3)=Y
47837 GOSUB 47866
47838 REM IF D=0 THEN THE PRECISION LIMIT OF THE COMPUTER HAS BEEN REACHED
47839 IF D=0 THEN RETURN
47840 REM UPDATE X(I)
47841 GOSUB 47856
47842 REM CHECK FOR CONVERGENCE
47843 N=N+1
47844 IF N>=M THEN RETURN
47845 IF ABS(Y(3)-Y(2))<E THEN RETURN
47846 REM TRY ANOTHER ITERATION
47847 GOTO 47822
47848 REM FIND THE MAGNITUDE OF THE GRADIENT
47849 D=0
47850 FOR I=1 TO L
47851 D=D+D(I)*D(I)
47852 NEXT I
47853 D=SQR(D)
47854 RETURN
47855 REM UPDATE THE X(I)
47856 FOR I=1 TO L
47857 REM SAVE OLD VALUES
47858 X1(I)=X(I)
47859 X(I)=X(I)+K*D(I)/D
47860 NEXT I
47861 GOSUB 44300
47862 Y(3)=Y
47863 RETURN
47864 REM FINITE DIFFERENCES SUBROUTINE FOR THE D(I) APPROXIMATION
47865 REM LOOK AHEAD ONE HALF INTERVAL
47866 FOR I=1 TO L
47867 REM SAVE X(I)
47868 A=X(I)
47869 REM FIND INCREMENT
47870 B=D(I)*K/(2*D)
47871 REM MOVE INCREMENT IN X(I)
```

```
47872 X(I)=X(I)+B
47873 REM OBTAIN Y
47874 GOSUB 44300
47875 REM GUARD AGAINST DIVIDE BY ZERO NEAR MAXIMUM
47876 IF B=0 THEN B=1E-11
47877 REM UPDATE D(I)
47878 D(I)=(Y-Y(3))/B
47879 REM GUARD AGAINST LOCKED UP DERIVATIVE
47880 IF D(I)=0 THEN D(I)=1E-05
47881 REM RESTORE X(I) AND Y
47882 X(I)=A
47883 Y=Y(3)
47884 NEXT I
47885 REM OBTAIN D
47886 REM GOSUB 47849
47887 RETURN
```

Index

acceleration:
   Aitken, 363, 378
   formula, 363
Aitken:
   acceleration, 363, 378
   Δ^2 method, 363
   Steffenson iteration, 373, 378
Akima, semi-spline interpolation, 286
algebra, matrix, 31
algorithms:
   comparison of, 486
   CORDIC, 231
      general, 233
      modified, 236
   Horner's, 155
   Marquart's, 491
alternating-sign series, 94
analytic functions, 396
approximations, piece-wise, 266
arcsine by recursion, 199
arcsinh x, 120
arctangent by recursion, 194
arithmogeometrical mean, 194
ascending power series, 102
asymptotic series, 12, 13, 15, 102
Bairstow's method, 480
Bessel function, 93, 103
   by recursion, 209
binary fraction, 180
Binet function, 119
Birge-Vieta relations, 315
bisection, 320, 398
   method, 398
Cauchy-Riemann equations, 397, 445
Chebyshev:
   polynomials, 91, 123, 212
   series, 124
      truncated, 12
chi-square:
   distribution, 108
   probability density function, 109
   statistic, 108
circumference of an ellipse, 202
coefficient:
   matrix, 33
   vector, 32
collocation polynomial, 14
comparisons:
   of algorithms, 486
   numerical, 360

complementary error function, 162
complex:
   conjugate pairs, 315, 473
   domain, 395
      Newton's method in the, 423
   functions, 393
   roots of polynomials, 393, 453, 473, 480
   series, 169
      summation, 453
convergence, 330, 333, 346, 363
   factor, 365
   quadratic, 346, 373
coordinate rotational digital computer (CORDIC) algorithm, 231
   general, 233
   modified, 236
correlations, linear, 7
cumulative-distribution function, 82, 307
curvature, 347, 361
   minimization, 286
curve-fitting, 285
curves:
   french, 266
   ship-builder's, 266
decomposition, 130, 180, 235
deflation, 454
derivatives:
   finite difference, 500
   from tables, 291
   steepest descent with approximate, 499
   steepest descent with functional, 492
Descartes rules, 315
distributions:
   chi-square, 108
   cumulative, 82, 307
   normal, probability density, 162
divided differences, 274
   Newton, interpolation, 273
division, synthetic, 381, 454
domain, complex, 395
double-precision, 670
   arithmetic, 132
double roots, 323
economization, 16, 89, 123, 130
ellipse, circumference of, 202
elliptic integrals by recursion, 202
equations, Cauchy-Riemann, 397, 445
erf(x), 119
error function, 32, 119, 307
errors:
   integration, 299

minimizing the maximum, 269
min-max, 503
relative, 17
round-off, 15, 34, 65, 95, 155, 175, 381, 494
truncation, 9, 236, 275
estimates, unbiased, 8, 17, 76, 87
execution speeds, 4
expansion point shifting, 155
exponential function, 234, 669
extrapolation, Richardson's, 297
factors, convergence, 365
false-position method, 351
fitting:
  curvature, 285
  min-max polynomial, 502
  parabolic, 24, 428
fixed points, 328, 363
FOR/NEXT loops, 668
formula:
  acceleration, 363
  Gauss recursion, 195
  half-angle, 243
  Heron's, 178
  quadratic, 468
Forsythe polynomials, 56
fractions, binary, 180
french curves, 266
functions:
  analytic, 396
  Bessel, 93, 103
    by recursion, 209
  Binet, 119
  chi-square probability density, 109
  complementary error, 162
  complex, 393
  cumulative-distribution, 82, 307
  error, 32, 119, 307
  exponential, 234, 669
  gamma, 108, 118
  objective, 489
  normal probability distribution, 306
  roots of, 313
  square root, 668
  TAB, 668
  trigonometric, 233, 669
    hyperbolic, 249
    inverse, 245
    inverse hyperbolic, 250, 257
  Weibull cumulative-distribution, 75
gamma function, 108, 118
Gauss recursion formula, 195
gaussian, 78
Gibbs' phenomenon, 11
half-angle formulae, 243
Hermite polynomials, 225
Heron's formula, 178
Hessian matrix, 491
Horner's:
  algorithm, 155

  rule, 93
hyperbolic trigonometric functions, 249
  inverse, 250, 257
infinite products, 166
integration:
  elliptic integrals, by recursion, 202
  error, 299
  table, 296
  trapezoidal, 297
interpolation, 404
  Akima semi-spline, 286
  inverse, 312
  Lagrange, 267, 291
  Newton divided-differences, 273
  semi-spline, 285
  sin x cubic, 269
interval:
  halving, 320, 494
  searching, 320, 398
inverse:
  hyperbolic trigonometric functions, 250, 257
  interpolation, 312
  trigonometric functions, 245
isochronous, periods, 202
iteration, 65, 175, 280
  Aitken-Steffenson, 373, 378
  general, relation, 344
  Newton, 345, 480
  Newton-Raphson, 190, 491
  regression, 65
  roots by, 177
  tangent, 190
Lagrange interpolation, 267, 291
Laguerre polynomials, 221
Laurent series, 82
least-squares:
  first-order, 16
  multidimensional, 45
  Nth-order, 31
  parametric, 75
  second-order, 24
Legendre polynomials, 212, 216
likelihood, 8
linear correlation, 7
linearization, 75
Lin's method, 472
ln $x!$, 109
logarithms:
  natural, 241, 669
    by recursion, 206
  Taylor series expansion, 243
loops, FOR/NEXT, 668
Maclaurin series, 9, 89, 246
mapping, 398
Marquart's algorithm, 491
matrices:
  algebra, 31
  coefficient, 33
  Hessian, 491

maxima and minima, 335
maximum-likelihood estimators, 8
mean, arithmogeometrical, 394
methods:
   Aitken Δ^2, 363
   Bairstow's, 480
   bisection, 398
   false-position, 351
   Lin's, 472
   Mueller's, 428, 437, 444
   Newton's, 345-346, 382
     in the complex domain, 423
   secant, 346
   secant and false-position, 351
minimization:
   curvature, 286
   maximum error, 269
   standard deviation, 501
min-max:
   error, 503
   near, 503
   polynomials, 12, 91, 125
     fitting, 502
   principle, 280
   steepest descent and, 502
Mueller's method:
   in one dimension, 428
   in the complex plane, 444
   in two dimensions, 437
multidimensional least squares, 45
natural logarithm, 241, 669
   by recursion, 206
Newton:
   divided-differences interpolation, 273
   iteration, 345, 480
     Raphson, 190, 491
   method, 345-346, 382
     in the complex domain, 423
normal distribution probability density, 162
normal probability distribution function, 306
numerical comparisons, 360
optimization, 76, 489
orthogonal polynomials, 56, 215
   coefficients, 215
parabolic fitting, 24, 428
periods:
   of a real pendulum, 202
   isochronous, 202
phenomenon, Gibbs', 11
points, fixed, 328, 363
poles, 323
polynomials:
   Chebyshev, 91, 123, 212
   collocation, 14
   complex roots of, 393, 453, 473, 480
   Forsythe, 56
   Hermite, 225
   inversion, 148
   Laguerre, 221

Legendre, 212, 216
min-max, 12, 91, 125
   fitting, 502
orthogonal, 56, 215
   coefficients, 215
rational, 159
reversion, 142
roots of, 315
Taylor series, 396
precision, double, 132
products, infinite, 166
quadratic:
   convergence, 346, 373
   formula, 468
random number generators, 667
rational polynomials, 159
recursion, 57, 94, 125, 175, 215, 243, 307
   arcsine by, 199
   arctangent by, 194
   Bessel functions by, 209
   elliptic integrals by, 202
   Gauss, formula, 195
   natural logarithm by, 206
   recessive, 210, 229
regression, iterated, 65
regula-falsi, 351, 404
   modified, 352, 363
relations:
   Birge-Vieta, 315
   general iteration, 344
Richardson's extrapolation, 297
roots, 313, 315, 393
   by iteration, 177
   double, 323
   of functions, 313
   of polynomials, 315
     complex, 393, 453, 473, 480
   multiple, 379, 400, 439
   removal of, 381, 453
round-off error, 15, 34, 65, 95, 155, 175, 381, 494
rules:
   Descartes, 315
   Horner's, 93
secant method, 346
   and false-position, 351
semi-spline interpolation, 285
   Akima, 286
series:
   alternating-sign, 94
   ascending power, 102
   asymptotic, 12-13, 15, 102
   Chebyshev, 124
     truncated, 12
   complex, 169
     summation, 453
   Laurent, 82
   Maclaurin, 9, 89, 246
   Taylor, 8, 16, 93, 169
     expansion, 8, 273

logarithmic, expansion, 243
  polynomial, 396
  truncated, 9, 275
shipbuilder's curve, 266
simultaneous zeroes, 445
sin x, 669
  cubic interpolation, 269
  table spacings, 280
speeds, execution, 4
standard deviation, 15, 17
  minimizing the, 501
statistics, chi-square, 108
steepest descent, 489
  and min-max, 502
  with approximate derivatives, 499
  with functional derivatives, 492
substitution, successive, 281, 328, 332
summations, complex-series, 453
synthetic division, 381, 454
TAB function, 688
table:
  derivatives, 291
  integration, 296
  sin x, spacings, 280
tangent iteration, 190
Taylor series, 8, 16, 93, 169
  expansion, 8, 273
    logarithmic, 243
  polynomial, 396
  truncated, 9, 275
telescoping, 125
trapezoidal integration, 297
trigonometric functions, 233, 669
  hyperbolic, 249
    inverse, 250, 257
  inverse, 245
truncation:
  Chebyshev series, 12
  error, 9, 236, 275
  Taylor series, 9, 275
unbiased estimates, 8, 17, 76, 87
uniqueness, 268
variables, reserved, 4
variance, 8
vectors:
  coefficient, 32
  rotation of, 233
Weibull cumulative-distribution function, 75
zeroes, simultaneous, 445

Text set in Paladium Medium
by BYTE Publications

Edited by Blaise Liffick and Bruce Roberts

Design and Production Supervision
by Ellen Klempner

Production and Editing by Peg Clement

Production by Mike Lonsky

Typeset by Donna Sweeney

Printed and bound by Halliday
Lithograph Corporation

The programs presented in this book are available on cassette and disk for most personal computers. For information on availability, contact your local computer store or:

DYNACOMP Inc
6 Rippingale Rd
Pittsford NY 14534
Phone: (716) 586-7579